# Lecture Notes in Bioinformatics     **9656**

Subseries of Lecture Notes in Computer Science

More information about this series at http://www.springer.com/series/5381

Francisco Ortuño · Ignacio Rojas (Eds.)

# Bioinformatics and Biomedical Engineering

4th International Conference, IWBBIO 2016
Granada, Spain, April 20–22, 2016
Proceedings

 Springer

*Editors*
Francisco Ortuño
Universidad de Granada
Granada
Spain

Ignacio Rojas
Universidad de Granada
Granada
Spain

ISSN 0302-9743                    ISSN 1611-3349   (electronic)
Lecture Notes in Bioinformatics
ISBN 978-3-319-31743-4           ISBN 978-3-319-31744-1   (eBook)
DOI 10.1007/978-3-319-31744-1

Library of Congress Control Number: 2016934189

LNCS Sublibrary: SL8 – Bioinformatics

Printed on acid-free paper

This Springer imprint is published by Springer Nature
The registered company is Springer International Publishing AG Switzerland

# Preface

We are proud to present the papers selected for the fourth edition of the IWBBIO conference "International Work-Conference on Bioinformatics and Biomedical Engineering" held in Granada (Spain) during April 20–22, 2016.

IWBBIO 2016 sought to provide a discussion forum for scientists, engineers, educators, and students on the latest ideas and realizations in the foundations, theory, models, and applications for interdisciplinary and multidisciplinary research encompassing the disciplines of computer science, mathematics, statistics, biology, bioinformatics, and biomedicine.

The aims of IWBBIO 2016 were to create a friendly environment that could lead to the establishment or strengthening of scientific collaborations and exchanges among attendees, and, therefore, IWBBIO 2016 solicited high-quality original research papers (including significant work-in-progress) on any aspect of bioinformatics, biomedicine, and biomedical engineering.

New computational techniques and methods in machine learning, data mining, text analysis, pattern recognition, data integration, genomics and evolution, next-generation sequencing data, protein and RNA structure, protein function and proteomics, medical informatics and translational bioinformatics, computational systems biology, modelling and simulation, and their application in the life science domain, biomedicine, and biomedical engineering were especially encouraged. The list of topics in the successive Call for Papers also evolved, resulting in the following list for the present edition:

1. **Computational proteomics**. Analysis of protein–protein interactions. Protein structure modelling. Analysis of protein functionality. Quantitative proteomics and PTMs. Clinical proteomics. Protein annotation. Data mining in proteomics.
2. **Next-generation sequencing and sequence analysis**. De novo sequencing, re-sequencing, and assembly. Expression estimation. Alternative splicing discovery. Pathway analysis. Chip-seq and RNA-Seq analysis. Metagenomics. SNPs prediction.
3. **High performance in bioinformatics**. Parallelization for biomedical analysis. Biomedical and biological databases. Data mining and biological text processing. Large-scale biomedical data integration. Biological and medical ontologies. Novel architecture and technologies (GPU, P2P, Grid, etc.) for Bioinformatics.
4. **Biomedicine**. Biomedical computing. Personalized medicine. Nanomedicine. Medical education. Collaborative medicine. Biomedical signal analysis. Biomedicine in industry and society. Electrotherapy and radiotherapy.
5. **Biomedical engineering**. Computer-assisted surgery. Therapeutic engineering. Interactive 3D modelling. Clinical engineering. Telemedicine. Biosensors and data acquisition. Intelligent instrumentation. Patient monitoring. Biomedical robotics. Bio-nanotechnology. Genetic engineering.
6. **Computational systems for modelling biological processes**. Inference of biological networks. Machine learning in bioinformatics. Classification for biomedical

data. Microarray data analysis. Simulation and visualization of biological systems. Molecular evolution and phylogenetic modelling.

7. **Health care and diseases**. Computational support for clinical decisions. Image visualization and signal analysis. Disease control and diagnosis. Genome-phenome analysis. Biomarker identification. Drug design. Computational immunology.

8. **E-health**. E-health technology and devices. E-health information processing. Telemedicine/e-health application and services. Medical image processing. Video techniques for medical images. Integration of classical medicine and e-health.

After a careful peer-review and evaluation process (286 submission were submitted and each submission was reviewed by at least three, and on average 3.2, Program Committee members or additional reviewer), 69 papers were accepted to be included in the LNBI proceedings.

During IWBBIO 2016, several special sessions were carried out. Special sessions are a very useful tool with which to complement the regular program with new and emerging topics of particular interest for the participating community. Special sessions that emphasize multi-disciplinary and transversal aspects as well as cutting-edge topics are especially encouraged and welcomed, and in this edition of IWBBIO 2016 they were the following:

1. **SS1: Tools for Next-Generation Sequencing Data Analysis.** Next-generation sequencing (NGS) is a broadly used term to describe the most recent sequencing technologies, including Illumina, Roche/454, Ion Torrent, SOLiD, and Pacific Biosciences. These technologies allow for the quick and cost-effective sequencing of DNA and RNA, opening new ways for the study of genomics, transcriptomics, gene expression, and systems biology, among others.

    The continuous improvements in these technologies (longer read length, fewer base calling errors, greater throughput, etc.) and the broad application of NGS in many research fields are producing a continuous increase of data requiring improved bioinformatics tools. Therefore, we invite authors to submit original research, pipelines, and review articles on topics helping in the study of NGS data, such as (but not limited to):

    (a) Tools for data pre-processing (quality control and filtering)
    (b) Tools for sequence mapping
    (c) Tools for de novo assembly
    (d) Tools for quality check of sequence assembling
    (e) Tools for the analysis of genomic data: identification and annotation of genomic variants (variant calling, variant annotation)
    (f) Tools for functional annotation: identification of domains, orthologues, genetic markers, controlled vocabulary (GO, KEGG, InterPro, etc.)
    (g) Tools for biological enrichment in non-model organisms
    (h) Tools for the analysis of transcriptomic data: RNA-Seq analyses (quantification, normalization, filtering, differential expression) and distinguishing transcripts, alleles, and paralogs
    (i) Tools for Chip-Seq data

(j) Tools for big-data analyses

(k) Tools for handling and editing complex workflows and pipelines

Organizers: Prof. M. Gonzalo Claros Diaz, PhD, Department of Molecular Biology and Biochemistry, University of Malaga (Spain)

Dr. Javier Perez Florido, PhD, Genomics and Bioinformatics Platform of Andalusia (GBPA), Seville, (Spain).

2. **SS2: Fundamentals of Biological Dynamics.**
   Motivation: For a proper description of the dynamic phenomenon it is necessary to discriminate between the trajectory of the evolution toward the limit set and the properties of the limit set. The limit set is the final state that may be dynamic, i.e., oscillating or following the trajectory of deterministic chaos. Concrete examples relevant to the proposed session are, for example, the living cycle of one living organism from the connection of gametes to death — this is the trajectory of the evolution of a partly discrete self-organized system — or behavior of the unconstrained fish shoal in free space — this is possible for assessing a limit set. Structured systems on the trajectory to the limit set prevail in nature.
   Objectives: The objective of the session is to gather researchers active in discrete dynamic systems research, i.e., multilevel cellular automata or agent-based models, researchers in qualitative dynamics, and the relevant experimentalists, namely, behavioral science researchers, cell dynamic researchers etc. Experimenters on relevant model systems such as chemical self-organization are also welcome. Such meetings occur only seldom and are never balanced in attendance; often producers of primary datasets do not have a sufficient audience.
   The session should contribute to answering following questions:

   (a) Which primary time and space element is predicted by the dynamic model?
   (b) Which are the experimental limits of behavioral and cell dynamic experiment and how well do they reproduce model predictions?

   Only with the merging of these two aspects can an answer on the proper setting and interpretation of biological measurements be provided.
   Organizer: Prof. Dalibor Stys, PhD, Head of Laboratory of Experimental Complex Systems, Institute of Complex Systems, University of South Bohemia (Czech Republic).

3. **SS3: Maximization of Information Extraction from the Experiments in the Biological Systems.**
   Interpretation of experimental results depends on the proper evaluation of the measured data, their comparison, and their classification. The experimental set-up, conditions, and measurement device attributes are also of importance. The significance of the interpretation could be optimized via experimental design. The collection of all settings requires the approach of catalogization and protocol databases. In this way, there could be an emphasis on the requirements necessary to obtain the best possible information, in the maximal relevant content, so as to assure nonbiased interpretation and knowledge extraction. There are, of course, specific details presented in bioinformatical cases. On the other hand, there are

already solved methods in different fields, which could serve in several bioinformatic cases as well.

In this special section we describe broad examples from experimental design through information tuning and data standardization to real implementation using optical sensors, image processing and analysis, and distributed knowledge database. The aim of this section is to present the potential increase of data interpretation and related methods.

Organizer: Dr. Jan Urban, PhD, Head of Laboratory of Signal and Image Processing, Institute of Complex Systems, University of South Bohemia (Czech Republic)

4. **SS4: Using Quantitative Systems Pharmacology for Novel Drug Discovery: A "Systems-level" Perspective to Reduce Toxicity and Increase the Therapeutic Effect of Drugs towards 'precision medicine'.**

Over the past three decades, the predominant paradigm in drug discovery was designing selective ligands for a specific target to avoid unwanted side effects. However, in the last 5 years, the aim has shifted to take into account the biological network in which they interact. Quantitative and systems pharmacology (QSP) is a new paradigm that aims to understand how drugs modulate cellular networks in space and time, in order to predict drug targets and their role in human pathophysiology.

The goal of this symposium is to go through the current computational and experimental QSP approaches such as polypharmacology techniques combined with systems biology information and consider the use of new tools and ideas in a wider systems-level context in order to design new drugs with improved efficacy and fewer unwanted off-target effects.

The use of network biology produces valuable information such as new indications for approved drugs, drug–drug interactions, proteins–drug side effects and pathways–gene associations. However, we are still far from the aim of QSP, both because of the huge effort needed to model precisely biological network models and the limited accuracy that we are able to reach with those. Hence, moving from "one molecule for one target to give one therapeutic effect" to the "big systems-based picture" seems obvious moving forward although whether our current tools are sufficient for such a step is still under debate. We will discuss all these issues in this symposium.

Organizer: Dr. Violeta I. Perez-Nueno, PhD, Inria Nancy Grand Est (LORIA), Villers les Nancy (France).

5. **SS5: High-Performance Computing in Bioinformatics, Computational Biology, and Computational Chemistry.**

The goal of this special session is to explore the use of emerging parallel computing architectures as well as high-performance computing systems (supercomputers, clusters, grids) for the simulation of relevant biological systems in the context of structural bioinformatics, computational biology, and computational chemistry. We welcome papers, not submitted elsewhere for review, with a focus on topics of interest on:

(a) Programming models and systems for multicore, manycore, and clusters of multicore/manycore
(b) Parallel stochastic simulation
(c) Biological and numerical parallel computing
(d) Parallel and distributed architectures
(e) Emerging processing architectures (e.g., GPUs, Intel Xeon Phi, FPGAs, mixed CPU-GPU or CPU-FPGA, etc.)
(f) Parallel model-checking techniques
(g) Parallel algorithms for biological analysis
(h) Cluster and grid deployment for systems biology
(i) Soft computing algorithms in bioinformatics, computational biology, and computational chemistry
(j) Application of HPC developments in structural bioinformatics, computational biology, and computational chemistry
(k) Biological and chemical databases for big data management
(l) Automated verification in computational biology
(m) System infrastructure for high-throughput analysis
(n) Biomolecular structure prediction and large molecular systems
(o) Molecular visualization of biological big data
(p) Large-scale proteomics

Organizers: Dr. Horacio Perez-Sanchez
Dr. Jose M. Cecilia, Universidad Catolica San Antonio de Murcia (UCAM), (Spain).

6. **SS6: Advances in Drug Discovery.**
We welcome research papers, not submitted elsewhere for review, with a focus on topics of interest ranging from but not limited to:

(a) Target identification and validation
(b) Computational chemistry: modelling biological processes through quantum chemistry and hybrid QM:MM approaches, interpreting ligand–protein binding sites with novel non-covalent interactions analysis, applying density functional theory (DFT) for predicting drug reactivity towards protein fragments and DNA, computing excited states to disclose the in vivo action of DNA/sensitizers and related molecules.
(c) Chemoinformatics: methodological basis and applications to drug discovery of: QSAR, Docking, CoMFA-like methods, high-performance computing (HPC), cloud computing, biostatistics, artificial intelligence (AI), machine learning (ML), and bio-inspired algorithms like artificial neural networks (ANN), genetic algorithms, or swarm intelligence
(d) Multidimensional QSAR. Applications and recent advances in QSAR concepts exploring higher dimensions related to their benefit in the drug-discovery process
(e) Bioinformatics biosystems: methodological basis and applications to drug design, target or biomarkers discovery of: alignment tools, pathway analysis, complex networks, non-linear methods, microarray analysis, software and Web servers

(f) High-throughput screening (HTS) of drugs; fragment-based drug discovery; combinatorial chemistry and synthesis

Organizers: Dr. Horacio Perez-Sanchez, Dr. Alfonso Perez-Garrido
Dr. Jose Pedro Ceron, Universidad Catolica San Antonio de Murcia (UCAM), (Spain).

7. **SS7: Human Behavior Monitoring, Analysis, and Understanding (HBMAU).**
Most prevalent diseases are partly caused or aggravated by lifestyle choices that people make in their everyday life. Accordingly, there is an urgent need for mechanisms to automatically and autonomously identify and eventually change people's unhealthy behaviors. This special session aims at showcasing the latest achievements in the monitoring, analysis, and understanding of human behavior through smart technologies. We welcome novel, innovative, and exciting contributions in areas including but not limited to:

(a) Applications for cognitive and physical health
(b) Ambient-assisted living applications
(c) Smart coaching systems
(d) Mobile social networks
(e) Behavioral grouping
(f) Participatory sensing (crowd-sensing)
(g) Behavior analysis for alerts and recommendations
(h) Behavior sensing and personalization of mobile devices
(i) Context-awareness and semantic modeling
(j) Affective computing (emotion recognition)
(k) Activity and action recognition
(l) User mobility modeling (location tracking)
(m) User-centric computing
(n) Benchmarking, databases, and simulation tools

Organizers: Dr. Oresti Banos, Biomedical Signals and Systems Group (BSS), University of Twente (UT) - Enschede, (The Netherlands)
Dr. Hector Pomares, University of Granada (Spain).

8. **SS8: Pattern Recognition and Machine Learning in the -omics Sciences.**
The biological sciences are quickly becoming data-centered sciences and, as a result, they are, more than ever, one of the key targets for data scientists, particularly from the viewpoints of pattern recognition (PR) and machine learning (ML). Nowhere else this is more evident than in the -omics sciences, spanning the fields of genomics, proteomics, metabolomics, and transcriptomics. In all of these, the advances in data acquisition are creating a true deluge of information and, with it, new challenging scenarios related to data management in all forms, compounded by privacy and legal issues.
In this scenario, PR and ML methods hold the promise of opening new avenues for transforming raw and unstructured -omics information into usable knowledge for the biomedical and bioinformatics domains.
Topics that are of interest to this session include (but are not necessarily limited to):

(a) Applications of PR and ML methods to problems in the -omics sciences (including genomics, proteomics, metabolomics, and transcriptomics)
(b) Development of novel PR and ML methods suitable for application to -omics problems
(c) PR- and ML-based methods for decision support in the -omics sciences
(d) PR and ML for data processing, preparation, and transformation in -omics
(e) Scalable PR and ML models for big data problems in -omics, including online and stream learning, as well as parallel and high-performance computing approaches

More information is available at
http://www.cs.upc.edu/~avellido/research/conferences/IWBBIO16-PRML-Omics.html
Main Organizer: Alfredo Vellido, PhD, Department of Computer Science, Universitat Politecnica de Catalunya, BarcelonaTECH (UPC), Barcelona (Spain).
Co-organizers: Jesus Giraldo, PhD, Institut de Neurociencies and Unitat de Bioestadistica, Universitat Autonoma de Barcelona (UAB), Cerdanyola del Valles, Barcelona (Spain).
Rene Alquezar, PhD, Department of Computer Science, Universitat Politecnica de Catalunya, BarcelonaTECH (UPC), Barcelona (Spain).

9. **SS9: Resources for Bioinformatics.**
From dealing with the complex storage and accessibility requirements, to the challenge of organizing and extracting meaningful information from big data, resources have a key role in today's research workflows. Bioinformatics support involves very different domains: from the creation of standards and rules to enable data exchange and representation to the maintenance of sustainable and efficient hardware and software infrastructures where data can be hosted and analyzed. Although recognized by the research community, this support role often does not find a devoted discussion forum in scientific events.
In this special session we want to provide an opportunity to share the latest ideas and developments on how different resources provide service to bioinformatics research. We welcome contributions focused on all aspects of bioinformatics and biomedicine support, roughly divided into the following topics:

(a) Data providers: Creating and maintaining databases hosting organized data enabling research involves some serious issues, from raw data hosting to curation and organization of the information to enable users' accessibility.
(b) Tool development: Analyzing and integrating data needs for the creation of tools and algorithms that help users make sense of the current overflow of information.
(c) Computing and infrastructure: Data organization and analysis is not possible without specific infrastructure and technical support enabling storage and accessibility, while ensuring sustainability and scalability.
(d) Standards and integration: Information integration is one of the most demanding challenges we face in bioinformatics research. Developing

standards and common guidelines is the way forward to eliminate current barriers for data exchange and re-use.

Organizers: Dr. Pablo Porras Millan, PhD, European Molecular Biology Laboratory, European Bioinformatics Institute (EMBL-EBI) (UK)
Dr. Rafael Jimenez, PhD, ELIXIR Chief Technical Officer (UK).

10. **SS10: Assistive Technology for People with Neuromotor Disorders.**
The motivation for this special session arises from the limitations caused by neuromotor disorders in the fundamental areas of life: mobility, communication, manipulation, orientation, and cognition. This special session collects publications about the newest advances and trends in assistive technologies for people with motor disorders. The term motor disorder includes pathologies with some common patterns such as spinal cord injury, cerebral palsy, and stroke.

According to the Association for the Advancement of Assistive technology in Europe (AAATE), assistive technology (AT) is any item, piece of equipment, or product system, whether acquired commercially off the shelf, modified, or customized, that is used to increase, maintain, or improve the functional capabilities of children with disabilities. AT is used in the context of the whole life of a person with a disability, to function as a part of an individual's total system of support, and not just in one role, setting, or functional application.

AT frequently uses methods and knowledge developed in other technological (robotics, informatics, telecommunications, artificial intelligence, etc.,) or humanistic (psychology, pedagogy, sociology, etc.,) fields. The main objective of AT (equipments and services) is to contribute to a better quality of life of the many persons affected by disabilities worldwide, through the integration of technological aspects in equipment, services, and contexts.

These technologies must be adapted to the special features of the users affected by disabilities. Human–computer interaction, for instance, can contribute significantly to AT. Rehabilitation robotics is also an emerging field focused on recovering physical and cognitive skills.

Tha main topics of this special session are:

(a) Mobility and manipulation technical aids
(b) Human–computer interaction
(c) Rehabilitation robotics
(d) Physiological sensors
(e) Biofeedback and virtual reality

Organizers: Dr. Rafael Raya, PhD, Department of Information Technologies and Researcher, Bioengineering Laboratory, Universidad CEU San Pablo (Spain).
Dr. Gabriel Caffarena, PhD, Department of Information Technologies and Researcher, Bioengineering Laboratory, Universidad CEU San Pablo (Spain).

11. **SS11: Toward an Effective Telemedicine: An Interdisciplinary Approach.**
In the last 20 years, many resources have been spent in experimentation and marketing of telemedicine systems, but — as pointed by several researchers - no

real product has been fully realized — neither in developed nor in underdeveloped countries. Many factors could be detected:

(a)  Lack of a decision support system in analyzing collected data
(b)  The difficulty of using the specific monitoring devices;
(c)  The caution of patients and/or doctors toward e-health or telemedicine systems
(d)  The passive role imposed to the patient by the majority of experimented systems
(e)  The limits of profit-driven outcome measures
(f)  A lack of involvement of patients and their families as well as an absence of research on the consequences in the patient's life

The constant improvement of ICT tools should be taken into account: at-home and mobile monitoring are both possible; virtual visits can be seen as a new way of performing an easier and more accepted style of patient–doctor communication (which is the basis of a new active role of patients in monitoring symptoms and evolution of the disease). The sharing of this new approach could be extended from patients to healthy people, obtaining tools for a real preventive medicine: A large amount of data could be gained, stored, and analyzed outside the sanitary structures, contributing to a low-cost approach to health.

The goal of this session is to bring together interdisciplinary experts to develop (discuss about) these topics:

(a)  Decision support systems for the analysis of collected data
(b)  Customized monitoring based on the acuteness of the disease
(c)  Integration of collected data with e-health systems
(d)  Attitudes of doctors and sanitary staff
(e)  Patient–doctor communication
(f)  Involvement of patients and of their relatives and care-givers
(g)  Digital divide as an obstacle/hindrance
(h)  Alternative measurements on the effectiveness of telemedicine (quality of life of patients and care-givers, etc.)
(i)  Mobile vs. home monitoring (sensors, signal transmissions, etc.)

Organizer: Maria Francesca Romano, Institute of Economics, Scuola Superiore Sant Anna, Pisa (Italy)

12. **SS12: Medical Planning: Sterilization Department Design.**
Since the sterilization department is one of the most sophisticated and complex sections inside health-care facilities and hospitals, creation of this department is a very important task. Sterilization departments should satisfy special requirements and conditions to comply with infection control. These conditions reflect the department space area, department location inside the hospital, functional relation to the operating rooms department and intensive care unit, as well as devices capacities. In this section, we will introduce the minimum requirements and needs from location selection, minimum required area, device capacities, and how to comply with the quality and infection control. These requirements have been

applied to many hospitals and centers and we have recorded some improvements and enhancements for work flow and infection control.

Organizers: Dr. Khaled S. Ahmed, PhD, Department of Bio-Electronics, Modern University for Technology and Information (Egypt).

13. **SS13: Biological Sequence Modeling with Computational Algorithms as HMMs.**
Sequence modeling is one of the most important problems in bioinformatics. In the sequential data modeling, the computational algorithms can be used for modeling and comparing biological sequences. Hidden Markov models (HMMs) have been widely used to find similarity between sequences, since the performance of HMMs is suitable for the handling of sequence patterns with various lengths. Some of the most important topics in this session are:

(a) Modeling of biological sequences in bioinformatics
(b) The application of hidden Markov models (HMM)
(c) HMM in modeling of sequential data
(d) The advantages of HMM in biological sequence modeling compare with other algorithms
(e) The new algorithms of training HMM
(f) Gene sequence modeling with HMM

Organizer: Dr. Mohamad Soruri, PhD, Ferdows Faculty of Engineering, University of Birjand, (Iran).

14. **SS14: Information Security Optimization and Big Data in Bioinformatics and Biomedical Engineering (ISOBD).**
The challenge of handling big data is considered one of the main hot topics that concerns most organizations especially in the medical sector. Nowadays, the usage of information systems and strategic tools has been applied to various fields. Through time a vast amount of data is generated. Although there are several data processing tools to extract knowledge and various visualization tools to present these data, the overarching challenge of the current big data paradigm is to encourage researchers to develop novel data processing and computational algorithms. Furthermore, securing these data is still a challenge that leads to competitions aimed at developing and applying a number of new promising security algorithms.
The aim of this special session is to bring scientists, researcher scholars, and students from academia and industry to present recent and ongoing research activities related to the advances in securing data and applying optimization techniques, especially in medical applications as well as to allow the exchange and sharing of ideas and algorithm experiences. Topics of interest include but are not limited to:

(a) Big data analytics modeling
(b) Bioinformatics and biomedical applications
(c) Data mining and knowledge representation
(d) Machine learning for medical data analysis
(e) Information security and biometrics

(f) Supervised and unsupervised learning of big data
(g) IOT applications
(h) Optimization algorithms

More information is available at:
Information Security Optimization and Big Data (ISOBD)
Organizers: Dr. Kareem Kamal A. Ghany, PhD, Faculty of Computers Information, Beni-Suef University, Egypt. Director of Intelligent Systems and Informatics Research Lab.
Eng. Hossam M. Zawbaa, Research Assistant, Faculty of Mathematics and Computer Science, Babes-Bolyai University, Romania.
Eng. Heba Ayeldeen, Assistant Lecturer, Akhbar El-Yom Academy, Egypt.

**Plenary Talks**

In this edition of IWBBIO, we were honored to have the following invited speakers:

(a) Prof. Rita Casadio, Group Leader of the Bologna Biocomputing Unit, University of Bologna (UNIBO), Bologna (Italy)
(b) Prof. Ferran Sanz, Director Research Programme on Biomedical Informatics (GRIB), Hospital del Mar Medical Research Institute (IMIM), Department of Health and Life Sciences, Universitat Pompeu Fabra (Spain)
(c) Dr. Andreas Bender, Department of Chemistry, Centre for Molecular Informatics, Group Leader: Bender Group, University of Cambridge (UK)

It is important to note, that for the sake of consistency and readability the presented papers are classified under 12 chapters including contributions on:

(a) Bioinformatics for healthcare and diseases
(b) Biomedical image analysis
(c) Biomedical signal analysis
(d) Computational systems for modelling biological processes
(e) E-health
(f) Tools for next-generation sequencing data analysis
(g) Assistive technology for people with neuromotor disorders
(h) Fundamentals of biological dynamics and maximization of information extraction from experiments in biological systems
(i) High-performance computing in bioinformatics, computational biology, and computational chemistry
(j) Human behavior monitoring, analysis, and understanding
(k) Pattern recognition and machine learning in the -omics sciences
(l) Resources for bioinformatics

This fourth edition of IWBBIO was organized by the Universidad de Granada together with the Spanish Chapter of the IEEE Computational Intelligence Society. We wish to thank to our main sponsor BioMed Central, and the Faculty of Science, Department of Computer Architecture and Computer Technology and the CITIC-UGR of the University of Granada for their support and grants. We also wish to thank the Editors-in-Chief of different international journals for their interest in publishing special issues of the best papers of IWBBIO.

We would also like to express our gratitude to the members of the different committees for their support, collaboration, and good work. We especially thank the Local Committee, Program Committee, the reviewers, and special session organizers. Finally, we want to thank Springer, and in particular Alfred Hofmann and Anna Kramer for their continuous support and cooperation.

April 2016                                                    Francisco Ortuño
                                                                  Ignacio Rojas

# Organization

## Program Committee

| | |
|---|---|
| Miguel A. Andrade | University of Mainz, Germany |
| Jesus S. Aguilar | University of Pablo Olavide, Spain |
| Carlos Alberola | Universidad de Valladolid, Spain |
| Hesham H. Ali | University of Nebraska at Omaha, USA |
| René Alquézar | Universitat Politecnica de Catalunya, BarcelonaTECH (UPC), Spain |
| Rui Carlos Alves | University of Lleida, Spain |
| Eduardo Andrés León | Spanish National Cancer Center (CNIO), Spain |
| Antonia Aránega | University of Granada, Spain |
| Saúl Ares | Spanish National Center for Biotechnology (CNB), Spain |
| Ruben Armañanzas | Universidad Politecnica de Madrid, Spain |
| Joel P. Arrais | University of Coimbra, Portugal |
| O. Bamidele Awojoyogbe | Federal University of Technology, Minna, Nigeria |
| Jaume Bacardit | University of Newcastle, UK |
| Hazem Bahig | University of Haíl, Saudi Arabia |
| Pedro Ballester | Computational Biology and Drug Design, Inserm, France |
| Oresti Baños | Ubiquitous Computing Lab (UCLab), Kyung Hee University, South Korea |
| Ugo Bastolla | Center of Molecular Biology Severo Ochoa, Spain |
| Steffanny A. Bennett | University of Ottawa, Canada |
| Alfredo Benso | Politecnico di Torino, Italy |
| Armando Blanco | University of Granada, Spain |
| Ignacio Blanquer | Universidad Politecnica de Valencia, Spain |
| Giorgio Buttazzo | TeCIP Institute, Scuola Superiore Sant'Anna, Italy |
| Gabriel Caffarena | Technical University of Madrid, Spain |
| Mario Cannataro | University Magna Græcia of Catanzaro, Italy |
| Carlos Cano | University of Granada, Spain |
| Jose María Carazo | Spanish National Center for Biotechnology (CNB), Spain |
| Rita Casadio | University of Bologna, Italy |
| Jose M. Cecilia | Universidad Catolica San Antonio de Murcia (UCAM), Spain |
| M. Gonzalo Claros | University of Malaga, Spain |
| Darrell Conklin | UPV/EHU, Spain |
| Clare Coveney | Nottingham Trent University, UK |

| | |
|---|---|
| Miguel Damas | University of Granada, Spain |
| Guillermo de La Calle | Politechnical University of Madrid, Spain |
| Javier De Las Rivas | CSIC, Bioinformatics and Functional Genomics Research Group, CIC, USAL, Cancer Research Center Salamanca, Spain |
| Joaquin Dopazo | Centro de Investigacion Principe Felipe (CIPF) Valencia, Spain |
| Werner Dubitzky | University of Ulster, UK |
| Khaled El-Sayed | Modern University for Technology and Information |
| Christian Exposito | ICAR-CNR, Italy |
| Afshin Fassihi | Universidad Catolica San Antonio de Murcia (UCAM), Spain |
| Jose Jesús Fernandez | University of Almeria, Spain |
| Jean-Fred Fontaine | Max Delbrück Center for Molecular Medicine (MDC), Germany |
| Xiaoyong Fu | Case Western Reserve University, USA |
| Razvan Ghinea | University of Granada, Spain |
| Humberto Gonzalez | University of the Basque Country (UPV/EHU), Spain |
| Daniel Gonzalez Peña | Bioinformatics and Evolutionary Computing, Spain |
| Concettina Guerra | College of Computing, Georgia Tech, USA |
| Christophe Guyeux | IUT Belfort-Montbeliard, France |
| Michael Hackenberg | University of Granada, Spain |
| Luis Javier Herrera | University of Granada, Spain |
| Lynette Hirschman | Mitre Corp., USA |
| Michelle Hussain | University of Salford, UK |
| Andy Jenkinson | European Bioinformatics Institute (EBI), UK |
| Craig E. Kapfer | Research Computing, KAUST, Kingdom of Saudi Arabia |
| Narsis Aftab Kiani | Computational Medicine Unit, Karolinska Institute, Sweden |
| Ekaterina Kldiashvili | New Vision University/Georgian Telemedicine Union, Georgia |
| Tomas Koutny | University of West Bohemia, Czech Republic |
| Natalio Krasnogor | University of Newcastle, UK |
| Marija Krstic-Demonacos | University of Salford, UK |
| Sajeesh Kumar | Health Science Center, University of Tennessee, USA |
| Pedro Larrañaga | Universidad Politecnica de Madrid, Spain |
| Jose Luis Lavin | cicBIO Gune |
| Rémi Longuespée | University of Liège, Belgium |
| Miguel Angel Lopez Gordo | Universidad de Cadiz, Spain |
| Ernesto Lowy | European Bioinformatics Institute (EBI) |
| Natividad Martinez | Reutlingen University, Germany |
| Francisco Martinez Alvarez | University of Seville, Spain |
| Marco Masseroli | Politechnical University of Milan, Italy |
| Roderik Melnik | Wilfrid Laurier University, Canada |
| Jordi Mestres | Pompeu Fabra University, Spain |

| | |
|---|---|
| Federico Moran | Complutense University of Madrid, Spain |
| Antonio Morreale | Repsol, Spain |
| Walter N. Moss | Yale University and Howard Hughes Medical Institute, USA |
| Cristian R. Munteanu | University of Coruna (UDC), Spain |
| Enrique Muro | Johannes Gutenberg University, Institute of Molecular Biology, Germany |
| Jorge A. Naranjo | New York University Abu Dhabi, UAE |
| Isabel A. Nepomuceno | Universidad de Sevilla, Spain |
| Michael Ng | Hong Kong Baptist University, SAR China |
| Baldomero Oliva | GRIP, Universidad Pompeu Fabra, Spain |
| Jose Luis Oliveira | University of Aveiro, IEETA, Portugal |
| Jose L. Oliver | University of Granada, Spain |
| Juan Antonio Ortega | University of Seville, Spain |
| Julio Ortega | CITIC-UGR, University of Granada, Spain |
| Francisco Ortuno | CITIC-UGR, University of Granada, Spain |
| Paolo Paradisi | Istituto di Scienza e Tecnologie dell Informazione, CNR, National Research Council of Italy |
| Alejandro Pazos | University of Coruna (UDC), Spain |
| Alexandre Perera | Politechnical University of Catalunya, Spain |
| Javier Perez Florido | Genomics and Bioinformatics Platform of Andalusia, Spain |
| María Del Mar Pérez Gómez | University of Granada, Spain |
| Violeta I. Pérez Nueno | Harmonic Pharma, France |
| Horacio Pérez Sánchez | Universidad Catolica San Antonio de Murcia (UCAM), Spain |
| Antonio Pinti | Valenciennes University, France |
| Alberto Policriti | University of Udine, Italy |
| Héctor Pomares | CITIC-UGR, University of Granada, Spain |
| Alberto Prieto | CITIC-UGR, University of Granada, Spain |
| Carlos Puntonet | CITIC-UGR, University of Granada, Spain |
| Omer F. Rana | Cardiff University, UK |
| Jairo Rocha | University of the Balearic Islands, Spain |
| Fernando Rojas | CITIC-UGR, University of Granada, Spain |
| Ignacio Rojas | CITIC-UGR, University of Granada, Spain |
| Maria Francesca Romano | Institute of Economics, Scuola Superiore Sant'Anna, Italy |
| Gregorio Rubio | Institute of Multidisciplinary Mathematics, UPV, Spain |
| Antonio Rueda | Genomics and Bioinformatics Platform of Andalusia, Spain |
| Michael Sadovsky | Siberian Federal University, Russia |
| Yvan Saeys | VIB - Ghent University, Belgium |
| Maria Jose Saez | University of Granada, Spain |
| José Salavert | European Bioinformatics Institute (EBI), UK |
| Vicky Schneider | The Genome Analysis Centre, UK |

Jean-Marc Schwartz        University of Manchester, UK
Ralf Seepold              Ubiquitous Computing Lab, HTWG Konstanz,
                            Germany
Jose Antonio Seoane       University of Bristol, UK
Istvan Simon              Institute of Enzymology, Research Centre for Natural
                            Sciences, Hungary
Richard Sinnott           National e-Science Centre, University of Glasgow, UK
Mohammad Soruri           University of Birjand, Iran
Yoshiyuki Suzuki          Tokyo Metropolitan Institute of Medical Science, Japan
Li Teng                   University of Iowa, UK
Pedro Tomas               INESC-ID, Portugal
Carolina Torres           University of Granada, Spain
Oswaldo Trelles           University of Malaga, Spain
Paolo Trunfio             University of Calabria, Italy
Olga Valenzuela           University of Granada, Spain
Alfredo Vellido           Universidad Politecnica de Catalunya (UPC), Spain
Renato Umeton             CytoSolve Inc., USA
Jan Urban                 University of South Bohemia, Czech Republic

## Additional Reviewers

Abellán, José L.                    Langa, Jorge
Aguiar-Pulido, Vanessa              Llanes Castro, Antonio
Alquezar, Rene                      Martin-Guerrero, Jose D.
Belanche, Lluís                     Martínez España, Raquel
Bueno-Crespo, Andrés                Minguillon, Jesus
Caballero, Alberto                  Mir, Arnau
Císař, Petr                         Olier, Iván
Falgueras Cano, Juan                Ortega-Martorell, Sandra
Fernández-Pascual, Ricardo          Ribas, Vicent
Fontaine, Jean-Fred                 Saberioon, Mohammadmehdi
Garcia-Gomez, Juan M.               Smith, Peter
Garcia-Rodriguez, Jose              Teruel, Antonio
Guillen, Alberto                    Tosi, Alessandra
Gómez Luna, Juan                    Vilamala, Albert
Imbernón, Baldomero

# Contents

**Biomedical Image Analysis**

## Computational Systems for Modelling Biological Processes

## eHealth

## Tools for Next Generation Sequencing Data Analysis

## Assistive Technology for People with Neuromotor Disorders

## Human Behavior Monitoring, Analysis and Understanding

## Pattern Recognition and Machine Learning in the -omics Sciences

## Resources for Bioinformatics

# Bioinformatics for Healthcare
and Diseases

# The Complexity of Some Pattern Problems in the Logical Analysis of Large Genomic Data Sets

Giuseppe Lancia[(⊠)] and Paolo Serafini

Department of Mathematics and Computer Science, University of Udine, Udine, Italy
giuseppe.lancia@uniud.it

**Abstract.** Many biomedical experiments produce large data sets in the form of binary matrices, with features labeling the columns and individuals (samples) associated to the rows. An important case is when the rows are also labeled into two groups, namely the *positive* (or healthy) and the *negative* (or diseased) samples. The *Logical Analysis of Data* (LAD) is a procedure aimed at identifying relevant features and building boolean formulas (rules) which can be used to classify new samples as positive or negative. These rules are said to explain the data set. Each rule can be represented by a string over {0,1,-}, called a *pattern*. A data set can be explained by alternative sets of patterns, and many computational problems arise related to the choice of a particular set of patterns for a given instance. In this paper we study the computational complexity of these pattern problems and show that they are, in general, very hard. We give an integer programming formulation for the problem of determining if two sets of patterns are equivalent. We also prove computational complexity results which imply that there should be no simple ILP model for finding a minimal set of patterns explaining a given data set.

## 1 Introduction

The recent advances in technology together with the massive use of computers have made modern molecular biology a science where the huge size of the input and/or output data of most experiments is by itself a serious computational problem. Very large data sets can be found everywhere in computational molecular biology. For instance, the European Molecular Biology Laboratory Nucleotide Sequence Database has grown exponentially in the past ten years, and, currently, the archive comprises around two billion records covering almost two trillion base pairs [2]. A similar growth, albeit slower, has interested the Protein Data Bank in which the number of protein structures deposited (each of which is a large data set by itself) is today around 100,000 [3]. Microarray experiments produce GigaBytes of information about the expression of hundreds of thousands of genes in hundreds of individuals at once [6]. These are only a few examples of data sets and experiments where the size of the information hinders the analyis and interpretation of the information itself.

© Springer International Publishing Switzerland 2016
F. Ortuño and I. Rojas (Eds.): IWBBIO 2016, LNBI 9656, pp. 3–12, 2016.
DOI: 10.1007/978-3-319-31744-1_1

In many biological experiments, the data have the form of a two dimensional array, where the rows correspond to *individuals* and the columns are associated to some biological *features*. This is the case, for instance of *microarrays*, bi-dimensional grids in which each column is associated to a gene, and each row to some tissue cells [17]. Each entry $(i, j)$ of the grid contains a number measuring the level of expression of gene $j$ in cells $i$. The dimension of the data sets makes imperative the use of computer algorithms in order to understand what is relevant information and what, on the other hand, is background noise and extra information of little practical use.

**Classification and Feature Selection.** In many experiments the rows are partitioned in classes, where each class contains cells that are "of the same type" but different from the other classes. A typical and prominent example is the case of two classes in the study of a disease: the *healthy* cells vs the *infected* cells.

When the individuals are divided into two or more classes associated with their characteristics, it is of interest to identify some rules that put the values of the features of an individual in relation with its class. Such rules can be used to try and understand the hidden underlying properties defining each class (e.g., one may discover a certain combination of some over-expressed genes with other under-expressed is common to all infected individuals while it is not found in any of the healthy ones). These type of rules can then be used to classify new individuals, i.e., to predict the class of an individual of unknown class. The general goal in data-analysis studies for large and complex data sets is to identify the relevant features (i.e., to perform *Feature Selection*) as well as their mutual relations in order to build simple (hopefully) rules that allow to discriminate between individuals that belong to different classes (i.e., to perform *Classification*).

Feature Selection and Classification techniques have been extensively used in the field of bioinformatics [4, 8, 9, 12, 15]. Since both problems are computationally hard, many times they are studied separately but this is not the best approach. For instance, some optimization approaches to feature selection are meant at minimizing the number of features (see, for example, the barcoding problem [14]), but this objective may not be appropriate when the true goal is that of using the selected features for classification purposes. For one, a classifier based only on only a few features may be not very robust. But the main weakness of this type of approach is that the number of features becomes more important than their mutual relations: it could happen, e.g., that classification should be based on three biologically correlated features, but this will never be discovered if there is a way to use two other, perhaps unrelated, features in their place.

**Logical Analysis of Data.** Consider a data set consisting in a binary matrix of $m$ rows and $n$ columns, in which $m' < m$ rows are labeled as *positive* while the remaining rows are labeled as *negative*. This is a typical data set for many biology experiments in which the rows are associated, e.g., to healthy and diseased samples. If the data comes from sequencing, the columns would be associated to nucleotides of a reference sequence, and a 0 in a row would mean "no mutation" while a 1 would mean "a mutation occurred". If the data comes from a microarray experiment, the values 0 and 1 would be related to the level of expression of each gene, with 0 meaning "normal" and 1 meaning "abnormal" [11, 16].

A technique known as Logical Analysis of Data (LAD [1, 7, 13]) has been often applied to data sets of this type with the objective of discovering a set of simple boolean formulas (or "rules") that can be used to classify new binary vectors $(b_1, \ldots, b_n)$. Each rule specifies the values that some bits must have in order to classify a vector as positive or negative. For instance, a "positive rule" could be

$$(b_2 = 0) \wedge (b_5 = 1) \wedge (b_9 = 1)$$

meaning that any vector that has a 0 in the 2nd component, and a 1 in the 5th and 9th component is classified as positive. Similarly, "negative rules" will specify how to classify a vector as negative. The goal of LAD is to derive positive and negative rules from the data in such a way that (i) each positive row satisfies at least one of the positive rules, while no negative row does and (ii) each negative row satisfies at least one of the negative rules, while no positive row does.

The techinque has been successfully applied to many biomedical contexts, such as the detection of rules that can be related to ovarian cancer, breast cancer and lymphoma [13]. The application of LAD to bioinformatics/biomedical data is based on the solution of optimization problems corresponding to large set covering instances.

**Boolean Formulas as Patterns.** Let us define an $n$-pattern (or simply a pattern) $p$ as a string of $n$ symbols where each symbol can take the values 0, 1 or $-$, i.e.,

$$p_i \in \{0, 1, -\}, \qquad i = 1, \ldots n.$$

We say that a pattern $p$ covers, or generates, a binary string $b = (b_1, \ldots, b_n)$ if $b_k = p_k$ for each $k$ such that $p_k \in \{0, 1\}$. Let us denote the set of all binary strings generated by the pattern $p$ by

$$S(p) = \{s \in \{0, 1\}^n : s_i = p_i \text{ if } p_i \in \{0, 1\}\}$$

Given a set $P$ of patterns the set $S(P)$ generated by $P$ is the set

$$S(P) = \bigcup_{p \in P} S(p)$$

The condition of being covered by a pattern can be interpreted as a boolean formula akin to the positive and negative rules of LAD, namely,

$$\bigwedge_{k: p_k \in \{0, 1\}} (b_k = p_k)$$

For instance, if $n = 10$, the rule $(b_2 = 0) \wedge (b_5 = 1) \wedge (b_9 = 1)$ is represented by the pattern

$$\texttt{-0--1---1-}$$

In view of the equivalence of rules and patterns, we can consider LAD as the problem of finding positive and negative patterns in order to explain a set of positive and negative vectors.

For a given instance of LAD, there might be many alternative sets of patterns explaining the data. A widely accepted objective, when more theories can be adopted to explain a phenomenon, is to choose the simplest one (a principle known as *Okkam's razor*). In this case, the simplest solution would be a set with a minimum number of patterns. Finding a min-size set of patterns is computationally demanding (see in [18] a branch-and-price procedure), and so also heuristic procedures have been proposed and used [5]. Some interesting computational problems emerge, such as

1. SIMPLE PATTERN FEASIBILITY: given a set $A$ of strings and a set $P$ of patterns, $A = S(P)$?
2. SIMPLE PATTERN MINIMALITY: given a set $A$ of strings and a constant $K$, does there exist a set of patterns $P$ such that $A = S(P)$ and $|P| \leq K$?
3. PATTERN EQUIVALENCE: given two sets $P$, $P'$ of patterns, are they equivalent, i.e., $S(P) = S(P')$?
4. PATTERN MINIMALITY: given a set $P$ of patterns and a constant $K < |P|$, does there exist an equivalent set of patterns $Q$ such that $|Q| \leq K$?
5. FULL PATTERN COVERAGE: given a set $P$ of patterns, $S(P) = \{0,1\}^n$?
6. PARTIAL PATTERN COVERAGE: given a set $P$ of patterns, does there exist a string $s \in \{0,1\}^n$ such that $s \notin S(P)$?

All these questions are motivated when LAD is used for biomedical analysis. For instance, one might be interested in knowing if a solution could be improved upon, or if two solutions obtained by different heuristics are equivalent. Furthermore, consider this situation: Given a set of patterns $P$ and a set of strings $Y$, find a set of patterns $Q$ such that $S(Q) = S(P) \cup Y$. This is a problem that arises when one wants to update a classifier in light of new data that has become available. Notice how this problem is substantially different from the "regular" LAD problem: in LAD, we are given a set of *strings* and we want to find a set of patterns that generate them. Here, we are given a set of *patterns* (and strings) and we want an equivalent set. The study of this problem motivates both questions 2. and 3. above, since we would certainly look for a set $Q$ such that $|Q| < |P| + |Y|$. Problems 5. and 6. are complement of each other and their interest is more theoretical in studying the other problems.

**Paper Contribution and Organization.** In this paper we study the computational complexity of pattern problems arising in the logical analysis of bidimensional data with positive/negative samples (such as biomedical data coming from microarray experiments). We show that these problems are, in general, quite complex. We give an integer programming formulation for PATTERN EQUIVALENCE. Formulating an ILP model for PATTERN MINIMALITY seems quite challenging due to the computational complexity results of the next section.

## 2   Computational Complexity Results

**Proposition 1.** *SIMPLE PATTERN FEASIBILITY is polynomial.*

**Proof:** For each $a \in A$ and each $p \in P$ we may easily check if $a$ is generated by $p$. Hence in time $O(n\,|A|\,|P|)$ we may decide whether $A \subseteq S(P)$ or not. In order to decide whether $S(P) \subseteq A$ or not, we start generating all strings in $S(P)$. For each generated string $s$, deciding whether $s \in A$ or not can be done in time $O(n\,|A|)$. We don't need to generate more than $|A| \cdot |P|$ strings. If a generated string does not belong to $A$ we have found that $S(P) \not\subseteq A$. Otherwise each generated string belongs to $A$. Since a string can be generated by more than one pattern, i.e., in the worst case by $|P|$ patterns, no more than $|A| \cdot |P|$ generated strings can belong to $A$. If there are more strings to be generated, these do not belong to $A$. Hence we may decide whether $S(P) \subseteq A$ or not in time $O(n\,|A|^2\,|P|)$. ∎

**Proposition 2.** *SIMPLE PATTERN MINIMALITY is NP-complete.*

**Proof:** It suffices to observe that SIMPLE PATTERN MINIMALITY is exactly MINIMUM DISJUNCTIVE NORMAL FORM (see [10] p. 261), which we repeat here for the sake of completeness. A set $U = \{u_1, u_2, \ldots, u_n\}$ of variables, a set $A \subseteq \{T, F\}^n$ of "truth assigments", and a positive integer $K$ are given. We ask whether there exists a disjunctive normal form expression $E$ over $U$, having no more than $K$ disjuncts, such that $E$ is true for precisely those truth assigments in $A$, and no others.

We may replace the variables $u_i$ with the symbols in the strings, each element $a \in A$ is a particular binary string, each disjunct is a pattern with a 1 for each variable, a 0 for each negated variable and a $-$ for each variable not present in the disjunct. ∎

**Proposition 3.** *PARTIAL PATTERN COVERAGE is NP-complete.*

**Proof:** The proof goes via a transformation from SAT. Given an instance of SAT with $n$ literals and $m$ clauses we generate a pattern from each clause as follows: if the literal $x_i$ is present in the clause we set $p_i = 0$, if $\neg x_i$ is present we set $p_i = 1$, and if neither $x_i$ nor $\neg x_i$ are present we set $p_i = -$.

Now assume SAT is satisfiable and let $x$ be the truth assignment that makes the instance satisfiable. Then we form the string $s_i = 1$ if $x_i = $ TRUE and $s_i = 0$ if $x_i = $ FALSE. Since at least one of the literals of each clause must be true, then, for each $p \in P$, at least one of the symbols $s_i$ corresponding to 0, 1 positions of $p$ must be different from $p_i$ and therefore cannot be in $S(p)$. Conversely, given a string not in $S(P)$ we may just reverse the reasoning and find a truth assignment that makes the SAT instance satisfiable.

PARTIAL PATTERN COVERAGE is also in NP since it suffices to show a string not in $S(P)$ and the verification that it does not belong to any $S(p)$ takes polynomial time. ∎

Since PARTIAL PATTERN COVERAGE is complement to FULL PAT-
TERN COVERAGE we clearly have:

**Corollary 1.** *FULL PATTERN COVERAGE is co-NP-complete.*    ∎

**Proposition 4.** *PATTERN EQUIVALENCE is co-NP-complete.*

**Proof:** We transform FULL PATTERN COVERAGE into PATTERN EQUIV-
ALENCE, by taking $P'$ the single pattern $\{- - \ldots -\}$ generating $\{0,1\}^n$. Furthermore for a no-instance there exists a string $s \in S(P)$ and $s \notin S(P')$, or vice
versa, and this string is a succinct certificate.    ∎

**Proposition 5.** *PATTERN MINIMALITY is co-NP-hard.*

**Proof:** The proof goes via a transformation from FULL PATTERN COVER-
AGE. Given an instance of FULL PATTERN COVERAGE we define a similar
instance of PATTERN MINIMALITY by choosing $K = 1$. Note that we may
freely assume that for each $i$, the $p_i$'s, $p \in P$, are not all equal, otherwise we may
drop each position with all symbols equal and reduce the instance to an equivalent
one. Hence the only pattern that can be equivalent to $P$ is $\{- - \ldots -\}$.    ∎

Since SIMPLE PATTERN MINIMALITY is a particular case of PATTERN
MINIMALITY we have by Proposition 2:

**Proposition 6.** *PATTERN MINIMALITY is NP-hard.*    ∎

By putting together Propositions 5 and 6 we see that it is very unilkely that
PATTERN MINIMALITY is in NP or in coNP.

Our main problem is finding a smaller pattern set generating the same strings
of a given pattern set. The previous results show that, even if we have available
the two pattern sets, the simple check that they generate the same set of binary
strings is NP-hard. This observation seems to rule out ILP models in which
patterns are represented in some way and the check is left to a polynomial
system of equalities/inequalities on integer variables.

## 3   ILP Models

Two sets $P$ and $Q$ of patterns are given. Consider the following models

$$
\begin{aligned}
v = \min \quad & \sum_{q \in Q} z_q \\
& \sum_{i:q_i=0} x_i + \sum_{i:q_i=1} (1 - x_i) \geq 1 - z_q && q \in Q \\
& y_p \leq 1 - x_i && i : p_i = 0, \quad p \in P \quad (1) \\
& y_p \leq x_i && i : p_i = 1, \quad p \in P \\
& \sum_{p \in P} y_p \geq 1 \\
& x_i \in \{0,1\}, y_p \in \{0,1\}, z_q \geq 0 \text{ integer}
\end{aligned}
$$

$$w = \min \quad \sum_{p \in P} y_p$$

$$\sum_{i:p_i=0} x_i + \sum_{i:p_i=1} (1 - x_i) \geq 1 - y_p \qquad p \in P$$

$$z_q \leq 1 - x_i \qquad\qquad\qquad i : q_i = 0, \quad q \in Q \qquad (2)$$

$$z_q \leq x_i \qquad\qquad\qquad\quad i : q_i = 1, \quad q \in Q$$

$$\sum_{q \in Q} z_q \geq 1$$

$$x_i \in \{0,1\}, z_q \in \{0,1\}, y_p \geq 0 \text{ integer}$$

In the sequel $\subset$ means strict inclusion.

**Proposition 7**

- $S(P) \subset S(Q)$ *if and only if* $v > 0$ *and* $w = 0$;
- $S(Q) \subset S(P)$ *if and only if* $w > 0$ *and* $v = 0$;
- $S(Q) = S(P)$ *if and only if* $v > 0$ *and* $w > 0$.

**Proof:** It is enough to show that $S(P) \subseteq S(Q)$ if and only if $v > 0$. If $y_p = 1$ then $x$ is generated by $p \in P$. The constraint $\sum_{p \in P} y_p \geq 1$ forces $x$ to be generated by at least one pattern in $P$. Hence feasible $x$ are in $S(P)$. Consider any $x \in \{0,1\}^n$. If $x$ is generated by $q \in Q$ then $z_q = 1$. If $x$ is not generated by $q \in Q$ then $z_q = 0$ is feasible (along with possible integer values $z_q \geq 1$). The objective forces $z_q$ to be zero in this case.

Hence $v = 0$ if and only if $x \in S(P)$ and $x \notin S(Q)$. If $v > 0$, for any pattern $x \in S(P)$ we have that $x \in S(Q)$, i.e., $S(P) \subseteq S(Q)$. ∎

Note that, if $S(P) \nsubseteq S(Q)$, i.e., when $v = 0$, model (1) exhibits also a string $x$ in $S(P)$ that is not in $S(Q)$, whereas if $S(P) \subseteq S(Q)$, i.e., when $v > 0$, model (1) exhibits also a string $x$ in $S(P)$ that is necessarily also in $S(Q)$. Similarly if $S(Q) \nsubseteq S(P)$, i.e., when $w = 0$, model (2) exhibits also a string $x$ in $S(Q)$ that is not in $S(P)$, whereas if $S(Q) \subseteq S(P)$, i.e., when $w > 0$, model (2) exhibits also a string $x$ in $S(Q)$ that is necessarily also in $S(P)$.

We may further distinguish the case $w = 0$, $v = 0$, via the following model

$$\hat{w} = \min \quad w_p + w_q$$

$$y_p \leq 1 - x_i \qquad\qquad i : p_i = 0, \quad p \in P$$

$$y_p \leq x_i \qquad\qquad\quad i : p_i = 1, \quad p \in P$$

$$\sum_{p \in P} y_p \geq 1 - w_p$$

$$z_q \leq 1 - x_i \qquad\qquad i : q_i = 0, \quad q \in Q \qquad (3)$$

$$z_q \leq x_i \qquad\qquad\quad i : q_i = 1, \quad q \in Q$$

$$\sum_{q \in Q} z_q \geq 1 - w_q$$

**Proposition 8.** $S(P)$ and $S(Q)$ are disjoint if and only if $\hat{w} > 0$.     ■

If $S(P)$ and $S(Q)$ are not disjoint then the model exhibits a string $x$ shared by both sets. Note that the problems FULL PATTERN COVERAGE and PARTIAL PATTERN COVERAGE are solved by

$$u = \min \quad \sum_{p \in P} y_p$$

$$\sum_{i:p_i=0} x_i + \sum_{i:p_i=1} (1 - x_i) \geq 1 - y_p \qquad p \in P$$

$$x_i \in \{0, 1\}, y_p \geq 0 \text{ integer}$$

**Proposition 9.** $S(P) = \{0, 1\}^n$ if and only if $u > 0$.     ■

As a simple example of the previous results suppose we are given the two following sets of patterns. These are fictitious instances and $P$ has been obtained from $Q$ by simply replacing some "0" or some "1"with a "-" so that we know in advance that $S(Q) \subset S(P)$.

$$Q = \begin{bmatrix} 1\,1\,-\,-\,1\,1\,1\,1\,-\,-\,1\,1\,-\,-\,1\,1\,1\,1\,-\,-\,1\,1\,-\,-\,1\,1\,1\,1\,-\,-\,0\,1\,-\,-\,1\,1\,1\,1\,-\,- \\ 1\,1\,-\,1\,0\,1\,1\,1\,1\,-\,1\,1\,-\,1\,0\,1\,1\,1\,1\,-\,1\,1\,-\,-\,1\,1\,1\,1\,-\,-\,1\,1\,-\,-\,1\,0\,1\,1\,-\,- \\ 1\,0\,-\,1\,-\,-\,1\,-\,0\,1\,1\,0\,-\,1\,-\,-\,1\,-\,0\,1\,1\,1\,-\,-\,1\,1\,1\,1\,-\,-\,1\,1\,-\,-\,1\,1\,1\,1\,-\,- \\ 1\,0\,1\,0\,1\,-\,0\,-\,1\,1\,1\,0\,1\,0\,1\,-\,0\,-\,1\,1\,1\,1\,-\,-\,1\,1\,0\,1\,-\,-\,1\,1\,-\,-\,1\,1\,1\,1\,-\,- \\ 1\,1\,1\,1\,-\,1\,-\,1\,-\,1\,1\,1\,1\,1\,-\,1\,-\,1\,-\,1\,1\,1\,-\,-\,1\,1\,1\,1\,-\,-\,1\,1\,-\,-\,0\,1\,1\,1\,-\,- \\ 0\,0\,-\,1\,0\,0\,0\,-\,0\,-\,0\,0\,-\,1\,0\,0\,0\,-\,0\,-\,0\,0\,-\,1\,0\,0\,0\,-\,0\,-\,0\,0\,-\,1\,0\,0\,0\,-\,0\,- \\ 1\,0\,-\,1\,-\,-\,0\,-\,0\,0\,1\,0\,-\,1\,-\,-\,0\,-\,0\,0\,0\,-\,1\,0\,1\,-\,0\,-\,0\,1\,0\,-\,1\,0\,1\,-\,0\,-\,0\,1 \\ 0\,1\,0\,0\,-\,0\,-\,1\,-\,-\,0\,1\,0\,0\,-\,0\,-\,1\,-\,-\,0\,-\,1\,0\,1\,-\,0\,-\,0\,1\,0\,-\,1\,0\,1\,-\,0\,-\,0\,1 \\ 0\,-\,1\,0\,1\,-\,0\,-\,0\,1\,0\,-\,1\,0\,1\,-\,0\,-\,0\,1\,0\,-\,1\,0\,1\,-\,0\,-\,0\,1\,0\,-\,1\,0\,1\,-\,0\,-\,0\,1 \\ 1\,0\,-\,1\,-\,-\,0\,-\,0\,0\,1\,0\,-\,1\,-\,-\,0\,-\,0\,0\,0\,0\,-\,1\,0\,0\,0\,-\,0\,-\,0\,0\,-\,1\,0\,0\,0\,-\,0\,- \end{bmatrix}$$

$$P = \begin{bmatrix} 1\,1\,-\,-\,1\,1\,1\,1\,-\,-\,1\,1\,-\,-\,1\,1\,1\,1\,-\,-\,1\,1\,-\,-\,1\,-\,1\,1\,-\,-\,0\,1\,-\,-\,1\,1\,1\,1\,-\,- \\ 1\,1\,-\,1\,0\,1\,1\,1\,1\,-\,1\,1\,-\,1\,0\,1\,1\,1\,1\,-\,1\,1\,-\,-\,1\,1\,1\,1\,-\,-\,1\,1\,-\,-\,1\,0\,1\,-\,-\,- \\ 1\,0\,-\,1\,-\,-\,1\,-\,0\,-\,1\,0\,-\,1\,-\,-\,1\,-\,0\,1\,1\,1\,-\,-\,1\,1\,1\,1\,-\,-\,1\,1\,-\,-\,1\,1\,1\,1\,-\,- \\ 1\,0\,1\,0\,1\,-\,0\,-\,1\,1\,-\,0\,1\,0\,1\,-\,0\,-\,1\,1\,1\,1\,-\,-\,1\,1\,0\,1\,-\,-\,1\,1\,-\,-\,1\,1\,1\,1\,-\,- \\ 1\,1\,1\,1\,-\,1\,-\,1\,-\,1\,1\,1\,1\,1\,-\,1\,-\,1\,-\,1\,1\,1\,-\,-\,1\,1\,1\,1\,-\,-\,1\,1\,-\,-\,0\,-\,1\,1\,-\,- \\ 0\,0\,-\,1\,0\,0\,0\,-\,0\,-\,0\,0\,-\,1\,0\,0\,0\,-\,0\,-\,0\,0\,-\,1\,0\,0\,0\,-\,0\,-\,0\,0\,-\,1\,0\,0\,0\,-\,0\,- \\ 1\,0\,-\,1\,-\,-\,0\,-\,0\,0\,1\,0\,-\,-\,-\,-\,0\,-\,0\,0\,0\,-\,1\,0\,1\,-\,0\,-\,0\,1\,0\,-\,1\,0\,1\,-\,0\,-\,0\,1 \\ 0\,1\,0\,0\,-\,0\,-\,1\,-\,-\,0\,1\,0\,0\,-\,0\,-\,-\,-\,-\,0\,-\,1\,0\,1\,-\,0\,-\,0\,1\,0\,-\,1\,0\,1\,-\,0\,-\,0\,1 \\ 0\,-\,1\,0\,1\,-\,0\,1\,0\,-\,-\,0\,1\,-\,0\,-\,0\,1\,0\,-\,0\,1\,0\,-\,1\,0\,1\,-\,0\,-\,0\,1\,0\,-\,1\,0\,1\,-\,0\,-\,0\,1 \\ 1\,0\,-\,-\,-\,-\,0\,-\,0\,0\,1\,0\,-\,1\,-\,-\,0\,-\,0\,0\,0\,0\,-\,1\,0\,0\,0\,-\,0\,-\,0\,0\,-\,1\,0\,0\,0\,-\,0\,- \end{bmatrix}$$

We run in sequence (1) and (2) and obtain $v = 0$ and $w > 0$, that implies $S(Q) \subset S(P)$, according to Proposition 7. In this case there is no need of running (3). The two programs take less than one second on a Mac OS personal computer. As a byproduct we obtain from (1) and (2) respectively the strings

$$x^1 = 10110001001010000100001010010100101000101000101$$
$$x^2 = 11001111001100111100111011110101101111101$$

One can easily check that $x^1 \in S(P)$ and $x^1 \notin S(Q)$ and also that $x^2 \in S(Q) \subset S(P)$.

Now suppose the first symbol of the first pattern in $P$ is changed to "0". We run (1) and (2) and obtain $v = 0$ and $w = 0$, so that we know that neither $S(P) \subset S(Q)$ nor $S(Q) \subset S(P)$. We have to run (3) and obtain $\hat{w} = 0$ so that we know that the two sets are not disjoint according to Proposition 8. Also in this the running time is negligible. As byproducts we obtain respectively

$$x^1 = 1\,1\,1\,0\,1\,1\,1\,1\,0\,0\,1\,1\,1\,0\,1\,1\,1\,1\,0\,0\,1\,1\,0\,0\,1\,1\,1\,1\,0\,0\,0\,1\,0\,0\,1\,1\,1\,1\,0\,0$$
$$x^2 = 1\,0\,1\,1\,0\,0\,0\,1\,0\,0\,1\,0\,1\,0\,0\,0\,0\,1\,0\,0\,0\,0\,1\,0\,1\,0\,0\,1\,0\,1\,0\,0\,1\,0\,1\,0\,0\,1\,0\,1$$
$$x^3 = 0\,0\,1\,1\,0\,0\,0\,1\,0\,0\,0\,0\,1\,1\,0\,0\,0\,1\,0\,0\,0\,0\,1\,1\,0\,0\,0\,1\,0\,1\,0\,0\,1\,1\,0\,0\,0\,1\,0\,1$$

One can easily check that $x^1 \in S(P)$ and $x^1 \notin S(Q)$, that $x^2 \in S(Q)$ and $x^2 \notin S(P)$ and that $x^3 \in S(P) \cap S(Q)$.

## 4 Conclusions

The use of software tools for feature selection and LAD poses some challenging problems with respect to the solutions (sets of patterns) that migh be obtained. In this paper we have studied the computational complexity of these problems and have show that they are, in general, very hard. One consequence of our complexity results is that there should be no simple ILP model for finding a minimal set of patterns explaining a given data set. On the other hand, we have given integer programming formulations for the problem of determining if two sets of patterns are equivalent. A line of future research would be to try and use these models in order to solve the PATTERN MINIMALITY problem.

## References

1. Alexe, G., Alexe, S., Bonates, T.O., Kogan, A.: Logical analysis of data - the vision of Peter L. Hammer. Ann. Math. Artif. Intell. **49**, 265–312 (2007)
2. Baker, W., van den Broek, A., Camon, E., Hingamp, P., Sterk, P., Stoesser, G., Tuli, M.A.: The EMBL nucleotide sequence database. Nucleic Acid Res. **28**, 19–23 (2000)
3. Berman, H.M., Westbrook, J., Feng, Z., Gilliland, G., Bhat, T.N., Weissig, H., Shindyalov, I.N., Bourne, P.E.: The protein data bank. Nucleic Acid Res. **28**, 235–242 (2000)
4. Bertolazzi, P., Felici, G., Festa, P., Lancia, G.: Logic classification and feature selection for biomedical data. Comput. Math. Appl. **55**, 889–899 (2008)
5. Boros, E., Hammer, P., Ibaraki, T., Kogan, A., Mayoraz, E., Muchnik, I.: An implementation of logical analysis of data. IEEE Trans. Knowl. Data Eng. **12**, 292–306 (2000)
6. Chee, M., Yang, R., Hubbell, E., Berno, A., Huang, X.C., Stern, D., Winkler, J., Lockhart, D.J., Morris, M.S., Fodor, S.P.A.: Accessing genetic information with high-density DNA arrays. Science **274**, 610–614 (1996)
7. Chikalov, I., Lozin, V., Lozina, I., Moshkov, M., Son Nguyen, H., Skowron, A., Zielosko, B.:Logical analysis of data: theory, methodology and applications. In: Three Approaches to Data Analysis, pp. 147–192. Springer (2013)

8. Dash, M., Liu, H.: Feature selection for classification. Intell. Data Anal. **1**, 131–156 (1997)
9. Felici, G., de Angelis, V., Mancinelli, G.: Feature selection for data mining. In: Felici, G., Triantaphyllou, E. (eds.) Data Mining and Knowledge Discovery Approaches Based on Rule Induction Techniques, pp. 227–252. Springer (2006)
10. Garey, M.R., Johnson, D.S.: Computers and Intractability: A Guide to the Theory of NP-Completeness. W.H. Freeman and Company, San Francisco (1979)
11. Golub, T.R., Slonim, D.K., Tamayo, P., Huard, C., Gaasenbeek, M., Mesirov, J.P., Coller, H., Loh, M.L., Downing, J.R., Caligiuri, M.A., Bloomfield, C.D., Lander, E.S.: Molecular classification of cancer: class discovery and class prediction by gene expression monitoring. Science **286**, 531–537 (1999)
12. Guyon, I., Weston, J., Barnhill, S., Vapnik, V.: Gene selection for cancer classification using support vector machines. Mach. Learn. **46**, 389–422 (2002)
13. Hammer, P., Bonates, T.: Logical analysis of data: from combinatorial optimization to medical applications. RUTCOR Research Report, 10-05, Rutgers University, NJ (2005)
14. Hebert, P.D.N., Cywinska, A., Ball, S.L., de Waard, J.R.: Biological identifications through DNA barcodes. Proc. R. Soc. Lond. B **270**, 313–321 (2003)
15. Hu, H., Li, J., Plank, A., Wang, H., Daggard, G.: A comparative study of classification methods for microarray data analysis. In: Proceedings of the fifth Australasian Conference on Data Mining and Analytics, vol. 61, pp. 33–37, Sydney, Australia (2006)
16. Li, T., Zhang, C., Ogihara, M.: A comparative study of feature selection and multiclass classification methods for tissue classification based on gene expression. Bioinformatics **20**, 2429–2437 (2004)
17. Montgomery, D., Undem, B.L.: CombiMatrix' customizable DNA microarrays-Tutoial: In situ computer-aided synthesis of custom oligo microarrays. Genetic Eng. News **22**, 32–33 (2002). Drug Discovery
18. Serafini, P.: Classifying negative and positive points by optimal box clustering. Discrete Appl. Math. **165**(270–282), 10 (2014)

# Development of a Handheld Side-Stream Breath Analyser for Point of Care Metabolic Rate Measurement

T.A. Vincent[1], A. Wilson[2,3], J.G. Hattersley[3], M.J. Chappell[1], and J.W. Gardner[1(✉)]

[1] School of Engineering, University of Warwick, Coventry, UK
J.W.Gardner@warwick.ac.uk
[2] Department of Physics, University of Warwick, Coventry, UK
[3] Human Metabolic Research Unit, University Hospitals Coventry and Warwickshire NHS Trust, Coventry, UK

**Abstract.** A novel handheld side-stream breath analyser has been developed. The low-cost device offers breath-by-breath measurements of $O_2$, $CO_2$, temperature, relative humidity and gas flow rate. Metabolic rate can be calculated from the inspired and expired gas concentrations; a knowledge of this over a 24 h period can guide calorific intake. The analyser provides easy-to-read results on either a laptop or smart phone. Results for the $O_2$ and $CO_2$ sensors demonstrate the device's potential for metabolic rate breath analysis. The $O_2$ sensor is not able to follow changes in $O_2$ concentration during the breathing cycle; however, the newly developed affordable and low-power consumption $CO_2$ sensor performs comparably to a bulky, high power consumption commercial device.

**Keywords:** Breath analysis · Metabolic rate · Energy expenditure · Smart phone sensing · Handheld breath analyser

## 1 Introduction

The measurement of energy expenditure (EE) through indirect whole body calorimetry is considered a 'gold standard' for human metabolism measurement [1]. In this system, subjects are studied in a sealed room where the difference in the concentrations of $O_2$ and $CO_2$ entering and leaving the room are measured, from which the metabolic rate is determined. Besides the enormous cost involved in constructing and maintaining calorimetry rooms, the measurements are limited to small groups of volunteers and patients. In order to enable a wider study population, a novel handheld analyser that can measure exhaled and inhaled $O_2$ and $CO_2$ has been developed. The low cost analyser permits real-time breath-by-breath data logging of the gas contents during an exhalation. The device has been designed considering the needs of patients, where the analysis of metabolic rate in a free-living environment is desired, without the need for a trained practitioner to make the measurements. The low power consumption of the device enables the measured $O_2$ and $CO_2$ content of breath samples to be recorded either on a smart phone or on a laptop computer, as required.

© Springer International Publishing Switzerland 2016
F. Ortuño and I. Rojas (Eds.): IWBBIO 2016, LNBI 9656, pp. 13–21, 2016.
DOI: 10.1007/978-3-319-31744-1_2

The total EE of a human being can be categorized into three components: resting metabolic rate (RMR), thermal effect of food (TEF) and physical activity [2]. Whilst respiratory chamber EE measurements give accurate and reproducible measurements [3], the subject is confined to a small room with limited ability to undertake activities of daily living (ADL). It has been reported that the energy expended in physical work is the most variable component of daily EE, where it can contribute between 15 and 30 % of the total EE [4]. RMR contributes between 60 and 75 % [4], where the remaining 10 % is attributed to TEF. The device that has been developed offers the capabilities to take multiple metabolic rate measurements throughout a 24 h period.

EE has been shown to vary by the time of day, where an increased EE is observed after eating or exercise [5]. Through monitoring of EE at regular intervals over a period of one day, and then repeating this over many days and weeks, we aim to investigate how EE varies in a free-living environment. The doubly-labelled water technique has been used to research daily EE in free-living human beings previously [6]; however, this technique does not give a dynamic measurement of energy expenditure but merely a measure of total energy over the observation. Additionally, DLW, as whole-body calorimeters, requires specialised knowledge, equipment and a non-trivial protocol. Hand-held calorimetry based on breath analysis, gives a uniquely low cost, non-invasive and low-risk measuring device suitable for use in an EE population monitoring programme. Metabolic carts provide a portable means of investigating a patient's metabolism, only a limited number of subjects can be analysed with such instruments, in part due to limited availability, long warm-up times and necessary calibration procedures [7]. Predictive equations are often used to calculate calorific requirements, in particular in intensive care units where correct energy intake is critical to recovery, however underlying assumptions may result in large errors [8].

Whole body calorimeters and metabolic carts are expensive and complex to use. Whilst domestic devices are available the quality and accuracy of the gas sensors in these are not at the level required for clinical quality measurements. A handheld calorimeter incorporating innovative sensor developments negate the need for long warm-up times associated with traditional sensors and low manufacturing costs means an affordable, simple-to-use precision hand-held indirect calorimeter is now possible.

## 2   Methods

To enable measurements to be taken periodically over a normal day for a sustained period, the analyser must be compact and portable whilst being able to deliver accurate and reproducible measurements. A side-stream system was selected for its reduced size and to provide the gas sensors with a constant and much reduced gas flow rate when compared to the main-stream flow. In our device, a sample of both exhaled and inhaled gas is extracted from the main-stream exhalation at a rate of 150 ml/min. A block diagram of the system is shown in Fig. 1. Metabolic rate can be determined from the rate at which $O_2$ is consumed, $\dot{V}_{O_2}$, and $CO_2$ is produced, $\dot{V}_{CO_2}$, through published equations, e.g. the abbreviated Weir equation (1) [9].

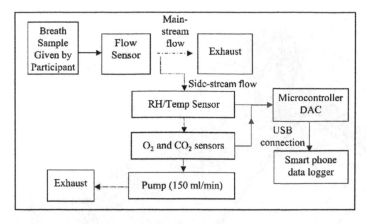

**Fig. 1.** Functional block diagram of side-stream sampling system, showing sensors included and the data-logging procedure.

$$EEtotal[kcal.] = 3.9(\dot{V}_{O_2}) - 1.1(\dot{V}_{CO_2}) \qquad (1)$$

Sensors are included for carbon dioxide ($CO_2$), oxygen ($O_2$), temperature, relative humidity (RH) and flow. An electrochemical sensor is used for $O_2$ monitoring (City Technology, MOX-20, UK) and a SFM3000 (Sensirion, Switzerland) for flow rate measurement. A ChipCap2 sensor (GE, USA) provides both temperature and RH measurements in the side-stream tubing. The data are logged via a microcontroller (PJRC, Teensy 3.2, USA) on either an Android smartphone or a Microsoft Windows laptop computer. An acetal shell was used to house the sensors, with 3D printed parts used where necessary, to meet the manufacturing tolerances required for our sensors.

The completed unit is shown in Fig. 2, with data logging performed on a smartphone. A novel non-dispersive infrared (NDIR) MEMS based sensor was developed for the breath analyser. Low cost commercial devices failed to provide the necessary response times and the compact low-power design required for a portable breath-by-breath sensor. The power consumption of the newly developed system was measured to peak at around 100 mW. Experimental data collected previously from respiratory chambers indicated the $O_2$ content on an exhalation needed to be measured to an accuracy of 0.52 % and the $CO_2$ content to 1.25 %, in order to obtain EE to an accuracy of 1 %. To put this requirement into perspective the maximum change in energy expenditure associated with the digestion of food (TEF) is between 30 and 70 kJ/h on a resting metabolic rate (RMR) of around 400 kJ/h [10].

A LabVIEW Virtual Instrument (National Instruments, v2014) was developed as an interface to visualise the outputs from the breath analyser on any Windows computer (Windows 7 used in the experiments presented, as shown in Fig. 3). The $O_2$, RH and temperature sensors were analogue inputs, where a raw voltage was recorded and converted into quantity. The flow sensor provided a digital output, which was converted into a flow value in litres per minute. The $CO_2$ sensor output is a sinusoidal waveform, the amplitude of which corresponds to the $CO_2$ concentration in the sensor

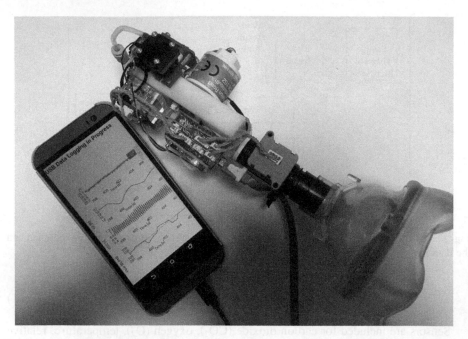

**Fig. 2.** Side-stream sampling system with smart phone data logging system. Software shown on screen demonstrating ambient conditions, with $CO_2$, $O_2$, temperature, RH and flow plots.

chamber. Further post-processing is required to convert the voltage recorded to the gas concentration desired than can currently be performed in real time. Thus only the raw voltage output is plotted on the readout graphs.

The microcontroller was selected to allow up to 16-bit analogue to digital conversion and the necessary $I^2C$ and SPI connections for the gas sensors. The data logging rate is selectable, where a 200 Hz rate provides a suitable balance between file storage size and capture speed. A serial output from the microcontroller, and simple USB connection, enables the output data to also be displayed on a smart phone. The low-power nature of the system enables it to be powered from either a laptop or smart phone alone. Software written in Python (v2.7) permitted displaying and logging of the results on an Android operating system (v5). Future generations of the system will allow the data to be logged wirelessly, to permit Apple iPhone mobile phones to be compatible with our system.

The current Android smartphone application plots the recorded data in real time, including conversion from raw data to a physical quantity, with the exception of the $CO_2$ readout. A screen print of the application is shown in Fig. 3(a), after a subject has exhaled and inhaled twice through the device. The current generation application allows limited analysis of the recorded data, extracted from the logged data files. Two basic features are extracted, the minimum value and peak value over the course of one measurement, as shown in Fig. 3(b). The flow sensor output provides an indication of when a subject is exhaling or inhaling. Current work includes validation of a complete breath sample, by the flow rate recorded through the flow sensor. The flow rate can be

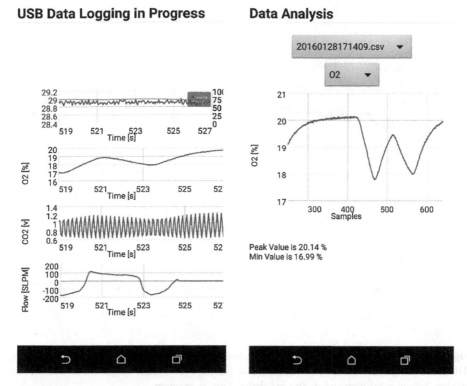

**Fig. 3.** (a) Screen print showing the data logging application on an Android smartphone, logging temperature, relative humidity, $O_2$, $CO_2$ and flow rate; (b) Peak value and minimum value are extracted from the measurement data in a separate data analysis section of the smartphone application.

used to determine the baseline readings and used to indicate the end of an exhalation, when the gas contents of exhaled breath is of interest for our metabolic rate calculations.

Our system has been trialed on three volunteers at the University of Warwick. Data from one subject are presented from experiments where subjects were asked to breathe through the device for a period of 1 min (see Fig. 4). During the experiment, subjects were requested to maintain a fixed, relaxed breathing pattern and regulate their breathing to have a breath-to-breath period of either ten seconds or six seconds, whilst maintaining a ratio of 1:1 between the inhalation period and the exhalation period. The six second cycle provides a similar respiratory rate to that of a resting adult, where 10 to 12 breaths per minute are considered normal for an awake adult [11].

The $CO_2$ sensor was calibrated against a commercial breath-by-breath sensor, for the 10 s breathing cycle provided by the first subject. The commercial sensor was connected to the output of the side-stream chamber, prior to the pump. The $O_2$ sensor was calibrated on a gas testing bench, using known values of gas concentrations. The gas concentrations were generated using a mixture of nitrogen and synthetic air. These calibration values were used for subsequent tests.

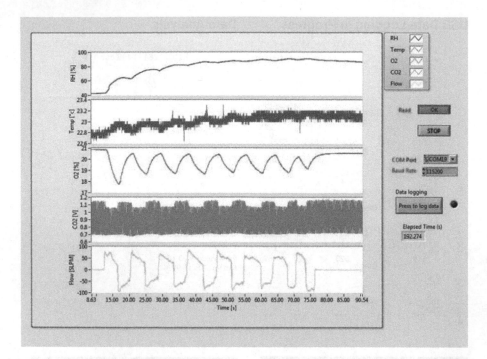

**Fig. 4.** LabVIEW VI logging system (for Windows computers) showing $O_2$, $CO_2$, temperature, RH and flow graphs for a 90 s experimental period with 8 exhalations. Note that during inhalation flow is positive and during exhalation flow is negative.

## 3 Results

The results obtained were consistent and Figs. 5 and 6 show an example of the results for one subject with a breath period of 6 s. A 90 s section of $CO_2$ data containing 10 breaths from one subject is shown in Fig. 5. The oxygen sensor output is shown in Fig. 6.

The $CO_2$ sensor captures the exhalations well, demonstrating a similar performance to that of a commercial device. The $t_{90}$ response time (time for the response to reach 90 % of its final value) was on average ∼1.7 s for the research sensor. The commercial device offers a $t_{10} - t_{90}$ response time of around 100 ms. Comparison between the commercial sensor and research device yielded a variance of 0.29 % for a 95 % confidence interval. The mean recorded $CO_2$ value across the 10 breath samples was 4.70 %, which is within the expected range for normal subjects of 4–5 %.

The research $CO_2$ sensor is driven at a rate of 5 Hz, which limited the number of measurements that could be taken per breath. A faster drive rate, perhaps up to 100 Hz, would enable a greater number of measurements to be taken per breath. The sensor output does not plateau for all the 10 sample breaths. This outcome is perhaps in part due to the limited measurement rate. The dimensions of the SOI CMOS IR emitter are related to the response time of the device. The larger emitter size chosen in these

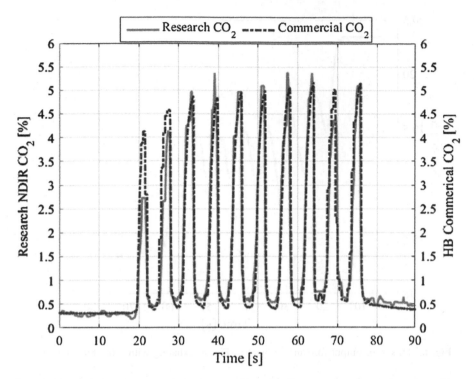

**Fig. 5.** $CO_2$ sensor output compared to commercial device, for 10 breaths provided by a subject over a period of 1 min.

experiments provides greater emissivity for a higher detector response, but prevents the faster drive signals desired for reduction in noise and greater measurement rate.

The $O_2$ sensor does not perform to its specification for breath-by-breath measurements. The baseline for room air is $\sim 21\ \%\ O_2$. The sensor does not return to this value when the subject is inhaling. The sensor could be affected by the elevated humidity level in an exhalation (peak of measured $\sim 80\ \%$ in the side-stream after an exhalation). Furthermore, the minimum $O_2$ concentration in normal subjects is expected to be in the range of $\sim 16$–$17\ \%$ in exhaled breath; a 4–5 % decrease compared to inhaled air.

A complementary response would be expected from the $O_2$ and $CO_2$ sensors to a number of exhales and inhales from a healthy subject. The response of the $CO_2$ sensor, demonstrating a peak to a maximum concentration (average amplitude of 4.1 %), indicates the exhaled breath gases are reaching the sensors located in the side-stream. The $O_2$ sensor does not recover to the baseline throughout the inhalation phase of any of the 10 sample breaths. The mean value during inhalation is 19.5 %, where the mean value during exhalation is only 0.9 % lower. The cause of this inadequate response could perhaps be due to the sensors detection principle. The electrochemical device measures $O_2$ concentration through a reaction inside a gel solution. The reaction rate could become slower as the device ages, due to the layers becoming blocked or from contamination.

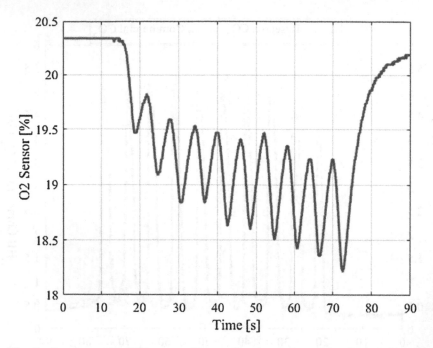

**Fig. 6.** $O_2$ sensor output for same subject as Fig. 4, exhaling with a 6 s breath cycle.

The smartphone logging application is a desired feature of a hand-held breath analyser, to enable easy integration into daily living as a portable unit. For the case of a personal breath analyser, when measurements are taken for a matter of minutes over the course of one day, a smartphone is ideal for metabolism analysis. However, in this prototyping phase, a computer was often used to log the measurement data. The current breath analyser unit connects to the data logging device by a USB power, which also serves as the power supply. Smart phones have only one USB port for connections to external devices, thus only one breath analyser can be connected. Furthermore, the phone cannot be charged simultaneously. Perhaps a future device, with wireless integration, will provide a solution to these limitations, which are not present on a traditional computer logging system.

The android application was used to log data at a rate of 200 Hz. The current measurement program is limited in functionality, due to the amount of data recorded. It was found no more than 10 s of data could be displayed on the screen for all five sensors. Therefore, any further data processing was completed after the data had been logged into file storage.

## 4    Conclusions

A microcontroller based portable breath analyser has been developed. In terms of gas sensors, a novel research $CO_2$ sensor has been compared to a commercial $CO_2$ sensor, where it demonstrated similar performance. The performance of the $O_2$ sensor requires further investigation. Whilst the specification for the device suggests that it should be able to follow breath-by-breath changes in $O_2$ the measured data did not demonstrate the expected change in $O_2$ concentration during exhalation (observed maximum $\sim 2.5$ %, expected 4–5 %). This could be due to elevated humidity levels in the side-stream section ($\sim 80$ %).

The side-stream system demonstrated a promising performance. It was powered by a smart phone alone, with peak consumption $\sim 100$ mW. In the future we aim to develop a wireless sensor system, which will allow measurements on other models of smart phone as the current system is compatible only with Android phone technology.

# References

1. Whybrow, S., Ritz, P., Horgan, G.W., Stubbs, R.J.: An evaluation of the IDEEA™ activity monitor for estimating energy expenditure. Br. J. Nutr. **109**, 173–183 (2013)
2. Sims, E.A., Danforth, E.: Expenditure and storage of energy in man. J. Clin. Invest. **79**, 1019–1025 (1987)
3. de Jonge, L., Nguyen, T., Smith, S.R., Zachwieja, J.J., Roy, H.J., Bray, G.A.: Prediction of energy expenditure in a whole body indirect calorimeter at both low and high levels of physical activity. Int. J. Obes. Relat. Metab. Disord. **25**, 929–34 (2001)
4. Poehlman, E.T.: A review: exercise and its influence on resting energy metabolism in man. Med. Sci. Sports Exerc. **21**, 515–525 (1989)
5. Ravussin, E., Lillioja, S., Anderson, T.E., Christin, L., Bogardus, C.: Determinants of 24-hour energy expenditure in man. Methods and results using a respiratory chamber. J. Clin. Invest. **78**, 1568–1578 (1986)
6. Buchowski, M.S.: Doubly labeled water is a validated and verified reference standard in nutrition research. J. Nutr. **144**, 573–574 (2014)
7. Boullata, J., Williams, J., Cottrell, F., Hudson, L., Compher, C.: Accurate determination of energy needs in hospitalized patients. J. Am. Diet. Assoc. **107**, 393–401 (2007)
8. Singer, P., Doig, G.S., Pichard, C.: The truth about nutrition in the ICU. Intensive Care Med. **40**, 252–255 (2014)
9. Weir, J.B.: New methods for calculating metabolic rate with special reference to protein metabolism. J. Physiol. **109**, 1–9 (1949)
10. Reed, G., Hill, J.O.J.: Measuring the thermic effect of food. Am. J. Clin. Nutr. **63**, 164–169 (1996)
11. Littleton, S.W.: Impact of obesity on respiratory function. Respirology **17**, 43–49 (2012)

# Confluence of Genes Related to the Combined Etiology DOISm (Diabetes, Obesity, Inflammation and Metabolic Syndrome) in Dissecting Nutritional Phenotypes

Ana Paula Moreira Bezerra[1], Samara Cardoso Silva-Santiago[2],
José Francisco Diogo Da Silva Jr.[1], Emanuel Diego S. Penha[1],
Monalisa M. Silveira[1], Myrna S. Ramos[1], Mônica M. Silva[3],
Ana Carolina L. Pacheco[3], and Diana Magalhaes Oliveira[1,2(✉)]

[1] Mestrado Acadêmico em Nutrição e Saúde, Centro de Ciências da Saúde (CCS),
Universidade Estadual Do Ceará (UECE), Campus do Itaperi, Fortaleza,
CE 60740-000, Brazil
[2] Programa de Pós-Graduação em Biotecnologia (RENORBIO),
Universidade Estadual Do Ceará (UECE), Campus do Itaperi, Fortaleza,
CE 60740-000, Brazil
diana.magalhaes@uece.br
[3] Bacharelado Em Nutrição, Universidade Federal Do Piauí (UFPI),
Campus Senador Helvídio Nunes de Barros, Picos, PI, Brazil

**Abstract.** The term DOISm (Diabetes, Obesity, Inflammation and metabolic Syndrome) describes a confluence of comorbidities specifying these disease phenotypes. Recent studies using genome-wide association analysis have identified genes and variations that correlate human phenotype within phenotype prediction programs. Benefiting from such post-genomics outcomes, we catalogued genes that have been associated with each of the four conditions before searching for confluence of any two or three conditions, and the confluence of genes concomitantly involved in all phenotypes. Bioinformatics analyses were performed using multi-relational data mining techniques to cover sequence, structure and functional/clinical features. We used high-confidence predictions for gene functional classification analyses for better phenotyping DOISm confluence. Our curated panel of 1439 DOISm genes and a subset of 217 confluent genes represents a platform to assist in dissecting complex nutritional phenotypes. Our repertoire of human genes likely to be involved in DOISm is an attempt to guide further subtyping of complex phenotypes.

**Keywords:** Diabetes · Obesity · Inflammation · Metabolic syndrome · DOISm · Nutrigenomics · Core-meta analysis · Bioinformatics · Complex phenotyping · Analysis association

## 1 Introduction

The term "diabesity" [4] was proposed to group together two associated disorders: type 2 diabetes mellitus (T2DM) and obesity. The term "diabesin" was

© Springer International Publishing Switzerland 2016
F. Ortuño and I. Rojas (Eds.): IWBBIO 2016, LNBI 9656, pp. 22–37, 2016.
DOI: 10.1007/978-3-319-31744-1_3

later used to describe concomitant morbid conditions of obesity, T2DM, insulin resistance (IR), metabolic syndrome (metS), and inflammation response [10]. In order to establish an abbreviation that better identifies the confluence of such comorbidities, we coined the term DOISm (Diabetes, Obesity, Inflammation and metabolic Syndrome) to join such multifactorial phenotypes in a so-called combined etiology [9]. The clinical confluence characteristic of DOISm can be described as the following: first, metS raises the risk for development of T2DM; then, the obesity-inflammation linkage arises from the circumstance that pro-inflammatory cytokines increase in obesity, while visceral obesity relates to metS. As recognized, obesity leads to a pro-inflammatory condition starting at highly metabolic (such as adipocytes, hepatocytes, or muscle cells) and/or immune cells, which release cytokines, such as TNF-$\alpha$, IL-6, and adiponectin [37]. It has been hypothesized [19] that obesity-driven inflammation is the basis for complications such as hypertension (HTN), atherosclerosis, dyslipidemia, IR, and T2DM, all of them features of metS.

Because DOISm conditions are all determined by both genetic and environmental factors, a better understanding of their combined etiology will require a careful investigation of gene-environment interactions. The importance of DOISm in Nutrition Sciences has gained attention with the growing interest in chronic, non-communicable diseases (NCDs) [29,45,52]. NCDs are the number one cause of death and disability in the world, being the biggest global killers today [57]. The four major NCDs responsible for almost 2/3 of deaths annually are cardiovascular diseases (CVDs), cancers, chronic respiratory diseases and diabetes [57]. Obesity-related disorders are also increasingly relevant in this regard because T2DM often results from excess body weight and physical inactivity [57], merging with metS to build the framework for the DOISm combined etiology.

We must recall that obesity can be characterized as an inflammatory condition [52]. Obesity causes a non-specific, general state of chronic inflammation, also referred as para-inflammation, which contributes to several systemic metabolic dysfunctions easily associated with obesogenic disorders [20] and metS [45,48]. In fact, metS is a direct risk factor for both T2DM [26] and CVDs [30] and is a related co-factor for inflammation response [19,20] and other NCDs [45]. Para-inflammation is likely to be a key factor in several NCDs [19]. Recently, microRNAs and adipokines have been implicated in senescence, inflammation, CVDs [14] and T2DM [5]. Be affecting endothelial activity and causing atherosclerosis, adipokines help to maintain low-grade inflammation, which connects morbid conditions of metS, obesity and CVDs [55]. Various aspects contribute to metS, such as genetic background, diet and lifestyle [3].

Genetic analysis of such complex diseases (multifactorial phenotypes, indeed in DOISm) are difficult due to several inherent properties, which might affect their progression, including but not limited to gene expression and polymorphic variants, chromosome anomalies, diet and environmental differences [15,18], as well as epigenetic changes, such as DNA methylation [41]. Attempting to identify genetic and epigenetic markers directly linked to DOISm is a good alternative to unravel candidate genes. Aiming to identify an exhaustive panel of human genes implicated in DOISm, we have attempted to catalogue all genes associated with each one of the four conditions before searching for a confluence of any two and three conditions, plus the full confluence of genes concomitantly involved in the combined etiology. Bioinformatics analyses of such DOISm confluent genes have been performed through data mining techniques covering sequence, structure and functional/clinical features in a core meta-analysis approach [35].

Intrinsic variability within DOISm confluent genes might serve as prospective indicators for establishing potential risk haplotypes, which can be exploited for the fine mapping of disease genes. A multitude of genetic association studies investigating the influence of common polymorphisms on disease susceptibility has been published to date [35], and they have identified several candidate genes that are likely associated with one, two or even three of the four DOISm conditions [2,27,31,40,43,54]. However, no screening has been performed, to our knowledge, to encompass all four-disease phenotypes concomitantly, besides genetic and genomic features, such as epigenetic changes.

One key epigenetic phenomenon, the genomic imprinting [39], is a remarkable regulatory process in which a relatively small number of human genes (approximately 120 reported by Silva-Santiago [51]) undergo a monoallelic expression where only one copy of the gene is active—the second is turned off [42]. Many of the imprinted genes have been recognized as important players of major gene-environment interactions, appearing to be highly interconnected through interactions mediated by proteins, RNA, and DNA [56]. As recently posed [56], these interactions in imprinted genes often favor the evolution of genetic coadaptation, where beneficially interacting alleles evolve to become co-inherited. Imprinted genes are, thus, good candidates for searching for a variation basis in nutritional phenotypes, as we have previously demonstrated by examining 24 (twenty-four) human imprinted genes [10], which have now been added to 1416 (non-imprinted) other genes to be studied as putative candidates in the confluence of DOISm combined etiology.

Therefore, using a semantic, multi-step approach for comparing clinical and genetic/epigenetic data, candidate disease-gene associations could be further identified in a platform to assist in dissecting complex nutritional phenotypes, which would be especially useful for discerning potential pathways to be used in future gene prioritization approaches.

## 2 Methods

### 2.1 Ethics Statement

All experimental designs, data and procedures were exclusively in silico and in agreement with the guidelines of the Open Data Commons (http://opendatacommons.org/guide/), as well as the Bermuda principles and the Fort Lauderdale agreement (http://www.geno-me.gov/10506537). Pertinent human, mouse and other mammalian biological samples also cited (including DNA, RNA and protein sequences, structures and interactions) were exactly as provided by their original data sources and corroborated with STREGA (STrengthening the REporting of Genetic Association Studies), as posed by Little et al. [38]. We have thus performed systematic core meta-analyses, derived from Genome-Wide Association Studies (GWAS) and/or candidate gene studies, according to Lill and Bertram [35]. This research is part of a project approved by the Research Ethics Board (Comissão de Ética em Pesquisa) of Universidade Estadual do Ceará (UECE), under the process number 09231334-5/2010.

## 2.2   Data Sources and Collection, with Detailed Annotations

Bioinformatics analyses involved exhaustive human genome and literature searches were performed using queries from a table of keywords such as "diabetes", "diabetes mellitus", T2DM", "obesity", "inflammation", "para-inflammation", inflammasome", "metabolic syndrome", "diabesity", and "DNA methylation". A catalogue, including gene names, was compiled based on analyses of these reviews and their cited references. For each DOISm or predicted-to-be DOISm-related gene, we extracted annotations, references, and sequences by performing comprehensive searches in relevant databases (DBs). Gene features, such as chromosome localization (genome coordinates), size in kb, number of exons/introns and transcripts, among others, which could not be found in existing DBs, were obtained by analyzing related publications. The compiled annotations were formalized and imported into a MySQL DB with tables comprising genes for all four morbid conditions of DOISm. Each table was assigned general annotation fields, as well as specific annotation fields for its corresponding category (Supplementary Tables S1-S4). Gene annotations were also obtained by mutual cross-referencing literature and DBs queries.

A collection of pertinent human genes was mainly extracted from research papers, while specific datasets were downloaded from all available DBs and repositories, primarily: (i) T-HOD (http://bws.iis.sinica.edu.tw/THOD/) [15]; (ii) Rat Genome Database (http://rgd.mcw.edu/); (iii) HGNC (HUGO Gene Nomenclature Committee) at the European Bioinformatics Institute (http://www.genenames.org/); (iv) T2D-Db (http://t2ddb.ibab.ac.in) [1]; (v); T2D Genetic Association Database (T2DGADB - http://t2db.khu.ac.kr:8080/index.jsp) [36]; and (vi) T2D@ZJU – (http://tcm.zju.edu.cn/t2d) [58]. Additional data were collected from the Online Mendelian Inheritance in Man (OMIM) (http://www.ncbi.nlm.nih.gov/omim/), NHGRI-GWAS Catalog (http://www.genome.gov/gwastudies/), GeneCards® (http://www.genecards.org/); and Diseasome (http://wifo5-03.informatik.uni-mannheim.de/diseasome/). All data were collected in an annotation-rich genomic context and/or in a gene-centric view to benefit from many relevant annotations and discriminations among DOISm candidate genes. Both validated and predicted data were collected and imported into our datasets. Sequences were downloaded from the human genome (GRCh37.p13) assembly using the UCSC Genome Browser, with its updated UCSC Genes track GRCh37/hg19, which has 82,960 total transcripts and a total number of 31,848 canonical human genes. The UCSC Genes track includes both protein-coding genes and non-coding RNA genes, being a set of gene predictions based on data from Ref-Seq, GenBank, CCDS, Rfam, and tRNA Genes track.

Annotations for diseases and gene associations were manually compiled from the literature and from ClinVar DB release 20130226 (www.ncbi.nlm.nih. gov/clinvar/), while phenotype association data were extracted from OMIM (http://omim.org/) for variation Phenotype/Feature schema. Ensembl Variation stores areas of the genome that differ among individual genomes ("variants", such as copy number variants or CNVs) and, where available, associated disease and phenotype information. Additionally, data from pertinent gene association studies, such as GIANT (Genetic Investigation of ANthropometric Traits) and MAGIC (Meta-Analyses of Glucose and Insulin-related traits Consortium), were added after Ensembl Variation DB and before being imported into our datasets. Furthermore, we benefited from the data provided by major research initiatives, such as International 1q Consortium, DIAGRAM (DIAbetes Genetics Replication and Meta-analysis) consortium and Metabochip [43].

## 2.3    Data Management Operations (Data Sharing, Extraction, and Verification)

The Vertebrate and Genome Annotation (VEGA), Human Vega release 53 (http://vega.sanger.ac.uk/Homo_sapiens/Info/) was used for gene structures presented in the merged human geneset shown in Ensembl release 73. This corresponds to Gencode release 18 (http://www.gencodegenes.org/). All annotations are from the Havana group (Human and Vertebrate Analysis and Annotation) at the Wellcome Trust Sanger Institute and GENCODE, after the ENCODE pilot project [28]. To visualize genomic data, specifically for detailed features of gene names, genes involved with disease, loci mutations and variable regions, we used either VEGA or Artemis browsers (http://sanger.ac.uk/resources/ software/artemis). In addition, Cytoscape 3.0 (www.cytoscape.org) [49] was used for network visualization and analysis, while illustrations were built using Venn's and/or Euler's diagrams.

## 2.4    *In Silico* Prediction of DOISm Confluence Genes

Similar to previous work [9,51], we performed in silico surveys with multi-relational data mining (MRDM) approaches to sequence pattern recognition and computer-assisted methods employing training algorithms [12,17] to mine the human genome for candidate DOISm genes and their features. We extended our search to epigenomic features, such as DNA methylation status (the chemical marks on the genome that control gene expression in response to environmental factors), nucleosome positioning, chromatin states, histone modifications (methylation and acetylation) and nutritional phenotypes [2]. Phenotypes result from the expression of genes as well as the influence of environmental factors such as diet, lifestyle, and environmental exposure and interactions between the two. Phenotyping is the measurement of a composite of individual characteristics. Thus, we used high-confidence predictions in the human genome to perform Gene Ontology (GO) (www.geneontology.org), gene functional classi-fication/clustering analyses using AmiGO v1.8 search engine in order to help better phenotype DOISm individual entities and their confluence. We searched public DBs and unpublished resources for each transcript associated with a lead DOISm-related gene polymorphism (or variant), and then, we identified the respective lead SNPs and then estimated the linkage disequilibrium (LD) between them, using 1000 Genomes data (www.1000genomes.org/data) to assess coincidence of the signals, as described in Morris et al. [43].

To set standards for each primary phenotype comprising DOISm, we used the National Library of Medicine (NLM)/Medline "Medical Subject Head-ings" (MeSH) Major Topics and Hierarchy. We started with Nutritional and Metabolic Diseases, following Metabolic Diseases and Nutrition Disorders, then Glucose/Lipid Metabolism Disorders and Overnutrition, and finally: (i) Diabetes Mellitus, Type 2 (68003924[uid]); (ii) Obesity, Abdominal; and (iii) Metabolic Syndrome X. Regarding Inflammation, we initially used the MeSH term Patho-logic Processes, but no MeSH term has been yet assigned to para-inflammation or inflammasome.

All policies that apply to the data stored in the European Genome-phenome Archive (EGA - www.ebi.ac.uk/ega), available at the European Bioinformatics Institute (EBI), were accommodated.

## 3 Results

### 3.1 The (up-to-date) Full Repertoire of DOISm-Related 1439 Candidates Genes

Previous works [9,11] have dealt with DOISm-related genes in a less detailed manner or on a smaller scale. Now, using a large-scale, MRDM gathering approach, we have selected a greater number of potential DOISm genes with significant clinical relevance to determine which ones clustered with known comorbidity-regulated genes (Fig. 1).

From our list, we chose to verify and further characterize *in silico* a few of them that could be subsequently tested for associations with each comorbidity in DOISm, as indicated, by combining several studies of genome-wide data integra-tion and meta-analyses, phenotypic data from multiple cohorts, SNP and CNV detection and analyses, and genetic refinement through re-sequencing and fine-mapping, among other approaches. To determine where our candidate genes (or their functional products) are expected to be located within a given human tar-get cell/tissue, we mined the literature for subcellular distribution and variation in a transcript-specific manner.

By identifying a large number of studies for each gene in a given cell or tissue localization, we could also begin to systematically match the cell or tissue profiles for each comorbidity in DOISm, if clear distinctions were identified. For instance, if we use pancreatic beta cells as a target, we can seek experimental evidence to support functional roles such as the FKN/CX3CR1 system in modulation of $\beta$-cell insulin secretory function [34] or microRNA (miRNA) sequence variants ("isomiRs") acting as regulatory hubs in a T2DM gene network [5]. Such assignment of gene-to-cell associations could enable the identification of underlying causes, such as differences in levels of protein expression in the cells or the presence of different transcripts in tissues affected by DOISm diseases. In addition, we employed pertinent SNP and CNV profiles to GWAS data, provided these observational studies correlated with the subjects' genomes for relevant disease phenotypic traits [13], particularly in DOISm.

We have compiled over 1439 unique human genes that were unambiguously associated with at least one of the four morbid conditions of DOISm, comprising a full repertoire to be mined (Dataset DS1). These genes were divided into subgroups (Supplementary Tables S1-S4) to comprise more amenable lists, displaying several gene features (gene product and architecture, sequence length and assembly, SNP content, CNVs frequency, etc.). We are aware that, for several complications that compromise detection of genetic associations, our repertoire is only a starting point for scrutiny large-scale variant association signals reflecting causal alleles and greater effects.

Through this exhaustive, curated panel of human genes implicated in DOISm (Fig. 1), we were able to catalogue the occurrence of genes involved with these phenotypes by identifying underlying genetic variation in order to assist in subtyping nutritional disease phenotypes in the combined etiology. Out of 1439 human genes related to DOISm (Dataset DS1), we found 769 potentially related to T2DM (Table S1) and 1059 potentially related to obesity (Table S2), whereas 288 were potentially related to inflammation (Table S3) and only 139 potentially were related to metS (Table S4).

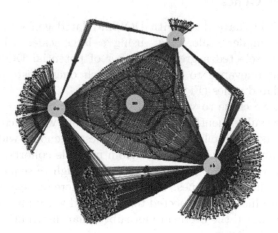

**Fig. 1.** Network diagram illustrating the relationship of 1439 genes associated with the combined DOISm (Diabetes, Obesity, Inflammation and metabolic Syndrome) etiology. The projection attempts to integrate the global 1439-gene dataset into functional annotations of one of each of the four DOISm comorbidities. Cytoscape Apps were used to perform advanced analysis and modeling to establish powerful visual mappings across these data.

Restricting our data to the DOI (diabetes, obesity and inflammation) gene confluence, we found only 78 human genes concomitantly related to these three disease-phenotypes (Table S6) but not directly associated with metS. This is an intriguing aspect to consider if we assume metS as the most natural confluence of DOISm comorbidities. Adding the 139 genes related to metS (Table S4) to this restricted group, we yielded 217 genes potentially confluent for any given two comorbidities and at the same time related to metS. The latter poses an interesting point due to estimating a proportional DOISm risk in groups undergoing any two comorbidities.

## 4  Discussion

### 4.1  DOISm Confluence: 217 Genes Likely to be Concomitantly Associated with all Four Comorbidities

Our curated panel of DOISm genes (Table S5) represents a platform to assist in subtyping complex phenotypes, such as the two distinct subtypes of obesity, referred to as "metabolically healthy" (MHO) and "metabolically unhealthy" obesity (MUHO), as theoretically modeled by a previous study [2]. MHO and MUHO may represent distinct subtypes of obesity, genetically predetermined to confer differing metabolic and CVD risks. Few research studies have yet considered in greater depth this phenotypic heterogeneity of obesity, and our work aims to fill in these gaps. Examples of such DOISm confluent genes are cholecystokinin (CCK), cadherin 18 (CDH18), myosin X (MYO10), the anchor protein 6 of AMPK (AKAP6), the neuronal PAS domain protein 3 (NPAS3), the fatty acid binding protein 3 (*fabp3*), and GPRC5B, among others [44,59]. For instance, the gene product of GPRC5B is a transmembrane lipoprotein with signaling function in adipose tissue, suggesting that it might be a key factor in adipose signaling pathways, which, in turn, is thought to be associated with obesity after DMT2 dietary complications [33]. Such an example illustrates how the novel approach of gene confluence might increase understanding of the fine-tuning regulation believed to underlie the DOISm combined etiology.

Regarding fractalkine (FKN), a chemokine that mediates cell-to-cell adhesion, inflammation and communication in hepatocytes [34,52], its receptor CX3CR1 (which is also known as GPR13) is thought to evoke FKN gene modulation of monocyte attachment to adipocytes. Some FKN SNPs were recently associated with an increased incidence of metS and T2DM [34], providing additional reasons to include this gene as a convincing candidate for DOISm confluence.

Obesity is associated with gene modifications involved in lipid metabolism and energy expenditure, whereas weight loss, such as that induced by gastric bypass surgery, can normalize expression of these important genes. A recent report identified two genes – PGC-1$\alpha$ and PDK4 – where changes in DNA methylation took place in obesity and were restored to normal after weight loss [6]. Cholecystokinin (CCK), a gastrointestinal hormone with intrinsic pancreatic and gall bladder functional co-associations [25], is another gene product for which nervous system communication in satiety switch is potentially valuable for DOISm confluence. CCK has been shown to interact with calcineurin, which activates the transcription factors NFAT1-3, the stimulation of which cause pancreatic hypertrophy. All these gene products are likely to be involved in severe epigenetic alterations due to their diet-environmental influences. Therefore, the inclusion of such genes in a DOISm-related repertoire, from an analytical standpoint, reinforces the notion that epigenetic alterations induced by weight loss are another key factor of interest for the management of both obesity and metabolic changes that follows, in other words, the DOISm combined etiology.

## 4.2   Inflammatory Roles of the Imprinted DOISm-Related Genes: Towards Epigenome-Wide Association Studies (EWAS)

It is becoming increasingly evident that obesity and diabetes are linked to many pro-inflammatory and other epigenetic conditions. The origin of phenotypic variation remains one of the main challenges of contemporary biology. Because the epigenetic state of an organism (or 'epigenome') incorporates a landscape of complex and plastic molecular events that may underlie the'missing link' that integrates genotype with phenotype [22], we sought to address the epigenomic influences on DOISm through genomic imprinting and the human imprintome [10,51]. Genomic imprinting alters gene expression without altering DNA sequence. Several nutritional diseases are caused by environmental and physical imprinting disorders [7,23,46], and here, we focus on diabetes, obesity and CVDs because there remains a knowledge gap on how dosage changes or the absence of genomic imprinting might induce these disease phenotypes [7].

As human complex disease, such as those comprising DOISm, we are still unaware of the proportion of phenotypic variation attributable to inter-individual epigenomic variation [47]. This setback might begin to be elucidated by large-scale, systematic epigenomic equivalents of GWAS – the so-called epigenome-wide association studies (EWAS). EWAS mainly consists of profiling DNAm patterns as a function of time, tissues and developmental stages. Furthermore, EWAS associations can be causal as well as consequential for the phenotype in question [47].

Assuming imprinted genes are a fairly good source of epigenetic traceable records and are likely amenable for future EWAS approach [47], we have investigated the human imprintome [51] for relevant features within the DOISm context. Consider, for example, the KLF14 and GNAS genes, as seen on Table S7 with twenty-four imprinted genes predicted to be DOISm related (Fig. 2). KLF14 and GNAS have maternal and paternal expression, respectively. We found a total number of 167 and 1117 SNPs, 142 and 25 SNP flanking genes, 0 and 951 SNPs within introns, and 25 and 141 SNPs within exons for each gene, respectively (Table S8). Such high polymorphic imprinted genes, associated with a recent report on KLF4 [16,53], can be a new, unanticipated genetic roadmap for exploring obesity and T2DM with the additional feature of tracing parental origin.

**Fig. 2.** Ideogram illustrating distribution and localization of a few genes associated with the DOISm (Diabetes, Obesity, Inflammation and metabolic Syndrome) combined etiology. Each gene is represented by their symbol and relative position on its respective autosome (no distribution shown on sexual chromosomes).

Twenty-four genes are written in red, nineteen imprinted and five predicted to be imprinted in the human genome, according to Silva-Santiago et al. [51].

Another point of interest is the inflammation response coupled with imprinted (NLRP2) and predicted-to-be imprinted (TMEM88 and VENTX) genes, as listed on Table S7. NLRP2 is a member of the NOD-like receptor family (NLR - nucleotide-binding domain and leucine-rich repeat protein), containing the *pyrin* 2-protein domain, 17 transcripts and several aliases ("NACHT, *leucine rich repeat and PYD containing* 2", NALP2). NLRPs have gained importance as major members of a protein family – the NALPs – implicated in caspase-1 activation by Toll-like receptors (TLRs) during the cell response to both infection and inflammation [8,50].

One of the most recently identified signaling pathways is the "inflammasome," which activation appears to affect many metabolic disorders, is a multi-protein complex composed of NLRP3 (nucleotide-binding domain and leucine-rich repeat protein 3), ASC (apoptosis-associated speck-like protein containing a caspase recruitment domain, CARD), and procaspase-1 (reviewed by Benetti *et al.* [8]

and Schroder *et al.* [8]. The assembly of inflammasomes relies on NALP proteins to activate inflammatory caspases [8]; hence, their strategic situation fosters our interest. The assembly of different inflammasomes elicits a common downstream cascade, namely the activation of inflammatory caspases, which all have a CARD, followed by a domain containing the catalytic residue cysteine. The main substrates of caspases are cytokines (such as pro-IL-1$\beta$, pro-IL-18 and eventually pro-IL-33), which are cleaved to produce their active and secreted form. IL-18, a gene linked to diseases with a sex-specific prevalence, is subject to complex, regional, and sex-specific parental effects in the brain [24], another probable case of interaction with imprinting balance.

In addition, IL-1$\beta$ secretion by the NLRP3 inflammasome [8,53] is triggered by high extracellular glucose levels in beta-cells. We must recall that elevated IL-1$\beta$ is a risk factor for T2DM, contributing to IR. Thus, by functioning as a sensor for metabolic stress [8], such as hyperglycemia, the NLRP3 inflammasome likely contributes to the pathogenesis of both obesity and T2DM. As an imprinted member of the NLR family, NLRP2 offers novel insights into DOISm gene confluence due to its potential as a future target for EWAS and an important parameter in either NLRP1 or NLRP3 activation. Moreover, growing evidence has suggested the convergence of a combination of events, escalating the pro-inflammatory state [40], which is likely relevant in the DOISm context.

# 5 Concluding Remarks

In this work, we have performed in silico surveys on the characterization of heterogeneity concerning 217 relevant genes that might be related to a confluence of DOISm-combined etiology. Extending the idea [13] regarding "the end of the beginning in the search for obesity predisposition genes," we support the same view; it will be a matter of time until the wider characterization of DOISm-related genes and their functional roles will shorten the development of new drugs, targeted treatment and diagnostic tools for this combined etiology. Nevertheless, much work is needed to transfer nutrigenomics knowledge [32] from bench to bedside (or to table, if we may dare). Large-scale systems biology [46] meta-analyses of GWAS and EWAS data can improve interpretations of genetic variations and genetic risk factors [21,48]. Here, we have demonstrated how a simple, stand-alone bioinformatics framework built upon four complex phenotypes (DOISm) may serve as a guide for further study of DOISm comorbidities, which are rapidly rising globally, driven by population growth and ageing, in addition to sedentary lifestyles and other factors [21,32,46,48]. We must prepare to detect DOISm early and manage it properly if we hope to avoid its deleterious sequelae.

# 6 Authors' Contributions

The study was designed by DMO, APMB and EDSP. SCS-S was responsible for seeking legal and institutional authorizations. Data were acquired by APMB,

EDSP, MMS, ACLP, and JFDSJr. All authors were responsible for analyses and interpretation of the data. DMO and APMB drafted the manuscript. DMO, APMB, EDSP and JFDSJr wrote the final version of manuscript. All authors reviewed and approved the final version of the manuscript.

**Acknowledgements.** The authors are funded by grants and fellowships from the following Brazilian agencies: CNPq (Conselho Nacional de Desenvolvimento Científico e Tecnológico), CAPES (Coordenação de Aperfeiçoamento de Pessoal de Nível Superior) and FUNCAP (Fundação Cearense de Apoio ao Desenvolvimento Científico e Tecnológico).

# References

1. Agrawal, S., Dimitrova, N., Nathan, P., Udayakumar, K., Lakshmi, S.S., Sriram, S., Manjusha, N., Sengupta, U.: T2D-Db: an integrated platform to study the molecular basis of Type 2 diabetes. BMC Genomics **9**, 320 (2008)
2. Alam, I., Ng, T.P., Larbi, A.: Does inflammation determine whether obesity is metabolically healthy or unhealthy? The aging perspective. Mediators Inflamm. **2012**, 1–15 (2012)
3. Alberti, K.G.M.M., Eckel, R.H., Grundy, S.M., Zimmet, P.Z., Cleeman, J.I., Donato, K.A., Fruchart, J.C., James, W.P.T., Loria, C.M., Smith, S.C.: Harmonizing the metabolic syndrome. Circulation **120**(16), 1640–1645 (2009)
4. Astrup, A., Finer, N.: Redefining type 2 diabetes: 'diabesity' or 'obesity' dependent diabetes mellitus'? Obesity Rev. **1**(2), 57–59 (2000)
5. Baran-Gale, J., Fannin, E.E., Kurtz, C.L., Sethupathy, P.: Beta cell 5'-shifted isomiRs are candidate regulatory hubs in type 2 diabetes. PloS One **8**(9), e73240 (2013)
6. Barres, R., Kirchner, H., Rasmussen, M., Yan, J., Kantor, F.R., Krook, A., Näslund, E., Zierath, J.R.: Weight loss after gastric bypass surgery in human obesity remodels promoter methylation. Cell Rep. **3**(4), 1020–1027 (2013)
7. Bartolomei, M.S., Ferguson-Smith, A.C.: Mammalian genomic imprinting. Cold Spring Harb. Perspect. Biol. **3**(7), a002592 (2011)
8. Benetti, E., Chiazza, F., Patel, N.S.A., Collino, M.: The NLRP3 inflammasome as a novel player of the intercellular crosstalk in metabolic disorders. Mediators Inflamm. **2013**, Article ID 678627 (2013)
9. Bezerra, A.P.M., Martin, C.P., Silva-Santiago, S.C., Farias, K.M., Penha, E.D.S., Oliveira, D.M.: Predictive analyses of human genes concomitantly involved in the etiology of Diabetes, Obesity, Inflammation & Metabolic Syndrome - Sm. In: 7th Congress of the International Society of Nutrigenetics/Nutrigenomics (ISNN), Quebec City, October 5–8, J. Nutrigenet. Nutrigenom. **6**(4–5), 201–255(253) (2013)
10. Bezerra, A.P.M., Silva-Santiago, S.C., Oliveira, D.M., Vasconcelos, E.J.R., Silva, M.M., Pacheco, A.C.L.: In silico analyses of 24 human imprinted genes associated to diabesin triad (diabetes, obesity and inflammation): In search for a potential diabesin risk. J. Nutrigenet. Nutrigenom. **5**(4–5), 222 (2012)

11. Bezerra, A.P.M., Silva-Santiago, S.C., Vasconcelos, E.J.R., Pacheco, A.C.L., Silva, M.M., Oliveira, D.M.: In silico analyses of human genes involved in the diabesin triad (diabetes, obesity, inflammation), their association to imprinted genes. In: Proceedings of the 6th Congress of the International Society of Nutrigenetics/Nutrigenomics (ISNN), São Paulo, Brazil, vol. 5, pp. 171–302 (2012). J. Nutrigenetics Nutrigenomics (2012)
12. Brideau, C.M., Eilertson, K.E., Hagarman, J.A., Bustamante, C.D., Soloway, P.D.: Successful computational prediction of novel imprinted genes from epigenomic features. Mol. Cell. Biol. **30**(13), 3357–3370 (2010)
13. Choquet, H., Meyre, D.: Molecular basis of obesity: current status and future prospects. Curr. Genomics **12**, 169–179 (2011)
14. Corsten, M.F., Papageorgiou, A., Verhesen, W., Carai, P., Lindow, M., Obad, S., Summer, G., Coort, S.L.M., Hazebroek, M., Van Leeuwen, R., Gijbels, M.J.J., Wijnands, E., Biessen, E.A.L., De Winther, M.P.J., Stassen, F.R.M., Carmeliet, P., Kauppinen, S., Schroen, B., Heymans, S.: MicroRNA profiling identifies MicroRNA-155 as an Adverse Mediator of Cardiac Injury and Dysfunction during Acute Viral MyoCarditis. Circul. Res. **111**(4), 415–425 (2012)
15. Dai, H.-J., Wu, J.C.-Y., Tsai, R.T.-H., Pan, W.-H., Hsu, W.-L.: T-HOD: a literature-based candidate gene database for hypertension, obesity and diabetes. Database **2013**, bas061, January 2013
16. Ding, Q., Gupta, R.M., Raghavan, A., Musunuru, K.: Abstract 70: KLF14 is a novel regulator of human metabolism. Arterioscler. Thromb. Vasc. Biol. **34**(Suppl 1), A70–A70 (2014)
17. Diniz, M.C., Pacheco, A.C.L., Girão, K.T., Araujo, F.F., Walter, C.A., Oliveira, D.M.: The tetratricopeptide repeats (TPR)-like superfamily of proteins in Leishmania spp., as revealed by multi-relational data mining. Pattern Recogn. Lett. **31**(14), 2178–2189 (2010)
18. Doris, P.A.: Hypertension genetics, single nucleotide polymorphisms, and the common disease: common variant hypothesis. Hypertension **39**(2), 323–331 (2002)
19. Emanuela, F., Grazia, M., De Marco, R., Paola, L.M., Giorgio, F., Marco, B.: Inflammation as a link between obesity and metabolic syndrome. J. Nutr. Metab. **2012**, Article ID 476380, January 2012
20. Fantuzzi, G.: Adipose tissue, adipokines, and inflammation. J. Allergy Clin. Immunol. **115**(5), 911–919 (2005)
21. Fenech, M., El Sohemy, A., Cahill, L., Ferguson, L.R., French, T.-A.C., Tai, E.S., Milner, J., Koh, W.-P., Xie, L., Zucker, M., Buckley, M., Cosgrove, L., Lockett, T., Fung, K.Y.C., Head, R.: Nutrigenetics and nutrigenomics: viewpoints on the current status and applications in nutrition research and practice. J. Nutrigenet. Nutrigenom. **4**(2), 69–89 (2011)
22. Finer, S., Holland, M.L., Nanty, L., Rakyan, V.K.: The hunt for the epiallele. Environ. Mol. Mutagen. **52**, 1–11 (2011)
23. Girardot, M., Cavaillé, J., Feil, R.: Small regulatory RNAs controlled by genomic imprinting and their contribution to human disease. Epigenetics **7**(12), 1341–1348 (2012)
24. Gregg, C., Zhang, J., Butler, J.E., Haig, D., Dulac, C.: Sex-specific parent-of-origin allelic expression in the mouse brain. Science **329**(5992), 682–685 (2010)
25. Gurda, G.T., Guo, L., Lee, S.-H., Molkentin, J.D., Williams, J.A.: Cholecystokinin activates pancreatic calcineurin-NFAT signaling in vitro and in vivo. Mol, Biol. Cell **19**(1), 198–206 (2008)
26. Haffner, S., Taegtmeyer, H.: Epidemic obesity and the metabolic syndrome. Circulation **108**(13), 1541–1545 (2003)

27. Hale, P.J., López-Yunez, A.M., Chen, J.Y.: Genome-wide meta-analysis of genetic susceptible genes for Type 2 Diabetes. BMC Syst. Biol. **6**(Suppl 3), S16 (2012)
28. Harrow, J., Frankish, A., Gonzalez, J.M., Tapanari, E., Diekhans, M., Kokocinski, F., Aken, B.L., Barrell, D., Zadissa, A., Searle, S., Barnes, I., Bignell, A., Boychenko, V., Hunt, T., Kay, M., Mukherjee, G., Rajan, J., Despacio-Reyes, G., Saunders, G., Steward, C., Harte, R., Lin, M., Howald, C., Tanzer, A., Derrien, T., Chrast, J., Walters, N., Balasubramanian, S., Pei, B., Tress, M., Rodriguez, J.M., Ezkurdia, I., Van Baren, J., Brent, M., Haussler, D., Kellis, M., Valencia, A., Reymond, A., Gerstein, M., Guigó, R., Hubbard, T.J.: GENCODE: the reference human genome annotation for the ENCODE project. Genome Res. **22**(9), 1760–1774 (2012)
29. Horton, R.: Non-communicable diseases: 2015 to 2025. The Lancet **381**(9866), 509–510 (2013)
30. Isomaa, B.O., Almgren, P., Tuomi, T.: Cardiovascular morbidity and mortality associated with the metabolic syndrome. Diabetes Care **24**(4), 683–689 (2001)
31. Jain, P., Vig, S., Datta, M., Jindel, D., Mathur, A.K., Mathur, S.K., Sharma, A.: Systems biology approach reveals genome to phenome correlation in type 2 diabetes. PloS One **8**(1), e53522 (2013)
32. Kang, J.X.: Nutrigenomics and systems biology. J. Nutrigenet. Nutrigenom. **5**(6), I–II (2012)
33. Kim, Y.-J., Sano, T., Nabetani, T., Asano, Y., Hirabayashi, Y.: GPRC5B activates obesity-associated inflammatory signaling in adipocytes. Sci. Signal. **5**(251), ra85 (2012)
34. Lee, Y.S., Morinaga, H., Kim, J.J., Lagakos, W., Taylor, S., Keshwani, M., Perkins, G., Dong, H., Kayali, A.G., Sweet, I.R., Olefsky, J.: The fractalkine/CX3CR1 system regulates β cell function and insulin secretion. Cell **153**(2), 413–425 (2013)
35. Lill, C.M., Bertram, L.: Developing the "next generation" of genetic association databases for complex diseases. Hum. Mutat. **33**(9), 1366–1372 (2012)
36. Lim, J.E., Hong, K.-W., Jin, H.-S., Kim, Y.S., Park, H.K., Oh, B.: Type 2 diabetes genetic association database manually curated for the study design and odds ratio. BMC Med. Inf. Dec. Mak. **10**(1), 76 (2010)
37. Lin, Z., Tian, H., Lam, K.S.L., Lin, S., Hoo, R.C.L., Konishi, M., Itoh, N., Wang, Y., Bornstein, S.R., Xu, A., Li, X.: Adiponectin mediates the metabolic effects of FGF21 on glucose homeostasis and insulin sensitivity in mice. Cell Metabol. **17**(5), 779–789 (2013)
38. Little, J., Higgins, J.P.T., Ioannidis, J.P.A., Moher, D., Gagnon, F., Von Elm, E., Khoury, M.J., Cohen, B., Davey-Smith, G., Grimshaw, J., Scheet, P., Gwinn, M., Williamson, R.E., Zou, G.Y., Hutchings, K., Johnson, C.Y., Tait, V., Wiens, M., Golding, J., Van Duijn, C., McLaughlin, J., Paterson, A., Wells, G., Fortier, I., Freedman, M., Zecevic, M., King, R., Infante-Rivard, C., Stewart, A., Birkett, N.: STrengthening the REporting of genetic association studies (STREGA)- An extension of the STROBE statement. Genet. Epidemiol. **33**(7), 581–598 (2009)
39. Macdonald, W.A.: Epigenetic mechanisms of genomic imprinting: common themes in the regulation of imprinted regions in mammals, plants, and insects. Genet. Res. Int. **2012**, Article ID 585024 (2012)
40. McArdle, M.A., Finucane, O.M., Connaughton, R.M., McMorrow, A.M., Roche, H.M.: Mechanisms of obesity-induced inflammation and insulin resistance: insights into the emerging role of nutritional strategies. Front. Endocrinol. **4**, 1–23 (2013)
41. Moore, L.D., Le, T., Fan, G.: DNA methylation and its basic function. Neuropsychopharmacol. Rev. **38**(1), 23–38 (2013)

42. Moore, T., Haig, D.: Genomic imprinting in mammalian development: a parental tug-of-war. Trends Genet. **7**(2), 45–49 (1991)
43. Morris, A.P., Voight, B.F., Teslovich, T.M., Ferreira, T., Segrè, A.V., Steinthorsdottir, V., Strawbridge, R.J., Khan, H., Grallert, H., Mahajan, A., Prokopenko, I., Kang, H.M., Dina, C., Esko, T., Fraser, R.M., Kanoni, S., Kumar, A., Lagou, V., Langenberg, C., Luan, J., Lindgren, C.M., Müller-Nurasyid, M., Pechlivanis, S., Rayner, N.W., Scott, L.J., Wiltshire, S., Yengo, L., Kinnunen, L., Rossin, E.J., Raychaudhuri, S., Johnson, A.D., Dimas, A.S., Loos, R.J.F., Vedantam, S., Chen, H., Florez, J.C., Fox, C., Liu, C.-T., Rybin, D., Couper, D.J., Kao, W.H.L., Li, M., Cornelis, M.C., Kraft, P., Sun, Q., van Dam, R.M., Stringham, H.M., Chines, P.S., Fischer, K., Fontanillas, P., Holmen, O.L., Hunt, S.E., Jackson, A.U., Kong, A., Lawrence, R., Meyer, J., Perry, J.R.B., Platou, C.G.P., Potter, S., Rehnberg, E., Robertson, N., Sivapalaratnam, S., Stančáková, A., Stirrups, K., Thorleifsson, G., Tikkanen, E., Wood, A.R., Almgren, P., Atalay, M., Benediktsson, R., Bonnycastle, L.L., Burtt, N., Carey, J., Charpentier, G., Crenshaw, A.T., Doney, A.S.F., Dorkhan, M., Edkins, S., Emilsson, V., Eury, E., Forsen, T., Gertow, K., Gigante, B., Grant, G.B., Groves, C.J., Guiducci, C., Herder, C., Hreidarsson, A.B., Hui, J., James, A., Jonsson, A., Rathmann, W., Klopp, N., Kravic, J., Krjutškov, K., Langford, C., Leander, K., Lindholm, E., Lobbens, S., Männistö, S., Mirza, G., Mühleisen, T.W., Musk, B., Parkin, M., Rallidis, L., Saramies, J., Sennblad, B., Shah, S., Sigurðsson, G., Silveira, A., Steinbach, G., Thorand, B., Trakalo, J., Veglia, F., Wennauer, R., Winckler, W., Zabaneh, D., Campbell, H., van Duijn, C., Uitterlinden, A.G., Hofman, A., Sijbrands, E., Abecasis, G.R., Owen, K.R., Zeggini, E., Trip, M.D., Forouhi, N.G., Syvänen, A.-C., Eriksson, J.G., Peltonen, L., Nöthen, M.M., Balkau, B., Palmer, C.N.A., Lyssenko, V., Tuomi, T., Isomaa, B., Hunter, D.J., Qi, L., Shuldiner, A.R., Roden, M., Barroso, I., Wilsgaard, T., Beilby, J., Hovingh, K., Price, J.F., Wilson, J.F., Rauramaa, R., Lakka, T.A., Lind, L., Dedoussis, G., Njølstad, I., Pedersen, N.L., Khaw, K.-T., Wareham, N.J., Keinanen-Kiukaanniemi, S.M., Saaristo, T.E., Korpi-Hyövälti, E., Saltevo, J., Laakso, M., Kuusisto, J., Metspalu, A., Collins, F.S., Mohlke, K.L., Bergman, R.N., Tuomilehto, J., Boehm, B.O., Gieger, C., Hveem, K., Cauchi, S., Froguel, P., Baldassarre, D., Tremoli, E., Humphries, S.E., Saleheen, D., Danesh, J., Ingelsson, E., Ripatti, S., Salomaa, V., Erbel, R., Jöckel, K.-H., Moebus, S., Peters, A., Illig, T., de Faire, U., Hamsten, A., Morris, A.D., Donnelly, P.J., Frayling, T.M., Hattersley, A.T., Boerwinkle, E., Melander, O., Kathiresan, S., Nilsson, P.M., Deloukas, P., Thorsteinsdottir, U., Groop, L.C., Stefansson, K., Hu, F., Pankow, J.S., Dupuis, J., Meigs, J.B., Altshuler, D., Boehnke, M., McCarthy, M.I.: Large-scale association analysis provides insights into the genetic architecture and pathophysiology of type 2 diabetes. Nat. Genet. **44**(9), 981–990 (2012)
44. Ng, M.C.Y., Tam, C.H.T., So, W.Y., Ho, J.S.K., Chan, A.W., Lee, H.M., Wang, Y., Lam, V.K.L., Chan, J.C.N., Ma, R.C.W.: Implication of genetic variants near NEGR1, SEC16B, TMEM18, ETV5/DGKG, GNPDA2, LIN7C/BDNF, MTCH2, BCDIN3D/FAIM2, SH2B1, FTO, MC4R, and KCTD15 with obesity and type 2 diabetes in 7705 Chinese. J. Clin. Endocrinol. Metabol. **95**(5), 2418–2425 (2010)
45. Ouchi, N., Parker, J.L., Lugus, J.J., Walsh, K.: Adipokines in inflammation and metabolic disease. Nat. Rev. Immunol. **11**(2), 85–97 (2011)
46. Phillips, C.M.: Nutrigenetics and metabolic disease: current status and implications for personalised nutrition. Nutrients **5**(1), 32–57 (2013)
47. Rakyan, V.K., Down, T.A., Balding, D.J., Beck, S.: Epigenome-wide association studies for common human diseases. Nat. Rev. Genet. **12**(8), 529–541 (2011)

48. Richardson, V.R., Smith, K.A., Carter, A.M.: Adipose tissue inflammation: Feeding the development of type 2 diabetes mellitus. Immunobiology **218**(12), 1497–1504 (2013)
49. Saito, R., Smoot, M.E., Ono, K., Ruscheinski, J., Wang, P.-L., Lotia, S., Pico, A.R., Bader, G.D., Ideker, T.: A travel guide to Cytoscape plugins. Nat. Methods **9**(11), 1069–1076 (2012)
50. Schroder, K., Tschopp, J.: The Inflammasomes. Cell **140**(6), 821–832 (2010)
51. Silva-Santiago, S.C., Pacheco, C., Rocha, T.C.L., Brasil, S.M.V., Pacheco, A.C.L., Silva, M.M., Araújo, F.F., de Vasconcelos, E.J.R., de Oliveira, D.M.: The linked human imprintome v1.0: over 120 genes confirmed as imprinted impose a major review on previous censuses. Int. J. Data Mining Bioinform. **10**(3), 329–356 (2014)
52. Sirois-Gagnon, D., Chamberland, A., Perron, S., Brisson, D., Gaudet, D., Laprise, C.: Association of common polymorphisms in the fractalkine receptor (CX3CR1) with obesity. Obesity **19**(1), 222–227 (2011)
53. Small, K.S., Hedman, A.K., Grundberg, E., Nica, A.C., Thorleifsson, G., Kong, A., Thorsteindottir, U., Shin, S.-Y., Richards, H.B., Soranzo, N., Ahmadi, K.R., Lindgren, C.M., Stefansson, K., Dermitzakis, E.T., Deloukas, P., Spector, T.D., McCarthy, M.I.: Identification of an imprinted master trans regulator at the KLF14 locus related to multiple metabolic phenotypes. Nat. Genet. **43**(6), 561–564 (2011)
54. Szalowska, E., Dijkstra, M., Elferink, M.G., Weening, D., de Vries, M., Bruinenberg, M., Hoek, A., Roelofsen, H., Groothuis, G.M., Vonk, R.J.: Comparative analysis of the human hepatic and adipose tissue transcriptomes during LPS-induced inflammation leads to the identification of differential biological pathways and candidate biomarkers. BMC Med. Genomics **4**(1), 71 (2011)
55. Vandanmagsar, B., Youm, Y.-H., Ravussin, A., Galgani, J.E., Stadler, K., Mynatt, R.L., Ravussin, E., Stephens, J.M., Dixit, V.D.: The NLRP3 inflammasome instigates obesity-induced inflammation and insulin resistance. Nat. Med. **17**(2), 179–188 (2011)
56. Jason, B.: Wolf: Evolution of genomic imprinting as a coordinator of coadapted gene expression. PNAS **110**(13), 5085–5090 (2013)
57. World Health Organization. World health statistics 2015 (2015)
58. Yang, Z., Yang, J., Liu, W., Wu, L., Xing, L., Wang, Y., Fan, X., Cheng, Y.: T2D@ZJU: a knowledgebase integrating heterogeneous connections associated with type 2 diabetes mellitus. Database **2013**, bat052 (2013)
59. Zhang, Y., Kent, J.W., Lee, A., Cerjak, D., Ali, O., Diasio, R., Olivier, M., Blangero, J., Carless, M.A., Kissebah, A.H.: Fatty acid binding protein 3 (fabp3) is associated with insulin, lipids and cardiovascular phenotypes of the metabolic syndrome through epigenetic modifications in a Northern European family population. BMC Med. Genomics **6**(9), January 2013

# Studying the Herd Immunity Effect
# of the Varicella Vaccine in the Community
# of Valencia, Spain

A. Díez-Gandía[1], R.-J. Villanueva[2(✉)], J.-A. Moraño[2], L. Acedo[2],
J. Mollar[1], and J. Díez-Domingo[3]

[1] Hospital La Fe, Valencia, Spain
[2] Instituto de Matemática Multidisciplinar,
Universitat Politecnica de Valencia, Valencia, Spain
rjvillan@imm.upv.es
[3] FISABIO-Salud Pública, Valencia, Spain

**Abstract.** In 2013, the Spanish Agency of Medicines blocked the varicella vaccine distribution based on a partial coverage of the vaccine that
hypothetically could induce an increase of cases in adults: as non vaccinated children reach adulthood without having had contact with the
virus in case the herd immunity stopped the virus circulation, or a hypothetical loss of vaccine protection in the long run. Also, this measure
wanted to avoid increasing the number of cases of herpes zoster in adults.
In this paper we develop a mathematical model to study the transmission dynamics of varicella in order to assess the impact of the partial
coverage of the vaccination program. This is of paramount importance
because, from the Public Health point of view, the herd immunity may
be an undesirable effect of the partial vaccination due to that varicella
and/or herpes zoster in adults use to be severe.

**Keywords:** Varicella Zoster Virus (VZV) · Mathematical model · Herd
immunity

## 1 Introduction

Chickenpox is a highly contagious disease caused by the Varicella Zoster Virus
(VZV) [1–3]. This virus is the responsible for both Chickenpox and Herpes Zoster
and it has a very high prevalence in populations all around the world. Chickenpox affects mostly children and, in most cases, it is a benign infection, and
occasionally may complicate and with a low death rate. However, it is more
severe in adults.

The implementation of vaccine programs with a live attenuated OKA–varicella virus strain has decreased the impact of the disease by a factor 10 in the USA
and reduced the mortality. In 2004 varicella vaccine was licensed in Spain. In two
of the autonomous communities, Madrid and Navarra, a universal vaccination

© Springer International Publishing Switzerland 2016
F. Ortuño and I. Rojas (Eds.): IWBBIO 2016, LNBI 9656, pp. 38–46, 2016.
DOI: 10.1007/978-3-319-31744-1_4

of infants was started. The benefits of the programs are reported. In Madrid, in spite of the clear benefits, the vaccine was withdrawn for political reasons.

In the rest of the Spanish autonomous communities, the program included vaccinating 11–12 years individuals that had no history of varicella or varicella vaccination, with the objective of decreasing varicella cases in adults.

Apart from that program, paediatricians recommended the vaccine (Spanish Paediatric Association recommendation), and parents paid for a two-dose schedule at 12 months and 3 years of age. Vaccination coverage in the young ages reached 50–70 % depending on the Communities.

In 2013, the Spanish Agency of Medicines blocked the vaccine distribution based on a partial coverage of the vaccine that hypothetically could induce an increase of cases in adults, as non vaccinated children reach adulthood without having had contact with the virus (in case the herd immunity stopped the virus circulation), or a hypothetical loss of vaccine protection in the long run. Also, this measure wanted to avoid increasing the number of cases of herpes zoster in adults. Now, the only existing vaccination program is the vaccination of 11–12 years old individuals that had no history of varicella with a coverage of 90 %.

Therefore we aimed to develop a mathematical model to study the transmission dynamics of varicella that allow to assess the impact of this partial coverage of the vaccination program. Note that, from the Public Health point of view, the herd immunity may be an undesirable effect of the partial vaccination because varicella and/or herpes zoster in adults use to be severe.

The paper is organized as follows. In Sect. 2, we present the data of varicella incidence. In Sect. 3, we build the model to study the transmission dynamics of varicella per age group. Also, we calibrate the model and obtain interesting data about the disease from the results of the model. In Sect. 4, we simulate some vaccination strategies in order to assess the impact of their partial coverage and find out if there is herd immunity. Finally, we finish the paper with some conclusion.

## 2   Data

The data on the number of infected individuals per week was obtained from a report from Royal College of General Practitioners entitled: *New RSC Communicable and Respiratory Disease Report for England & Wales, 36/2014* [4]. They averaged over a period of 10 years prior to the introduction of the vaccine to get the data in Fig. 1.

We are going to assume these figures for the Community of Valencia (Spain). In fact, they are very similar. An example is the varicella report bulletin of the Community of Valencia 2012 [7] (Fig. 2). But the similarity can be checked against other years [5, 6, 8, 9].

## 3   Mathematical Modeling

The basic model we propose is an $S - L_1 - L_2 - I - R - L_1$, i.e., a Susceptible-Infected-Recovered or SIR model but with two intermediate latent stages and

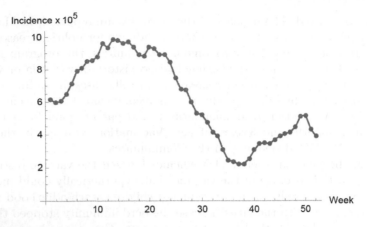

**Fig. 1.** Reported individuals infected with varicella per $10^5$ as a function of the week averaged over a period of ten years in UK. The epidemiological behavior of the disease is clearly seasonal.

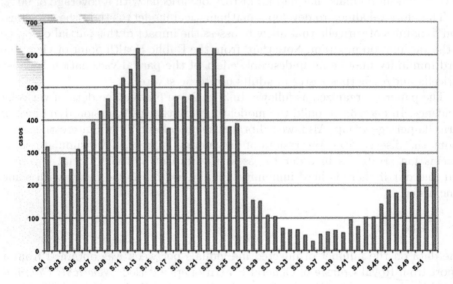

**Fig. 2.** Weekly reported cases in the Community of Valencia in 2012 [7]. Note the similarity with the UK cases.

the possibility to become latent after being recovered. Every individual has an assigned state with respect to the virus:

– Immune: a child under six months of age who is immune to the disease as a consequence of the maternal protection.
– Susceptible: a healthy individual who could become infected.
– Latent 1 week: an infected individual during the first week. In this state the individual cannot be the source of contagion. This is an incubation period [2].

- Latent 2 week: the same as above but during the second week.
- Infectious: After the two latent states the infected individual develops the disease and the characteristic pustules and he/she becomes contagious. This state lasts another week.
- Recovered: These individuals have developed the disease and, after clearing the infection, they have almost total lifelong immunity. Reinfection may occur in some cases but it is very rare.

Ours is also an age-group model in which we take into account the following age-groups:

- $G(1)$: 0–6 months.
- $G(2)$: 6–12 months.
- $G(3)$: 1–3 years.
- $G(4)$: 3–12 years.
- $G(5)$: older than 12 years.

As we are interested in a model with constant population the death rate for age-group $G(5)$ is chosen accordingly.

In order to consider the vaccination, we are going to include the vaccinated states into the model. Then, the model that describes the weekly transmission dynamics of varicella in the Community of Valencia is given by the following system of difference equations ($t$ in weeks):

$$Imm(t+1) = Imm(t) + \mu - c_1 Imm(t) - d_1 Imm(t),$$

$$
\begin{aligned}
S_2(t+1) &= S_2(t) + c_1 Imm(t) - c_2 S_2(t) - d_2 S_2(t) - \alpha_2 S_2(t) I(t),\\
L_2^{1w}(t+1) &= -c_2 L_2^{1w}(t) - d_2 L_2^{1w}(t) + \nu_2 R_2(t) + \alpha_2 S_2(t) I(t),\\
L_2^{2w}(t+1) &= L_2^{1w}(t) - c_2 L_2^{2w}(t) - d_2 L_2^{2w}(t),\\
I_2(t+1) &= L_2^{2w}(t) - c_2 I_2(t) - d_2 I_2(t),\\
R_2(t+1) &= I_2(t) + R_2(t) - \nu R_2(t) - c_2 R_2(t) - d_2 R_2(t),
\end{aligned}
$$

$$
\begin{aligned}
S_3(t+1) &= S_3(t) + (1-\rho_1)c_2 S_2(t) - c_3 S_3(t) - d_3 S_3(t) - \alpha_3 S_3(t) I(t),\\
L_3^{1w}(t+1) &= c_2 L_2^{1w}(t) - c_3 L_3^{1w}(t) - d_3 L_3^{1w}(t) + \nu_3 R_3(t) + \alpha_3 S_3(t) I(t),\\
L_3^{2w}(t+1) &= L_3^{1w}(t) + c_2 L_2^{2w}(t) - c_3 L_3^{2w}(t) - d_3 L_3^{2w}(t),\\
I_3(t+1) &= L_3^{2w}(t) + c_2 I_2(t) - c_3 I_3(t) - d_3 I_3(t),\\
R_3(t+1) &= I_3(t) + R_3(t) - \nu_3 R_3(t) + c_2 R_2(t) - c_3 R_3(t) - d_3 R_3(t),\\
V_3(t+1) &= V_3(t) + \rho_1 c_2 S_2(t) - c_3 V_3(t) - d_3 V_3(t),
\end{aligned}
$$

$$
\begin{aligned}
S_4(t+1) &= S_4(t) + c_3 S_3(t) - c_4 S_4(t) - d_4 S_4(t) - \alpha_4 S_4(t) I(t),\\
L_4^{1w}(t+1) &= c_3 L_3^{1w}(t) - c_3 L_4^{1w}(t) - d_4 L_4^{1w}(t) + \nu_4 R_4(t) + \alpha_4 S_4(t) I(t),\\
L_4^{2w}(t+1) &= L_4^{1w}(t) + c_3 L_3^{2w}(t) - c_4 L_4^{2w}(t) - d_4 L_4^{2w}(t),\\
I_4(t+1) &= L_4^{2w}(t) + c_3 I_3(t) - c_4 I_4(t) - d_4 I_4(t),\\
R_4(t+1) &= I_4(t) + R_4(t) - \nu_4 R_4(t) + c_3 R_3(t) - c_4 R_4(t) - d_4 R_4(t),\\
V_4(t+1) &= V_4(t) + c_3 V_3(t) - c_4 V_4(t) - d_4 V_4(t),
\end{aligned}
$$

$$
\begin{aligned}
S_5(t+1) &= S_5(t) + (1-\rho_2)c_4 S_4(t) - d_5 S_5(t) - \alpha_5 S_5(t) I(t),\\
L_5^{1w}(t+1) &= c_4 L_4^{1w}(t) - d_5 L_5^{1w}(t) + \nu_5 R_5(t) + \alpha_5 S_5(t) I(t),\\
L_5^{2w}(t+1) &= L_5^{1w}(t) + c_4 L_4^{2w}(t) - d_5 L_5^{2w}(t),\\
I_5(t+1) &= L_5^{2w}(t) + c_4 I_4(t) - d_5 I_5(t),\\
R_5(t+1) &= I_5(t) + R_5(t) - \nu_5 R_5(t) + c_4 R_4(t) - d_5 R_5(t),\\
V_5(t+1) &= V_5(t) + \rho_2 c_4 S_4(t) + c_4 V_4(t) - d_5 V_5(t),
\end{aligned}
$$

$$(1)$$

where Imm = immune, $S_i$ = susceptible, $L_i^{1w}$ = latent of 1st week, $L_i^{2w}$ = latent 2nd week, $I_i$ = infected, $R_i$ = recovered, for $i = 2, 3, 4, 5$, $V_i$ = vaccinated for $i = 3, 4, 5$ and $I(t) = \beta(I_2(t) + I_3(t) + I_4(t) + I_5(t))/5\,129\,266$.

The underlying demography in the model follows data for the Community of Valencia, Spain [10]. Therefore, the total constant population is 5 129 266, the cohort of newborns is 47 574 persons each year and

- The constant populations per age group are $G(1) = 22\,945.5$, $G(2) = 22\,945.5$, $G(3) = 103\,401$, $G(4) = 475\,759$, $G(5) = 4\,504\,215$, respectively.
- The weekly birth rate is $\mu = 914.885$.
- The weekly death rates per age group are $d_1 = 4.31161 \times 10^{-7}$, $d_2 = 5.62384 \times 10^{-8}$, $d_3 = 8.24829 \times 10^{-8}$, $d_4 = 1.45095 \times 10^{-6}$, $d_5 = 0.00020296$, respectively.
- The weekly growth rates from $G(i)$ to $G(i+1)$ are $c_1 = 0.0398716$, $c_2 = 0.0398716$, $c_3 = 0.00884774$, $c_4 = 0.00192151$, respectively.

Also, as we describe above, the transitions from $L_i^{1w}$ to $L_i^{2w}$, from $L_i^{2w}$ to $I_i$ and from $I_i$ to $R_i$, $i = 2, 3, 4, 5$ are equal to 1 week.

The structure and flow diagram of the above compartmental model with vaccination (1) is given in Fig. 3.

In Fig. 3, the black lines determine the transitions between disease states, ages, births and deaths. The red lines determine the transitions for vaccinated individuals. On the one hand, 1-year-old susceptible individuals get vaccinate and transit to box $V_3$. Here we are assuming that all the children vaccinated at 1 year of age, will be vaccinated again at 3 years of age. On the other hand, 12-year-old susceptible individuals get vaccinated and transit to box $V_5$. In this model we consider that, once a child is vaccinated at 1 and 3 years old or at 12 years old, he/she gets a permanent immunization.

Our model is stated as an age-structure model of a system of difference equations where we are going to consider that the transmission is not only depending on the age group, but also on the week of the year. Other models have been used to study the transmission dynamics of varicella using systems of differential equations and systems of partial derivatives equations. See, for instance [3,11].

Assuming a base case consisting of no vaccination, we calibrate the model using data in Fig. 1 and percentages for the number of cases of varicella in Spain according to infected percentages per age group appearing in [12]:

- 0–1 year: 3 %,
- 1–4 years: 48 %,
- 5–9 years: 32 %,
- 10–14 years: 6 %,
- older than 15 years: 11 %.

To calibrate the model, we built it and implemented the appropriate procedures using *Mathematica* [13]. From the calibrated model we also get the following information:

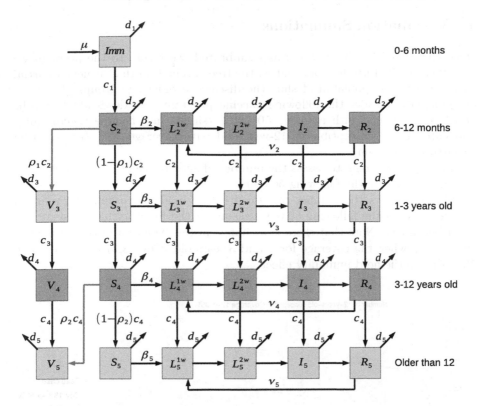

**Fig. 3.** Age-structured compartmental model for varicella infection with vaccination. $d_i$, $i = 1, \ldots, 5$ are the death rates, $\mu$ is the birth rate, $\beta_i$, $i = 2, \ldots, 5$ are the infection rates, $\nu_i$, $i = 2, \ldots, 5$ are the reinfection rates, $\rho_1$ and $\rho_2$, are the vaccination rates and $c_i$, $i = 1, \ldots, 4$ are the transition rates between age-groups chosen in such a way that the total population remains constant. Imm = immune, $S_i$ = susceptible, $L_i^{1w}$ = latent of 1st week, $L_i^{2w}$ = latent 2nd week, $I_i$ = infected, $R_i$ = recovered, $V_i$ = vaccinated (Color figure online).

– Susceptible individuals are 7.3 % of the total population.
– There is an average of 15 infectious individuals every 10 000.
– Recovered account for 92.70 % of the total population.
– Around 54.75 % people have got infected with 5 years old or younger.
– Around 1300 persons are re-infected each year.
– 98.22 % of the cohort is infected every year.
– 6200 children are still susceptible with age 12 years (around 13 % of the cohort). These individuals should be vaccinated under the current strategy.

All these above data contribute to the validation of the model because the calibration is correct and the model satisfies other facts appearing in the literature.

## 4    Vaccination Simulations

Once the model has been stated and calibrated, we return to the problem we described in the Introduction, that is, the free vaccination that achieved partial coverage was the potential of shift the disease to older age groups. Thus, we are going to consider the following vaccine program: two doses at 12 months and 3 years of age, with coverage 50–100% (simulating the free vaccination), and catch up of susceptible at 12-years-old with coverage 90 % (the existing vaccination program).

Then, we are going to assess the number of susceptible at 12-years-old individuals per year in the following 50 years.

Figure 4 shows the number of susceptible individuals reaching 12 years of age per year in the following 50 years assuming the free vaccination strategies described above. Note that there is a change in the evolution of the number of susceptible when the coverage for 1 and 3 years old (Cov 1&3) is greater than 70 % due to the herd immunity effect.

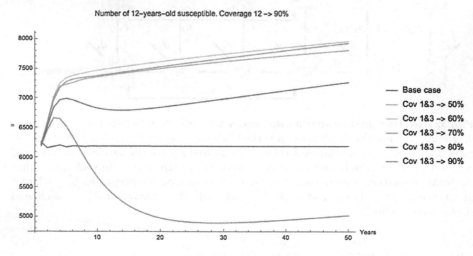

**Fig. 4.** Number susceptible 12-years-old children per year in each one of the next 50 years. Note that there is a trend change when the Cov 1&3 is greater than 70 %.

This effect can also be seen in Fig. 5 where the percentage of 3-years-old susceptible that are susceptible at 12-years-old are drawn. Note that the differences are increasing as the partial coverage increases and the jumps are remarkable when the coverage is greater than 70 %. It is clear that the number of susceptible decreases as the coverage increases, however these non-vaccinated susceptible are better protected against infection because the herd immunity.

With vaccination coverage in children 1–3 years old less than 70 %, the percentage of non vaccinated individuals that reach susceptible at 12 years of age over the next 50 years is low (less than 50 %) and the main effect of the vaccine

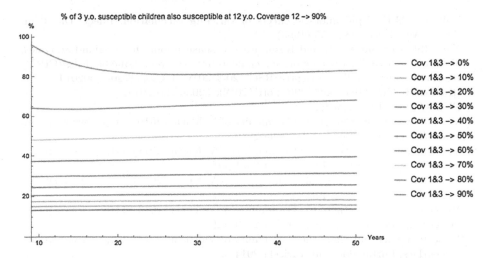

**Fig. 5.** Percentage of 3-years-old susceptible that are susceptible at 12-years-old. Note that the differences are increasing as the partial coverage increases and the jumps are remarkable when the coverage is greater than 70 %. It is clear that the number of susceptible decreases as the coverage increases, however these non-vaccinated susceptible are better protected against infection because the herd immunity.

is the direct protection. However, for coverage 80–90%, the herd immunity effect produces a remarkable increase in the percentage of susceptible individuals.

## 5   Conclusion

Varicella vaccine is highly efficient in the prevention of varicella. The free vaccination program, where parents pay for the vaccine, does not show a shift in the disease incidence always that the vaccination at 12 years of age is maintained and the coverage of the free vaccination is 70 % or less.

Important herd immunity is shown when the vaccination coverage at 1 and 3 years of age is higher than 70 %. Below this the main effect of the vaccine is the direct protection to the vaccinated.

Lack of herd immunity implies that the virus is freely circulating, therefore it is expected that the contact of adults with the virus persists and the potential for a herpes zoster increase in the adults will be minimal.

## References

1. Lin, F., Hadler, J.L.: Epidemiology of primary varicella and herpes zoster hospitalizations: the pre-varicella vaccine era. J. Infect. Dis. **181**(6), 1897–1905 (2000)
2. Lenne, X., Díez Domingo, J., Gil, A., Ridao, M., Lluch, J.A., Dervaux, B.: Economic evaluation of varicella vaccination in Spain-Results from a dynamic model. Vaccine **24**(47–48), 6980–6989 (2006). http://dx.doi.org/10.1016/j.vaccine.2006.04.051

3. Halloran, M.E.: Epidemiologic effects of varicella vaccination. Infect. Dis. Clin. North. Am. **10**(3), 631–655 (1996)
4. New RSC Communicable and Respiratory Disease Report for England & Wales, 36/2014.    http://www.rcgp.org.uk/clinical-and-research/~/media/Files/CIRC/ Research-and-Surveillance-Centre/RSC%20-%20Week%2036%20-%202014/ RSC%20RCGP%20Weekly%20Report%20Wk%2036%20_2014_ FINAL%20VERSION.ashx
5. Varicella report bulletin of the Community of Valencia (2010). http://www.sp.san. gva.es/DgspPortal/docs/Inf_Varicela_2010.pdf
6. Varicella report bulletin of the Community of Valencia (2011). http://www.sp.san. gva.es/DgspPortal/docs/Inf_Varicela_2011.pdf
7. Varicella report bulletin of the Community of Valencia (2012). http://www.sp.san. gva.es/DgspPortal/docs/Inf_Varicela_2012.pdf
8. Varicella report bulletin of the Community of Valencia (2013). http://www.sp.san. gva.es/DgspPortal/docs/Inf_Varicela_2013.pdf
9. Varicella report bulletin of the Community of Valencia (2014). http://www.sp.san. gva.es/DgspPortal/docs/Inf_Varicela_2014.pdf
10. Statistic site of the Valencian Government. http://www.ive.es
11. Poletti, P., Melegaro, A., Ajelli, M., del Fava, E., Guzzetta, G., Faustini, L., Scalia Tomba, G., Lopalco, P., Rizzo, C., Merler, S., Manfredi, P.: Perspectives on the impact of varicella immunization on herpes zoster. A model-based evaluation from three European Countries. PLoS ONE **8**(4), e60732 (2013). http://dx.doi.org/ 10.1371/journal.pone.0060732
12. Peña-Rey, I., Martínez de Aragón, M.V., Villaverde, A., Terres, M., Alcalde, E., Suárez, B.: Epidemiología de la Varicela en España en los períodos de pre- y post-vacunación. Rev. Esp. de Salud Pública **83**, 711–724 (2009). http://scielo.isciii.es/ pdf/resp/v83n5/original3.pdf
13. Wolfram Research, Inc., Mathematica, Version 10.1, Champaign, IL (2015). http:// www.wolfram.com/mathematica

# Angel: Towards a Multi-level Method for the Analysis of Variants in Individual Genomes

Ginés Almagro-Hernández[1], Francisco García-Sánchez[1],
María Eugenia de la Morena-Barrio[2(✉)], Javier Corral[2],
and Jesualdo Tomás Fernández-Breis[1]

[1] Facultad de Informática, Universidad de Murcia,
IMIB-Arrixaca, CP 30100 Murcia, Spain
{gines.almagro, frgarcia, jfernand}@um.es
[2] Centro de Hemodonación, IMIB-Arrixaca, CP 30003 Murcia, Spain
{mm113317, jcc}@um.es

**Abstract.** Genomic medicine pursues to develop methods for improving early diagnosis processes, the efficiency of treatments and facilitating the discovery of new therapies, and mainly searches for associations between the genotype of individuals and their phenotypical features. The huge genomic variability is a major difficulty for developing effective computational methods, since the correlation of a locus and a phenotype does not necessarily mean causality. Hence, methods for genome-based diagnosis need to take into account the complexity of the genomic background and the biological networks involved in the manifestation of phenotypes and disorders.

We describe a method for analysing the variants identified in the genome of human individuals, sequenced using Next-Generation Sequencing techniques, and such analysis is based on the existing knowledge about the genes, pathways and phenotypes. This method is capable of generating quantitative scores at the levels of gene, pathway and phenotype, which represent the degree of functional disorder of the corresponding gene or pathway, and the level of contribution to development of a specific phenotype of the genomic variant. The validation experiments performed with exomes of patients with "Congenital Disorder of Glycosylation, Type IA" (CDG1A) have shown positive results.

**Keywords:** Genomic medicine · Bioinformatics · Genome analysis · Variant analysis

## 1 Introduction

Genomic medicine is based on the analysis and interpretation of the human genome with clinical objectives, through the identification and annotation of genomic variations that affect the genes that are responsible of a given phenotype, either for monogenic or multigenic disorders. Personalized medicine pursues finding the association between the genotype of an individual and its phenotypical features [1] with the aim of improving early diagnosis processes, the efficiency of treatments and facilitating the discovery of new therapies. All these aspects would contribute to reduce the economic

© Springer International Publishing Switzerland 2016
F. Ortuño and I. Rojas (Eds.): IWBBIO 2016, LNBI 9656, pp. 47–58, 2016.
DOI: 10.1007/978-3-319-31744-1_5

cost and the amount of erroneous treatments assigned to patients, with the corresponding negative impact on their health.

Due to the huge genomic variability that exists, one of the main challenges for genomic medicine is to identify which genomic variations associated with a concrete phenotype are the real cause or predispose to develop the phenotype and which do not have a real influence [2]. This difficulty holds at both individual and population level [3]. It should be noted that the correlation of a given locus and a phenotype does not mean causality [2]. In most situations, the progress of science over time has changed the hypotheses about the types of variations and the types of disorders or phenotypes they cause [4]. Another challenge for genomic medicine is the complexity of the operation and regulation of genomic mechanisms, based on complex networks for more than 22,000 gene products with functional RNA, epigenetic factors and metabolic factors.

The analysis of the state of the arts reveals that methods used in genomic medicine for the identification of genomic variations can be classified in (1) biochemical techniques, like G-banding or allele-specific PCR [5], (2) sequencing techniques, like Sanger [6] (which provide higher reliability at a higher cost and analyzing a few locus per experiment) or Next Generation Sequencing (NGS) [7] (which permit massive experiments on complete genomes, exomes or specific genes panels). The progressive and quick evolution of the NGS technology permits its usage in the search for genotype-phenotype relations. The genomic variations are analyzed by means of studies of cohorts of patients who present the phenotype of interest, Genome Wide Association Studies (GWAS) or studies of the segregation of haplotypes at family level. These types of studies have difficulties to deal with situations like incomplete penetrance, pleiotropic effect of genes, linkage disequilibrium, genetic drift, development of de novo variations, variations of low frequency in the population, allelic heterogeneity and population bias, among others.

The goal of our work is to develop a method for a precise quantification of the genotype-phenotype relation from the genomic variations identified by means of NGS technologies. We expect our method to achieve a high degree of reliability and confidence. This method analyses the genome of a person and focuses on the functional alteration of the gene products, which would permit to avoid the aforementioned limitations of the state of the art methods for identifying the relation variation-phenotype. Our method will be based on a multi-layer and multi-model analysis: genes, pathways and phenotypes. This layered structure intends to represent the functional and structural complexity of the genome in a machine-friendly way. The method will combine the evaluation that will be performed at each layer, which will permit to quantify the functional implications of the variations for each gene, pathways or phenotype.

## 2   Methods

### 2.1   Variant Model

**Representation and Annotation.** We use VCF files, which contain the genomic variations obtained by using NGS technologies. We apply filters on the variations based on their quality score, (i.e., p-value <= 0.01, Phred score[1] >= 20). The genomic coordinates are converted to the version of reference genome we are using (GRCh38 [8]). The contents of this file are used to obtain or calculate the attributes associated with variants in our model: chromosome (Chr), variant start position (Pos), cross-reference (ID), variant reference allele sequence (seq_Ref), variant alternative allele sequence (seq_Alt), genotype code (Genotype_cod), variant type (Type_var), localisation length (Length_loc), consequence length (Length_cons) and consequence position (Pos_cons). All these attributes are stored in a tabular file. The field "strand" is not included because all variants are represented in the forward strand.

## 2.2 Gene Model

**Representation.** A *gene_model* is generated for each gene analysed, including curated information that is automatically collected from the following bioinformatic resources: Ensembl R.81 [9] RefSeqGene R.71 [10], CCDS R.18 [11], UNIPROT [12] and INTERPRO v.52 [13].

The *gene_model* covers both non-coding (introns, UTR5', UTR3', promoter, splicing sites and regulatory elements) and coding region (CDS). Its canonical transcript is considered to be its only coding product. The *gene_model* consists of hierarchical levels of regions, subregions and Basics (see Fig. 1). These levels are processed and analysed separately, but the model takes into account the overall impact regarding the gene functionality in which they are or on which they act. One region can be considered as a sequence of nucleotides (non-coding regions) or amino acids (coding region). Each region can be divided into a limited number of non-overlapping subregions of variable length (subset from one nucleotide/amino acid to the total length of the region). Also, one subregion can be divided into a limited number of non-overlapping Basics of variable length (subset from one nucleotide/amino acid to the total length of the subregion). There are two types of subregions and Basics: (1) "Defined", which means explicitly defined for that particular gene, and there might be several ones in a single region or subregion; and (2) "Trivial", that includes all the

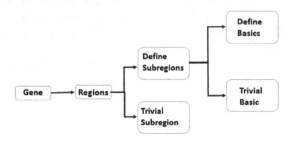

**Fig. 1.** Architecture of gene_model in levels

---

[1] $Q = -10 \ \log_{10} P$; P is the base-calling error probabilities.

nucleotides or amino acids not included in any defined subregion or Basic. A Basic has no further divisions. A key feature of each Basic is that all its nucleotides/amino acids have the same functional value, so it is considered the fundamental unit for calculating the *gene_score*, as it will be described later. The regions, subregions and Basics permit to increase the sensitivity and accuracy of the method.

**Scoring.** Each *gene_model* is given a score, which depends on three types of properties:

(1) Functional scores: Each region, subregion or Basic is defined by at least three attributes, start, end and *funct_score*. The latter indicates the degree of gene functional disorder to which it belongs or on which it acts, in only one allele, if this region, subregion or Basic was entirely deleted. The functional role of a nucleotide or amino acid is calculated by dividing this parameter by the length of the region, subregion o Basic to which it belongs.

(2) Variant Type Consequence (*var_type_conseq*): This property is used in the non-coding regions. It is the correction factor of the effect of the genomic variant on the Basic according to the type of genomic variant. We are currently analyzing single nucleotide variants, substitutions, deletions, insertions and indels.

(3) Consequence Type Score (*conseq_type_score*): This property is used in the coding regions. It is the correction factor of the effect caused by the genomic variant on each affected amino acid. The types of consequences analyzed in this method are: synonymous, missense, nonsense, frameshift, deletion, insertion, lost_stop and lost_start.

The values for the different properties can be estimated from scientific literature, using quantitative or statistical methods such as the conservation rate of nucleotides, the similarity of physic-chemical properties between amino acids, logistic regression, etc., or by experimental evidence.

**Analysis.** This method implements the gene alteration function (Eq. 1). Its domain is the set of all the genes analysed (Xgene) and its range is the set of all possible values of the *gene_score* (Eq. 2). This score indicates that the degree of the total functional disorder of gene and it is calculated on the basis of *gene_model* and the genomic variants that affect it. Its value is the sum of all the *region_score* of both the non-coding regions and coding region (CDS). The calculation of *region_score* will be described later. If the *gene_score* is larger than 1 then it is standardized at 1. The score is a number in the range [0,1], where, 0 means "no functional disorder" and 1 means that there is "total functional disorder".

Its value depends on the three types of properties introduced in the Scoring section. Next, we describe how the method obtains the scores for such types of properties.

$$Fg(x) : Xgene \rightarrow gene\_score \tag{1}$$

$$gene\_score = [\Sigma\ non\text{-}coding\_region\_score] + cds\_score \tag{2}$$

*Variants Classification.* Firstly, all the genomic variants of a gene are selected. One genomic variant affects a gene when the intervals defined between there is an overlap of their genomic coordinates. Then, the genomic variants are classified by region, sub-region and Basic.

*Non-coding Regions.* For each non-coding regions, the analysis is initiated at the Basic level. The consequence score (*var_conseq_score*) of each genomic variant associated with the Basic is calculated. This score is the total number of affected nucleotides by it on both alleles at gene, according to its genotype and the correction factor *var_type_conseq* previously defined. Then, the score of each Basic is obtained (*basic_score*) by taking into account all the affected nucleotides by all the genomic variants, the functional role of each of them determined by the *basic_funct_score* (see *funct_score*) and the gene alleles number. This score indicates the degree of the functional disorder caused by all the genomic variants associated with this Basic in the gene. Its numerical value range falls in [0,1].

This method takes into account that a single genomic variant within a Basic may alter fully the functionality of the gene that contains it. Subsequently, the subregion score (*subregion_score*) is calculated as the sum of the *basic_score* of its defined and trivial Basics. This score indicates the degree of functional disorder caused by all the genomic variants associated with this subregion in the gene.

In case a genomic variant alters fully the functionality of the gene, its *basic_score* would be 1. If the *subregion_score* value is larger than 1 then it is standardized at 1. Therefore, its numerical value range falls in [0,1]. Finally, the value of region score (*region_score*) (Eq. 3) is calculated as the sum of the *subregion_score* of its defined and trivial subregions. This score indicates the degree of functional disorder caused by all the genomic variants associated with this region in the gene. As in the case of *subregion_score*, if the *region_score* value is larger than 1 then it is standardized at 1. Therefore, its numerical value range falls in [0,1].

$$\text{region\_score} = \Sigma \text{ subregion\_score} \qquad (3)$$

*Coding Region.* For each genomic variant in the coding region, the affected amino acids are determined, which permits to obtain the amino acid alteration score (*aa_alteration_score*). This score indicates the degree of gene functional disorder in case of deleting it, and it depends on the parameter *basic_funct_score* of the Basic to which it belongs and on the correction factors *conseq_type_score* that have been previously defined. Next, the genomic variant consequence score (*var_conseq_score*) is calculated as the sum of all the *aa_alteration_score*. This score indicates the degree of functional disorder caused by one genomic variant associated with this region in the gene. This parameter takes into account both the gene alleles number (average) and the genomic variant genotype (total), similar to non-coding regions. An amino acid affected by one genomic variant may alter fully the functionality of the gene that contains it, that is, its *aa_alteration_score* would be 1. If the *var_conseq_score* value is larger than 1 then it is standardized at 1. Therefore, its numerical value range falls in [0,1]. Finally, the value of coding region score (*cds_score*) (Eq. 4) is calculated as the sum of all the *var_conseq_score*. This score indicates the degree of functional disorder caused by all the genomic variants associated with this region in the gene. If the *cds_score* is larger than 1, then it is standardized at 1. Therefore, its numerical value range falls in [0,1].

$$cds\_score = \Sigma var\_conseq\_score \qquad (4)$$

## 2.3   Pathway Model

**Representation.** A *pathway_model* is generated for each pathway included in the analysis. To date, the information about the pathway is manually collected from resources such as Reactome V.53 [14] and KEGG R.75.0 [15].

A *pathway_model* contains a set of pathways or/and genes, which are structured in a series of levels. There is no limit in the number of levels but the number of levels may affect the sensitivity and accuracy of the method. (see Fig. 2) shows an example of a *pathway_model* with four levels. The pathways included as *pathway_model* elements are usually biosynthesis or degradation pathways of cofactors, substrates, proteins or/and regulatory metabolites, etc., which are involved in the functionality of this pathway. The same gene may appear in the model as many times as independent reactions are catalysed by it. However, the gene appears only once in case of homopolymer forms of the same locus. One gene may appear as element of different pathways, too. The genes playing a regulatory role in a specific reaction can also be included in the model.

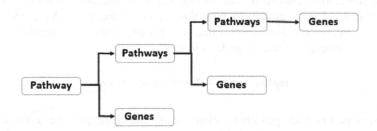

**Fig. 2.** Architecture of the pathway_model in levels, with 4 levels

**Scoring.** Each element of a *pathway_model* has a functional score (*functional_score*). This score indicates the degree of the functional disorder caused by that element if it is deleted. This *functional_score* depends on the corresponding *gene_score* or *pathway_score*. The functional scores can be estimated from scientific literature, using quantitative or statistical methods, or by experimental evidence.

**Analysis.** This method implements the pathway alteration function (Eq. 5). The domain is the set of the pathways analysed (Xpath) and its range is the set of all possible values of *pathway_score* (Eq. 6). This score indicates the degree of the functional disorder according to the genomic background analysed for the pathway. This score is calculated on the basis of the *pathway_model*, as the sum of the corresponding *pathway_score* and *gene_score* of all its elements, which were previously calculated and weighted according to their *functional_score*. If the *pathway_score* is larger than 1 then it is standardized at 1. Therefore, its numerical value range falls in the range [0,1] where 0 means "no pathway functional disorder" and 1 means "pathway total functional disorder".

$$Fp(x) : Xpath \rightarrow pathway\_score \qquad (5)$$

$$Pathway\_score = \Sigma(element\_score \times functional\_score) \qquad (6)$$

## 2.4   Phenotype Model

**Representation.** A *phenotype_model* is generated for each phenotype included in the analysis. The information of each phenotype is currently manually retrieved from resource such as OMIM [16] and Orphadata [17]. One *phenotype_model* contains a set of phenotypes, pathways and genes, which are directly or indirectly related to the phenotype. These elements are structured in levels as described in Fig. 3. This method assumes an inheritance model for the phenotypes, because it takes into account the genomic variant genotype in the calculation of *gene_score*.

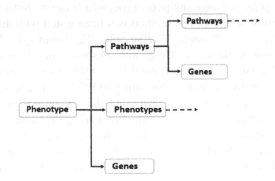

**Fig. 3.** Architecture of the phenotype_model in levels

**Scoring.** Each element of a *phenotype_model* has a contribution score (*contrib_score*). This score indicates the contribution of this element to the development of phenotype if this element was fully no functional. The contribution scores can be estimated from scientific literature, using quantitative or statistical methods, or by experimental evidence.

**Analysis.** This method implements the phenotype contribution function (Eq. 7). Its domain is the set of the phenotypes analysed (Xphen) and its range is the set of all possible values of *phenotype_score* (Eq. 8). This score indicates the contribution of all the elements of this phenotype according to the genomic background analysed. This score is calculated as the sum of the corresponding *phenotype_score*, *pathway_score* and *gene_score* of all its elements, which have been previously calculated and weighted according to their *contrib_score*. If the *phenotype_score* is larger than 1 then it is standardized at 1. Therefore, its numerical value range falls in the range [0,1], where 0 means "the phenotype would not be developed", and 1 means "the phenotype would surely be developed".

$$Fph(x) : Xfen \rightarrow phenotype\_score \tag{7}$$

$$phenotype\_score = \Sigma(element\_score \times contrib\_score) \tag{8}$$

## 3 Results

The method described in this paper has been implemented in a software tool implemented in Perl. The input data are the human genome genomic variants in VCF files, which are analyzed, as well as the parameters needed for calculating the *gene_model*, *pathway_model* and *phenotype_model*. The outcomes are presented in a comprehensive report sorted by genes. This report shows the molecular basis responsible for the phenotype under study. A summary report is obtained, and it contains the scores calculated for each gene, pathway and phenotype, which can be further analyzed using biostatistics or machine learning. The method has been tested with the exomes of five individuals: "Sample_1" to "Sample_5". "Sample_1", "Sample_2", "Sample_3" and "Sample_4" have been diagnosed as "Congenital Disorders of Glycosylation Type IA" (CDG1A), while "Sample_5" does not have this disorder. Table 1 lists all models used in this test. The *gene_model* are named according to HGNC nomenclature. Table 2 lists the default parameters used in this test. Provided that the aim of this validation is to allow comparisons between all samples, we are giving the same weight to all the elements used for obtaining the scores for *pathway_model* and *phenotype_model*. A scale based on the materiality of each region or subregion was applied for the different functional scores of *gene_model*. For this study, defined Basics were not created. A relative scale based on biological function impairment of a nucleotide or amino acid was defined for *var_type_conseq* and *conseq_type_score* respectively. For example, a single nucleotide deletion completely alters its function, but a substitution generally involves a minor functional alteration (we do not take account commonalities between their physico-chemical properties, etc.).

Sample_1 presents a *phenotype_score* = 1 (see Table 3). This score is due mainly to one homozygous deletion of one nucleotide in the coding region within the gene ALG9. This deletion leads to a frameshift + lost_stop. The gene ALG9 catalyses two independent reactions in the pathway_bios_llo, so its *gene_score* is added twice. Sample_2 presents a *phenotype_score* = 0.181. This score is due mainly to one compound heterozygous deletion of one nucleotide and SNV in the coding region within the gene RPN2. These mutations lead to a frameshift + lost_stop and a

**Table 1.** Gene_model, pathway_model and phenotype_model for validation

| Model | Model name and description |
|---|---|
| Gene_model | "MVD", "DHDDS", "SRD5A3", "DOLK", "DOLPP1", "GFPT1", "GFPT2", "GNPNAT1", "PGM3", "UAP1", "MPI", "PMM1", "PMM2", "GMPPA", "GMPPB", "DPM1", "DPM2", "DPM3", "ALG5", "DPAGT1","ALG13", "ALG14", "ALG1", "ALG2", "ALG11", "RFT1", "ALG3", "MPDU1", "ALG9", "ALG12", "ALG6", "ALG8", "ALG10", "ALG10B", "DAD1", "DDOST", "RPN1", "RPN2", "OST4", "TUSC3", "MAGT1", "STT3A", "STT3B" |
| Synthesis of dolichyl-phosphate-glucose | pathway_synt_dol_p_gluc = {1:ALG5 } |
| Synthesis of dolichyl-phosphate mannose | pathway_synt_dol_p_man = {1:DPM1, 1:DPM2, 1:DPM3 } |
| Synthesis of GDP-mannose | pathway_synt_gdp_man = {1:MPI, 1:PMM1, 1:PMM2, 1:GMPPA, 1:GMPPB } |
| Synthesis of UDP-N-acetyl-glucosamine | pathway_synt_udp_ac_glucamin = {1:GFPT1, 1:GFPT2, 1:GNPNAT1, 1:PGM3, 1:UAP1 } |
| Synthesis of dolichyl-phosphate | pathway_synt_dol_p = {1:MVD, 1:DHDDS, 1:SRD5A3, 1:DOLK, 1:DOLPP1 } |
| Biosynthesis of the N-glycan precursor LLO | pathway_bios_llo = {1:pathway_synt_dol_p, 1:pathway_synt_udp_ac_glucamin, 1:pathway_synt_gdp_man, 1:pathway_synt_dol_p_man, 1:pathway_synt_dol_p_gluc, 1:DPAGT1, 1:ALG13, 1: ALG14, 1:ALG1, 1:ALG2, 1:ALG11, 1:RFT1, 1:ALG3, 1: ALG9, 1:ALG9, 1:ALG12, 1:MPDU1, 1:ALG6, 1:ALG8, 1: ALG10, 1:ALG10B } |
| Transfer to a nascent protein | pathway_transf_llo = {1:DAD1, 1:DDOST, 1:RPN1, 1:RPN2, 1:OST4, 1:TUSC3, 1:MAGT1, 1:STT3A, 1:STT3B } |
| CDGIA | phentype_CDGIa = {1:pathway_bios_llo, 1: pathway_transf_llo} |

synonymous amino acid change respectively. Sample_3 presents a *phenotype_score* = 0.074. This score is due mainly to one homozygous SNV in the coding region within the gene DDOST. These mutations lead to a missense amino acid change. Sample_4 presents a phenotype_score = 0.206. This score is due mainly to the same mutation as Sample_2. Furthermore, Sample_5 presents a *phenotype_score* = 7.646e-04, which means that the contribution of the genomic background of Sample_1 to the development of CDGIA is four orders of magnitude larger than the genomic background of Sample_5, and that CDGIA should have been developed in Sample_1. Furthermore, the contribution of both Sample_3 and Sample_4 is three orders of magnitude larger, while the contribution of Sample_2 is two orders of magnitude larger.

**Table 2.** Default values for the parameters

| Model | Parameter | Value |
|---|---|---|
| pathway_model | functional_score | 1.00 |
| phenotype_model | contrib_score | 1.00 |
| utr5',utr3', exons,splicing_sites, promoter | region_funct_score | 1.00 |
| introns | region_funct_score | 0.60 |
| regulatory elements | region_funct_score | 0.25 |
| promoter_core, cds_domains | subregion_funct_score | 1.00 |
| SNV, sustitution | var_type_conseq | 0.25 |
| deletion | var_type_conseq | 1.00 |
| insertion | var_type_conseq | 0.50 |
| indel | var_type_conseq | 0.75 |
| synonymous | conseq_type_score | 0.10 |
| missense | conseq_type_score | 0.75 |
| deletion | conseq_type_score | 1.00 |
| insertion | conseq_type_score | 0.50 |
| frameshift | conseq_type_score | 0.75 |

**Table 3.** Scores of the main models for the five samples.

| | CDG1A | Model | Parameter | Score |
|---|---|---|---|---|
| Sample_1 | + | phenotype_CDGIa | phenotype_score | **1.00** |
| | | pathway_bios_llo | pathway_score | 1.00 |
| | | pathway_transf_llo | pathway_score | 0.022 |
| | | gene_model (ALG9) | gene_score | 1.00 |
| Sample_2 | + | phenotype_CDGIa | phenotype_score | **0.181** |
| | | pathway_bios_llo | pathway_score | 0.052 |
| | | pathway_transf_llo | pathway_score | 0.129 |
| | | gene_model (RPN2) | gene_score | 0.110 |
| Sample_3 | + | phenotype_CDGIa | phenotype_score | **0.074** |
| | | pathway_bios_llo | pathway_score | 0.053 |
| | | pathway_transf_llo | pathway_score | 0.021 |
| | | gene_model (DDOST) | gene_score | 0.019 |
| Sample_4 | + | phenotype_CDGIa | phenotype_score | **0.206** |
| | | pathway_bios_llo | pathway_score | 0.075 |
| | | pathway_transf_llo | pathway_score | 0.131 |
| | | gene_model (RPN2) | gene_score | 0.110 |
| Sample_5 | – | phenotype_CDGIa | phenotype_score | **7.646e-04** |
| | | pathway_bios_llo | pathway_score | 4.438e-04 |
| | | pathway_transf_llo | pathway_score | 3.208e-04 |

# 4 Discussion

This method addresses human genome analysis challenges from a multi-scale point of view, through the integration of several levels: from the upper level or phenotype level (monogenic, complex, cancer) to the molecular level. In this method the gene level is the unit scale. This method can analyse the following types of genomic variants: SNV, substitutions, insertions, deletions and indels. It can also analyse inherited and de novo mutations, independently of the sample source (somatic or germ line). This method is independent from genomic variants and polymorphisms classification mistakes. It analyses the coding (canonical transcript) and non-coding regions of each gene. It is based on the idea of expert systems and can increase its sensitivity and accuracy through feedback. This method allows analysing a single gene, a specific genes panel, an exome or the whole genome. To measure both the functional alteration of a gene or pathway and the likelihood of developing a phenotype through its genomic background, the method provides the *gene_score*, *pathway_score* and *phenotype_score* as vectors for further analysis with biostatistics methods or machine learning. This method has all these features compared to the current genomic variants annotation tools, which are limited to analyse and annotate each genomic variant separately, without any kind of integration process between them. These tools usually analyse only the coding regions and SNVs. Their annotation process involves to create a list of the genomic variants with their consequence over the genetic product (synonymous, missense, nonsense, etc.) and to determine their clinical significance based on information provided by previous works.

# 5 Conclusions

The field of computational genomic analysis needs to find the breakeven between the simplicity of the models and the algorithms used and the generation of consistent, reliable, realistic results. We believe that the analysis method described in this work is a good starting point for achieving that breakeven. The preliminary results obtained with the method are promising since they are in line with the clinical situation of the patients analysed, although we do not aim to make any claim about the generalisation of the results with such a small number of samples analysed in this paper. However, the results have been obtained using a very basic model, without including subregions or the defined basic ones in the *gene_model*, and without including formation and degradation reactions of substrates, cofactors and other involved metabolites. The method has some limitations such as the non-analysis of structural variations, alternative splicing or functional RNA, functional complementarity between gene products or information specific on the tissues or developmental stage of the phenotype, among others. The evolution of NGS technologies will also provide more reliable and complete data about the genomic variations of a person. The analysis of this information supported by domain knowledge expressed in ontologies [18], and the use of statistical models like logistic regression or Bayesian networks will permit to optimize the model, including the weights of the parameters used. This will permit to increase the sensitivity and specificity of the bioinformatics tools applied to genomic medicine.

**Acknowledgments.** This work has been supported by the Ministerio de Economía y Competitividad and the FEDER programme through grant TIN2014-53749-C2-2-R2, and by the Ministerio de Educación, Cultura y Deportes through grant FPU14/06303.

# References

1. Offit, K.: Personalized medicine: new genomics, old lessons. Hum. Genet. **130**(1), 3–14 (2011)
2. MacArthur, D., Manolio, T., Dimmock, D., Rehm, H., Shendure, J., Abecasis, G., Adams, D., Altman, R., Antonarakis, S., Ashley, E., Barrett, J., Biesecker, L., Conrad, D., Cooper, G., Cox, N., Daly, M., Gerstein, M., Goldstein, D., Hirschhorn, J., Leal, S., Pennacchio, L., Stamatoyannopoulos, J., Sunyaev, S., Valle, D., Voight, B., Winckler, W., Gunter, C.: Guidelines for investigating causality of sequence variants in human disease. Nature **508** (7497), 469–476 (2014)
3. The 1000 genomes project consortium. http://www.1000genomes.org
4. Marian, A.: Molecular genetic studies of complex phenotypes. Transl. Res. **159**(2), 64–79 (2012)
5. Myakishev, M., Khripin, Y., Hu, S., Hamer, D.: High-throughput SNP genotyping by allele-specific PCR with universal energy-transfer-labeled primers. Genome Res. **11**(1), 163–169 (2001)
6. Sanger, F., Nicklen, S., Coulson, A.: DNA sequencing with chain-terminating inhibitors. Proc. Natl. Acad. Sci. U.S.A. **74**(12), 5463–5467 (1977)
7. Soon, W., Hariharan, M., Snyder, M.: High-throughput sequencing for biology and medicine. Mol. Syst. Biol. **9**, 640–640 (2013)
8. Genome reference consortium. http://www.ncbi.nlm.nih.gov/projects/genome/assembly/grc
9. The Ensembl project. http://www.ensembl.org
10. RefSeqGene. http://www.ncbi.nlm.nih.gov/refseq/rsg
11. The consensus CDS protein set. http://www.ncbi.nlm.nih.gov/CCDS/CcdsBrowse.cgi
12. UniProt consortium. http://www.uniprot.org
13. INTERPRO protein sequence analysis and classification. http://www.ebi.ac.uk/interpro
14. REACTOME a curated pathway database. http://www.reactome.org
15. KEGG pathway database. http://www.genome.jp/kegg/pathway.html
16. Online mendelian inheritance in man. http://www.omim.org
17. Free access data from orphanet. http://www.orphadata.org
18. The OBO foundry. http://www.obofoundry.org

# Transcriptome-Based Identification of a Seed Olive Legumin (11S Globulin). Characterization of Subunits, 3D Modelling and Molecular Assessment of Allergenicity

Adoración Zafra[1], José Carlos Jimenez-Lopez[1], Rosario Carmona[1],
Gonzalo Claros[2], and Juan de Dios Alché[1(✉)]

[1] Plant Reproductive Biology Laboratory, Department of Biochemistry,
Cell and Molecular Biology of Plants Estación Experimental Del Zaidín, CSIC, Granada, Spain
{dori.zafra,josecarlos.jimenez,juandedios.alche}@eez.csic.es
[2] Department of Molecular Biology and Biochemistry, University of Málaga, Malaga, Spain
{rosariocarmona,claros}@uma.es

**Abstract.** Seed storage proteins (SSPs) are fundamental molecules for seed germination as an important source of carbon and nitrogen. Among the main four protein families that integrate SSPs, legumins (11S globulins) are widely distributed in dicots, and represent the major contribution to the pool of seed proteins in olive. In the present study, we have used an olive seed transcriptome generated de novo by 454/Roche Titanium+ sequencing to identify a broad panel of 11S protein sequences. Among these identified legumin sequences, five were selected using their presence within the output results from the BLASTP alongside the whole NCBI database, and their clustering with previously-characterized 11S sequences in the phylogenetic analysis as the criteria. The selected sequences were identified as corresponding to the isoform 2 of the 11S protein precursor, and one of the sequences was used for further analysis. Individual acidic and basic subunits within this sequence were recognised, 3D-modeled and assessed as regard to their potential molecular interaction by docking methods. Furthermore, T-cell epitopes were forecast by using predictive software in order to evaluate the putative implications of the olive 11S proteins in food allergy. The potential use of this protein highly present in the olive seed as a food source is discussed.

**Keywords:** Electrostatic potential · Globulin · Legumin · Molecular modelling · Olive · T-cell epitopes · Transcriptome · 11S

## 1 Introduction

The olive tree (*Olea europaea* L.) is agriculturally and economically very important, mainly in the Mediterranean area where it represents 95 % of the total world production of green and black table olives (16 million tons) and olive oil (2.7 million tons). Among Mediterranean countries, Spain is the major producer of olive oil (44 % of the world production) [1, 2]. Olive oil is unavoidably rooted to a genuine diet that has been associated to protective features against cardiovascular disease and cancer [3–7].

© Springer International Publishing Switzerland 2016
F. Ortuño and I. Rojas (Eds.): IWBBIO 2016, LNBI 9656, pp. 59–70, 2016.
DOI: 10.1007/978-3-319-31744-1_6

The use of new olive by-products and specific procedures envisaged to improve economic returns from olive drupes is currently under development. Among these, the use of olive stones and seeds is an alternative with a high potential.

Seed proteins are synthesized at high levels and stored in specific tissues (endosperm and cotyledon) at certain stages of development [8], particularly during the maturation process. After their synthesis, seed storage proteins (SSPs) are deposited in membranous specialized organelles namely protein bodies (PBs) inside of cotyledon and the endosperm cells [9]. They are further mobilized and hydrolysed during seed germination in order to supply amino acids used as nitrogen, carbon and sulphur source [10].

On the basis of their solubility properties [11], SSPs are classified in four major families: albumins, globulins (legumin and vicilin), prolamins, and glutelins.

Legumin and vicilin globulins belong to the cupin superfamily, mainly characterized by displaying a β-barrel structural feature [12], with sedimentation coefficients of 11S and 7S, respectively. Legumins and vicilins are originated from a common ancestral gen [13], which is under positive selection for 11S globulins in dicot plants only [14]. Thus, 11S globulins are widely distributed in dicots [15], and represent the most abundant protein form (70 % of the total seed nitrogen) in dicot seeds [8]. 11S proteins are normally in an oligomeric state, composed of up to six subunits situated at the vertices of an octahedron, with a symmetry 32 [16]. Each subunit is composed of an acidic and a basic chain derived from a single precursor of 50–60 kDa [17–20].

According to canonical synthesis and maturation pathways, their integrating polypeptides enter the endoplasmic reticulum (ER) because of the presence of an N-terminal signal peptide [21], where inter-chain disulphide bridges are established between both the basic and the acidic subunits [22], followed by a cleavage step directed by a asparaginyl endopeptidase [22–24] to finally yield the mature hexameric 11S protein integrated by α and β subunits associated into trimmers, through covalent and non-covalent interactions [25, 26]. This process (11S globulins endoproteolytic cleavage into two subunits linked by a disulphide bridge) is evolutionarily conserved in seeds of conifers, monocots and dicots [22].

After a desiccation period in seeds, SSPs are mobilized under seed hydration by *de novo* synthesis of cysteine proteinases [27, 28] required for the breakdown of peptides, allowing seed germination and seedling growth [29].

The 11S proteins have been widely studied in numerous monocot and dicot plant species [14]. However only few studies about 11S and other SSPs are available as regard to the olive tree [9, 30–33]. Based on proteomic studies, it was proposed the presence of 3 precursors of the 11S protein, obtained from combinations of 3 α subunits and 2 β subunits [9]. However, the identification of the nucleotide and amino acid sequences of one of these subunits is reported here for the first time. We have also performed *in silico* analysis of the structure of one of these precursors by homology modelling and molecular docking analysis to show the α/β subunits interacting surface, and the identification of T-cell epitopes.

# 2 Materials and Methods

## 2.1 *De Novo* Construction and Functional Annotation of an Olive Seed Transcriptome

For the construction of the transcriptome, mature seeds of the olive cultivar 'Picual' were used. Samples were thoroughly grinded with a pistil and liquid $N_2$ followed by the extraction of the total RNA as RNeasy Plant Mini Kit (Qiagen) manual instructions recommends. RNA integrity was checked by formaldehyde gel analysis [34]. The mRNAs were purified using the Oligotex mRNA mini kit (Qiagen). The concentration and quality of the mRNAs were determined by the Ribogreen method (Quant-it Ribo-Green RNA Reagent and kit) and the Agilent RNA 6000 Pico assay chip (Bioanalizer 2100). The isolated mRNAs were subjected to 454/Roche Titanium+ sequencing. Reads were processed, assembled and the functional annotation was performed as described by Zafra *et al.* [35].

## 2.2 Alignment and *in Silico* Analysis of the Sequences

Different strategies were defined to select 11S transcripts. First, semantic and name searches were carried out in the generated annotations using key words. BLASTP searches of known, conserved sequences for 11S were also performed in the annotated database. Finally, annotations were manually screened for specific sequences selected from well-established bibliography resources.

Partial amino acid sequences from olive 11S proteins were compiled and aligned using Clustal Omega software (http://www.ebi.ac.uk/Tools/msa/clustalo/). Alignments visualization was performed by using Bioedit V7.0.5.3 (http://www.mbio.ncsu.edu/bioedit/bioedit.html). Paired sequence (overlapping) reads were matched and joined to complete full-length sequences. Various mismatches in the overlapping sequence reads were allow in order to obtain these full sequences. Protein sequences derived from these final alignments were further analyzed for the presence of putative functional (biologically meaningful) motifs by the ScanProsite program (http://www.expasy.org/tools/scanprosite), and to confirm the identity of the final (mature) sequences. Protparam tool (http://web.expasy.org/protparam/) was used to identify theoretical pI and molecular weight.

## 2.3 Phylogenetic Tree

Phylogenetic tree was constructed with the aid of the software Seaview [36] using the maximum likelihood (PhyML) method implemented with the LG model of the most probable amino acid substitution calculated by the ProtTest 2.4 server [37]. The branch support was estimated by bootstrap resampling with 100 replications.

Based on the results of the phylogenetic tree, five sequences among all identified legumin sequences in the seed transcriptome were selected. As the criteria for selection, we designated their presence within the output results from the BLASTP (E-value = 0.0 to e-14) alongside the whole NCBI database, and their clustering with previously characterized 11S sequences in the phylogenetic analysis.

## 2.4    3D-Modeling

The sequences of two subunits taking part of one of the five 11S globulin precursors selected as indicated above (ID: 002407), were subjected to 3D homology modeling (http://swissmodel.expasy.org/workspace/) [38–41] by using 3ehk, 2e9q, 1od5, 3c3v, 3 kgl and 2d5f protein templates (alpha subunit) and 1fxz, 2e9q, 1od5, 3c3v, 3qac and 3 kgl protein templates (beta subunit) available in PDB (http://www.pdb.org). Models were visualized by PyMol software (https://www.pymol.org/).

## 2.5    Identification of T-cell Epitopes

Identification of linear T-cell epitopes was performed as described by Jiménez-López *et al.* [42].

# 3    Results

## 3.1    Searching the Olive Seed Transcriptome for 11S Sequences

A total of five inputs corresponding to 11S seed storage proteins were obtained from the database generated (Table 1).

**Table 1.** Selected sequences from the olive seed transcriptome identified as 11S seed storage proteins. Lengths and identity of the coding proteins, as well as those of the alpha and beta subunits are indicated, and compared to other species.

| Olive sequences | | | Identity with other species | | |
|---|---|---|---|---|---|
| Identifier | Length (aa) | Description (α/ß-subunits) | Percent | Accession number | Description |
| 040125 | 253 | α: lack N-terminal | 51 % | NP_001291336.1 | 11S globulin seed storage protein 2 precursor [*Sesamum indicum*] |
| | | ß: lack C-terminal | | | |
| 067540 | 165 | α: not sequenced | 56 % | XP_012831308.1 | 11S globulin seed storage protein 2-like [*Erythranthe guttatus*] |
| | | ß: lack N-terminal | | | |
| 012745 | 291 | α: lack N-terminal | 70 % | NP_001291336.1 | 11S globulin seed storage protein 2 precursor [*Sesamum indicum*] |
| | | ß: complete | | | |
| 002407 | 435 | α: complete | 67 % | NP_001291336.1 | 11S globulin seed storage protein 2 precursor [*Sesamum indicum*] |
| | | ß: complete | | | |
| 052095 | 52 | α: lack N-terminal and C-terminal | 67 % | XP_009761684.1 | 11S globulin seed storage protein 2-like [*Nicotiana sylvestris*] |
| | | ß: not sequenced | | | |

## 3.2   Alignment of the NGS-Retrieved Sequences of 11S Proteins to the NCBI Database and Phylogenetic Analysis

BLASTP query of the individual sequences and ScanProsite analysis of motifs confirmed all 11S sequences as seed storage proteins. These sequences exhibited high identity to the 11S globulin precursor 2 (Table 1). Alignment of the sequences with high identity allowed us to identify the presence of both conserved and hyper-variable regions [43] (Fig. 1). Four cysteines involved in the establishment of disulphide bridges between subunits were also identified. From these, C14 and C47 were implicated in intra-chain disulphide bonds [25], while C90 and C251 were involved in the disulphide bridge between the α and ß subunits [44].

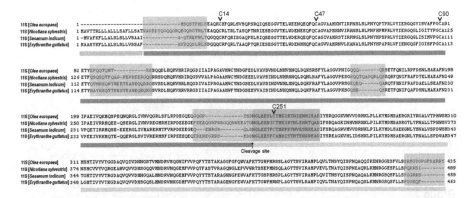

**Fig. 1.** Sequences alignment of the olive 11S (sequence id: 002407) and the 11S isoform 2 from tobacco (accession number: XP_009761684.1), sesame (accession number: NP_001291336.1) and common yellow monkey flower (accession number: XP_012831308.1). Hyper-variable regions are highlighted with gray boxes. The consensus pattern was located within a green box. The signal peptide and the acid/basic subunits (α/ß) are underlined in orange/red/blue colours, respectively. The four cysteine residues involved in disulphide bridges are pointed out with black arrows (Color figure online).

A total of 39 sequences were used to build a phylogenetic tree, which included a selection of the 11S sequences retrieved from the olive seed transcriptome, as well as the most relevant sequences (best BLAST scores) from other plant species. The olive retrieved sequences grouped together, and close to a cluster including the isoform 2 of the 11S precursor protein from other different plant species (named as "Group 1") (Fig. 2). Among such sequences, we found some from plants members of the following orders: Solanales, Gentianales, Sapindales, Ericales, Cariophyllales, Myrtales, Cucurbitales, Malgiphiales, Caryophyliales, Asterales, Rosales, and Lamiales, these later displaying the larger similarity with the olive sequences.

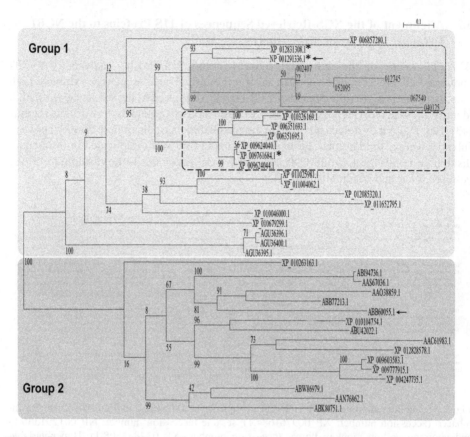

**Fig. 2.** Phylogenetic analysis of olive 11S seed storage protein sequences. Olive sequences obtained by NGS and most relevant orthologous identified in the NCBI database were used. Tree is divided in two big clusters, named Group 1 and Group 2. Group 2 contains sequences less-related to olive from a variety of plant species. Sequences from group 1 included all retrieved olive sequences (dark grey box). Sequences from Lamiales/Solanales are framed with doted and dashed lines. Asterisks (*) highlight the sequences from tobacco, sesame and common yellow monkey flower used for the alignment in Fig. 1. Arrows point out two sequences from group 1 and 2 corresponding to 11S isoform 2 and 4 from the same species (sesame). Measurements of support for the nodes are represented as percentages.

## 3.3 3D Modeling, Docking Analysis and Identification of T-cell Epitopes in the Olive 11S Globulin Subunits α and β

The olive 11S sequence identified as 002407 was used for homology modeling. Two models, built individually for the α subunit (Fig. 3A) and the β subunit (Fig. 3B), showed a β-barrel structure integrated by 11 antiparallel β-strands, which has been described as an specific feature of the cupin superfamily.

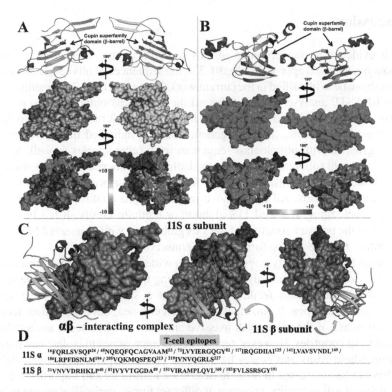

**Fig. 3.** Structural analysis of the olive (Olea europaea L.) 11S protein. Three-dimensional structure of olive 11S protein corresponding to the α (A) and ß (B) subunits. Structures were depicted as a cartoon diagram. α-helices, β-sheets and coils are depicted in red, yellow and green, respectively. Two views rotated 180 degrees round the x-axis are provided for both subunits of surface (green and blue) and electrostatic potential representation. The electrostatic potential surface colors are clamped at red (−10) or blue (+10). (C) Representation of the 11S α/β – interacting complexes. Three views rotated 30 and 45 degrees around the x-axis are provided (α in purple and ß as a cartoon diagram). (D) T-cell epitopes identified for the 11S individual α and ß subunits. Position of the epitope in each subunit is numbered at the beginning and end of the sequence (Color figure online).

The analysis of the surface electrostatic potential revealed several prominent charged residues, with more than half of the side exhibiting extensive number of negative residues in the α-subunit (Fig. 3A), and positive residues in the β-subunit (Fig. 3B). Docking analysis revealed predicted features for the heterodimeric association of α/β subunits (Fig. 3C). The models generated are in good agreement with the models present in the PDB database, corresponding to ortologous forms from other species. Both, the IE and IA faces, as well as the covalent and non covalent interactions between subunits were also predicted to occur properly in the olive forms.

The identification of T-cell epitopes in olive 11S (Fig. 3D) showed a large difference between both subunits in the number of epitopes predicted (eight and four for α and ß subunits, respectively).

# 4    Discussion

Although evidence of the presence of 11S proteins in the olive seed is available throughout biochemical approaches [9, 30–33], the sequences of olive seed 11S globulin transcripts have been identified in the current work for the first time. The results retrieved from the BLASTP analyses revealed that all the sequences obtained from olive transcriptomic approaches shared high identity to the isoform 2 of these proteins in other plant species. Furthermore, the phylogenetic analysis indicated that olive isoform 2 sequences grouped with orthologous sequences from other species as well. Structural analysis of both α/β subunits conforming the isoform 2 sequence showed a comparable 2-D distribution of elements, overall 3D folding and surface and charges to other described cupin proteins. All these above results lead to confirm the identity of the isoform 2 as belonging to a seed 11S globulin or legumin family. It has been largely described that the primary structure of the 11S globulin is well-conserved [45, 46]. Our results confirm that the 11S isoform 2 is also conserved in the olive.

Such high level of identity, together with the wide presence of these proteins in the olive seed represent convincing arguments favouring the potential use of olive seeds for animal and/or even human feeding purposes, in a similar way to other legumin containing plants like pea, lupin, chicken pea, pigeon pea, almond, coffee, lentil and others. In the case of the olive seed, massive amounts of olive seeds are available in those producing countries. However, technological approaches to isolate and purify the seeds from whole fruits and stones have yet to be developed. Moreover, additional studies have to be conducted in order to determine additional and important aspects of these proteins like digestibility, presence of different forms, variability among cultivars, or allergenicity. Knowledge of the corresponding transcript sequences is of particular interest as a way to determine these aspects, or to generate recombinant proteins able to easy the design and improve the reliability of numerous assays.

Food allergy is a growing world concern. Seed proteins are of special relevance since a high percentage of allergies diagnosed in atopic patients are caused by proteins present in seeds (i.e., sesame, mustard, grains as cereals), legumes (i.e., soybean, peanut, lupin) or nuts, (i.e., hazelnut and almond) almond [47–52]. Some of the proteins responsible for these effects are also responsible for cross-allergenicity (i.e., between tree nuts and peanut and mustard [50]), and include 11S and 7S SSPs. The identification of T-cell epitopes in the olive 11S, as reported here, revealed that these proteins might also putatively trigger allergy phenomena in humans. Moreover, the model proposed here indicates that alpha subunits from olive 11S may be more reactive than beta subunits, as a larger number of T-epitopes was predicted in the former. These analyses could help to identify hyper- and low- reactive forms of the protein, and for extension, forms with a higher or lower potential of cross-reactivity to other species. This knowledge would help to prevent allergy problems among the population providing that the olive seeds could become a source of proteins form human use, either on the whole, or as elaborated foods (i.e., flours, protein isolates, protein hydrolisates, etc.) [53].

# 5 Conflict of Interest

The authors confirm that this article content has no conflicts of interest.

**Acknowledgments.** This work was supported by ERDF-cofunded projects RTC-2015-4181-2 (MINECO), 201540E065 (CSIC) and P2010-AGR-6274 and P2011-CVI-7487 (Junta de Andalucía). A. Zafra thanks the Agrifood Campus of International Excellence ceiA3/Elayotecnia S.L. for grant funding. This research was also partially supported by the European Research Program MARIE CURIE (FP7-PEOPLE-2011-IOF), under the grant reference number PIOF-GA-2011-301550 to JCJ-L and JDA. JCJ-L thanks the Spanish Ministry of Economy and Competitiveness for the grant ref. number RYC-2014-16536 (Ramón y Cajal Research Program).

# References

1. FAO Water Development and Management Unit. http://www.fao.org/nr/water/cropinfo_olive.html
2. Sola-Guirado, R.R., Castro-García, S., Blanco-Roldán, G.L., Jiménez-Jiménez, F., Castillo-Ruiz, F.J., Gil-Ribes, J.A.: Traditional olive tree response to oil olive harvesting technologies. Biosyst. Eng. **118**, 186–193 (2014)
3. Esposito, K., Marfella, R., Ciotola, M., Di Palo, C., Giugliano, F., Giugliano, G., D´Armiento, M., D´Andrea, F., Giugliano, D.: Effect of a mediterranean-style diet on endothelial dysfunction and markers of vascular inflammation in the metabolic syndrome: a randomized trial. JAMA **292**, 1440–1446 (2004)
4. Covas, M.-I., Nyyssönen, K., Poulsen, H.E., Kaikkonen, J., Zunft, H.-J.F., Kiesewetter, H., Gaddi, A., de la Torre, R., Mursu, J., Bäumler, H., Nascetti, S., Salonen, J.T., Fitó, M., Virtanen, J., Marrugat, J.: The effect of polyphenols in olive oil on heart disease risk factorsa randomized trial. Ann. Intern. Med. **145**, 333–341 (2006)
5. Covas, M.: Olive oil and the cardiovascular system. Pharmacol. Res. **55**, 175–186 (2007)
6. Martín-Peláez, S., Mosele, J.I., Pizarro, N., Farràs, M., de la Torre, R., Subirana, I., Pérez-Cano, F.J., Castañer, O., Solà, R., Fernandez-Castillejo, S., Heredia, S., Farré, M., Motilva, M.J., Fitó, M.: Effect of virgin olive oil and thyme phenolic compounds on blood lipid profile: implications of human gut microbiota. Eur. J. Nutr. 1–13 (2015)
7. Davis, C., Bryan, J., Hodgson, J., Murphy, K.: Definition of the mediterranean diet: a literature review. Nutrients **7**, 9139–9153 (2015)
8. Shewry, P.R., Napier, J.A., Tatham, A.S.: Seed storage proteins: structures and biosynthesis. Plant Cell **7**, 945–956 (1995)
9. Alche, J., Jimenez-Lopez, J., Wang, W., Castro-Lopez, A., Rodriguez-Garcia, M.: Biochemical characterization and cellular localization of 11S type storage proteins in olive (Olea europaea L.) seeds. J. Agric. Food Chem. **54**, 5562–5570 (2006)
10. Shutov, A., Bäumlein, H., Blattner, F., Müntz, K.: Storage and mobilization as antagonistic functional constraints on seed storage globulin evolution. J. Exp. Bot. **54**, 1645–1654 (2003)
11. Osborne, T.B.: The Vegetable Proteins. Monographs on Biochemistry, 2nd edn, p. xiii+154. Longmans Green and Co, London (1924)
12. Dunwell, J.M.: Cupins: a new superfamily of functionally diverse proteins that include germins and plant storage proteins. Biotechnol. Genet. Eng. Rev. **15**, 1–32 (1998)
13. Shutov, A.D., Kakhovskaya, I.A., Braun, H., Baumlein, H., Müntz, K.: Legumin-like and vicilin-like seed storage proteins: evidence for a common single-domain ancestral gene. J. Mol. Evol. **41**, 1057–1069 (1995)

14. Li, C., Li, M., Dunwell, J.M., Zhang, Y.-M.: Gene duplication and an accelerated evolutionary rate in 11S globulin genes are associated with higher protein synthesis in dicots as compared to monocots. BMC Evol. Biol. **12**, 15 (2012)

15. Casey, R.: Distribution and some properties of seed globulins. In: Shewry, P.R., Casey, R. (eds.) Seed Proteins, pp. 159–169. Kluwer Academic Publishers, Dordrecht (1999)

16. Plietz, P., Damaschun, G., Zirwer, D., Gast, K., Scwenke, K., Prakash, V.: Shape and quaternary structure of a-globulin from sesame (Sesamun indicum L.) seed as revealed by small angle x-ray scattering and quasi-elastic light scattering. J. Biol. Chem. **261**, 12686–12691 (1986)

17. Lycett, G.W., Croy, R.R.D., Shirsat, A.H., Boulter, D.: The complete nucleotide sequence of a legumin gene from pea (Pisum sativum L.). Nucleic Acids Res. **12**, 4493–4506 (1984)

18. Momma, T., Negoro, T., Hirano, H., Matsumoto, A., Udaka, K., Fukazawa, C.: Glycinin A5A4B3 mRNA: cDNA cloning and nucleotide sequencing of a splitting storage protein subunit of soybean. Eur. J. Biochem. **149**, 491–496 (1985)

19. Chlan, C.A.: cDNA and amino acid sequences of members of the storage protein families. Plant Mol. Biol. **7**, 475–489 (1986)

20. Hayashi, M., Mori, H., Nishimura, M., Akazawa, T., Hara-Nishimura, I.: Nucleotide sequence of cloned cDNA coding for pumpkin 11-S globulin beta subunit. Eur. J. Biochem. **172**, 627–632 (1988)

21. Ohmiya, M., Hara, I., Matsubara, H.: Pumpkin (Cucurbita sp.) seed globulin IV. Terminal sequences of the acidic and basic peptide chains and identification of a pyroglutamyl peptide chain. Plant Cell Physiol. **21**, 157–167 (1980)

22. Dickinson, C.D., Hussein, E.H., Nielsen, N.C.: Role of posttranslational cleavage in glycinin assembly. Plant Cell **1**, 459–469 (1989)

23. Hara-Nishimura, I., Shimada, T., Hiraiwa, N., Nishimura, M.: Vacuolar processing enzyme responsible for maturation of seed proteins. J. Plant Physiol. **145**, 632–640 (1995)

24. Jung, R., Scott, M.P., Nam, Y.W., Beaman, T.W., Bassüner, R., Saalbach, I., Müntz, K., Nielsen, N.C.: The role of proteolysis in the processing and assembly of 11S seed globulins. Plant Cell **10**, 343–357 (1998)

25. Jung, R., Nam, Y.-W., Saalbach, I., Müntz, K., Nielsen, N.C.: Role of the sulfhydryl redox state and disulfide bonds in processing and assembly of 11 s seed globulins. Plant Cell **9**, 2037–2050 (1997)

26. Mills, E.N.C., Marigheto, N.A., Wellner, N., Fairhurst, S.A., Jenkins, J.A., Mann, R., Belton, P.S.: Thermally induced structural changes in glycinin the 11S globulin of soya bean (Glycine max)—an in situ spectroscopic study. Biochim. Biophys. Acta (BBA) Proteins and Proteomics **1648**, 105–114 (2003)

27. Baumgartner, B., Chrispeels, M.J.: Purification and characterization of vicilin peptidohydrolase, the major endopeptidase in the cotyledons of mung-bean seedlings. Eur. J. Biochem./FEBS **77**, 223–233 (1977)

28. Baumgartner, B., Tokyyasu, K., Chrispeels, M.: Localization of vicilin peptidohydrolase in the cotyledons of mung bean seedlings by immunofluorescence microscopy. J. Cell Biol. **79**, 10–19 (1978)

29. Müntz, K., Belozersky, M.A., Dunaevsky, Y.E., Schlereth, A., Tiedemann, J.: Stored proteinases and the initiation of storage protein mobilization in seeds during germination and seedling growth. J. Exp. Bot. **52**, 1741–1752 (2001)

30. Wang, W., Alche, J., Castro, A., Rodríguez-García, M.I.: Characterization of seed storage proteins and their synthesis during seed development in Olea europaea. Int. J. Dev. Biol. **45**, 63–64 (2001)

31. Jimenez-Lopez, J.C., Hernandez-Soriano, M.C.: Protein bodies in cotyledon cells exhibit differential patterns of legumin-like proteins mobilization during seedling germinating states. Am. J. Plant. Sci. **4**, 2444–2454 (2013)
32. Zienkiewicz, A., Zienkiewicz, K., Rejón, J.D., Alché, JdD, Castro, A.J., Rodríguez-García, M.I.: Olive seed protein bodies store degrading enzymes involved in mobilization of oil bodies. J. Exp. Bot. **65**, 103–115 (2014)
33. Jimenez-Lopez, J.C., Zienkiewicz, A., Zienkiewicz, K., Alché, J.D., Rodríguez-García, M.I.: Biogenesis of protein bodies during legumin accumulation in developing olive (Olea europaea L.) seed. Protoplasma 1–14 (2015)
34. Sambrook, J., Rusell, D.: Molecular Cloning: a laboratory manual. Cold Spring Harbor Laboratory Press, New York (2001)
35. Zafra, A., Carmona, R., Jimenez-Lopez, J., Pulido, A., Claros, M.G., de Dios Alché, J.: Identification of distinctive variants of the olive pollen allergen ole e 5 (cu,zn superoxide dismutase) throughout the analysis of the olive pollen transcriptome. In: Ortuño, F., Rojas, I. (eds.) IWBBIO 2015, Part I. LNCS, vol. 9043, pp. 460–470. Springer, Heidelberg (2015)
36. Gouy, M., Guindon, S., Gascuel, O.: SeaView version 4: a multiplatform graphical user interface for sequence alignment and phylogenetic tree building. Mol. Biol. Evol. **27**, 221–224 (2010)
37. Abascal, F., Zardoya, R., Posada, D.: ProtTest: selection of best-fit models of protein evolution. Bioinformatics **21**, 2104–2105 (2005)
38. Guex, N., Peitsch, M.C.: SWISS-MODEL and the Swiss-PdbViewer: an environment for comparative protein modeling. Electrophoresis **18**, 2714–2723 (1997)
39. Benkert, P., Biasini, M., Schwede, T.: Toward the estimation of the absolute quality of individual protein structure models. Bioinformatics **27**, 343–350 (2011)
40. Arnold, K., Bordoli, L., Kopp, J., Schwede, T.: The SWISS-MODEL workspace: a web-based environment for protein structure homology modelling. Bioinformatics **22**, 195–201 (2006)
41. Schwede, T., Kopp, J., Guex, N., Peitsch, M.C.: SWISS-MODEL: an automated protein homology-modeling server. Nucleic Acids Res. **31**, 3381–3385 (2003)
42. Jimenez-Lopez, J.C., Kotchoni, S.O., Hernandez-Soriano, M.C., Gachomo, E.W., Alché, J.D.: Structural functionality, catalytic mechanism modeling and molecular allergenicity of phenylcoumaran benzylic ether reductase, an olive pollen (Ole e 12) allergen. J. Comput. Aided Mol. Des. **27**, 873–8795 (2013)
43. Wright, D.J.: The seed globulins. In: Hudson, B.J.F. (ed.) Development in Food Proteins, p. 81. Elsevier, London (1897)
44. Shotwell, M.A., Afonso, C., Davies, E., Chesnut, R.S., Larkins, B.A.: Molecular characterization of oat seed globulins. Plant Physiol. **87**, 698–704 (1988)
45. Nielsen, N.C., Dickinson, C.D., Cho, T.J., Thanh, V.H., Scallon, B.J., Fischer, R.L., Sims, T.L., Drews, G.N., Goldberg, R.B.: Characterization of the glycinin gene family in soybean. Plant Cell **1**, 313–328 (1989)
46. Adachi, M., Takenaka, Y., Gidamis, A.B., Mikami, B., Utsumi, S.: Crystal structure of soybean proglycinin A1aB1b homotrimer. J. Mol. Biol. **305**, 291–305 (2001)
47. Koppelman, S.J., Knol, E.F., Vlooswijk, R.A.A., Wensing, M., Knulst, A.C., Hefle, S.L., Gruppen, H., Piersma, S.: Peanut allergen Ara h 3: isolation from peanuts and biochemical characterization. Allergy **58**, 1144–1151 (2003)
48. Beyer, K., Bardina, L., Grishina, G., Sampson, H.A.: Identification of sesame seed allergens by 2-dimensional proteomics and Edman sequencing: seed storage proteins as common food allergens. J. Allergy Clin. Immunol. **110**, 154–159 (2002)

49. Beyer, K., Grishina, G., Bardina, L., Grishin, A., Sampson, H.A.: Identification of an 11S globulin as a major hazelnut food allergen in hazelnut-induced systemic reactions. J. Allergy Clin. Immunol. **110**, 517–523 (2002)
50. Sirvent, S., Akotenou, M., Cuesta-Herranz, J., Vereda, A., Rodríguez, R., Villalba, M., Palomares, O.: The 11S globulin Sin a 2 from yellow mustard seeds shows IgE cross-reactivity with homologous counterparts from tree nuts and peanut. Clin. Transl. Allergy **2**, 23 (2012)
51. Schiller, D., Hellmuth, M., Gubesch, M., Ballmer-Weber, B., Bindslev-Jensen, C., Niggemann, B., Scibillia, J., Hanschmann, K.-M., Kühne, Y., Reuter, A., Vieths, S., Wangorsch, A., Holzhauser, T.: Soybean allergy: IgE epitopes of glycinin (Gly m 6), an important soybean allergen. Clin. Transl. Allergy **4**, O2 (2014)
52. Willison, L.N., Tripathi, P., Sharma, G., Teuber, S.S., Sathe, S.K., Roux, K.H.: Cloning, expression and patient IgE reactivity of recombinant Pru du 6, an 11S globulin from almond. Int. Arch. Allergy Immunol. **156**, 267–281 (2011)
53. Zafra, A., Zienkiewicz, A., Clemente, A., Al-lach, S., Fernandez-Márquez, A., Martín-Aznarte, I., Rueda, A., Salmerón, C., Jimenez-López, J.C., Castro, A.J., Rodríguez-García, M.I., Alché, J.D.D.: Aislados proteicos de semillas de olivo: composición química, propiedades funcionales y caracterización proteica. XVI Simposio Científico Técnico del Aceite de Oliva. 1–7 (2013)

# A Feature Selection Scheme for Accurate Identification of Alzheimer's Disease

Hao Shen[1,2,3], Wen Zhang[1,2,3], Peng Chen[4], Jun Zhang[5], Aiqin Fang[6],
and Bing Wang[1,2,3(✉)]

[1] School of Electronics and Information Engineering, Tongji University, Shanghai China
wangbing@ustc.edu
[2] The Advanced Research Institute of Intelligent Sensing Network,
Tongji University, Shanghai China
[3] The Key Laboratory of Embedded System and Service Computing,
Tongji University, Shanghai China
[4] Institute of Health Sciences, Anhui University, Hefei Anhui, China
[5] College of Electrical Engineering and Automation, Anhui University, Hefei Anhui, China
[6] Department of Chemistry, University of Louisville, Louisville, KY 40292, USA

**Abstract.** Effective biomarkers play important roles for accurate diagnosis of Alzheimer's Disease (AD), including its intermediate stage (i.e. mild cognitive impairment, MCI). In this paper, a new feature selection scheme was proposed to improve the identification AD and MCI from healthy controls (HC) by a support vector machine (SVM) based-classifier with recursive feature addition. Our method can find the significant features automatically, and the experiments in this work demonstrates that our scheme can achieve better classification performance based on a dataset with 103 subjects where three biomarkers, i.e., structural MR imaging (MRI), functional imaging PET, and cerebrospinal fluid(CSF), had been used. Our proposed method demonstrated its effectiveness in identifying AD from HC with an accuracy of 95.0 %, while only 89.3 % for the classifier without the step of feature selection. In addition, some features selected in this work had shown strong relation with AD by other previous studies, which can provide the support for the significance of our results.

**Keywords:** Mild cognitive impairment (MCI) · Alzheimer's disease (AD) · Support vector machine (SVM) · Feature selection (FS)

## 1 Introduction

Alzheimer's Disease (AD), a most frequent type of psychiatric or neurological disorders, is becoming a growing health problem. The main characteristic of this brain disease is to make people produce a gradual loss of cognitive functions. It is reported that an increasing number of elderly people are affected by this dementia from poles to poles every year [1, 2]. Obviously, reliable diagnosis of AD is not merely significant but also a challenging for subsequent treatments.

© Springer International Publishing Switzerland 2016
F. Ortuño and I. Rojas (Eds.): IWBBIO 2016, LNBI 9656, pp. 71–81, 2016.
DOI: 10.1007/978-3-319-31744-1_7

Even though definitive cure to date has not been found, it is well-known that AD is correlated with various aspects of neurodegeneration, such as structural atrophy [3–5], pathological amyloid depositions [4], decreased blood perfusion [6] and metabolic alteration [4, 7] in the brain. Some of these different modalities of biomarkers are allowed the non-invasive investigation of human brain by modern neuroimaging techniques, including poliodystrophia measured in structural magnetic resonance imaging (sMRI) [8, 9], glucose metabolism measured by 18-fluorodeoxyglucose positron emission tomography (FDG-PET) [8], cerebral blood flow measured in single photon emitting computer tomography (SPECT) [6], and amyloid burden in brain tissue measured through Pittsburgh Compound-B (PIB) [10]. In addition, some biological or genetic biomarkers also have been used for diagnosis of AD. For example, neurofibrillary tangle pathology effected by increased CSF total tau (t-tau) and tau hyperphosphorylated threonine 181 (p-tau), amyloid plaque pathology related to the decrease amyloid β (Aβ42), cognitive decline or conversion to AD reflected through the presence of the apolipoprotein E(APOE) ε4 allele [11–13].

In addition, for further accurate diagnosis of AD, machine learning and pattern recognition techniques are often proposed to distinguish those above-mentioned biomarkers from two groups of subjects, i.e. patients vs. controls [14]. These techniques can generalize a pattern from biomarkers (e.g. MRI individually [15, 16]) of AD training samples to recognize the state of a new test sample. Also, an appropriate combination of different biomarkers as features may have a better performance has been proved by many studies, such as combination of MRI and CSF [17, 18], or MRI, CSF and PET [17, 19, 20], or MRI, CSF, PET and APOE [21], cognitive scores. These works have achieved promising results, but few studies have previously investigated deeply a suitable features selection for these features of biomarkers and got a desired performance [18, 20]. Feature selection is a crucial technique which selects the most discriminative features and reduces the redundant features at the meantime in machine learning and pattern recognition field.

In this work, we investigated the role of feature selection for the diagnosis of AD, and we found a suitable selection strategy that can better predict the disease. Three biomarkers, i.e. MRI, PET and CSF, which suggested by a criterion of the National Institute of Neurological and Communicative Disorders and Stroke and the Alzheimer's disease and related Disorders Association are used as our baseline features, which our feature selection strategy will be implemented on [22, 23]. Then a support vector machine (SVM) algorithm based classifier will be trained on the selected features to predict AD from healthy controls (HCs). Here a 10-cross validation is adopted to performance evaluation for the predictor. The experimental results achieved in this study demonstrated the effectiveness of our proposed feature selection method in the diagnosis of AD.

## 2 Materials and Methods

### 2.1 Support Vector Machines

Support vector machine is a type of popular machine learning or pattern recognition method that proposed by Vapnik [24], which is featured by a low generalization error and computational overhead in practice. There are a lot of applications in different research studies because of its successful performance, including in neuroscience data analysis [14]. Here, we performed our work using LIBSVM package [25], with a linear kernel and a default value for the parameter $C$ (i.e. $C = 1$), for classification.

## 2.2 Feature Selection

As an important part for machine learning or pattern recognition, specially, in high dimension and small samples problems, a proper feature selection method can improve classification accurate by wiping out those redundant or irrelevant features. Whilst reducing the number of features means reducing the calculation effort and speeding up the learning process. And among previous studies, there is increasing interest in applying feature selection to enhance the performance of AD classification [26].

A recursive feature addition based-feature selection strategy which can combine with SVM method had been adopted in this work. For each subject in our dataset, there are 189 features which represent pre-selected ROIs based on prior knowledge. Before the features selected, a two-sample t-test was performed respectively for each feature in training set, and a feature array, F, can be obtained where all ROIs were ranked according to their p-value. Firstly, the top 1st ROI in F will be used as feature vector input into the SVM classification, and selected into S, the feature set selected. Then, the top 1st and 2nd ROIs were put to the SVM for classification, and if the performance is better than before, the 2nd feature will be selected into the set S, otherwise, not. In this way, a new feature selected sequentially from the array F will be combined with the feature in S, and put into the classifier. If the classification performance can be improved, this feature will be selected into S, otherwise wiped. Finally, a feature set S will be selected with the best combination of important ROIs. The workflow can be found in Fig. 1.

## 2.3 Validation

To assess the performance of a classifier in an objective way, a 10-fold cross-validation strategy is used to our work. The whole sample set was randomly divided into 10 subsets in the almost same size, each time one subset was selected as test samples and all the rest subjects were treated as training samples. For the sake of avoiding any tendency or bias affected by random partition, this process would be repeated for 10 times independently. Moreover, we measured the classification accuracy (i.e. the proportion of whole subjects correctly classified), sensitivity (i.e. the proportion of whole patients correctly classified) and specificity (i.e. the proportion of whole healthy controls correctly classified) as indicators of classifier's performance. Additionally, a normalization strategy (i.e. $f_i = (f_i - \min_{fi})/(\max_{fi} - \min_{fi})$) was performed for each feature $f_i$, where $\min_{fi}$ and $\max_{fi}$ are the minimum and maximum of the $i$th feature for whole training samples.

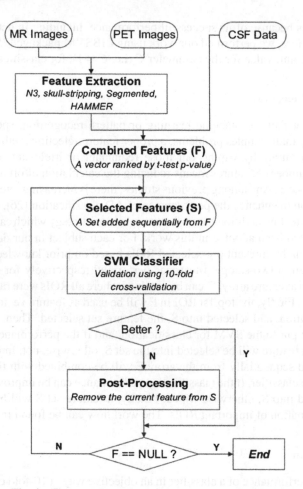

**Fig. 1.** The workflow of our proposed in training phase

# 3  Results

## 3.1  Subjects and Feature Extracting

In this work, 103 subjects between 55–90 years of age had been used which includes 51 AD patients and 52 healthy controls. In the data preprocessing step, a voxel-based workflow was adapted to all MRIs and then the volume of their ROIs region was calculated. After that, all PET images had been implemented by a rigid transformation and the average intensity of each ROI region would be measured as a feature. Also, corresponding CSF information for each subject were downloaded from the ADNI website. More detailed information can be found in Zhang's work [20]. Finally, each subject can be represented by 189 features as a vector, where 93 of them are corresponding for MRI, 93 for PET and 3 for CSF.

## 3.2 AD Classification and Feature Selection

Firstly, we test the performance of our classification model with feature selection for 103 subjects in ADNI, based on MRI, PET and CSF biomarkers. Table 1 shows the classification accuracy, sensitivity and specificity of the different methods. It should be noted that this table only shows the averaged results of 10 independent tests and the minimal and maximal values in brackets. Moreover, the corresponding ROC curves of different methods were plotted in Fig. 2.

It can be seen that our feature selection approach can achieve better performance than the methods without a feature selection for classifying ADs from Health controls. Feature selection for the combined features of MRI, PET and CSF biomarkers results in a classification accuracy of 95.0 % (sensitivity = 97.6 %, specificity = 92.7 % and

**Table 1.** Performance of single-model and multi-model AD classification.

|  | ACC (%) | SEN (%) | SPE (%) |
|---|---|---|---|
| MRI | 86.7(85.3–87.7) | 88.7(87.2–90.1) | 85.1(83.8–86.1) |
| MRI(FS) | 88.8(87.8–91.7) | 96.2(94.7–97.3) | 83.8(82.6–88.1) |
| PET | 85.0(84.1–85.9) | 84.9(83.8–86.2) | 85.3(84.4–86.5) |
| PET(FS) | 90.7(88.9–92.5) | 90.2(88.6–91.6) | 91.3(88.2–93.8) |
| Combined | 89.3(88.6–89.8) | 88.9(87.8–89.8) | 89.7(88.7–90.8) |
| Combined(FS*) | 96.3(95.1–97.1) | 97.6(96.1–98.0) | 95.2(94.2–96.2) |

*FS denotes model with Feature Selection

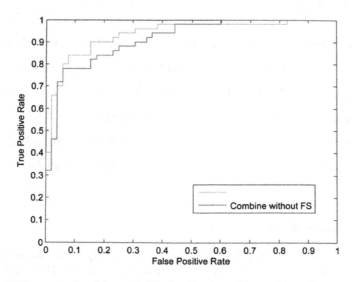

**Fig. 2.** ROC curves of multi-model AD classification with or without feature selection.

AUC = 0.9147) compared to 89.3 % for the model without feature selection (sensitivity = 88.9 %, specificity = 89.7 % and AUC = 0.9146) at baseline.

Additionally, we applied this feature selection method for only MRI measures or PET measures, and it can achieve a better performance, too. Such as MRI, Table 1 shows the results, although less pronounced for AD vs. HC model. For PET-based classifier, the improvement is also obvious. For CSF measures, we did not take any feature selection because the number of the features is very small.

### 3.3    The Selected ROIs Set

It can be found that the 10-cross validation strategy divided the original dataset into 10 subsets randomly, which will cause different features selected by our algorithm. Especially for those selected features have a low p-value in t-test, they did not have a stable perform every time. In this paper, we therefore gathered all ROIs selected in our 10 times experiment and set a present threshold value (5 in this work). If the present number of the ROI is greater than the threshold, the ROI was certainly selected. The results in Table 1 were calculated by using these features. Table 2 presents all the features which we finally used for training our model in AD classification.

**Table 2.** The selected brain regions for AD classification.

|  | Brain regions |
|---|---|
| MRI | Hippocampal formation right; hippocampal formation left; uncus left; inferior temporal gyrus right |
| PET | Precuneus right; precentral gyrus left; angular gyrus left; precuneus left; postcentral gyrus right |
| CSF | Aβ42; T-tau |

**Table 3.** Performance of multi-model ADclassification with differentfeature selectionschema

|  | ACC (%) | SEN (%) | SPE (%) |
|---|---|---|---|
| Forward sequence | 94.8(92.7–96.0) | 96.3(93.7–97.9) | 93.5(91.7–94.8) |
| Reverse sequence | 94.0(92.4–94.9) | 93.5(89.8–95.5) | 94.5(93.3–96.1) |
| Random sequence | 94.2(91.2-96.2) | 94.8(92.0–96.7) | 93.7(89.2–98.0) |
| Recursive feature addition | 96.3(95.1–97.1) | 97.6(96.1–98.0) | 95.2(94.2–96.2) |

### 3.4    The Sequence of Feature Selection

Our feature selection strategy picked up features sequentially from a fixed sequence to the training model if the added feature can improve the accuracy. This fixed sequence is a crucial point to our feature selection. To investigate the effect of different sequence on the performance of our classification model, we tested four selected strategies (forward sequence, reverse sequence, random sequence and the sequence sorted by p-value of t-test) for the combined features and compared their results in Table 3. It can be seen that the sequence sorted by p-value of t-test outperformed other three schemes, even though the other schemes can also improve the performance of classification model.

## 4   Discussion

In this paper, we proposed a strategy of feature selection to improve the identification of AD patients from healthy controls automatically, which adopts a scheme of recursive feature addition, and the results showed a better classification performance. The outcome on 202 subjects from ADNI database clearly demonstrated that our feature selection can improve the capability of classification model for AD, and achieve a higher accuracy (96.3 % for AD classification).

### 4.1   Data Fusion and Classification

For diagnosis of AD, different kinds of biomarkers may contain complementary information, which was already reported by many previous studies [13, 19, 20, 27]. Recently, several methods for combine different biomarkers have been proposed, where the different biomarkers had been combined with an ingenious kernel method and showed a good performance [19, 20]. However, many works just fused the different modalities into a longer vector [12, 21, 28], which might bring some noise or abundant information into the analysis. In this work, a feature selection strategy were proposed to investigate the significant ROIs to gain a better performance for AD classification.

From Table 1, the results proved our feature selection for combined vector can work well for AD classification. The approach was more robust and accurate with a high sensitivity and a low specificity. This characteristic of classification model may be advantageous for the purpose of diagnosis, because the cost for misclassified a patient into a healthy control is much higher than misclassified a healthy control into a patient. And we found this feature selection can also have a better performance for individual modality classification.

Furthermore, we compared our work with Zhang's method [20], where a kernel combination had been adopted to fuse MRI, PET and CSF data, and showed an accuracy (93.2 %) for AD classification. Its performance is inferior to our method, and the results can be seen in Table 4.

**Table 4.** Comparison of performance with Zhang's Method.

|                | ACC (%)          | SEN (%)          | SPE (%)          |
|----------------|------------------|------------------|------------------|
| Zhang's method | 93.2(89.0–96.5)  | 93.0(88.7–96.3)  | 93.3(89.1–96.6)  |
| Our method     | 96.3(95.1–97.1)  | 97.6(96.1–98.0)  | 95.2(94.2–96.2)  |

## 4.2   Feature Selection

As a selective method based a scheme of recursive feature addition, our proposed approach can select the most suitable feature set automatically to SVM model for training and utilize the whole information from the original data. As shown in Table 1, the performance of our classification model had been tested on two situations, i.e. with and without feature selection. It can be found that the accuracy of the model with feature selection is obviously higher than the one without feature selection, which implies that the original input data contain some noise information. Removing non-informative features and increasing the contrast between different groups indeed improve classification performance, as shown in this work.

Recently, many studies had applied feature selection for their classification work, especially when the dimensionality of input feature space is larger than the number of subjects, and t-test statistics has been widely applied in the classification when pre-selected regions were used as features [29, 30]. Compared to other feature selection approaches, the advantage of our method is its ability of automatically selecting the most suitable features for discriminate among the subjects, while several researches [18, 20] only selected top n (a fixed constant) features that sorted by p-values to their train model. However, the features with low p-value may support complementary information for AD diagnosis, just like the different modalities contain complementary information for classification. Therefore we did not select top n features, but a combination of features with the highest differential ability.

Furthermore, some of the brain regions we selected, including hippocampus, uncus, and temporal pole et al. had been proven to be related to AD in many previous studies [20, 31, 32]. For example, hippocampus plays an important role in the memory and cognition of human, and also, it is a crucial region related to spatial navigation and the consolidation of information from working memory to long-term memory. Hippocampal atrophy is one of the earliest signs for AD patient [33] and many MRI studies measured hippocampal volumes individually for AD classification [34, 35]. Both uncus and temporal pole, are a part of temporal lobe which is vital for visual memories and language comprehension [36–38].

## 5   Conclusion

In this paper, we proposed a strategy to optimize the prediction of AD diagnose by combining feature selection into the predictor. Our feature selection approach can pick up the most suitable features automatically into the training of SVM model. In this work, a 10-fold cross validation was implemented, and a small variance of prediction error can be achieved, which demonstrated the robustness of our method. Compared to other models with feature selection, the greatest advantage of our method is that the training model can replenish automatically the necessary features for a better prediction. Moreover, some features selected in our strategy had been reported by previous works that they are related to AD or other psychological diseases, which can prove the effectiveness of our model. Nevertheless, for the application of clinical diagnosis, only a good performance for AD classification is not enough. In the future, we will further investigate the multi-class classification of AD, and we will seek more effective methods in machine learning or pattern recognize for classification, i.e. ensemble learning, to address the problems in this work.

**Acknowledgments.** This work was supported by the National Natural Science Foundation of China (Nos. 61300058, 61472282 and 61374181), Anhui Provincial Natural Science Foundation (No. 1508085MF129). The authors give special thanks to Professor D.Q. Zhang in Nanjing University of Aeronautics and Astronautics for his work in data preprocessing, and the data support from ADNI.

# References

1. Ramaroson, H., Helmer, C., Barberger-Gateau, P., Letenneur, L., Dartigues, J.F.: Prevalence of dementia and Alzheimer's disease among subjects aged 75 years or over: updated results of the PAQUID cohort. Rev. Neurol. (Paris) **159**, 405–411 (2003)
2. Brookmeyer, R., Johnson, E., Ziegler-Graham, K., Arrighi, H.M.: Forecasting the global burden of Alzheimer's disease. Alzheimer's Dement. **3**, 186–191 (2007)
3. De Toledo-Morrell, L., Stoub, T.R., Bulgakova, M., Wilson, R.S., Bennett, D.A., Leurgans, S., et al.: MRI-derived entorhinal volume is a good predictor of conversion from MCI to AD. Neurobiol. Aging **25**, 1197–1203 (2004)
4. Nestor, P.J., Scheltens, P., Hodges, J.R.: Advances in the early detection of Alzheimer's disease. Nat. Med. **10**(suppl.), S34–S41 (2004). (Review)
5. Jack Jr., C.R., Shiung, M.M., Weigand, S.D., O'Brien, P.C., Gunter, J.L., Boeve, B.F., et al.: Brain atrophy rates predict subsequent clinical conversion in normal elderly and amnestic MCI. Neurology **65**, 1227–1231 (2005)
6. Ramírez, J., Górriz, J.M., Salas-Gonzalez, D., Romero, A., López, M., Álvarez, I., et al.: Computer-aided diagnosis of Alzheimer's type dementia combining support vector machines and discriminant set of features. Inf. Sci. **237**, 59–72 (2013)
7. Hoffman, J.M., Welsh-Bohmer, K.A., Hanson, M., Crain, B., Hulette, C., Earl, N., et al.: FDG PET imaging in patients with pathologically verified dementia. J. Nucl. Med. **41**, 1920–1928 (2000)
8. Ishii, K., Sasaki, H., Kono, A.K., Miyamoto, N., Fukuda, T., Mori, E.: Comparison of gray matter and metabolic reduction in mild Alzheimer's disease using FDG-PET and voxel-based morphometric MR studies. Eur. J. Nucl. Med. Mol. Imaging **32**, 959–963 (2005)
9. McEvoy, L.K., Fennema-Notestine, C., Roddey, J.C., Hagler Jr., D.J., Holland, D., Karow, D.S., Pung, C.J., Brewer, J.B., Dale, A.M.: Alzheimer disease: quantitative structural neuroimaging for detection and prediction of clinical and structural changes in mild cognitive impairment. Radiology **251**, 195–205 (2009)
10. Klunk, W.E., Engler, H., Nordberg, A., Wang, Y., Blomqvist, G., Holt, D.P., et al.: Imaging brain amyloid in Alzheimer's disease with Pittsburgh Compound-B. Ann. Neurol. **55**, 306–319 (2004)
11. Ji, Y., Permanne, B., Sigurdsson, E.M., Holtzman, D.M., Wisniewski, T.: Amyloid beta40/42 clearance across the blood-brain barrier following intra-ventricular injections in wild-type, apoE knock-out and human apoE3 or E4 expressing transgenic mice. J. Alzheimers Dis. **3**, 23–30 (2001)
12. Bouwman, F.H., Schoonenboom, S.N., van der Flier, W.M., van Elk, E.J., Kok, A., Barkhof, F., et al.: CSF biomarkers and medial temporal lobe atrophy predict dementia in mild cognitive impairment. Neurobiol. Aging **28**, 1070–1074 (2007)

13. Fjell, A.M., Walhovd, K.B., Fennema-Notestine, C., McEvoy, L.K., Hagler, D.J., Holland, D., et al.: CSF biomarkers in prediction of cerebral and clinical change in mild cognitive impairment and Alzheimer's disease. J. Neurosci. **30**, 2088–2101 (2010)
14. Orru, G., Pettersson-Yeo, W., Marquand, A.F., Sartori, G., Mechelli, A.: Using support vector machine to identify imaging biomarkers of neurological and psychiatric disease: a critical review. Neurosci. Biobehav. Rev. **36**, 1140–1152 (2012)
15. Vemuri, P., Gunter, J.L., Senjem, M.L., Whitwell, J.L., Kantarci, K., Knopman, D.S., et al.: Alzheimer's disease diagnosis in individual subjects using structural MR images: validation studies. Neuroimage **39**, 1186–1197 (2008)
16. Magnin, B., Mesrob, L., Kinkingnehun, S., Pelegrini-Issac, M., Colliot, O., Sarazin, M., et al.: Support vector machine-based classification of Alzheimer's disease from whole-brain anatomical MRI. Neuroradiology **51**, 73–83 (2009)
17. Kohannim, O., Hua, X., Hibar, D.P., Lee, S., Chou, Y.Y., Toga, A.W., et al.: Boosting power for clinical trials using classifiers based on multiple biomarkers. Neurobiol. Aging **31**, 1429–1442 (2010)
18. Westman, E., Muehlboeck, J.S., Simmons, A.: Combining MRI and CSF measures for classification of Alzheimer's disease and prediction of mild cognitive impairment conversion. Neuroimage **62**, 229–238 (2012)
19. Walhovd, K.B., Fjell, A.M., Brewer, J., McEvoy, L.K., Fennema-Notestine, C., Hagler Jr., D.J., et al.: Combining MR imaging, positron-emission tomography, and CSF biomarkers in the diagnosis and prognosis of Alzheimer disease. AJNR Am. J. Neuroradiol. **31**, 347–354 (2010)
20. Zhang, D., Wang, Y., Zhou, L., Yuan, H., Shen, D.: Multimodal classification of Alzheimer's disease and mild cognitive impairment. Neuroimage **55**, 856–867 (2011)
21. Hinrichs, C., Singh, V., Xu, G., Johnson, S.C.: Predictive markers for AD in a multi-modality framework: an analysis of MCI progression in the ADNI population. Neuroimage **55**, 574–589 (2011)
22. McKhann, G., Drachman, D., Folstein, M., Katzman, R., Price, D., Stadlan, E.M.: Clinical diagnosis of Alzheimer's disease: report of the NINCDS-ADRDA work group under the auspices of department of health and human services task force on Alzheimer's disease. Neurology **34**, 939–944 (1984)
23. McKhann, G.M., Knopman, D.S., Chertkow, H., Hyman, B.T., Jack Jr., C.R., Kawas, C.H., et al.: The diagnosis of dementia due to Alzheimer's disease: recommendations from the National Institute on Aging-Alzheimer's Association workgroups on diagnostic guidelines for Alzheimer's disease. Alzheimers Dement. **7**, 263–269 (2011)
24. Vapnik, V.: The Nature of Statistical Learning Theory. Springer, New York (1999)
25. Chang, C.-C., Lin, C.-J.: LIBSVM: a library for support vector machines. ACM Trans. Intell. Syst. Technol. **2**, 27:27–27:27 (2011)
26. Chu, C., Hsu, A.L., Chou, K.H., Bandettini, P., Lin, C.: Does feature selection improve classification accuracy? Impact of sample size and feature selection on classification using anatomical magnetic resonance images. Neuroimage **60**, 59–70 (2012)
27. Walhovd, K.B., Fjell, A.M., Dale, A.M., McEvoy, L.K., Brewer, J., Karow, D.S., et al.: Multi-modal imaging predicts memory performance in normal aging and cognitive decline. Neurobiol. Aging **31**, 1107–1121 (2010)
28. Fan, Y., Resnick, S.M., Wu, X., Davatzikos, C.: Structural and functional biomarkers of prodromal Alzheimer's disease: a high-dimensional pattern classification study. Neuroimage **41**, 277–285 (2008)

29. Fan, Y., Rao, H., Hurt, H., Giannetta, J., Korczykowski, M., Shera, D., et al.: Multivariate examination of brain abnormality using both structural and functional MRI. Neuroimage **36**, 1189–1199 (2007)
30. Costafreda, S.G., Chu, C., Ashburner, J., Fu, C.H.: Prognostic and diagnostic potential of the structural neuroanatomy of depression. PLoS ONE **4**, e6353 (2009)
31. Chetelat, G., Desgranges, B., De La Sayette, V., Viader, F., Eustache, F., Baron, J.-C.: Mapping gray matter loss with voxel-based morphometry in mild cognitive impairment. NeuroReport **13**, 1939–1943 (2002)
32. Misra, C., Fan, Y., Davatzikos, C.: Baseline and longitudinal patterns of brain atrophy in MCI patients, and their use in prediction of short-term conversion to AD: results from ADNI. Neuroimage **44**, 1415–1422 (2009)
33. Hampel, H., Burger, K., Teipel, S.J., Bokde, A.L., Zetterberg, H., Blennow, K.: Core candidate neurochemical and imaging biomarkers of Alzheimer's disease. Alzheimers Dement. **4**, 38–48 (2008)
34. Jack Jr., C.R., Petersen, R.C., Xu, Y.C., O'Brien, P.C., Smith, G.E., Ivnik, R.J., et al.: Prediction of AD with MRI-based hippocampal volume in mild cognitive impairment. Neurology **52**, 1397–1403 (1999)
35. Barnes, J., Scahill, R.I., Boyes, R.G., Frost, C., Lewis, E.B., Rossor, C.L., et al.: Differentiating AD from aging using semiautomated measurement of hippocampal atrophy rates. Neuroimage **23**, 574–581 (2004)
36. Edward, E.S., Stephen, M.K.: Cognitive Psychology: Mind and Brain, pp. 21, 194–199, 349. Prentice Hall, New Jersey (2007)
37. Arnold, S.E., Hyman, B.T., Van Hoesen, G.W.: Neuropathologic changes of the temporal pole in Alzheimer's disease and Pick's disease. Arch. Neurol. **51**, 145–150 (1994)
38. Yang, J., Pan, P., Song, W., Huang, R., Li, J., Chen, K., et al.: Voxelwise meta-analysis of gray matter anomalies in Alzheimer's disease and mild cognitive impairment using anatomic likelihood estimation. J. Neurol. Sci. **316**, 21–29 (2012)

# HYDROWEB, an Online Tool for the Calculation of Hydrodynamic Properties of Macromolecules

Horacio Pérez-Sánchez[1]([⊠]), Jorge Peña-García[1], Helena den-Haan[1],
Ricardo Rodríguez-Schmidt[1], José P. Cerón-Carrasco[1], Adriano N. Raposo[2],
Mounira Bouarkat[4], Sid Ahmed Sabeur[4], and Francisco Guillermo Díaz-Baños[3]

[1] Bioinformatics and High Performance Computing Research Group (BIO-HPC),
Computer Science Department, Universidad Católica San Antonio de Murcia
(UCAM), Campus de los Jerónimos, 30107 Murcia, Spain
hperez@ucam.edu
[2] Instituto de Telecomunicações (IT), Universidade da Beira Interior (UBI),
Av. Marquês D'Avila e Bolama, 6200-001 Covilhã, Portugal
[3] Department of Physical Chemistry, Faculty of Chemistry, University of Murcia,
Campus of Espinardo, 30071 Murcia, Spain
[4] Laboratoire d'Etude Physique des Matériaux, Faculté de Physique,
Université des Sciences et de la Technologie d'Oran (USTOMB),
BP 1505 El M'naouer, 31000 Oran, Algeria

**Abstract.** Calculation and prediction of hydrodynamic properties of biological and synthetic macromolecules through computational approaches is a technique that has experimented a great advance in the last decades. However, most of the hydrodynamics software was designed decades ago and it is rather complex to use for less computer experienced users. With this objective in mind we have developed HYDROWEB, a tool that easily allows to work with hydrodynamic models of macromolecules and the calculation of their properties (using the softwares HYDROPRO, HYDRO++ and SIMUFLEX) and convenient visualization of its results. The tool can be accessed at http://bio-hpc.eu/software/hydroweb/.

## 1 Introduction

As it is well known, the calculation of hydrodynamic properties of macromolecules in solution, e.g., diffusion coefficients, relaxation times, and intrinsic viscosity are determined by the shape and the size of biomolecules. Whether the macromolecules are flexible or rigid, different numerical models have been developed during the last four decades [1,3,7]. Although these models are now well established, the heavy numerical calculations underlying the estimation of hydrodynamic properties of macromolecules is still considered difficult and sometimes inaccessible for end-users. Nowadays, the widespread use of internet gave birth to a lot of web servers dedicated to calculations. In this framework, HYDROWEB has been developed to facilitate the estimation of hydrodynamic properties,

© Springer International Publishing Switzerland 2016
F. Ortuño and I. Rojas (Eds.): IWBBIO 2016, LNBI 9656, pp. 82–90, 2016.
DOI: 10.1007/978-3-319-31744-1_8

using the softwares developed by García de la Torre et al. HYDROPRO [12], HYDRO++ [5] and SIMUFLEX [4], through a user-friendly web interface. Hydrodynamic calculations can be submitted to the server, first, by uploading a PDB file for the desired macromolecule, then using a simple editing procedure which consists of introducing the main parameters of the simulation (temperature, solvent viscosity, molecular weight and solvent density). The bead-spring model replacing the repeating units (amino acid residues) in the macromolecule can also be tuned in the web interface by setting the position, radius and energy for each bead and spring of the chain. Once the The simulation is done, the results are sent automatically to the user via email.

## 2    Methods

### 2.1    Web Interface

Some of the most popular 3D molecular visualization and manipulation tools such as USCF Chimera [13], BALLView [11] and PyMol [15], are not web-based, i.e., they are desktop applications that need to be downloaded and installed in the user's computer, with the drawback of having different versions for different operating systems. As an attempt to provide cross-platform web-based solutions, some molecular visualization tools were developed using Java3D applets, as it was the case of Jmol [8]. The drawback of this approach was that many devices do not run Java. However, most devices do run Javascript, and this is why, more recently, some WebGL/HTML5 molecular visualization and manipulation tools like, for example, JSmol [9], Chemozart [10] and Chemdoodle [2] were presented. However, being more or less user-friendly for someone with less experience in the manipulation of 3D scenes, all the existing 3D chemical modeling tools, are atom-oriented, i.e., not exactly oriented for a coarse-grained hydrodynamic modeling of macromolecules like the one used by our system. Having this in mind, we developed an easy to use web-based 3D interface that allows anyone, even a less experienced user, to build 3D coarse-grained hydrodynamic models of macromolecules in just a few *drag-and-drop* moves using a general purpose web browser without installing any additional plug-in. Like some of the tools already mentioned, our web interface is based on WebGL and HTML5, the state-of-the-art in 3D technology for the web. More specifically, we use *Three.js* (http://threejs.org/) to handle the interactive 3D scene. This Javascript library was created to simplify the manipulation of WebGL 3D objects and scene rendering and interaction operations like, for example, the capture and handling of mouse events in the scene. To build the input components we used *gui.dat*, a lightweight graphical user interface for changing variables in JavaScript (https://code.google.com/p/dat-gui/). The choice of WebGL/HTML5 is due to the fact that, in theory, this technology is compatible with most of web browsers, including the mobile versions.

Our web interface is divided into three main parts: (a) the GUI; (b) the interactive 3D canvas; and (c) the output components.

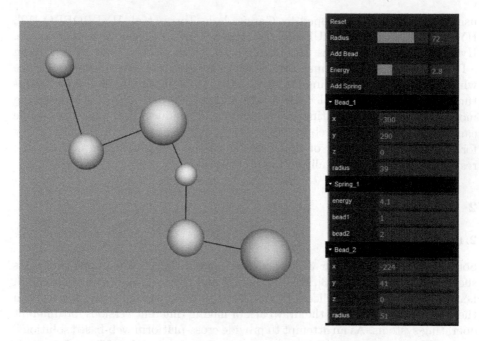

**Fig. 1.** 3D Interactive Canvas (left) and GUI (right).

**GUI:** The user only needs to on the *Add Bead* and *Add Spring* buttons in the GUI (see Fig. 1 (right)), to easily add beads and springs to the 3D interactive canvas. The size of the beads can be adjusted before adding it to the canvas just by using the *Radius* slider placed just above the *Add Bead* button. Analogously, the user can adjust the value of the energy of the spring using the *Energy* slider placed above the *Add Spring* button. In the current version of the web interface the user can add a spring between two beads and can not add a bead without adding a spring before it (except for the first bead). In this sense, when the user adds a bead or a spring to the canvas, a corresponding folder is automatically added to the GUI. Each bead folder contains its 3D coordinates and its radius. On its turn, each spring folder contains information about its energy and the beads that it connects to. To start a new model the user must click in the *Reset* button in the top of the GUI.

**Interactive 3D Canvas:** This is the core component of our web interface. As one can see in Fig. 1 (left), in our interactive canvas the beads are represented by 3D spheres and the springs are represented by black lines between the beads. The user can easily build an hydrodynamic molecular model just by dragging and dropping the beads around the 3D canvas. When the user adds a bead to the canvas it is placed on the center of the scene. Then, he just has to drag it to the wanted position. When the user displaces a bead its connecting springs are automatically adjusted and the bead's coordinates are updated in the corresponding folder in the GUI. In this sense, and for the sake of usability simplification, the user does not need to interact with the springs and, therefore, just the beads will react to the user's interaction.

**Output:** The final goal of our web interface is to be fully integrated within our system. Thus, the web interface just makes sense if, not only allows less experienced users to build coarse grained 3D hydrodynamic molecular models, but also generates the input data needed for the following stages of the calculation, i.e., a file containing the 3D coordinates of the beads (see Fig. 2 (left)), and a file containing the information about the springs, namely, the beads that each spring connects and the corresponding energy value (see Fig. 2 (right)). Our web interface can export this information to regular HTML components from where the user can copy the data, or directly generate the needed files in our server side.

```
-299.9918791057484 290.04739692544723 0
-223.750849056789 41.435342417920666 1.1368683772161603e-13
-4.9722410901491685 125.96344095048583 0
59.66689308180534 -23.20379175403533 0
58.009479385088454 -200.54705730273741 -1.1368683772161603e-13
285.0751558352938 -246.9546408108071 1.1368683772161603e-13
```

```
1 2 4.118495621930772
2 3 6.502887824101219
3 4 2.9262995208455482
4 5 5.527454650486035
5 6 2.8179180571105276
```

**Fig. 2.** HTML textareas containing the 3D coordinates of 6 beads (left) and the corresponding 5 springs data (right).

## 2.2 Underlying Calculations and Formulas

Hydrodynamic properties calculations implies running a molecular dynamics simulation that includes hydrodynamic interactions via an appropriate diffusion tensor $D$.

$$\mathbf{R_{ij}} = \sum_j \mathbf{D_{ij}f_j} + \sum_j \xi_{ij}\eta_{ij} \tag{1}$$

In the above equation, the diffusion tensor $D$ can be approximated by the Rotne-Prager tensor [6,14] with solvent viscosity $\zeta$. The force $\mathbf{f_j}$ is the total force acting on bead $j$ by the other beads within the chain. The solvent particles surrounding the chain are effectively replaced by the friction $\xi$ where it fullfills $\xi\xi^t = \mathbf{D}$. $\eta$ is a Gaussian distributed stochastic force acting on the beads.

Using the beads coordinates and radius, a serie of transformations are applied before obtaining the hydrodynamics properties. According to Garcia de la Torre et al. [3], they can be summarized as follow:

- First, the interaction tensor $\mathbf{T}$ can be obtained using Oseen formula [16] by the following equation:

$$\mathbf{T_{ij}} = (8\pi\eta_0\mathbf{R_{ij}})^{-1}(I + \mathbf{R_{ij}R_{ij}}/\mathbf{R_{ij}}^2) \tag{2}$$

Here, $\mathbf{R_{ij}}$ represents the distance vector between beads $i$ an $j$ and $\mathbf{I}$ is the unit tensor.

- A $3N \times 3N$ blocks supermatrix $\mathbf{B}$ is then derived from the interaction tensor $\mathbf{T}$ where:

$$\mathbf{B_{ij}} = \mathbf{T_{ij}} \quad if \quad i \neq j \tag{3}$$

and

$$\mathbf{B_{ii}} = \left(\frac{1}{\xi_i}\right)\mathbf{I} \tag{4}$$

$\xi_i$ is the friction coefficient of bead $i$ and can be obtained by the Stockes law:

$$\xi_i = 6\pi\zeta\sigma_i \tag{5}$$

where $\zeta$ and $\sigma_i$ are respectively the viscosity of the solvent and the radius of bead $i$.

- The supermatrix $\mathbf{B}$ is first inverted and then fractionned to $3 \times 3$ blocks to build up the components of the friction tensor $\mathbf{\Gamma}$ as follow:

$$\mathbf{C} = \mathbf{B}^{-1} \tag{6}$$

$$\mathbf{\Gamma}_{tt} = \sum_i \sum_j \mathbf{C}_{ij} \tag{7}$$

$$\mathbf{\Gamma}_{rr} = \sum_i \sum_j \mathbf{U}_i \mathbf{C}_{ij} \mathbf{U}_j \tag{8}$$

$$\mathbf{\Gamma}_{tr} = \sum_i \sum_j \mathbf{U}_i \mathbf{C}_{ij} \tag{9}$$

The subscripts $tt$, $rr$ and $tr$ are respectively the translational, rotational and tranlational-rotational coupling in the $\mathbf{\Gamma}$ tensor. $\mathbf{U}$ is a $3 \times 3$ matrix used in the calculation of the $\mathbf{\Gamma}$ tensor and is defined as:

$$\begin{pmatrix} 0 & -z_i & y_i \\ z_i & 0 & -x_i \\ -y_i & x_i & 0 \end{pmatrix} \tag{10}$$

- Finally, for a temperature $T$, the diffusion tensor $\mathbf{D}$ can be obtained through the generelized Einstein relationship $\mathbf{D} = kT\mathbf{T}^{-1}$.

Finding interesting properties from the diffusion tensor $\mathbf{D}$ is fairly straightforward. As the diffusion tensor $\mathbf{D}$ is partionned in four $3 \times 3$ blocks as follow

$$\begin{pmatrix} \mathbf{D}_{tt} & \mathbf{D}_{tr}^T \\ \mathbf{D}_{tr} & \mathbf{D}_{rr} \end{pmatrix} \tag{11}$$

The translational diffusion coefficient can be obtained from the trace of the tensor.

$$D_t = \frac{1}{3} Tr(\mathbf{D}_{tt}) \tag{12}$$

The friction coefficient is then derived from the diffusion coefficient.

$$f_t = \frac{kT}{D_t} \tag{13}$$

Estimating the eigenvalues $D_1, D_2, D_3$ of the matrix $\mathbf{D}_{rr}$ can also give access to five relaxation times $\tau_1, \tau_2, \tau_3, \tau_4$ and $\tau_5$ which determine the time or frequency dependence in dynamic electrooptical or spectroscopic properties. These relaxation times are obtained as follow:

$$\begin{cases} \tau_1 = 6D_r - 2\Delta \\ \tau_2 = 3(D_r + D_1) \\ \tau_3 = 3(D_r + D_2) \\ \tau_4 = 3(D_r + D_3) \\ \tau_5 = 6D_r + 2\Delta \end{cases} \tag{14}$$

where

$$D_r = \frac{1}{3}(D_1 + D_2 + D_3) \tag{15}$$

and

$$\Delta = \sqrt{D_1^2 + D_2^2 + D_3^2 - D_1 D_2 - D_1 D_3 - D_2 D_3} \tag{16}$$

For the intrinsic viscosity $[\eta]$, the Tsuda [17] expression is generally adopted in the calculations, it is defined as:

$$[\eta] = \frac{N_A \pi}{M} \left( \sum_i \sigma_i R_i^2 \right) \cdot$$

$$\left[ 1 + \frac{1}{\sum_i \sigma_i R_i^2} \frac{3}{4} \left( \sum_{i \neq} \sum_j \sigma_i \sigma_j \left( \frac{R_i R_j cos\alpha_{ij}}{R_{ij}} + \frac{4(R_i^2 + R_j^2)R_i R_j cos\alpha_{ij} - R_i^2 R_j^2(1 + 7cos^2\alpha_{ij})}{10R_{ij}^3} \right) \right) \right] \tag{17}$$

$N_A$ and $M$ in the expression of the intrinsic viscosity $v$ are respectively Avogadro's number and the macromolecule molecular weight.

## 3 Results

The above method has been tested for the case of recombinant human interferon gamma macromolecule, which is available in the RCSB Protein Data Bank server under the code "1HIG".

These files contain descriptions of the macromolecules at an atomic detail level, so that the corresponding option was chosen in the "Submit calculations" section in HYDROWEB site. Under this option, the "PDB code" suboption was chosen in order to let the server retrieve the corresponding structure file from the RCSB server.

As for the simulation parameters, the temperature was set to $T = 20\,^\circ\mathrm{C}$. The solvent viscosity was given as 0.01 Poise, and the solvent density as $1.0\,cm^3g^{-1}$. For the protein, we filled in a molecular weight of 16.6 KDa, and $0.4\,cm^3g^{-1}$ for the solute partial specific volume.

After the submission of this data, the server processed the request, resulting in the generation of a HYDROPRO calculation task with INDMODE = 1 due to the choice of an atomic detail level. The parameters supplied by the user where complemented with a default hydrodynamic bead radius of 2.84 Å, letting HYDROPRO decide on the radio of the shell model beads (NSIG = −1). As HYDROWEB is meant to provide hydrodynamic properties only, the non-hydrodynamic properties calculation mode of HYDROPRO was disabled by giving a value of zero to the parameters NS, NQ, NTRIALS and IFDIF.

Once the appropriate files for the calculation were generated, they were offloaded to a remote calculation relay, and added to the work queue. Once the simulation was processed, the results summarized in Table 1 were retrieved.

**Table 1.** List of hydrodynamic properties as retreived from HYDROWEB server.

| Hydrodynamic properties | values |
| --- | --- |
| Translational diffusion coefficient | 7.175E-07 $cm^2/s$ |
| Rotational diffusion coefficient | 6.024E+06 $s^-1$ |
| Relaxation time (1) | 3.354E-08 s |
| Relaxation time (2) | 3.305E-08 s |
| Relaxation time (3) | 3.085E-08 s |
| Relaxation time (4) | 2.855E-08 s |
| Relaxation time (5) | 2.831E-08 s |
| Intrinsic viscosity | 1.056E+01 $cm^3/g$ |

## 4    Conclusions

In this work (currently in progress) we have shown that performing hydrodynamic calculations for macromolecules can be a complicated process and that not all users are able to run them via command line. Trying to solve this problem, we have implemented a web based tool called HYDROWEB that eases the process of performing such calculations via a convenient web interface, using the softwares developed by García de la Torre et al. HYDROPRO [12], HYDRO++ [5] and SIMUFLEX [4]. The implementation of the tool and the way it works is explained in this paper.

**Acknowledgements.** This work was supported by the Fundación Séneca–Agencia de Ciencia y Tecnología de la Región de Murcia under Projects 18946/JLI/13 and 19499/PI/14, by the Nils Coordinated Mobility under grant 012-ABEL-CM-2014A, in part financed by the European Regional Development Fund (ERDF), and by TETRA-COM Technology Transfer in Computing Systems (funded by the European Union Commission Framework Programme 7 - Contract No. 609491), under Project ACD-DTHPC. Powered@NLHPC: This research was partially supported by the supercomputing infrastructure of the NLHPC (ECM-02). The authors also thankfully acknowledge the computer resources and the technical support provided by the Plataforma Andaluza de Bioinformática of the University of Málaga. This work was partially supported by the computing facilities of Extremadura Research Centre for Advanced Technologies (CETACIEMAT), funded by the European Regional Development Fund (ERDF). CETACIEMAT belongs to CIEMAT and the Government of Spain.

# References

1. Bloomfield, V.A., Dalton, W.O., Holde, K.E.V.: Frictional coefficients of multisubunit structures. II. Application to proteins and viruses. Biopolymers **5**, 149–159 (1967)
2. Burger, M.C.: Chemdoodle web components: HTML5 toolkit for chemical graphics, interfaces, and informatics. J. Cheminform **7**, 35 (2015)
3. Carrasco, B., García de la Torre, J.: Hydrodynamic properties of rigid particles: comparison of different modeling and computational procedures. Biophys. J. **75**, 3044–3057 (1999)
4. García de la Torre, J., Hernández Cifre, J.G., Ortega Retuerta, A., Rodriguez Schmidt, R., Fernandes, M., Pérez Sánchez, H.E., Pamies, R.: Simuflex: algorithms and tools for simulation of the conformation and dynamics of flexible molecules and nanoparticles in dilute solution. J. Chem. Theory Comput. **5**(10), 2606–2618 (2009). PMID: 26631776
5. García de la Torre, J., del Rio Echenique, G., Ortega Retuerta, A.: Improved calculation of rotational diffusion and intrinsic viscosity of bead models for macromolecules and nanoparticles. J. Phys. Chem. B **111**(5), 955–961 (2007). PMID:17266248
6. Ermak, D.L., McCammon, J.A.: Brownian dynamics with hydrodynamic interactions. J. Chem. Phys. **69**, 1352 (1978)
7. Filson, D.P., Bloomfield, V.A.: Shell model calculations of rotational diffusion coefficients. Biochemistry **6**, 1650–1658 (1967)
8. Hanson, R.M.: Jmol - a paradigm shift in crystallographic visualization. J. Appl. Crystallogr. **43**(5 Part 2), 1250–1260 (2010)
9. Hanson, R.M., Prilusky, J., Renjian, Z., Nakane, T., Sussman, J.L.: Jsmol and the next-generation web-based representation of 3D molecular structure as applied to proteopedia. Isr. J. Chem. **53**(3–4), 207–216 (2013)
10. Mohebifar, M., Sajadi, F.: Chemozart: a web-based 3D molecular structure editor and visualizer. J. Cheminform **7**, 56 (2015)
11. Moll, A., Hildebrandt, A., Lenhof, H.-P., Kohlbacher, O.: BALLView: a tool for research and education in molecular modeling. Bioinformatics **22**(3), 365–366 (2006)
12. Ortega Retuerta, A., Amorós Cerdán, D., García de la Torre, J.: Prediction of hydrodynamic and other solution properties of rigid proteins from atomic- and residue-level models. Biophys. J. **101**(4), 892–898 (2011)

13. Pettersen, E.F., Goddard, T.D., Huang, C.C., Couch, G.S., Greenblatt, D.M., Meng, E.C., Ferrin, T.E.: UCSF chimera-a visualization system for exploratory research and analysis. J. Comput. Chem. **25**(13), 1605–1612 (2004)
14. Rotne, J., Prager, S.: Variational treatment of hydrodynamic interaction in polymers. J. Chem. Phys. **50**, 4831–4837 (1969)
15. Schrödinger, L.L.C.: The PyMOL molecular graphics system, version 1.3r1, August 2010
16. Teller, D.C., Swanson, E., de Haen, C.: The low Reynold number translational friction of ellipsoid, cylinders, dumbbells, and hollow spherical caps. Numerical testing of the validity of the modified Oseen tensor in computing the friction of objects modeled as beads on a shell. J. Chem. Phys. **68**, 5097–5102 (1978)
17. Tsuda, K.: Hydrodynamic properties of rigid complex molecules. Polym. J. **1**, 616–631 (1970)

# The Use of the Miyazawa-Jernigan Residue Contact Potential in Analyses of Molecular Interaction and Recognition with Complementary Peptides

Nikola Štambuk[1](✉), Paško Konjevoda[1], Zoran Manojlović[2], and Renata Novak Kujundžić[1]

[1] Ruđer Bošković Institute, Zagreb, Croatia
{stambuk,pkonjev,rnovak}@irb.hr
[2] Croatian Institute for Toxicology and Anti-Doping, Zagreb, Croatia
zoran.manojlovic@antidoping-hzta.hr

**Abstract.** The classic results by Biro, Blalock and Root-Bernstein link genetic code nucleotide patterns to amino acid properties, protein structure and interaction. This study explores the use of the Miyazawa-Jernigan residue contact potential in analyses of protein interaction and recognition between sense and complementary (antisense) peptides. We show that Miyazawa-Jernigan residue contact energies, derived from 3D data, define the recognition rules of peptide-peptide interaction based on the complementary coding of DNA and RNA sequences. The model is strongly correlated with several other chemoinformatic scales often used for the determination of protein antigenic sites and transmembrane regions (Parker et al. $r = 0.94$; Rose et al. $r = -0.92$; Manavalan-Ponnuswamy $r = -0.92$; Cornette et al. $r = -0.91$; Kolaskar-Tongaonkar $r = -0.91$; Grantham $r = 0.90$; White-Wimley (octanol) $r = -0.88$; Kyte-Doolittle $r = -0.85$). The algorithms presented have important biomedical and proteomic applications related to modulation of the peptide-receptor function and epitope-paratope interaction, the design of lead compounds and the development of new immunochemical assays and diagnostic procedures.

**Keywords:** Contact · Potential · Protein-protein interaction · Peptide · Binding

## 1 Introduction

A large body of experimental evidence over the last two decades supports the thesis that sense and complementary (antisense) peptides interact with increased probability [1–4]. Genetic coding and the possible interaction of complementary peptides, specified by the sense and antisense mRNA sequences, was first discussed by Biro, Mekler and Idlis, Root-Bernstein, Blalock, and others [5–8]. Critical examination of the concept has confirmed the relevance and applicability of this method to *in vitro* and *in vivo* research [9–12]. Over the past decade, antisense peptide-based modeling has become a valuable procedure for the design of new bioactive peptides and antibodies [13–15]. It is also a useful tool for the efficient modeling and investigation of ligand-receptor interactions in medicinal chemistry and immunochemistry [1, 12, 16].

© Springer International Publishing Switzerland 2016
F. Ortuño and I. Rojas (Eds.): IWBBIO 2016, LNBI 9656, pp. 91–102, 2016.
DOI: 10.1007/978-3-319-31744-1_9

Sixty-four codons of the genetic code consist of three nucleotide bases. Sixty-one are codons for 20 amino acids and three are stop signals (Table 1). The complementary (antisense) peptide sequence is obtained from the sense mRNA by transcribing uracil (U) into its complement adenine (A) and cytosine (C) into its complement guanine (G), or *vice versa* (Table 1). With respect to the standard genetic coding sense and complementary peptides are characterized by: (1) a tendency for the antisense peptide to show opposite polarity patterns compared to those of the sense peptide structure, and (2) a different number of antisense peptides depending on the direction of the mRNA transcription (from left to right or *vice versa*; 3' → 5'/5' → 3') [1, 3, 7].

These opposite polarity patterns shown by complementary peptides arise from the importance of the second base for peptide structure and complementary binding, i.e., from the fact that the second base of the genetic code table specifies nonpolar (U) and polar (A) amino acids (Table 1) [1, 3, 7]. Antisense codons correspond to tRNA anticodons, and they are described by reversed 5' and 3' end directions compared to mRNA [4, 10].

**Table 1.** A. The standard genetic code table with 3-letter nucleotide coding that specifies amino acids, and stop codons for protein synthesis. B. Complementary antisense amino acid transcription of the genetic code table in both directions: from left to right (3' → 5'), and the reverse, from right to left (5' → 3').

## A.

protein structure
complementary peptide binding

| First (5') letter | Second letter | | | | Third (3') letter |
|---|---|---|---|---|---|
| | **U** | **C** | **A** | **G** | |
| **U** | F | S | Y | C | **U** |
| | F | S | Y | C | **C** |
| | L | S | stop | stop | **A** |
| | L | S | stop | W | **G** |
| **C** | L | P | H | R | **U** |
| | L | P | H | R | **C** |
| | L | P | Q | R | **A** |
| | L | P | Q | R | **G** |
| **A** | I | T | N | S | **U** |
| | I | T | N | S | **C** |
| | I | T | K | R | **A** |
| | M | T | K | R | **G** |
| **G** | V | A | D | G | **U** |
| | V | A | D | G | **C** |
| | V | A | E | G | **A** |
| | V | A | E | G | **G** |

## B.

| Sense 5'→3' | Antisense 3'→5' | Antisense 5'→3' |
|---|---|---|
| F | K | K, E |
| L | D, E, N | E, Q, K |
| I | Y | N, D, Y |
| M | Y | H |
| V | H, Q | H, D, N, Y |
| S | S, R | G, R, T, A |
| P | G | G, W, R |
| T | W, C | G, S, C, R |
| A | R | R, G, S, C |
| Y | M, I | I, V |
| H | V | V, M |
| Q | V | L |
| N | L | I, V |
| K | F | F, L |
| D | L | I, V |
| E | L | L, F |
| C | T | T, A |
| W | T | P |
| R | A, S | A, S, P, T |
| G | P | P, S, T, A |

## 2    Results and Discussion

### 2.1    The Concept of Complementary Peptides

The codons for sense amino acids could be transcribed into complementary pairs in two different directions – from the left to the right side of the genetic code Table ($1^{st}2^{nd}3^{rd}$ base), and the reverse, i.e., from the right to left side ($3^{rd}2^{nd}1^{st}$ base, Table 1) [1, 4, 14]. Sense mRNA codons transcribed from the left to the right side of the genetic code table define the translation of 27 possible complementary pairs for 20 sense amino acids (3' → 5' antisense direction, Table 1A and B). The transcription of sense mRNA codons from the right to the left side of the genetic code table defines the translation of 52 possible complementary pairs for 20 sense amino acids (5' → 3' antisense direction, Table 1A and B). Consequently, for the sense peptide sequence of length $n$ the number of complementary peptides ($N$) translated in the direction 3' → 5' is approximately the square root of the number of complementary peptides translated in the 5' → 3' direction, i.e., $N_{3'\to5'} \approx \sqrt{N_{5'\to3'}}$. This leads to the significantly fewer antisense peptides when the 3' → 5' direction algorithm of complementary peptide binding is applied [1, 4].

Complementary peptide binding algorithms for the antisense peptide design in both the 3' → 5' and the 5' → 3' direction were introduced in the 1980s by Root-Bernstein and Blalock et al. [1, 8, 9]. Over the last three decades, their efficiency has been experimentally verified for different ligand-receptor systems [1, 2, 5, 6]. However, until this study, the Kyte-Doolittle scale was the only support for this important property of the genetic code – that codons for the hydrophilic amino acids are complemented by codons for the hydrophobic amino acids and *vice versa*, while the codons for those that are neutral pair with each other [3, 4, 7, 9, 14].

### 2.2    New Data Supporting the Concept of Complementary Peptide Interactions

In this study we applied and tested the Miyazawa-Jernigan residue contact potential in analyses of molecular interaction and recognition in complementary peptides [17, 18]. The weight or the energy value of each possible complementary amino acid pair, in Fig. 1A and B, was calculated as the absolute difference between amino acid pairwise contact energies by Miyazawa and Jernigan method A [17]. The energy values were derived from 3D data based on a large set of PDB structures [17]. The values of the relative partition energies obtained by the Miyazawa-Jernigan model for the sense-antisense amino acid pairs form two distinct clusters in the 3' → 5' and the 5' → 3' antisense pairing direction (Fig. 1A and B; PAST software version 2.17c, Paired group algorithm, Euclidean similarity measure). Strong correlations of residue potentials in the 3' → 5' ($r = 0.90$) and in 5' → 3' directions ($r = 0.90$) confirm that the Miyazawa-Jernigan method supports the complementary peptides concept. The relative partition energies (or hydrophobic energies) of amino acid residues calculated by the Miyazawa-Jernigan model are highly correlated with those of the Kyte-Doolittle hydrophobicity scale ($r = -0.85$) [18–20]. Consequently, amino acid pairwise contact

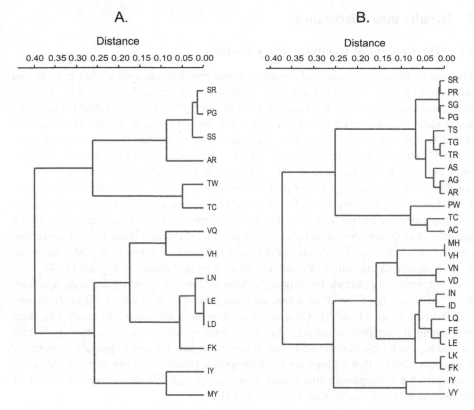

**Fig. 1.** The values of amino acid relative partition energies obtained using the Miyazawa-Jernigan method [17, 18] form two distinct clusters in the 3' → 5' (A) and 5' → 3' directions (B), and reconstruct the complementary genetic code patterns presented in Table 1.

energies by Miyazawa and Jernigan derived from 3D data (Method A [17]) both reconstruct the algorithms for complementary peptide binding, and support the theoretical framework developed by Root-Bernstein and Blalock et al. [1, 4, 8, 9].

In addition to the standard Kyte-Doolittle hydrophobicity scale, the Miyazawa-Jernigan relative partition (hydrophobic) energies of the complementary amino acids used in this study correlate strongly with several other bioinformatic scales often used for the determination of protein antigenic sites and transmembrane regions, e.g.:

1. Parker et al. $r = 0.94$ [21],
2. Rose et al. $r = -0.92$ [22],
3. Manavalan-Ponnuswamy $r = -0.92$ [23],
4. Cornette et al. $r = -0.91$ [24],
5. Kolaskar-Tongaonkar $r = -0.91$ [25],
6. Grantham $r = 0.90$ [26], and
7. White-Wimley (octanol) $r = -0.88$ [27].

The fact that hydrophobic amino acids are complemented by hydrophilic residues, and neutral amino acids by neutral residues, suggests that the driving force behind this type of molecular recognition is the minimization of free energy [28]. The algorithm of complementary peptide interaction implies a high probability of interaction for 26 out of a total of 190 possible different amino acid pairs (and one additional SS pair) in the 3' → 5' direction, and 52 out of a total of 190 possible different amino acid pairs in the 5' → 3' direction [1, 4, 26].

It is important to note that the critical distance, defined as less than 6.5 Å, for efficient modeling with the Miyazawa-Jernigan method satisfies the recommended criteria for an accurate protein-protein interaction algorithm, which include different bonding types (e.g., ionic or electrostatic bonds, hydrogen bonds, Van der Waals and hydphobic interactions, or multiple bonding types) [18, 29, 30]. The Miyazawa-Jernigan contact potential is a widely used knowledge-based potential for globular proteins, derived by using the quasi-chemical approximation from databases of proteins having known structures [31].

The use of this method in the context of sense-antisense peptide analysis could be of importance for further comparative investigations of the protein structure and molecular recognition [17, 31, 32]. Antisense (complementary) peptide technology can be also used for pharmacophore modeling [12, 33]. The binding pocket of a protein receptor is usually defined by the residues that have at least one heavy atom, i.e., an atom other than hydrogen, within a distance of 5 Å of a heavy atom of the ligand [12, 34]. This is within the critical distance of 6.5 Å needed for efficient modeling with the Miyazawa-Jernigan method [17, 18].

## 2.3   Miyazawa-Jernigan Model and Complementary Clustering of Amino Acids

The clustering of amino acids into hydrophilic, hydrophobic and neutral residues was reported by different authors [3, 4, 7, 19, 23, 35]. However, until this study (based on the Miyazawa-Jernigan model), the Kyte and Doolittle scale was the only method clearly confirming this type of clustering for the pairs of amino acids coded by the complementary mRNA codons (Table 1) [3, 4, 7, 9, 14, 19].

As shown in Figs. 2 and 3, the Kyte and Doolittle scale and the Miyazawa-Jernigan method permit the analysis of surface-exposed and transmembrane protein regions based on full scale (20 amino acids) or three cluster approximations.

The three clusters of amino acids defined by partition around medoids for the Kyte-Doolittle and Miyazawa-Jernigan methods are the nonpolar (I, L, F, V, M, C), neutral (T, G, Y, H, A, S, W, P) and polar (D, E, N, Q, K, R) groups of amino acid residues (Fig. 2). Representative medoid values for normalised Kyte-Doolittle and Miyazawa-Jernigan clusters are I (1, 0.89), T (0.42, 0.24) and E (0.11, 0.09), respectively, where polar values → 0, nonpolar values → 1 (Fig. 2; R statistical software, version 3.12). For the three normalised amino acid clusters of Fig. 2, the correlation between Kyte-Doolittle and Miyazawa-Jernigan values is strong ($r = 0.85$). Individual proteins retain a strong correlation between the moving average values of both methods. This is illustrated in Fig. 3, using the Erythropoietin example (sliding block of 9 residues, $r = 0.97$).

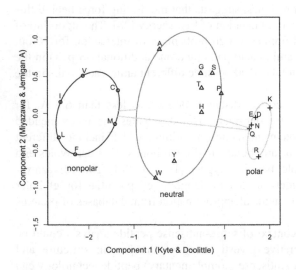

| amino acid | Kyte & Doolittle | Miyazawa & Jernigan A |
|---|---|---|
| F | 0.81 | 0.98 |
| L | 0.92 | 1.00 |
| I | 1.00 | 0.89 |
| M | 0.71 | 0.72 |
| V | 0.97 | 0.72 |
| S | 0.41 | 0.15 |
| P | 0.32 | 0.17 |
| T | 0.42 | 0.24 |
| A | 0.70 | 0.31 |
| C | 0.78 | 0.61 |
| W | 0.40 | 0.70 |
| R | 0.00 | 0.19 |
| G | 0.46 | 0.20 |
| Y | 0.36 | 0.57 |
| H | 0.36 | 0.31 |
| Q | 0.11 | 0.15 |
| N | 0.11 | 0.13 |
| K | 0.07 | 0.00 |
| D | 0.11 | 0.09 |
| E | 0.11 | 0.09 |

**Fig. 2.** The three groups of amino acids defined by partition around medoids for the Kyte-Doolittle and Miyazawa-Jernigan methods (left), and their normalised values (right).

## 2.4 Erythropoietin Quantification Using Immunoassay Based on Complementary Peptide Binding

The complementary peptide recognition concept may be applied as a simple three-step method for the design of an antisense peptide targeting receptor-binding site [1]. Firstly, the receptor binding region of the molecule (epitope) is used as a template for the antisense peptide modeling. Secondly, mRNA transcription of the human epitope sequence in the 3' → 5' direction and computational screening of potential human paratope structures with BLAST are used to design complementary (antisense) peptide ligands [1]. Thirdly, sense-antisense (epitope-paratope) peptide binding and affinity are evaluated to select potential lead compounds for further diagnostic or therapeutic purposes. Recently, this method was applied for the modeling of the central binding region P2 of the human Erythropoietin (EPO) molecule presented in Fig. 3 [1]. It was shown that, using fluorescence spectroscopy and microscale thermophoresis, the carboxyl-terminal domain peptide LKLYTGEACRTGDR (aa 153–166) binds antisense peptide DFDIWPLRTAWPLS with the affinities ($K_d$) of 850 and 816 µM, respectively [1].

In this study antisense EPO peptide DFDIWPLRTAWPLS, designed by Štambuk et al. [1], was used for two-step nanoparticle-based immunoassays. Following the coating of carboxyl magnetic particles with this EPO (P2) antisense, we used Magnetic Particle Enzyme Immunoassays (MPEIA) to quantify the levels of the hormone and its derivative Darbepoetin alfa.

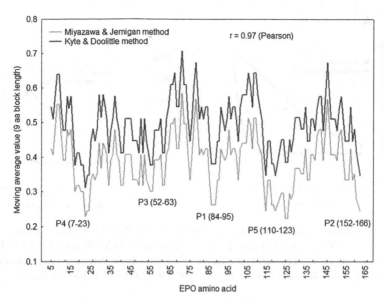

**Fig. 3.** The plotting of moving average values of human Erythropoietin (EPO) for the Kyte-Doolittle and Miyazawa-Jernigan methods. P1-P5 are experimentally verified epitopes [1].

**Coating of Carboxyl Magnetic Particles with Peptides.** SPHERO™ carboxyl magnetic particles (1.0–1.4 μm) were coated with either a test peptide (antisense peptide DFDIWPLRTAWPLS, > 97 % purity, $M_W$ 1716.50, GenScript, Piscataway, NJ, USA; directed to EPO region P2, aa 153–166, Fig. 3) or control peptide (EPO unrelated peptide YGGFM [33]) using a one-step, 1-ethyl-3-(3-dimethylaminopropyl) carbodiimide (EDC) coupling method, as suggested by the manufacturer. Briefly, reaction mixtures were composed of a 2 mL sodium acetate buffer (0.01 M, pH 5.0), 1 mg of peptide, 0.2 ml of 2.5 % w/v carboxyl magnetic particles and 10 mg of EDC.

The reactions proceeded in glass reaction tubes over two hours at room temperature with occasional vortexing. The tubes were centrifuged at 3000 × g for 15 min, supernatant was carefully discarded, and the pellets were twice washed in 4 mL isotonic buffered saline (IBS), followed by centrifugation. After washing, the magnetic particles were re-suspended in IBS to obtain a 0.125 % w/v suspension.

**Magnetic Particle Enzyme Immunoassays (MPEIA).** Two fold serial dilutions (1000 ng – 8 ng) of two commercial preparations of erythropoietin Recombinant Human Erythropoietin-alpha (EPO-α, ProSpec, East Brunswick, NJ, USA; Fig. 4A) and Darbepoetin alfa (Aranesp®, Amgen Europe B.V., Breda, NL; Fig. 4B) were made in 96-well microtiter plates to generate standard curves and compare the MPEIA using antibodies to the MPEIA using biotinylated peptides. A suspension (0.125 % w/v) of carboxyl magnetic particles, coated with either a test or a control peptide, was added (25 μL/well). Plates were incubated at room temperature for 30 min and washed with PBS-Tween three times. The washing solution was removed each time using the Spherotech UltraMag Separator (Sperotech, Inc., Lake Forest, IL, USA).

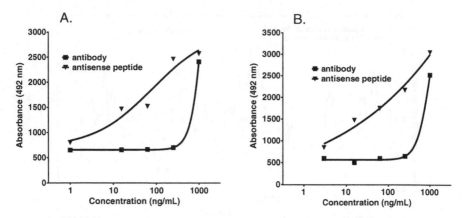

**Fig. 4.** A. Detection of Recombinant Human Erythropoietin-alpha by Magnetic Particle Ezyme Immunoassay (MPEIA), using antibody and antisense peptide RAQDLSIDELRFLR-Lys (Biotin). B. Detection of Darbepoetin alfa (Aranesp®) by Magnetic Particle Ezyme Immunoassay (MPEIA), using antibody and antisense peptide RAQDLSIDELRFLR-Lys(Biotin).

For the Magnetic Particle Enzyme Immunoassay (MPEIA) using antibody detection, 100 µL of Mouse Anti-Human Erythropoietin antibody (Raybiotech, Inc., Norcross, GA, USA), directed to EPO region P4 within aa 1–26 (Fig. 3) and diluted in PBS (1:40000), was added to each well and incubated at room temperature for 30 min. The plate was washed, 100 µL of HRP-conjugated goat anti-mouse IgG (Raybiotech, Inc., Norcross, GA, USA) diluted in PBS (1:5000) was then added to each well and incubated for 30 min at room temperature. Following the washing procedure, SIGMA-*FAST*™ OPD peroxidase substrate (Sigma-Aldrich, St. Louis, MO, USA) was added to the wells and incubated in the dark for 30 min prior to stopping the reaction with 3 M HCl. The plate was read at 492 nm. Regression analyses and plotting were made using GraphPad Prism 5 software (Fig. 4A and B).

For the Magnetic Particle Enzyme Immunoassay (MPEIA) using biotinylated peptide, 100 µL of biotinylated antisense peptide RAQDLSIDELRFLR-Lys(Biotin) (99.1 % purity, $M_W$ 2086.43, GenScript, Piscataway, NJ, USA) directed to EPO region P4 (SRVLERYLLEAKEA, Fig. 3, aa 9–22) in PBS (125 ng/µL) were added to the wells and incubated at room temperature for 30 min. The plates were washed and 100 µL of anti-biotin-peroxidase antibody produced in goats (Sigma-Aldrich, St. Louis, MO, USA), diluted in PBS (1:2000), were added to the wells. The plate was incubated for 30 min at room temperature, then washed; SIGMA*FAST*™ OPD peroxidase substrate was then added to the wells. After 30 min incubation in the dark, the reaction was stopped with 3 M HCl. The absorbances were read at 492 nm. The data plots and regression analyses were made using GraphPad Prism 5 software (Fig. 4A and B).

**Antisense Peptides and Magnetic Particle Enzyme Immunoassay.** When coated to magnetic nanoparticles, the antisense peptide DFDIWPLRTAWPLS captures EPO at its receptor binding site (P2, Fig. 3). This permits the detection of unmodified and modified

EPO molecules by Magnetic Particle Enzyme Immunoassay (MPEIA). The detection of different EPO derivatives bound to antisense-coated nanoparticles may be performed by MPEIA using secondary antibody to EPO P4 epitope (Figs. 3 and 4A and B).

Alternatively, EPO derivatives could be detected by MPEIA using biotinylated antisense peptide RAQDLSIDELRFLR-Lys (Biotin) directed to EPO region P4 (Fig. 3), instead of the secondary antibody (Fig. 4A and B). The quantification of unmodified recombinant human Erythropoietin-alpha, and Darbepoetin alfa modified with two additional sialic acid-containing carbohydrate chains (aa 30 and 88), was equally successful in both approaches, even in the low concentration range (Fig. 4A and B).

This result implies that antisense peptides applied to specific regions of the protein molecule could be used for the development of new immunochemical methods that avoid the binding problems caused by glycosylation or sequence modification [1]. The concept of complementary peptide binding may be also used for the magnetic separation of molecules and cells from different body fluids, e.g., using Sepmag® technology and immunopurification [36–39].

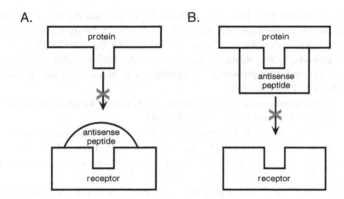

**Fig. 5.** Modulation of receptors (A) and proteins (B) using complementary (antisense) peptides.

## 3  Conclusion

The use of Miyazawa-Jernigan model in the context of sense-antisense peptide analysis could be of importance for further comparative investigations of the protein structure and molecular recognition. Complementary peptides can be used to derive bioactive peptide fragments, and to modulate the activity of receptors and related proteins (Fig. 5) [1, 6, 12, 13].

The important applications of this technology are vaccine research and the modulation of peptide molecules in medicinal chemistry and immunochemistry research [1, 6, 12, 13]. It may prove to be a useful tool for virtual screening and the selection of sense-antisense peptide pairs, supplementing and/or optimizing conventional epitope mapping methods and immunoassays [1, 40].

The method may easily be adapted for high-throughput screening with the technique of microscale thermophoresis, which provides a quick, inexpensive and simple method for binding detection and the quantification of affinity [1, 12]. Selected complementary peptides with acceptable affinity are potential lead compounds for the further development of diagnostic and therapeutic substances.

**Abbreviations**

Amino acids: Alanine (A), Arginine (R), Asparagine (N), Aspartate (D), Cysteine (C), Glutamate (E), Glutamine (Q), Glycine (G), Histidine (H), Isoleucine (I), Leucine (L), Lysine (K), Methionine (M), Phenylalanine (F), Proline (P), Serine (S), Threonine (T), Tryptophan (W), Tyrosine (Y), Valine (V).

**Acknowledgments.** The support of the Croatian Institute for Toxicology and Anti-Doping, and the Croatian Ministry of Science, Education and Sports is gratefully acknowledged (grant No. 098-0982929-2524).

# References

1. Štambuk, N., Manojlović, Z., Turčić, P., Martinić, R., Konjevoda, P., Weitner, T., Wardega, P., Gabričević, M.: A simple three-step method for design and affinity testing of new antisense peptides: an example of erythropoietin. Int. J. Mol. Sci. **15**, 9209–9223 (2014)
2. Heal, J.R., Roberts, G.W., Raynes, J.G., Bhakoo, A., Miller, A.D.: Specific interactions between sense and complementary peptides: the basis for the proteomic code. ChemBioChem **3**, 136–151 (2002)
3. Biro, J.C.: The proteomic code: a molecular recognition code for proteins. Theor. Biol. Med. Model. **4**, 45 (2007). doi:10.1186/1742-4682-4-45
4. Štambuk, N., Konjevoda, P., Boban-Blagaić, A., Pokrić, B.: Molecular recognition theory of the complementary (antisense) peptide interactions. Theory Biosci. **123**, 265–275 (2005)
5. Siemion, I.Z., Cebrat, M., Kluczyk, A.: The problem of amino acid complementarity and antisense peptides. Curr. Protein Pept. Sci. **5**, 507–527 (2004)
6. Miller, A.D.: Sense-antisense (complementary) peptide interactions and the proteomic code; potential opportunities in biology and pharmaceutical science. Expert Opin. Biol. Ther. **15**, 245–267 (2015)
7. Blalock, J.E.: Genetic origin of protein shape and interaction rules. Nat. Med. **1**, 876–878 (1995)
8. Root-Bernstein, R.S.: Amino acid pairing. J. Theor. Biol. **94**, 885–894 (1982)
9. Blalock, J.E., Bost, K.L.: Binding of peptides that are specified by complementary RNAs. Biochem. J. **234**, 679–683 (1986)
10. Štambuk, N.: On the genetic origin of complementary protein coding. Croat. Chem. Acta **71**, 573–589 (1998)
11. Root-Bernstein, R.S.: Peptide self-aggregation and peptide complementarity as bases for the evolution of peptide receptors: a review. J. Mol. Recognit. **18**, 40–49 (2005)
12. Turčić, P., Štambuk, N., Konjevoda, P., Kelava, T., Gabričević, M., Stojković, R., Aralica, G.: Modulation of $\gamma_2$-MSH hepatoprotection by antisense peptides and melanocortin subtype 3 and 4 receptor antagonists. Med. Chem. **11**, 286–925 (2015)

13. Root-Bernstein, R.: How to make a non-antigenic protein (auto) antigenic: molecular complementarity alters antigen processing and activates adaptive-innate immunity synergy. Anticancer Agents Med. Chem. **15**, 1242–1259 (2015)
14. Houra, K., Turčić, P., Gabričević, M., Weitner, T., Konjevoda, P., Štambuk, N.: Interaction of α-melanocortin and its pentapeptide antisense LVKAT: effects on hepatoprotection in male CBA mice. Molecules **16**, 7331–7343 (2011)
15. McGuire, K.L., Holmes, D.S.: Role of complementary proteins in autoimmunity: an old idea re-emerges with new twists. Trends Immunol. **26**, 367–372 (2005)
16. Štambuk, N., Konjevoda, P., Gotovac, N.: A new rule-based system for the construction and structural characterization of artificial proteins. In: Stavrinides, S.G., Banerjee, S., Caglar, S.H., Ozer, M. (eds.) Chaos and Complex Systems: Proceedings of the 4th International Interdisciplinary Chaos Symposium, pp. 95–103. Springer, Berlin (2013)
17. Miyazawa, S., Jernigan, R.L.: Self-consistent estimation of inter-residue protein contact energies based on an equilibrium mixture approximation of residues. Proteins. **34**, 49–68 (1999)
18. Miyazawa, S., Jernigan, R.L.: Estimation of effective inter-residue contact energies from protein crystal structures: quasi-chemical approximation. Macromolecules **18**, 534–552 (1985)
19. Kyte, J., Doolittle, R.F.: A simple method for displaying the hydropathic character of a protein. J. Mol. Biol. **157**, 105–132 (1982)
20. Pokarowski, P., Kloczkowski, A., Jernigan, R.L., Kothari, N.S., Pokarowska, M., Kolinski, A.: Inferring ideal amino acid interaction forms from statistical protein contact potentials. Proteins **59**, 49–57 (2005)
21. Parker, J.M.R., Guo, D., Hodges, R.S.: New hydrophilicity scale derived from high-performance liquid chromatography peptide retention data: correlation of predicted surface residues with antigenicity and X-Ray-derived accessible sites. Biochemistry **25**, 5425–5432 (1986)
22. Rose, G.D., Geselowitz, A.R., Lesser, G.J., Lee, R.H., Zehfus, M.H.: Hydrophobicity of amino acid residues in globular proteins. Science **229**, 834–838 (1985)
23. Manavalan, P., Ponnuswamy, P.K.: Hydrophobic character of amino acid residues in globular proteins. Nature **275**, 673–674 (1978)
24. Cornette, J.L., Cease, K.B., Margalit, H., Spouge, J.L., Berzofsky, J.A., DeLisi, C.: Hydrophobicity scales and computational techniques for detecting amphipathic structures in proteins. J. Mol. Biol. **195**, 659–685 (1987)
25. Kolaskar, A.S., Tongaonkar, P.C.: A Semi-empirical method for prediction of antigenic determinants on protein antigens. FEBS. **276**, 172–174 (1990)
26. Grantham, R.: Amino acid difference formula to help explain protein evolution. Science **185**, 862–864 (1974)
27. White, S.H., Wimley, W.C.: Membrane protein folding and stability: physical principles. Annu. Rev. Biophys. Biomol. Struct. **28**, 319–365 (1999)
28. Kastritis, P.L., Bonvin, A.M.J.J.: On the binding affinity of macromolecular interactions: daring to ask why proteins interact. Interface **10**, 20120835 (2012). doi:10.1098/rsif.2012.0835
29. Mihel, J., Šikić, M., Tomić, S., Jeren, B., Vlahoviček, K.: PSAIA – protein structure and interaction analyzer. BMC Struct. Biol. **8**, 21 (2008). doi:10.1186/1472-6807-8-21
30. Singh, G., Dhole, K., Pai, P.P., Mondal, S.: SPRINGS: prediction of protein-protein interaction sites using artificial neural networks. J. Proteomics Computational. Biol. **1**, 1–7 (2014)

31. Leelananda, S.P., Feng, Y., Gniewek, P., Kloczkowski, A., Jernigan, R.L.: Statistical contact potentials in protein coarse-grained modeling: from pair to multi-body potentials. In: Kolinski, A. (ed.) Multiscale Approaches to Protein Modeling, pp. 127–157. Springer, New York (2011)

32. Tsurui, H., Takahashi, T., Matsuda, Y., Lin, Q., Sato-Hayashizaki, A., Hirose, S.: Exhaustive characterization of TCR-pMHC binding energy estimated by the string model and Miyazawa-Jernigan matrix. Gen. Med. **2**, 126 (2013). doi:10.4172/2327-5146.1000126

33. Martinić, R., Šošić, H., Turčić, P., Konjevoda, P., Fučić, A., Stojković, R., Aralica, G., Gabričević, M., Weitner, T., Štambuk, N.: Hepatoprotective effects of met-enkephalin on acetaminophen-induced liver lesions in male cba mice. Molecules **19**, 11833–11845 (2014)

34. Chou, K.C., Wei, D.Q., Zhong, W.Z.: Binding mechanism of coronavirus main proteinase with ligands and its implication to drug design against SARS. Biochem. Biophys. Res. Commun. **308**, 148–151 (2003)

35. Wolfenden, R.V., Cullis, P.M., Southgate, C.C.F.: Water, protein folding, and the genetic code. Science **206**, 575–577 (1979)

36. Spherotech, Inc.: SpheroTechnical Notes #10 - Magnetic Particle Enzyme Immunoassay (MPEIA) Test Procedure. http://www.spherotech.com/tech_SpheroTech_Note_10.html

37. Kala, M., Bajaj, K., Sinha, S.: Magnetic bead enzyme-linked immunosorbent assay (ELISA) detects antigen-specific binding by phage-displayed scFv antibodies that are not detected with conventional ELISA. Anal. Biochem. **254**, 263–266 (1997)

38. Sepmag Systems.: The Basic Guide to Magnetic Bead Cell Separation. http://www.sepmag.eu/free-basic-guide-magnetic-bead-cell-separation

39. Thevis, M., Kuuranne, T., Geyer, H., Schänzer, W.: Annual banned-substance review: analytical approaches in human sports drug testing. Drug. Test. Anal. **5**, 1–19 (2013)

40. Morris, G.E.: Epitope Mapping: B-Cell Epitopes. Encyclopedia of Life Sciences (eLS). Wiley, Chichester (2007). doi:10.1002/9780470015902.a0002624.pub2. http://www.els.net

# Comparative Analysis of microRNA-Target Gene Interaction Prediction Algorithms - The Attempt to Compare the Results of Three Algorithms

Anna Krawczyk[✉] and Joanna Polańska

Data Mining Group, Institute of Automatic Control Silesian University of Technology Gliwice, ul. Akademicka 16, 44-100 Gliwice, Poland
{anna.krawczyk,joanna.polanska}@polsl.pl

**Abstract.** MicroRNAs are non-coding, small molecules (21–25 nucleotides). They regulate gene expression by downregulation of the target gene or translational repression. What is more, they are involved in cancer growth. Nowadays, we can observe a sustainable growth and development of computational target prediction programs. The plethora of prediction algorithms cause the problem with a choice of the one algorithm, that may give satisfactory results– the possible binding site of the microRNA to the target gene. What is crucial is that the result is considered as satisfactory one, when it is statistically significant and there is a high probability that a specific gene is the target of real microRNA. In order to compare the results obtained from different algorithms we have to define one probability space for each of them. We performed a proper statistical test (Fisher's exact test) to ensure that we can juxtapose the results from three different algorithms which take into account different aspects of binding microRNA to the target gene. The conclusion of our work is the suggestion of the way in which one can juxtapose the results from algorithms based on different methods of prediction the possible miRNA-target gene interactions.

**Keywords:** MicroRNA · Target gene · P-value integration · MicroRNA-target prediction algorithms

## 1 Introduction

MicroRNAs (miRNAs) are small (21–25 nucleotides), non-coding, single-stranded RNA molecules [2]. What is more, they are well conserved in many species, both in plants and animals. The first miRNA was identified in Caenorhabditis elegans in 1993 [12].

MicroRNA molecules regulate gene expression post-transcriptionally [12] by pairing with their targets' transcripts [2]. Pairing is based on Watson-Crick complementarity. For instance, when they bind to the genes which are responsible for coding proteins, the result of such a binding is downregulation of the target gene or even repression [2]. What is more, miRNAs regulate the key processes, such as: cell cycle, proliferation and apoptosis.

© Springer International Publishing Switzerland 2016
F. Ortuño and I. Rojas (Eds.): IWBBIO 2016, LNBI 9656, pp. 103–112, 2016.
DOI: 10.1007/978-3-319-31744-1_10

In the nucleus, by the enzyme Drosha, miRNAs are transcribed from pri-miRNAs into stem-loop pre-miRNAs. Then, by Exportin-5, they are transported into cytoplasm, where Dicer processed them into single-stranded miRNA molecule. Mature miRNA enters RISC complex and binds to the 3'-UTR of the target mRNA [2].

MicroRNAs can act as an oncogenes, as well as suppressors. What is worth saying is that 50 % of microRNAs genes are in the genome fragile sites. The increase of miRNA expression level could take place as a result of the microRNA locus amplification. When the target of miRNA is tumor suppressors mRNA gene, the result could be the decrease of the target expression. This can cause the proliferation increase. On the other hand, the decreased expression of miRNA (for instance by deletion of the microRNA locus) could result in increased level of a target gene expression [13].

In these days, many computational approaches for microRNAs-target interaction identification were developed. It implies that there is a huge problem to decide which of these bioinformatics tool use in order to obtain significant results.

## 2    Methods

Nowadays, there are many online algorithms that can be used to define the interactions between miRNAs and genes but in order to obtain results, we used three algorithms: miRWalk v. 2.0, DIANA-microT v. 5.0 and Target Scan v. 7.0. The reason why we have chosen these very three algorithms (miRWalk, DIANA-microT and TargetScan) is that we performed a thorough research on many online accessible tools for miRNA-target gene interactions annotation and what has turned out is that only these mentioned three algorithms are based on comprehensive knowledge – taking into account both experimental and bioinformatical approach. Furthermore, each of these algorithms is based on another meaningful attitude towards prediction the binding site of miRNA molecule to the target gene. What is also worth saying is that miRWalk is the most impartial algorithm due to its capacity to give as a result the comparison of results obtained by 12 different microRNA-gene prediction programs [1]. Otherwise, all of them are still developing.

### 2.1    miRWalk

It is a freely accessible database, which provides the huge collection of experimentally verified microRNA-target interactions and also of the predicted ones for human, mouse and rat. What is more, one of the feature that distinguish mentioned algorithm from DIANA-microT-CDS and TargetScan is that the binding sites could be in such regions: promoter, 3'-UTR, 5'-UTR and CDS. MiRWalk searches possible binding sites by walking the complete gene sequence with a heptamer seed (7 nucleotides) of miRNA, from position 1 to 6 [1, 2].

Using Poisson distribution, the probability distribution of random matches of a subsequence (from the 5' end of miRNA sequence) in a complete gene sequence is calculated. Default p-value: 0.05 [1, 2].

Detailed p-value calculation [1, 2]:

$$P(X = K) = \frac{e^{-\lambda} \times \lambda^K}{K!} \tag{1}$$

$$\lambda = n \times P^L \tag{2}$$

n - length of analyzed sequence
P - probability of nucleotides
L - length of subsequence

Therefore,

$$P(X \geq 1) = 1 - P(X = 0) = 1 - (\frac{e^{-\lambda} \times \lambda^K}{K!}) = 1 - e^{-\lambda} \tag{3}$$

## 2.2   DIANA - MicroT

MicroT is the algorithm that takes into account two binding sites: 3'-UTR and CDS region. Unlike of the miRWalk algorithm, it is based on thermodynamics, as it estimates the free binding energy between the miRNA and 3'-UTR sequence [3].

MicroRNA Recognition Element (MRE) those UTR sites that have 6- (with G:U wobble base pair), 7-, 8- or 9-nucleotides long base pairing (Watson-Crick base pairing) with the microRNA. Algorithm starts from position 1 or 2 from the 5' end of the microRNA [4].

MiTG score – weighted sum of the scores of all identified MREs on the 3'-UTR, higher value refers to the higher possibility of correct prediction [4]. We used the default threshold value: 7.3 [3]. Weights were estimated for each binding category (7mer, 8mer and 9mer). The estimation of weights takes into account the slope of a fitted line, where mentioned fitting was based on linear least squares approximation [3].

Score indicates the contribution of each binding site to the overall miTG score [3] (Table 1).

$$w = \frac{slope\ for\ binding\ category}{slope\ for\ 9mer} \tag{4}$$

**Table 1.**  Weights for specific binding category.

| Binding category | Weights |
|---|---|
| 9mer | 1.00 |
| 8mer | 0.75 |
| 7mer | 0.50 |
| other | 0.25 |

## 2.3  TargetScan

This algorithm searches for possible binding sites of microRNA to target gene by looking for the presence of 6-, 7- and 8mer in the seed region of each microRNA (position 2–7 of mature microRNA) in the 3'-UTR binding site [5]. Predictions are evaluated on the basis of cumulative weighted context++ scores of the sites [5] and probability of conserved targeting - PCT [6]. The context++ score is the sum of the contribution of following features [5]:

- site type
- supplementary pairing – supplementary pairing at the microRNA 3' end
- local AU – AU content near the site
- minimum distance – minimum distance of site from stop codon or polyadenylation site
- sRNA1A – identity of nucleotide 'A' at position 1 of the sRNA
- sRNA1C – identity of nucleotide 'C' at position 1 of the sRNA
- sRNA1G – identity of nucleotide 'G' at position 1 of the sRNA
- sRNA8A - identity of nucleotide 'A' at position 8 of the sRNA
- sRNA8C - identity of nucleotide 'C' at position 8 of the sRNA
- sRNA8G - identity of nucleotide 'G' at position 8 of the sRNA
- site8A
- site8C
- site8G
- 3' UTR length
- SA (predicted structural accessibility) – probability that a 14 nt segment centered on the match to sRNA positions 7 and 8 is unpaired
- ORF length
- ORF 8mer count– numer of 8mer sites in the ORF
- 3' UTR offset 6mer count – number of offset-6mer sites in the 3'-UTR
- TA (target site abundance) – number of sites in all annotated 3'-UTRs
- SPS (predicted seed-pairing stability) – predicted thermodynamic stability of seed pairing
- $P_{CT}$ (probability of conserved targeting) – probability of site conservation, controlling for dinucleotide evolution and site context. Signal-to-background-ratio is converted to a $P_{CT}$ (Table 2).

**Table 2.** The comparison of algorithms.

| Algorithm | miRBase version | 3'-UTR | 5'-UTR | CDS | Promoter | Method |
|---|---|---|---|---|---|---|
| miRWalk | 20 | + | + | + | + | Seed complementarity |
| DIANA-microT-CDS | 18 | + | – | + | – | Thermodynamics |
| TargetScan | 21 | + | – | – | – | Seed complementarity |

$$P_{CT} = \frac{\frac{S}{B} - 1}{\frac{S}{B}} \tag{5}$$

Aggregate PCT:

$$1 - \left( \left(1 - P_{CT}\right)_{site1} \times \left(1 - P_{CT}\right)_{site2} \times \left(1 - P_{CT}\right)_{site3} \cdots \right) [6] \tag{6}$$

As can be seen above, each algorithm provides another measure of the hit significance. From miRWalk algorithm what we obtain is p-value, whilst the result from DIANA-microT is miTG score and from TargetScan is aggregate PCT. It implies the fact that only miRWalk and TargetScan use probability distribution to measure the miRNA-target interaction significance hit, whereas the third mentioned algorithm uses score calculation. What is worth noticing is that both miRWalk and TargetScan calculate the probability distribution, but the first algorithm follows the Poisson distribution, whereas the second – Bayesian estimation. Referring to this, we cannot compare this two algorithms in a simple way.

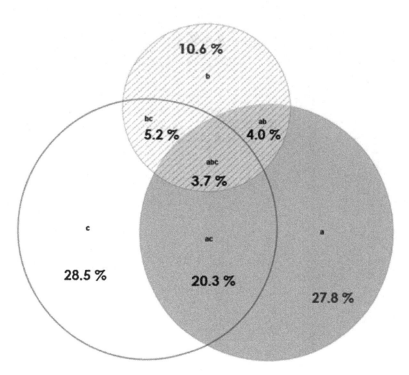

**Fig. 1.** Venn diagram of the results obtained using three algorithms for microRNA hsa-let-7a-5p [10].

## 3 Results

For comparison we have chosen three microRNAs: hsa-let-7a-5p, hsa-mir-21 and hsa-miR-17. The first one has oncogenes (RAS, HMGA2) as its targets, as well as genes which are responsible for cell cycle, proliferation and apoptosis [8]. Whilst the second one has tumor suppressors as its targets and what is more, it inhibits the expression of phosphatases what has an influence on signaling pathways [9]. The target of hsa-miR-17-5p is E2F1 – the transcription factor that regulates the cell cycle (transition from G1 to S phase) [15] (Fig. 1).

Number of predicted target genes for hsa-let-7a-5p:

- **a** - miRWalk: 7863
- **b** - DIANA-microT: 2997
- **c** - TargetScan: 8060

As can be observed, TargetScan provides the bigger number of results and the largest intersection of results is miRWalk-TargetScan. Whereas in juxtaposition with miRWalk, we can see that the difference between size of result sets is not significant. The reason why the results number obtained from these two algorithms is a little bit different is probably the fact that in miRWalk algorithm the user can set the p-value cutoff, whilst in the TargetScan there is no such a possibility. On the other hand, what is crucial is that

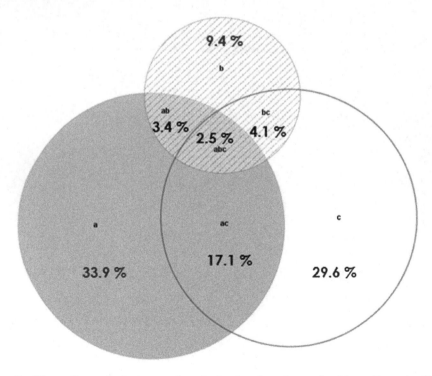

**Fig. 2.** Venn diagram of the results obtained using three algorithms for microRNA hsa-miR-21 [10].

each of three mentioned algorithms is based on different method (seed complementarity/ thermodynamics). DIANA-microT and TargetScan takes into account also the conservation across species (Fig. 2).

Number of predicted target genes for hsa-mir-21:

- **a** - miRWalk: 18892
- **b** - DIANA-microT: 5250
- **c** - TargetScan: 16510 (Fig. 3)

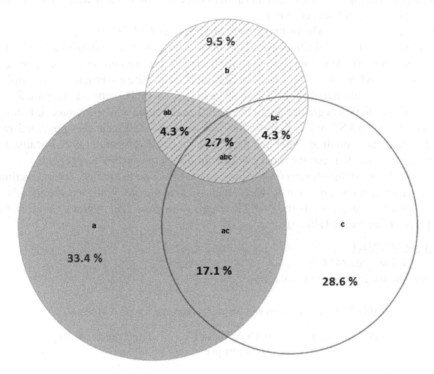

**Fig. 3.** Venn diagram of the results obtained using three algorithms for microRNA hsa-miR-17 [10].

Number of predicted target genes for hsa-miR-17-3p:

- **a** - miRWalk: 33474
- **b** - DIANA-microT: 9573
- **c** - TargetScan: 28733

As we can notice, the vast majority of microRNAs targets are predicted by miRWalk. By using DIANA-microT the obtained number of results are small in comparison to the other algorithms. As it was for hsa-let-7a-5p and hsa-mir-21 the intersection is the largest for miRWalk and TargetScan.

We would like to know whether the gene is the target of miRNA. We use three algorithms: miRWalk, DIANA-microT-CDS and TargetScan to find the miRNA-target gene interaction. As the matter of fact, we obtained plethora of results but each tool provides the different ones, so according to this fact, we have to join them into one probability space.

We do not know which algorithm is better so we have to compare results from each and choose the best one. The best one is the result which is statistically significant and the most probable to be the real target of microRNA.

We conducted the Fisher's exact test in order to combine p-values obtained from our own calculations, which were done using the Poisson distribution. These 3 studies were done in different methodologies but all of them were done in order to assess whether the gene is the microRNA target or not.

Calculations were made for two genes: MAP3K1 and YOD1.

For gene MAP3K1 DIANA-microT points out the possible binding site in 3'-UTR as a 6mer (score: 0.00457) and 9mer (score: 0.11652), but we choose for further calculations the second one due to the lower score value, because the lower score value implies the higher contribution to the miTG score. The higher the miTG score is, the probability of accurate prediction is greater. For gene YOD1 in the same way the choice of binding site length for DIANA-microT algorithm was made. TargetScan also predicted two possible lengths of binding: 7mer and 8mer, but we chose the greater one. Because we are looking for the longest, the most possible binding site miRNA-target gene.

We use Poisson distribution for p-values calculation due to the fact, the interactions miRNA-target are the rare events. $\lambda$ was calculated as in the miRWalk algorithms, where the length of the sequence is the 3'-UTR length for a gene. The lower p-value is, the most significant hit is (Table 3).

Gene MAP3K1
3'-UTR length: 2431 bp
The obtained p-values are as follows:

**Table 3.** P-values calculated for each algorithm for gene MAP3K1.

| miRWalk (11 nt) | DIANA-microT-CDS (9 nt) | TargetScan (8 nt) |
|---|---|---|
| 0.00060 | 0.00927 | 0.03709 |

All obtained results are statistically significant.

Gene YOD1
3'-UTR length: 5257 bp (Table 4)

**Table 4.** P-values calculated for each algorithm for gene YOD1.

| miRWalk (12 nt) | DIANA-microT-CDS (9 nt) | TargetScan (8 nt) |
|---|---|---|
| 0.00031 | 0.02005 | 0.08022 |

Only the result obtained from TargetScan is not a statistically significant one, due to the fact that p-value is greater than 0.05.

Null hypothesis $H_0$: A gene is not the microRNA target.
Alternative hypothesis $H_A$: A gene is the microRNA target.

$$F = -2 \sum_{i=1}^{k} \log(p_i) \sim \chi^2_{2k} \tag{7}$$

There are 3 tests, so the test has 6 degrees of freedom. Critical value for chi-squared distribution for 8 degrees of freedom and at a significance level at 0.05 is 12.5916.

Fisher's Combined Probability Test [14] value for gene MAP3K1 is equal to 30.7879, whereas for gene YOD1 is equal to 29.0229. Both p-values are greater than critical value, so we can reject the null hypothesis and say that the gene MAP3K1 and YOD1 are the microRNA targets.

## 4 Discussion

Our thorough comparison of those three prediction algorithms shows that one can use all of them. Until our analysis was done one was not able to say whether the results obtained from the one algorithm are better than those obtained from another one. Now it is possible to use them and what is worth to underline– we know if obtained result is statistical significant or not, what implies, that it is possible to choose the best result. Each of mentioned algorithm takes into account another aspect of predicting the binding site of microRNA to the target gene, so it was so crucial to decide which of them provides the most probable results. The conjunction of the algorithms were essential to decide whether the results are possible to be the real target of microRNA.

Newly developed research provides more information on miRNA-target gene interactions and their influence on gene expression, so all algorithms should be actualized and should take into account the results of the newest research.

**Acknowledgements.** We thank Dr. Harsh Dweep of the University of Heidelberg for comprehensive explanation of calculating the p-value in miRWalk algorithm. This work was financed by BK/227/Rau-1/2015/t.10 and BioRadInt Harmonia 4 grant DEC-2013/08/M/ST6/00924. Calculations were carried out using infrastructure of GeCONiI (POIG. 02.03.01-24-099/13).

## References

1. Dweep, H., et al.: miRWalk2.0: a comprehensive atlas of microRNA-target interactions. Nat. Methods **12**(8), 697 (2015)
2. Dweep, H., et al.: miRWalk - Database: prediction of possible miRNA binding sites by "walking" the genes of three genomes. J. Biomed. Inform. **44**, 839–847 (2011)
3. Maragkakis, M., et al.: Accurate microRNA target prediction correlates with protein repression levels. BMC Bioinform. **10**, 295 (2009)
4. Maragkakis, M., et al.: DIANA-microT web server: elucidating microRNA functions through target prediction. Nucleic Acids Research W273-W276 (2009)

5. Lewis, B.P., et al.: Conserved seed pairing, often flanked by adenosines, indicates that thousands of human genes are MicroRNA targets. Cell **120**, 15–20 (2004)
6. Agarwal, V., et al.: Predicting effective microRNA target sites in mammalian mRNAs. Elife **4**, e05005 (2015)
7. Friedman, R.C., et al.: Most mammalian mRNAs are conserved targets of microRNAs. Genome Res. **19**, 92–105 (2009)
8. Mayr, C., et al.: Disrupting the Pairing Between let-7 and Hmga2 Enhances Oncogenic Transformation. Science **315**(5818), 1576–1579 (2007)
9. Papagiannakopoulos, T., et al.: MicroRNA-21 targets a network of key tumor-suppressive pathways in glioblastoma cells. Cancer Res. **68**(19), 8164–8172 (2008)
10. Micallef, L., Rodgers, P.: eulerAPE: Drawing Area-Proportional 3-Venn Diagrams Using Ellipses. PLoS ONE **9**, e101717 (2014)
11. Bartel, D.P.: MicroRNAs: Genomics, Biogenesis, Mechanism, and Function. Cell **116**, 281–297 (2004)
12. Paraskevopoulou, Maria D., et al.: DIANA-microT web server v5.0: service integration into miRNA functional analysis workflows. Nucleic Acids Res. **41**, W169–W173 (2013)
13. Majorek, K., Krzyżosiak, W.J.: Rola mikroRNA w patogeneie, diagnostyce i terapii nowotworów. Współczesna Onkologia **10**(8), 359–366 (2006)
14. Fisher, R.A.: Statistical Methods for Research Workers, 4th edn. Oliver and Boyd, Edinburgh (1932)
15. Stahlhut Espinosa, C.E., Slack, F.J.: The role of microRNAs in cancer. Yale J Biol Med. **79**(3–4), 131–140 (2006)

# A Novel Divisive iK-Means Algorithm with Region-Driven Feature Selection as a Tool for Automated Detection of Tumour Heterogeneity in MALDI IMS Experiments

Grzegorz Mrukwa[1], Grzegorz Drazek[1], Monika Pietrowska[2], Piotr Widlak[2], and Joanna Polanska[1(✉)]

[1] Silesian University of Technology, Data Mining Group, Gliwice, Poland
{grzegorz.mrukwa,grzegorz.drazek,joanna.polanska}@polsl.pl
[2] Maria Sklodowska-Curie Memorial Cancer Center and Institute of Oncology, Gliwice, Poland
{m_pietrowska,widlak}@io.gliwice.pl

**Abstract.** Due to the constantly increasing cancer incidence rates and varying levels of effectiveness of the utilised therapeutic approaches, obtaining a clear understanding of the underlying phenomena is of the utmost importance. The problem is tackled by numerous research groups worldwide, utilising a number of molecular biology quantification techniques. MALDI-IMS (Matrix-Assisted Laser Desorption Ionization – Image Mass Spectrometry) is a quantification technique that brings together MALDI spectroscopy with tissue imaging by multiple applications of the laser beam to a raster of points on the surface of the analysed tissue. The application of MALDI-IMS in cancer research allows for the spatial identification of molecular profiles and their heterogeneity within the tumour, but leads to the creation of highly complicated datasets of great volume. Extraction of relevant information from such datasets relies on the design of appropriate algorithms and using them as the base to construct efficient data mining tools. Existing computational tools for MALDI-IMS exhibit numerous shortcomings and limited utility and cannot be used for fully automated discovery of heterogeneity in tumour samples. We developed a novel signal analysis pipeline including signal pre-processing, spectrum modelling and intelligent spectra clustering with region-driven feature selection to efficiently analyse that data. The idea of combining divisive iK-means algorithm with peptide abundance variance based dimension reduction performed independently for each analysed sub-region allowed for discovery of squamous cell carcinoma and keratinized stratified squamous epithelium together with stratified squamous epithelium within an exemplary head and neck tumour tissue.

**Keywords:** MALDI IMS · Clustering · K-means · Tumour heterogeneity

## 1 Introduction

Matrix-assisted laser desorption ionization (MALDI) imaging mass spectrometry (IMS) is a powerful tool for investigating the spatial distribution of proteins and small molecules (metabolites) within biological samples. This technique combines the potential of

© Springer International Publishing Switzerland 2016
F. Ortuño and I. Rojas (Eds.): IWBBIO 2016, LNBI 9656, pp. 113–124, 2016.
DOI: 10.1007/978-3-319-31744-1_11

MALDI-ToF mass spectrometry with the ability to scan series of pixels across the surface of tissues, which generates multiplex space-correlated mass spectra (Fig. 1). Numerical processing of the data allows the visualization of specific molecular species and the assessment of their correlation with morphological image of the sample. Hence, MALDI-IMS is a multiplex untargeted analysis that enables the characterization of tissue regions based on their endogenous biomolecular content. The major advantage of this technique is its potential to define tissue regions based on their molecular profiles independently of their histological and morphological characteristics given by traditional tools. MALDI-IMS can detect potential cancer foci within histologically normal tissue, reveal intra-tumour heterogeneity or altered protein/metabolite levels at tumour/ normal tissue interface zones. Hence, MALDI-IMS is an emerging technology that could fill gaps in the knowledge of the spatial nature of cancer, allowing for the better understanding of connections between structure and molecular processes. Application of MALDI-IMS leads to the generation of datasets of significant complexity and massive volumes and retrieval of useful information is highly dependent on the development of appropriate algorithms, incorporated into bioinformatics tools, for processing and analysing the data. The development of such tools is the main goal of the work.

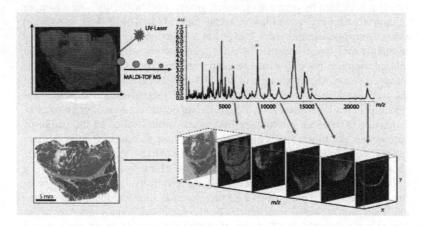

**Fig. 1.** The key idea of MALDI mass spectrometry imaging [13]

## 2    State-of-the-Art in IMS Data Analysis

Existing computational tools for MALDI-IMS exhibit numerous shortcomings and limited utility. Due to the lack of appropriate, specialized compression tools, there are serious problems with the storage and transfer of MALDI-IMS data. A serious limitation of the existing data analysis systems is also their insufficient variety of statistical tests and data mining algorithms implemented. Although standard statistical analysis methods are incorporated in some of the commercial and shareware systems, none of them provides comprehensive inter- and intra-sample analysis techniques.

Software suites from spectrometer sellers include an application to load the instrument-specific IMS data, reconstruct an image, and offer post-processing tools to navigate through the data, enhance images, and help with data interpretation: FlexImaging (Bruker, Billerica, MA, USA), ImageQuest (ThermoFisher Scientific, Bremen, Germany), and Tissueview (AB SCIEX, Foster City, CA, USA). The functionality and flexibility of such commercial software is unsatisfactory. MSiReader is an open source program built in the MATLAB platform to process, analyse, and visualize imaging data. The most popular, commercial SCiLS Lab (SCiLS, Bremen, Germany) is a program for visualization, analysis and interpretation of MALDI-IMS data. All of the above software uses local maximum based peak identification or applies simple modelling of signal fragments, and does not offer an extensive spectrum modelling, which might be crucial for final discoveries. The systems were originally designed to support identification of tissue functional structure and in majority use Principal Component Analysis and/or spatial k-means clustering to achieve the goal. Despite this the discovery of heterogeneity within tumour tissue remains unsolved.

While performing literature search one can find a few groups of scientific publications devoted to the development of the algorithms suitable for extensive IMS data analysis. Combination of PCA and hierarchical clustering allowed separation of gastric cancer foci from non-malignant gastric mucosa [5]. Six different methods of unsupervised analysis were tested in dataset generated by MALDI-IMS for myxofibrosarcoma samples [8]. Semi-supervised segmentation of MALDI-IMS data based on spatial k-means clustering on PCA component heat maps allowed to reveal distinct sub-regions of laryngeal cancer [1]. The results of unsupervised segmentation of HNC performed for lipidome imaging by MALDI FT-ICR was recently published by Krasny et al. [10]. Authors used SCiLS Lab pipeline for OMP based peak picking [6] and bisecting k-means segmentation of MALDI spectra proceeded by spatial denoising.

# 3 Material

## 3.1 Clinical Material

Tissue material was collected from a patient who underwent surgery due to oral cavity squamous cell carcinoma, cancer stage T4N2M0. Tissue specimen containing tumour and surrounding tissues was H&E stained and evaluated by an experienced pathologist in fresh postoperative material (Fig. 2). The study was approved by the appropriate Bioethical Committee, and performed in accordance with national guidelines.

## 3.2 Sample Preparation for MALDI IMS

Frozen tissue section was plated onto indium tin oxide-coated conductive slide, dried, washed twice in ethanol, dried again and coated with a solution of trypsin using an automatic spraying device (ImagePrep, Bruker Daltonik – standard matrix coating program). Optical images were registered before matrix deposition. Tissue section was subjected to peptide imaging with the use of MALDI ToF ultrafleXtreme mass spectrometer (Bruker Daltonik) equipped with a smartbeam II™ laser operating at 1 kHz

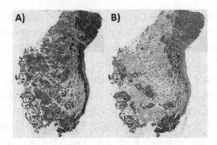

**Fig. 2.** The H&E stained fresh section (panel A) and results of pathologists' analysis (panel B). The following colour code was used for region definitions: red – tumour area, blue – healthy epithelium, green – saliva, yellow – muscle (Color figure online).

repetition rate. Ions were accelerated at 25 kV with PIE time of 100 ns. Spectra were acquired in positive reflectron mode within 800–4000 m/z and externally calibrated with Bruker's Peptide Calibration Standard II. A raster width of 100 μm was applied, 400 spectra were collected and averaged from each ablation point. The primary dataset consisted of 9,492 averaged spectra with 109,568 mass channels [m/z].

### 3.3 Spectra Pre-processing and Modelling

The following spectrum pre-processing steps were applied: spectrum resampling (to unify mass channels across dataset), baseline identification and removal, TIC normalization, and Fast Fourier Transform based alignment to the average spectrum [14]. The Gaussian mixture model (GMM) approach [12] was used for spectra modelling and peak detection (Fig. 3). That step reduces significantly the data dimension with almost no information loss. The peptide abundance was estimated by performing convolution of the adequate GMM component and the individual spectrum, and subsequently calculating the area below the obtained curve. The neighbouring peaks, modelling the right

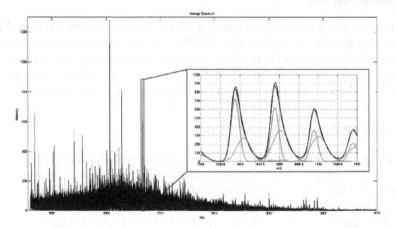

**Fig. 3.** The final result of spectra pre-processing and modelling.

skewness of spectral peaks [9], were identified and merged by summing their estimated abundance. The location of the dominant component was set as the peptide ion m/z value. The additional reduction of dimensionality was obtained by filtering out the peptides with abundance below the noise threshold. The signal denoising procedure requires the data-driven estimation of noise level. The GMM approach presented in [11] was used for that purposes.

## 4   Spectra Clustering

A flowchart of the proposed spectra clustering algorithm is presented in Fig. 4. The procedure, named [4] divisive iK-means algorithm with region-driven feature selection, consists of recursive sub-region splitting performed in reduced domain independently customized for every sub-region to be split. Both its elements, the iterative character of the developed algorithm and the sub-region dependent reduction of feature domain help in discovery of hidden secondary tissue structure.

The primary tissue structure usually predominates its molecular image and the heterogeneous tumour regions can be overlooked. Additionally, performing clustering in high dimensional space with only small fraction of features being significantly different among sub-regions does not ensure satisfactory results. The need for selection of informative feature has been already reported by Friedman and Meulman [7]. Although reducing the feature domain by optimizing performance index composed of both cluster

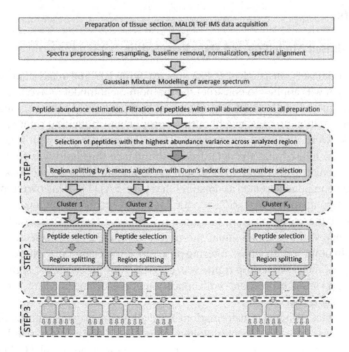

**Fig. 4.** The diagram showing core elements of divisive iK-means algorithm with region-driven feature selection.

separation related component and feature weighting component, the COSA algorithm is not used in an iterative manner and it also does not give passable results in case of MALDI imaging of tumour samples.

The developed algorithm starts with k-means clustering of all modelled spectra from the sample. After the first split, the k-means algorithm is applied independently to each sub-region obtained in the antecedent split. The splitting is then continued until the specified number of recursions is reached or the size of obtained cluster is less than a priori assumed value. After testing several distance metrics, Pearson's correlation coefficient was chosen due to its best performance in capturing spectral similarity. The number of clusters at each splitting was not predefined, k-means clustering was performed for 2–10 clusters and the Dunn index was used for the selection of the optimal number of clusters [2, 3].

Since the final result of k-means partitioning strongly depends on the initial configuration, we developed a novel procedure for setting highly effective initial partitions. Rigorous numerical evaluation demonstrated its predominance over standard approaches. The procedure does not require repetitions to protect against hitting local optima. First, the linear regression model is built using the most locally informative features for a given subset of spectra from a sub-region. The most distant data point, defined as the spectrum with the highest residuum, is chosen as the initial center of the first cluster. The remaining K-1 initial centers are chosen sequentially in such a way that the minimal distance from the new center to all of the centers found yet has to be maximal.

## 5    Results

### 5.1    GMM-Based Modelling and Data Processing

The obtained dataset consisted of 9,492 spectra with 109,568 mass channels [m/z]. GMM approach was applied to construct the mathematical model of average spectrum. The complete model contained 6,714 Gaussian components. At the beginning of the iterative segmentation process, neighbouring components modelling right-skewness of spectral peaks were identified and merged. Then the components with relatively low abundance were filtered out; the data-driven abundance threshold was found through modelling abundance distribution as a sum of Gaussian-shaped functions, with the smallest mean component treated as noise-related (Fig. 5A) [11]. This two steps reduced the number of components from 6,714 to 3,671. During each sub-region splitting step, independent of the recursion step, the most informative features (i.e. these with the highest variance within the individual sub-region of interest) were selected out of the set of 3,671. In uninformative peak filtering procedure, the Gaussian mixture component with the highest mean value (top right) was chosen from the model of signal variance distribution and variance threshold (with condition of preserving at least 5 % of features) was calculated (Fig. 5B).

It is crucial that all signal transformations used allow for easy recognition of peptides as they preserve information about biological dependencies (in contrary to PCA-based methods, for example), while reducing disk space necessary from about 8 GB per sample to about 200 MB (per sample).

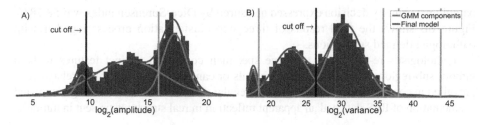

**Fig. 5.** The histogram based filtration of (A) low abundance, and (B) uninformative peaks. From all potential thresholds (magenta) present on (B), the highest preserving at least 5 % of features is chosen (black, indicated by arrow) (Color figure online).

### 5.2 Divisive IK-Means with Region-Driven Feature Selection

In the first step of clusterization process, based on GMM filtration procedure, 1,892 features were chosen as possibly differentiating across whole sample. Two regions were found (Fig. 6A, upper plot) - one corresponding to epithelium and tumour marked by pathologist (997 data points/spectra, black colour), another to the rest of the sample (7,008 points, grey colour). Dice-Sorensen similarity index was 78.82 % while comparing to pathologist decision. For both regions further analysis was conducted independently. 119 and 74 features were chosen for both regions respectively. Figure 7 shows graphically, which features were chosen at particular step. Focusing back on epithelium and tumour cluster found, segmentation into three sub-regions was selected as optimal in the meaning of Dunn's index. There were 443, 234 and 320 data points/ spectra assigned to regions assumed as tumour, epithelium and part of epithelium that is characterized by prominent keratinization. Cancer region found similarity with

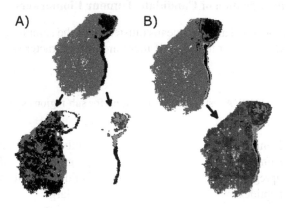

**Fig. 6.** Results of first two steps of divisive iK-means based clusterization algorithm with region-driven feature selection. Panel (A) demonstrates the first split and the second splits of primary sub-regions. Panel (B) presents the visualization of final results of two-step divisive iK-means algorithm. Red line shows border of tumour region, blue line – healthy epithelium, as marked by experienced pathologist during the visual inspection of H&E stained specimen (Color figure online).

pathologist primary decision expressed measured by Dice-Sorensen index was 58.78 %. Figure 6B shows the final results of three-step clusterization process together with pathologist defined sub-regions.

Pathologist's re-inspection of tissue specimen confirmed distinct features in both tumour sub-regions: foci of actual cancer cells or cancer microenvironment-related cells prevailed in corresponding areas. Hence, molecular differences detected during automated segmentation of IMS data had an apparent reflection in real structures present in tumour.

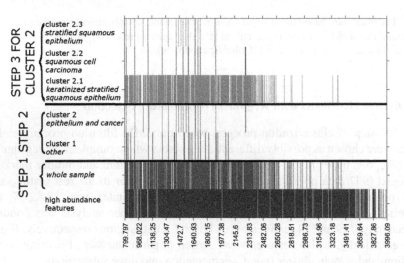

**Fig. 7.** The location of spectral features active during particular step of divisive iK-means clusterization algorithm.

### 5.3 Statistical Identification of Candidate Tumour Biomarkers

Using the partition obtained after iK-means clustering of the sample, it was possible to use permutation ANOVA test to check, if there are any characteristic peptides up- or

**Table 1.** Permutation ANOVA test results in searching for sub-regions signature (significance level $\alpha = 2.5 \cdot 10^{-6}$).

| Tissue region | Number of significantly upregulated peptides | Number of significantly downregulated peptides | Number of upregulated peptides with at least large effect size | Number of downregulated peptides with at least large effect size |
|---|---|---|---|---|
| Tumour | 512 | 74 | 127 | 43 |
| Epithelium | 229 | 154 | 103 | 72 |
| Parakeratosis | 48 | 197 | 15 | 19 |

downregulated for specific sub-regions such as tumour or epithelium. P value less than $2.5 \cdot 10^{-6}$ was assumed statistically significant (Table 1). Among them there were peptides with Cohen's D large or very large effect size. Peptide abundance heat maps of exemplary candidate biomarkers chosen from the ones mentioned in Table 1 are presented on Fig. 8.

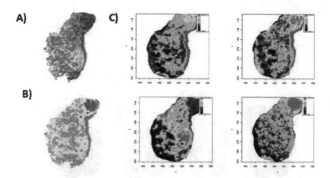

**Fig. 8.** The spatial distribution of chosen biomarkers: (A) original H&E stained specimen, (B) tissue sub-regions found in two-step clusterization, (C) heat maps of four chosen peptide biomarkers specific for particular tissue regions: epithelium (upper left); tumour and epithelium (upper right); muscle (bottom left); and tumour only (bottom right).

The peptides forming the obtained signatures of different sub-regions were hypothetically annotated with tryptic peptides identified by LC-MS/MS in the same tissue preparations. GO terms related to negative regulation of apoptosis, cell motility and protein folding were associated with proteins which fragments were putatively upregulated in tumour area, while GO terms related to canonical glucose metabolism were associated with downregulated ones. Peptides upregulated in tumour are exemplified by the component m/z = 2172.08 (putatively fragment of pyruvate kinase - enzyme involved in the Warburg effect).

## 6    Discussion

### 6.1    PCA-Based Approach

Since the most popular method for dimension reduction and pattern recognition within MALDI IMS data is PCA transformation, we have also performed it on our exemplary data - 9,492 spectra with 109,568 mass channels each (raw MALDI data). Consecutive PCA components variance explanation was summed in order to obtain cumulative distribution of variance across the features of transformed data. To find components with the highest input to variance explanation, a piecewise-linear approximating model was built and the crossing point was considered as discriminating (Fig. 9).

Twelve PCA components were taken as the most sub-region specific ones. Discretized heat maps of chosen PCA components are presented on Fig. 10 together with pathologist decision based tumour and epithelium regions marked with red and blue

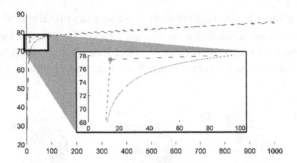

**Fig. 9.** The cumulative explained variance versus number of PCA components

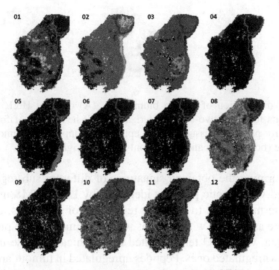

**Fig. 10.** Spatial distribution of discretized first 12 PCA components with tumour and healthy epithelium marked in red and blue colour (Color figure online).

colours. Advanced comparison study of obtained heat maps supported by pathologist expertise is necessary to identify different sub-regions within sample.

The tumour and epithelium sub-regions are seen as one cluster on heat map of second principal component, while parakeratosis of epithelium can be noticed on heat map of third and eighth principal component. The heterogeneity of tumour area is observed on the heat maps of second and eighth principal components.

## 6.2   Computational Load

Computational load of presented algorithm is much lower than for standard methods applied to this kind of data, as after pre-processing, all further analysis are carried in reduced space, which is less than 5 % of original data size. Whole pre-processing was carried on a computer with 16 GB of RAM, while further calculations were rather calculated on a PC with 8 GB, however even computer with just 4 GB of memory was sufficient in this case. Moreover, implementation was designed to be able to resume calculations even after forced stops by caching partial results on computer's hard disk. This allows for quick result recovery or recalculation on machine with worse parameters when primarily all of them were performed on bigger one. Clusterization time was measured for presented sample while computations were carried out on computer equipped with Intel(R) Core(TM) i3-2100 CPU @ 3.10 GHz 3.10 GHz and 5.00 GB of RAM: 671 s for first run, lower than 20 s in next runs (if properly set).

### 6.3 Summary and Conclusions

In contrast to many algorithms used nowadays, the proposed solution does not try to repeat classification made by pathologist, but is focused on the interdependencies actually present in the data. Therefore, some divergences appear in generated cluster assignment plots in comparison to direct inspection, however they are caused by strong differentiation of spectra among considered sub-regions. Further investigation revealed that real information was successfully found. Region corresponding to cancer and epithelium (obtained in first iteration) splits into three sub-regions with different properties. First sub-region revealed substantial presence of foci of squamous cell carcinoma, i.e., transformed cells derived from normal epithelium. The other tumour sub-region, showed molecular similarity with connective tissues and was markedly different from epithelial cells. Re-analysis of corresponding tissue by a pathologist revealed substantial contribution of inflammation-related cells and other features putatively related to cancer microenvironment. Hence, molecular differences detected during automated segmentation of IMS data had an apparent reflection in functional structures present in cancer area.

Quick review of PCA-based approach heat maps does not allow for straightforward conclusion, that region marked by pathologist is internally differentiated and complex. Moreover, among twelve PCA components representing highest variance of the sample, only few of them pass valuable information, while, still, it is hard to evaluate which peptides take part in each of the components.

We demonstrated that the developed divisive iK-means algorithm with region-driven feature selection allows for discovery of hidden structure within tumour tissue. Independent feature selection within each region to be split together with intelligent setting initial conditions for k-means algorithm are crucial for satisfactory result of clustering. Divisive nature of the proposed solution is necessary for automated distinguishing between tumour and healthy tissue.

**Acknowledgement.** The work was partially financed by NCN grant no. DEC2013/08/M/ ST6/924. The GeCONiI IT infrastructure (grant on POIG 02.03.01-24-099) *"Upper Silesian Centre for Computational Science and Engineering"* was used for performing calculations and numerical simulations.

# References

1. Alexandrov, T., Becker, M., Guntinas-Lichius, O., Ernst, G., et al.: MALDI-imaging segmentation is a powerful tool for spatial functional proteomic analysis of human larynx carcinoma. J. Cancer Res. Clin. Oncol. **139**, 85–95 (2013)
2. Bolshakova, N., Azuaje, F., Machaon, C.V.E.: Cluster validation for gene expression data. Bioinformatics **19**, 2494–2495 (2003)
3. Celebi, M.E., Kingravi, H.A.: Linear, deterministic, and order-invariant initialization methods for the k-means clustering algorithm. In: Celebi, M.E. (ed.) Partitional Clustering Algorithms, pp. 79–98. Springer, Switzerland (2014)
4. Chandan, K.R.: Bhanukiran, V: A survey of partitional and hierarchical clustering algorithms. In: Charu, C.A., Chandan, K.R. (eds.) Data Clustering: Algorithms and Applications. Chapman and Hall/CRC, Boca Raton (2013)
5. Deininger, S.O., Ebert, M.P., Futterer, A., Gerhard, M., et al.: MALDI imaging combined with hierarchical clustering as a new tool for the interpretation of complex human cancers. J. Proteome Res. **7**, 5230–5236 (2008)
6. Denis, L., Lorenz, D.A., Trede, D.: Greedy solution of ill-posed problems: error bounds and exact inversion. Inverse Probl. **25**(11), 115017 (2009)
7. Freidman, J.H., Meulman, J.J.: Clustering objects on subsets of variables (with discussion). J. Roy. Stat. Soc. B **66**, 815–849 (2004)
8. Jones, E.A., van Remoortere, A., van Zeijl, R.J.M., Hogendoorn, P.C.W., et al.: Multiple statistical analysis techniques corroborate intratumor heterogeneity in imaging mass spectrometry datasets of Myxofibrosarcoma. PLoS ONE **6**, e24913 (2011)
9. Kempka, M., Sjodahl, J., Bjork, A., Roeraade, J.: Improved method for peak picking in matrix-assisted laser desorption/ionization time-of-flight mass spectrometry. Rapid Commun. Mass Spectrom. **18**, 1208–12 (2004)
10. Krasny, L., Hoffmann, F., Ernst, G., Trede, D., et al.: Spatial segmentation of MALDI FT-ICR MSI data: a powerful tool to explore the head and neck tumor in situ lipidome. J. Am. Soc. Mass Spectrom. **26**, 36–43 (2015)
11. Marczyk, M., Jaksik, R., Polanski, A., Polanska, J.: Adaptive filtering of microarray expression data based on Gaussian mixture decomposition. BMC Bioinform. **14**, 101 (2013)
12. Polanski, A., Marczyk, M., Pietrowska, M., Widlak, P., Polanska, J.: Signal partitioning algorithm for highly efficient Gaussian mixture modeling in mass spectrometry. PLoS ONE **10**, e0134256 (2015)
13. Rauser, S., Höfler, H., Walch, A.: In-situ-Proteomanalyse von Geweben. Pathologe **S2**(30), 140–145 (2009)
14. Wong, J.W., Durante, C., Cartwright, H.M.: Application of fast Fourier transform cross-correlation for the alignment of large chromatographic and spectral datasets. Anal. Chem. **77**, 5655–5661 (2005)

# Multigene P-value Integration Based on SNPs Investigation for Seeking Radiosensitivity Signatures

Joanna Zyla[1(✉)], Christophe Badie[2], Ghazi Alsbeih[3],
and Joanna Polanska[1]

[1] Data Mining Group, Institute of Automatic Control, Silesian University
of Technology, Gliwice, ul. Akademicka 16, 44-100 Gliwice, Poland
{joanna.zyla,joanna.polanska}@polsl.pl
[2] Centre for Radiation, Chemical and Environmental Hazards,
Public Health England, Didcot, Chilton, Oxfordshire OX11 ORQ, UK
christophe.badie@phe.gov.uk
[3] Radiation Biology Section, Biomedical Physics Department,
King Faisal Specialist Hospital and Research Centre,
Riyadh 11211, Kingdom of Saudi Arabia
galsbeih@kfshrc.edu.sa

**Abstract.** Dysregulation of apoptosis is a key attribute of cancer, especially the one induced by p53 expression disruption. Radiotherapy, sometimes supported by chemotherapy and/or pre-surgery, is recommended in majority of cases, but despite of the very well defined treatment protocols and high quality irradiation procedure, the huge dispersion in response to the radiotherapy is observed among cancer patients. Patient radiosensitivity, according to up-to-date knowledge, is at least partially responsible for different reactions to ionising radiation. Here we concentrate on investigation of single nucleotide polymorphisms (SNP) which can possibly explain the radiation response phenomena. To reach this goal dependent and independent methods of p-value integrations are presented and compared. Both statistical and molecular function domains are used in comparison study. We propose a novel method of p-value integration which includes the control of gene expression trend and introduces the adaptive significance level. What is more the multigene approach is proposed in contrary to classical single gene investigation. As a result, set of statistically significant polymorphisms was obtained, among which some were identified as possible deleterious for KRAS signalling pathway.

**Keywords:** P-value integration · Single nucleotide polymorphism · Radiosensitivity

## 1 Introduction

Apoptosis is a process of programmed cell death, which defectiveness can lead to many diseases, usually by uncontrolled cell proliferation and dysregulation of p53 gene. Cancer is one of them [1]. One of the treatment method of cancer is radiotherapy,

© Springer International Publishing Switzerland 2016
F. Ortuño and I. Rojas (Eds.): IWBBIO 2016, LNBI 9656, pp. 125–134, 2016.
DOI: 10.1007/978-3-319-31744-1_12

which lead to many side effects in patients [2] and is source of high radiation dose. The individual reaction to the harmful effect of ionising radiation is called radiosensitivity, and as phenomena is observe from single cell to whole organism. In presented study two proapoptotic genes PUMA (p53 upregulated modulator of apoptosis) and FDXR (Ferredoxin Reductase) will be investigated, where both of them have been identified as possible candidates to describe radiosensitivity process [3]. As a base of the presented work single nucleotide polymorphism will be investigated. This will allow for individual understanding of investigated phenomena represented as a point mutation. To achieve this goal modelling of genotype-phenotype interaction accompanied with expression trend control will be introduced. While to find integral part of relevant outcomes between two genes, method of p-value integration will be used. The integration methods are mostly known from meta-analysis. However, in the recent years' integration methods base on applied statistics become more popular in genomic and transcriptomic studies [4, 5] also their nature in pure statistical way were widely disused [6, 7]. The main advantage of integration methods is possibility to join several studies or outcomes and give one statistical answer. Also they allow to point out results significant in one group but in parallel not significant in other, which normal want be taken into consideration. In presented study we proposed novel application of dependent and independent methods of p-value integration dedicated to multigene integration with expression trend control based on single nucleotide polymorphism information which is in contrary to most popular single gene investigation. To sum up main biological goal of this study is to propose group of candidate SNPs which can be responsible for disruption of apoptosis pathway related to radiation. This in feature can lead to development of new treatment individual to patient genome profile. From the other hand the new application of p-value integration is proposed adjusted by expression change trend control.

## 2  Materials

The population under investigation is composed of 44 unrelated individuals (unR) all healthy Caucasians. From each of individuals, two types of information's were collected. First one was the result of genotyping of 567,096 polymorphisms by Axiom GW Human hg36.1 arrays. All SNPs were marked to the newest genome 38 by NCBI annotation with the usage of internal SNPlab software. Quality control step was performed and one of SNP was removed from analysis due to missing genotypes for 91 % of individuals. The second dataset includes qPCR measurements for BBC3 (better known as PUMA) and FDXR gene, taken in two conditions: (1) in normal conditions - non irradiated cells and (2) just after the irradiation with a single dose of 2 Gy. Both genes are proapoptotic and have been reported as highly relevant to understanding of radiosensitivity phenomena [8]. The standardized fold change was calculated for all investigated genes. The irradiation was performed at room temperature with an A.G.O. HS X-ray system (Aldermaston, Reading, UK) (output 13 mA, 250 kV peak, 0.5 Gy/min for doses 0.5 4 Gy and 0.2 mA 4.9 mGy/min for doses up to 100 mGy). The T-lymphocyte cultures were used and prepared using the method described previously [9, 10].

# 3 Methods

## 3.1 Preliminary Statistical Analysis

Each polymorphism-gene interaction was investigated in three models: genotype (each allele gives different signal of expression), dominant (no mutation versus single/double mutation) and recessive (no mutation and single mutation vs. double mutation) all models were disused before [11]. For each model the normality of distribution was checked by Shapiro-Wilk test. Further homogeneity of variance was assessed by Bartlett test for genotype model and F test for dominant and recessive model. Based on results from normality and variance homogeneity assessments best test to check mean equivalence was chosen (ANOVA or Kruskal-Wallis for genotype model, T-test/Welch test or Mann-Whitney with normal approximation test for dominant and recessive models). All tests for mean equivalence in dominant and recessive models were performed as left and right sided, which will represent the trend of expression change by alleles (right side – down trend, left side – up trend).

## 3.2 Integration of Probabilities

To obtain group of significant polymorphism to investigated phenomena the p-values from previous step were integrated to each gene-SNP interactions and each model including consistence of trend (up with up, down with down). Three methods of probability integration were tested, where each of it have assumption about consistence in hypothesis of tests, however all of them represent different assumptions about dependency between integrated p-values. First one is classic Fisher method (Eq. 1), which assumes independency of integrated p-values [12].

$$F = -2 \sum_{i=1}^{k} \log(p_i) \sim \chi^2_{2k} \tag{1}$$

Second tested method of integration assumes dependency between integrated p-values and was proposed by Brown [13] (Eq. 2).

$$K = -2 \sum_{i=1}^{k} \log(p_i) \sim c\chi^2_f \tag{2}$$

The methods assume that the first two momentums of Fisher method and Brown method are equal, which allows to estimation of degrees of freedom "f" and scale parameter "c" (Eq. 3).

$$c = \frac{Var[F]}{2E[F]} \quad \text{and} \quad f = 2\frac{E[F]^2}{Var[F]} \tag{3}$$

In dependent integration the expected value is the same like for Fisher methods (Eq. 4) however, the variance has to be adjusted by covariance between random variable (Eq. 5) (in Fisher method cov = 0 from independence assumption). The covariance estimation was calculated by exact formulas proposed by Kost and McDermott [14], which based on the correlation coefficient between variables. In presented study the relation was estimated by Pearson correlation between qPCR results for FDXR and PUMA and overall p-value may be computed as $P(\chi_f > K/c)$.

$$E[F] = 2k \tag{4}$$

$$Var[F] = 4k + 2 \sum_{i<j} \text{cov}(-2\log(p_i), -2\log(p_j)) \tag{5}$$

Last method was proposed by Rüschendorf [15] and proved by Vovk [16] for scaling parameter equal two (Eq. 6). This method allows for integrating p-values without making any assumptions about their dependence structure.

$$2\bar{p} = \frac{1}{N}(p_1 + \cdots + p_N) \tag{6}$$

All presented methods were applied to each model-trend interaction from one gene to corresponding model-trend interaction in other. This allows for establishing the assumption about consistence of test hypothesis.

### 3.3    Significant Polymorphism Selection

To select statistically significant polymorphism, the best model-trend interaction was chosen as minimum from five obtained p-values (Eq. 7).

$$\min\{p\text{-value}_{genotype}, \ p\text{-value}_{dominant\_up}, \ p\text{-value}_{dominant\_down}, p\text{-value}_{recessive\_up}, p\text{-value}_{recessive\_down}\} \tag{7}$$

Further, to each method significance level alpha was adapted and individually set by the same integration method where the standard 0.05 level was used as probability to integration – scheme at Fig. 1. For Fisher method new alpha = 0.0175, for Vovk alpha = 0.1 where for Brown method there is individual alpha level for each SNP due to changing degrees of freedom for SNPs which is needed to estimate the covariance by Kost and McDermott method. The integration methods will be compared to restricted approach where for both gene p-values obtained in Sect. 3.1 have to be lower than 0.05 ($p_{FDXR} < 0.05$ and $p_{PUMA} < 0.05$).

### 3.4    *In Silico* Comparison

Final, group of statistically significant SNPs were investigated by their possible distribution of signal cascades. For this purpose, the Gene Set Enrichment Analysis was performed by procedure dedicated to SNP investigation proposed by Serge et al. with

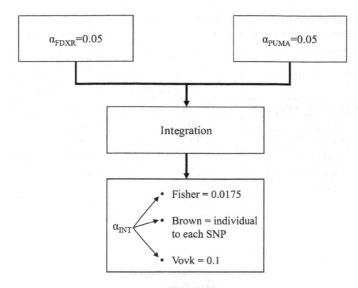

**Fig. 1.** The general idea of adaptive alpha level for integration method.

including Sidak correction for linkage disequilibrium [17]. As a pathway collection Hallmark gene set was used from MSigDB resources [18].

## 4  Results

Standardized fold change was calculated for qPCR measurements and taken as main signal in presented work. According to description in Sect. 3.1 to each SNP the p-values were obtained for both investigated genes (FDXR and PUMA) and for all possible model-trend interactions to assess the impact of allele change to signal expression. For both genes majority of SNPs represent normal distribution (94.5 % for FDXR and 93.5 % for PUMA). Homogeneous variances have 86.1 % of SNPs for FDXR and 84.8 % for PUMA. Nevertheless, the best test was established for each polymorphism. In the, next step three method of p-value integration were applied and compared to restricted approach – Table 1.

As can be observed in Table 1 from three investigated methods of integration Vovk method gives the most restricted results very close to standard approach in such analysis here represented as restricted approach. For Fisher and Brown method the results give similar amount of significant polymorphism to adaptive alpha level. To show overlap of SNP in model trend exemplary for Brown methods the Venn diagram of relevant polymorphism was constructed with the usage of InteractiVenn software [19] – Fig. 2.

The figure shows that coverage between SNPs appears only in polymorphisms with consistent trend but different model of gene-SNP interaction. This indicate the fact about existing additive model in data however, the model was not investigated in presented research due to small sample size. Also it allows to choose the minim p-value

**Table 1.** Results of p-value integration for three tested methods and restricted approach. The grey colour represents polymorphisms of interest with adaptive alpha level to each method.

| Method | Cut off | Model of interaction with trend | | | | |
|---|---|---|---|---|---|---|
| | | Genotype | Dominant | | Recessive | |
| | | | DOWN | UP | DOWN | UP |
| Restricted | p<0.05 both genes | 1 900 | 4 581 | 4 219 | 4 849 | 5 052 |
| Fisher | p<0.05 | 13 004 | 23 926 | 24 221 | 27 618 | 26 652 |
| | p<0.0175 | 5 786 | 11 573 | 11 445 | 13 168 | 12 743 |
| Brown | p<0.05 | 7 855 | 15 164 | 15 166 | 17 311 | 16 664 |
| | p<adaptive alpha | 5 778 | 11 563 | 11 432 | 13 160 | 12 727 |
| Vovk | p<0.05 | 1 226 | 2 870 | 2 654 | 3 041 | 3 178 |
| | p<0.1 | 3 254 | 7 277 | 6 987 | 8 047 | 8 058 |

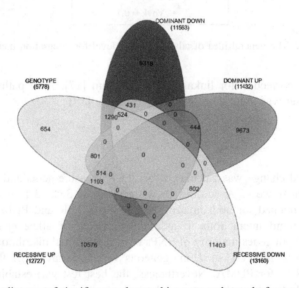

**Fig. 2.** Venn diagram of significant polymorphisms to each trend of expression change.

from all 5 possibilities as representative to each SNP-model-trend interaction, while there is no trend coverage. The results for this stage are presented in Table 2.

As can be seen in Table 2 the difference in final statistically significant polymorphism between Fisher and Brown methods is only 47 SNPs. From all three methods Vovk one is the most restricted giving 37 % less SNPs compering to two other. However, there is large difference in case of standard 0.05 significance level which show that adaptive alpha is able to equate dependent and independent method. To present distribution of integration p-value versus initial the heatmaps for best model were created – Fig. 3.

**Table 2.** Statistically significant results for best SNP-model-trend interactions for three investigated methods of integration.

| Method | Cut off | Best model of interaction | | | TOTAL |
|--------|---------|----------|----------|-----------|-------|
|        |         | Genotype | Dominant | Recessive |       |
| Fisher | p < 0.05 | 2 223 | 44 450 | 24 714 | 97 233 |
|        | p < 0.0175 | 1 108 | 21 847 | 24 714 | 47 669 |
| Brown  | p < 0.05 | 1 451 | 28 542 | 32 164 | 62 139 |
|        | p < adaptive alpha | 1 107 | 21 824 | 24 691 | 47 622 |
| Vovk   | p < 0.05 | 331 | 5 391 | 6 088 | 11 810 |
|        | p < 0.1 | 755 | 13 684 | 15 552 | 29 991 |

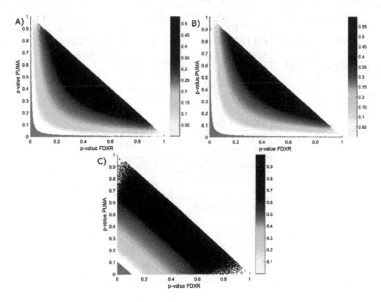

**Fig. 3.** Illustrations of integration mechanism; grey colour on left bottom in each plot represent significant polymorphisms; other colours represent integrated p-value level according to the scale in the bar; x and y axis represent p-values for FDXR and PUMA (respectively) gene before integration. The A) panel represents dependent type of integration, B) panel represents independent type of integration, C) panel represents Vovk integration.

As can be seen in Fig. 3 the independent Fisher method of integration gives slightly lower level of integrated p-value comparing to Brown. In case of Vovk methods it is unable to get polymorphism highly significant in one gene and on the boarder of significance to other gene, where this property is its leading power of integration, and is clearly observed in case of Fisher and Brown integration. As a last step all polymorphisms were in silico investigated by Gene Set Enrichment methods dedicated for SNPs called MAGENTA (Meta-Analysis Gene-set Enrichment of variaNT Associations). The method bases on global cut of level usually set as 5th percentile of all p-values. In this case of this study the adaptive alpha level was set as cut off in MAGENTA-GSEA, where for Brown method the average of all alpha level was taken into account. This allows for marking the significant polymorphism form integration as those which can possibly disturb the signal cascade. In next step to each Gene Set (in this case gene set from Hallmark collection) the number of genes below threshold are counted and further by permutation of gene set the p-value is established – more details in [17]. The results for significant gene sets are present in Table 3.

**Table 3.** The results of MAGENTA-GSEA in silico analysis for three investigated integration methods. The grey colour represents common gene sets between Brown and Fisher method, while the bold italic represents common genes set between Fisher and Vovk method.

| Brown method of integration | | Fisher method of integration | | Vovk method of integration | |
|---|---|---|---|---|---|
| Gene Set Name | p-value | Gene Set Name | p-value | Gene Set Name | p-value |
| COAGULATION | 0.0017 | *WNT BETA CATENIN SIGNALING* | *0.0013* | *WNT BETA CATENIN SIGNALING* | *0.0416* |
| UV RESPONSE UP | 0.0102 | HYPOXIA | 0.0015 | | |
| KRAS SIGNALING UP | 0.0372 | CHOLESTEROL HOMEOSTASIS | 0.0119 | | |
| IL2 STAT5 SIGNALING | 0.0439 | IL2 STAT5 SIGNALING | 0.0149 | | |
| | | MTORC1 SIGNALING | 0.0152 | | |
| | | COAGULATION | 0.0163 | | |
| | | UV RESPONSE DN | 0.0206 | | |
| | | UV RESPONSE UP | 0.0236 | | |
| | | MYOGENESIS | 0.0264 | | |
| | | ANGIOGENESIS | 0.0342 | | |
| | | ANDROGEN RESPONSE | 0.0352 | | |

From the obtain results we can observe that Fisher method detect the larger amount of possibly disturbed signal cascades which also covers in finding of the most conservative Vovk method. The Brown method detects 4 gene sets where only one was not pointed by Fisher integration. Considering this pathway, the KRAS (rat sarcoma viral oncogene homolog) signalling and mutation in KRAS are common ad well known in colorectal cancer and non-small cell lung cancer (NSCLC) [20]. It is also related to radiosensitivity specially with combination of mutation in MET1 which could contribute to radioresistance in NSCLC patients [21]. In case of apoptosis KRAS is part of KRAS-mediated apoptosis resistance process mostly observed in colorectal cancer patients, where also possible treatment strategies were proposed [22]. Most pathways detected in Fisher method are related with apoptosis however only WNT signalling pathway and Hypoxia have linkage for both radiosensitivity and apoptosis [23, 24]. The results of *in silico* validation show main problem of integration based on SNPs over genes. The difference in no. of significant SNPs between Fisher and Brown methods can be treated as "by chance" and if those polymorphisms will be analysed individual the difference would not be observed. However, the distribution of p-values is different between those two methods which have an impact to GSEA analysis. In presented paper the dependency between variables is observed in biological nature of investigated process, where both genes come from apoptosis pathway. This indicated the correctness of using Brown method in such kind of problems.

# 5 Conclusions

In presented work the strategy for integrating p-values for multi SNP multigene analysis was proposed. The developed p-value integration pipeline includes trend control of gene expression response combined with different SNP-gene interaction models performed in domain of two chosen genes. In contrary to authors' previous works, where p-value integration of the same SNP-gene interaction between two datasets was done, the proposed multigene approach supports search for radiosensitivity biomarkers in more complex way, which might be crucial for understanding of the radiosensitivity phenomena. The correction after dependency/correlation between expression of analysed genes is necessary in the process of global biomarker discovery. The results obtained for exemplary two genes co-occurring in apoptosis pathway demonstrate the importance of that step. The set of candidate polymorphisms disturbing the oncogenic KRAS signalling pathway was discovered. The final step, combining both - multigene and multi datasets p-value integration would allow for in silico validation of those polymorphisms.

**Acknowledgement.** This work was funded by NCN grant HARMONIA 4 no. DEC-2013/ 08/M/ST6/00924 (JZ, CB, JP). Calculations were carried out using infrastructure of GeCONiI (POIG.02.03.01-24-099/13).

# References

1. Pan, H., Yin, C., Van Dyke, T.: Apoptosis and cancer mechanisms. Cancer Surv. **29**, 305–327 (1996)
2. West, C.M., Barnett, G.C.: Genetics and genomics of radiotherapy toxicity: towards prediction. Genome Med. **3**(8), 52 (2011)
3. Bunz, F.: Principles of Cancer Genetics. Springer, The Netherland (2007)
4. Papiez, A., Kabacik, S., Badie, C., Bouffler, S., Polanska, J.: Statistical integration of p-values for enhancing discovery of radiotoxicity gene signatures. In: Ortuño, F., Rojas, I. (eds.) IWBBIO 2015, Part I. LNCS, vol. 9043, pp. 503–513. Springer, Heidelberg (2015)
5. Zaykin, D.V., Zhivotovsky, L.A., Czika, W., Shao, S., Wolfinger, R.D.: Combining p-values in large-scale genomics experiments. Pharm. Stat. **6**(3), 217–226 (2007)
6. Whitlock, M.C.: Combining probability from independent tests: the weighted Z-method is superior to Fisher's approach. J. Evol. Biol. **18**(5), 1368–1373 (2005)
7. Chen, Z.: Is the weighted z-test the best method for combining probabilities from independent tests? J. Evol. Biol. **24**(4), 926–930 (2011)

8. Budworth, H., et al.: DNA repair and cell cycle biomarkers of radiation exposure and inflammation stress in human blood. PLoS ONE **7**(11), e48619 (2012)
9. O'Donovan, M.R., et al.: Extended-term cultures of human T-lymphocytes: a practical alternative to primary human lymphocytes for use in genotoxicity testing. Mutagenesis **10**, 189–201 (1995)
10. Finnon, P., et al.: Evidence for significant heritability of apoptotic and cell cycle responses to ionising radiation. Hum. Genet. **123**(5), 485–493 (2008)
11. Zyla, J., Badie, C., Alsbeih, G., Polanska, J.: Modelling of genetic interactions in GWAS reveals more complex relations between genotype and phenotype. In: Proceedings of the International Conference on Bioinformatics Models, Methods and Algorithms (BIOSTEC 2014), pp. 204–208 (2014). ISBN 978-989-758-012-3
12. Fisher, R.A.: Statistical Methods for Research Workers, 4th edn. Oliver and Boyd, Edinburgh (1932)
13. Brown, M.B.: 400: A method for combining non-independent, one-sided tests of significance. Biometrics **31**, 987–992 (1975)
14. Kost, J.T., McDermott, M.P.: Combining dependent p-values. Stat. Probab. Lett. **60**(2), 183–190 (2002)
15. Rüschendorf, L.: Random variables with maximum sums. Adv. Appl. Probab. **14**, 623–632 (1982)
16. Vovk, V.: Combining p-values via averaging (2012). arXiv preprint arXiv:1212.4966
17. Segrè, A.V., Groop, L., Mootha, V.K., Daly, M.J., Altshuler, D., Diagram Consortium, Magic Investigators: Common inherited variation in mitochondrial genes is not enriched for associations with type 2 diabetes or related glycemic traits. PLoS Genet, **6**(8), e1001058 (2010)
18. Subramanian, A., et al.: Gene set enrichment analysis: a knowledge-based approach for interpreting genome-wide expression profiles. PNAS **102**(43), 15545–15550 (2005)
19. Heberle, H., et al.: InteractiVenn: a web-based tool for the analysis of sets through Venn diagrams. BMC Bioinform. **16**(1), 169 (2015)
20. Jančík, S., Drábek, J., Radzioch, D., Hajdúch, M.: Clinical relevance of KRAS in human cancers. BioMed Res. Int. (2010). doi:10.1155/2010/150960
21. Chakrabarti, G.: Mutant KRAS associated malic enzyme 1 expression is a predictive marker for radiation therapy response in non-small cell lung cancer. Radiat. Oncol. **10**(1), 1 (2015)
22. Okamoto, K., Zaanan, A., Kawakami, H., Huang, S., Sinicrope, F.A.: Reversal of mutant KRAS-mediated apoptosis resistance by concurrent Noxa/Bik induction and Bcl-2/Bcl-xL antagonism in colon cancer cells. Mol. Cancer Res. **13**(4), 659–669 (2015)
23. Wang, G., et al.: LincRNA-p21 enhances the sensitivity of radiotherapy for human colorectal cancer by targeting the Wnt/β-catenin signaling pathway. Oncol. Rep. **31**(4), 1839–1845 (2014)
24. Leszczynska, K.B., et al.: Hypoxia-induced p53 modulates both apoptosis and radiosensitivity via AKT. J. Clin. Invest. **125**(6), 2385–2398 (2015)

# Epithelial-Mesenchymal Transition Regulatory Network-Based Feature Selection in Lung Cancer Prognosis Prediction

Borong Shao[1,2]([✉]) and Tim Conrad[1,2]

[1] Department of Mathematics and Computer Science, Freie Universität Berlin,
Berlin, Germany
borong.shao@fu-berlin.de, tconrad@math.fu-berlin.de
[2] Zuse Institute Berlin, Berlin, Germany
{shao,conrad}@zib.de

**Abstract.** Feature selection technique is often applied in identifying cancer prognosis biomarkers. However, many feature selection methods are prone to over-fitting or poor biological interpretation when applied on biological high-dimensional data. Network-based feature selection and data integration approaches are proposed to identify more robust biomarkers. We conducted experiments to investigate the advantages of the two approaches using epithelial mesenchymal transition regulatory network, which is demonstrated as highly relevant to cancer prognosis. We obtained data from The Cancer Genome Atlas. Prognosis prediction was made using Support Vector Machine. Under our experimental settings, the results showed that network-based features gave significantly more accurate predictions than individual molecular features, and features selected from integrated data (RNA-Seq and micro-RNA data) gave significantly more accurate predictions than features selected from single source data (RNA-Seq data). Our study indicated that biological network-based feature transformation and data integration are two useful approaches to identify robust cancer biomarkers.

**Keywords:** Cancer prognosis prediction · Epithelial mesenchymal transition · Feature selection · Data integration · Network motif

## 1 Introduction

Reliable prognosis prediction is desirable in cancer treatment. It can help stratify patients and determine the most appropriate therapy [1,2]. With the availability of high-throughput molecular data, researchers are looking for molecular patterns that can differentiate different prognosis groups. Due to the high dimensionality of data, feature selection is usually a necessary step before building predictive models [3]. There are mainly three types of feature selection methods: filter, wrapper and hybrid methods [3,4]. The filter methods select features only based on the data while the wrapper and hybrid methods involve iteratively

© Springer International Publishing Switzerland 2016
F. Ortuño and I. Rojas (Eds.): IWBBIO 2016, LNBI 9656, pp. 135–146, 2016.
DOI: 10.1007/978-3-319-31744-1_13

evaluating the selected features with a model, for example, a classifier. However, because of the complexity of biological data, feature selection is not an easy task. Firstly, since all the three types of feature selection methods are based on statistics and machine learning, seldom do these methods involve relevant biological principles or knowledge while working with biological data. Secondly, regarding the features selected using a single dataset, it is hard for the features to achieve high prediction accuracy on this dataset and other independent datasets [5,6].

To improve the quality of features, biological domain knowledge, especially networks, such as PPI (protein-protein interaction) network are more often referred to during feature selection to give more robust and generalizable features. A few studies [7–9] have successfully used PPI network to obtain more robust feature sets. These results indicate the advantage of using biological domain knowledge in feature selection. However, since the PPI network is large and usually the entire network is applied, it is hard to tell which parts or which properties of the network contribute to the predictive capability. The answers to this question are necessary to figure out how to take advantage of biological network in feature selection. Another approach to improve the quality of features is to use multiple data sources, such as gene expression, DNA methylation, miRNA expression, etc. [10]. Regarding the specific task of prognosis prediction, several studies obtained higher prediction accuracy by integrating different sources of data [11,12]. However, due to the complexity of data, there has been no single best biological data integration methodology. In addition, depending on the experimental settings and the data sources, the results obtained from different data integration studies vary [11,13].

To investigate the effects of features, networks, and data integration approaches on the prediction accuracy of the selected features, it is useful to design a set of experiments based on a specific biological network. This can reduce the effects of potential confounding variables in the experiment and lead to more reproducible and interpretable results. In this study, we chose Epithelial Mesenchymal Transition (EMT) regulatory network. Our motivation is that EMT is a key process in cancer metastasis and it has been demonstrated to be highly relevant to cancer prognosis [14–16]. We obtained RNA-Seq, miRNA expression data, and clinical data for Lung Adenocarcinoma (LUAD) cancer from The Cancer Genome Atlas (TCGA) database and mapped the RNA-Seq and miRNA data to the EMT network. Then we performed feature selection in comparative settings: (1) select features within EMT network or select features from the entire dataset, (2) select single molecules as features or select network motifs as features, (3) select features only from RNA-Seq data or select features from integrated RNA-Seq and miRNA data. Then we took the selected feature sets as input and used Support Vector Machine (SVM) to predict patient prognosis (either long survival or short survival). We acknowledge that the choice of classifier affects the prediction results to a certain extent. However, since the focus of our study is the comparison of different feature sets, we chose one representative and commonly applied classifier - SVM as a benchmark.

From the experimental results we observed: (1) EMT molecules have significant predictive power in the setting of integrated data, (2) EMT network motif features achieved significantly higher prediction accuracy than molecular features, and (3) features selected from integrated data achieved significantly higher prediction accuracy than features selected from RNA-Seq data. Our study indicated that biological network-based feature transformation and data integration are useful approaches to select cancer biomarkers.

## 2  Hypotheses

We proposed three hypotheses:

1. EMT network molecules have significant predictive power in prognosis prediction.
2. The transformation of molecular features to network motif features using EMT network structure can significantly improve prognosis prediction accuracy.
3. Features selected from integrated data have significantly higher prediction accuracy than features selected from RNA-Seq data.

## 3  Methods

In this section we described our work flow and the design of the experiment in details.

### 3.1  Generate EMT Network from Literature

We searched for and adopted the most recent and comprehensive review papers regarding EMT regulatory network to manually construct the network. The main literatures we referred to are [17–20]. When necessary, we also referred to the references of these papers. In the network, a node is either a gene or a miRNA. The other types of molecules such as proteins are matched to the corresponding genes. The interactions between the nodes are represented as directed edges.

### 3.2  Data Acquisition and Processing

**Download the Data.** We obtained data from TCGA (The Cancer Genome Atlas) using *TCGA-Assembler* software [21]. This software is able to download all available samples of a specified cancer type and specified data sources from TCGA data store. We downloaded level-3 RNA-Seq data, miRNA expression data, and clinical data of Lung Adenocarcinoma (LUAD) cancer. We chose LUAD cancer type because it has more available samples.

**Keep the Samples that Have Survival Information.** For the aims of the experiment, we only keep the samples that have survival information. In the downloaded TCGA clinical information data, there exist three fields: *vital_status*, *days_to_last_followup*, and *days_to_death*. *vital_status* has two alternative values: *dead* and *alive*. We calculated the average *days_to_death* value among the samples whose *vital_status* values are *dead*. Then we dropped the samples whose *vital_status* values are *alive* but with *days_to_last_followup* values less than the average *days_to_death* value. As in these cases, the measurements did not sufficiently reflect the survival information of the patients.

**Merge RNA-Seq and miRNA Expression Data.** The aforementioned *TCGA-Assembler* software is able to merge the downloaded different data sources to a single file. In this file, each record is a molecular feature and each column is a sample. If a sample does not have all types of data sources being integrated, the sample is not included in the merged data. We used *TCGA-Assembler* to merge RNA-Seq and miRNA data.

**Data Normalization.** We removed all-zeros features and applied z-score normalization on the datasets. For merged data, since the measurements of RNA-Seq features and miRNA features are on different scales, we normalized them separately and then combined them.

### 3.3    Select Six Feature Sets from Each Dataset

We selected six feature sets from RNA-Seq data and six feature sets from merged data. Since the procedure is similar with both datasets, below we took RNA-Seq data to illustrate.

- Feature set 1. We selected the $N$ molecular features that are within the EMT network and applied Lasso[1] [22] on the $N$ features.
- Feature set 2. We applied Lasso on all RNA-Seq features.
- Feature set 3. We firstly identified all $M$ three nodes network motifs in the EMT network using *FANMOD* software [23]. We then calculated the activity score vector of each motif using the approach proposed in [9]. The equations we adopted from this paper were given below. To explain, we calculated the mutual information $MI_i$ of each component gene $i$ and the class label $c$. $z_i$ is the normalized expression values of gene $i$. Then we calculated the weight $w_i$ of each gene and the score of the motif $A$ $score_A$. After obtaining the scores of all $M$ network motifs, we applied Lasso on the motif features (scores).

---

[1] We applied the *lasso* function implemented in *MATLAB R2015a* to select the feature set that has the minimum mean squared error.

$$MI_i(x;y) = \sum_{x \in z_i} \sum_{y \in c} p(x,y) log \frac{p(x,y)}{p(x)p(y)} \tag{1}$$

$$w_i = \frac{MI_i(x;y)}{\sum_{i=1}^{M} MI_i(x;y)} \tag{2}$$

$$score_A = \frac{1}{\sqrt{\sum_{i=1}^{M} w_i^2}} \sum_{i=1}^{M} w_i z_i \tag{3}$$

– Feature set 4. We selected $N$ random features from the RNA-Seq dataset and applied Lasso on the $N$ random features.
– Feature set 5. We used the same $N$ random features from the previous step and calculated $M$ motif features (scores) using the same network structure and same equations as above. Then we applied Lasso on the $M$ motif features.
– Feature set 6. We randomly selected $M$ motifs from the dataset and calculated the motif scores using the same equations above. Then we applied Lasso on the $M$ motif features.

Note that before selecting feature set 1 and 3 from RNA-Seq data, we removed the miRNA nodes and their associated interactions from the EMT network. Above all, the motivation of obtaining the six feature sets was to provide experimental and control groups to test the hypotheses.

### 3.4  Compare the Prediction Accuracy of the Six Feature Sets

We took the selected feature sets as the input to Support Vector Machine (SVM) classifier to predict prognosis. We used LibSVM implementation [24] with Radial Basis Function (RBF) kernel (*degree* $= 3$) and cost parameter $C = 1$. We performed 10-fold cross-validation and 10-times repetitions. For each feature set and dataset, we measured the classification accuracy using AUC (the area under the receiver operating characteristic curve).

## 4  Results

### 4.1  EMT Network

The molecules in the network and their interactions were given in the appendix. There are totally 140 nodes and 1947 motifs in the EMT network. When miRNA molecules and their associated interactions are removed, there are 116 nodes and 1477 motifs in the network.

### 4.2  Data

The descriptions of the datasets were given in Table 1.

**Table 1.** Datasets description

| Datasets | Long survival samples | Short survival samples | No. features |
|---|---|---|---|
| LUAD RNA-Seq dataset | 61 (>1400 days) | 67 (<700 days) | 19993 |
| LUAD merged dataset | 48 (>1400 days) | 60 (<700 days) | 20777 |

### 4.3 Feature Selection and Classification

We selected six feature sets from RNA-Seq dataset and six feature sets from merged dataset. The classification accuracy (AUCs) of the twelve feature sets were given in Table 2.

**Table 2.** Classification accuracy of different feature sets using SVM classifier

| Datasets | Feature set 1 | Feature set 2 | Feature set 3 | Feature set 4 | Feature set 5 | Feature set 6 |
|---|---|---|---|---|---|---|
| RNA-Seq data | 0.70 | 0.81 | 0.77 | 0.73 | 0.72 | 0.91 |
| Merged data | 0.73 | 0.95 | 0.81 | 0.67 | 0.70 | 0.95 |

## 5 Discussion

To test the hypotheses, we performed pairwise Wilcoxon signed-rank test on the AUCs of the feature sets. To test hypothesis 1, we compared AUCs of feature set 1 with AUCs of feature set 4. To test hypothesis 2, we compared the AUCs of feature set 1 with AUCs of feature set 3 and compared the AUCs of feature set 4 with AUCs of feature set 5. To test hypothesis 3, we compared the AUCs of feature set 1, 2, and 3 from RNA-Seq data with the AUCs of feature set 1, 2, and 3 from merged data correspondingly. The results were given in Table 3.

**Table 3.** Hypotheses testing using pairwise Wilcoxon signed-rank test

| Hypotheses | Feature set | RNA-Seq data | Merged data |
|---|---|---|---|
| | | AUC (p-value) | |
| Hypothesis 1 | 1 and 4 | 0.0969 | **0.0243** |
| Hypothesis 2 | 1 and 3 | **0.0058** | **0.0059** |
| | 4 and 5 | 0.7665 | 0.1829 |
| Hypothesis 3 | 1 | 0.1913 | |
| | 2 | **0.0059** | |
| | 3 | **0.0091** | |

*The transformation of molecular features to biological network-based features significantly improved prediction accuracy.* Under the settings of our experiment, EMT features (feature set 1) gave significantly higher prediction accuracy than random selected features (feature set 4) in the case of merged dataset. EMT network motif features (feature set 3) gave significantly higher prediction accuracy than EMT molecular features (feature set 1) on both datasets. In contrast, network motif features from random network (feature set 5) did not perform better than the random molecular features (feature set 4). This indicated that the transformation of molecular features to network motif features based on a real biological network can increase the predictive capability of features.

In addition to the three hypotheses, we observed that in the case of RNA-Seq dataset, features selected from 1477 random network motifs (feature set 6) achieved significantly higher prediction accuracy than features selected from the entire dataset (feature set 2). With merged dataset, feature set 6 was able to achieve equivalent prediction accuracy as feature set 2. Note that these random network motifs were randomly sampled from the entire datasets without any network structure and they had only about one fifth feature dimension of the entire dataset. This further indicated that feature transformation from molecular features to network-based features is a useful approach to identify caner prognosis biomarkers.

*Features selected from integrated data have significantly higher prediction accuracy than features selected from RNA-Seq data.* Under the settings of our experiment, four out of six feature sets selected from merged data gave significantly higher prediction accuracy than the corresponding feature sets selected from RNA-Seq data. The observation is consistent with state-of-the-art research, where integrated data were used [11, 25] to achieve higher prognosis prediction accuracy. Compared with single-data-source approach, integrating multiple data sources can compensate for unreliable information in any single data source and allow for more comprehensive modeling of complex phenotypes [10].

To the best of our knowledge, this is the first study that employed a concrete regulatory network (EMT) to investigate the advantages of network-based feature transformation and data integration approaches in cancer prognosis biomarker identification. We mapped RNA-Seq data and integrated data to the EMT network and performed network-based feature selection. The experimental results indicated that biological network-based feature selection and data integration are useful approaches to identify more robust and more biologically relevant cancer biomarkers.

**Acknowledgment.** This study was funded by the German Ministry of Research and Education (BMBF) Project Grant 3FO18501 (Forschungscampus MODAL).

# Appendix

EMT network interactions were given in Table 4.

**Table 4.** EMT regulatory network

| | |
|---|---|
| SNAI1 repress CDH1 | GSK3$\beta$ phosphorylate SNAI1 |
| SNAI2 repress CDH1 | SNAI1 degraded_by $\beta$TRCP1 |
| ZEB1 repress CDH1 | TNF$\alpha$ stabilize SNAI1 |
| ZEB2 repress CDH1 | NFKB1 cooperate TNF$\alpha$ |
| TCF3 repress CDH1 | NFKB1 cooperate CSN2 |
| KLF8 repress CDH1 | NFKB2 cooperate TNF$\alpha$ |
| Brachyury repress CDH1 | NFKB2 cooperate CSN2 |
| SNAI1 repress MUC1 | CSN2 disrupt GSK3$\beta$ |
| SNAI1 repress cytokeratin_18 | MDM2 degrade SNAI1 |
| SNAI1 repress claudin_IMP | MDM2 degrade SNAI2 |
| SNAI1 repress occludin | FBXL14 degrade SNAI1 |
| SNAI1 incude fibronectin | FBXL14 degrade SNAI2 |
| SNAI1 induce LEF1 | PPA1 degrade SNAI1 |
| SNAI1 induce ZEB1 | PPA1 degrade SNAI2 |
| ILK1_kinase induce SNAI1 | PPA1 degrade TWIST1 |
| ILK1_kinase induce ZEB1 | PPA1 degrade ZEB2 |
| ZEB1 repress MUC1 | PPA2 degrade SNAI1 |
| SNAI1 associate CDH1 | PPA2 degrade SNAI2 |
| DNMT1 interact SNAI1 | PPA2 degrade TWIST1 |
| SNAI1 induce CDH1 | PPA2 degrade ZEB2 |
| SNAI1 recruit HDAC1 | PAK1 stabilize SNAI1 |
| SNAI1 recruit HDAC2 | ATM stabilize SNAI1 |
| HDAC1 deacetylate CDH1 | LOXL2 stabilize SNAI1 |
| HDAC2 deacetylate CDH1 | O-GlcNAc stabilize SNAI1 |
| HDAC3 deacetylate CDH1 | MMP3 increase SNAI1 |
| SIRT deacetylate CDH1 | LATS2 stabilize SNAI1 |
| EZH2 methyltransferase CDH1 | PAK1 nucleus_localize SNAI1 |
| SUZ12 methyltransferase CDH1 | LATS2 nucleus_localize SNAI1 |
| SUV39H1 methyltransferase CDH1 | LIV1 nucleus_import SNAI1 |
| G9a methyltransferase CDH1 | LATS2 attenuate TAZ |
| ZEB1 interact BRG1 | Scribble repress TAZ |
| EZH2 cooperate TWIST1 | PRKD1 nucleus_export SNAI1 |
| BMI1 cooperate TWIST1 | ZEB1 interact RASGRP3 |
| LOXL2 repress CDH1 | ZEB1 activate RRAS |
| HDAC3 cooperate SNAI1 | ZEB2 attenuate CDH1 |
| HIF1a direct_target HDAC3 | $\beta$-catenin activate TCF7 |
| HDAC3 interact WDR5 | $\beta$-catenin activate TCF7L1 |
| WDR5 activate HMT | $\beta$-catenin activate TCF7L2 |
| ZEB1 repress miR-203 | $\beta$-catenin activate LEF1 |

*(Continued)*

**Table 4.** (*Continued*)

| | |
|---|---|
| ZEB1 repress miR-141 | TGF$\beta$1 interact TGF$\beta$R2 |
| ZEB1 repress miR-200a | TGF$\beta$R2 interact TGF$\beta$R1 |
| ZEB1 repress miR-200b | TGF$\beta$R2 phosphorylate PAR6 |
| ZEB1 repress miR-200c | PAR6 interact SMURF1 |
| ZEB1 repress miR-429 | SMURF1 degrade RhoA |
| miR-200b target SUZ12 | miR-155 inhibit RhoA |
| miR-200c target BMI1 | TGF$\beta$R1 phosphorylate SMAD2 |
| miR-203 target BMI1 | TGF$\beta$R1 phosphorylate SMAD3 |
| miR-183 target BMI1 | SMAD2 bind SMAD4 |
| miR128-1 target BMI1 | SMAD3 bind SMAD4 |
| miR128-2 target BMI1 | SMAD3 activate HMGA2 |
| miR-200a repress ZEB1 | SMAD4 activate HMGA2 |
| miR-200b repress ZEB1 | HMGA2 regulates SNAI1 |
| miR-200a repress ZEB2 | SMAD3 interact ETS1 |
| miR-200b repress ZEB2 | SMAD1 increase GSC |
| SNAI1 recruit SUZ12 | ETS1 increase ZEB1 |
| SNAI1 recruit EZH2 | miR-192 inhibit ZEB2 |
| SUZ12 repress CDH1 | SMAD2 inhibit ID2 |
| miR-200b inhibit SUZ12 | SMAD3 inhibit ID2 |
| MIR101-1 repress EZH2 | ID2 inhibit E2A |
| MIR101-2 repress EZH2 | E2A increase SNAI1 |
| p53 induce miR-34a | GSC increase FOXC2 |
| p53 induce miR-34b | FOXC2 increase MMP-2 |
| p53 induce miR-34c | FOXC2 increase MMP-9 |
| p53 activate miR-200c | miR-200b inhibit JAG1 |
| miR-34a repress SNAI1 | miR-200b inhibit JAG2 |
| miR-34b repress SNAI1 | JAG1 activate NOTCH |
| miR-34c repress SNAI1 | JAG2 activate NOTCH |
| miR-34a repress SNAI2 | NICD increase SNAI1 |
| miR-34a repress ZEB1 | NICD increase SNAI2 |
| miR-34a repress BMI1 | NICD increase LOXL2 |
| miR-34a repress CD44 | NICD activate NFKB1 |
| miR-34a repress CD133 | NICD activate NFKB2 |
| miR-34a repress OLFM4 | EGF phosphorylate EGFR |
| miR-34a repress c-MYC | EGFR activate JAK |
| SNAI1 repress miR-34a | EGFR binds GRB2 |
| SNAI1 repress miR-34b | EGFR activate PI3K |
| SNAI1 repress miR-34c | HRAS interact Raf |

(*Continued*)

**Table 4.** (*Continued*)

| | |
|---|---|
| ZEB1 repress miR-34a | KRAS interact Raf |
| ZEB1 repress miR-34b | Raf activate MEK1 |
| ZEB1 repress miR-34c | Raf activate MEK2 |
| ZEB1 repress miR-141 | JAK phosphorylate STAT3 |
| ZEB1 repress miR-200c | MEK1 activate ERK1 |
| miR-205 repress ZEB1 | MEK1 activate ERK2 |
| miR-205 repress ZEB2 | MEK2 activate ERK1 |
| miR130b repress ZEB1 | MEK2 activate ERK2 |
| miR92a repress CDH1 | PI3K activate AKT2 |
| SAM68 attenuate SRSF1 | STAT3 increase TWIST1 |
| ESRP1 regulate FGFR2 | ERK1 increase ZEB1 |
| ESRP2 regulate FGFR2 | ERK1 increase SNAI1 |
| ESRP1 regulate CTNND1 | ERK1 increase SNAI2 |
| ESRP2 regulate CTNND1 | ERK2 increase ZEB1 |
| AKT2 increase SNAI1 | ERK2 increase SNAI1 |
| CTNND1 stabilize CDH1 | ERK2 increase SNAI2 |
| CTNND1 attenuate RhoA | AKT2 increase SNAI2 |
| ESRP1 regulate CD44 | ZEB1 increase N-cadherin |
| ESRP2 regulate CD44 | SNAI1 increase N-cadherin |
| ESRP1 regulate ENAH | SNAI2 increase N-cadherin |
| ESRP2 regulate ENAH | HGF phosphorylate c-Met |
| SNAI1 repress ESRP1 | c-Met interact GAB1 |
| ZEB1 repress ESRP1 | c-Met interact GRB2 |
| ZEB2 repress ESRP1 | GAB1 interact SHP2 |
| SNAI1 repress ESRP2 | GRB2 interact SOS1 |
| ZEB1 repress ESRP2 | SOS1 interact HRAS |
| ZEB2 repress ESRP2 | SOS1 interact KRAS |
| YB1 increase SNAI1 | ERK1 increase EGR1 |
| YB1 increase ZEB2 | ERK2 increase EGR1 |
| YB1 increase LEF1 | EGR1 increase SNAI1 |
| YB1 increase TWIST1 | SNAI1 increase Claudin-3 |
| HNRNPE1 inhibit EEF1A1 | PDGFA increase SNAI2 |
| TGF$\beta$1 phosphorylation HNRNPE1 | FGF activate FGFR2 |
| TGF$\beta$1 increase ILEI | GAB1 interact PI3K |
| TGF$\beta$1 increase DAB2 | AKT2 inhibit GSK3$\beta$ |
| WT1 activate SRPK1 | GSK3$\beta$ phosphorylate $\beta$-catenin |
| SRPK1 phosphorylate SRSF1 | GSK3$\beta$ inhibit SNAI1 |
| SRPK1 nucleus_localize SRSF1 | GLI1 induce SNAI1 |
| CK1 phosphorylate SNAI1 | wnt1 inhibit GSK3$\beta$ |

# References

1. Ludwig, J.A., Weinstein, J.N.: Biomarkers in cancer staging, prognosis and treatment selection. Nat. Rev. cancer **5**(11), 845–856 (2005)
2. Hanash, S.M., Pitteri, S.J., Faca, V.M.: Mining the plasma proteome for cancer biomarkers. Nature **452**(7187), 571–579 (2008)
3. Saeys, Y., Inza, I., Larraaga, P.: A review of feature selection techniques in bioinformatics. Bioinformatics **23**(19), 2507–2517 (2007)
4. Guyon, I., Elisseeff, A.: An introduction to variable and feature selection. J. Mach. Learn. Res. **3**, 1157–1182 (2003)
5. Thousands of Samples are Needed to Generate a Robust Gene List for Predicting Outcome in Cancer, vol. 103. National Academy Sciences (2006)
6. Haury, A.-C., Gestraud, P., Vert, J.-P.: The influence of feature selection methods on accuracy, stability and interpretability of molecular signatures. PloS One **6**(12), e28210 (2011)
7. Patel, V.N., Gokulrangan, G., Chowdhury, S.A., Chen, Y., Sloan, A.E., Koyutrk, M., Barnholtz-Sloan, J., Chance, M.R.: Network signatures of survival in glioblastoma multiforme. PLoS Comput. Biol. **9**(9), e1003237 (2013)
8. Dao, P., Colak, R., Salari, R., Moser, F., Davicioni, E., Schönhuth, A., Ester, M.: Inferring cancer subnetwork markers using density-constrained biclustering. Bioinformatics **26**(18), i625–i631 (2010)
9. Clarke, R., Ressom, H.W., Zhang, Y., Xuan, J.: Module-based breast cancer classification. Int. J. Data Min. Bioinform. **7**, 284–302 (2013)
10. Holzinger, E.R., Li, R., Pendergrass, S.A., Kim, D., Ritchie, M.D.: Methods of integrating data to uncover genotype-phenotype interactions. Nat. Rev. Genet. **16**, 85–97 (2015)
11. Kim, D., Shin, H., Song, Y.S., Kim, J.H.: Synergistic effect of different levels of genomic data for cancer clinical outcome prediction. J. Biomed. Inform. **45**(6), 1191–1198 (2012)
12. Huang, H.-L., Wu, Y.-C., Su, L.-J., Huang, Y.-J., Charoenkwan, P., Chen, W.-Li., Lee, H.-C., Chu, W.C.-C., Ho, S.-Y.: Discovery of prognostic biomarkers for predicting lung cancer metastasis using microarray and survival data. BMC Bioinform. **16**(1) (2015)
13. Zhao, Q., Shi, X., Xie, Y., Huang, J., Shia, B.C., Ma, S.: Combining multidimensional genomic measurements for predicting cancer prognosis: observations from TCGA. Briefings Bioinform. **16**(2), 291–303 (2015)
14. Schliekelman, M.J., Taguchi, A., Zhu, J., Dai, X., Rodriguez, J., Celiktas, M., Zhang, Q., Chin, A., Wong, C.-H., Wang, H., et al.: Molecular portraits of epithelial, mesenchymal, and hybrid states in lung adenocarcinoma and their relevance to survival. Cancer Res. **75**(9), 1789–1800 (2015)
15. Chaffer, C.L., Weinberg, R.A.: A perspective on cancer cell metastasis. Science **331**(6024), 1559–1564 (2011)
16. Elsevier. EMT as the Ultimate Survival Mechanism of Cancer Cells, vol. 22 (2012)
17. Derynck, R., Lamouille, S., Xu, J.: Molecular mechanisms of epithelial-mesenchymal transition. Nat. Rev. Mol. Cell Biol. **15**, 178–196 (2014)
18. Kalluri, R., Weinberg, R.A.: The basics of epithelial-mesenchymal transition. J. Clin. Invest. **119**(6), 1420–1428 (2009)

19. Amin, E.M., Oltean, S., Hua, J., Gammons, M.V.R., Hamdollah-Zadeh, M., Welsh, G.I., Cheung, M.-K., Ni, L., Kase, S., Rennel, E.S., Symonds, K.E., Nowak, D.G., Royer-Pokora, B., Saleem, M.A., Hagiwara, M., Schumacher, V.A., Harper, S.J., Hinton, D.R., Bates, D.O., Ladomery, M.R.: WT1 mutants reveal SRPK1 to be a downstream angiogenesis target by altering VEGF splicing. Cancer Cell **20**(6), 768–780 (2011)
20. Berx, G., De Craene, B.: Regulatory networks defining EMT during cancer initiation and progression. Nat. Rev. Cancer **13**(6), 97–110 (2013)
21. Ji, Y., Zhu, Y., Qiu, P.: TCGA-Assembler: open-source software for retrieving and processing TCGA data. Nat. Methods **11**, 599–600 (2014)
22. Tibshirani, R.: Regression shrinkage and selection via the lasso. J. Roy. Stat. Soc.: Ser. B (Methodol.) **58**, 267–288 (1996)
23. Wernicke, S., Rasche, F.: FANMOD: a tool for fast network motif detection. Bioinformatics **22**(9), 1152–1153 (2006)
24. Chang, C.-C., Lin, C.-J.: LIBSVM: a library for support vector machines. ACM Trans. Intell. Syst. Technol. (TIST) **2**(3), 27 (2011)
25. World Scientific. Integrative Network Analysis to Identify Aberrant Pathway Networks in Ovarian Cancer (2012)

# Biomedical Image Analysis

# Tracking a Real Liver Using a Virtual Liver and an Experimental Evaluation with Kinect v2

Hiroshi Noborio[1]([✉]), Kaoru Watanabe[1], Masahiro Yagi[1], Yasuhiro Ida[1],
Shigeki Nankaku[1], Katsuhiko Onishi[1], Masanao Koeda[2], Masanori Kon[2],
Kosuke Matsui[2], and Masaki Kaibori[2]

[1] Department of Computer Science, Osaka Electro-Communication University, Osaka, Japan
{nobori,nankaku,onishi}@isc.osakac.ac.jp,
watanabe@wtlab.osakac.ac.jp
[2] Medical School, Kansai Medical University, Osaka, Japan
koeda@isc.osakac.ac.jp, {kon,matsuik,kaibori}@hirakata.kmu.ac.jp

**Abstract.** In this study, we propose a smart transcription algorithm for translation and/or rotation motions. This algorithm has two phases: calculating the differences between real and virtual 2D depth images, and searching the motion space defined by three translation and three rotation degrees of freedom based on the depth differences. One depth image is captured for a real liver using a Kinect v2 depth camera and another depth image is obtained for a virtual liver (a polyhedron in stereo-lithography (STL) format by z-buffering with a graphics processing unit). The STL data are converted from Digital Imaging and Communication in Medicine (DICOM) data, where the DICOM data are captured from a patient's liver using magnetic resonance imaging and/or a computed tomography scanner. In this study, we evaluated the motion precision of our proposed algorithm based on several experiments based using a Kinect v2 depth camera.

**Keywords:** Depth image · Graphics processing unit · Parallel processing · Randomized steepest descent method · Z-buffering

## 1 Introduction

In the last decade, many researchers have designed fast and robust approaches for surface registration [1–7]. However, they employed three-dimensional (3D) point-point matching and irregular (x, y, z-axes) 3D matching is time consuming. Therefore, these methods are not suitable for surgical navigation in real-time. To overcome this drawback, we propose another approach where the movements of a real liver in an operating room are replicated by the movements of a virtual liver simulated on a PC. This motion transfer function is important for constructing a surgical navigation system.

Our algorithm involves two types of parallel processing: depth difference calculation for all of the pixels by z-buffering with a graphics processing unit (GPU) and selecting the best neighbor from a large number using the multicores on a GPU. In the former procedure, we use two-dimensional (2D) depth-depth matching between a real depth image in the real world and its virtual depth image, which is calculated automatically by z-buffering in the virtual world. This type of matching comprises regularly arranged

© Springer International Publishing Switzerland 2016
F. Ortuño and I. Rojas (Eds.): IWBBIO 2016, LNBI 9656, pp. 149–162, 2016.
DOI: 10.1007/978-3-319-31744-1_14

one-dimensional (z-axis = depth) matching. Therefore, all of the matches for a vast number of pixels can be calculated in parallel by z-buffering on the GPU. In the latter procedure, we perform a rapid search for the best neighbor, where the difference is the minimum/medium/average/maximum compared with a large number of neighbors based on six degrees of freedom (DOFs) translation/rotation movements in space in parallel using the multicores on the GPU. Thus, we successively find a position/posture that is most likely to exist based on various differences between the depth image captured by a Kinect v2 from the real world and that in the z-buffer captured by the GPU from the virtual world. To eliminate local minima, we use two types of randomization processes in the steepest descent method.

A practical system for surgical navigation requires several basic functions. Therefore, we are now developing several types of software, hardware, sensing equipment, and control schema. Our navigation system mainly comprises the following software/hardware/sensing items.

1. Initial position/orientation adjustment for virtual and real livers [8]: In our study, a 3D computer generated (CG) virtual environment controlled by OpenGL on the GPU is theoretically adjusted for a 3D real camera environment, which is controlled by Kinect SDK and Kinect Studio API. Thus, the virtual 3D coordinate system is roughly consistent with the real 3D world. Furthermore, a user can appreciate differences in depth images captured of real and virtual livers in 3D real and virtual environments according to various color types based on the Kinect v1 [8]. While watching the colors, a user can select several parameters related to the 3D CG virtual environment according to the consistency between the real liver in the camera coordinate system and the virtual liver in the graphics coordinate system. In this image-based position/orientation adjustment system, the CG virtual world captured artificially by the z-buffer of the GPU should agree with the real world captured by a real depth camera [8].

2. Tracking a master virtual moving liver with its slave virtual liver (theoretical evaluation) [9–11]: One year ago, we proposed the following new framework using a 2D depth-depth matching method based on a real depth image and a virtual depth image (the Z-buffer) on a GPU. We then evaluated the algorithm based on simulations of master-slave livers supported by a GPU in a virtual environment [9]. Furthermore, we investigated the performance of the simulated master-slave livers by varying the search space from small to large [10], which comprised three DOFs for translation and three DOFs for rotation, as well as by dividing the search space into three translational DOFs and one rotational DOF [11]. Moreover, we have developed a randomized steepest descendent algorithm, which can be replaced in a flexible manner with another optimal or near-optimal search algorithm.

3. Tracking a real moving liver with a virtual liver (experimental evaluation) [12]: Using the Kinect v1, we developed a trial version for real liver motion tracking using a virtual liver. Unfortunately, compared with Kinect v2, the image resolution and depth resolution are lower with Kinect v1. In addition, there are few software functions for controlling the 3D virtual environment. Thus, we are developing a new

version based on Kinect v2. In the new version, we have added some original functions to evaluate precise overlapping between a real and virtual liver during a surgical operation based on the video and the graph, as described in this study.

4. Position/orientation calibration for a real liver and scalpel [13, 14]: This is a technique for identifying the position/orientation of a real liver and scalpel in the real world. In our method, we use a Micron tracker 3 (a position tracking system supported by various types of differential check patterns, which has high precision of about 70 μm). If and only if the positions/orientations are recognized precisely in the real world, they are expressed in the virtual world.

5. Calculation of the Euclidean distance between the scalpel tip and the virtual liver and its blood vessels [15, 16]: We developed a fast algorithm based on parallel processing by a GPU to calculate the Euclidean distance between a scalpel tip and cancer areas and/or blood vessels. This function allows a doctor to reduce their errors, such as injuries to blood vessels during surgical operations.

6. Fast cutting of a virtual liver is equivalent to cutting a real liver [17, 18]: In this stage, when a doctor operates on a human organ in a real environment, the operation and cutting of the virtual liver are synchronized in our system based on augmented reality (AR). However, cutting a virtual liver expressed by a polyhedron in stereo-lithography (STL) format is generally time consuming. Therefore, in the proposed method, we operate on a polyhedral liver in STL format by Z-buffering and by using multicore parallel processing on the GPU during surgical navigation under the control of AR.

In this study, we developed our new algorithm using Kinect v2, which has several original functions that allow precise evaluation in the real world. A replica of a real liver was printed using a 3D printer and the virtual liver was modeled by STL. The 3D plastic liver was then moved by a human according to several different translations and rotations. Finally, to consider the positional precision, we checked the performance by tracking a real 3D liver with a 3D virtual liver using our 2D depth-depth surface matching algorithm. Thus, we demonstrated that our algorithm can potentially be employed for translational and rotational liver tracking during surgical navigation.

The remainder of this paper is organized as follows. In Sect. 2, we describe the real and virtual livers. We explain our real-virtual 2D depth-depth matching algorithm in Sect. 2.1. In Sect. 3, we present experimental results obtained when a real liver moved along the X, Y, and Z axes, respectively, as well as experimental results obtained when a real liver moved around the Z-axis. Finally, we give our conclusions in Sect. 4.

## 2   Liver Modeling

Using the proposed algorithm, the liver of a patient is captured as Digital Imaging and Communication in Medicine (DICOM) data by magnetic resonance imaging (MRI) or computed tomography (CT), and the DICOM data are then converted into a polyhedron in the STL format. STL is employed to maintain the visual quality and to rapidly calculate a depth image using the Z-buffer of the GPU (Fig. 1). STL is used to model a virtual liver. We constructed a plastic replica of a real liver using a 3D printer based on the STL

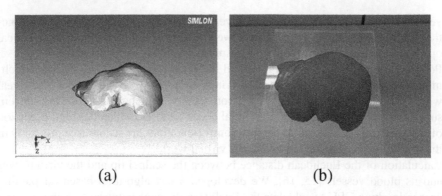

Fig. 1. (a) A polyhedral liver in STL format and (b) the 3D printed plastic liver.

data, although this replica did not possess any elastic or visco-elastic properties. Thus, we checked the transcription of all motions excluding deformation of the liver.

## 2.1 Real-Virtual Depth-Depth Matching Algorithm

Next, we define the real and virtual 3D environments (camera and CG coordinate systems) based on adjustments of the parameters shown in Table 1. Using these parameters, we can specify the CG environment with Open GL, as follows.

```
// programming codes
        const int imageWidth = 512;
        const int imageHeight = 424;
        const float cameraFovy = 60;
        const float cameraAspect = float(imageWidth)/float(imageHeight);
        const float cameraZNear = 0.5;
        const float cameraZFar = 4.5;
        const glm::mat4 cameraProjMat = glm::perspective(cameraFovy,
        cameraAspect, cameraZNear, cameraZFar);
// perspective view
```

First, the replica plastic liver and the virtual STL liver were set in the same position/posture using our image-based initial position/orientation adjustment system [1].

Next, our real-virtual liver tracking algorithm was performed according to the following five steps.

[Step 1] A human user who acted as a doctor translationally or rotationally operated on the replica plastic liver printed using a 3D printer.

[Step 2] The replica plastic liver was captured as a set of rectangular parallelepipeds by the Kinect v2 depth camera, which is widely used for human motion capture.

[Step 3] The virtual liver STL was moved into the best position/posture from among 24 that neighbored the present position/posture (Fig. 2).

**Table 1.** Comparison of Kinect v1 and v2.

|  |  | Kinect v1 | Kinect v2 |
|---|---|---|---|
| Color | Resolution | 640 × 480 | 1920 × 1080 |
|  | fps | 30 fps | 30 fps *3 |
| Depth | Resolution | 320 × 240 | 512 × 424 |
|  | fps | 30 fps | 30 fps |
| Range of Depth |  | 0.8–4.0 m(Near Mode > 0.4 m) (Extended Depth > 10.0 m) | 0.5–8.0 m |
| Range of Detection |  | 0.8–4.0 m(Near Mode 0.4–3.0 m) | 0.5–4.5 m |
| Angle Depth | Horizontal | 57° | 70° |
|  | Vertical | 43° | 60° |

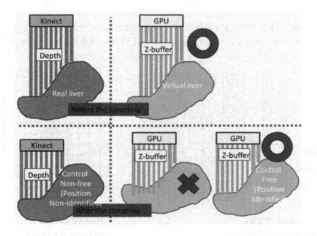

**Fig. 2.** The position/orientation of the virtual liver relative to that of the real liver can be matched better by using real and virtual depth images.

The virtual liver was moved in motion space with six DOFs (X, Y, and Z translational DOFs, and roll, pitch, and yaw rotational DOFs). In our randomized steepest descendent algorithm, we always selected one of three motions, such as positive two steps, positive one step, stop, negative one step, or negative two steps along each DOF. Therefore, there were 24 (= 6*2*2) candidate neighbors for the current candidate (Fig. 3).

**(3-1)** For each position/posture candidate, the depth image for the virtual liver STL was captured by z-buffering in the GPU.

**(3-2)** For each position/posture candidate, we compared a real depth image obtained by the Kinect v2 depth camera and a virtual depth image prepared in the z-buffer of the GPU. In general, the camera coordinate system (camera parameters) is known for Kinect v2. Therefore, we adjusted the CG coordinate system (CG parameters) to those of the Kinect coordinate system. The rendering resolution for CG was set as 512 × 424 pixels, which is that of Kinect v2 (Fig. 4).

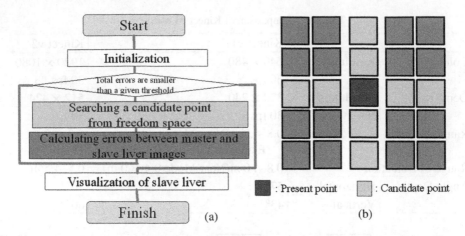

**Fig. 3.** (a) Flowchart illustrating our depth-depth matching algorithm, and (b) 24 neighboring candidates with six dimensions, i.e., three translational DOFs and three rotational DOFs.

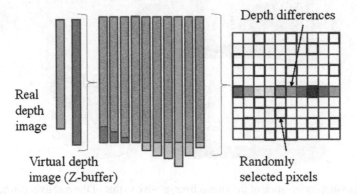

**Fig. 4.** Summed differences between the real and virtual depth images based on pixels selected randomly in parallel on the GPU.

**(3-3)** We selected 20 pairs of depth images and we selected 100 random pixels in each image. Next, we determined their average based on a histogram of the depth differences among the 100 pixels. We also selected the median based on the histogram of the averages for all 20 images.

**[Step 4]** The PC moved the virtual liver STL (a truly virtual liver) according to the selected position/posture.

**[Step 5]** Return to [Step 1].

# 3   Experimental Results

In this section, we describe the results obtained when a human operated on a liver replica printed using a 3D printer (colored in red in the images), where we checked whether the virtual liver (colored in blue) matched precisely with the real liver replica.

First, we compared the virtual 3D environment (CG coordinate system) with the real 3D environment (camera coordinate system). Next, we overlapped the replica liver with the virtual liver in 3D space by matching various parameters of the real and virtual depth images in 2D space [1].

Using various windows, we determined positional errors in the X, Y, and Z coordinates, as well as orientation errors in the pitch, yaw, and roll coordinates. In each of the left-hand windows, a 3D virtual liver colored in blue translationally or rotationally tracked the 3D replica liver colored in red, which moved along the X-axis. In the right-hand bottom window, the overlapping area in 3D XYZ space was colored as green pixels.

By watching a color image (green: XYZ matched area; yellow: XY matched area; red: replica liver 2D projection area; blue: virtual liver 2D projection area) that comprised pixel differences in 2D depth images, a user overlapped the 2D projections of the replica and virtual livers on the XY plane by combining the X and/or Y movements. In the first stage, the yellow region was increased in the 2D color image by the user. Next, the 3D replica and virtual livers were overlapped along or around the Z-axis by combining translation and/or rotation movements along and/or around the Z-axis. In the second stage, the green region was increased in the 2D color image by the user.

As shown in Fig. 5 and Table 2, the 3D virtual liver translationally tracked the 3D replica liver, which moved along the X-axis at a speed of 15 cm/10 s. The 3D virtual liver was colored in yellow and the 3D replica liver was colored in red. In Fig. 5(a)–(c), we present three stroboscope shots from the initial stage, middle stage, and the final stage, respectively. In this experiment, the overlapping area in 3D XYZ space was colored with a set of blue pixels. In Table 2, we show the overlapping ratio for the replica liver and the virtual liver. The overlapping ratio was defined precisely as: the number

**Table 2.** Overlapping ratio (the number of 2D green pixels overlapping in the 3D real and virtual livers/the number of 2D blue pixels projected from a 3D virtual liver) for a translational movement along the X axis at a velocity of 15 cm/10 s.

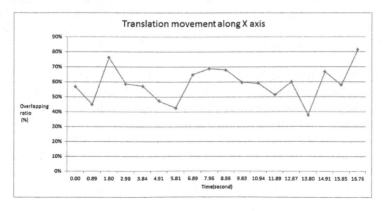

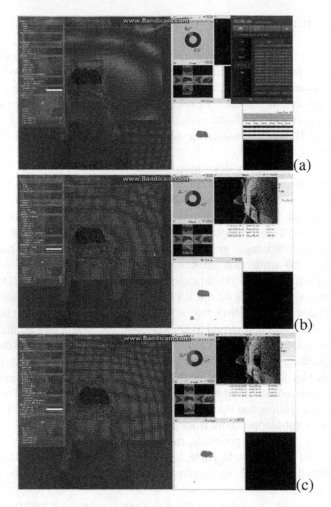

**Fig. 5.** (a),(b),(c) Three successive stroboscope shots showing a 3D real liver replica moving along the X axis at a speed of 15 cm/10 s. In each of the left-hand windows, a 3D virtual liver colored in blue translationally tracks the 3D real liver colored in red. In the right-hand bottom window, the overlapping area in 3D XYZ space is colored in green. In addition, in the right-hand upper windows, the overlapping ratio is indicated by the blue arc in the circular graph, while six side projections are presented for the real and virtual livers (Color figure online).

of 2D green pixels where the 3D replica liver overlaps with the 3D virtual liver/the number of 2D blue pixels projected from the 3D virtual liver. As shown in Fig. 5 and Table 2, the replica and virtual livers generally overlapped in the 3D space, so our algorithm is of practical use during navigation in liver surgery.

In Fig. 6 and Table 3, a 3D virtual liver translationally tracks the 3D replica liver, which moves along the X-axis at a speed of 15 cm/20 s, where only the speed was changed. The 3D virtual liver was colored in yellow and the 3D replica liver was colored in red. In Fig. 6(a)–(c), we present three stroboscope shots from the initial, middle, and final stages, respectively. The overlapping area in 3D XYZ space was also colored with a set of blue pixels. In Table 3, we show the overlapping ratio between the replica liver and the virtual liver. As shown in Fig. 6 and Table 3, the overlapping ratio was better when the replica liver moved slowly. Thus, our algorithm is more suitable for slower navigation during liver surgery.

When the 3D virtual liver translationally tracked the 3D replica liver as it moved along the Y or Z axes with rapid or slow translational motions, the overlapping area and ratio were generally maintained in the 3D XYZ space. Therefore, in general, the overlapping precision was lower when the translational movement speed of the replica liver was faster.

**Table 3.** Overlapping ratio (the number of 2D green pixels overlapping in the 3D real and virtual livers/the number of 2D blue pixels projected from a 3D virtual liver) for translational movement along the X axis at a velocity of 15 cm/20 s.

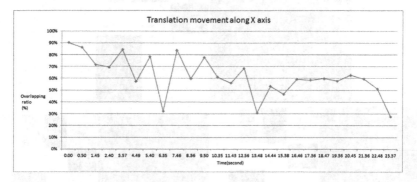

**Table 4.** Overlapping ratio (the number of 2D green pixels overlapping between 3D real and virtual livers/the number of 2D blue pixels projected from a 3D virtual liver) for rotational movement around the Z axis at a rate of 45°/10 s.

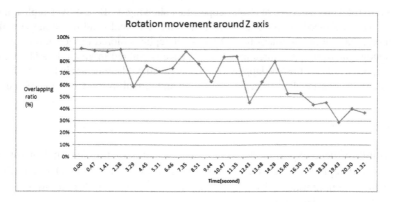

Finally, Fig. 7, Table 4, Fig. 8, and Table 5 show the 3D virtual liver rotationally tracking the 3D replica liver as it moved around the Z axis at rates of 45°/20 s and 45°/10 s, respectively. The 3D virtual liver was colored in blue and the 3D replica liver was colored in red. Figures 7(a)–(c) and 8(a)–(c) present three stroboscope shots from the initial, middle, and final stages, respectively. The overlapping area in 3D XYZ space was colored with a set of blue pixels. We also present the overlapping ratio between the replica liver and the virtual liver. As shown in Fig. 7, Table 4, Fig. 8, and Table 5, the overlapping ratio was better when the replica liver moved slowly. Thus, our proposed algorithm is more suitable for slower navigation during liver surgery.

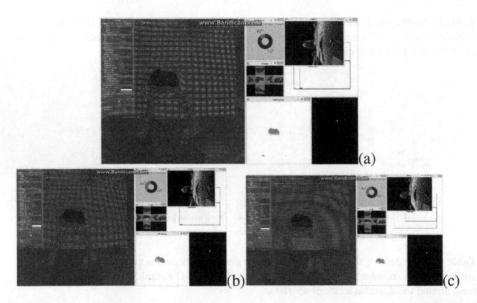

**Fig. 6.** (a),(b),(c) Three successive stroboscope shots showing a 3D real liver replica moving along the X axis at a speed of 15 cm/20 s. In each of the left-hand windows, a 3D virtual liver colored in blue translationally tracks the 3D real liver replica colored in red. In the right-hand bottom window, the overlapping area in 3D XYZ space is colored in green. In addition, in the right-hand upper windows, the overlapping ratio is indicated by the blue arc in the circular graph and six side projections are also presented for the real and virtual livers (Color figure online).

According to these experiments, under various types of translational or rotational movements, the replica and virtual livers generally overlapped in 3D space, thereby demonstrating the potential utility of the proposed 2D depth-depth matching algorithm based on a Kinect v2 depth camera.

**Table 5.** Overlapping ratio (the number of 2D green pixels overlapping between the 3D replica and virtual livers/the number of 2D blue pixels projected from a 3D virtual liver) with rotational movement around the Z-axis at a rate of 45°/20 s.

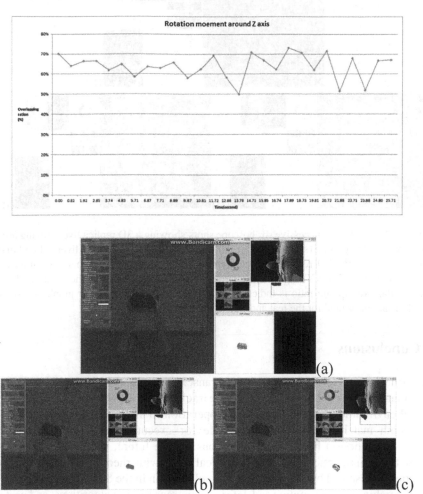

**Fig. 7.** (a),(b),(c) Three successive stroboscope shots showing a 3D replica liver moving around the Z-axis at a rate of 45°/10 s. In each of the left-hand windows, a 3D virtual liver colored in blue rotationally tracks the 3D replica liver colored in red. In the right-hand bottom window, the overlapping area in 3D XYZ space is colored in green. In addition, in the right-hand upper windows, the overlapping ratio is indicated by the blue arc in the circular graph and six side projections are presented for the replica and virtual livers (Color figure online).

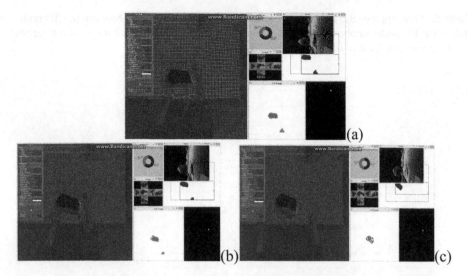

**Fig. 8.** (a),(b),(c) Three successive stroboscope shots showing a 3D replica liver moving around the Z-axis at a rate of 45°/20 s. In each of the left-hand windows, a 3D virtual liver colored in blue rotationally tracks the 3D replica liver colored in red. In the right-hand bottom window, the overlapping area in 3D XYZ space is colored in green. In addition, in the right-hand upper windows, the overlapping ratio is indicated by the blue arc in the circular graph and six side projections are presented for the real and virtual livers (Color figure online).

## 4    Conclusions

In this study, we developed a new motion transcription algorithm using Kinect v2 and we evaluated the overlapping precision while tracking a replica liver using a virtual liver in a real environment similar to a surgical operation room. We found that our fast 2D depth-depth matching algorithm is suitable for executing motion tracking during surgical navigation in human liver operations. In the future, we will develop a new deformation transcription algorithm for a real liver with rheological or visco-elastic properties represented by a virtual liver as a polyhedron in the STL format. In addition, we will combine our new motion and deformation transcription algorithms, and perform experimental evaluations with various depth cameras.

**Acknowledgments.** This study was supported partly by 2014 Grants-in-Aid for Scientific Research (No. 26289069) from the Ministry of Education, Culture, Sports, Science, and Technology, Japan. Further support was provided by the 2014 Cooperation Research Fund from the Graduate School at Osaka Electro-Communication University.

## References

1. Besl, P.J., McKay, N.D.: A method for registration of 3-D Shapes. IEEE Trans. Pattern Anal. Mach. Intell. **14**(2), 239–256 (1992)

2. Zhang, Z.: Iterative point matching for registration of free-form surfaces. J. Comput. Vis. **13**(2), 119–152 (1994)
3. Granger, S., Pennec, X.: Multi-scale EM-ICP: a fast and robust approach for surface registration. In: Heyden, A., Sparr, G., Nielsen, M., Johansen, P. (eds.) ECCV 2002, Part IV. LNCS, vol. 2353, pp. 418–432. Springer, Heidelberg (2002)
4. Liu, Y.: Automatic registration of overlapping 3D point clouds using closest points. J. Image Vis. Comput. **24**(7), 762–781 (2006)
5. Salvi, J., Matabosch, C., Fofi, D., Forest, J.: A review of recent range image registration methods with accuracy evaluation. J. Image Vis. Comput. **25**, 578–596 (2007)
6. Rusu, R.B., Cousins, S.: 3D is here: Point Cloud Library (PCL): In: IEEE International Conference on Robotics and Automation, pp. 1–4 (2011)
7. Wu, Y.F., Wang, W., Lu, K.Q., Wei, Y.D., Chen, Z.C.: A new method for registration of 3D point sets with low overlapping ratios. In: 13th CIRP Conference on Computer Aided Tolerancing, pp. 202–206 (2015)
8. Noborio, H., Watanabe, K., Yagi, M., Ida, Y., Onishi, K., Koeda, M., Nankaku, S., Matsui, K., Kon, M., Kaibori, M.: Image-based initial position/orientation adjustment system between real and virtual livers. Jurnal Teknologi, Med. Eng. **77**(6), 41–45 (2015). Penerbit UTM Press
9. Noborio, H., Onishi, K., Koeda, M., Mizushino, K., Yagi, M., Kaibori, M., Kwon, M.: Motion transcription algorithm by matching corresponding depth image and Z-buffer. In: 10th Anniversary Asian Conference on Computer Aided Surgery, pp. 60–61. Kyusyu University, Japan, June 2014
10. Watanabe, K., Yagi, M., Ota, K., Onishi, K., Koeda, M., Nankaku, S., Noborio, H., Kon, M., Matsui, K., Kaibori, M.: Parameter identification of depth-depth-matching algorithm for liver following. Jurnal Teknologi, Med. Eng. **77**(6), 35–39 (2015). Penerbit UTM Press
11. Watanabe, K., Yagi, M., Shintani, A., Nankaku, S., Onishi, K., Koeda, M., Noborio, H., Kon, M., Matsui, K., Kaibori, M.: A new 2D depth-depth matching algorithm whose translation and rotation freedoms are separated. In: International Conference on Intelligent Informatics and Biomedical Sciences, Okinawa Japan, pp. 271–278, November 2015
12. Noborio, H., Watanabe, K., Yagi, M., Ida, Y., Nankaku, S., Onishi, K., Koeda, H., Kon, M., Matsui, K., Kaibori, M.: Experimental results of 2D depth–depth matching algorithm based on depth camera Kinect v1. J. Bioinform. Neurosci. **1**(1), 38–44 (2015). ISSN: 2188-8116
13. Koeda, M., Tsukushi, A., Noborio, H., Onishi, K., Mizushino, K., Kunii, T., Watanabe, K., Kaibori, M., Matsui, K., Kwon, M.: Depth camera calibration and knife tip position estimation for liver surgery support system. In: 17th International Conference on Human-Computer Interaction, Los Angeles, CA, USA, pp. 496–502 August 2015
14. Doi, M., Yano, D., Koeda, M., Noborio, H., Onishi, K., Kayaki, M., Mizushino, K., Matsui, K., Kaibori, M.: Knife tip position estimation using multiple markers for liver surgery support. In: 6th International Conference on Advanced Mechatronics (ICAM 2015), pp. 74–75. Nishiwaseda Campus of Waseda University, Tokyo Japan, 1A2-08, 5–8 December 2015
15. Noborio, H., Kunii, T., Mizushino, K.: GPU-based shortest distance algorithm for liver surgery navigation. In: 10th Anniversary Asian Conference on Computer Aided Surgery, pp. 42–43. Kyusyu University, Japan, June 2014
16. Noborio, H., Kunii, T., Mizushino, K.: Comparison of GPU-based and CPU-based algorithms for determining the minimum distance between a CUSA scalpel and blood vessels. In: The 7th International Conference on Bioinformatics Models, Methods and Algorithms, Roma, Italy (to appear)

17. Onishi, K., Noborio, H., Koeda, M., Watanabe, K., Mizushino, K., Kunii, T., Kaibori, M., Matsui, K., Kwon, M.: Virtual liver surgical simulator by using Z-buffer for object deformation. In: 17th International Conference on Human-Computer Interaction (HCII 2015), Los Angeles, CA, USA, pp. 345–351 August 2015
18. Noborio, H., Onishi, K., Koeda, M., Mizushino, K., Kunii, T., Kaibori, M., Kon, M., Chen, Y.-W.: Fast surgical algorithm for cutting with liver standard triangulation language format using Z-Buffers in graphics processing unit. In: Fujie, M. (ed.) Computer Aided Surgery, pp. 127–140. Springer, Japan (2016). doi:10.1007/978-4-431-55810-1_11

# Thermal Imaging-Based Muscular Activity in the Biomechanical Study of Surgeons

Ramon Sancibrian[1(✉)], Maria C. Gutierrez-Diez[2], Carlos Redondo-Figuero[3],
Esther G. Sarabia[4], Maria A. Benito-Gonzalez[5], and Jose C. Manuel-Palazuelos[6]

[1] Department of Structural and Mechanical Engineering,
University of Cantabria, Santander, Spain
sancibrr@unican.es
[2] Department of Medicine and Psychiatry, University of Cantabria, Santander, Spain
mariacruz.gutierrez@unican.es
[3] Department of Medical and Surgical Sciences, University of Cantabria, Santander, Spain
carlos.redondo@unican.es
[4] Electronic Technology and Automatic Systems Department,
University of Cantabria, Santander, Spain
esther.gonzalezs@unican.es
[5] Department of Nursery, Marques de Valdecilla University Hospital, Santander, Spain
abenito@hvvaldecilla.es
[6] Surgical Unit of the Valdecilla Virtual Hospital, Santander, Spain
jcpalazuelos@hvvaldecilla.es

**Abstract.** The use of minimally invasive surgery has introduced many modifications in surgical procedures. Despite the advantages that this kind of surgery provides, surgeons have to confront many ergonomic problems during their interventions. In fact, the poor ergonomic characteristics of the workplace reduce the efficiency of the interventions and produce undesirable effects such as physical fatigue or musculoskeletal injuries. Electromyography has been used traditionally for measurement of the muscular effort in the workplace. However, in recent studies thermal imaging has been highlighted as a valuable alternative in the determination of muscular activity. One of the main advantages of using thermal imaging is that there is no necessary to foresee the muscular groups activated in the performance of surgery. In this paper thermal imaging is used to evaluate the muscular effort of surgeons and the results are compared with electromyography. The paper shows the features of this technique and the relationship with electromyography.

**Keywords:** Muscular effort · Electromyography · Thermal imaging · Ergonomics · Biomechanics

## 1 Introduction

The use of minimally invasive surgery (MIS) procedures has led to a great advance in the performance of surgical interventions. Twenty years since the first use of these surgical techniques, a retrospective view shows several advantages for patients.

© Springer International Publishing Switzerland 2016
F. Ortuño and I. Rojas (Eds.): IWBBIO 2016, LNBI 9656, pp. 163–174, 2016.
DOI: 10.1007/978-3-319-31744-1_15

Indeed, they need less recovery time, their risk of infections is reduced and there are aesthetic benefits [1, 2]. However, this kind of surgery has significantly changed the working life of surgeons [3, 4]. In fact, from the perspective of the surgeon MIS does not only imply adaptation to new technologies but also involves important restrictions on how they perform their work. The biomechanics of surgeons reveals that laparoscopic surgery strongly restricts the degrees of freedom of forearms, wrist and hands [5, 6]. Surgeons cannot move freely as they do in open surgery because they work with special instruments which limit the access to the patient's organs obliging them to adopt awkward postures. The literature shows that several factors influence these ergonomic deficiencies [7, 8]. For instance, the layout of the surgical room plays an important role because it determines the range of motion of the surgical team. However, the main issues influencing the ergonomic features of the workplace can be focused on three main factors: the height of the worktable, the position of the monitors [3] and the design of the handles of the surgical instruments [5]. Manufacturers of surgical instruments try to design the best ergonomic handles for their surgical instruments. Nevertheless, the design of the handles has to accommodate many constraints making it difficult to find satisfactory solutions.

Due to the long duration of surgical interventions, the lack of ergonomic characteristics produces physical fatigue in the surgeon. This fact not only has negative effects on the surgeon but also on the patient. Indeed, considering the long duration of surgical interventions, a tired surgeon is more likely to make mistakes. In the long term, the poor ergonomics produces injuries and musculoskeletal disorders (MSDs), reducing the capacity and efficiency of the surgeons [9, 10]. In order to avoid all these problems, it is necessary to develop accurate methodologies to assess the ergonomic features of new designs.

Surface electromyography (sEMG) is considered the best technology to evaluate the muscular effort in surgeons [11, 12]. This is not an invasive procedure because the electrodes are situated on the skin [13, 14]. However, cables are always present to transport the signal from the surgeon's body to the analyzer. The presence of cables is uncomfortable for the surgeons and could change the way in which they work during the tests. On the other hand, the location of the sEMG electrodes needs to be foreseen before making the measurements [15]. In other words, the analyst should know which muscles will be involved in the performance of the work before carrying out the test. In this way they can locate the electrodes on the correct muscular group to collect the data. Those muscles which are not considered important would not be measured. Obviously, this fact could lead to overlooking important muscular activity which has not been foreseen.

The technological evolution of infrared (IR) technology has provided an important tool for ergonomic research [16, 17]. Thermal imaging provides accurate information about the variation in temperature on the skin surface. The increase in the muscle temperature is due to the heat production in the metabolic process that takes place during the dynamic effort [18–20]. Thus, the modification of the temperature can be correlated with the muscular activity and therefore used to obtain information about the effort necessary to complete surgical tasks [21].

This research work delves into the use of thermal imaging for the determination of the muscular activity in surgeons. The objective is to propose an alternative to EMG in the measurement of the muscular effort during the performance of surgery. Using the results obtained from the tests, this paper compares the two technologies and determines the influence and relationship of the muscular effort on temperature.

## 2   Materials and Methods

### 2.1   Participants and Inclusion Criteria

Ten surgeons (6 males and 4 females) volunteered to participate in the experiment, which was carried out in the Laboratory of Human Factors and Ergonomics of the Valdecilla Virtual Hospital in Santander, Spain. The age of the participants ranged from 25 to 35 years old and their experience in surgery was from 1 to 5 years. All participants were healthy and they did not mention any musculoskeletal disorder. Pregnant and left-handed surgeons were excluded from the experiment. The survey was approved by the Institutional Review Board (IRB) of Cantabria. The participants were informed about the objective and conditions of the experiment and all of them gave their informed consent before starting the tests.

### 2.2   Design of the Tests

The experiment consists of two tasks which were carried out by each volunteer taking part in the survey. The first task (T1) was a controlled test in which the forearm muscles of the surgeon were fatigued. The main objective of this test is to monitor all parameters and to know when exactly the participant is fatigued. For this, a handgrip is used and the participant is requested to exert dynamic hand contractions until the forearm muscles are exhausted. The duration of the controlled test depends on the volunteer's strength but in all cases was around one minute or less. The second part of the experiment is a surgical task (T2) where the surgeon has to perform an activity during twelve minutes in an endotrainer (see Fig. 1). The objective is to simulate the actual conditions in an operating room reproducing the movements of the forearm, wrist and hand. The design of the tests pays special attention to the muscular effort necessary to complete the tasks. The surgical activity was designed to activate the forearm muscles in a similar way to that used in real surgical operations.

All participants stood on a height-adjustable platform which was used to obtain an angle of 110 degrees between the arm and forearm. In this way, the effect of the position of the forearm did not have influence on the results of the experiment.

During the tests the EMG values and infrared imaging were recorded. In order to avoid the influence of room temperature on the results of the experiment, the environmental conditions of the laboratory were controlled during the performance of the test and the temperature was constant at 24.5 °C ($\pm 3$ %). The participants rested in the laboratory for at least 15 min before starting the tests in order to acclimatize them to the laboratory conditions.

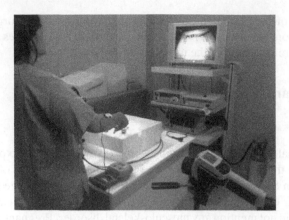

**Fig. 1.** Participant performing the surgical task (T2).

### 2.3 Measuring EMG

The EMG signal was registered with surface electrodes (B&L Engineering, USA). The electrodes were located on the skin to measure the activity of the muscular groups which are the extensor digitorum and flexor carpi radialis. The analysis of EMG was carried out with Matlab™, R2014b (Mathwork, USA). A passband filter was applied to the raw signal to obtain the information in the frequency range from 20 Hz to 250 Hz. A notch filter was applied to the electromyograms in order to remove the spike signal generated at 50 Hz from electrical noise.

The analysis of the EMG signal involves the parameters commonly used in investigating muscle activity. However, it is necessary to consider that in the case of the surgeon's work, the dynamic contraction of the muscles is more important than sustained contraction [22]. Normally, time-domain and frequency domain are used to obtain EMG information. Parameters such as root mean square (RMS) and zero crossing per second are commonly used in the time-domain. RMS (%) was used to obtain the force exerted by means of the comparison with the maximum voluntary contraction (MVC). However, time-domain parameters are normally used for the determination of the force exerted during the activity [23]. On the other hand, the frequency-domain is used to obtain information about muscle fatigue because it is related with the reduction of muscular activity when the surgeon becomes tired. Several parameters are defined in the literature for the study of muscular fatigue in the frequency-domain but two of them can be highlighted as the most important. They are the median frequency (MDF) and the mean frequency (MNF) [24].

In this work MDF and MNF are processed from EMG signals which were recorded from the forearms of the surgeons. The MDF parameter can be defined as the frequency value which divides the power spectrum into two equal regions. It is expressed as follows,

$$\sum_{j=1}^{m} P_j = \sum_{j=m}^{N} P_j = \frac{1}{2} \sum_{j=1}^{m} P_j \qquad (1)$$

Where $N$ is the size of frequency bin, $P_j$ is the power spectrum for the $jth$ value of the frequency, and the parameter $m$, once obtained, is the value MDF. On the other hand, the MNF is the average frequency that can be computed as follows,

$$MNF = \frac{\sum_{j=1}^{N} f_j P_j}{\sum_{j=1}^{N} P_j} \qquad (2)$$

where $f_j$ is the frequency at $j$.

During the tests the raw EMG signal was recorded and filtered. The analysis of the signal requires the RMS values to be obtained in order to calculate the magnitude of the force applied during the test. The muscular fatigue is obtained using the Fast Fourier Transform (FFT) in the standard way but applied consecutively in windows over the whole signal. In order to avoid the leakage problem, Hanning was applied to the Short-time Fourier Transform (STFT). The result of the sliding window technique is the variation of MDF and MNF with time. The slopes of the regression lines for MDF and MNF values are used as a fatigue index (FI).

### 2.4  Measuring Thermal Parameters

An IR camera (ThermalCAM® B2, Flir™, Sweden) was used to measure the temperature distribution and variation of muscles. The spectral range is 7.5 to 13 $\mu$m, the type of detector was focal plane array (FPA) uncooled microbolometer and the resolution was $320 \times 240$ pixels. The temperature range was from $-20\ °C$ to $+100\ °C$. The sensitivity of the camera was less than $0.1\ °C$. The emissivity of the human body surface was selected as 0.98.

The IR camera was located in front of the forearm of the participant and separated 0.5 m from the body's surface as is shown in Fig. 1. Thermograms were taken at different instants of the experiments in order to obtain the temperature variation. The measurement of temperature consisted of identifying the hottest zones of the forearm during the evolution of the tests. For this, an algorithm was developed to analyze the increment of temperature in each pixel of the forearm. That is,

$$\Delta T_i = T_{ki} - T_{0i}; \quad i = 1, 2, \ldots np \qquad (3)$$

where $np$ stands for the total number of pixels and $T_{ki}$ and $T_{0i}$ are temperatures obtained in the instant $k$ and at the beginning the task, respectively. The maximum value given by Eq. (3) gives the maximum increment of temperature along the forearm. The algorithm also identifies which is the hottest zone and relates it to the muscular group affected by the increment in temperature.

## 3  Results

### 3.1  Demographic Results

The demographic characteristics of the participants who took part in the experiment are presented in Table 1. Except the weight, there are no significant differences between

**Table 1.** Mean and standard deviation of the demographic values of the participants

|  | Age (years) | Height (m) | Weight (kg) | Forearm length (cm) | Hand size (cm) | Skinfold (mm) |
|---|---|---|---|---|---|---|
| Male | 28.2 (3.71) | 1.8 (0.06) | 81.2 (15.92) | 26.8 (2.15) | 22.8 (0.12) | 4.8 (0.69) |
| Female | 24.8 (4.07) | 1.6 (0.07) | 61.4 (13.46) | 25.2 (2.19) | 20.1 (0.13) | 4.9 (0.92) |
| Group | 26.5 (4.10) | 1.7 (0.10) | 71.3 (17.45) | 26.0 (2.24) | 21.4 (0.18) | 4.8 (0.78) |

male and female variables. The results shown in this table could be useful to be compared with other demographic data from other countries.

### 3.2 EMG Results

The measurements obtained from the EMG electrodes were recorded for each participant in the two tasks in the experiment. An example of EMG graphical results obtained during the first task is given in Fig. 2. In this case the participant was exhausted after 40 s of using the handgrip. Figure 2a and b shows the raw and filtered signal, respectively. Figure 2c shows the values of the MNF during the task and Fig. 2d displays the evolution of the MDF. In these figures the lines depict the linear regression of the MNF and MDF, which indicate the fatigue of the participant. The negative slopes of Fig. 2c and d shows that in both cases, (MNF and MDF) the effect of physical fatigue is reflected from the beginning of the experiment to the end.

Finally, Fig. 2e gives the spectrogram showing how the spectrum of frequencies varies with time. It is expected that when the participant becomes tired the range of frequencies is reduced as well as the amplitude of the signal. In the spectrogram, it is clear that this happens in the last eight seconds where the intensity of the predominant frequencies decreases.

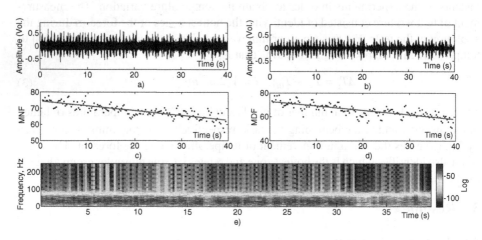

**Fig. 2.** EMG data collection recorded from the task (T1).

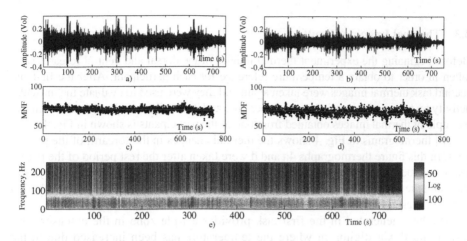

**Fig. 3.** EMG data collection obtained from one participant during the surgical task (T2).

Figure 3 shows the same results as Fig. 2, but in this case they correspond to the surgical task (T2). As is expected the slope of the regression lines is negative but with lower intensity than in the first task. That means that the physical fatigue of the participant is less because the surgical task requires less muscular effort to be completed. The comparison of Fig. 3c and d shows that the dispersion of the MNF values is lower than MDF, and this behavior is found in all participants. The spectrogram also shows that the variation of the frequencies involved in the task is lower. The RMS (%) values are measured as a percentage of the MVC in order to obtain relative values with respect to the maximum effort exerted by each participant. The values of MNF and MDF are shown using the slope of the regression lines. In this way the more fatigued a participant is, the more negative the index obtained. Indeed, in the last 100 s the amplitude of the frequencies involved is strongly diminished.

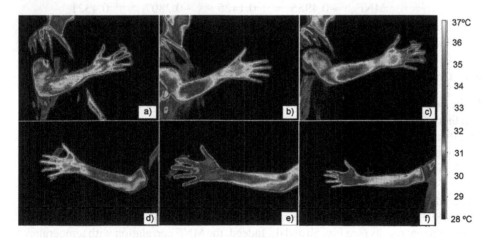

**Fig. 4.** Thermographs obtained at the beginning and end of the tasks.

## 3.3    Thermal Imaging Results

Before beginning the experiment thermal images of the dorsal and ventral forearm were taken in order to obtain a temperature reference for each participant. After the first and second task thermal images were taken again and they were used to evaluate the muscular activity by means of the variation in temperature and its distribution along the forearm. An example of thermal images obtained from one of the participants is shown in Fig. 4.

All thermograms in Fig. 4 shows the thermal changes in the forearm of the participant. In this figure thermographs 4a and d were taken after the rest period of the participant, just before starting the first task. Figure 4b and e were taken immediately after finishing the first task. Finally, Fig. 4c and d corresponds to the temperature distribution of the participant after the surgical task.

The heat generated in the first task provides a wide zone in the extensor digitorum and flexor digitorum where the temperature has been increased due to the muscular activity. The images in Fig. 4c and f clearly shows an increment of the temperature in the performance of the surgical task, but in this case it is less important than in the first task.

**Table 2.** Pearson's correlation and significance between EMG and fatigue index (FI).

| Parameters | | T1 | | T2 | |
|---|---|---|---|---|---|
| | | Correlation | $p$ - value | Correlation | $p$ - value |
| Dorsal | RMS | 0.9669 | 0.0407 | 0.7893 | 0.0066 |
| | MNF | −0.4986 | 0.1424 | −0.5000 | 0.1411 |
| | MDF | −0.8401 | 0.0023 | −0.7568 | 0.0113 |
| Ventral | RMS | 0.8006 | 0.0054 | 0.8844 | 0.0007 |
| | MNF | −0.4985 | 0.1425 | −0.2807 | 0.4321 |
| | MDF | −0.2313 | 0.5202 | −0.1500 | 0.6791 |

## 3.4    Correlation Between Thermal Parameters and EMG

In the next paragraphs the outcome of the first and second test are graphically shown together with the linear regression, Pearson's correlation and statistical significance, $p$. Significant relationship was considered when $p < 0.05$.

Figure 5 shows the results obtained in comparing EMG and thermal parameters for the dorsal part of the forearm and Table 2 shows the Pearson's correlation and significance. Figure 5a and c shows the relationship between temperature and the RMS (%) obtained from the EMG readings. The Pearson's correlations for these values are 0.9669 and 0.7893, respectively, which are significant ($p < 0.05$). However, MNF and MDF show different results (see Fig. 5b and d). Indeed, the MNF correlation with temperature is not significant whereas MDF shows a good correlation with a $p$ value of 0.0023.

Figure 6 shows the relationship of the EMG and temperature in the ventral part of the forearm. In this case, again the relationship of RMS values with temperature presents a good correlation which is significant. However, MNF and MDF show poor correlation and significance.

## 4    Discussion

This paper shows that the muscular effort in the forearm of surgeons performing laparoscopic surgery generates thermal effects on the skin. Thus, by means of infrared imaging, it is possible to monitor the changing temperature during the working hours. Furthermore, the statistical analysis showed that there is a significant correlation between temperature and muscular activity in the forearm.

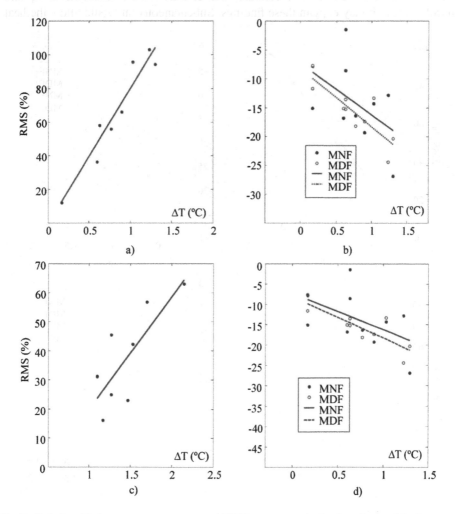

**Fig. 5.** Relationship between temperature and EMG parameters in the dorsal part of the forearm.

The thermographs show that the temperature is clearly increased in the muscular groups with more activity during the tests. However, the Pearson's correlation is only significant when temperature and RMS values obtained from EMG are compared. In fact, the results obtained do not permit a clear relationship to be established between temperature and muscle fatigue. In other words, the temperature is related with the instant effort but not with the cumulative effort.

In this study MNF and MDF have been used as the references in the measurement of muscle fatigue obtained from EMG. However, it is necessary to take into account that the accuracy of these indexes is not always reliable and many authors recognize the lack of precision under certain circumstances [12, 14]. More research is necessary in order to find a significant correlation of the thermal readings and muscular fatigue.

The dispersion of the results obtained is important in all cases and it may be due to several factors. Differences in vasculature, skin characteristics, subcutaneous fat and muscular strength may explain these findings. Subcutaneous fat would allow the heat

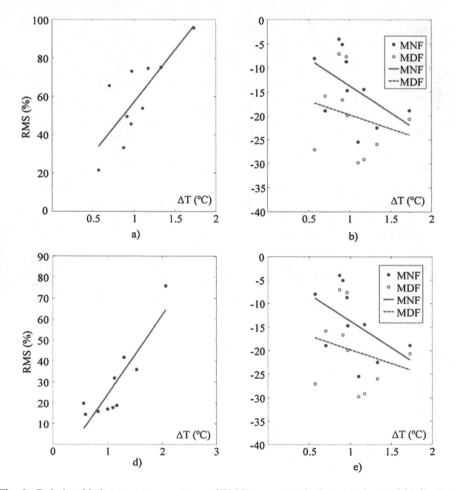

**Fig. 6.** Relationship between temperature and EMG parameters in the ventral part of the forearm.

transmission to be reduced because fat acts as a thermal insulator in the human body. The results obtained in this study were compared with the skinfold values obtained from each participant. However, these results were not significant because the sample size was not large enough. Further research is necessary about how subcutaneous fat influences the increment of temperature due to the muscular effort.

Qualitative information obtained from thermal imaging can be useful for the identification of muscular groups taking part in the surgical work. However, it is clear that further research is necessary to explore the possibilities of using thermal imaging as a quantitative technique for the assessment of muscular activity. For example, it would be interesting to use spectral analysis in the data obtained from the thermal outcomes. This information could be correlated with the spectral analysis obtained from EMG data.

# References

1. Berguer, R., Forkey, D.L., Smith, W.D.: Ergonomic problems associated with laparoscopic surgery. Surg. Endosc. **13**, 466–468 (1999)
2. Fan, Y., Kong, G., Meng, Y., Tan, S., Wei, K., Zang, Q., Jin, J.: Comparative assessment of surgeons' task performance and surgical ergonomics associated with conventional and modified flank positions: a simulation study. Surg. Endosc. **28**, 3249–3256 (2014)
3. Manasnayakorn, S., Cuschieri, A., Hanna, G.B.: Ergonomic assessment of optimum operating table height for hand-assisted laparoscopic surgery. Surg. Endosc. **23**, 783–789 (2009)
4. McDonald, M.E., Ramirez, P.T., Munsell, M.F., Greer, M., Burke, W.M., Naumann, W.T., Frumovitz, M.: Physician pain and discomfort during minimally invasive gynecologic cancer surgery. Gynecol. Oncol. **134**, 243–247 (2014)
5. Sancibrian, R., Gutierrez-Diez, M.C., Torre-Ferrero, C., Benito-Gonzalez, M.A., Redondo-Figuero, C., Manuel-Palazuelos, J.C.: Design and evaluation of a new ergonomic handle for instruments in minimally invasive surgery. J. Surg. Res. **188**, 88–89 (2014)
6. Szeto, G.P.Y., Ho, P., Ting, A.C.W., Poon, J.T.C., Tsang, R.C.T., Cheng, S.W.: A study of surgeons' postural muscle activity during open, laparoscopic, and endovascular surgery. Surg. Endosc. **24**, 1712–1721 (2010)
7. Szeto, G.P.Y., Cheng, S.W.K., Poon, J.T.C., Ting, A.C.W., Tsang, R.C.C., Ho, P.: Surgeons' static posture and movement repetitions in open and laparoscopic surgery. J. Surg. Res. **172**, E19–E21 (2012)
8. Tang, B., Hou, S., Cuschieri, A.: Ergonomics of and technologies for single-port laparoscopic surgery. Minim. Invasiv. Ther. **21**, 46–54 (2012)
9. Ridtitid, W., Cote, G.A., Leung, W., Buschbacher, R., Lynch, S., Fogel, E.L., Watkins, J.L., Lehman, G.A., Sherman, S., McHenry, L.: Prevalence and risk factors for musculoskeletal injuries related to endoscopy. Gastrointest. Endosc. **81**, 294–302 (2015)
10. Quick, N.E., Gillette, J.C., Shapiro, R., Adrales, G.L., Gerlach, D., Park, A.E.: The effect of using laparoscopic instruments on muscle activation patterns during minimally invasive surgical training procedures. Surg. Endosc. **17**, 462–465 (2003)
11. Roman-Liu, D., Bartuzi, P.: The influence of wrist posture on the time and frequency EMG signal measures of forearm muscles. Gait Posture **37**, 340–344 (2013)
12. Cifrek, M., Medved, V., Tonkovic, S., Ostojic, S.: Surface EMG based muscle fatigue evaluation in biomechanics. Clin. Biomech. **24**, 327–340 (2009)
13. Elfving, B., Nemth, G., Arvidsson, I., Lamontagne, M.: Reliability of EMG spectral parameters in repeated measurements of back muscle fatigue. J. Electromyogr. Kines. **9**, 235–243 (1999)

14. Frahm, K.S., Jensen, M.B., Farina, D., Andersen, O.K.: Surface EMG crosstalk during phasic involuntary muscle activation in the nociceptive withdrawal reflex. Muscle Nerve **46**, 228–236 (2012)

15. Takala, E., Toivonen, R.: Placement of forearm surface EMG electrodes in the assessment of hand loading in manual task. Ergonomics **56**, 1159–1166 (2013)

16. Bartuzi, P., Roman-Liu, D., Wisniewski, T.: The influence of fatigue on muscle temperature. Int. J. Occup. Saf. Ergon. **18**, 233–243 (2012)

17. Bertmaring, I., Babski-Reeves, K., Nussbaum, M.A.: Infrared imaging of the anterior deltoid during overhead static exertions. Ergonomics **51**, 1606–1619 (2008)

18. Brioschi, M.L., Okimoto, M.L.L.R., Vargas, J.V.C.: The utilization of infrared imaging for occupational disease study in industrial work. Work-J. Prev. Assess. Rehabil. **41**, 503–509 (2012)

19. Govindu, N.K., Babski-Reeves, K.: Reliability of thermal readings of the skin surface over the anterior deltoid during intermittent, overhead tapping tasks. Int. J. Ind. Ergonom. **42**, 136–142 (2012)

20. Merla, A., Mattei, P.A., Donato, L., Romani, G.L.: Thermal imaging of cutaneous temperature modifications in runners during graded exercise. Ann. Biomed. Eng. **38**, 158–163 (2010)

21. Sormunen, E., Rissanen, S., Oksa, J., Pienimaki, T., Remes, J., Rintamäki, H.: Muscular activity and thermal responses in men and women during repetitive work in cold environments. Ergonomics **52**, 964–976 (2009)

22. Rodriguez-Falces, J., Izquierdo, M., Gonzalez-Izal, M., Place, N.: Comparison of the power spectral changes of the voluntary surface electromyogram and M wave during intermittent maximal voluntary contractions. Eur. J. Appl. Physiol. **114**, 1943–1954 (2014)

23. Bonato, P., Gagliati, G., Knaflitz, M.: Analysis of myoelectric signals recorded during dynamic contractions. IEEE Eng. Med. Biol. Mag. **15**, 102–111 (1996)

24. Bonato, P., Roy, S.H., Knaflitz, M., De Luca, C.J.: Time-frequency parameters of the surface myoelectric signal for assessing muscle fatigue during cyclic dynamic contractions. IEEE Trans. Biomed. Eng. **48**, 745–753 (2001)

# FCM-Based Method for MRI Segmentation of Anatomical Structure

Pinti Antonio[1,2(✉)]

[1] University of Lille Nord de France, UVHC, 59313 Valenciennes, France
antonio.pinti@univ-valenciennes.fr
[2] EA 4708, I3MTO, CHRO, 1 rue Porte Madeleine, 45032 Orléans, France

**Abstract.** Fuzzy C-means (FCM) has been widely applied to segmentation of medical images, especially MRI images for identifying living organs and supporting medical diagnosis. However, in practice, this method is too sensitive to image noises. Then, many methods have been proposed to improve the objective function of FCM by adding a penalty term to it. One drawback of these methods is that they can determine neither the appropriate size of observation window for each pixel of interest for incorporating spatial information, nor the suitable importance coefficient of the penalty term. Moreover, the modification of the objective function of FCM often causes additional complex derivations. In this paper, we develop a new FCM-based method for medical MRI image segmentation. This method permits to dynamically determine the optimal size of observation window for each pixel of interest without adding any penalty term. Moreover, a n-dimensional feature vector including both local and global spatial information between neighboring pixels is generated to describe each pixel in the objective function. And specialized a priori knowledge is integrated into the segmentation procedure in order to control the application of FCM for tissue classification of thigh. The effectiveness and the robustness of the proposed method have been validated by real MRI image of thigh.

**Keywords:** Magnetic Resonance Imaging (MRI) · Fuzzy C-means (FCM) · Image segmentation · Spatial constraint · Expert knowledge · Medical imaging

## 1 Introduction

Magnetic resonance imaging (MRI) allows us to explore the living organs in a non-invasive way. And it has advantages over computerized tomography (CT) and ultra-sound image because it does not involve ionizing radiation [1]. The analysis of MRI images can be used in the quantification of human body composition [2], such as the quantification of muscle/fat ratio [3], and the assessment for the variation of body fat content. All these works have important medical significance in human nutrition and muscle physiology, in the study of pathologic consequences of obesity, and in the study of diseases of muscle [4]. In the analysis of MRI images, image segmentation techniques play a key role [5]. It aims at partitioning an image into a number of non-overlapping, constituent regions which are homogeneous with respect to some characteristics such as grey level or texture [6].

© Springer International Publishing Switzerland 2016
F. Ortuño and I. Rojas (Eds.): IWBBIO 2016, LNBI 9656, pp. 175–183, 2016.
DOI: 10.1007/978-3-319-31744-1_16

In this paper, we develop a new FCM-based image segmentation method to solve the previous drawbacks, it includes three novelties:

- We propose an algorithm to define the window size in a dynamical way for extracting appropriate global spatial information from images.
- Moreover, we use the standard FCM objective function without adding any penalty term. This can effectively reduce the difficulties related to the selection of coefficient of penalty term and complex derivations.
- Also, we use a n-dimensional feature vector to describe each pixel. This vector includes grey level and both local and global spatial information on neighboring pixels.

In the other hand, as the FCM clustering algorithm will be recurrently applied to medical image segmentation and the feature vector does not include any specialized information on medical images, a priori knowledge concerning the tissue geometric structure [7] of thigh is used to control the application of the FCM algorithm during the segmentation procedure for tissue classification of thigh. This treatment can provide additional information to the segmentation procedure so that the results can be more accurate.

The organization of this paper is as follows. Section 2 presents the standard FCM method and describes the proposed FCM-based method for image segmentation. Section 3 shows the experimental results of the proposed segmentation method applied to a real MRI medical image. The experimental results of tissue classification and the analysis of related results are given in Sect. 4. A conclusion is given in Sect. 5.

## 2   Image Segmentation

### 2.1   Classical FCM Method

Let $X = (x_1, x_2, \ldots x_n)$ represent one image with n pixels which will be segmented into c clusters or classes. $x_i$ denotes one pixel characterized by a multidimensional feature vector. The FCM method is an iterative optimization which minimizes the objective function defined below:

$$J = \sum_{j=1}^{n} \sum_{i=1}^{c} u_{ij}^{m} \left\| x_j - v_i \right\|^2 \tag{1}$$

$u_{ij}$ represents the adhesion of the pixel $x_j$ to the $i$th cluster. $v_i$ is the center of the $i$th cluster. $\|*\|$ is the Euclidean distance. m is a constant controlling the fuzzy degree in the clustering process, and m = 2 is usually used to facilitate the calculation. $\{u_{ij}\} = U$ represents one clustering matrix which satisfies:

$$U \in \left\{ u_{ik} \in [0, 1] \mid \sum_{i=1}^{c} u_{ik} = 1, \forall k, and, 0 < \sum_{k=1}^{n} u_{ik} < n, \forall i \right\} \tag{2}$$

The function (1) is minimized when high adhesion degrees are assigned to pixels close to related cluster centers, and low adhesion degrees to pixels far from related cluster centers.

The adhesion function represents the level of belonging of one pixel to a specific cluster. In FCM method, it only depends on the distance between one pixel and each cluster center. The adhesion function and the cluster center are updated using (3) and (4):

$$u_{ij} = \frac{1}{\sum_{k=1}^{c} \left( \frac{\|x_j - v_i\|}{\|x_j - v_k\|} \right)^{\frac{2}{m-1}}} \tag{3}$$

$$v_i = \frac{\sum_{j=1}^{n} u_{ij}^m x_j}{\sum_{j=1}^{n} u_{ij}^m} \tag{4}$$

When $\|U^{r+1} - U^r\| < \varepsilon$, stop the iteration.

The fuzzy c-means algorithm can be summarized in the following steps:

1. Fix c, m, $\varepsilon$, and initialize memberships $u_{ij}^0$
2. Update cluster centers $v_i^t$ using Eq. (4).
3. Update membership degrees $u_{ij}^t$ using Eq. (3).
4. Repeat Steps 2 and 3 until the following termination criterion is satisfied:

$$\left\| u_{ij}^{t+1} - u_{ij}^t \right\| < \varepsilon \tag{5}$$

(Note: In the definition of classical FCM, shown in Eq. (1), $x_j$ and $v_i$ are n-dimension vectors. However, in the application of FCM for image segmentation [1–3, 9, 10], $x_j$ and $v_i$ are always one dimension. And the grey levels of pixels in images are chosen as $x_j$.)

## 2.2 Our New FCM-Based Method for Image Segmentation

In this part, we firstly introduce our algorithm to define the window size in a dynamical way for extracting appropriate global spatial information from images for each pixel. Then we explain how to generate the n-dimensional feature vector to describe each pixel. This vector includes grey level and both local and global spatial information on neighboring pixels. Then, we give the final algorithm.

### 2.2.1 Algorithm for Dynamically Selecting the Size of Observation Window for Each Pixel

For each pixel of the image, we define an observation window centered on it and then calculate the mean value of all Euclidean distances of grey levels between the central pixel and its neighboring pixels inside this window. On the image, we move this window from left to right and from top to bottom in order to extract spatial information of objects. The size of this window can be $3 \times 3, 5 \times 5, 7 \times 7$. It is formally expressed by:

$$M = \frac{\sum_{i=1}^{N} (\|G_{(0)} - G_{(i)}\|)}{N} \tag{6}$$

where M is the mean value of the Euclidean distances, N the number of neighboring pixels inside the window, $G_{(0)}$ the grey level of the central pixel, $G_{(i)}$ the grey level of $i$th neighbor of the central pixel (i=1,2,...,N).

By calculating the mean values of the Euclidean distances for different sized windows, we select the window corresponding to the smallest averaged Euclidean distance. The uniformity of grey levels in this window is the highest. Obviously, when moving the window on the image, the size of window can vary with the distribution of grey levels centered on the central pixel. In this way, each object, generally considered as a group of pixels having similar grey levels, can be extracted.

In practice, only the $3 \times 3$ window, $5 \times 5$ window, and $7 \times 7$ window are considered. The corresponding values of N are 8, 24 and 48.

### 2.2.2    Generation of Features for Each Pixel

For describing each pixel X, we generate the following feature vector:

$$X = \left(G_{(0)}, G_{(1)}, \dots .G_{(8)}, W_{(1)}, W_{(2)}, \dots .W_{(c)}, M, P_{(1)}, P_{(2)}, \dots .P_{(c)}\right) \tag{7}$$

where $G_{(0)}$ is the grey level of the central pixel, $G_{(1)}, \dots ..G_{(8)}$ the grey levels of its eight neighbors in a $3 \times 3$ window. They represent the local distribution of grey levels around the central pixel.

$$W_{(k)} = \frac{\sum_{i=1}^{8} (1 - u_{ik})}{8} \tag{8}$$

$u_{ik}(i = 1, 2, \dots .8)$ is the fuzzy adhesion that $i$th neighbor of central pixel belongs to class k (k=1,...,c). So $W_{(k)}$ is the average of the fuzzy adhesions that the eight neighbors of the central pixel do not belong to class k. It represents the local similarity of the 8 neighboring pixels related to one specific class k.

M is the smallest mean value of Euclidean distances defined in Sect. 2.2.

$P_{(k)}$ is the average of the fuzzy adhesions that the neighbors of the central pixel inside the adaptive window do not belong to class k. It represents the global similarity of all the neighboring pixels in the window related to class k. We calculate $P_{(k)}$ according to the same principle as $W_{(k)}$, N = 8, 24, or 48.

$$P_{(k)} = \frac{\sum_{i=1}^{N} (1 - u_{ik})}{N} \tag{9}$$

$(G_{(1)}, \ldots ..G_{(8)}, W_{(1)}, \ldots ..W_{(c)})$ represents the local spatial information in the nearest neighboring area of the central pixel. This information is more detailed than any other observation windows. $(M, P_{(1)}, \ldots ..P_{(c)})$ represents the spatial information in a larger and more adaptive neighboring area of the central pixel. Thus, the feature vector includes not only local but also general spatial information around the central pixel. The spatial distribution of grey levels of objects on the image can be effectively taken into account in this feature vector.

### 2.2.3    Our New FCM-Based Algorithm

The proposed fuzzy-based algorithm can be summarized in the following steps:

1.  Fix c, m, $\varepsilon$, and initialize memberships $u_{ij}^0$
2.  For each pixel, calculate the **n-dimensional** feature vector using Eq. (7).
3.  Update cluster centers $v_i^t$ using Eq. (4).
4.  Update membership degrees $u_{ij}^t$ using Eq. (3).
5.  Repeat Step 3 and Step 4 until the following termination criterion is satisfied:

$$\left\| u_{ij}^{t+1} - u_{ij}^t \right\| < \varepsilon \tag{10}$$

## 3    Experimental Results for Classifying Tissues of Thigh

In this section, we show the experimental results of the knowledge guided segmentation for classifying tissues of thigh [16]. We set the parameters as follows: $m = 2$, $\varepsilon = 0.001$.

To compare the performance of dynamically selected window, $3 \times 3$ window, $5 \times 5$ window, and $7 \times 7$ window, we test them with 52 images of thigh. Figure 1 present the comparison of classification results by the four methods and the reference classification result which is achieved manually by a doctor in Hospital of Lille (France). For each result image, grey levels are set in the following way:

Background: 20; Adipose tissue: 100; Spongy bone: 150; Cortical bone: 200; Muscle: 250;

Original images are shown in Fig. 1(a), corresponding reference classification are shown in Fig. 1(b) and the classification results by the four methods are shown in Fig. 1(c–f). Table 1 gives quantitative comparison corresponding to Fig. 1(a) using four methods for adipose tissue, cortical bone, muscle and spongy bone, where the quantitative comparison is calculated by:

$$s = \frac{A_{eg} \cap A_{fg}}{A_{eg} \cup A_{fg}} \tag{11}$$

where $A_{eg}$ represents the set of pixels belonging to the $g$th class obtained by the $e$th method, and $A_{fg}$ the set of pixels belonging to $g$th class obtained by $f$th method. We see

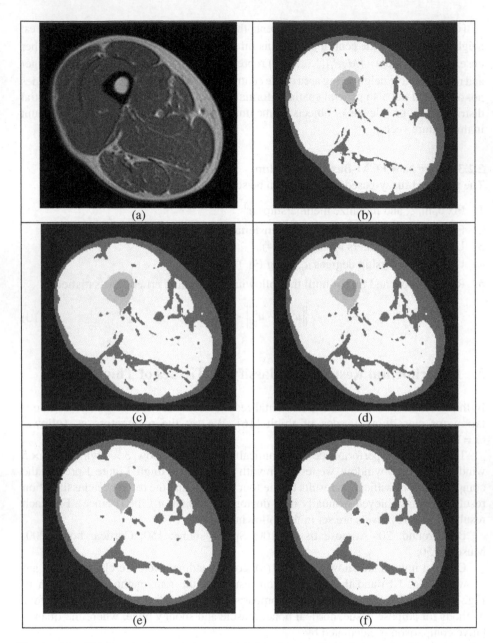

**Fig. 1.** Classification results on one image. (a) Original image. (b) Reference classification result. (c) Result of dynamically selected window. (d) Result of $3 \times 3$ window. (e) Result of $5 \times 5$ window. (f) Result of $7 \times 7$ window

that the method of dynamically selected window size shows the best classification performance of all the methods.

**Table 1.** Quantitative comparison corresponding to Fig. 1(a) using four methods for adipose tissue, cortical bone, muscle and spongy bone.

|        | Adipose tissue | Cortical bone | Muscle | Spongy bone |
| ------ | -------------- | ------------- | ------ | ----------- |
| **DS** | 0.8100         | 0.7520        | 0.9133 | 0.8032      |
| 3 × 3  | 0.8078         | 0.7509        | 0.9130 | 0.8030      |
| 5 × 5  | 0.7392         | 0.6573        | 0.8837 | 0.7335      |
| 7 × 7  | 0.6669         | 0.5599        | 0.8468 | 0.6848      |

DS - Dynamically Selected

In order to show the effects of window size, we generate a synthetic image from the original image. In this new image, we assign 30 to pixels in which 3 × 3 windows are selected using the proposed method. In the same way, we assign 130 and 230 to pixels of 5 × 5 windows and 7 × 7 windows respectively.

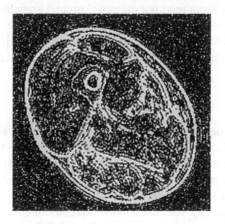

**Fig. 2.** Synthetic image showing the range of window size.

We can see in Fig. 2 that in the homogeneous area, such as background and muscle, 3 × 3 windows are always selected. In the inhomogeneous area, such as the border of two tissues, 7 × 7 windows are always selected. In fact, the selection of window size is strongly related to local continuity on grey level.

## 4   Discussion

The first novelty of our method is that we select the window size in a dynamical way for extracting wide spatial information. It means that we can get the best segmentation result in one time. However, in other methods [6, 8, 9], they always usc a 3 × 3 window. They don't give any explication why they select this window size and they don't give any proof that this window size can give best segmentation result among 3 × 3 window, 5 × 5 window, 7 × 7 window.

The second novelty of our method is that we don't add any penalty terms to standard FCM. However, in other FCM-based methods [5, 6, 8, 9, 11–14], a penalty term is always added. And there is always a coefficient for this penalty term. The disadvantage of all these methods is that they can't determine the best value for the coefficient in a dynamical way. Thus, they have to calculate several times using different values for the coefficient then select the value which gives best result.

The third novelty of our method is that we use a n-dimension vector to describe each pixel in the image. Nevertheless, in other FCM-based methods, each pixel in the image is described by a one-dimension vector, the grey level of this pixel.

In addition, we incorporate a priori knowledge to guide the segmentation so that we can get the tissue classification. By our algorithm, we use four steps to separate the five classes: background, spongy bone, cortical bone, adipose tissue and muscle. The result is satisfactory. Compared to our result, muscle and adipose tissue can't be separated in [15], cortical bone and spongy bone can't be separated in [4], and adipose tissue and spongy bone can't be separated in [2]. The reason why they fail to classify all the four tissues (muscle, cortical bone, spongy bone, and adipose tissue) is that they just use the segmentation method without incorporating background knowledge.

Another advantage of our method is that it's very simple. In other FCM-based methods, complex derivations in the modification to objective function are often involved. However, in our method, we use the original FCM method only.

## 5   Conclusion

In this paper, we firstly develop a FCM-based method for image segmentation. It shows good results on a real image [16]. However, it fails to classify the five classes (background, spongy bone, cortical bone, adipose tissue, and muscle) in MRI image of thigh. Thus, we incorporate a priori knowledge to guide the segmentation so that we get the classification of each class in the end.

The ideas of dynamically selected window size, n-dimension vector to describe a pixel, and without adding penalty term to standard FCM can be used in other FCM-based method to reduce the algorithm complexity and running time, as well as to avoid selecting the best coefficient for the penalty term.

In the future, by the obtained classification results, we will work on the quantification of the muscle/fat ratio, assessment of the muscle/fat temporal variation, and measurement of the volume of muscle, fat and bone in human legs. We will also consider selecting other features to describe each pixel in the image. In addition, we will analyze the performance when we use kernel-based distance [17, 18] to replace the Euclidean distance in standard FCM method. Another work may be the development of an automatic and intelligent system for tissue classification.

# References

1. Seidell, J.C., Bakker, C.J.C., Van der Kooy, K.: Imaging techniques for measuring adipose-tissue distribution-a comparison between computed tomography and 1.5-T magnetic resonance. Am. J. Clin. Nutr. **51**, 953–957 (1990)
2. Schaefer, G., Hassanien, A., Jiang, J.: Computational Intelligence in Medical Imaging. CRC Press, Taylor & Francis group, London (2009)
3. Imamoglu, N., Gomez-Tames, J., Siyu, H., Dong-Yun, G., Kita, K., Wenwei, Y.: Unsupervised muscle region extraction by fuzzy decision based saliency feature integration on thigh MRI for 3D modeling. In: IEEE International Conference on Machine Vision Applications, IAPR, Japan, pp. 150–153 (2015)
4. Barra, V., Boire, J.V.: Segmentation of fat and muscle from MR images of the thigh by a possibilistic clustering algorithm. Comput. Methods Programs Biomed. **68**, 185–193 (2002)
5. Liew, A.W., Yan, H.: An adaptive spatial fuzzy clustering algorithm for 3-D MR image segmentation. IEEE Trans. Med. Imaging **22**(9), 1063–1075 (2003)
6. Zhang, D., Chen, S.: A novel kernelized fuzzy C-means algorithm with application in medical image segmentation. Artif. Intell. Med. **32**, 37–50 (2004)
7. Adhikaria, S.K., Sing, J.K., Basu, D.K., Nasipuri, M.: Conditional spatial fuzzy C-means clustering algorithm for segmentation of MRI images. Appl. Soft Comput. **34**, 758–769 (2015)
8. Chen, S., Zhang, D.: Robust image segmentation using FCM with spatial constraints based on new kernel-induced distance measure. IEEE Trans. Syst. Man Cybern. B Cybern. **34**(4), 1907–1916 (2004)
9. Ahmed, M.N., Yamany, S.M., Mohamed, N., Farag, A.A., Moriarty, T.: A modified fuzzy C-means algorithm for bias field estimation and segmentation of MRI data. IEEE Trans. Med. Imaging **21**(3), 193–199 (2002)
10. Pham, D.L., Prince, J.L.: An adaptive fuzzy C-means algorithm for image segmentation in the presence of intensity inhomogeneity. Pattern Recogn. Lett. **20**, 57–68 (1999)
11. Chen, J.H., Chen, C.S.: Fuzzy kernel perceptron. IEEE Trans. Neural Netw. **13**, 1364–1373 (2002)
12. Ahmed, M.N., Yamany, S.M., Mohamed, N., Farag, A.A., Moriarty, T.: A modified fuzzy C-means algorithm for bias field estimation and segmentation of MRI data. IEEE Trans. Med. Imaging **21**, 193–199 (2002)
13. Tolias, Y.A., Panas, S.M.: Image segmentation by a fuzzy clustering algorithm using adaptive spatially constrained membership functions. IEEE Trans. Syst. Man Cybern. A **28**, 359–369 (1998)
14. Tolias, Y.A., Panas, S.M.: On applying spatial constraints in fuzzy image clustering using a fuzzy rule-based system. IEEE Sig. Process. Lett. **5**, 245–247 (1998)
15. Douglas, T.S., Solomonidis, S.E., Lee, V.S.P., Spence, W.D., Sandham, W.A., Hadley, D.M.: Automatic segmentation of magnetic resonance images of the trans-femoral residual limb. Med. Eng. Phys. **20**, 756–763 (1998)
16. Kang, H.: Contribution to automatic corporal tissue classification by integrating qualitative medical knowledge: application to the analysis of musculo skeletal diseases and disabilities from MRI sequences. Ph.D thesis, Universite de Valenciennes et du Hainaut-Cambresis, France (2009)
17. Muller, K.R., Mika, S., Ratsch, G., Tsuda, K., Scholkopf, B.: An introduction to kernel-based learning algorithms. IEEE Trans. Neural Netw. **12**(2), 181–201 (2001)
18. Girolami, M.: Mercer kernel-based clustering in feature space. IEEE Trans. Neural Netw. **13**, 780–784 (2002)

# An Automated Tensorial Classification Procedure for Left Ventricular Hypertrophic Cardiomyopathy

Santiago Sanz-Estébanez[1(✉)], Javier Royuela-del-Val[1],
Susana Merino-Caviedes[1], Ana Revilla-Orodea[2], Teresa Sevilla[2],
Lucilio Cordero-Grande[3], Marcos Martín-Fernández[1],
and Carlos Alberola-López[1]

[1] Laboratorio de Procesado de Imagen, Universidad de Valladolid, Valladolid, Spain
{ssanest,jroyval,smercav}@lpi.tel.uva.es, {marcma,caralb}@tel.uva.es
[2] I. de Ciencias Del Corazón, Hospital Clínico Universitario, Valladolid, Spain
[3] Department Biomedical Engineering, King's College, London, UK
lucilio.cordero_grande@kcl.ac.uk

**Abstract.** Cardiovascular diseases are the leading cause of death globally. Therefore, classification tools play a major role in prevention and treatment of these diseases. Statistical learning theory applied to magnetic resonance imaging has led to the diagnosis of a variety of cardiomyopathies states. We propose a two-stage classification scheme capable of distinguishing between heterogeneous groups of hypertrophic cardiomyopathies and healthy patients. A multimodal processing pipeline is employed to estimate robust tensorial descriptors of myocardial mechanical properties for both short-axis and long-axis magnetic resonance tagged images using the least absolute deviation method. A homomorphic filtering procedure is used to align the cine segmentations to the tagged sequence and provides 3D tensor information in meaningful areas. Results have shown that the proposed pipeline provides tensorial measurements on which classifiers for the study of hypertrophic cardiomyopathies can be built with acceptable performance even for reduced samples sets.

**Keywords:** Magnetic resonance tagging · Fuzzy clustering · Support vector machines · Homomorphic filtering · Harmonic phase · Hypertrophic cardiomyopathy · Least absolute deviation

## 1 Introduction

Classifications of heart muscle diseases have proved to be exceedingly complex and in many respects contradictory. Cardiomyopathies are an important and complex group of heart muscle diseases with multiple etiologies and heterogeneous phenotypic expression. Therefore, systematic classifications have traditionally been viewed as useful exercises that promote greater understanding of these diseases [1].

© Springer International Publishing Switzerland 2016
F. Ortuño and I. Rojas (Eds.): IWBBIO 2016, LNBI 9656, pp. 184–195, 2016.
DOI: 10.1007/978-3-319-31744-1_17

Hypertrophic cardiomyopathy (HCM) is very common and can affect people of any age. About one out of every 500 people has HCM. Men and women are equally affected [2]. HCM is a common cause of sudden cardiac arrest in young people, including young athletes. Etiological factors are of great importance in cardiovascular disease detection [3]; specifically, genetic studies have been conducted in order to classify cardiomyopathies and to assess patients predisposition to suffer HCM [4,5].

Imaging techniques provide essential descriptors for the study and classification of cardiomyopathies and, from them, cardiac magnetic resonance (MR) is increasingly becoming the standard technique as it provides information to assess the myocardial morphology, function and structure. Its use is especially relevant for quantitative analysis of myocardial motion, the anomalies of which are directly related with impaired cardiac function. From the set of MR acquisition techniques, MR-Tagging has become the reference modality for evaluating strain evolution in the human heart. This modality is based on the generation of a set of saturated magnetization planes on the imaged volume which may be subsequently tracked throughout the cardiac cycle [6], from which the cardiac function can be assessed. Harmonic Phase (HARP) based methods [7] are capable of reconstructing deformation fields accurately grounded on the assumption of constant local phase, which turns out to be more reliable than a constant pixel brightness assumption.

Global image-derived measures have been reported, such as the global longitudinal strain [8], which turns out to be an interesting tool that correlates with the global presence of fibrosis. A relation between extent of cardiac muscle cell disorganization and left ventricular wall thickness has also been established [9]; however, imaging studies focused on the characterization and classification of the nature of HCM are relatively scarce.

In order to provide greater understanding about these factors, comparative regional studies have been carried out in [10] for athletes, controls and HCM patients; the authors reveal a significant reduction in the diagonal components of the strain in HCM patients and athletes, but this reduction was not associated to any particular segment and it was even present in non-fibrotic regions. In [11] a local analysis is performed fusing the information of MR-Cine and Late Enhancement MR to provide more insight into the mechanical properties of the fibrotic tissue in HCM. Automated classifiers have been developed in [12] and [13] using global biomarkers derived from MR-Cine and electrocardiogram, respectively. Although the authors report noticeable prognostic values for the identification of different cardiovascular diseases, only global measures were used; however, local measurements may be of additional utility for the characterization of the fibrosis that accompanies primary/secondary HCM.

In this paper we propose an automated processing pipeline to classify heterogeneous groups of HCM from myocardial functional descriptors obtained out of the deformation gradient tensor estimated from MR-Tagging images by means of a robust reconstruction method. We have applied different machine learning methods (supervised and unsupervised) using a sequential methodology that adapts to the characteristics of the subjects at every stage.

## 2    Materials and Methods

### 2.1    Materials

For the validation of the proposed approach on real data, we have performed cardiac studies in a population of 47 subjects, 23 of which are affected by primary HCM, 10 are affected by secondary HCM and the control group consists of 14 healthy volunteers.

A short axis (SA) MR-Tagging dataset, providing a coverage from apex to base, is acquired for each patient using a MR Complementary SPAtial Modulation of Magnetization (CSPAMM) SENSitivity Encoding (SENSE) Turbo Field Echo sequence on a Philips Achieva 3 T scanner. Additionally, we have also acquired a SENSE balanced Turbo Field Echo SA MR-Cine sequence at the same spatial location for each patient, where the myocardium has been manually segmented at the end diastole (ED) phase. Segmentations are also used to define a region of interest on which to compute meaningful measures of the strain. Long axis (LA) MR-Tagging datasets and the corresponding MR-Cine sequence have also been acquired following the aforementioned acquisition protocol in each case. Additional details on these sequences are included in Table 1.

**Table 1.** Details on the sequences of MR images used in the paper. $\Delta_p$ : Spatial Resolution (mm). $\Delta_l$ : Slice Thickness (mm). $N_p$: Number of pixels along each direction. $N_t$: Number of Temporal Phases. $N_s$: Number of slices. $T_R$ : Repetition Time (ms). $T_E$: Echo Time (ms). $\alpha$: Flip Angle ($°$).

|  | $\Delta_p$ | $\Delta_l$ | $N_p$ | $N_t$ | $N_s$ | $T_R$ | $T_E$ | $\alpha$ |
|---|---|---|---|---|---|---|---|---|
| MR-Tagging SA | 1.21–1.32 | 10 | 256–432 | 16–25 | 10–15 | 2.798–6.154 | 1.046–3.575 | 7–25 |
| MR-Cine SA | 0.96–1.18 | 8–10 | 240–320 | 30 | 10–15 | 2.902–3.918 | 1.454–2.222 | 45 |
| MR-Tagging LA | 1.21–1.34 | 10 | 240–340 | 15–27 | 1–3 | 2.903–4.507 | 1.097–2.897 | 10–45 |
| MR-Cine LA | 0.98–1.25 | 8–10 | 256–448 | 30 | 1–3 | 2.858–3.529 | 1.251–2.132 | 45 |

### 2.2    Methods

The processing pipeline is divided in the following steps:

**Alignment.** An alignment stage is performed with the purpose of mapping the MR-Cine segmentations provided by the cardiologist onto the MR-Tagging sequence. First, the temporal correspondence of the MR-Tagging images with the MR-Cine sequence is established by means of the DICOM timestamps; then, an affine registration method is performed to align MR-Tagging and MR-Cine images at the correct time instant. To that end, the MR-Tagging sequence is detagged following an homomorphic filtering procedure [14] to improve registration performance. Similar detagging procedures have been reported in [15] for global measures estimation. The proposed dettaging method is described below.

Let us assume a simple case in which an image $I(\mathbf{x})$ consists of an anatomical image $I_0(\mathbf{x})$(low-pass) multiplied by a tag pattern $f(\mathbf{x}; \mathbf{g})$, where the gradient directions are given by $\mathbf{g} = [g_1, g_2]$. Our purpose is to estimate $I_0(\mathbf{x})$ from the final image $I(\mathbf{x})$. Our method is based on the assumption that the anatomical image shows a low variability, i.e., it can be considered a low pass signal. This principle is not always fulfilled as boundaries present sharp intensity changes. Consequently, we calculate the local mean of the image in order to alleviate the power smearing caused by the abrupt myocardium background transition and next, we separate anatomical and tag signals by applying the logarithm:

$$\log(I_n(\mathbf{x})) = \log(\overline{I}(\mathbf{x})) \approx \log I_0(\mathbf{x}) + \log f(\mathbf{x}; \mathbf{g}). \tag{1}$$

where $\overline{I}(\mathbf{x})$ denotes the local mean. The tag pattern term $\log f(\mathbf{x}; \mathbf{g})$ has its energy localized at specific frequencies, while the term $\log I_0(\mathbf{x})$ is a low frequency signal. We can suppress the residual influence of the tag pattern using a notch filter on $\log(I_n(\mathbf{x}))$. The position of the spectral peaks can be easily estimated from the information obtained from the DICOM headers, so we have resorted to an isotropic Gaussian filter with radius $r$ for every spectral peak detected. The filter bandwidth is normalized with respect to the wave number $k$ of the applied modulation by using the parameter $\mu = r/k = 0.3$ according to [16].

The whole filtering pipeline is depicted in Fig. 1:

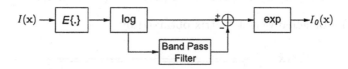

**Fig. 1.** Pipeline of homomorphic filtering procedure.

**Estimation of the Material Deformation Gradient Tensor.** The estimation technique is based on the extraction of the local phase of the grid pattern according to the method presented in [16]. A windowed Fourier Transform (WFT) is applied to the image at the end-systole (ES) phase. The WFT provides a representation of the image spectrum in the surroundings of each pixel of the original image, so HARP Band Pass filtering techniques can be directly applied on the spatially localized spectrum of the image. The complex image can be reconstructed in the spatial domain by using an inverse discrete WFT (IWFT) prior to extract the phase.

As stated in [7], 3D HARP motion reconstruction using the SPAMM technique requires a minimum of 3 linearly independent wave vectors. In this paper we extend the aforementioned HARP methodology by allowing the application of a set of 4 wave vector (performing this estimation technique on CSPAMM MR-Tagging LA and SA images), so 3D deformations can be robustly recovered, at least at the intersection points of both axes, applying the methodology presented in [17].

We can arrange the set of the four given wave vectors $\mathbf{k}_i^T$ in matrix form by:

$$\mathbf{K} = \left[\mathbf{k}_{1,SA}^T, \mathbf{k}_{2,SA}^T, \mathbf{k}_{1,LA}^T, \mathbf{k}_{2,LA}^T\right]^T \tag{2}$$

The material deformation gradient tensor $\mathbf{F}(\mathbf{x})$ is related to the gradient of the phase image $\phi_i(\mathbf{x})$ as stated in [7]. Rearranging the gradient of the phase images in matrix form as:

$$\mathbf{Y}(\mathbf{x}) = \left[\frac{\partial^*\phi_{1,SA}}{\partial \mathbf{x}^T}(\mathbf{x}), \frac{\partial^*\phi_{2,SA}}{\partial \mathbf{x}^T}(\mathbf{x}), \frac{\partial^*\phi_{1,LA}}{\partial \mathbf{x}^T}(\mathbf{x}), \frac{\partial^*\phi_{2,LA}}{\partial \mathbf{x}^T}(\mathbf{x})\right]^T, \tag{3}$$

we obtain the material deformation gradient $\mathbf{F}(\mathbf{x})$ from:

$$\mathbf{K} = \mathbf{Y}(\mathbf{x})\mathbf{F}(\mathbf{x}). \tag{4}$$

In order to estimate $\mathbf{F}(\mathbf{x})$, one could resort to the Least Squares (LS) method. However, bearing in mind that HARP-like methods suffer phase interferences (especially within the vicinity of boundaries), which give rise to outliers, our proposal is to resort to the Least Absolute Deviation (LAD) method, due to its robustness [17]. Hence, the reconstruction is performed iteratively by:

$$\mathbf{F}_{l+1}(\mathbf{x}) = (\mathbf{Y}^T(\mathbf{x})\mathbf{W}_l(\mathbf{x})\mathbf{Y}(\mathbf{x}))^{-1}\mathbf{Y}^T(\mathbf{x})\mathbf{W}_l(\mathbf{x})\mathbf{K}, \tag{5}$$

with $\mathbf{W}_l(\mathbf{x})$ a diagonal weight matrix obtained by:

$$W_l^{jj}(\mathbf{x}) = \frac{1}{\sqrt{\sum_{h=1}^{3}\left(K^{jh} - \sum_{g=1}^{3} Y^{jg}(\mathbf{x})F_l^{gh}(\mathbf{x})\right)^2}} \tag{6}$$

and initially establishing $\mathbf{F}_0(\mathbf{x}) = \mathbf{I}$, with $\mathbf{I}$ the identity matrix.

Once $\mathbf{F}(\mathbf{x})$ is reconstructed, the Green-Lagrange strain tensor can be easily obtained as: $\mathbf{E}(\mathbf{x}) = 1/2(\mathbf{F}^T(\mathbf{x})\mathbf{F}(\mathbf{x}) - \mathbf{I})$.

**Classification.** Mechanical descriptors are extracted from the aforementioned tensors. We considered the projected components of the strain tensor on the usual radial-circumferential-longitudinal $\{r, c, l\}$ space. These components will be referred to as $E^{ab}$, where $a$ and $b$ will be components of this space. In all the cases, all the features will be averaged within three areas in the heart along its LA; specifically, we will consider components at the base, mid-ventricle and apex. On the other hand, we will also calculate the curl of the deformation field as a measure of rotation. We will generically refer to the twist as the unsigned difference between two components that measure rotation [18], one at the base, the second one at the apex. When we refer to the rotation modulus, the specific component will be the unsigned average of the component in the mid-ventricle area. Additionally, since some rotation-related components have opposite directions in apex and base, we will consider the location of the zero crossing for these components.

Once the principal heart motion descriptors have been calculated, a normalization stage [19] is applied in order to diminish the influence of possible outliers. A sigmoidal function is used to this end; data are mapped on the interval $(0, 1)$.

Our purpose is to classify a sample into one of three classes, namely, control, primary HCM and secondary HCM. Since secondary HCM patients have subtle differences with respect to the other two classes, we have resorted to a two-stage classification procedure (see Fig. 2). At the first stage we grossly pursue to distinguish between controls and primary HCM. Secondary HCM patients will fall on either of the two mentioned partitions. As for the second stage the purpose is to tell apart controls and secondary HCM, on one hand, and primary and secondary HCM, on the other hand. Notice that primary HCM patients incorrectly classified as a control in the first stage (or vice versa) will not be correctly classified; however, as we will describe later, this situation hardly takes place.

Optimal feature selection and classification are intimately related; those features whose classification figures are the highest are considered as the optimal features for that classification stage. To minimize overtraining, despite the reduced sample size, a Leave-10-out cross validation procedure has been used and data samples have been randomized; the proportions of control/primary/secondary have been kept unaltered along trials. For every trial we record the feature set with best performance; after one hundred trials we can identify the feature vector that has been selected the most; this will be the chosen one for the classification stage. As for finding classification performance, the procedure is repeated with a new randomized Leave-10-out procedure using the aforementioned selected feature vector as the input. In all the cases, 3-component feature vectors have been tested. Higher dimensionality has been tried but with no improve in performance.

The procedure described above has been carried out using Fuzzy c-Means (FCM) [20] and Support Vector Machines [21] both with quadratic (SVMq) and Gaussian (SVMg) kernels [22]. We have also tested a combination of them at different stages. As for the fuzzy clustering, the method requires that a threshold is set for the membership grade of each partition to carry out a final hard threshold. This value is another parameter that has to be optimized.

As for the first stage, the selected features are the ones that provide the highest accuracy involving only healthy and primary HCM. The modulus of $E^{cc}$ and curl as well as the twist of the curl of the 2D deformation field [18] have proven to provide a good separability between these groups for each one of the tested classification techniques.

Then, for both output partitions of the first stage, another classification is established with the purpose of detecting secondary HCM cases in groups of healthy volunteers and primary HCM, respectively. The objective of this stage is to increase the classifier sensitivity with respect to secondary HCM cases; if several feature vectors share the maximum sensitivity with respect to this group, we select the feature vector with maximum specificity (i.e., with maximum

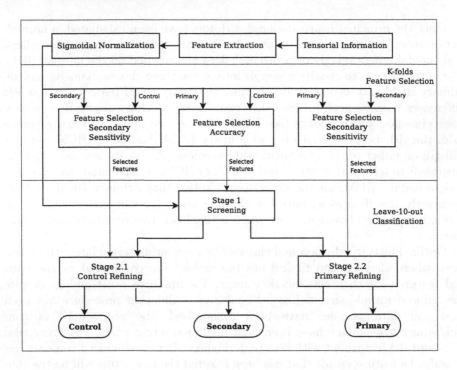

**Fig. 2.** Pipeline of the feature selection and classification stages.

sensitivity with respect to the other class, controls in classifier 2.1 and primaries in classifier 2.2, following the notation in Fig. 2).

It turns out that the selected feature vector for classifier 2.2 consists of $E^{lc}$ and $E^{rc}$ components in mid-ventricular slices and $E^{ll}$ over basal slices for the FCM analysis; for both SVM analyses, zero crossing of $E^{rc}$ component as well as the curl of the 2D deformation field substitute the components $E^{ll}$ and $E^{rc}$. On the other hand, for the 2.1 classifier, higher variability was found. Modulus of $E^{cc}$ and $E^{rr}$ components in the mid-ventricular area as well as the twisting obtained from eigenvector shift [18] were used in the SVMg analysis, while for SVMq the latter was replaced by the twisting given by the principal directions tangent angle [18]. As for FCM analysis, $E^{cc}$ modulus, curl zero crossing and its modulus have been accounted for.

## 3   Results

### 3.1   Alignment

For the homomorphic filtering procedure, the validation is performed in terms of contour overlapping, measured by means of the Dice Coefficient [23]. Different affine registration procedures have been performed in order to align MR-Cine image at ED phase with its corresponding in the original MR-Tagging and the filtered sequences. The resulting transformation is applied to the MR-Tagging segmentation at ED phase and the Dice coefficient is calculated between the latter and the original MR-Cine segmentation. Boxplot diagrams of the Dice coefficient distributions obtained with the three similarity metrics compared are displayed in Fig. 3.

Mann-Whitney U-tests have been used to determine whether significant improvements exist in the medians of the Dice coefficient distributions when carrying out the homomorphic filtering stage; three different metrics have used, the results of which are, $p = 0.0154$ for cross-correlation [24], $p = 2.03 \times 10^{-11}$ for mutual information and $p = 0.0134$ for sum of squares; consequently, in the three cases significant differences have been observed.

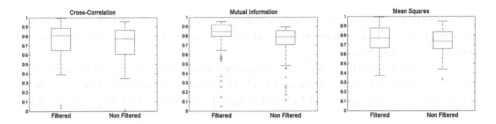

**Fig. 3.** Boxplot diagrams of the Dice Coefficient for different registration methods with and without homomorphic filtering.

### 3.2 Tensorial Estimation

For each patient described in Sect. 2.1, 3D deformation gradient tensors are obtained from the LS and the LAD methods; the 2D components of these tensors (i.e., excluding the five terms related to the longitudinal component) should ideally be equal to the components of the 2D tensor obtained directly from the MR-Tagging SA images. So, since in the 2D case four parameters are estimated out of two orientations and in 3D nine are estimated out of four, we take the 2D tensor as the reference and measure the quality of the estimator in 3D as the similarity between its 2D components with those of the 2D tensor. Similarity is calculated in terms of the Frobenius norm difference ($FND$) [17] between the 2D tensor and its correspondence in the 3D tensor.

Figure 4 shows the distribution of the $FND$ indexed for each of the methods used for the 3D tensor estimation (left) as well as the corresponding $FND$ Cumulative Distribution Function (CDF) (right). Significant differences have been found between medians of both error distributions by means of the Mann-Whitney U-test ($p = 8.94 \times 10^{-22}$).

### 3.3 Classification

Finally, we show the performance of the classification algorithm in Table 2 by means of normalized confusion matrices. Each column of the matrix represents

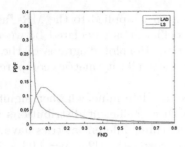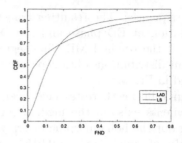

**Fig. 4.** FND and CDF of the tensors obtained by either LAD or LS reconstruction method with respect to the 2D tensor reconstructed with minimal wave vectors.

the instances in the predicted class, while each row represents the instances in the actual class. This allows more detailed analysis than mere proportion of correct guesses (accuracy).

From Table 2 global accuracy as well as measures of specificity and sensitivity can be obtained for each group and each classification method. The approach referred to as *mixed* consists of FCM in stages 1 and 2.2 and SVMg in stage 2.1.

**Table 2.** Confusion matrices for the FCM, SVMq, SVMg and *mixed* classifers.

|     | FCM | | | SVMq | | | SVMg | | | Mixed | | |
| --- | --- | --- | --- | --- | --- | --- | --- | --- | --- | --- | --- | --- |
|     | Con | Sec | Pri | Con | Sec | Pri | Con | Sec | Pri | Con | Sec | Pri |
| Con | 0.245 | 0.055 | 0 | 0.225 | 0.072 | 0.003 | 0.215 | 0.085 | 0 | 0.239 | 0.061 | 0 |
| Sec | 0.051 | 0.125 | 0.024 | 0.063 | 0.136 | 0.001 | 0.08 | 0.119 | 0.001 | 0.036 | 0.147 | 0.017 |
| Pri | 0.012 | 0.016 | 0.472 | 0 | 0.148 | 0.352 | 0.004 | 0.069 | 0.427 | 0 | 0.034 | 0.466 |

## 4   Discussion

In Sect. 3.1 we have described an alignment stage aimed at mapping the MR-Cine segmentations provided by cardiologist onto the MR-Tagging sequence. The segmentations are used to define a region of interest on which to compute meaningful measures of the tensor. This stage could be performed for every cardiac phase as MR-Cine sequence has been preprocessed by means of a groupwise elastic registration procedure in order to propagate the ED segmentations along the whole cardiac cycle.

Figure 3 shows an increase in the overlapping of the segmentations when the homomorphic filter is applied to the image independently of the metric used in the registration procedure; specifically, the improvement in performance caused by the homomorphic filter is higher than the improvement obtained by switching the registration metric. Therefore, we can conclude that a preprocessing stage is relevant in order to perform precise registration over MR-Tagging images.

Although in this paper the ultimate goal of the alignment stage is the projection of segmentations, this stage can be also applied to other more ambitious objectives, such as a multimodal scheme of material point tracking along the cardiac cycle or finding the association between gadolinium accumulation in late enhancement images and local mechanical abnormalities in the myocardium. We have taken some steps in this latter direction [11, 25].

As for the tensor estimation technique, a better response in terms of robustness is observed in Fig. 4 for the LAD estimator with respect to LS. The results conclusively support the hypothesis that the LAD estimator is better suited for this reconstruction problem, where the main source of inconsistencies seems to be the presence of phase interferences as opposed to the presence of noise in the measurements. Specifically, the CDF curve is left skewed although heavier tails are observed, i.e., the majority of the estimations are more accurate than for the LS method, at the expense of the onset of larger errors whenever estimations are inaccurate.

From the results in Sect. 3.3 we can see that the methodology here presented seems effective in classifying HCM patients out of tensorial descriptors obtained from MR-Tagging sequences. Better sensitivity figures are observed for both control and primary HCM patients with respect to the secondary patients, for which performance is clearly lower (specially for SVMq), possibly due to the small data sample included in the study for this group. A poor sensitivity to primary HCM is observed when using SVMq, resulting in the lowest global accuracy. A quadratic kernel seems inadequate for this problem, possibly due to the multiple states present in primary HCM. On the other hand, SVMg provides best performance in detecting secondary HCM. Consequently, we have resorted to a *mixed* approach to take advantage of the accuracy in primary HCM detection of FCM analysis and the secondary HCM sensitivity shown by SVMg, obtaining sensitivity figures higher than 70 % for each group (especifically, 80 % for control, 73 %for secondary patients and 93 % for primary patients). It is worth mentioning that no primaries as classified as controls and vice versa; therefore, the pipeline proposed seems a proper screening tool.

## 5   Conclusion

A processing pipeline for the tensorial classification of hypertrophic cardiomyopathies is presented which builds upon a robust 3D tensor estimation technique from SA and LA MR-Tagging sequences and a novel homomorphic filtering preprocessing step. This filtering method has significantly improved the accuracy of the alignment and it paves the way to construct multimodal processing schemes in which different modalities can be accurately dealt with, as it would be the case for MR-Cine, MR-Tagging and Late Enhancement MR.

A comparative study in terms of robustness provided by LAD and LS estimators in real datasets has also been carried out, supporting the hypothesis that LAD estimator is worth taking for an overdetermined reconstruction problem in tagging images.

For the classifier itself, we have compared three different classification methods used in machine learning, namely, FCM and SVM with quadratic and Gaussian kernels. We have shown that the fuzzy approach provides better global accuracy results although it is not suitable for small data sets, while SVM do. Consequently, we have resorted to a mixed approach that takes advantage of both techniques obtaining high rates in global accuracy with more balanced sensitivities of each class with respect to those obtained with a unique classifier in the sequential procedure.

Although our classifier is designed for HCM patients, it can be easily tuned for other cardiovascular diseases as long as appropiate biomarkers are available; these biomarkers could be derived from different technologies.

**Acknowledgments.** This work was partially supported by the Spanish Ministerio de Ciencia e Innovacion under Research Grant TEC2013-44194-P, the Spanish Ministerio de Ciencia e Innovacion and the European Regional Development Fund (ERDF-FEDER) under Research Grant TEC2014-57428-R and the Spanish Junta de Castilla y Leon under Grant VA136U13.

# References

1. Baron, B.J.: The 2006 american heart association classification of cardiomyopathies is the gold standard. Circ. Heart Fail. **1**, 72–76 (2008)
2. Hypertrophic cardiomyopathy. http://www.heart.org/HEARTORG/Conditions/More/Cardiomyopathy/Hypertrophic-Cardiomyopathy_UCM_444317_Article.jsp#.Vkyih3YvdD8. Accessed 18 November 2015
3. Karamitsos, T.D., Neubauer, S.: The interplay between cardiac strain and fibrosis in non-ischaemic cardiomyopathies: insights from cardiovascular magnetic resonance. Eur. J. Heart Fail **13**, 927–928 (2011)
4. Richard, P., Charron, P., Carrier, L., Ledeuil, C., Cheav, T., Pichereau, C., Benaiche, A., Isnard, R., Dubourg, O., Burban, M., Gueffet, J.P., Millaire, A., Desnos, M., Schwartz, K., Hainque, B., Komajda, M.: Hypertrophic cardiomyopathy: distribution of disease genes, spectrum of mutations, and implications for a molecular diagnosis strategy. Circulation **107**, 2227–2232 (2003)
5. Braunwald, E., Seidman, C.E., Sigwart, U.: Contemporary evaluation and management of hypertrophic cardiomyopathy. Circulation **106**, 1312–1316 (2002)
6. Shehata, M.L., Cheng, S., Osman, N.F., Bluemke, D.A., Lima, J.A.: Myocardial tissue tagging with cardiovascular magnetic resonance. J. Cardiovasc. Magn. Reson. **11**, 55 (2009)
7. Osman, N.F., McVeigh, E.R., Prince, J.L.: Imaging heart motion using harmonic phase MRI. IEEE Trans. Med. Imaging **19**(3), 186–202 (2000)
8. Shimon, A., Reisner, M.D., Lysyansky, P., Agmon, Y., Mutlak, D., Lessick, J., Friedman, Z.: Global longitudinal strain: a novel index of left ventricular systolic function. J. Am. Soc. Echocardiogr. **17**, 630–633 (2000)
9. Maron, B.J., Wolfson, J.K., Roberts, W.C.: Relation between extent of cardiac muscle cell disorganization and left ventricular wall thickness in hypertrophic cardiomyopathy. Am. J. Cardiol. **70**, 785–790 (1992)

10. Piella, G., De Craene, M., Bijnens, B.H., Tobon-Gómez, C., Huguet, M., Avegliano, G., Frangi, A.F.: Characterizing myocardial deformation in patients with left ventricular hypertrophy of different etiologies using the strain distribution obtained by magnetic resonance imaging. Rev. Esp. Cardiol. **63**, 1281–1291 (2010)
11. Cordero-Grande, L., Sevilla, T., Revilla, A., Martín-Fernández, M., Alberola-López, C.: Assessment of the fibrotic myocardial tissue mechanics by image processing. in: Computing in Cardiology Conference, pp. 635–638 (2013)
12. Gopalakrishnan, V., Menon, P.G., Madan, S.: cMRI-BED: A novel informatics framework for cardiac MRI biomarker extraction and discovery applied to pediatric cardiomyopathy classificationn. In: 2nd International Work-Conference on Bioinformatics and Biomedical Engineering, Granada, Spain (2014)
13. Rahman, Q.A., Tereshchenko, L.G., Kongkatong, M., Abraham, T., Abraham, M.R., Shatkay, H.: Utilizing ECG-based heartbeat classification for hypertrophic cardiomyopathy identification. IEEE Trans. Nanobiosci. **14**, 505–512 (2015)
14. Aja-Fernández, S., Pieçiak, T., Vegas-Sánchez-Ferrero, G.: Spatially variant noise estimation in MRI: a homomorphic approach. Med. Image Anal. **20**, 184–197 (2014)
15. Makram, A., Khalifa, A., El-Rewaidy, H., Fahmy, A., Ibrahim, E.S.H.: Assessment of global cardiac function from tagged magnetic resonance images. Comparison with cine MRI. In: 23rd Proceedings of the International Society on Magnetic Resonance in Medicine, Toronto, Canada, vol. 23, p. 4472 (2015)
16. Cordero-Grande, L., Vegas-Sánchez-Ferrero, G., Casaseca-de-la-Higuera, P., Alberola-López, C.: Improving harmonic phase imaging by the windowed Fourier transform. In: 8th IEEE International Symposium on Biomedical Imaging: From Nano to Macro, Chicago, USA, pp. 520–523 (2011)
17. Cordero-Grander, L., Royuela-del-Val, J., Sanz-Estébanez, S., Martín-Fernández, M., Alberola-López, C.: Multi-oriented windowed harmonic phase reconstruction for robust cardiac strain imaging. Med. Image Anal. **19**, 1–11 (2015)
18. Fung, Y.C.: Foundations of Solid Mechanics. Prentice-Hall, Englewood Cliffs (1965)
19. Theodoridis, S., Koutroumbas, K.: Pattern Recognition. Academic Press, San Diego (1999)
20. Bezdec, J.C.: Pattern Recognition with Fuzzy Objective Function Algorithms. Plenum Press, New York (1981)
21. Cortes, C., Vapnik, V.: Support-vector networks. Mach. Learn. **20**, 273–297 (1995)
22. Vert, J.P., Tsuda, K., Schölkopf, B.: A primer on kernel methods. In: Kernel Methods in Computational Biology, pp. 1–42 (2004)
23. Dice, L.R.: Measures of the amount of ecologic association between species. Ecology **26**, 297–302 (1945)
24. Avants, B.B., Epstein, C.L., Grossman, M., Geel, J.C.: Symmetric diffeomorphic image registration with cross-correlation: evaluating automated labeling of elderly and neurodegenerative brain. Med. Image Anal. **12**, 26–41 (2008)
25. Sanz-Estébanez, S., Merino-Caviedes, S., Sevilla, T., Revilla-Orodea, A., Martín-Fernández, M., Alberola-López, C.: Cardiac strain assessment for fibrotic myocardial tissue detection in left ventricular hypertrophic cardiomyopathy. In: Congreso Anual de la Sociedad Española de Ingeniería Biomédica, Madrid, Spain, pp. 10–13 (2015)

# Depth Image Matching Algorithm for Deforming and Cutting a Virtual Liver via Its Real Liver Image Captured Using Kinect v2

Hiroshi Noborio[1]([✉]), Kaoru Watanabe[1], Masahiro Yagi[1], Kentaro Takamoto[1], Shigeki Nankaku[1], Katsuhiko Onishi[1], Masanao Koeda[2], Masanori Kon[2], Kosuke Matsui[2], and Masaki Kaibori[2]

[1] Department of Computer Science, Osaka Electro-Communication University, Osaka, Japan
{Nobori,nankaku,onishi}@isc.osakac.ac.jp,
watanabe@wtlab.osakac.ac.jp
[2] Medical School, Kansai Medical University, Osaka, Japan
koeda@isc.osakac.ac.jp, {kon,matsuik,kaibori}@hirakata.kmu.ac.jp

**Abstract.** In this paper, we propose a smart deforming and/or cutting transcription algorithm for rheology objects such as human livers. Moreover, evaluation of performance and shape precision under the proposed algorithm are experimentally verified by deforming a real clay liver and/or cutting a gel block prepared at human body temperature. First, we capture the image of the liver of a patient by digital imaging and communication in medicine (DICOM) generated by magnetic resonance imaging (MRI) and/or computed tomography (CT) scanner. Then, the DICOM data is segmented and converted into four types of stereolithography (STL) polyhedra, which correspond to the whole liver and three blood vessels. Second, we easily overlap the virtual and real liver images in our mixed reality (MR) surgical navigation system using our initial position/orientation/ shape adjustment system that uses color images to differentiate between real and virtual depth images. After overlapping, as long as the real liver is deformed and/ or cut by a human (doctor), the liver is constantly captured by Kinect v2. Subsequently, by using the real depth image captured in real time, many vertices around the virtual polyhedral liver in STL format are pushed/pulled by viscoelastic elements called the Kelvin–Voigt materials located on the vertices. Finally, after determining the displacements of the vertices, we obtain an adequately shaped STL. The vertex position required for fixing the shape is calculated using the Runge–Kutta method.

**Keywords:** Depth camera image · Z-buffering · Kelvin–voigt material · Runge–kutta method · GPU · Parallel processing

## 1 Introduction

Recently, many surgical navigation systems have been proposed [1–7]. In almost all the navigators, 3D mechanical or 2D non-mechanical probes with ultrasonic sensors have been used. However, in general, because the image resolution of an ultrasonic sensor

© Springer International Publishing Switzerland 2016
F. Ortuño and I. Rojas (Eds.): IWBBIO 2016, LNBI 9656, pp. 196–205, 2016.
DOI: 10.1007/978-3-319-31744-1_18

is not accurate, we cannot detect the position, orientation, and shape of a real liver precisely for manipulating it in surgical navigation systems. In few research studies, translational and rotational motions and deformation of a real liver have been captured using the stereo vision of 3D camera, with and without markers. If some artificial markers are located on the real liver of a patient, the position, orientation, and shape of the real liver can be precisely determined. However, the liver can get severely damaged if the markers are peeled off from the liver after the surgical operation. In addition, the markers cannot often be detected because they get soiled because of materials and body fluids. In addition, the calculation of the position, orientation, and shape of the liver is immensely time consuming [8–14].

To eliminate these drawbacks, we use a smart depth camera to capture a depth image for manipulating a real liver in the exact manner. The advantage of using the depth image for a real liver is that it is easy to compare this image with the virtual depth image that is efficiently calculated by z-buffering used in the graphics processing unit (GPU) against its virtual liver. In our research, the STL is originally converted from the DICOM data captured for a patient, which is generated by magnetic resonance imaging (MRI) and/or a computed tomography (CT) scanner. In DICOM, several types of veins, arteries, and gate vein bloods and cancer organizations are segmented.

First, a 3D CG virtual environment controlled by OpenGL on a GPU is theoretically adjusted for a 3D real camera environment controlled by Kinect SDK and Kinect Studio API. Subsequently, in order to overlap the image of a real liver with its virtual image, we use our initial position/orientation/shape adjustment system [15]. This algorithm matches depth-depth images in the same camera and CG coordinate systems. In this system, a human can differentiate between the depth images captured for real and virtual livers in 3D real and virtual environments based on several colors (e.g., both depths coincide with each other at blue pixel in XYZ spaces) based on Kinect v1. By observing the colors corresponding to depth differences in the image, a doctor can comfortably determine the position and orientation of the virtual liver in the 3D CG virtual environment, which coincides with the position and orientation of the real liver in the camera coordinate system. A real liver can be practically captured using a real depth camera by employing an image-based position/orientation adjustment system such that it coincides with the virtual liver artificially captured by the z-buffer used in the GPU.

Simultaneously, we design several types of motion transcription algorithms by matching real and virtual 2D depth images and searching the space defined by three translational and three rotational degrees of freedom based on their depth differences [16–19]. The depth image for the real liver is always captured by the depth cameras Kinect v1 and v2, and the other depth image obtained for its virtual liver (a polyhedron with STL format) is captured by z-buffering in the GPU. Moreover, we evaluate the performances (motion precision and calculation time) of the algorithms ascertained in several experiments based on the images captured by depth cameras Kinect v1 and v2.

The evaluation of performances and shape precisions of all the algorithms was carried out on a 3D printed plastic liver based on the STL model converted from DICOM. Clearly, because the human liver is a rheology object, it is always deformed and/or cut by doctors during surgical operations. Therefore, in this paper, we propose a smart deforming/cutting transcription algorithm for a virtual liver replicated from a real liver.

Moreover, the performances and precisions of the proposed algorithm were evaluated using a clay block possessing plastic property and a gel block possessing viscoelastic property.

In our algorithm, we first capture an image of the real liver of the patient using a CT scan in order to obtain the polyhedron with STL. Second, using the liver STL, we construct a real clay gel block shaped from the STL and a gel liver is cut from the middle portion in advance. Then, the virtual and the real liver images are overlapped in our surgical navigation system based on color image matching under mixed reality (MR) using the previously explained initial position/orientation/shape adjustment software in Kinect v2. Further, the real liver is manually deformed and/or cut and the deformation is captured by Kinect v2. Third, by using the real depth image in real time, many vertices around the virtual liver with the STL format are pushed/pulled using viscoelastic elements such as Kelvin–Voigt materials located on the vertices. Finally, we demonstrate adequate deformation in the virtual liver with the STL format. The vertex positions were determined by the Runge-Kutta method.

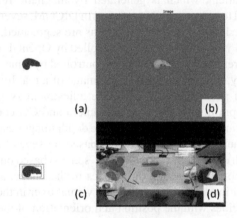

**Fig. 1.** (a) 2D virtual depth image generated by z-buffer in GPU for a 3D virtual liver, (b) 3D virtual liver represented by polyhedron with STL format, (c) 2D real depth image captured by depth camera Kinect v2 for a 3D real liver, (d) 3D real liver represented by a gel block with human body temperature.

In realistic surgical operation, as long as a virtual liver consists of four types of stereolithography (STL) polyhedra, which correspond to the whole liver and three blood vessels. Therefore, before a doctor cuts a real liver, he understands some risks from its virtual liver in advance. As a result, if and only if real and virtual livers coincide with each other, the real-time animation reduces the possibility of mistakes made by the doctor, such as blood vessel injuries in surgery performed using mixed reality.

In this paper, Sect. 2 explains the real–virtual 2D depth-depth matching algorithm for deforming and/or cutting a virtual liver with STL format. Section 3 provides several experimental results of when the clay liver is deformed and the gel liver is cut. Finally, Sect. 4 concludes our research.

## 2    Deforming and Cutting Transcription Algorithm Based on Real-Virtual Depth-Depth Matching

In this section, we explain the algorithm to replicate a real liver to design its virtual liver in mixed reality for deforming and cutting a rheology object. In our algorithm, we first overlap the real and virtual livers in real and virtual environments by comparing the real and virtual depth images, respectively. A real depth image is always captured by the depth camera Kinect v2 and the virtual depth image is always captured by the z-buffering in the GPU (Fig. 1). Moreover, in our deforming and cutting transcription algorithm, we consider matching real depth and virtual depth images. The difference is that all the vertices of the polyhedron with STL format (virtual liver) were either pushed or pulled by the depth differences in the corresponding pixels of real and virtual depth images. The pushing or pulling is calculated by Kelvin-Voigt materials located on the vertices, and simultaneously, the positions of all vertices are calculated by the Runge-Kutta method in real time.

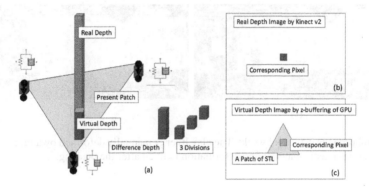

**Fig. 2.**   (a) Pushing/pulling of three vertices of the corresponding patch via Kelvin-Voigt material and Runge-Kutta method. (b) A real depth image by Kinect v2 and its corresponding pixel. (c) A virtual depth image by z-buffering and its corresponding pixel.

### 2.1    Real-Virtual Depth-Depth Matching Algorithm

Our algorithm consists of the following four steps. The real clay or gel liver and its virtual liver with the STL format are initially set at the same position/posture using the image-based initial position/orientation adjustment system in MR [15].

[**Step 1**] The clay or gel liver is manually deformed or cut. Simultaneously, a real depth image of the clay or gel liver is captured by Kinect v2 (Fig. 1).
[**Step 2**] By using z-buffering in the GPU, we obtained a virtual depth image of the virtual liver with STL format (Fig. 1).
[**Step 3**] By considering the depth difference between each corresponding pixel in the real and virtual depth images, three vertices of a patch corresponding to a pixel are pulled or pushed (Fig. 2). These pulling or pushing operations at all vertices are processed in parallel by multiple GPU cores. Subsequently, we summed up the depth

differences of all pixels on the patch and converted one-third of the sum into a corresponding force at each vertex of the patch (Fig. 2). The force at all vertices is calculated in parallel by multiple GPU cores.

**[Step 4]** The Kelvin-Voigt material at each vertex is calculated by the force (Fig. 2), and the moving displacement is determined by the Runge-Kutta method during a given time interval. Here, we can choose the parameters of the elastic and viscous elements of the Kelvin–Voigt materials and the time steps of the Runge-Kutta method.

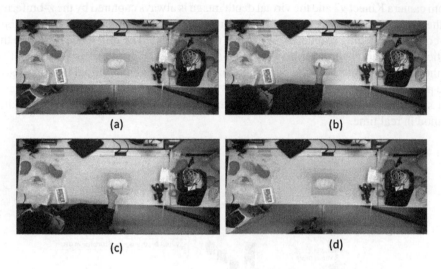

(a)                                              (b)

(c)                                              (d)

**Fig. 3.** Stroboscopes in the real world. The clay liver is manually deformed by a finger. Because the clay liver has plastic material, the deformation is fixed after the pushing.

## 3    Experimental Results

When the real clay liver is pushed manually and the gel liver is cut by a user, we check whether the virtual liver is also deformed in the exact manner.

First, we compare the virtual 3D environment (CG coordinate system) with the real 3D one (camera coordinate system). Then, we overlap the real liver by its virtual one in 3D space by observing the status between the real and virtual depth images in 2D space [15]. For this purpose, several kinds of windows were employed to find position errors in X, Y, Z and the orientation errors in pitch, yaw, and roll coordinate systems. While observing a color image (green: XYZ coincidence, yellow: XY coincidence, red: real liver 2D projection, blue: virtual liver 2D projection), which consists of pixel differences in 2D depth images, the doctor first overlaps the 2D projections of real and virtual livers on the XY plane by mixing X and/or Y movements. Then, the 3D real and virtual livers are overlapped along or around the Z-axis by mixing translational and/or rotational motions along and/or around the Z-axis, respectively.

Furthermore, the clay liver, which possesses plastic properties, is pushed by a finger as shown in Fig. 3. According to the deformation, the shape of the virtual polyhedral liver with STL format is slightly altered as shown in Fig. 4. Simultaneously, the overlapping ratio between real and virtual livers is altered when the liver is deformed as shown in Table 1. As can be observed from Fig. 4 and Table 1, the shape transcription algorithm replicates the shape of the real clay liver to that of its virtual liver adequately.

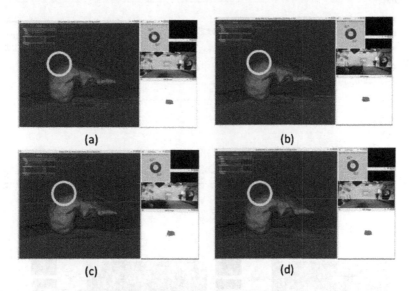

**Fig. 4.** Stroboscopes in the virtual world. The upper left portion of virtual liver is concaved. In the bottom right, red, blue, yellow, and green pixels are indicated as the projection pixels of 3D real, virtual, XY overlapping, and XYZ overlapping livers. In the upper right, the circular blue graph represents the overlapping ratio (green pixel area)/(blue pixel area) × 100. (Color figure online)

Second, the human cuts a gel liver possessing viscoelastic property by pulling it with both hands as shown in Fig. 5. We prepared a section in the gel liver in advance. According to the cutting operation stated previously, the shape of the virtual polyhedral liver with STL format is completely altered around the upper left portion as illustrated in Fig. 6. Simultaneously, the overlapping ratio between the real and virtual 3D livers is also altered as shown in Table 2. From Fig. 6 and Table 2, we can observe that the shape transcription algorithm replicates the shape of the real gel liver to that of its virtual liver adequately.

In the deforming and cutting operations shown above, which are the main operations in the surgical navigation system, real and virtual livers were almost overlapped in the 3D real and virtual space. All experimental results were processed in real time by the parallel processing of the GPU. As a result, the effectiveness of our proposed deforming and cutting copy algorithm based on 2D depth-depth matching using the Kinect v2 depth camera was ascertained.

**Table 1.** Overlapping ratio (green pixel area)/(blue pixel area) × 100 during the operation described in real world as shown in Fig. 3 and the virtual world as shown in Fig. 4.

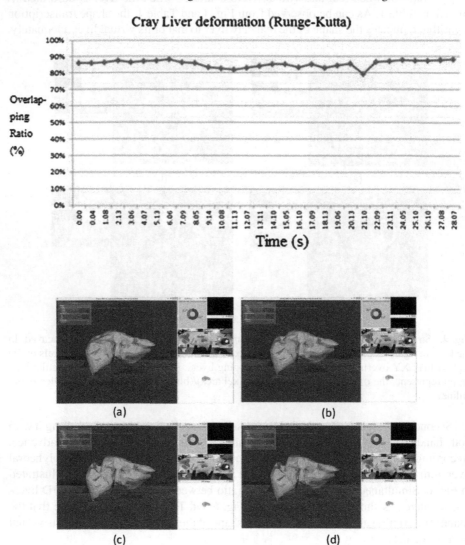

**Fig. 5.** Stroboscopes in the real world. The human cuts the gel liver by pulling it with both hands. Because the gel liver is made of viscoelastic material, the deformation is backed after the cutting.

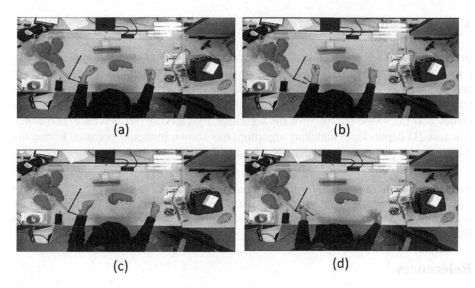

(a)                                              (b)

(c)                                              (d)

**Fig. 6.** Stroboscopes in the virtual world. The upper left portion of the virtual liver is cut deeply. In the bottom right, red, blue, yellow, and green pixels indicate the projection pixels of 3D real, virtual, XY overlapping, and XYZ overlapping livers. In the upper right, the circular blue graph represents the overlapping ratio (green pixel area)/(blue pixel area) × 100. (Color figure online)

**Table 2.** Overlapping ratio (green pixel area)/(blue pixel area) × 100 during the operation described in the real world as shown in Fig. 5 and the virtual world as shown in Fig. 6.

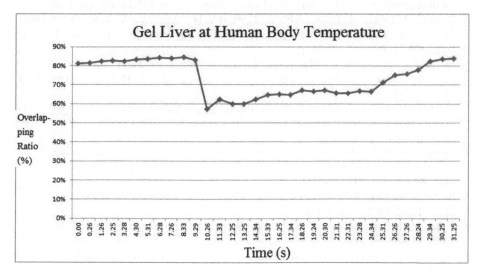

## 4　Conclusions

In this study, we designed a fast shape transcription algorithm to replicate a real liver to form a virtual liver by matching the real and virtual depth images. The real depth image was captured using the Kinect v2 depth camera and the virtual depth image was simultaneously captured by the z-buffering in the GPU. Subsequently, we evaluated the efficiency and the shape precision of the algorithm by carrying out several experiments. Our fast 2D depth-depth matching algorithm has shown immense potential to execute the shape transcription of a human liver in surgical navigation systems.

**Acknowledgment.** This study is supported in parts by 2014 Grants-in-aid for Scientific Research (No. 26289069) from the Ministry of Education, Culture, Sports, Science and Technology, Japan. It is also supported in parts by 2014 Cooperation research fund of graduate school in Osaka Electro-Communication University.

## References

1. Peterhans, M., vom Berg, A., Dagon, B., Inderbitzin, D., Baur, C., Candinas, D., Weber, S.: A navigation system for open liver surgery: design, workflow and first clinical applications. Int J. Med. Robot. **7**(1), 7–16 (Mar 2011). doi:10.1002/rcs.360. Epub 29 October 2010
2. Nicolas, C.B., Francesco, V., François, P., Christian, T., Matteo, F., Kate, G., Pietro, E.M., Matthias, P., Stefan, W., Philippe, M.: Augmented environments for the targeting of hepatic lesions during image-guided robotic liver surgery. J. Surg. Res. **184**(2), 825–831 (2013)
3. Satou, S., Aoki, T., Kaneko, J., Sakamoto, Y., Hasegawa, K., Sugawara, Y., Arai, O., Mitake, T., Miura, K., Kokudo, N.: Initial experience of intraoperative three-dimensional navigation for liver resection using real-time virtual sonography. Surgery **155**(2), 255–262 (2014)
4. Pessaux, P., Diana, M., Soler, L., Piardi, T., Mutter, D., Marescaux, J.: Towards cybernetic surgery: robotic and augmented reality-assisted liver segmentectomy. Langenbecks Arch. Surg. **400**(3), 381–385 (2015)
5. Morita, Y., Takanishi, K., Matsumoto, J.: A new simple navigation for anatomic liver resection under intraoperative real-time ultrasound guidance. Hepatogastroenterology **61**(34), 1734–1738 (2014)
6. Mahmud, N., Cohen, J., Tsourides, K., Berzin, T.M.: Computer vision and augmented reality in gastrointestinal endoscopy. Gastroenterol Rep. (Oxf) **3**(3), 179–184 (2015). doi:10.1093/gastro/gov027. Epub 1 July 2015
7. Chen, X.-P., Zhang, W.-D., Wang, D., Cui, W., Yu, Y.-L.: Image classification of liver cancer surrounding right hepatic pedicle and its guide to precise liver resection. Int. J. Clin. Exp. Med. **8**(7), 11093–11100 (2015)
8. Besl, P.J., McKay, N.D.: A method for registration of 3-D shapes. IEEE Trans. Pattern Anal. Mach. Intell. **14**(2), 239–256 (1992)
9. Zhang, Z.: Iterative point matching for registration of free-form surfaces. Int. J. Comput. Vis. **13**(2), 119–152 (1994)
10. Granger, S., Pennec, X.: Multi-scale EM-ICP: a fast and robust approach for surface registration. In: 7th European Conference on Computer Vision, vol. 4, pp. 69–73 (2002)
11. Liu, Y.: Automatic registration of overlapping 3D point clouds using closest points: J. Image Vis. Comput. **24**(7), 762–778 (2006)

12. Salvi, J., Matabosch, C., Fofi, D., Forest, J.: A review of recent range image registration methods with accuracy evaluation. J. Image Vis. Comput. **25**, 578–596 (2007)
13. Rusu, R.B., Cousins, S.: 3D is here: point cloud library (PCL). In: IEEE International Conference on Robotics and Automation, pp. 1–4 (2011)
14. Wu, Y.F., Wang, W., Lu, K.Q., Wei, Y.D., Chen, Z.C.: A new method for registration of 3D point sets with low overlapping ratios. In: 13th CIRP Conference on Computer Aided Tolerancing, pp. 202–206 (2015)
15. Noborio, H., Watanabe, K., Yagi, M., Ida, Y., Onishi, K., Koeda, M., Nankaku, S., Matsui, K., Kon, M., Kaibori, M.: Image-based initial position/orientation adjustment system between real and virtual livers. Jurnal Teknologi Med. Eng. **77**(6), 41–45 (2015). doi:10.11113/jt.v77.6225. Penerbit UTM Press
16. Noborio, H., Onishi, K., Koeda, M., Mizushino, K., Yagi, M., Kaibori, M., Kon, M.: Motion transcription algorithm by matching corresponding depth image and Z-buffer. In: 10th Anniversary Asian Conference on Computer Aided Surgery, pp. 60–61. Kyusyu University, Japan (2014)
17. Watanabe, K., Yagi, M., Ota, K., Onishi, K., Koeda, M., Nankaku, S., Noborio, H., Kon, M., Matsui, K., Kaibori, M.: Parameter identification of depth-depth-matching algorithm for liver following. Jurnal Teknologi Med. Eng. **77**(6), 35–39 (2015). doi:10.11113/jt.v77.6224. Penerbit UTM Press
18. Watanabe, K., Yagi, M., Shintani, A., Nankaku, S., Onishi, K., Koeda, M., Noborio, H., Kon, M., Matsui, K., Kaibori, M.: A new 2D depth-depth matching algorithm whose translation and rotation freedoms are separated. In: International Conference on Intelligent Informatics and Biomedical Sciences (ICIIBMS2015), Track 3: Bioinformatics, Medical Imaging and Neuroscience, pp. 271–278. Okinawa Institute of Science and Technology Graduate University (OIST), Okinawa, 28–30 November 2015
19. Noborio, H., Watanabe, K., Yagi, M., Ida, Y., Nankaku, S., Onishi, K., Koeda, H., Kon, M., Matsui, K., Kaibori, M.: Experimental results of 2D depth-depth matching algorithm based on depth camera Kinect v1. J. Bioinform. Neurosci **1**(1), 38–44 (2015). ISSN:2188–8116

# Optic Disc Segmentation with Kapur-ScPSO Based Cascade Multithresholding

Hasan Koyuncu[✉] and Rahime Ceylan

Electrical and Electronics Engineering, Selcuk University, Konya, Turkey
{hasankoyuncu, rpektatli}@selcuk.edu.tr

**Abstract.** The detection of significant retinal regions (segmentation) constitutes an indispensible need for computer aided diagnosis of retinal based diseases. At this point, image segmentation algorithm is wanted to be quick in order
to spare time for feature selection and classification parts. In this paper, we deal
with the fast and accurate segmentation process of optic discs in retinal images.
For this purpose, a cascade multithresholding (CMT) process is proposed by a
novel optimization algorithm (Scout Particle Swarm Optimization) and an
efficient cost function (Kapur).

Scout Particle Swarm Optimization (ScPSO) is originated from Particle Swarm
Optimization (PSO) and improves standard PSO by using a necessary part taken
from Artificial Bee Colony (ABC) Optimization. In other words, the most
important handicap of PSO (regeneration of useless particles) is eliminated via the
formation of ScPSO that can be obtained by adding the scout bee phase from ABC
into standard PSO. In this study, this novel method (ScPSO) constitutes the
optimization part of multithresholding process. Kapur function is preferred as
being the cost function to be used in ScPSO, since Kapur provides low standard
deviations on output of optimization based multithresholding techniques in literature. In this manner, a well-combined structure (Kapur-ScPSO) is generated
for cascade multithresholding. Optic disc images taken from DRIVE database are
used for statistical and visual comparison. As a result, Kapur-ScPSO based CMT
can define the optic disc quickly (7–8 s) with the rates of 77.08 % precision,
57.89 % overlap and 95.59 % accuracy.

**Keywords:** Scout particle swarm optimization · Kapur · Optic disc
segmentation · Multithresholding

## 1 Introduction

Optic Disc (OD) or Optic Nerve Head (ONH) is the exit for ganglion cell axons leaving
the eye and this situation brings into prominence to optic disc about its vitality for a
qualified vision. There are no rods or cones overlying the optic disc and it corresponds
to a physiological blind spot in each eye.

Detection of optic disc has an importance on definition of some diseases (like as
glaucoma, diabetic retinopathy) and on identification of some abnormalities (like as
neovascularisation) [1]. For this purpose, remarkable segmentation algorithms have
been generated in literature.

© Springer International Publishing Switzerland 2016
F. Ortuño and I. Rojas (Eds.): IWBBIO 2016, LNBI 9656, pp. 206–215, 2016.
DOI: 10.1007/978-3-319-31744-1_19

Singh et al. [2] designed an efficient method of optic disc segmentation by using Region Growing Technique. In their study, the least required computation time is around 16 s. Marin et al. [3] proposed a 2-step automatic thresholding procedure. In their approach, the first step is attendant for finding a pixel within or near enough the OD and the second step is for discarding the darkest values of a neighbourhood window centred on it. In this method, threshold algorithm is the Otsu and the end of segmentation is performed by circular Hough transform. Besides, their average computation time is approximately 5.425 s for an image. Welfer et al. [4] achieved the segmentation of optic disc via an adaptive morphological approach in which watershed transform, image reconstruction and basic mathematical rationale are settled. In [4], obtainment of optic disc is 12.17 s for an image from DIARETDB1 database. Also, accuracy and overlap rates are respectively 100 % and 41.47 % for images taken from DRIVE database. Muramatsu et al. [5] compared Active Contour Model (ACM), Fuzzy C-Means (FCM) and Artificial Neural Network (ANN) on ONH segmentation. According to their results, ANN and ACM have better segmentation performance than FCM structure. Esmaeili et al. [6] used Curvelet transform and Deformable Variational Level Set Model on segmentation of optic disc. Mendonca et al. [7] achieved the localization of optic disc by combination of vascular and intensity information. The computation time of their proposed method is 8 s on average. Hsiao et al. [8] used a Supervised Gradient Vector Flow snake (SGVF snake) model for contour segmentation. Their results showed that proposed structure has a good performance on localization, but contour of optic disc obtained by their proposed method is not superior on reflecting the real contour. Zhou et al. [9] used Mean Shift based Gradient Vector Flow (MSGVF) for segmentation process. Time consumption of MSGVF is 22 s for only an image. Kumara and Meegama [10] designed an active contour-based segmentation for removal of optic disc. As well as the active contour model, morphological operators and circle detection process are used in their method which performs 90 % accuracy on segmentation of OD. Issac et al. [11] realized OD and optic cup segmentation by an adaptive threshold based algorithm that achieves to 92.06 % accuracy. Mohamed et al. [12] used an approach based on Local Binary Patterns which obtains 89.51 % accuracy and 73.90 % precision on segmentation of OD.

As seen in literature studies, techniques can achieve segmentation of OD which could be realized as more accurate in less time than the available ones. For this purpose, we designed a cascade multithresholding approach which uses Scout Particle Swarm Optimization and Kapur function. So, Kapur-ScPSO based CMT is obtained which divides the optic disc by multithresholding. For performance evaluation, proposed CMT is tested on 20 optic disc images taken from DRIVE database [17].

# 2 Methods

## 2.1 Scout Particle Swarm Optimization

Scout Particle Swarm Optimization (ScPSO) is a recent & promising algorithm designed by Koyuncu and Ceylan [13]. This novel optimization algorithm combines the Particle Swarm Optimization (PSO) with the scout bee phase of Artificial Bee

Colony Optimization (ABC) in order to generate a PSO variant which eliminates the most important handicap of PSO. As known from general PSO algorithm, there is a need of reproduction of ineffective particles which cannot improve its position along an iteration number defined by user. This iteration number is known as parameter 'limit' in ABC algorithm.

In PSO algorithm, Eq. (1) symbolizes the velocity formula while Eq. (2) stands for the position equation. In ABC algorithm, Eq. (3) represents the scout bee phase. In ScPSO algorithm, all processes in PSO are available and Eq. (1–2) are used for the consistent update of particle values. On the other hand, scout bee phase is added to the end of algorithm so as to regenerate ineffective particles. In other words, the particle which cannot improve its $p_{best}$ value along iteration number equal to 'limit', is regenerated according to Eq. (3) in ScPSO algorithm.

$$V_i(t+1) = \omega V_i(t) + c_1 r_1 (X_{pbest(i)}(t) - X_i(t)) + c_2 r_2 (X_{gbest}(t) - X_i(t)) \qquad (1)$$

$$X_i(t+1) = X_i(t) + V_i(t+1) \qquad (2)$$

$$x_m = l_i + rand(0,1) * (u_i - l_i) \qquad (3)$$

Pseduocode of ScPSO algorithm is shown in Table 1. As seen in Table 1, ScPSO algorithm can be divided into two basic parts: Initialization and loop parts. Also, loop part can be separated into four subparts: calculation of fitness, update of $p_{best}$ and $g_{best}$ values, update of positions, update of inertia weight.

For a detailed explanation about PSO, ABC and ScPSO algorithms, please see [13] and [14].

**Table 1.** Pseudocode for ScPSO algorithm

| |
|---|
| Initialize all particles within the user defined boundaries |
| (The first best position (Pbest) values are equal to position of particles) |
| **-Define a limit value within the range [1, (maximum iteration number-1)]** |
| While (iteration number < maximum iteration number) |
| -Calculate fitness according to cost function for all particles |
| -Update best position values according to fitness values for all particles |
| -Choose the best Pbest vector as being Gbest (the vector achieved to minimum cost) |
| -Calculate new positions according to Eq.(1) and Eq.(2) |
| -If a variable inertia weight is used, change it in accordance with the used rule |
| -Control all particles which exceed the parameter 'limit', then regenerate the useless particle(s) according to Eq.(3) |
| End |

## 2.2  Kapur's Entropy Criterion

Kapur entropy criterion was designed by Kapur et al. [15] for thresholding process. According to Kapur function, an image (*I*) brings about a 2D grayscale intensity function, containing *n* pixels with gray levels from *0* to *L-1* [16].

Let assume that the number of pixels with gray level *i* is denoted by $h(i)$, then the probability of gray level *i* in *I* can be defined as $p_i = h(i)/n$. In here, $p(i)$ indicates the probability of occurrences of gray level *i* in the image *I*. The subdivision of an image into $k + 1$ class can be considered as a k-dimensional optimization problem and solution of this problem means the obtainment of *k* optimal thresholds $(t_0, t_1, \ldots, t_{k-1})$. At this point, the optimal thresholds are achieved by maximizing the objective function in Eq. (4). In Eq. (4) [16], entropies are symbolized as $H_i$ which are calculated according to Eq. (5) and $w_i$ values stands for the distribution of gray levels' probabilities to be used at the calculation of entropies.

$$f(t_0, t_1, \ldots, t_{k-1}) = \sum_{i=0}^{k} (H_i) \tag{4}$$

$$
\begin{aligned}
H_0 &= -\sum_{i=0}^{t_0-1} \frac{p_i}{w_0} \ln \frac{p_i}{w_0}, & w_0 &= \sum_{i=0}^{t_0-1} p_i \\
H_1 &= -\sum_{i=t_0}^{t_1-1} \frac{p_i}{w_1} \ln \frac{p_i}{w_1}, & w_1 &= \sum_{i=t_0}^{t_1-1} p_i \\
&\qquad\qquad \vdots \\
H_k &= -\sum_{i=t_{k-1}}^{L-1} \frac{p_i}{w_k} \ln \frac{p_i}{w_k}, & w_k &= \sum_{i=t_{k-1}}^{L-1} p_i
\end{aligned}
\tag{5}
$$

In optimization based multithresholding, optimization algorithm tries to find optimal threshold values $(t_0, t_1, \ldots, t_{k-1})$ which maximizes the objective value of Kapur.

## 3  Experimentals

### 3.1  Cascade Multithresholding Procedure

In this study, cascade multithresholding is used for the obtainment of optic disc. With this aim, we apply the cascade Kapur-ScPSO to images. Segmentation procedure is shown in Fig. 1.

As known from literature studies, retinal images seem formidable for solving it by simple approaches. Also, these images cannot be segmented by only one thresholding process. So, the idea of Kapur-ScPSO based CMT is produced in this way. The basic idea of the first multithresholding in cascade process is to render the optic disc from background as much as possible. Thus, second multithresholding will remove the irrelevant parts stated out of optic disc.

CMT basically includes the proposed optimization based multithresholding technique (Kapur-ScPSO). Additionly, it operates two important tools of image analysis (sharpening and filling) helping to obtain more clear results.

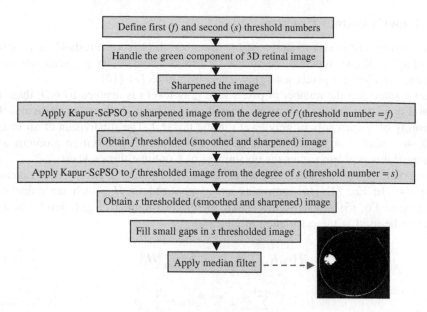

**Fig. 1.** CMT procedure

## 3.2    Trials and Results

In this paper, all experiments are realized with a personal computer (2.60 GHz CPU & 6 GB RAM) running Windows 8 system. During operation of Kapur-ScPSO in CMT, population size and maximum iteration number were settled to 100. Limit value was fixed to 10, $c_1/c_2$ was 2/2 and inertia weight was adjusted as a parabolic descent movement (0.9 → 0.4). Flow of processes in CMT is shown in Fig. 2 by using 09_test image from DRIVE database.

For 09_test image, the best segmentation was obtained, once the first and second threshold numbers (f&s) in Kapur-ScPSO are respectively 5&4.

As seen in Fig. 2, original image is transformed to grayscale and sharpened at the first step for rendering the optic disc clearly. At the second step, 5 thresholded image obtained by first multithresholding is formed in order to remove vessels and other parts in image. At the third step, optic disc is obtained by second multithresholding, but a little bit in piece. At the fourth step, small gaps are filled and at the last step, a median filter is applied to image for the achievement of optic disc more clear.

Through flow of processes, it's obvious that first multithresholding used for highlighting the optic disc enables us to obtain a more clear result by the second multithresholding. Besides, computation time of cascade multithresholding takes only 7–8 s on average. For a visual comparison, results of proposed method on 5 images are presented in Fig. 3.

As seen in Fig. 3, Kapur-ScPSO based CMT can reveal the optic disc obviously and it's seen that proposed method obtains the optic disc with a little noise, since the original images own some dark regions inside. So, these dark regions affect the optimal threshold values. Also vessels cause a negative effect on the obtainment of thresholds too.

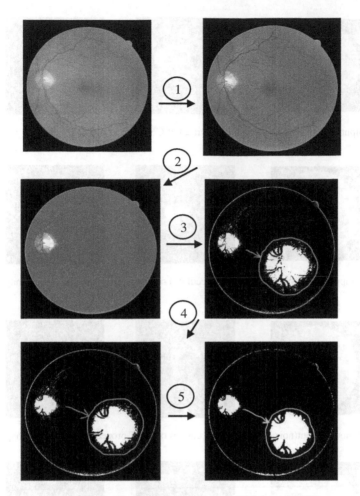

**Fig. 2.** Flow of processes in Kapur-ScPSO based CMT

By two-step multithresholding, these handicaps can be minimized. In Table 2, optimum threshold pairs are shown for every image.

According to Table 2, the most repeated threshold pair is 5&4 with repeat number of 6, while the second pair is 5&3 with the repetition of 4. In all trials, "5" is repeated 10 times as being the first threshold number and "4" is repeated 7 times as being second threshold. So, we can infer that the choice of 5&4 is generally effective on segmentation of optic disc from retinal images for cascade multithresholding. In Table 3, proposed method is compared with the recent promising algorithms on computation time.

As seen in Table 3, proposed method exposes a superior performance on computation time. Thus it's obvious that Kapur-ScPSO is not only successful on segmentation process, but also owns a low operation time on the achievement of segmentation.

In Table 4, our method is compared with literature studies via three metrics.

For a detailed explanation of these metrics, please see [4, 10, 16] and [17].

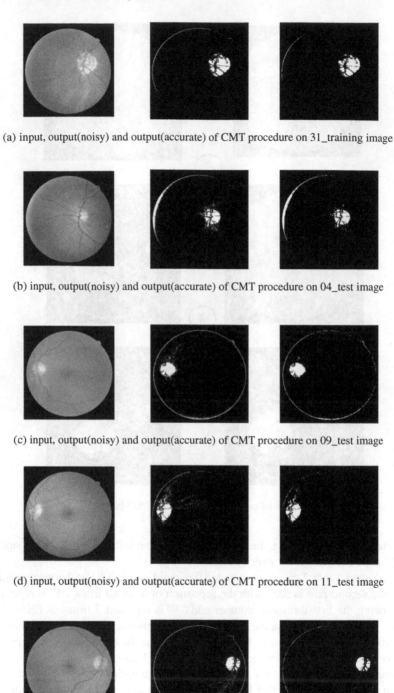

(a) input, output(noisy) and output(accurate) of CMT procedure on 31_training image

(b) input, output(noisy) and output(accurate) of CMT procedure on 04_test image

(c) input, output(noisy) and output(accurate) of CMT procedure on 09_test image

(d) input, output(noisy) and output(accurate) of CMT procedure on 11_test image

(e) input, output(noisy) and output(accurate) of CMT procedure on 16_test image

**Fig. 3.** Results of Kapur-ScPSO based CMT on 5 retinal images

**Table 2.** First and second threshold numbers in trials

| Image | Threshold number (First & second) | Image | Threshold number (First & second) |
|---|---|---|---|
| 25_training | 5 & 3 | 37_training | 4 & 1 |
| 26_training | 5 & 4 | 38_training | 5 & 4 |
| 27_training | 4 & 3 | 40_training | 3 & 2 |
| 28_training | 5 & 3 | 02_test | 8 & 2 |
| 29_training | 4 & 2 | 04_test | 5 & 4 |
| 31_training | 12 & 17 | 07_test | 3 & 4 |
| 33_training | 5 & 4 | 09_test | 5 & 4 |
| 34_training | 7 & 2 | 10_test | 5 & 4 |
| 35_training | 2 & 1 | 11_test | 5 & 3 |
| 36_training | 2 & 1 | 16_test | 5 & 3 |

**Table 3.** Comparison of proposed method with literature studies by means of computation time

| Algorithm | Database/no of OD images | Year | Computation time per image |
|---|---|---|---|
| Region Growing Technique [2] | —/8 image | 2014 | 16 s |
| 2-step automatic thresholding procedure [3] | MESSIDOR/2948 images | 2015 | 5.425 s |
| Adaptive morphological approach [4] | DIARETDB1/89 images | 2010 | 12.17 s |
| Combination of vascular and intensity information [7] | DRIVE/40 images | 2013 | 8 s |
| MSGVF [9] | DRIVE/40 images | 2013 | 22 s |
| Kapur-ScPSO based CMT | DRIVE/20 images | 2015 | 7–8 s |

**Table 4.** Comparison of proposed method with literature studies by means of 3 metrics

| Study | Database/OD images | Overlap | Precision | Accuracy |
|---|---|---|---|---|
| Welfer et al. [4] | DRIVE/40 images | 41.47 | — | 100 |
| Kumara and Meegama [10] | —/130 images | — | — | 90 |
| Issac et al. [11] | —/63 images | — | — | 92.06 |
| Mohamed et al. [12] | RIM-One/12 images | — | 73.90 | 89.51 |
| Kapur-ScPSO based CMT | DRIVE/20 images | 57.89 | 77.08 | 95.59 |

## 4   Conclusions

In this paper, a novel segmentation approach is proposed for optic disc segmentation. This approach is a cascade multithresholding procedure including a novel optimization algorithm and Kapur function. Also, performance analysis of this method is realized by using computation time, precision, overlap and accuracy metrics.

As seen in statistical results, ScPSO-Kapur based CMT obtains a high accuracy rate on OD segmentation. When proposed method is compared with the Adaptive Morphological Approach (AMA) of Welfer et al. [4], it's seen that accuracy of our method is 4.41 % less than AMA. However, proposed method overrides AMA with a 16.42 % difference in terms of overlap ratio which specifies the performance of accuracy on only OD area. Furthermore, proposed method achieves to 77.08 % precision which is 3.18 % higher than Local Binary Patterns (LBP) based approach of Mohamed et al. [12]. Also, ScPSO-Kapur based CMT obtains higher accuracy (9.74 %) than LBP based approach. Besides, proposed method achieves to better accuracy (95.59 %) than Active Contour-Based Segmentation [10] and Adaptive Threshold based Algorithm [11].

According to visual results, it's seen that proposed algorithm obtains OD successfully, but it needs a supporter technique which achieves to only optic disc area without any noise or without other parts of retina. From the computation time comparison, proposed method usually exhibits a prior performance to others by achieving the segmentation in only 7–8 s for an image.

In future work, we want to improve the CMT algorithm which obtains better segmentation performance without any noise. Furthermore, an adaptive CMT algorithm will be proposed which can be applied to segmentation processes on different disciplines meaning that multithresholding pairs of this technique will change according to the used datasets.

**Acknowledgement.** This work is supported by the Coordinatorship of Selcuk University's Scientific Research Projects.

# References

1. Niemeijer, M., Abràmoff, M.D., Van Ginneken, B.: Fast detection of the optic disc and fovea in color fundus photographs. Med. Image Anal. 13(6), 859–870 (2009)
2. Singh, A., Dutta, M.K., Parthasarathi, M., Burget, R., Riha, K.: An efficient automatic method of optic disc segmentation using region growing technique in retinal images. In: 2014 IEEE International Conference on Contemporary Computing and Informatics (IC3I), pp. 480–484. IEEE Press, New York (2014)
3. Marin, D., Gegundez-Arias, M.E., Suero, A., Bravo, J.M.: Obtaining optic disc center and pixel region by automatic thresholding methods on morphologically processed fundus images. Comput. Meth. Prog. Bio. 118(2), 173–185 (2015)
4. Welfer, D., Scharcanski, J., Kitamura, C.M., Dal Pizzol, M.M., Ludwig, L.W., Marinho, D. R.: Segmentation of the optic disk in color eye fundus images using an adaptive morphological approach. Comput. Biol. Med. 40(2), 124–137 (2010)
5. Muramatsu, C., Nakagawa, T., Sawada, A., Hatanaka, Y., Hara, T., Yamamoto, T., Fujita, H.: Automated segmentation of optic disc region on retinal fundus photographs: comparison of contour modeling and pixel classification methods. Comput. Meth. Prog. Bio. 101(1), 23–32 (2011)
6. Esmaeili, M., Rabbani, H., Dehnavi, A.M.: Automatic optic disk boundary extraction by the use of curvelet transform and deformable variational level set model. Pattern Recogn. 45(7), 2832–2842 (2012)

7. Mendonca, A.M., Sousa, A., Mendonca, L., Campilho, A.: Automatic localization of the optic disc by combining vascular and intensity information. Comput. Med. Imag. Grap. **37**(5), 409–417 (2013)
8. Hsiao, H.K., Liu, C.C., Yu, C.Y., Kuo, S.W., Yu, S.S.: A novel optic disc detection scheme on retinal images. Expert Syst. Appl. **39**(12), 10600–10606 (2012)
9. Zhou, H., Li, X., Schaefer, G., Celebi, M.E., Miller, P.: Mean shift based gradient vector flow for image segmentation. Comput. Vis. Image Und. **117**(9), 1004–1016 (2013)
10. Kumara, M.R.S.P., Meegama, R.G.N.: Active contour-based segmentation and removal of optic disk from retinal images. In: 2013 International Conference on Advances in ICT for Emerging Regions (ICTer), pp. 15–20. IEEE Press, New York (2013)
11. Issac, A., Parthasarthi, M., Dutta, M. K.: An adaptive threshold based algorithm for optic disc and cup segmentation in fundus images. In: 2nd International Conference on Signal Processing and Integrated Networks (SPIN), pp. 143–147. IEEE, New York (2015)
12. Mohamed, N. A., Zulkifley, M. A., Hussain, A.: On analyzing various density functions of local binary patterns for optic disc segmentation. In: 2015 IEEE Symposium on Computer Applications & Industrial Electronics (ISCAIE), pp. 37–41. IEEE, New York (2015)
13. Koyuncu H., Ceylan R.: Scout particle swarm optimization. In: Proceedings of 6th European Conference of the IFMBE, pp. 82–85. Springer, Switzerland (2015)
14. Karaboga, D., Akay, B.: A comparative study of artificial bee colony algorithm. Appl. Math. Comput. **214**, 108–132 (2009)
15. Kapur, J.N., Sahoo, P.K., Wong, A.K.: A new method for gray-level picture thresholding using the entropy of the histogram. Comput. Vision Graph. **29**(3), 273–285 (1985)
16. Tuba, M.: Multilevel image thresholding by nature-inspired algorithms: a short review. Comput. Sci. J. Moldova. **22**(3), 318–338 (2014)
17. Staal, J.J., Abramoff, M.D., Niemeijer, M., Viergever, M.A., van Ginneken, B.: Ridge based vessel segmentation in color images of the retina. IEEE T. Med. Imaging. **23**, 501–509 (2004)

# Biomedical Signal Analysis

# Uncertainty in 1D and 3D Models of a Fiber Stimulated by an External Electrode

Wanda Krassowska Neu[(✉)]

Department of Biomedical Engineering, Duke University,
Box 90281, Durham, NC 27708, USA
wanda.neu@duke.edu
http://bme.duke.edu/faculty/wanda-neu

**Abstract.** One-dimensional (1D) cable model is used to study electrical excitation of nerves and muscle fibers, and to aid in the design of electrical therapies. However, approximations inherent in the cable model limit its validity. More realistic three-dimensional (3D) fiber models have been advocated but they require long computational times. This study investigates whether better accuracy of 3D models is worth the cost by computing the probability $p$ that the difference between outputs from 3D and 1D models could have arisen from uncertainties in parameter values. The results are summarized in contour maps of probability $p$ in the space of fiber-electrode distances and stimulus durations. The cable model is considered valid where $p > 0.05$. This region of validity depends on uncertainties in the parameters. In particular, the uncertainties must exceed 0.05 (0.02) of the nominal parameter values for the cable model to be valid in the regions where retinal (cochlear) implants operate.

**Keywords:** Electric potential · Stimulation · Passive fiber · Cable model · Uncertainty

## 1 Introduction

Success of electrical treatments involving nerves and skeletal muscle fibers, such as cochlear and retinal implants, depends strongly on the electrodes configuration and the stimulation protocol. Thus, planning electrical therapies or designing prosthetic devices would benefit from accurate information on where the stimulation takes place and which cells are stimulated or inhibited by the pulse. Some of this information can come from experiments, but in many cases experimental measurements are not feasible because of difficult access and small scale. Models are free from these limitations and, by computing spatial distribution and temporal evolution of the transmembrane potential, they can predict the cellular response to stimuli. Thus, models have been used extensively to elucidate the effect of electrical stimuli on nerves and muscle fibers, and to aid in the design of electrical therapies.

Currently, the standard modeling approach is based on the cable equation, which is a 1D approximation of a nerve or muscle fiber [9]. The 1D cable model

© Springer International Publishing Switzerland 2016
F. Ortuño and I. Rojas (Eds.): IWBBIO 2016, LNBI 9656, pp. 219–229, 2016.
DOI: 10.1007/978-3-319-31744-1_20

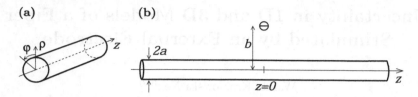

**Fig. 1.** (a) Sketch of a 3D fiber in cylindrical coordinates. (b) Stimulation of the fiber; ⊖ marks the position of the point electrode (cathode).

is so popular because its numerical solution is easy to implement and runs fast, so it can be used repeatedly in optimization studies. Thus, this model has been applied not only to the electrical stimulation of straight unmyelinated axons and skeletal muscle fibers, such as illustrated in Fig. 1, but also to myelinated axons, neurons of more complex geometry, and to the magnetic stimulation of neurons. Most importantly, the cable equation is used in all simulation packages for modeling neurons and their networks, such as NEURON or GENESIS [2].

However, the 1D cable equation ignores the presence of the stimulated fiber and its effects on the distribution of the potential established by the electrodes [4,17]. The importance of these effects is illustrated in Fig. 2, which compares the time evolution of the transmembrane potential in a 3D fiber ($\Phi_m$) and a 1D cable ($V_m$).[1] During the first microsecond the 3D fiber is polarized, with $\Phi_m$ on proximal and distal sides having opposite polarities (Fig. 2a, c). In contrast, the transmembrane potential $V_m$, computed from the 1D cable model, does not reproduce this transverse polarization because potential $V_m$ is an average of $\Phi_m$ over the fiber circumference. At steady state (Fig. 2b, d), the average transmembrane potential $V_m$ is approximately 30 times larger than the polarization of the fiber, so the differences between the 1D and 3D models are small. However, the average transmembrane potential takes over thousand times longer to develop than the transverse polarization. Consequently, one expects to find large discrepancies between 1D and 3D models during the first millisecond of the stimulus. Indeed, recent studies have shown that the cable model approximates transmembrane potential with an error below 5 % when the stimuli are longer than approximately 1 ms [16] and when the fiber-electrode distance is 0.2–4 mm [19].

Since many current therapies and prosthetic devices may operate outside these conditions, the use of 3D fiber models has been advocated and several such models have appeared in the literature [5,10,13,15,21,23]. However, 3D models are costly. First, one may need to write a dedicated solver because the boundary value problem governing the 3D fiber is nonstandard: time dependence appears not in the differential equation but in the continuity-of-current condition on the membrane. That precludes the use of most commercial software for solving partial differential equations, with the notable exception of COMSOL Multiphysics (COMSOL AB, Stockholm, Sweden). Second, a large number of

---

[1] In this paper, transmembrane potential is denoted by two symbols: $\Phi_m(\varphi, z, t)$ in the 3D model and $V_m(z, t)$ in the 1D model.

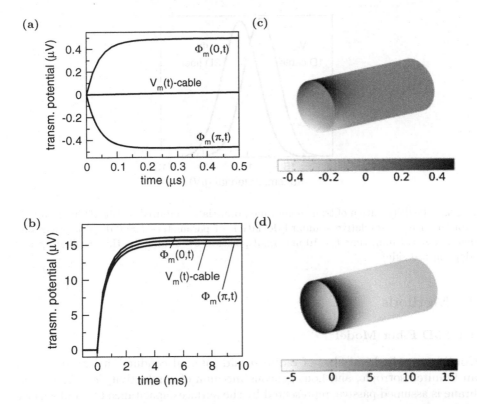

**Fig. 2.** (a-b)Time evolution of the transmembrane potential in a 3D fiber model $(\Phi_m)$ and a 1D cable model $(V_m)$ during the first 0.5 $\mu$s of the simulation and the entire 10-ms simulation. Included are $\Phi_m$ at $\varphi = 0$ (depolarized) and $\varphi = \pi$ (hyperpolarized) sides; axial position $z = 0$. (c-d) $\Phi_m$ over half fiber length $(z \geq 0)$ at $t = 0.5$ $\mu$s and 10 ms. In all panels, the fiber-electrode distance is 1 mm.

degrees of freedom is required to model a long and thin fiber, which results in long run times. It is understandable that the neuroscience community has been slow to adopt 3D models for practical use.

This study addresses the question that has not been asked before: given uncertainty in model parameters, is better accuracy of the 3D models worth the cost? Parameters of the models are known only approximately: measurement errors and intrinsic variability of biological quantities introduce some amount of uncertainty. This uncertainty propagates to the output of the model so the computed transmembrane potentials, $\Phi_m$ and $V_m$, should be viewed as random variables described by distributions. Figure 3 shows that under some experimental conditions, there is a large overlap between distributions of $\Phi_m$ and $V_m$, and that the difference between 3D and 1D models' results may not be significant compared to the width of the distributions. Therefore, present study revisits the question of the validity of the 1D cable model by taking into account uncertainties in the results introduced by uncertainties in the model parameters.

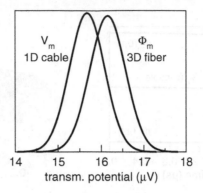

**Fig. 3.** The distribution of transmembrane potentials computed by the 3D fiber and 1D cable models. The relative standard deviation of parameters $f = 0.01$, fiber-electrode distance $b = 1$ mm, time $t = 10$ ms, axial position $z = 0$ and, for the 3D fiber, $\varphi = 0$ (depolarized side).

## 2    Methods

### 2.1    3D Fiber Model

Consider a cylindrical fiber of radius $a$ and internal conductivity $\sigma_i$, placed in an infinite, isotropic, and homogeneous medium of conductivity $\sigma_e$. The membrane is assumed passive, represented by the surface capacitance $C_m$ and surface resistance $R_m$, and the rest potential is zero. The fiber is stimulated by an extracellular electrode (Fig. 1b).

Potential $\Phi_i$ inside the fiber is governed by Laplace's equation and potential $\Phi_e$ outside the fiber is governed by Poisson's equation,

$$\nabla^2 \Phi_i = 0 \text{ in } \rho < a \quad \text{and} \quad \nabla^2 \Phi_e = -\frac{I_s}{\sigma_e}\,\delta(\rho - b, \varphi, z)\,u(t) \text{ in } \rho > a. \quad (1)$$

Using cylindrical coordinates $\rho, \varphi$ and $z$ defined in Fig. 1a, $\delta(\rho - b, \varphi, z)$ indicates that the current $I_s$ is delivered by a point electrode located at the distance $b$ from the fiber (Fig. 1b); $u(t)$ indicates that the current is a step function in time. Extracellular potential $\Phi_e$ decays to zero away from the electrode. Current density is continuous across the membrane:

$$\sigma_i \frac{\partial \Phi_i}{\partial \rho} = \sigma_e \frac{\partial \Phi_e}{\partial \rho} = -C_m \frac{\partial \Phi_m}{\partial t} - \frac{\Phi_m}{R_m} \text{ on } \rho = a. \quad (2)$$

As an initial condition, we assume that at $t = 0$, $\Phi_i$ and $\Phi_e$ are equal to zero.

Potentials $\Phi_i(\rho, \varphi, z, t)$ and $\Phi_e(\rho, \varphi, z, t)$ are determined from the boundary value problem (1-2) using separation of variables [16]. Knowing $\Phi_i$ and $\Phi_e$, the transmembrane potential $\Phi_m(\varphi, z, t) = \Phi_i(a, \varphi, z, t) - \Phi_e(a, \varphi, z, t)$ is expressed in the following form:

$$\Phi_m(\varphi, z, t) = -\sum_{n=0}^{\infty} \left[ \int_0^{\infty} \gamma(n, k) \frac{I_n - \frac{I_n'}{K_n'} K_n}{1 + \frac{1}{kR_m} \left( \frac{1}{\sigma_i} \frac{I_n}{I_n'} - \frac{1}{\sigma_e} \frac{K_n}{K_n'} \right)} \left( 1 - e^{-t/\tau(n,k)} \right) \right.$$

$$\left. \cos(n\varphi) \cos(kz) dk \right], \quad (3)$$

where $I_n$ and $K_n$ are modified Bessel functions and $I_n'$ and $K_n'$ are their derivatives with respect to the argument, evaluated at $\rho = a$. The effect of the stimulus is expressed by $\gamma$,

$$\gamma(n, k) = \frac{I_s u(t)}{2\pi^2 \sigma_e} \epsilon_n K_n(kb) \quad \text{with } \epsilon_n = 1 \text{ for } n = 0 \text{ and } 2 \text{ for } n > 0, \quad (4)$$

and the reciprocal of the time constant $\tau(n, k)$ is

$$\frac{1}{\tau(n, k)} = \frac{1}{R_m C_m} + \frac{1}{\frac{C_m}{k} \left( \frac{1}{\sigma_i} \frac{I_n}{I_n'} - \frac{1}{\sigma_e} \frac{K_n}{K_n'} \right)}. \quad (5)$$

## 2.2  1D Cable Model

Under the assumption that the intracellular potential $\Phi_i$ has a constant value over any cross-section of the fiber, the boundary value problem (1-2) can be reduced to a 1D cable equation [9]:

$$\lambda^2 \frac{\partial^2 V_m}{\partial z^2} - \tau \frac{\partial V_m}{\partial t} - V_m = F(z, t), \quad (6)$$

where $V_m(z, t)$ is the transmembrane potential, $\lambda = \sqrt{a R_m \sigma_i / 2}$ is the length constant of the fiber, and $\tau = R_m C_m$ is the time constant of the membrane. The effect of the stimulus is represented by the activating function $F(z, t)$ on right-hand side of (6) [14, 18]. For a fiber stimulated with an external point source (Fig. 1b), the activating function is:

$$F(z, t) = -\frac{\lambda^2 I_s u(t)}{4\pi \sigma_e} \left( \frac{3z^2}{R^5} - \frac{1}{R^3} \right), \quad (7)$$

with $R = \sqrt{z^2 + b^2}$.

The transmembrane potential $V_m$ can be computed as a convolution of the activating function $F$ and the fundamental solution $G$ of the cable equation:

$$V_m(z, t) = \int_{-\infty}^{\infty} F(x, t) G(x - z, t) \, dx. \quad (8)$$

The fundamental solution $G$ of (6) is [9]:

$$G(z, t) = -\frac{1}{4\lambda} \left[ e^{-|z|/\lambda} \operatorname{erfc} \left( \frac{|z|}{2\lambda} \sqrt{\frac{\tau}{t}} - \sqrt{\frac{t}{\tau}} \right) - e^{|z|/\lambda} \operatorname{erfc} \left( \frac{|z|}{2\lambda} \sqrt{\frac{\tau}{t}} + \sqrt{\frac{t}{\tau}} \right) \right],$$
$$(9)$$

where erfc is the complementary error function.

## 2.3  Uncertainty of Models' Predictions

Both 3D and 1D models depend on seven parameters, $P_i$, $i = 1,...7$, whose nominal values $\overline{P_i}$ are given in Table 1. Values of these parameters are assumed to be normally distributed around $\overline{P_i}$ with a standard deviation (SD) $\sigma(P_i) = f\overline{P_i}$. The relative standard deviation $f$ is assumed the same for each parameter and it varies from 0.01 to 0.2.

**Table 1.** Nominal values of the parameters

| Fiber radius | $a$ | 3 | $\mu$m |
|---|---|---|---|
| Surface capacitance | $C_m$ | 0.01 | F/m$^2$ |
| Surface resistance | $R_m$ | 0.1476 | $\Omega$m$^2$ |
| Intracellular conductivity | $\sigma_i$ | 2.82 | S/m |
| Extracellular conductivity | $\sigma_e$ | 1 | S/m |
| Fiber-electrode distance | $b$ | 4.5 $\mu$m to 10 cm | |
| Stimulating current | $I_s$ | -1 | $\mu$A |

The uncertainty in the transmembrane potential computed by the models is evaluated from the Gaussian error propagation formula [22]. For the 3D fiber,

$$\sigma(\Phi_m) = \sqrt{\sum_{i=1}^{7} \left[\frac{\partial \Phi_m}{\partial P_i}\sigma(P_i)\right]^2}, \tag{10}$$

where $\sigma(\Phi_m)$ represents the standard deviation characterizing the distribution of $\Phi_m$. Partial derivatives of $\Phi_m$ are computed by differentiating formula (3) with respect to each of the seven parameters listed in Table 1. The same approach is used to evaluate $\sigma(V_m)$ of the 1D cable model.

To ascertain whether potentials computed by 3D and 1D models are sufficiently different to overcome the uncertainties introduced by the parameters, we compute the z-score of the difference between their mean values $\overline{\Phi}_m$ and $\overline{V}_m$ [11],

$$Z = \frac{\overline{\Phi}_m - \overline{V}_m}{\sqrt{[\sigma(\Phi_m)]^2 + [\sigma(V_m)]^2}}. \tag{11}$$

Assuming normal distribution of the difference between potentials, $Z$ is used to compute the probability $p$ that uncertainties in parameters could cause a difference between potentials at least as large as $\overline{\Phi}_m - \overline{V}_m$. We assume that $p$ must exceed some value, say $p > 0.05$, for the 1D cable model to be considered valid.

The comparison between $\Phi_m$ and $V_m$ can be performed for any value of time $t$, axial position $z$ and, in the 3D model, angle $\varphi$. Here, we compare the maximum values of potentials (at $z = 0$ and $\varphi = 0$) as they correlate to the physiologically important threshold for excitation. The comparison is carried out for several values of time $t$ (0.1 $\mu$s to 0.1 s) and several distances $b$ between the fiber and the electrode (4.5 $\mu$m to 10 cm).

# 3   Results

## 3.1   Significance of the Difference Between Models' Predictions

The significance of the difference between models' predictions is summarized in Fig. 4a by plotting probability $p$ as a function of fiber-electrode distance and time. In this map, the relative SD of the parameters is small, $f = 0.01$. The white region of the map has $p < 0.05$, indicating that the difference between models is large compared to uncertainties in their outputs. The shaded region has $p > 0.05$, indicating that the difference between models is small compared to uncertainties in their outputs. In this shaded region, the simpler 1D model can be considered valid. Thus, for $f = 0.01$, the cable equation can be considered valid when the fiber-electrode distance is 0.18–10 mm and stimulus duration is above 7.2 ms. For stimuli between 0.18–7.2 ms, the range of fiber-electrode distances with the $p > 0.05$ shrinks, and it disappears completely for stimuli shorter than 0.18 ms.

As expected, the region of validity of the 1D cable model expands as the uncertainty in parameters grows. Figure 4b shows $p = 0.05$ contours for $f$ varying between 0.005 and 0.2. For the largest value of $f$, the cable model can be considered valid throughout most of the region, except for short stimuli and large fiber-electrode distances in the lower right corner of the map.

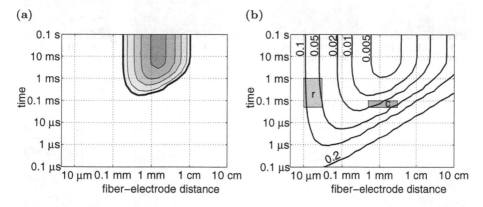

**Fig. 4.** Significance of the difference between the maxima of $\Phi_m$ and $V_m$ plotted as a function of the fiber-electrode distance and time. (a) Contours of probability $p = 0.05$ (heavy line), 0.1, 0.2, 0.3, and 0.4; relative SD $f = 0.01$. The white region has $p < 0.05$, indicating that here potentials of the 1D cable should beconsidered *different* than those of the 3D fiber. (b) The $p = 0.05$ contours for values of relative SD shown by the labels. Shaded areas show regions in which retinal (r) and cochlear (c) implants operate.

## 3.2    Contribution of Individual Parameters to Total Uncertainty

Figure 5 shows relative contributions of individual parameters to the uncertainty of the potentials. These contributions are dependent on time, with some parameters being more important for short pulses, while others for long pulses. The dependence on time changes with the fiber-electrode distance, as illustrated for four distances, 10 $\mu$m, 100 $\mu$m, 1 mm, and 1 cm. Moreover, the contribution of parameters is different in 3D and 1D models, especially for longer distances. Overall, the largest effect is due to the fiber-electrode distance $b$, stimulating current $I_s$ and, for short distances, extracellular conductivity $\sigma_e$. The least important parameters are $C_m$ and $R_m$, the surface capacitance and resistance of the membrane.

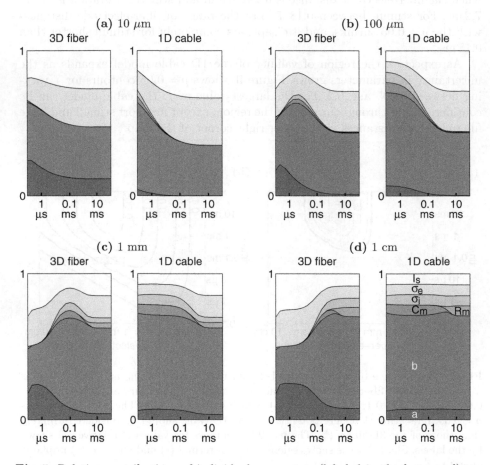

**Fig. 5.** Relative contributions of individual parameters (labeled in the last panel) to the SD of the transmembrane potentials, plotted as functions of time. (a-d) Four representative fiber-electrode distances. Relative SD $f = 0.01$.

# 4 Discussion

To assess the effect of approximations inherent in the 1D cable equation, two previous studies have investigated differences between transmembrane potentials computed with the 3D fiber and 1D cable models [16, 19]. These studies determined that the cable equation approximates the transmembrane potential of the 3D model with an error below 5 % when the fiber-electrode distance is 0.2–4 mm and stimulus duration is above 3.3 ms. For stimuli between 0.43–3.3 ms, the range of fiber-electrode distances with an error below 5 % shrinks, and it disappears completely for stimuli shorter than 0.43 ms.

In contrast, our study determines the validity of the 1D cable model by computing the probability $p$ that the difference between results from 3D and 1D models caused by uncertainties in parameters can be at least as large as the difference between the means. The region of validity of the 1D cable model, defined as the region with $p > 0.05$, is similar to the region determined by previous studies when the uncertainties in the parameters are small, 0.01 of their nominal value. However, the region of validity expands dramatically as the relative uncertainties in parameters increase to 0.2.

To put these findings in perspective: For retinal implants, stimulus phase durations range from 50 $\mu$s to 1 ms [1, 20] and the electrodes can be as close as 20–30 $\mu$m from their target ganglion cells and even closer to the layer of passing axons [6, 7]. For cochlear implants, stimulus phase durations are typically 50–100 $\mu$s [3] and the upper limit of the distance is determined by the height and width of scala tympani, 1–3 mm [8]. These regions, shaded in Fig. 4b, show that for retinal implants, the predictions the 1D cable are acceptable when the relative SD of parameters is above 0.05; for cochlear implants, the relative SD should be above 0.02. Therefore, the decision whether or not the 1D cable model can be used in a specific practical application depends on the uncertainty of model parameters.

This result points to the need for better estimates for the uncertainties in individual parameters. The preliminary study reported here assumed that all parameters were normally distributed with the same relative SD, which varied from 0.005 to 0.2. Consequently, the region of validity of 1D cable model depended on the assumed relative SD. A follow-up study is planned that will investigate uncertainties in individual parameters and that will compute probability $p$ from the true distribution of the difference between models' results instead of assuming the normal distribution. These improvements will lead to a realistic assessment of the region where the 1D cable model is valid for a specific application.

Assessing uncertainties in the parameters will be guided by the analysis of the contributions of different parameters to the total uncertainty of the potential. The follow-up study will concentrate on the most important parameters (fiber-electrode distance $b$, the stimulating current $I_s$ and the extracellular conductivity $\sigma_e$) and possibly ignore the least important ones ($C_m$ and $R_m$, the surface capacitance and resistance of the membrane). The two most important parameters, $b$ and $I_s$, are the ones that are determined by the experimenter, so in principle their uncertainty can be measured and, if desired, controlled.

The significant effect of $b$ and $I_s$ on the uncertainty of potentials calls attention to the importance of modeling the electrodes: their geometry and function should be as realistic as possible. Our study represented the electrode by a point source of current, which is rarely a good approximation for practical applications. Investigating the effect of the electrode geometry is outside the scope of this work but it should be explored in the future. It may turn out that modeling the electrodes has bigger effect on the potentials than modeling the fiber itself.

In our study, the uncertainty assessment has been performed on a straight cylindrical fiber, representing an unmyelinated axon or a skeletal muscle fiber with passive membrane. These restrictions were dictated by the use of analytical solutions of the 3D fiber and 1D cable models. In practical applications, 1D cable models are also used to represent myelinated axons, as well as neurons that bend, terminate, or include the soma and dendrites. In addition, the fiber membrane is usually assumed excitable, to allow modeling firing and propagation of an action potential. The uncertainty assessment should be expanded to those more realistic conditions. However, it would require a major change in methodology. With complex geometry of the fiber and with nonlinear model of the membrane, solutions to 3D fiber and 1D cable would need to be determined numerically. Consequently, the uncertainty of the results could no longer be assessed by the Gaussian error propagation formula; one needs to use a Monte Carlo or a spectral projection method [12]. These methods require repeated runs of the models to determine the probability distribution of their output and they can incorporate only a limited number of parameters in the analysis. Thus, the results on the contribution of individual parameters, obtained here with the analytical solutions, can guide future work involving more realistic models.

# References

1. de Balthasar, C., Patel, S., Roy, A., Freda, R., Greenwald, S., Horsager, A., Mahadevappa, M., Yanai, D., McMahon, M.J., Humayun, M.S., Greenberg, R.J., Weiland, J.D., Fine, I.: Factors affecting perceptual thresholds in epiretinal prostheses. Invest. Ophthalmol. Vis. Sci. **49**, 2303–2314 (2008)
2. Brette, R., Rudolph, M., Carnevale, T., Hines, M., Beeman, D., Bower, J.M., Diesmann, M., Morrison, A., Goodman, P.H., Harris Jr., F.C., Zirpe, M., Natschlger, T., Pecevski, D., Ermentrout, B., Djurfeldt, M., Lansner, A., Rochel, O., Vieville, T., Muller, E., Davison, A.P., Boustani, S.E., Destexhe, A.: Simulation of networks of spiking neurons: a review of tools and strategies. Comput. Neurosci. **23**, 349–398 (2007)
3. Chatterjee, M., Kulkarni, A.M.: Sensitivity to pulse phase duration in cochlear implant listeners: effects of stimulation mode. J. Acoust. Soc. Am. **136**, 829–840 (2014)
4. Clark, J., Plonsey, R.: A mathematical evaluation of the core conductor model. Biophys. J. **6**, 95–112 (1966)
5. Cranford, J.P., Kim, B.J., Krassowska Neu, W.: Asymptotic model of electrical stimulation of nerve fibers. Med. Biol. Eng. Comput. **50**, 243–251 (2012)

6. Eiber, C.D., Lovell, N.H., Suaning, G.J.: Attaining higher resolution visual prosthetics: a review of the factors and limitations. J. Neural Eng. **10**, 1–17 (2013). article 011002
7. Greenberg, R.J., Velte, T.J., Humayun, M.S., Scarlatis, G.N., de Juan Jr., E.: A computational model of electrical stimulation of the retinal ganglion cell. IEEE Trans. Biomed. Eng. **46**, 505–514 (1999)
8. Hatsushika, S., Shepherd, R., Tong, Y., Clark, G., Funasaka, S.: Dimensions of the scala tympani in the human and cat with reference to cochlear implants. Ann. Otol. Rhinol. Laryngol. **99**, 871–876 (1990)
9. Hodgkin, A.L., Rushton, W.A.H.: The electrical constants of a crustacean nerve fibre. Proc. Roy. Soc. B **133**, 444–479 (1946)
10. Joucla, S., Gliere, A., Yvert, B.: Current approaches to model extracellular electrical neural microstimulation. Front. Comput. Neurosci. **8**, 1–12 (2014). article 13
11. Kirkwood, B.R., Sterne, J.A.C.: Essential Medical Statistics, 2nd edn. Blackwell Science, Malden (2003)
12. Le Maître, O., Knio, O.: Spectral Methods for Uncertainty Quantification: With Applications to Computational Fluid Dynamics. Springer, Dordrecht-Heidelberg-London-New York (2010)
13. Leon, L.J., Hogues, H., Roberge, F.A.: A model study of extracellular stimulation of cardiac cells. IEEE Trans. Biomed. Eng. **40**, 1307–1319 (1993)
14. McNeal, D.R.: Analysis of a model for excitation of myelinated nerve. IEEE Trans. Biomed. Eng. **23**, 329–337 (1976)
15. Meffin, H., Tahayori, B., Grayden, D.B., Burkitt, A.N.: Modeling extracellular electrical stimulation: I. Derivation and interpretation of neurite equations. J. Neural Eng. **9**, 1–17 (2012). article 065005
16. Krassowska Neu, W.: Analytical solution for time-dependent potentials in a fiber stimulated by an external electrode. Med. Biol. Eng. Comput. (accepted)
17. Pickard, W.: A contribution to the electromagnetic theory of the unmyelinated axon. Math. Biosc. **2**, 111–121 (1968)
18. Rattay, F.: Analysis of models for external stimulation of axons. IEEE Trans. Biomed. Eng. **33**, 974–977 (1986)
19. Schnabel, V., Struijk, J.J.: Evaluation of the cable model for electrical stimulation of unmyelinated nerve fibers. IEEE Trans. Biomed. Eng. **48**, 1027–1033 (2001)
20. Sekirnjak, C., Hottowy, P., Sher, A., Dabrowski, W., Litke, A.M., Chichilnisky, E.J.: High-resolution electrical stimulation of primate retina for epiretinal implant design. J. Neurosc. **28**, 4446–4456 (2008)
21. Stickler, Y., Martinek, J., Rattay, F.: Modeling needle stimulation of denervated muscle fibers: voltage-distance relations and fiber polarization effects. IEEE Trans. Biomed. Eng. **56**, 2396–2403 (2009)
22. Taylor, J.R.: An Introduction to Error Analysis. University Science Books, Mill Valley, CA (1982)
23. Yu, H., Liu, X., Zheng, C., Wang, Y.: A novel model for excitation of peripheral nerves stimulated by a transverse electric field. Int. J. Wavelets Multiresolut. Inf. Process. **5**, 187–196 (2007)

# A Comparison of Feature Ranking and Rank Aggregation Techniques in Automatic Sleep Stage Classification Based on Polysomnographic Signals

Shirin Najdi[1,2(✉)], Ali Abdollahi Gharbali[1,2], and José Manuel Fonseca[1,2]

[1] Computational Intelligence Group of CTS/UNINOVA, Caparica, Portugal
[2] Faculdade de Ciências e Tecnologia, Universidade Nova de Lisboa Campus da Caparica,
Quinta da Torre, Monte de Caparica, Portugal
{s.najdi,a.gharbali}@campus.fct.unl.pt,
jmf@uninova.pt

**Abstract.** Sleep quality is one of the most important measures of healthy life, especially considering the huge number of sleep-related disorders. Identifying sleep stages using multi-channel recordings like polysomnographic (PSG) signals is an effective way of assessing sleep quality. However, manual sleep stage classification is time-consuming, tedious and highly subjective. To overcome this, automatic sleep classification was proposed, in which pre-processing, feature extraction and classification are the three main steps. Since the classification accuracy is deeply affected by the features selection, in this paper several feature selection methods as well as rank aggregation methods are compared. Feature selection methods are evaluated by three criteria: accuracy, stability and similarity. For classification two different classifiers ($k$-nearest neighbor and multi-layer feedforward neural network) were utilized. Simulation results show that MRMR-MID achieves highest classification performance while Fisher method provides the most stable rankings.

**Keywords:** Sleep stage classification · Feature selection · Rank aggregation · Feature ranking · Polysomnography · Biomedical signal processing

## 1 Introduction

Sleep disorders constitute a global epidemic affecting considerable number of the world's population. In addition to several physical and mental problems, sleep disorders can impose considerable costs to the national health care budget. Identification of sleep stages is the first step for diagnosing sleep related disorders and it is currently done through the visual inspection by an expert of polysomnographic (PSG) signals recorded from a subject's sleep session based on the rules of the American Academy of Sleep Medicine (AASM) [1]. According to AASM sleep is divided into five stages: *wake* (W), *stage 1* (S1), *stage 2* (S2), *slow wave stage* (SWS) and *rapid eye movement* (REM). Since manual identification of sleep stages is a time consuming, costly and subjective task, many different automatic sleep stage classification algorithms were proposed.

© Springer International Publishing Switzerland 2016
F. Ortuño and I. Rojas (Eds.): IWBBIO 2016, LNBI 9656, pp. 230–241, 2016.
DOI: 10.1007/978-3-319-31744-1_21

The common approach in automatic sleep stage classification consists of three main steps: pre-processing, feature extraction and classification. In the feature extraction stage, several linear and non-linear features are extracted in different time and frequency domains from PSG signals. Nevertheless some of these features may be redundant or irrelevant increasing the complexity of the model without any real benefit. Feature selection (FS) has been an important research topic for the researchers in the data mining and machine learning areas for a long time [2, 3]. However, in the sleep stage classification context, few studies applied FS on the extracted features before feeding them to the classifiers. In general, FS algorithms used in these studies are divided into three main categories: filter methods (such as Fisher score or the minimum Redundancy Maximum Relevance method (mRMR) [4, 5]), wrapper methods [6] and embedded methods [7–10]. Filter methods are simple and fast techniques which perform FS independently of the classifier by considering some intrinsic characteristics of the data providing a rank or a score for each feature. Low-ranked or low scored features are removed experimentally or according to a predefined threshold. The wrapper and the embedded methods, on the other hand, select a subset of features considering feature dependencies and the classification algorithm with higher computational complexity compared to the filter methods [11].

Different feature ranking techniques may produce different rankings according to their specific criteria for assessing features and there is no universal ranking algorithm that considers all the measures. Therefore, motivated by ensemble methods in supervised learning [12], rank aggregation methods are devised to combine different feature ranking methods and achieve more stable ranked feature lists and similar or even higher classification performance [3, 13]. In order to perform ensemble FS, one needs to decide on the method to aggregate the results from different ranking methods. There are many rank aggregation approaches from the very simple ones to some more complex [14]. To the best of our knowledge, there are no studies done on FS based on rank aggregation methods in the sleep stage classification area. In this paper different feature ranking and rank aggregation methods were compared within the sleep stage classification context. The main contributions of this paper are listed below:

1. A comprehensive feature set extracted from PSG signals including Itakura spectral distance (ISD) [15],
2. Similarity and stability comparison of different feature ranking and rank aggregation methods,
3. Classification performance comparison of different feature ranking and rank aggregation methods.

The paper is organized as follows: Sect. 2 explains the database and the information regarding the methods which are used. Section 3 gives a detailed overview of how experimental results compare the performance of FS methods. Section 4 finalizes the paper with the conclusions and future work directions.

## 2   Materials and Methods

In this section the sleep stage classification methodology used in this work is described in detail. Figure 1 shows the block diagram of the proposed algorithm for comparing the FS methods.

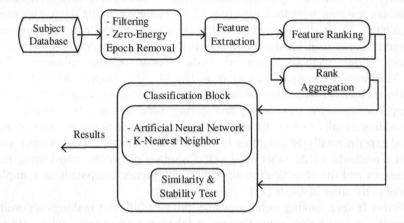

**Fig. 1.** Block diagram for comparison of FS methods

### 2.1   Data

The data used in this study was obtained from the Physionet Sleep-EDF Expanded Database [16]. This database is a collection of 22 polysomnographic (PSG) recordings including, EEG (Fpz-Cz and Pz-Oz), EOG (horizontal), submental chin EMG, together with the corresponding hypnograms. Except for slight difficulty in falling asleep, subjects were healthy without any sleep related medication. The data was segmented into 30-second epochs and all epochs were scored according to R&K guidelines for human sleep staging. In this paper Pz-Oz channel EEG recording together with submental chin EMG and horizontal EOG each sampled at 100 Hz were used in the evaluations.

### 2.2   Pre-processing

In order to guarantee the reliability of biomedical signal analysis, artifact free data is necessary. In this study for reducing the artifact, first the epochs with zero-energy were automatically detected and eliminated. Then EEG and EOG data were band-pass filtered with a low cut-off frequency of 0.3 Hz and a high cut-off frequency of 35 Hz according to the AASM manual for the sleep scoring [1]. For higher accuracy and less distortion, a wavelet packet (WP) decomposition and reconstruction method was used for filtering [17].

## 2.3 Feature Extraction and Normalization

In order to explore the information contained in PSG recordings, a set of features were extracted from EEG, horizontal EOG and submental chin EMG of each subject. This feature set includes 49 features which can be categorized into time, frequency, joint time-frequency domain, entropy-based and nonlinear types. These features, as summarized in Table 1, were extracted from 30-second signal segments.

In order to analyze the stochastic nature of EEG, we chose the wavelet packet analysis since it provides a valuable joint time-frequency domain analysis. In clinical applications, four main brain rhythms are associated with different states of sleep, including Delta (0–3.99 Hz), Theta (4–7.99 Hz), Alpha (8–13 Hz) and Beta (>13 Hz) [1]. According to the scheme proposed in [18], a WP tree with 7 decomposition levels was suitable to estimate the necessary frequency bands of EEG rhythms with adequate accuracy. After estimating the EEG frequency bands, features F13 to F26 were extracted from the corresponding WP coefficients according to Table 1.

The Itakura Spectral Distance (ISD) is broadly used in speech processing applications to measure the distance (similarity) between 2 auto regressive coefficients (AR) processes [19, 20]. ISD was also used in automatic sleep classification to find the relation between EEG and EOG signals during different epochs of sleep stages over the night [15]. In this paper ISD of sleep stages of EEG was measured. In order to calculate the distances the AR coefficients were extracted from 50 % of the wake epochs of each subject. Then, by getting the mean of the AR coefficients a representative model of wake epoch was generated and the ISD between this model and the W (remaining 50 %), S1, S2, SWS and REM epochs were calculated.

To avoid that features with greater numeric values dominate those with smaller numeric values affecting the classification performance, each feature $(x_{ij})$ is independently scaled to have zero mean and unit variance using the following equation:

$$x'_{ij} = \frac{x_{ij} - \bar{\mathbf{x}}_i}{\sigma_{\mathbf{x}_i}} \tag{1}$$

in which $\mathbf{x}_i$ is a vector of each independent feature, $\bar{\mathbf{x}}_i$ and $\sigma_{\mathbf{x}_i}$ are the mean and the standard deviation of each independent feature vector.

## 2.4 Feature Ranking Methods

In this paper, to select a subset of features containing most of the information of the original feature set, we used seven different feature ranking methods: ReliefF, Minimum Redundancy-Maximum Relevance (MRMR-MID and MRMR-MIQ), Fisher Score, Chi-Square (CHI), Information Gain (IG) and Conditional Mutual Information Maximization (CMIM). We have also implemented two different rank aggregation methods, Borda and Robust Rank Aggregation (RRA), to evaluate their ability to produce better feature rankings compared to feature ranking methods. A brief description of the used feature ranking methods is provided below:

**Table 1.** Summary of the features extracted from PSG recordings.

| Signal | Category | Feature name |
|--------|----------|--------------|
| EEG | Time domain (F1 to F12) | Statistical Features (Minimum Value, Maximum Value, Arithmetic Mean, Standard Deviation, Variance, Skewness, Kurtosis, Median), Zero-crossing Rate, Hjorth Parameters (Activity, Mobility and Complexity) [5] |
| | Time-frequency domain (F13 to F26) | Features Extracted from Wavelet Packet Coefficients including Energy of $\alpha, \delta, \beta1, \beta2, \theta$ and Spindle bands, Total Energy of all bands, Energy ratio of ($\frac{\alpha}{\delta+\theta}, \frac{\delta}{\alpha+\theta}, \frac{\theta}{\alpha+\delta}, \frac{\delta}{\theta}, \frac{\alpha}{\theta}$), Statistical Features (mean and standard deviation of coefficients in all of the bands) |
| | Entropy (F27 to F30) | Spectral Entropy, Rényi Entropy, Approximate Entropy, Permutation Entropy [5] |
| | Non-linear (F31 to F36) | Petrosian Fractal Dimension, Mean Teager Energy, Mean Energy, Mean Curve Length, Hurst Exponent [5], ISD |
| EOG | Time domain (F37 to F41) | Mean, Maximum, Standard Deviation, Skewness, Kurtosis [21] |
| | Non-linear (F42) | Energy [21] |
| EMG | Frequency domain (F43 to F46) | Total Power in the EMG Frequency Spectrum, Statistical Features of EMG Frequency Spectrum (Maximum, Mean, Standard Deviation) [21] |
| | Non-linear (F47 to F49) | Energy, Ratio of the EMG Signal Energy for the Current Epoch and Previous Epoch, Ratio of the EMG Signal Energy for the Current Epoch and Next Epoch [21] |

**ReliefF**: In 1992, Kira and Rendell [22] proposed an instance based method, Relief, for estimating the quality of features. In this method for a randomly selected sample two nearest neighbors were considered: one from the same class (nearest *hit*) and the other from different class (nearest *miss*). The quality estimation value for each feature is updated according to the randomly selected sample's distance from the nearest hit and miss. The Relief method is restricted to two-class problems and is highly sensitive to noisy and incomplete data. An extension of Relief, called ReliefF [23], was proposed improving the original method by estimating the probabilities more reliably and extending the algorithm to multi-class problems. The ReliefF algorithm uses $k$-nearest hits and $k$-nearest misses for updating the quality estimation for each feature.

**Minimum Redundancy-Maximum Relevance**: MRMR [24] is a FS method which selects a subset of features with maximum relevance for the target class and at the same time minimum redundancy between the selected features. In MRMR method the redundancy (R) and relevance (D) are expressed in terms of mutual information. In order to select the final feature set, an objective function $\varphi(D, R)$ is maximized. The $\varphi(D, R)$ can be defined either as the mutual information difference (MID), *D-R*, or the mutual information quotient (MIQ), *D/R*.

**Fisher Score**: This method is one of the efficient and the most widely used feature ranking methods. The key idea is to find a subset of feature matrix with maximum distance between the data points from different classes and minimum distance between data points of the same class in the feature space [25].

**Chi-Square**: CHI is another very common class sensitive feature selection method which ranks the features according to their CHI statistics without taking into account the interactions between features. Originally proposed for categorical data, this method was later extended to the continuous data [26]. For calculating the CHI statistics of each feature, the range of the numerical feature should be discretized into intervals.

**Information Gain**: Ross Quinlan proposed an algorithm for generating decision trees from a set of training data [27]. In this algorithm, IG is the measure for selecting the effective feature at each node. Generally IG can be described as the change in the marginal entropy of a feature set taking into account the conditional entropy of that feature set with the given class set.

**Conditional Mutual Information Maximization**: This method [28] is based on mutual information in such a way that all the selected features are informative and have two-by-two weak dependency. A feature is added to the selected feature subset if it carries information about the specific class and this information is not caught by any other previously selected feature.

**Borda**: The Borda algorithm is a feature aggregation method that ranks each feature based on its mean position in the different ranking methods considered, i.e.

$$Borda(f_i) = \sum_{j=1}^{N} \pi_j(f_i) \tag{2}$$

where $\pi_j(f_i)$ is the rank of the feature $f_i$ in the ranking method $\pi_j$. The feature with the highest Borda rank is considered the best.

**Robust Rank Aggregation**: This method proposed by Kolde et al. [29] is another rank aggregation method that compares the results from several feature ranking methods with a randomly ranked feature list. The RRA first looks how a specific feature is ranked by the various methods and lists the corresponding values in a so-called *rank order*, from best to worst. It is clear that if a feature has high quality, the dominance of ranks in the rank order will be towards smaller numbers. The probability of random list producing better ranking than the values seen in the actual rank order for that specific feature is determined. The features with the small probability are selected as better ones [30].

## 2.5    Classification

For discriminating between five sleep stages *W, S1, S2, SWS* and *REM*, in this study we selected two simple and widely used classifiers: $k$-nearest neighbor *($k$-NN)* and Multi-layer Feedforward Neural Network (MLFN). In this paper, by selecting $k = 1$, nearest neighbor was utilized. The Nearest Neighbor classifier is the simplest nonparametric classifier and assigns a pattern to a specific class based on its nearest neighbor's class. Despite of its simplicity, Duda et al. in [31] have proved that if one has fairly large collection of data, the error bound for nearest neighbor rule is quite tight, i.e. equal or less than twice the Bayes error. On the other hand, neural networks have been proved to be a very powerful computing paradigm which can learn from examples and gener-alize to the cases never seen before. Neural networks have been successfully applied to a wide range of data mining applications including classification [32].

# 3    Experimental Results and Discussion

To evaluate the feature selection methods six subjects were selected from the database described in Sect. 2.1. In the filtering stage of PSG signals, Daubechies order 20 (db20) was used as mother wavelet. After feature extraction and normalization, the feature set was fed into seven feature ranking methods. In order to combine the resulting ranked feature lists, Borda and RRA used the outputs of these seven methods, producing two additional ranked list of features. In the classification stage, the Euclidean distance was chosen as the distance metrics for the nearest neighbor classifier. In addition to the nearest neighbor classifier, a MLFN with 12 neurons and sigmoid transfer function was also used in our simulations. The Levenberg-Marquardt training algorithm was preferred for minimizing the cost function because of its fast and stable convergence. For training the classifiers, unlike the more conventional approaches in the literature, which imports all the existing epochs to the classifier, we used a quantity of epochs selected out of each subject. In this method, selected epochs from each subject have two characteristics. Firstly, the quantity of total epochs are the same for all the subjects, secondly, the selected number of epochs for each stage is dependent to the occurrence number of that stage for each subject. This method is suitable for large databases helping on the compu-tational complexity reduction of the classifier training stage.

## 3.1    Performance Assessment

In this paper three main criteria including stability, accuracy and similarity are consid-ered for evaluating and comparing different FS techniques. Stability of a FS method is defined as its sensitivity to variations in training set. Unstable FS may lead to inferior classification performance. In this work, in order to assess the stability of feature rank-ings, produced by different methods, a similarity based approach proposed by Kalousis et al. [33] was used.

In our simulations for each feature selection method $N = 50$ subsets were generated by bootstrapping. The stability of each method was evaluated as a function of the number of selected features ($d$) in which $d = 1, 3, 5\dots 29$. According to Fig. 2 while Fisher method shows the highest stability, CMIM method comes out to be the least stable one. Also, the stability of CHI and IG methods seems very convergent.

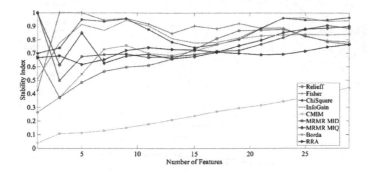

**Fig. 2.** Stability measure of each feature selection method

In this work, the overall accuracy value was calculated as the ratio of truly classified epochs to the total number of epochs [34]. In order to estimate the generalization ability of the classifier, *repeated random sub-sampling validation* with 200 runs was applied. Figure 3 shows the accuracy of the classifiers with respect to the number of selected features.

As Fig. 3 shows starting from one feature, each additional feature typically leads to an increment in the classification accuracy. However, at some point, the increment of the classification accuracy for each additional feature is not significant leading to an elbow in the graph. Inspired by the "elbow" point in the cost-benefit curves, in this work we used the Kneedle algorithm proposed in [35] for determining the optimal feature number which provides a satisfactory trade-off between selected number of features and classification accuracy.

Table 2 illustrates the top 10 features selected by each method. As it can be seen, F36 (which stands for ISD) appears always in the top 10. It means that ISD is a preferable feature for all the selection methods. Also, noticeable differences can be observed in the ranking of the selected features. Furthermore, the optimum number of features for each method, which is selected by the Kneedle algorithm, is shown in Table 2. For MLFN and *k*-NN classifiers, a slight difference exists in the optimum number. Considering the maximum accuracy that the methods reach in their optimum points, the MRMR-MID method outperforms all the others.

The CMIM method for both the classifiers reaches its best accuracy in the first 3 features. Considering Fig. 3, its accuracy is equal or less than the MRMR-MID method's accuracy at that point. Unanticipatedly, none of the aggregation methods outperformed the rest of the feature ranking methods. One possible reason for this is that the aggregation methods, especially Borda, is affected by the performance of all the methods from best to worst.

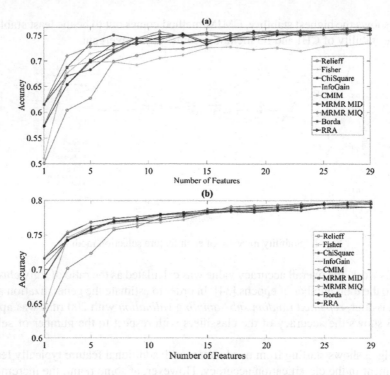

**Fig. 3.** Classification accuracy for different feature selection methods. (a) *k*-NN classifier, (b) MLFN classifier

**Table 2.** Top 10 features selected by each method and the corresponding optimum number selected by Kneedle algorithm.

|  | Relieff | Fisher | CHI | IG | CMIM | MRMR-MID | MRMR-MIQ | Borda | RRA |
|---|---|---|---|---|---|---|---|---|---|
| | F28 | **F36** | F35 | F9 | F15 | F35 | F35 | **F36** | **F36** |
| | **F36** | F35 | F9 | F35 | **F36** | F39 | F42 | F35 | F35 |
| | F7 | F31 | F11 | F11 | F9 | **F36** | F15 | F9 | F9 |
| | F49 | F9 | F31 | F31 | F8 | F22 | **F36** | F31 | F31 |
| | F41 | F29 | **F36** | **F36** | F1 | F15 | F22 | F22 | F27 |
| | F27 | F11 | F27 | F4 | F34 | F31 | F23 | F27 | F22 |
| | F20 | F25 | F26 | F27 | F35 | F29 | F31 | F29 | F17 |
| | F23 | F27 | F4 | F26 | F28 | F23 | F38 | F11 | F29 |
| | F6 | F12 | F25 | F25 | F6 | F9 | F29 | F15 | F11 |
| | F22 | F22 | F14 | F29 | F48 | F38 | F9 | F20 | F20 |
| **MLFN** | 7 | 5 | 7 | 7 | **3** | 5 | 5 | 5 | 7 |
| ***k*-NN** | 7 | 5 | 9 | 9 | **3** | 7 | 11 | 9 | 7 |

**Table 3.** Similarity of the feature selection technique

| | ReliefF | Fisher | CHI | IG | CMIM | MRMR-MID | MRMR-MIQ | Borda | RRA |
|---|---|---|---|---|---|---|---|---|---|
| **ReliefF** | 1 | 0.26 | 0.18 | 0.18 | 0.35 | 0.40 | 0.40 | 0.31 | 0.31 |
| **Fisher** | | 1 | 0.58 | 0.52 | **0.11** | 0.58 | 0.65 | 0.72 | 0.65 |
| **CHI** | | | 1 | **0.90** | 0.15 | 0.35 | 0.35 | 0.52 | 0.52 |
| **IG** | | | | 1 | 0.18 | 0.35 | 0.35 | 0.46 | 0.46 |
| **CMIM** | | | | | 1 | 0.22 | 0.22 | 0.22 | 0.22 |
| **MRMR-MID** | | | | | | 1 | **0.90** | 0.72 | 0.65 |
| **MRMR-MIQ** | | | | | | | 1 | 0.72 | 0.65 |
| **Borda** | | | | | | | | 1 | 0.72 |
| **RRA** | | | | | | | | | 1 |

The stability measure used for assessing the internal stability of a FS technique can also be used in a different context to assess the similarity of different FS techniques. The similarity measure provides information about the consistency and diversity of different FS algorithms [33]. According to Table 3 CHI and IG pair and MRMR-MID and MRMR-MIQ pair generate highly similar results. The similarity of MRMR methods can be explained by their likeness of theoretical background. On the contrary, CMIM and Fisher methods give the most dissimilar results. The average similarity of Borda and RRA methods are approximately 0.5 with the other methods. Regarding the aggregation characteristics it was predictable.

## 4 Conclusions and Future Work

Feature selection based on filtering techniques has several advantages such as being fast, easily scalable to high-dimensional datasets, decreasing computational complexity and working independently of the classifiers. In this paper we compared a group of seven feature ranking methods. Also, rank aggregation methods are believed to be robust through the broad variety of classifiers and produce comparable classification accuracy to the individual feature selection methods. So, two rank aggregation methods were applied to sleep stage classification problem. The Physionet Sleep-EDF Expanded Database was used to assess the effect of these methods on the classification accuracy of k-NN and MLFN. The stability and the similarity of the different FS methods were also studied. The results indicate that the MRMR-MID method slightly outperforms the other FS methods from the accuracy point of view. The Fisher method produces the most stable results. Although aggregation methods are known to generate more stable and better accuracy, in our simulations their performance was in the average level. Future steps could involve verifying the results with different databases as well as applying more rank aggregation methods.

**Acknowledgment.** This work was partially funded by FCT Strategic Program UID/EEA/ 00066/203 of UNINOVA, CTS and INCENTIVO/EEI/UI0066/2014 of UNINOVA.

# References

1. The AASM Manual for the Scoring of Sleep and Associated Events-Rules, Terminology and Technical Specifications. http://www.aasmnet.org/scoringmanual/
2. Abeel, T., Helleputte, T., Van de Peer, Y., Dupont, P., Saeys, Y.: Robust biomarker identification for cancer diagnosis with ensemble feature selection methods. Bioinformatics **26**, 392–398 (2010)
3. Prati, R.C.: Combining feature ranking algorithms through rank aggregation. In: Proceedings of the International Joint Conference on Neural Networks, pp. 10–15 (2012)
4. Herrera, L.J., Fernandes, C.M., Mora, A.M., Migotina, D., Largo, R., Guillen, A., Rosa, A.C.: Combination of heterogeneous EEG feature extraction methods and stacked sequential learning for sleep stage classification. Int. J. Neural Syst. **23**, 1350012 (2013)
5. Şen, B., Peker, M., Çavuşoğlu, A., Çelebi, F.V.: A comparative study on classification of sleep stage based on EEG signals using feature selection and classification algorithms. J. Med. Syst. **38**, 18 (2014)
6. Koley, B., Dey, D.: An ensemble system for automatic sleep stage classification using single channel EEG signal. Comput. Biol. Med. **42**, 1186–1195 (2012)
7. Zoubek, L., Charbonnier, S., Lesecq, S., Buguet, A., Chapotot, F.: Feature selection for sleep/ wake stages classification using data driven methods. Biomed. Signal Process. Control **2**, 171–179 (2007)
8. Özşen, S.: Classification of sleep stages using class-dependent sequential feature selection and artificial neural network. Neural Comput. Appl. **23**, 1239–1250 (2013)
9. Adnane, M., Jiang, Z., Yan, Z.: Sleep–wake stages classification and sleep efficiency estimation using single-lead electrocardiogram. Expert Syst. Appl. **39**, 1401–1413 (2012)
10. Lajnef, T., Chaibi, S., Ruby, P., Aguera, P.-E., Eichenlaub, J.-B., Samet, M., Kachouri, A., Jerbi, K.: Learning machines and sleeping brains: automatic sleep stage classification using decision-tree multi-class support vector machines. J. Neurosci. Methods **250**, 94–105 (2015)
11. Saeys, Y., Inza, I., Larrañaga, P.: A review of feature selection techniques in bioinformatics. Bioinformatics **23**, 2507–2517 (2007)
12. Seni, G., Elder, J.: Ensemble Methods in Data Mining: Improving Accuracy Through Combining Predictions. Morgan & Claypool Publishers (2010)
13. Dwork, C., Kumar, R., Naor, M., Sivakumar, D.: Rank aggregation methods for the Web. In: Proceedings of the 10th International Conference on World Wide Web, pp. 613–622 (2001)
14. Lin, S.: Rank aggregation methods. Wiley Interdisc. Rev. Comput. Stat. **2**, 555–570 (2010)
15. Estrada, E., Nava, P., Nazeran, H., Behbehani, K., Burk, J., Lucas, E.: Itakura Distance: a useful similarity measure between EEG and EOG signals in computer-aided classification of sleep stages. In: Conference Proceedings, IEEE Engineering Medical Biology Society, vol. 2, pp. 1189–1192 (2005)
16. The Sleep-EDF Database [Expanded]. http://www.physionet.org/physiobank/database/ sleep-edfx/

17. Wiltschko, A.B., Gage, G.J., Berke, J.D.: Wavelet filtering before spike detection preserves waveform shape and enhances single-unit discrimination. J. Neurosci. Methods **173**, 34–40 (2008)
18. Ebrahimi, F., Mikaeili, M., Estrada, E., Nazeran, H.: Automatic sleep stage classification based on EEG signals by using neural networks and wavelet packet coefficients. In: Conference of the Proceedings, IEEE Engineering in Medicine and Biology Society, pp. 1151–1154 (2008)
19. Zhang, Y., Zhao, Y.: Real and imaginary modulation spectral subtraction for speech enhancement. Speech Commun. **55**, 509–522 (2013)
20. Taşmaz, H., Erçelebi, E.: Speech enhancement based on undecimated wavelet packet-perceptual filterbanks and MMSE–STSA estimation in various noise environments. Digit. Signal Process. **18**, 797–812 (2008)
21. Dursun, M., Gunes, S., Ozsen, S., Yosunkaya, S.: Comparison of artificial immune clustering with fuzzy C-means clustering in the sleep stage classification problem. In: International Symposium on Innovations in Intelligent Systems and Applications, pp. 1–4 (2012)
22. Kira, K., Rendell, L.: The feature selection problem: traditional methods and a new algorithm. In: AAAI, pp. 129–134 (1992)
23. Robnik-Šikonja, M., Kononenko, I.: Theoretical and empirical analysis of ReliefF and RReliefF. Mach. Learn. **53**, 23–69 (2003)
24. Ding, C., Peng, H.: Minimum redundancy feature selection from microarray gene expression data. J. Bioinform. Comput. Biol. **3**, 185–205 (2005)
25. Gu, Q., Li, Z., Han, J.: Generalized Fisher Score for Feature Selection. CoRR. abs/1202.3725 (2012)
26. Liu, H.L.H., Setiono, R.: Chi2: feature selection and discretization of numeric attributes. In: Proceedings of 7th IEEE International Conference on Tools with Artificial Intelligence, pp. 5–8 (1995)
27. Quinlan, J.R.: C4.5: Programs for Machine Learning. Morgan Kaufmann Publishers (1993)
28. Fleuret, F.: Fast binary feature selection with conditional mutual information. J. Mach. Learn. Res. **5**, 1531–1555 (2004)
29. Kolde, R., Laur, S., Adler, P., Vilo, J.: Robust rank aggregation for gene list integration and meta-analysis. Bioinformatics **28**, 573–580 (2012)
30. Wald, R., Khoshgoftaar, T.M., Dittman, D.: Mean aggregation versus robust rank aggregation for ensemble gene selection. In: 2012 11th International Conference on Machine Learning and Applications, pp. 63–69 (2012)
31. Duda, R.O., Hart, P.E., Stork, D.G.: Pattern Classification. Wiley, Hoboken (2012)
32. Zhang, G., Patuwo, B.E., Hu, M.Y.: Forecasting with artificial neural networks: the state of the art. Int. J. Forecast. **14**(1), 35–62 (1998)
33. Kalousis, A., Prados, J., Hilario, M.: Stability of feature selection algorithms: a study on high-dimensional spaces. Knowl. Inf. Syst. **12**, 95–116 (2007)
34. Imtiaz, S.A., Rodriguez-Villegas, E.: Recommendations for performance assessment of automatic sleep staging algorithms. In: Conference Proceedings: Annual International Conference of the IEEE Engineering in Medicine and Biology Society, IEEE Engineering in Medicine and Biology Society, Annual Conference, pp. 5044–5047 (2014)
35. Satopaa, V., Albrecht, J., Irwin, D., Raghavan, B.: Finding a "Kneedle" in a haystack: detecting knee points in system behavior. In: 2011 31st International Conference on Distributed Computing Systems Workshops, pp. 166–171 (2011)

# Improved Dynamic Time Warping for Abnormality Detection in ECG Time Series

Imen Boulnemour[✉], Bachir Boucheham, and Slimane Benloucif

Département d'Informatique, Université 20 août 1955 de Skikda,
BP 26, DZ21000 Skikda, Algeria
boulnemourimen@hotmail.fr, boucheham_bachir@yahoo.fr,
slimane.bcf@hotmail.com

**Abstract.** Abnormality detection in ECG time series is very important for cardiologists to detect automatically heart diseases. In this study, we propose a novel algorithm that compare and align efficiently quasi periodic time series. We apply this algorithm to detect exactly in the ECG, where the anomaly is. For this purpose, we use a normal (healthy) ECG segment and we compare it with another ECG segment. Our algorithm is an improvement of the famous dynamic time warping algorithm, called Improved Dynamic Time Warping (I-DTW). Indeed, the alignment of quasi-periodic time series, such as those representing the ECG signal is impossible to achieve with the DTW, especially when the segment of ECGs are of different lengths and composed of different number of periods each. The tests were performed on ECG time series, selected from the public database of the "Massachusetts Institute of Technology - Beth Israel Hospital (MIT-BIH)". The results show that the proposed method outperforms the famous DTW method in terms of alignment accuracy and that it can be a good method for abnormalities detection in ECGs time series.

**Keywords:** Time series · DTW · ECG · Abnormality detection

## 1 Introduction

A time series is a continuation of couples $<(v_1,t_1),(v_2,t_2),\ldots,(v_i,t_i),\ldots>$ where $v_i$ is a value or a vector of values taken at a moment $t_i$. This notation can be abbreviated in $<v_1, v_2,\ldots, v_i,\ldots>$ when the reference to the time does not need to be clarified [1].

This study is dedicated to quasi-periodic time series and especially to electrocardiogram (ECG). Quasi-periodic time series are concatenations of quasi-similar patterns called pseudo periods (periods for short) [2]. Figure 1 presents the characteristics of an electrocardiogram composed of two quasi-regulars periods reflecting two heart cycles. Each cycle is itself composed of three consecutive basic patterns: P wave, QRS complex and T wave.

Alignment of time series and detection of anomalies in this series pose difficult problems. These problems are: difference in time axis scales (Fig. 2a), difference in amplitude axis scales (Fig. 2b), shift on the amplitude axis (Fig. 2c) between the series, shift on the time axis (Fig. 2d). In addition, ECG traces are particularly characterized

© Springer International Publishing Switzerland 2016
F. Ortuño and I. Rojas (Eds.): IWBBIO 2016, LNBI 9656, pp. 242–253, 2016.
DOI: 10.1007/978-3-319-31744-1_22

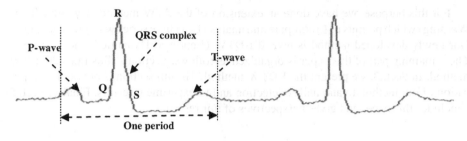

**Fig. 1.** A typical ECG segment with two periods.

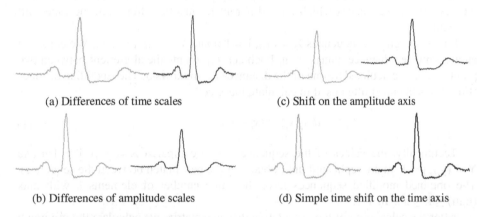

(a) Differences of time scales

(b) Differences of amplitude scales

(c) Shift on the amplitude axis

(d) Simple time shift on the time axis

**Fig. 2.** Difficulties encountered in time series alignment. (a) Different time scaling, (b) Different amplitude scaling, (c) Shift on the amplitude axis, (d) Shift on the time axis

by significant noises due to the transpiration of the patient and also by local morphological changes due to the activity of the patient.

There are various applications related to time series, we cite, ECG frame classification by DTW [3], Classification and identification of ECGs of patients with heart diseases, e.g. [4]. Annam, et al. and all, use K-medoids Clustering with DTW to identify the abnormalities in ECG heart beats through Clustering and Validation by using the QRS complexes of ECG, [5]. The detection of abnormalities in ECG by aligning time series with the SEA method (Shape Exchange Algorithm) of Boucheham [6] and the detection of abnormal pattern in activities of daily living using sequence alignment method [7].

The major methods for abnormality detection depend on extracting the characteristics of the ECG with special algorithms. This study aims to detect the abnormalities with a simple method without extraction of the ECG characteristics which allows having a small computation cost.

For this purpose, we have done an extension of the DTW method (Dynamic Time Warping) which permits us to compare and analyse ECG time series with great elegance. Our newly developed method is named I-DTW (Improved Dynamic Time Warping). The remaining part of this paper is organized as follows. Section 2 illustrates the DTW method. In Sect. 3, we present the I-DTW method. The 4th section is devoted to application of our method to anomalies detection and to experiment results. Finally, Sect. 5 concludes this paper and gives perspectives of our research.

## 2    Dynamic Time Warping (DTW)

The DTW method consists in establishing a non linear alignment of sequences to allow the mapping of sequences which have different lengths or suffer from the time-shift problem.

First, to align two sequences $X = (x_i)$, $i = 1{:}n$ and $Y = (y_j)$, $j = 1{:}m$, We construct an accumulated distance matrix m*n, Each cell represents the alignment between two points $x_i$ and $y_j$ calculated by one of the distances of Minkowski. The Euclidian distance (Eq. (1)) is the most often used to calculate these cells.

$$d(x_i, y_j) = (x_i - y_j)^2 \tag{1}$$

Technically, the values of the sequences to be compared $X = (x_i)$, $i = 1{:}n$ and $Y = (y_j)$, $j = 1{:}$ m are replicated until obtaining the best match between the sequences. The obtained modified sequences have the same number of elements k with max $(n,m) \leq k \leq n + m + 1$.

After the calculation of the cumulative distance matrix, we calculate the minimum warping path $W$ and we obtain the sequences X' and Y'. The values of cells are calculated as follows:

1- First cell:

$$M[1, 1] = [x_1 - y_1]$$

2- First line:

$$M[1,j] = [x_j - y_1] + M[1, j - 1]$$

3- First column:

$$M[i, 1] = [x_1 - y_i] + M[i - 1, 1]$$

$$M[i,j] = [x_j - y_i] + \min(M[i - 1, j - 1],$$
4- All the other elements: $M[i - 1, j], M[i, j - 1])$

There are several constraints on the warping path we cite:

**Boundary Conditions.** w1 = (1,1) and wK = (m,n), to force the end points in each series to match.

**Continuity.** Given wk = (a,b) then wk-1 = (a',b') where a–a' ≤ 1 and b-b' ≤ 1, to oblige the warping path to go to the adjacent cells.

**Monotony.** Given wk = (a,b) then wk-1 = (a',b') where a–a' ≥ 0 et b-b' ≥ 0 to force the points of W to be spaced monotonically in time. There are several warping path which satisfy these conditions, but we choose the one that minimizes the warping cost:

$$DTW(X, Y) = \min\left\{ \sqrt{\sum_{k=1}^{K} w_k} \Big/ K \right\} \qquad (2)$$

The kth of W is defined as $w_k = (i,j)_k$ such that:

$$W = w_1, w_2, .., w_k.., w_K \qquad (3)$$

Equation 4 illustrates the dynamic programming method used for the calculating of the optimal path which represents the DTW distance. Let D, this cumulative distance until $w_k$ cell.

$$D(x_i, y_j) = d(i,j) + \min\{D(i - 1,j); D(i,j - 1); D(i - 1,j - 1)\} \qquad (4)$$

# 3   Improved Dynamic Time Warping

Most of the methods that try to improve the DTW focus on the acceleration of its execution time [8, 9]. However, the alignment quality of the DTW poses many problems [2, 10, 11], especially when it comes to align quasi-periodic time series containing each a different number of quasi-similar phases. To overcome this problem we propose the new I-DTW method which is capable to align efficiently ECG time series. We apply this method to detect abnormalities in ECG time series.

**Step 1: Extracting the ECG segments**

- In this step, the normal ECG segment N and the input ECG segment R are fed to the algorithm. These can be specified from the MIT-BIH ECG database. We take in consideration the manual annotation of the ECGs taken by cardiologists to precise exactly segments containing the different abnormalities.

**Step 2: Finding the optimal warping path by the DTW**

- The algorithm DTW is executed to stretch the two segments N and R and to make the first mapping between them by the replication of certain values which are close in both time series.

**Step 3: Sorting on the magnitude**

- The sorting of the stretched segments, on the coordinates of their magnitude indexes is established to give them a stable signature. The result of this step is a matrix for each segment containing the magnitudes sorted in ascendant or descendent order with their equivalents temporal coordinates (not sorted). These sorted magnitudes represent the stable signatures of the segments N and R.

**Step 4: Signatures exchange**

- This step is divided into two parts. The first one consists in making the exchange of signatures between both segments. That is N will receive the coordinates of the sorted magnitudes (signature) of R and keep its own time indexes not sorted and vice versa for R. The second part consists in restoring the normal size of N and R after the exchange of signature which requires that the time series are of equal length. Thus, the temporal indexes replied in every new series must be deleted with their equivalent magnitudes. This part does not affect the results of alignment; it just allows having a clearer visual inspection of traces.

**Step 5: Sorting on the time index (Reconstruction)**

- Signatures represent well the characteristics of the time series but they do not consider the shifts and the differences of timescales. The sorting of N and R magnitudes and time indexes) on their respective temporal indexes will reconstructed them. The result of this operation for the abnormal segment R is a new segment $R_{Rec}$ having the magnitudes of the segment N and vice versa for N.

**Step 6: Comparison**

- In this step, the correlation factor (Eq. 5) is computed for the couple $(R, R_{Rec})$. The closer the value to 1, the more there is no abnormality in the segments.

$$Corr(X, Y) = \frac{cov(R, R_{Rec})^2}{var(R) \cdot var(R_{Rec})} \tag{5}$$

**Step 7: *Decision***

- We use a threshold value on the correlation factor. T is used for decision as follows:

  *If Corr(R,R$_{Rec}$) < T then* 'There-is-change'
  *Else* 'no-change-detected'

- The visual inspection for alignment is used as subjective criteria

  The proposed algorithm is described in the following diagram (Fig. 3):

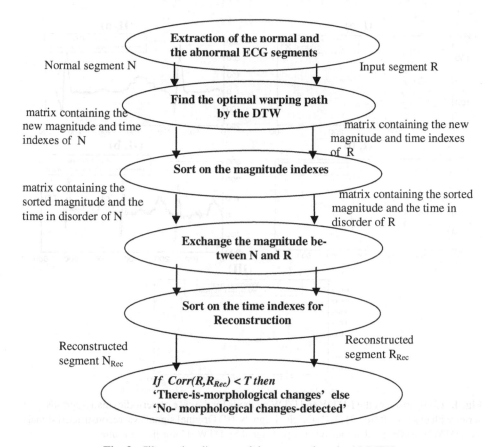

**Fig. 3.**   Illustrative diagram of the proposed method I-DTW

# 4   Experimental Results

## 4.1   Illustrative Example

In the example shown in Fig. 4, we refer to the time series rec. 215 of the MIT-BIH ECG database. Figure 4I shows the original traces X and Y. Figure 4II illustrates the alignment done by our method I-DTW for the time series X and its reconstructed time series (Fig. 4II,a) and for the time series Y and its reconstructed time series (Fig. 4II,b). Figure 4III illustrates the alignment done by the DTW method for the time series X and Y. Note that the two traces are from the same person ECG and that they don't contain abnormalities. Note also that they are of different lengths and have different number of periods each (two periods for X and three periods for Y). The segment (X) was purposely taking at the beginning of the recording and the segment (Y) at the end of the recording to illustrate the phase shift problem in quasi periodic time series.

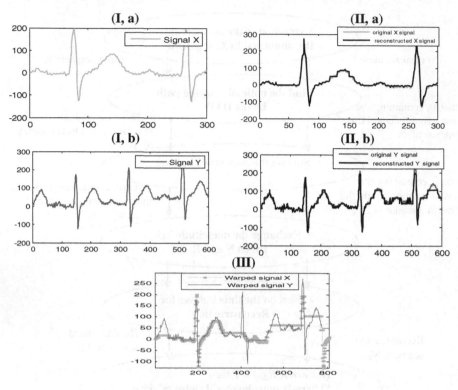

**Fig. 4.** Comparison of the I-DTW and DTW methods for the case of periodic- many-periods time series with phase shift (rec.215). (I) Original signals. (II) Original signals vs. reconstructed signals by I-DTW. (III) Original signals vs. warped signal by DTW (Color figure online).

Formulas are applied to the two traces (X and Y) with their respective reconstructed traces ($X_{Rec}$ and $Y_{Rec}$). As well we will report in Table 1 the Corr(X,$X_{rec}$) and Corr(Y,$Y_{rec}$) for I-DTW methods and theCorr(X,Y) for the DTW method.

The visual inspection of Fig. 4II shows a perfect alignment for the I-DTW, to the point that it is difficult to distinguish between original and reconstructed time series. Figure 4III shows a bad alignment done by DTW.

Numerical results reported in Table 1 confirm that our method arrives to align these similar and healthy segments provided from the same person ECG, unlike the DTW which completely fails. Our conclusion is that the DTW method doesn't deal efficiently with quasi periodic time series and that it always gives a low correlation even if the segments are similar and do not contain abnormalities.

**Table 1.** I-DTW vs. DTW for the time series rec.215

| Method | Corr(X,Y) | Corr(X,$X_{rec}$) | Corr(Y,$Y_{rec}$) |
|--------|-----------|-------------------|-------------------|
| I-DTW  | /         | **0.9906**        | **0.9817**        |
| DTW    | 0.7582    | /                 | /                 |

## 4.2   Experimental Tests

To test our method "I-DTW", we use the MIT-BIH ECG database [12]. According to [12], the recordings were digitized at 360 samples per second per channel with 11-bit resolution over a 10 mV range. We select 6 records numbered as records 219, 223, 218, 104, 223 in the database. In each of the records, we select a segment having abnormal heartbeats, and the other having no abnormal heartbeats, and we perform the alignment between them. Figures 5, 6, 7, 8 and 9 illustrate the capacity of our method in detecting the morphology change in every ECG person by comparing his healthy ECG segment and his abnormal ECG segment. Table 2 demonstrates numerically the alignment accuracy of our method. We stress the need to set the threshold value $T$ of step 7 (Decision) in the proposed algorithm. Through intensive experiments, we found that correlation factors that are less than 0.950 indicate that there is a morphology change between the compared segments.

**Table 2.** Obtained results for I-DTW

| ECG Segments | Anomaly | Correlation |
|---|---|---|
| (219 V, 219) | Premature ventricular contraction | 0.947 |
| (223 A, 223) | Atrial premature beat | 0.884 |
| (228 I, 228) | Isolated QRS-like artefact | 0.944 |
| (104, 104 f) | Fusion of paced and normal beat | 0.903 |
| (223, 223 V) | Premature ventricular contraction | 0.906 |

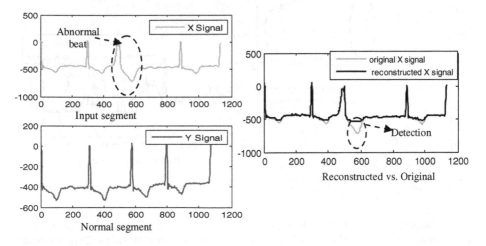

**Fig. 5.** Detection of morphology change by comparison of two ECG segments from the same person; in this case this is a PVC (Premature ventricular contraction), within record MITBIH #219 (Color figure online).

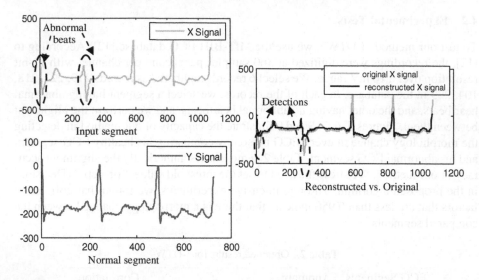

**Fig. 6.** Detection of morphology change by comparison of two ECG segments from the same person; in this case this is an Atrial premature beat, within record MITBIH #223 (Color figure online).

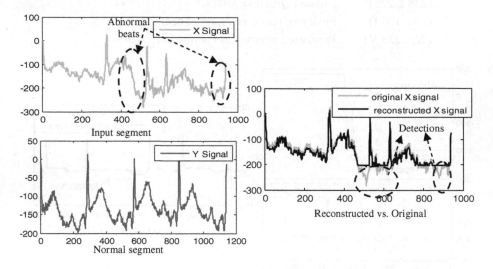

**Fig. 7.** Detection of morphology change by comparison of two ECG segments from the same person; in this case this is an Isolated QRS-like artefact, within record MITBIH #228 (Color figure online).

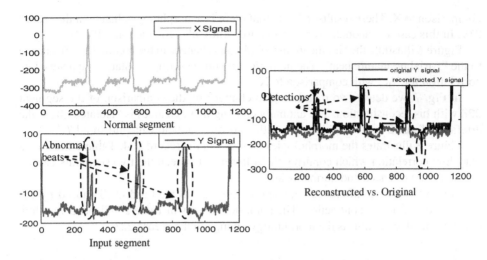

**Fig. 8.** Detection of morphology change by comparison of two ECG segments from the same person; in this case this is a Fusion of paced and normal beat, within record MITBIH #104 (Color figure online).

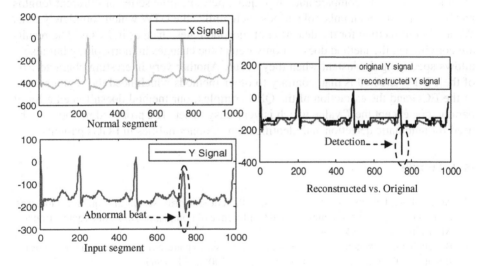

**Fig. 9.** Detection of morphology change by comparison of two ECG segments from the same person; in this case this is a Premature ventricular contraction, within record MITBIH #223 (Color figure online).

# 5    Results

Table 2 indicate the results of the comparison of the five used segments with their reconstructed segments. It indicates an average correlation, below the $T$ value for the segment 219. On the other hand, Fig. 5 shows significant difference in $X_{Rec}$ with comparison to X. These results indicate that there is a morphology change in the record 219. In this case the anomaly is a "Premature ventricular contraction" (PVC).

Figure 6 illustrate the significant morphologies changes in the record 223, this is due to the "Atrial premature beat". The numerical results confirm this state with a very low correlation for $X_{Rec}$ with comparison to X.

In Fig. 7, we detect two morphologies changes by the comparison of the segment 228 with his reconstruction. The anomaly causing these changes is "Isolated QRS-like artefact". Table 2 confirms this result with a correlation below the threshold $T$.

Figure 8 illustrates the morphologies changes in the record 104. Table 2 indicate a very low correlation which confirms the difference between the two segments caused by the fusion of paced and normal beat.

Figure 9 shows a difference between Y and $Y_{Rec}$ in the third QRS caused by the premature ventricular contraction. The numerical results in Table 2 indicate a very low correlation which confirms the morphology change in the record 223.

## 6    Conclusion

In this study, we propose a novel method I-DTW that improves the DTW method. Our method is capable to compare and align quasi periodic time series of different lengths and having a different number of periods each, unlike the DTW which completely fails. We apply our method for the detection of morphologies changes in ECGs. The results are convincing, the method does not only detect the changes in morphology but also it allows specifying the instant when they occur. Another very interesting characteristic of the I-DTW method is that, contrary to other methods which need the segmentation of the ECG and the extraction of the QRS complex, our method doesn't need a lot of parameters, except the $T$ value which is very easy to set. The proposed method can be used for automatic detection and identification of abnormalities in EGGs patients.

## References

1. Agrawal, R., Faloutso, C., Swami, A.N.: Efficient similarity search in sequence databases. In: Proceedings of the 4th International Conference of Foundations of Data Organization and Algorithms, pp. 69–84 (1993)
2. Boucheham, B.: Matching of quasi-periodic time series patterns by exchange of block-sorting signatures. Pattern Recogn. Lett. **29**, 501–514 (2008). (Elsevier)
3. Zhang, B., Kinsner, W., Huang, B.: Electrocardiogram data mining based on frame classification by dynamic time warping matching. Comput. Methods Biomech. Biomed. Eng. **12**(6), 701–707 (2009)
4. Vishwa, A., Sharma, A.: Arrhythmic ECG signal classification using machine learning techniques. IRACST- Int. J. Comput. Sci. Inf. Technol. Secur. (IJCSITS) **1**(2), 163–167 (2011)
5. Annam, J.R., Mittapalli, S.S., Bapi, R.S.: Time series clustering and analysis of ECG heartbeats using dynamic time warping. In: 2011 Annual IEEE India Conference (INDICON) (2011)

6. Boucheham, B.: Abnormality detection in electrocardiograms by time series alignment. Commun. Inf. Sci. Manag. Eng. (CISME) **1**(3), 7–11 (2011)
7. Jung, H.Y., Park, S.H., Park, S.J.: Detection abnormal pattern in activities of daily living using sequence alignment method. In: Proceedings of 30th Annual International Conference of the IEEE Engineering in Medicine and Biology Society, EMBS 2008
8. Junkui, L., Yuanzhen, W.: Early abandon to accelerate exact dynamic time warping. Int. Arab J. Inf. Technol. **6**(2), 144–152 (2009)
9. Salvador, S., Chan, P.: FastDTW: toward accurate dynamic time warping in linear time and space. J. Intell. Data Anal. **11**(5), 561–580 (2007)
10. Keogh, E., Pazzani, M.: Derivative dynamic time warping. In: Proceedings of the 2001 SIAM International Conference on Data Mining, Chicago (2001)
11. Boucheham, B.: Efficient matching of very complex time series. Int. J. Mach. Learn. Cybern. **4**(5), 537–550 (2013). Springer
12. MIT-BIH Arrhythmia Database. http://www.physionet.org/physiobank/database/mitdb/

# Hardware Accelerator to Compute the Minimum Embedding Dimension of ECG Records

Pablo Pérez-Tirador[1], Gabriel Caffarena[2(✉)], Constantino A. García[3],
Abraham Otero[2], Rafael Raya[2], and Rodrigo Garcia-Carmona[2]

[1] Department of Medical Physics and Biomedical Engineering,
University College London, London WC1E 6BT, UK
[2] University CEU-San Pablo, Urb. Monteprincipe, 28668 Madrid, Spain
gabriel.caffarena@ceu.es
[3] Centro Singular de Investigación en Tecnoloxías da Información (CiTIUS),
University of Santiago de Compostela, 15782 Santiago de Compostela, Spain
http://biolab.uspceu.com

**Abstract.** In this paper, a parallel hardware implementation to accelerate the computation of the minimum embedding dimension is presented. The estimation of the minimum embedding dimension is a time-consuming task necessary to start the non-linear analysis of biomedical signals. The design presented has as main goals maximum performance and reconfigurability. The design process is explained, giving details on the decisions taken to achieve massive parallelization, as well as the methodology used to reduce hardware usage while keeping a high mathematical accuracy. The results yield that hardware acceleration achieves a speedup of three orders of magnitude in comparison to a purely software approach.

## 1 Introduction

The heart rate regulation is one of the most complex systems in the human body. Heart rhythm is innervated by both the parasympathetic and sympathetic branches of the autonomic nervous system. At the same time, the autonomic nervous system is influenced by humoral effects, hemodynamic variables, respiratory rhythm and stroke volume, among others. Furthermore, there exist feedback loops among these mechanisms influencing each other in a nonlinear way. A consequence of these nonlinear interactions is that heart rate modulation cannot be fully understood by studying its components in isolation. Moreover, it has been argued that the heartbeat morphology is determined by nonlinear interactions of the heart tissues. Hence the interest in trying to quantify the complexity of both the heart rate and the beat morphology by using nonlinear statistics derived from the theory of dynamical system [1,2].

The Electrocardiogram (ECG) signal is a recording of the electrical activity of the heart. The ECG signal contains information about the beat morphology, which depends on the point of origin of the beat and on its propagation path through the myocardium. The ECG signal has already been subjected to a wide

© Springer International Publishing Switzerland 2016
F. Ortuño and I. Rojas (Eds.): IWBBIO 2016, LNBI 9656, pp. 254–265, 2016.
DOI: 10.1007/978-3-319-31744-1_23

variety of tests that showed that the dynamics of the heart is consistent with nonlinear dynamics or even chaos [2–4]. In this paper we focus on the non-linear analysis of the ECG to infer the non-linear behavior of the beat morphology.

The theory of dynamical systems relies on the concept of phase space reconstruction, which is a vector space where each point can be used to completely specify the state of a system. The analysis of the dynamics of the state of the system enables the extraction of features from the biomedical signal under study. Before starting the analysis it is necessary to obtain the *minimum embedding dimension*, which is the minimum dimension of the phase space that ensures that it is possible to fully describe the state of the system. In this work we focus on the technique presented by Cao [5] which keeps a good performance in the presence of noise.

Due to its computational complexity we propose in this paper the use of an FPGA-based accelerator for the computation of the minimum embedding dimension, aiming at introducing the non-linear analysis of ECG in the clinical routine. The design was performed according to the following goals:

– Maximum parallelization
– Use of fixed-point arithmetic to achieve maximum clock speed
– Run-time reconfiguration

The paper is divided as follows: In Sect. 2 the minimum embedding dimension algorithm is explained. Section 3 introduces FPGA computing. The next section deals with the hardware design, presenting the proposed hardware architecture. The results are in Sect. 5. And, finally, the conclusions are given in Sect. 6.

## 2  Embedding Dimension Algorithm

The starting point of the Cao's algorithm is a vector $\boldsymbol{x} = \{x_1, x_2, \ldots, x_N\}$ where the elements $x_i$ correspond to the sorted elements of a time series (e.g. ECG). Cao suggested to follow Takens' theorem and generate Takens vectors

$$\boldsymbol{y}_i(d) = \left( x_i, x_{i+\tau}, \ldots, x_{i+(d-1)\tau} \right), i = 1, 2, \ldots, N - (d-1)\tau \qquad (1)$$

where $d$ is the embedded dimension of the reconstruction and $\tau$ is a time delay. The value of $\tau$ can be computed by applying techniques based on the autocorrelation function or based on the mutual information.

Caos' method is based on computing the distance between the different Takens vectors with dimension $m$ by means of the infinite norm

$$\|\boldsymbol{y}_k(m) - \boldsymbol{y}_l(m)\| = \begin{array}{c} max \\ 0 \leq j \leq m-1 \end{array} |x_{k+j\tau} - x_{l+j\tau}|. \qquad (2)$$

Given a vector $\boldsymbol{y}_i(d)$ and its closest neighbor $\boldsymbol{y}_{n(i,d)}$, the parameter $a(i,d)$ is defined as

$$a(i,d) = \frac{\|\boldsymbol{y}_i(d+1) - \boldsymbol{y}_{n(i,d)}(d+1)\|}{\|\boldsymbol{y}_i(d) - \boldsymbol{y}_{n(i,d)}(d)\|}, \qquad (3)$$

and it is the ratio between the distance of $y_i(d)$ and its closest neighbor $y_{n(i,d)}(d)$ and the distance between the Takens vector that has as first element $x_i$ in the next dimension, and the vector in the next dimension that has as first element $x_{n(i,d)}$.

Equation 3 depends on the $i$-th vector and its neighbor. Cao suggests to compute $d_{min}$ by the mean of all $a(i,d)$, which only depends on $d$. For dimensions equal to or greater than the minimum dimension $d_{min}$ the closest neighbor to any Takens vector has always the same index. This situation can be checked through the parameters

$$E(d) = \frac{1}{N - d \cdot \tau} \sum_{i=1}^{N-d \cdot \tau} a(i, d) \tag{4}$$

$$E1(d) = \frac{E(d+1)}{E(d)}. \tag{5}$$

Equation 4 is the average of the $a(i, d,)$ of all vectors. Given a value $d_0$ that is the minimum value that makes $E1(d_0) = 1$, the minimum embedded dimension is $d_{min} = d_0 + 1$.

Another application of Cao's technique is to check whether the time series is deterministic or stochastic. This can be verified through the parameters:

$$E^*(d) = \frac{1}{N - d \cdot \tau} \sum_{i=1}^{N-d \cdot \tau} |x_{i+d \cdot \tau} - x_{n(i,d)+d \cdot \tau}| \tag{6}$$

$$E2(d) = \frac{E^*(d+1)}{E^*(d)}. \tag{7}$$

If the time series is stochastic $E2(d) = 1$ for all dimensions, while it will be different to one for some dimension otherwise.

In this work we have modified the original approach from Cao:

- It might be possible that for a given dimension $d$ two Takens vectors have exactly the same elements, that means that the distance between them is 0, so parameter $a(i, d)$ (Eq. 3) tends to infinite. Cao proposes using the second closest neighbor, however since this situation is extremely rare we have decided to take the vector out of the analysis.
- There is a temporal correlation between vectors with close indeces, so, in order to alleviate from this situation vectors that are close in time to each other are never compared. The so-called Theiler window is used [6]. Only vectors that are further than the distance specified by the Theiler window are used in the comparisons.

---

**Algorithm 1.** Computation of $E(d)$ and $E^*(d)$

---

**Input:** $\boldsymbol{x}$ $N$, $d$, $\tau$, $TH$
**Output:** $E(d)$, $E^*(d)$

1: $Takens = N - d \cdot \tau$ #No. of Takens vectors
2: # Generation of Takens vectors
3: **for** i=0 ... $Takens - 1$ **do**
4:     $\boldsymbol{y}_i(d) = [x_i x_i + \tau \ldots x_i + d \cdot \tau]$
5:     $\boldsymbol{y}_i(d+1) = [x_i x_i + \tau \ldots x_i + d \cdot \tau x_i + (d+1) \cdot \tau]$
6: **end for**
7: # Computation of $a(i, d)$
8: $sum_{E(d)} = sum_{E^*(d)} = 0$
9: $TakensValid = 0$
10: **for** $i = 0 \ldots Takens - 1$ **do**
11:     $D_1 = \infty$
12:     **for** $j = 0 \ldots Takens - 1$ **do**
13:         **if** $\neg\{i - TH \le j \le i + TH\}$ **then**
14:             **if** $||\boldsymbol{y}_i(d) - \boldsymbol{y}_j(d)|| < D_1$ **then**
15:                 $D_1 = ||\boldsymbol{y}_i(d+1) - \boldsymbol{y}_j(d+1)||$
16:                 $D_2 = ||\boldsymbol{y}_i(d) - \boldsymbol{y}_j(d)||$
17:                 $n(j, d) = k$
18:             **end if**
19:         **end if**
20:     **end for**
21:     **if** $D_2 > 0$ **then**
22:         $sum_{E(d)} + = D_1/D_2$
23:         $sum_{E^*(d)} + = ||x_{i+d\cdot\tau} - x_{n(j,d)+d\cdot\tau}||$
24:         $TakensValid + +$
25:     **end if**
26: **end for**
27: $E(d) = sum_{E(d)}/TakensValid$
28: $E^*(d) = sum_{E^*(d)}/TakensValid$

---

## 3    FPGA Computing

FPGA devices (Field-Programmable Gate Arrays) stand in a middle point between microprocessors and ASICs (Application-specific Integrated Circuits) in terms of performance, cost and development time. An FPGA is an electronic device composed of thousands of logic elements (LE) that can be configured to implement small logic function (i.e. 4:1 logic functions) and interconnected among each others. The configuration of both the logic and the connectivity is performed when the device is turned on. The FPGA is able to implement any digital system, given that there are enough available resources. As a main advantage they enable the design of optimum architectures designed to implement a specific algorithm. The design times are smaller than those of ASICs, while the performance is not as high as for those. They are specially suitable for exploiting temporal and spatial parallelization techniques and in many cases they are able to produce superior implementation than CPUs or GPUs. The main drawback is the high development time compared to those of programmed systems.

In the last decades, FPGAs has been extended to also include specific blocks for DSP computing (multipliers, MAC units, floating-point units, etc.), data storage (hundreds of customizable memories), high speed I/O transceivers, etc. [7].

The way to design a hardware accelerator with FPGAs follows the standard hardware design workflow [8]: specification of the system with a hardware description language (HDL), functional simulation, synthesis, place and route and temporal simulation. The last stage after the temporal simulation comprises the *programming* of the device, where the bitstream containing all the logic and connectivity configuration is uploaded onto the FPGA, and the final verification.

## 4    Hardware Design and Implementation

The main idea is to connect an FPGA device to a PC by means of a communications port (i.e. USB, PCIe, etc.) as depicted in Fig. 1-a. The goal of the hardware accelerator is to compute parameters $E(d)$ and $E^*(d)$ for a given $N-$element time series $x[n]$, a dimension $d$, a time delay $\tau$ and a Theiler Window $TH$. It must be stressed that the hardware system has *design parameters* and *configuration parameters*. The former specify the parameters for the computation of $E(d)$ and $E^*(d)$, while the latter affect the hardware architecture of the system and impose limits to the values of the configuration parameters. The design parameters are:

- $N_{max}$: Maximum number of elements of input vector $x$.
- $WL$: Wordlength of $x$.
- $FR$: Fractional bits at the output of the dividers.
- $d_{max}$: maximum dimension.
- $\tau_{max}$: maximum time delay.
- $TH_{max}$: maximum Theiler's window.
- $P$: number of processors.
- $pipeline_{div}$: pipeline levels of the dividers.

The configuration parameters are:

- $N$: Number of elements of input data $x$ $(N < N_{max})$.
- $d$: Current dimension $(d < d_{max})$.
- $\tau$: Current $\tau$ $(\tau < \tau_{max})$.
- $TH$: Current Theiler's window $(TH < TH_{max})$.

The design parameters are used by the FPGA compilation tool and if they are modified, the system must be reimplemented. The VHDL code generated is parametrizable, so it is possible to generate many versions of the systems by simply changing the value of the constants associated to each design parameter and, then, re-synthesizing the design. The configurable parameters are used in run-time to configure the behavior of the different hardware blocks. For instance, the hardware block that performs the computation of the distance is able to compare vectors with dimension $d_{max}$ but if $d < d_{max}$ then it performs the comparison between the involved Takens vectors using only $d$ elements.

The proposed processing goes as shown in Algorithm 2. The host sends the ECG data to the FPGA. This data transfer is only performed once, since the data is stored in the local memory of the FPGA board. Next, the host send the

---

**Algorithm 2.** Host-side computation of $E1(d)$ and $E2(d)$ with hybrid approach

---

1: Send vector $x$ to FPGA
2: Send parameters: $N, \tau, TH$
3: **for** $d = 1 \ldots d_{max}$ **do**
4:     Send parameter $d$
5:     Init FPGA computation
6:     Receive $E(d)$ and $E^*(d)$ from FPGA
7:     Store $E(d)$ and $E^*(d)$
8:     **if** $d > 1$ **then**
9:         $E1(d - 1) = E(d)/E(d - 1)$
10:        $E2(d - 1) = E^*(d)/E^*(d - 1)$
11:    **end if**
12: **end for**

---

parameters of the computation: $N$, $\tau$ and $TH$. From that moment, after sending the dimension $d$, it is possible to start the computation on the FPGA, and the host will receive parameters $E(d)$ and $E^*(d)$ after a few seconds. The loop traverses in increasing order the dimensions from 1 to $d_{max}$ and the parameters $E1(d)$ and $E2(d)$ are being computed. Note that this loop can be modified, for instance to have an exit condition, and also that other parameters, such as $\tau$ of $TH$, can also be changed in run-time.

The next subsections are devoted to the design of the system and its inner blocks.

### 4.1 Overall Design

Figure 1-b shows the main blocks in the hardware system. The block *TakensGen* is fed sequentially with the elements of the time series $x$ and it produces the Takens vector considering the parameters $d$ and $\tau$. This block is designed in a pipeline fashion so that, after an initial and constant delay, a new Takens vector is generated every clock cycle. The Takens vectors are fed to an array of processing elements (PE) that works in parallel and which compute the parameters $a(i, d)$

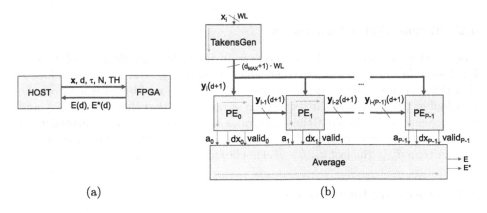

      (a)                                          (b)

**Fig. 1.** Hardware system: (a) basic connection to Host PC; (b) general diagram of hardware multiprocessor

**Algorithm 3.** FPGA-side computation of $E(d)$ and $E^*(d)$

```
 1: Receive x from Host
 2: for i = 0 … ⌈N/P⌉ − 1 do
 3:     #Parallel execution of loop
 4:     for p = 0 + iP … iP + P − 1 do
 5:         PE_p: Compute a_p = a(p, d)
 6:         PE_p: Compute dx_p = ||x_{p+d·τ} − x_{n(p,d)+d·τ}||
 7:         PE_p: valid_p = 1 if dx_p > 0
 8:         Average: Accumulate valid_p in Valid
 9:         Average: Accumulate a_p in A
10:         Average: Accumulate dx_p in Dx
11:     end for
12: end for
13: Average: E(d) = A/Valid
14: Average: E^*(d) = Dx/Valid
15: Send E(d) and E^*(d) to Host
```

($a_i$ in the figure) and $||x_{i+d·τ} − x_{n(i,d)+d·τ}||$ ($dx_i$ in the figure) for a given Takens vector. Again, these blocks compute in a pipeline fashion, so after all the Takens vectors are fed to the PEs and, after a constant delay, the parameters are fed to the block *Average*. This last block is in charge of accumulating the parameters produced by the PEs and to ultimately generate the parameters $E(d)$ and $E^*(d)$, so in terms of parallel computation is carrying out a reduction operation. The figure also displays the way that data propagates around the different blocks (dashed lines).

Algorithm 3 explains the overall behavior of the hardware system. Given that a PE is in charge of performing the required computations for a single Takens vector, and that there are $P$ processor, it is only possible to process the Takens vectors with indeces $0 … P − 1$ in parallel. In order to compute more Takens vector, it is necessary to read again the time series from the FPGA local memory. So it will be necessary to perform several rounds of memory reading and computation. In the first round, all the computations where the Takens vectors $y_0 … y_{P−1}$ are used as references are carried out, in the next round, the references are the Takens vectors $y_P … y_{2P−1}$, and so on.

### 4.2 Takens Vector Generation

This block produces the Takens vectors from the individual elements of $x$. This is achieved by chaining $d_{max}$ circular buffers of size $τ_{max} + 1$. The general idea is that the elements $x_i$ feeds the first buffers, while the second buffer is fed from the first buffer, the third from the second and so on. Given that the value of $d_{max}$ and $τ_{max}$ are expected to be small these memories can be efficiently implemented using the memory resources of the FPGA. The block considers that if the dimension is smaller than $d_{max}$ the last $d_{max} − d$ elements are set to zero.

### 4.3 Processing Elements $PE$

The inner blocks of the processing elements are depicted in Fig. 2-a. Note that the PEs receive the current Takens vector $y_i(d + 1)$ and the $p$−cycle delayed

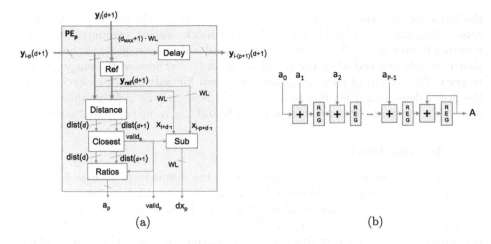

**Fig. 2.** Diagrams of: (a) processing element; (b) accumulator

Takens vector $y_{i-p}(d+1)$. The first input is used to obtain the reference vector and the delayed input is used in the comparisons with this reference. The block *Ref* stores the reference Takens vector that will be used during the first iteration of Algorithm 3. For the first round, $PE_0$ stores $y_0(d+1)$ in the first clock cycle. During the next cycles, it is possible to compare $y_0(d)$ with the remaining Takens vector of dimension $d$, as well as $y_0(d+1)$ with the rest of $(d+1)$-dimension vectors – note that the input vector has dimension $d+1$ so it is possible to easily obtain the $d$−element vector from that one. When $PE_1$ receives the first data of $y_{i-1}(d+1)$ it already has stored $y_1(d+1)$ since this vector arrived one clock cycle before, thus, it is possible to perform the comparison straight away. This situation is repeated for the remaining PEs. Note that the first PE to finish the processing is $PE_0$, one clock cycle later $PE_1$ will finish, two clock cycles later $PE_2$ will finish the computation and so son. This situation is beneficial for computing the averages of these data in the block *Average* as we will see in the next subsection.

The block *Ref* also prefetches the reference for the next memory reading round so that the reference vector is ready to use as soon as the next iteration starts. Remember, that $PE_p$ is in charge of processing all comparisons related to the Takens vector $y_p$ when $i = 0$ (see Algorithm 3). It will deal with $y_{p+2P}$ when $i = 1$, and son on. Thus, in the first round, vector $y_{p+P}$ is also stored so that it will be ready for the second iteration. In the second iteration, vector $y_{p+3P}$ is prefetched and stored. Summarizing, in each iteration the $PE$s posses the current reference and the next iteration's reference.

Once the reference vector is obtained, the block *Dist* computes the distances between the incoming Takens vectors and the reference vectors for dimensions $d$ and $d + 1$. At the same time, block *Sub* is performing the absolute value of the distance from Eq. 6. The distance $dist(d)$ is processed by the block *Closest* to determine if the last vector received is the closest neighbor so far. If this is the situation and the current vector is not the same as the reference (distance zero), then, the valid signal is active. The last block, *Ratios*, computes the ratio required to obtain $a(i, d)$. Both blocks *Ratio* and *Sub* has a register at the output that it is only enabled when the signal $valid_p$ is ON, otherwise the output is set to zero. The output of these registers are shown for only one clock cycle when the last Takens vector is processed, because this is just the right moment when we can ensure that the computation is finish and the value computed is valid.

### 4.4    *Average* Block

This last block is in charge of implementing the summation in Eqs. 4 and 6, as well as performing the divisions by the number of valid vectors, which in turn is computed with another chain of adders. Figure 2-b shows the way the accumulation is performed for Eq. 4. The accumulator adds all the $a(i, d)$ terms generated from the $PE$s, starting with the output of $PE_0$. Thus, each adder in the array of adders shown in the figure is adding two values and passing the result to the next adder. When the last processor generates the data, the adder connected to its output computes the summation of all the data generated during the current iteration. Eventually, there is an accumulator that adds all the partial summations of all iterations. This schema is used for the computation of signals $A$, $Dx$ and $Valid$, adding all signals $a_i$, $dx_i$ and $valid_i$. When the summations are done then, it is possible to perform the division by the number of valid vectors to compute $E(d)$ and $E^*(d)$.

### 4.5    Fixed-Point Optimization

The goal of fixed-point optimization is to reduce the number of bits used for each signal so that the resource usage is minimized and the mathematical precision of the implemented algorithm complies with a quality criterion. The decision taken for this system was to keep the maximum precision for all operations but the divisions. The divisions require an infinite precision, so it is necessary to truncate the number of bits required for the fractional part of the results.

A double-precision floating-point simulation of the algorithm was performed and the resulting $E(d)$ and $E^*(d)$ were used as reference. Then a fixed-point simulation of the algorithm with the output of the dividers truncated were performed. Several combinations of fractional number of bits were tried until a sensible error was obtained. As a final results, using 6 bits as fractional bits at the output of all divisions led to a maximum error in the computation of $E(d)$ and $E^*(d)$ around $10^{-4}$ with typical errors around $10^{-5}$.

As for the accumulator, the number of bits required for the chained adders, increases as long as the adders gets closer to the last $PE$ (check Fig. 2-b). We found analytically the number of bits for all adders and for the final accumulator that ensure that no overflow is produced. It must be stressed that the number of bits obtained is tailored for each individual adder, leading to an optimal implementation. Moreover, the computation of the wordlengths is integrated in the VHDL code, and it is a function of $WL$, $FR$ and $P$. Any change produced in any of these three design parameters will produce an implementation with the optimal wordlengths for all operators and signals in the design.

## 5   Results

In this section we present the implementation results. Firstly, the number of FPGA resources required for the different implementations is shown. Secondly, an estimation of the speedup achieved will be given.

The system was developed using the tool Quartus® II 13.1 from Altera [7]. The FPGA devices targeted were the Cyclone® IV EPCE115 and the Stratix® IV GX EP4SGX230KF40C2, thus covering low-end and mid-range FPGAs.

Both the area and performance results are computed for three FPGA implementations:

- 2, 8 and 35 processors on the Cyclone® device
- 2, 8 and 60 processors on the Stratix® device

The software version of the algorithm was run in a PC with an Intel® Core™ i5 (1.8 GHz and 4 GB of RAM). The speedup was computed using as baseline a single-thread execution of the CPU code.

Table 1 contains the area results for the implementation using the Cyclone IV device and the Stratix IV device, respectively. The results for a different number of processors are included. The last column shows the available resources for each device. The results yield that the resource usage scales almost linearly with the number of processors implemented. The maximum number of processors that the Cyclone IV device can hold is 35, while the other device is able to hold 60 processors.

The computing performance is limited by the clock frequency which is determined by the critical path (e.g. the slowest combinational path between two registers). The critical path was in the *Ratios* block due to the dividers. It was necessary to increase the pipeline levels of these components using values from 5 to 9 levels.

**Table 1.** FPGA resource usage

| Device | Resource type | No. of processors | | | Available resources |
|---|---|---|---|---|---|
| | | 2 | 8 | 35 | |
| Cyclone IV | Logic elements | 7, 294 | 15, 174 | 100, 826 | 114, 480 |
| | Memory bits | 3, 191 | 3, 172 | 3, 234 | 3, 981, 312 |
| | 9-bit multipliers | 4 | 4 | 4 | 532 |
| Stratix IV | Logic elements | 5, 257 | 17, 691 | 131, 558 | 182, 400 |
| | Memory bits | 3, 160 | 3, 072 | 3, 426 | 14, 625, 792 |
| | 18-bit DSP | 4 | 4 | 4 | 1288 |

**Table 2.** Computation performance of CPU and FPGA implementations

| Device | #Processors | $f_{ck}$ (MHz) | FPGA time (seconds) | CPU time (hours) | Speedup |
|---|---|---|---|---|---|
| Cyclone IV | 2 | | 4, 225 | | × 154 |
| | 8 | 50 | 1, 056 | | × 615 |
| | 35 | | 241.43 | 180, 5 | × 2692 |
| Stratix IV | 2 | | 2, 137 | | × 306 |
| | 8 | 100 | 528.13 | | × 1231 |
| | 60 | | 70.42 | | × 9230 |

The estimation of the speedup was performed by estimating the ratio of the execution times of the software implementation and the accelerated implementation. A 650, 000-sample ECG record was used as input. In fact, the execution times were estimated due to the long times required for the PC to carry out the processing. A single iteration of Algorithm 1 (lines 11–25 is run on the PC, which compares a single vector to the whole ECG record. This task takes less than a second. Given that the number of Takens vectors can be approximated by the size of the input vector, the execution time of a loop iteration was multiplied by the size of the ECG record, giving an estimated total execution of 180.5 hours. As for the execution time of the FPGA the latency generated for the pipelines is not considered, since it is negligible if compared to the total latency (e.g. 650, 000 clock cycles per round). If a single processor was available, then the comparison between all the vector would required $(65 \cdot 10^4)^2$ clock cycles. This number must be divided by the total number of processors, since they work in parallel and the latency introduced per processor is of one clock cycle and, again, its contribution is negligible. This leads to an FPGA execution time with $P$ processor of $(650, 000)^2/P$.

Table 2 shows the execution time times for the different implementations and the speedup achieved. The low-end FPGA, Cyclone IV, achieves a staggering result with a speedup around × 2, 700. It must be noted that even with a couple of processors the speedup is remarkable. The mid-range FPGA Stratix IV obtained a speedup greater than × 9, 200. The main reason for that is the higher density and the increased clock frequency.

The authors estimate the targeting high-end FPGAs, such as the Stratix 10, might lead to speedup around × 30, 000 (doubling both the number of processors and the clock frequency).

## 6   Conclusions

An FPGA-based accelerator for the computation of the minimum embedding dimension is presented. The results show that it is possible to accelerate the computation three orders of magnitude. This acceleration represent a very important step towards making nonlinear ECG analysis practical in the clinical routine: instead of requiring hours, or even days, of CPU time, now it can be performed in minutes, or even seconds.

As future lines we regard the inclusion of the communication controller to enable the data transfer between the FPGA and a host, as well as targeting high-end FPGA devices that might produce even better results. Also, the architecture presented has many common blocks with the one required for the computation of the Lyapunov exponents [9,10], so it can be extended to compute the minimum embedded dimension, as well as part of the posterior non-linear analysis of the time series. Moreover, slight changes in the presented architecture allow the analysis of the heart rate [1], and the study of electroencephalogram (EEG) recordings [11,12].

**Acknowledgments.** We thank Altera University Program for the support given to the Laboratory of Bioengineering, University CEU-San Pablo. This research was partially supported by the University CEU-San Pablo under project PPC12/2014.

# References

1. Voss, A., Schulz, S., Schroeder, R., Baumert, M., Caminal, P.: Methods derived from nonlinear dynamics for analysing heart rate variability. Philos. Trans. R. Soc. Lond. A: Math. Phys. Eng. Sci. **367**(1887), 277–296 (2009)
2. Foj, O., Holcik, J.: Applying nonlinear dynamics to ECG signal processing. IEEE Eng. Med. Biol. Mag. **17**(2), 96–101 (1998)
3. Kantz, H., Schreiber, T.: Human ECG: nonlinear deterministic versus stochastic aspects. In: IEE Proceedings- Science, Measurement and Technology, vol. 145, pp. 279–284. IET (1998)
4. Govindan, R., Narayanan, K., Gopinathan, M.: On the evidence of deterministic chaos in ECG: surrogate and predictability analysis. Chaos: Interdis. J. Nonlinear Sci. **8**(2), 495–502 (1998)
5. Cao, L.: Practical method for determining the minimum embedding dimension of a scalar time series. Phys. D **110**(1–2), 43–50 (1997)
6. Theiler, J.: Spurious dimension from correlation algorithms applied to limited time-series data. Phys. Rev. A **34**(3), 2427 (1986)
7. http://altera.com
8. Kilts, S.: Advanced FPGA Design: Architecture, Implementation, and Optimization. Wiley-IEEE Press, New Jersey (2007)
9. Perc, M.: Nonlinear time series analysis of the human electrocardiogram. Eur. J. Phys. **26**(5), 757 (2005)
10. Pereda, E., Gonzalez, J., Bhattacharya, J., Rial, R.: Nonlinear Analysis of Biomedical Data. University of La Laguna Press, Tenerife (2010)
11. Babloyantz, A.: Evidence of chaotic dynamics of brain activity during the sleep cycle. In: Mayer-Kress, G. (ed.) Dimensions and Entropies in Chaotic Systems. Springer Series in Synergetics, pp. 241–245. Springer, Heidelberg (1986)
12. Pradhan, N., Dutt, D.N.: A nonlinear perspective in understanding the neurodynamics of EEG. Comput. Biol. Med. **23**(6), 425–442 (1993)

# Low-Power, Low-Latency Hermite Polynomial Characterization of Heartbeats Using a Field-Programmable Gate Array

Kartik Lakhotia[1], Gabriel Caffarena[2], Alberto Gil[2], David G. Márquez[3], Abraham Otero[2], and Madhav P. Desai[1(✉)]

[1] Indian Institute of Technology (Bombay), Powai, Mumbai 400076, India
madhav@ee.iitb.ac.in
[2] Laboratory of Bioengineering, University CEU-San Pablo,
Boadilla Del Monte 28668, Spain
gabriel.caffarena@ceu.es
[3] Centro Singular de Investigación en Tecnoloxías da Información (CiTIUS),
Universidade de Santiago de Compostela, Santiago de Compostela 15782, Spain
david.gonzalez.marquez@usc.es

**Abstract.** The characterization of the heartbeat is one of the first and most important steps in the processing of the electrocardiogram (ECG) given that the results of the subsequent analysis depend on the outcome of this step. This characterization is computationally intensive, and both off-line and on-line (real-time) solutions to this problem are of great interest. Typically, one uses either multi-core processors or graphics processing units which can use a large number of parallel threads to reduce the computational time needed for the task. In this paper, we consider an alternative approach, based on the use of a dedicated hardware implementation (using a field-programmable gate-array (FPGA)) to solve a critical component of this problem, namely, the best-fit Hermite approximation of a heartbeat. The resulting hardware implementation is characterized using an off-the-shelf FPGA card. The single beat best-fit computation latency when using six Hermite basis polynomials is under $0.5\,ms$ with a power dissipation of $3.1\,W$, demonstrating the possibility of true real-time characterization of heartbeats for online patient monitoring.

## 1 Introduction

Cardiovascular disease is the number one cause of death worldwide [1]. The ECG is the main diagnostic tool of the pathologies that affect the heart. Visual inspection of the ECG is a tedious and time-consuming task due to the large amount of data; for example, a 24-h Holter recording can contain up to $100,000$ heartbeats per lead, and often multiple leads are used in each recording. Therefore, computational tools are needed to support this analysis. The first step in the analysis is the detection and characterization of the beats. This is a very important step because errors in this step invalidate the subsequent analysis.

© Springer International Publishing Switzerland 2016
F. Ortuño and I. Rojas (Eds.): IWBBIO 2016, LNBI 9656, pp. 266–276, 2016.
DOI: 10.1007/978-3-319-31744-1_24

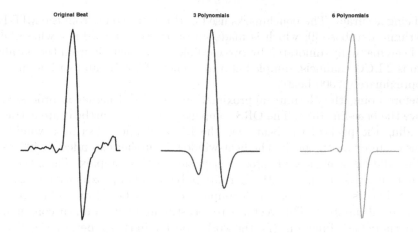

**Fig. 1.** Representation of heartbeat: left – original beat; center – 3-polynomial representation; right – 6-polynomial representation.

The characterization of the QRS complex by means of Hermite functions seems to be a reliable characterization technique of this part of the heartbeat [2]. The main advantages of the Hermite characterization are the low sensitivity to noise and artifacts, and the compactness of the representation (e.g. a 144-sample QRS can be reasonably characterized with 7 parameters [3]). These advantages have made the Hermite representation a very common tool for characterizing the morphology of the beats [2–5] (see Fig. 1).

ECG analysis using Hermite functions has a substantial amount of parallelism. Solutions to this problem have been investigated using processors (and multi-cores) and graphics processing units (GPU's). In this paper, we consider the alternative route of using an FPGA to implement the computations. In particular, our work is motivated by the potential of an FPGA (or eventually, a dedicated application-specific circuit) for low-latency energy efficient heart-beat analysis.

In generating the hardware for heart-beat analysis, we design the hardware starting from an algorithmic specification that is written in a high-level programming language (**C** in this case), which is then transformed to a circuit implementation using the AHIRV2 algorithm to hardware compilation tools [6–8]. Full double-precision floating point arithmetic is used in the hardware implementation. The resulting hardware is mapped to an FPGA card (the ML605 card from Xilinx, which uses a Virtex-6 FPGA). The circuit is then exercised through the PCI-express interface and used to classify beats. The round-trip latency of a single beat classification was found to be under $0.5\,ms$, and the power dissipation in the FPGA during beat processing was $3.1\,W$.

## 2    QRS Approximation by Means of Hermite Polynomials

The aim of using the Hermite approximation to estimate heartbeats is to reduce the number of dimensions required to carry out the ECG classification, without sacrificing accuracy. The benchmarks used in this work come from the MIT-BIH arrhythmia database [9] which is made up of 48 ECG recordings whose beats have been manually annotated by two cardiologists. Each file from the database contains 2 ECG channels, sampled at a frequency of 360 Hz and with a duration of approximately 2000 beats.

Before doing the Hermite approximation, the ECG signal is processed to remove the base-line drift. The QRS complexes for each heartbeat are extracted by finding the peak of the beat (e.g. the R wave) and selecting a window of 200 ms centered on the peak. The beat-window is further extended to 400 ms by padding 100-ms sequences of zeros at each side of the complex. Thus, the QRS beat data used as an input to the Hermite polynomial approximation consists of individual beats described as a 144-sample vector $x = \{x(t)\}$ of double precision floating point numbers. This vector is to be estimated with a linear combination of $N$ Hermite basis functions (for the work reported in this paper, we use $N = 6$).

The goal then is to find the best minimum-mean-square-error (MMSE) approximation to $\{x(t)\}$ as

$$\hat{x}(t) = \sum_{n=0}^{N-1} c_n(\sigma)\phi_n(t,\sigma), \tag{1}$$

with

$$\phi_n(t,\sigma) = \frac{1}{\sqrt{\sigma 2^n n! \sqrt{\pi}}} e^{-t^2/2\sigma^2} H_n(t/\sigma) \tag{2}$$

where $H_n(t/\sigma)$ is the $n^{th}$ Hermite polynomial. These polynomials can be computed recursively as

$$H_n(x) = 2x H_{n-1}(x) - 2(n-1)H_{n-2}(x), \tag{3}$$

where $H_0(x) = 1$ and $H_1(x) = 2x$. The parameter $\sigma$ is a time-scaling factor in the polynomials which needs to be chosen carefully. In [2] the maximum value of $\sigma$ for a given order $n$ is estimated. As the value of $n$ increases, the value of $\sigma_{MAX}$ decreases.

The Hermite polynomials are orthonormal. Thus, the optimal coefficients that minimize the estimation error for a given $\sigma$ are

$$c_n(\sigma) = \sum_t x(t) \cdot \phi_n(t,\sigma) \tag{4}$$

The best fit is calculated by comparing the MMSE approximation for each $\sigma$, and keeping the one with the smallest value. Once the best $\sigma$ and the corresponding fit coefficients $c = \{c_n(\sigma)\}$ ($n \in [0, N-1]$) are found for each heartbeat, it is possible to use only these figures to perform morphological classification of the heartbeats [2].

# 3   Beginning the FPGA Implementation: The Algorithm

The algorithm used in the FPGA implementation is illustrated in Fig. 2.

The implementation first receives the values of the Hermite polynomial basis functions, and stores them in distinct arrays in the hardware. In the current implementation, we use six arrays to store the basis functions for order $n = 0$ to $n = 5$. For each $n$, basis functions for ten different values of $\sigma$ are stored in the corresponding array. The values of $\sigma$ used range from $1/120$ to $1/90$.

```
void HermiteBestFit()
{
    receiveHermiteBasisFunctions();

    while(1)
    {
        receiveHeartBeat();
        innerProducts();
        findBestFit();
        reportResults();
    }
}
```

**Fig. 2.** High-level view of algorithm mapped to the FPGA

After this initialization step, the hardware executes a continuous loop. In the loop body, the hardware first listens for heart beats. When a complete heartbeat (144 samples) is received, the inner products of the heart-beat with all the basis functions are calculated in a double loop. After all inner products are calculated, the inner product coefficients are used to compute the best fit among the different values of $\sigma$. The best-fit $\sigma$ index and coefficients are then written out of the hardware.

The algorithm as described above is purely sequential and does not contain any explicit parallelization. The AHIRV2 compiler (described later in Sect. 4) is intelligent enough to extract parallelism from the two critical loops (in the inner-product and best-fit functions). Even with this simple coding of the hardware algorithm, we observe that excellent real-time performance is observed (in comparison with CPU/GPU implementations). Going further, it is possible to specify explicit parallelism by and exploit it by using multiple function units in hardware in order to reduce the processing latency. These investigations are currently in progress.

## 3.1   The Inner Product Loop

The inner product loop is shown in Fig. 3. The outer loop iterates over the samples, and the inner loop across the $\sigma$ values. There is a high-level of parallelism in

```
void innerProduct()
{
  int I;
  for (I=0; I < NSAMPLES; I++)
  { // outer-loop
      double x = inputData[I];
      for(SI = 0; SI < NSIGMAS; SI++)
      { // inner-loop
          int I0 = I + Offset[SI];
          double p0 = (x*hF0[I0]);
          double p1 = (x*hF1[I0]);
          double p2 = (x*hF2[I0]);
          double p3 = (x*hF3[I0]);
          double p4 = (x*hF4[I0]);
          double p5 = (x*hF5[I0]);
          dotP0[SI] += p0;
          dotP1[SI] += p1;
          dotP2[SI] += p2;
          dotP3[SI] += p3;
          dotP4[SI] += p4;
          dotP5[SI] += p5;
      }
  }
}
```

**Fig. 3.** Inner-product loop

the inner loop which can be further boosted by unrolling the outer loop. When translating this to hardware, the entire function uses one double-precision multiplier and one double-precision adder. Further note that the arrays $hFn$ and $dotPn$ are declared on a per-$n$ basis (for $n = 0$ to $n = 5$). This allows the arrays to be mapped to distinct memory spaces, thus increasing the memory access bandwidth in the hardware.

## 3.2    The Minimum-mean-square Loop

The MMSE calculation hardware uses the algorithm shown in Fig. 4. Note that the inner loop again has considerable parallelism. One pipelined double precision multiplier and one pipelined double precision adder are used to implement the loop. The arrays referred to in the loop $dotPn$ and $hFn$ are all implemented in disjoint memories to give high memory access bandwidth.

## 3.3    Optimizations: Loop Unrolling and Loop Pipelining

The current implementation uses a simple sequential specification. We have investigated the impact of further optimizations. In particular, we find that outer-loop unrolling (up to four) and inner-loop pipelining have substantial impact on the performance of the generated hardware.

```
void computeMSE()
{
  int I, SI;
  best_mse = 1.0e+20;
  best_sigma_index = -1;
  for (I=0; I<NSAMPLES; I=I+4)
  { // outer-loop
     for (SI=0; SI<NSIGMAS; SI++)
     { // inner-loop
        int fetchIndex0 = I + Offset[SI];
        double p0 = (dotP0[SI]*hF0[fetchIndex0]);
        double p1 = (dotP1[SI]*hF1[fetchIndex0]);
        double p2 = (dotP2[SI]*hF2[fetchIndex0]);
        double p3 = (dotP3[SI]*hF3[fetchIndex0]);
        double p4 = (dotP4[SI]*hF4[fetchIndex0]);
        double p5 = (dotP5[SI]*hF5[fetchIndex0]);
        double diff = (inputData[I]-
                     ((p0+p1) + (p2+p3) + (p4+p5)));
        err[SI] += (diff*diff);
     }
  }
  for (SI=0; SI<NSIGMAS; SI++)
  {
     if(err[SI] <  best_mse)
     {
        best_mse = err[SI];
        best_sigma_index = SI;
     }
  }
}
```

**Fig. 4.** MMSE calculation loop

If we look at the inner loop in Fig. 3, we observe that because of the accu-mulation of products in each loop iteration, the time interval between successive loop initiations is equal to the latency of the adder (because the next loop iter-ation needs to wait for the completion of the sum in the current iteration). This latency is around 20 cycles in our case, and becomes a performance limiter.

If the outer loop in the innner-product function (Sect. 3.1) is unrolled four times, the resulting code will have the form shown in Fig. 5. The amount of par-allelism in the inner loop increases by a considerable margin. The loop iteration initiation latency stays about the same, but the amount of computation done in each loop iteration is quadrupled, thus leading to higher performance. In Sect. 6, we provide actual measurements in hardware which support this observation.

A similar improvement is observed by unrolling the outer loop in the MMSE computation shown in Fig. 4.

```
void  innerProduct()
{
  for (I=0; I < NSAMPLES; I += 4)
  { // outer-loop
     double x0 = inputData[I];
     double x1 = inputData[I+1];
     double x2 = inputData[I+2];
     double x3 = inputData[I+3];
     for(SI = 0; SI < NSIGMAS; SI++)
     {
        int I0 = I + Offset[SI];
        I1 = I0+1; I2 = I0+2; I3 = I0+3;
        // compute inner product of
        // (x0,x1,x2,x3) with
        // (hFn[I0], hFn[I1], hFn[I2], hFn[I3])
        // for n=0,1,2,..5.
        //
        // accumulate into dotPn[SI].
        //
     }
  }
}
```

**Fig. 5.** Inner-product loop unrolled four times.

## 4    From Algorith-to-hardware Using AHIRV2, a C-2-VHDL Compiler

The AHIRV2 compiler tool-chain [6–8] provides a pathway from a C-program to actual synthesizable hardware. The tool-chain takes a description of an algorithm (described in C) and produces a VHDL logic circuit description which is equivalent to the algorithm.

The AHIRV2 compiler starts with a C program and produces VHDL. For the clang-2.8 compiler[1] is used as the C front-end and is used to emit LLVM byte-code[2], which is then transformed to VHDL using the following transformations:

1. The LLVM byte-code is translated to an internal intermediate format, which is itself a static-single assignment centric control-flow language (named **Aa**) which allows the description of parallelism using fork-join structures as well as arbitrary branching.
2. The **Aa** description is translated to a virtual circuit (the model is described in the next section). During this translation, the following major optimizations are performed: declared storage objects are partitioned into disjoint memory spaces using pointer reference analysis, and dependency analysis is used to

---

[1] www.clang.org.

[2] www.llvm.org.

generate appropriate sequencing of operations in order to maximize the parallelism. Inner loops in the **Aa** code are pipelined so that multiple iterations of a loop can be executed concurrently.

3. The virtual circuit is then translated to VHDL. At this point, decisions about operator sharing are taken. Concurrency analysis is used to determine if a shared hardware unit needs arbitration. Optimizations related to clock-frequency maximization are also carried out here. The generated VHDL uses a pre-designed library of useful operators ranging from multiplexors, arbiters to pipelined floating point arithmetic units (arbitrary precision arithmetic is supported, and in particular, there is full support for IEEE-754 single precision and double precision add/multiply with all rounding modes).

## 5    Hardware Implementation Details

The overall system has 3 major components:

- A host computer, which is used to calculate the Hermite basis functions, initialize the FPGA card, send beat data to the FPGA card and receive the best-fit coefficients from the FPGA card.
- The FPGA card, on which the best-fit algorithm is implemented. We use the Xilinx ML605 card which features a Virtex-6 FPGA and an 8-lane PCI express interface.
- The FPGA card driver, which is based on the RIFFA infrastructure [10].

The algorithm mapped to the FPGA is first described in a C program (code fragments described in Sects. 3, 3.1 and 3.2). The architecture of the hardware produced is shown in Fig. 6. In the initialization phase, the hardware-side listens on an input FIFO to acquire the Hermite polynomials (these are stored in six disjoint memories, one for each order $n = 0, 1, 2, \ldots 5$). After this step, the unit that receives the beat samples is triggered. This unit listens on the input FIFO and receives a 144 sample heartbeat (coded as 144 double-precision floating point numbers). After receiving the sample, it triggers the inner-product stage. In the inner-product stage, the hardware computes, for each $\sigma$ and $n$, an inner product of the received beat with the Hermite polynomials $\phi_n(t, \sigma)$. We are using ten values of $\sigma$ and 6 values of $n$. Thus, 60 inner-products are computed in this phase. The inner products are stored in ten disjoint memories, one for each $\sigma$. After this is done, the MMSE stage is triggered. In the MMSE stage, the inner-products are used to find the best fit $\sigma$. The computed best-fit coefficients are sent back to the host using the output FIFO. The hardware unit which listens for the next beat is then triggered (wait for the next beat).

The VHDL hardware for this design is generated using the AHIRV2 toolchain which was described in Sect. 4. The generated VHDL is instantiated in the FPGA together with the RIFFA wrappers, and the resulting design is synthesized and mapped to the Virtex-6 FPGA using the Xilinx ISE 14.3 toolset.

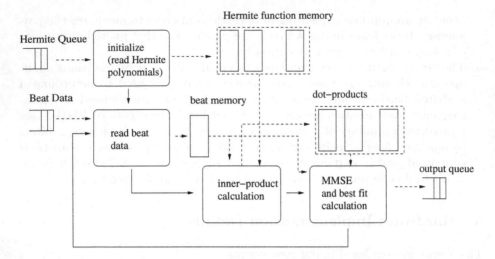

**Fig. 6.** Hardware architecture

## 6    Results

We measure the round-trip delay and FPGA core power consumption for processing one beat. The round-trip delay is the time interval between the beginning of transmission of beat-data from the host to the hardware and the beginning of reception of best fit coefficients from the hardware. The test feeds a single beat at a time to the FPGA and measures the latency.

In the implementation, the two outer-loops described in Sects. 3.1 and 3.2 were unrolled to different extents to see the impact of unrolling on the system performance. Three levels of unrolling were tried: one-way, two-way and four-way. The four-way unrolling gave the best performance, as expected. The results are summarized in Table 1 (the reported latency is the average value observed across 100 beats). The minimum latency achieved with four-way-unrolling was observed to be 0.39 ms (for the processing of a single 144-sample beat). The power dissipation values are measured using hardware monitoring while the beats are being processed, and represent peak power dissipation.

**Table 1.** Results: FPGA utilization and latency for different loop-unrolling levels

| Unroll-level | Slice LUT utilization | Slice register utilization | Avg. processing latency | FPGA core power consumption |
|---|---|---|---|---|
| 1-way | 56839 | 65995 | 1.39 ms | 2.75 W |
| 2-way | 65895 | 80709 | 0.80 ms | 2.88 W |
| 4-way | 84331 | 110165 | 0.44 ms | 3.09 W |

It is clear that FPGA can be used for *true* real-time processing since the computation time required to process a beat over two ECG leads (the MIT-BIH arrhythmia database has two leads) is less than 1 ms, which is much smaller than time-interval between actual heart beats (which is about 1 s). The observed power dissipation in the FPGA while the beat is being processed is 3.1 W. Hardware utilization in 4-way unrolled system is less than 55 % of the FPGA resource.

## 6.1 Comparison with GPU/CPU Implementations

Hermite basis fitting of heartbeats has been evaluated on GPU and CPU implementations as well. Such a study has been carried out in [11] using single precision floating point arithmetic. To summarize their results

- On a single CPU (an Intel-i7, 1.6 GHz core), Hermite polynomial fitting for a single beat sample-pair is accomplished in about $2\,ms$ using single-precision floating-point arithmetic.
- Using a GPU (NVIDIA TESLA C2050, 1.15 GHz), a substantial speed-up (relative to the single-CPU data) is observed when large chunks of data are processed: for example, the processing of a $10^6$ beat chunk is accomplished with an average processing latency per beat of 0.011 ms, using single-precision floating-point arithmetic. When processing a 10 beat chunk, the GPU manages an average processing latency per beat of 0.05 ms. When a single beat chunk is considered, the GPU latency is of the same order as that observed on the processor. Thus, as the number of beats being processed simultaneously is decreased, the GPU latency per beat increases.

In comparison, our FPGA implementation uses double precision floating point arithmetic, and achieves a latency of 1 ms for processing a pair of heartbeats, with a power dissipation of 3.1 W during the processing. This matches the CPU and GPU in term of the processing latency for a single pair of beats, while displaying a power consumption which is two orders of magnitude lower than the other two technologies. This demonstrates that custom hardware even on an FPGA is an attractive option for low energy real-time ECG classification applications in portable health monitoring devices.

## 7  Conclusions

In this paper, a solution to the problem of Hermite polynomial based heartbeat characterization using FPGAs is presented. We have mapped the problem to hardware using algorithm-to-hardware techniques (with the AHIRV2 tools). The Xilinx ML605 card with a Virtex-6 FPGA was used as the platform and the RIFFA host-interface was used to communicate with the FPGA card.

The time required to process a single heart-beat was less than 1 ms. This means that the implementation can be used for true online real-time beat processing. Further the power dissipation when the FPGA is actively processing the beat is 3.1 W. Assuming that the interval between two beats is 1 s, and assuming that the hardware is turned on only while the beat is actually being processed (which is less than 1 ms), we expect that the *average* power dissipation of hardware based characterization can be as small as 30 mW, which means that a custom hardware based portable device is practically feasible.

**Acknowledgments.** We thank Xilinx University Program for the support given. This research was partially supported by the University CEU-San Pablo under project PPC12/2014. David G. Márquez is funded by an FPU Grant from the Spanish Ministry of Education (MEC) (Ref. AP2012-5053).

# References

1. World Health Organization: Global Status Report on Noncommunicable Diseases. Accessed: September 27 (2010). http://whqlibdoc.who.int/publications/2011/9789240686458_eng.pdf?ua=1
2. Lagerholm, M., Peterson, C., Braccini, G., Edenbr, L., Sörnmo, L.: Clustering ECG complexes using Hermite functions and self-organizing maps. IEEE Trans. Biomed. Eng. **47**, 838–848 (2000)
3. Márquez, D.G., Otero, A., Félix, P., García, C.A.: On the accuracy of representing heartbeats with hermite basis functions. In: Biosignals Conference, pp. 338–341 (2013)
4. Braccini, G., Edenbrandt, L., Lagerholm, M., Peterson, C., Rauer, O., Rittner, R., Sörnmo, L.: Self-organizing maps and Hermite functions for classification of ECG complexes. Comput. Cardiol. **1997**, 425–428 (1997)
5. Linh, T.H., Osowski, S., Stodolski, M.: On-line heart beat recognition using Hermite polynomials and neuro-fuzzy network. IEEE Tran. Instrum. Measur. **52**(4), 1224–1231 (2003)
6. Sahasrabuddhe, S.D.: A competitive pathway from high-level programs to hardware. PH.D. thesis, IIT Bombay (2009)
7. Sahasrabudhe, S.D., Subramanian, S., Ghosh, K., Arya, K., Desai, M.P.: A c-to-rtl flow as an energy efficient alternative to the use of embedded processors in digital systems. DSD **2010**, 147–154 (2010)
8. Rinta-Aho, T., Karlstedt, M., Desai, M.: The clicktonetfpga tool-chain. In: USENIX ATC-2012, USENIX Association, Berkeley CA (2012)
9. Moody, G.B., Mark, R.G.: The impact of the MIT-BIH arrhythmia database. IEEE Eng. Med. Biol. Mag. **20**(3), 45–50 (2001)
10. Jacobsen, M., Kastner, R.: RIFFA 2.0: a reusable integration framework for FPGA accelerators. In: Field Programmable Logic and Applications, vol. 23, pp. 1–8 (2013)
11. Gil, A., Caffarena, G., Márquez, D., Otero, A.: Hermite polynomial characterization of heartbeats with graphics processing units. IWBBIO **2014**, 527–538 (2014)

# Assessing Parallel Heterogeneous Computer Architectures for Multiobjective Feature Selection on EEG Classification

Juan José Escobar, Julio Ortega[⊠], Jesús González,
and Miguel Damas

Department of Computer Architecture and Technology,
CITIC, University of Granada, Granada, Spain
rotty@correo.ugr.es,
{jortega,jesusgonzalez,mdamas}@ugr.es

**Abstract.** High-dimensional multi-objective optimization will open promising approaches to many applications on bioinformatics once efficient parallel procedures are available. These procedures have to take advantage of the present heterogeneous architectures comprising multicore CPUs and GPUs. In this paper, we describe and analyze several *OpenCL* implementations for an application comprising multiobjective feature selection for clustering in an EEG classification task on high-dimensional patterns. These implementation alternatives correspond to different uses of multicore CPU and GPU platforms to process irregular data codes. Depending on the dataset used, we have reached speedups of up to 14.9 and 17.2 with up to 24 threads for the implemented *OpenCL* CPU kernels and of up to 7.1 and 9.1 with up to 13 SMX processors and 256 local work-items for our OpenCL GPU kernels. Nevertheless, to provide this level of performance, careful considerations about the use of the memory hierarchy of the heterogeneous architecture and different strategies to cope with the irregularity of our target application have to be taken into account.

**Keywords:** Brain-Computer Interfaces (BCI) · EEG classification · Feature selection · GPU · Heterogeneous parallel architectures · Multiobjective optimization

## 1 Introduction

Multi-objective evolutionary optimization is useful in many classification and data mining applications on bioinformatics that frequently require the simultaneous optimization of several conflicting objectives [1, 2]. The goals of these applications deal with discovering useful models on large datasets, in many cases also involving high dimensional patterns, and comprising tasks such as classification, clustering, feature selection, association rule mining, etc. [1, 2]. Electroencephalogram (EEG) classification is a good example of such applications that deal with the processing of high-dimensional patterns. Indeed, an EEG is a multivariate signal whose classification has to cope with difficult problems. Some of them are related to the low signal-to-noise ratio in the EEG signals, the representation of time information in the features (as precisely brain patterns are

© Springer International Publishing Switzerland 2016
F. Ortuño and I. Rojas (Eds.): IWBBIO 2016, LNBI 9656, pp. 277–289, 2016.
DOI: 10.1007/978-3-319-31744-1_25

usually related to changes in time in the EEG signals), and the non-stationary nature of EEG signals. The solution to these problems usually implies the definition of high dimensionality feature vectors, despite the fact that few EEG patterns are available to determine the classifier parameters. This way, as in many other applications on bioinformatics that also imply high-dimensional pattern classification tasks, feature selection techniques should be applied to remove noisy or irrelevant features or to improve the learning accuracy and result comprehensibility whenever the number of features in the input patterns is higher than the number of available patterns. In this paper, different implementations for heterogeneous parallel architectures are proposed to accelerate feature selection implemented through multiobjective optimization in high dimensional spaces. The datasets used to evaluate our procedures correspond to BCI (Brain-Computer Interfaces) tasks [3].

Heterogeneous parallel architectures involving multiple general-purpose superscalar multicore CPUs and accelerators, mainly including GPUs, constitute the present mainstream approach to take advantage of technology improvements [4] and many previous works have reported high speedups on both multicore CPUs and GPUs. Precisely, GPUs offer the opportunity to take advantage of massive parallelism, sometimes with lower energy consumption [5]. Nevertheless, as it is pointed out in [6], there is a controversy about whether real applications can benefit from GPUs to get higher speedups than with multicore CPUs. As it is also argued in [6], the data location and the overhead to move data to where they are required, and back again if necessary, have to be taken into account whenever parallel efficient codes are to be developed. The purpose of this paper is to provide an insight into the issues to be taken into account to devise efficient parallel procedures for classification and optimization tasks on high-dimensional sets of patterns in these heterogeneous platforms. To do that, and as *OpenCL* is an open standard for writing programs to be executed across the considered heterogeneous computing platforms, we propose and analyze different *OpenCL* implementations of a multi-objective approach for feature selection in high-dimensional classification problems.

After this introduction, Sect. 2 describes the evolutionary multiobjective optimization approach to feature selection whose implementation we have parallelized by using either superscalar cores or GPUs. Section 3 analyzes the main issues to be taken into account in the development of efficient parallel codes for the present heterogeneous parallel platforms and gives the details of our proposed *OpenCL* multiobjective implementations along with the references to previous works on these issues. Finally, Sect. 4 describes the experimental results and compares the behavior of the considered alternatives, and Sect. 5 summarizes the conclusions.

## 2  Feature Selection as a Multiobjective Problem

The problem here addressed deals with feature selection in unsupervised classification. It is also considered that the patterns to be classified have a high number of features (components). Moreover, the number of available patterns to be classified could be even lower than the number of features, thus posing a curse of dimensionality [7] problem. In order to achieve an adequate performance for the classifier, the most relevant features have to be selected, thus contributing to (1) decrease the computational complexity of the classification procedure, (2) remove irrelevant/redundant features that would make the learning of the classifier more difficult, and (3) prevent the curse of dimensionality whenever there are many features and a low number of available patterns to be classified.

Nevertheless, finding the optimum set of features is an NP-hard problem and different metaheuristics such as simulated annealing, genetic algorithms, ant colony optimization, particle swarm optimization, etc., have been previously proposed [8].

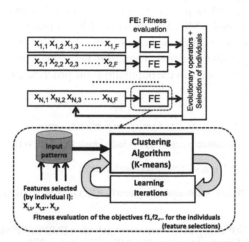

**Fig. 1.** Wrapper procedure for unsupervised feature selection (with K-means as clustering algorithm) by evolutionary multi-objective optimization

In the present paper, we apply multiobjective evolutionary computation to feature selection, as the performance of a classifier should be expressed not only through its accuracy for the set of patterns used for testing but also through other measures that quantify properties such as the generalization capability. This way, in order to search for the features that determine the best classifier performance once it is trained by using these features as pattern components, multiobjective optimization approaches such as [9–14] have been proposed for either supervised or unsupervised classification problems. The contributions of this paper deal with the parallel processing on multicore CPU and GPU architectures, of feature selection approached by multi-objective optimization in applications with a large number of features (decision variables in the multi-objective optimization problem). Some previous papers such as [15–18] have considered parallel processing as an interesting approach for feature selection. In [17], we have approached feature selection by using a parallel multiobjective procedure that is extended in [18] from a cooperative coevolutionary point of view.

Figure 1 describes our approach for feature selection. A multiobjective evolutionary procedure evolves a population of individuals that codify different feature selections. The fitness of each individual (a feature selection) is computed from the performance achieved by the well known K-means clustering algorithm applied to the patterns (whose components are the features codified by the corresponding individual of the population). To evaluate a set of features in unsupervised classification problems, clustering validation indexes (CVI) should be computed from the distribution of patterns in the space of the selected features. Several validation indices whose effectiveness depends on the characteristics of the application at hand have been proposed in the literature [19]. In this paper, we have used the intra-cluster and inter-cluster distances after applying a K-means clustering algorithm to the set of training patterns whose components correspond to the selected features.

Algorithm 1 corresponds to the pseudocode of an implementation of the procedure of Fig. 1. The *NSGAII_optimizer* procedure implements the evolutionary multi-objective feature selection based on the NSGA-II algorithm [20]. A population of N individuals is initialized (line 01) where each individual, S(i) (i = 1,.., N), corresponds to a selection among the set of F possible features. Then, the individuals are evaluated by calling N times (one time per individual in the population) the procedure *Evaluation* (lines 02 to 04 of this procedure).

```
NSGAII_Optimizer
01          Initialize a Population composed of N individuals P₀={S(i); i=1,…,N};
02          for i=1 to N
03                    (f1(S(i)),f2(S(i)))=Evaluation(S(i),DS);
04          end;
05          (P,f(P))=NSGAII_nondomination_sort(P₀,f(P₀));//f(P₀)≡(f1(S(i)),f2(S(i))(i=1,..,N)
06          t=1;
07          do
08                    (P',f(P'))=NSGAII_tournament_selection (P,f(P)); // P'={S'(j); j=1,…R}
09                    P''=Genetic_operators (P'); // P''={S''(j); j=1,…,V)}
10                    for i=1 to V
11                              (f1(S''(i)),f2(S''(i)))=Evaluation(S''(i),DS);
12                    end;
13                    (P*,f(P*))=NSGAII_nondomination_sort ( P, f(P);P'',f(P''));
14                    (P,f(P))=NSGAII_replace_chromosome ( P*,f(P*));
15                    t=t+1;
16          while stop criterion is not reached;
17          return(P,f(P));
18          end;


Evaluation(S(i),DS) // Dataset DS={D(k); k=1,…,M} (M training patterns of F components)
01          K=Kmeans(F(i),DS) // K={K(j); j=1,..,W}
02          f1(S(i))=intraclass(K,DS);
03          f2(S(i))=interclass(K,DS);
04          return (f1(S(i)),f2(S(i)));
05          end;
```

**Algorithm 1.** Pseudocode of the multi-objective feature selection procedure of Figure 1

The procedure *Evaluation* applies a K-means algorithm to the dataset, DS, including the training patterns (line 01 of *Evaluation*). Although the training patterns are vectors with F components (each corresponding to one of the possible features), the components used by the execution of the K-means algorithm in a given call of the *Evaluation* procedure are those chosen by the selection of features codified by the individual to be evaluated. The K-means algorithm has the following steps: (1) once the number of clusters, W, is set, generate W initial centroids (cluster centers); (2) assign each pattern to the nearest cluster according to its distances to the cluster centroids; (3) calculate the new cluster centroids; (4) if the end condition is not met (usually either changes are observed in the position of the centroids, or a maximum number of iterations have not been completed yet) repeat steps (2) and (3).

Once the K-means algorithm provides a set of K vectors corresponding to the centroids of the different clusters (as many as classes in the target problem), the intra-cluster and the inter-cluster costs are calculated (lines 02 and 03 of *Evaluation*, respectively) and returned to the procedure *NSGAII_optimizer*. Once the cost functions are evaluated for all the individuals in the population, they are ordered according to different non-dominance fronts used by the NSGA-II procedure. This is done by procedure *NSGAII_nondomination_sort* in line 05. Then, a given number of generations are implemented where selection, mutation and crossover evolutionary operators are applied (procedures of lines 08 and 09), along with calls to *Evaluation* procedure to determine the values of the cost functions for the new individuals (lines 10 to 12), another call to the non-domination sort procedure (line 13) applied to the union of the new generated individuals and the ones included in the previous population, and a procedure to select the population to be used in the next generation (line 14).

This way, the application here considered includes a multiobjective evolutionary computation algorithm and a clustering algorithm executed once for each individual in the population and generation of the evolutionary algorithm.

## 3 Multiobjective Feature Selection on Heterogeneous Platforms

In this section, we will describe the parallel implementations we have considered to take advantage of the present heterogeneous CPU-GPU architectures. In these architectures, the GPU plays the role of a coprocessor connected through a bus (usually a PCIe bus) to a host including multiple superscalar cores that share the main memory. The GPUs emerged at the end of the 1990 s fuelled by the demands of 3D graphic applications and have evolved towards parallel multi-threaded architectures with many cores that could be able to provide high speedups with lower costs as they devote more resources to data processing than to data caching and flow control than the superscalar cores. Nevertheless, during the execution of the programs, the data have to be transferred between the host memory and the GPU memory, and the latency and bandwidth of transferences between the host and the GPU affect the performance of the codes and should be taken into account for efficient programming of CPU-GPU platforms. The GPU has its own memory hierarchy which is usually deeper than in the multicore CPU host. The basic computing elements or cores of the GPU are the so called Stream Processors (SP). They do not contain instruction units and are able to execute scalar operations. Several SP along with one or several instruction units and a register file comprise a multiprocessor, also called Streaming Multiprocessor (SMX). A GPU can include several SMX and allows the simultaneous execution of the same program on different data (i.e. the SPMD model). This is done by using multiple threads always executed by the SMX according to an SIMT (Single Instruction Multiple Threads) model where all the threads on the SMX execute the same instruction per cycle (the SMX has only one program counter). The threads are organized within thread blocks in such a way that all the threads in a block are assigned to a single SMX. While the threads in a block are able to cooperate and share the instruction unit, the register file and some low latency memory, threads in different blocks can only communicate among themselves through the global memory.

In this paper we propose several *OpenCL* implementations for Algorithm 1. *OpenCL* [21] allows platform-independent parallel programming to develop applications built from a program that is executed in a host and also launches functions, called kernels, to other *OpenCL* devices that can be multi-core CPUs, or GPUs. In GPUs, the so called *work-items* are kernel instances, or threads, identified by indices of integer components in a space of one, two, or three dimensions. Several *work-items* can execute the same instruction over different data items according to a SIMD (Single Instruction Multiple Data) model, and they can be also organized as a *work-group*, in such a way that several *work-groups* can be executed according to an SPMD (Single Program Multiple Data) model. Our *OpenCL* implementations are built from a main program and two kernels that can be launched by that main program to either a set of superscalar CPU cores or a GPU. The main program reads the patterns included in the database and normalizes them (according to a logistic normalization) and implements the main steps of the evolutionary multiobjective optimization procedure, i.e. lines 05, 08, 09, 13, and 14 in the procedure *NSGAII_Optimizer* of Algorithm 1. The loops including the calls to the *Evaluation* procedure of lines 03 and 11 are implemented into two different kernels to be executed, respectively, in the GPU and in the CPUs (in order to compare the corresponding performances).

There are many details that have been taken into account to reach adequate and efficient implementations in the considered multicore CPU/GPU platform:

- A buffer is used to store the $N$ individuals of the population. The size of this buffer is the double of the memory required to store the $N$ individuals, because it is also used to store the offspring generated by the application of crossover and mutation operators to the parents. This way, an improvement in the memory accesses is expected through an *OpenCL* object that allows the asynchronous population reading or writing to the global memory.
- Before the first call to the *Evaluation* procedure, the main program chooses a set of initial centroids (as many as classes in the classification problem), which are randomly selected among the patterns included in the dataset (training patterns), as for example in [22]. After that, a barrier is set to synchronize the copy in the device of all the *OpenCL* objects previously used, the main program identifies the device where the kernel implementing the *Evaluation* function is going to be executed and launches the kernel once their arguments have being properly initialized.
- The kernels not only implement the *Evaluation* function, but also the loops including the calls to execute *Evaluation* for each individual in the population, i.e. lines 02 to 04 and 10 to 12 in the NSGAII_optimizer in Algorithm 1. This way, different *Evaluation* functions applied to different individuals can be executed in parallel either in the GPU (if the GPU kernel is launched) or using several cores in the host (if the CPU kernel is launched).

- The stop condition is determined by setting a maximum number of generations, and also by taking into account the improvement in the quality of the solutions between two consecutive generations. The quality of the solutions for this purpose is evaluated through the hypervolume of the first front of non-dominance which has been obtained from the procedure of Fonseca [23] by using the point (1,1) as reference (it has to be taken into account that the fitness values of the population are normalized).
- In the kernel to be executed in the host CPU, the input parameters are the population, the initial centroids, the normalized data set, and the indices of the individuals to be evaluated. The population and the indices of the individuals are stored in the *global memory* so that they could be accessed by all the computing units. The dataset and the initial centroids could be also accessed by all the computing units, but they can be stored in *constant memory,* as they do not change. Two auxiliary buffers are also stored in the local memory to keep information about the changes in the centroids along the K-means execution.
- The GPU kernel is quite similar to the CPU kernel with respect to the input parameters, although there are some differences related to the code distribution and execution. As the GPUs really have physically different memories, the data transference among buffers (after each K-means iteration) requires 3 N operations, thus having a complexity proportional to N. Moreover, in the GPU kernel the individuals are evaluated in parallel by the different SMX computing units (as in the CPU kernel they were evaluated by the different cores) but each computing unit includes several *work-items* that allow us to take advantage of the data parallelism present in the K-means algorithm and in the intra-cluster and inter-cluster computation. Nevertheless, it has to be taken into account that several barriers have to be inserted to synchronize these steps. These barriers would imply an efficiency loss.
- The kernels evaluate the population in parallel for each generation. This way, although the dataset and the initial centroids only have to be transferred to the GPU memory once, for each generation, the population to be evaluated has to be transferred from the main memory to the GPU memory, and the values of the cost functions for each population have to be sent back to the main memory from the GPU. This way, $(F \times M \times b + C \times M \times b)$ bytes have to be transferred once to the GPU, where $F$ is the number of features, $M$ is the number of patterns of the training set, $C$ is the number of clusters, and $b$ is the number of bytes per feature, $N \times F$ bits have to be transferred to the GPU once per generation ($N$ is the number of individuals in the population), and $N \times 2 \times b$ bytes (for two objectives) are transferred from the GPU memory to the host main memory.
- The different approaches for GPU-based implementations of evolutionary algorithms are analyzed in [24], and the issue of implementations of parallel metaheuristics on these multicore platforms has been surveyed in [25]. Nevertheless, works analyzing the effect in the parallel performances of heavy fitness functions requiring high-volume datasets, such as our target application, are less frequent. The paper by Pospichal et al. [26] describes a CUDA implementation of a parallel genetic algorithm based on an island model (parallel evolution of subpopulations that exchanges some individuals after some generations). With respect to the GPU implementations of evolutionary multi-objective procedures, the papers by Sharma et al. [27]

and Wong et al. [28] are good examples. The parallelization of the K-means algorithm on a GPU has been considered in many papers [29–33]. With respect to other *OpenCL* implementations [34] shows a genetic algorithm for feature selection in a biometric recognition application. Although it follows a quite similar strategy to the one considered in the present, it does not implement a multi-objective evolutionary algorithm.

## 4  Experimental Results

In this section, we analyze the speedup results obtained by our CPU and GPU *OpenCL* implementations using different resource configurations of our platform, including nodes with 32 GB of DDR3 memory and two Intel Xeon E5-2620 processors at 2.10 GHz comprising 24 threads (six cores per socket and two threads per core). It also includes a Tesla K20c with 5 GB of global memory, 208 GB/s as maximum memory bandwidth, and 2496 CUDA cores at 705.5 MHz, distributed into 13 stream multiprocessors (SMX) with 192 cores per SMX. We have used the data benchmark b480a extracted from the dataset recorded in the BCI Laboratory [35] at the University of Essex and used in [17, 18]. It includes 178 patterns with 480 features. We have also considered another larger data file built from the University of Essex dataset with 80 patterns and 2160 features per pattern. In the case of the dataset of 480 features, we have obtained speedups of up to 14.9 with up to 24 threads implemented on CPU kernels and up to 7.1 with our GPU kernels executed by up to 13 SMX processors. Whenever the dataset of 2160 features has been used, speedups of up to 17.2 and 9.1 have been respectively achieved by our CPU and GPU kernels. In what follows, we summarize the results of our experiments with the dataset of 480 features.

For each experimental condition, we have made several repetitions. Then, we apply a Kolmogorov-Smirnov test to the results of the repetitions in order to determine whether the data follow a standard normal distribution. According to the result of the Kolmogorov-Smirnov tests, we then apply either an ANOVA test if the data follow a normal distribution or a Kruskal-Wallis test if this is not the case. After analyzing the hypervolume results obtained in the execution of the different experiments, it has been observed that the differences in the obtained hypervolume values are not statistically significant with respect to the values obtained by the sequential procedures with the same conditions in the evolutionary algorithm (number of individuals in the population and generations), as it was be expected because our *OpenCL* implementations correspond to alternative parallel implementations that keep the behavior of the base sequential algorithm. Thus, it can be concluded that all the executions provide solutions with similar qualities and we will only consider the time behaviors of the different parallel implementations.

The experiments have been accomplished by considering different numbers of CPU cores and several SMX processors with different numbers of work-items in the GPU. Moreover, populations of 500 and 1000 individuals, and 20, 50, and 100 generations have been also evaluated. Table 1 provides the average execution times obtained by the same number of threads (8) in the CPU and GPU, and the best values provided by the

**Table 1.** Mean execution times for 8 and 24 CPU threads, and 8 and 13 SMX GPU processors (populations of 500 and 1000 individuals and 20, 50, and 100 generations).

|         | 500/20 | 500/50 | 500/100 | 1000/20 | 1000/50 | 1000/100 |
|---------|--------|--------|---------|---------|---------|----------|
| Seq.    | 331.84 | 823.38 | 1583.55 | 701.50  | 1745.80 | 3359.70  |
| CPU(8)  | 51.70  | 127.50 | 256.60  | 105.68  | 254.60  | 502.00   |
| GPU(8)  | 80.59  | 187.88 | 378.15  | 157.80  | 397.40  | 760.8    |
| CPU(24) | 25.47  | 57.23  | 114.30  | 49.64   | 117.10  | 237.00   |
| GPU(13) | 50.38  | 124.45 | 239.46  | 103.70  | 249.40  | 475.90   |

CPU threads in the node and by the GPU. The efficiencies for *OpenCL* CPU kernel are larger than 0.8 or quite close to this value. The efficiencies decrease, although very slightly, as the number of threads increases. The lower efficiencies observed for 24 threads (approximately 0.6) are related with the fact that we have only 12 cores, although each core implements simultaneous multithreading with two threads. The maximum mean efficiencies obtained by the *OpenCL* GPU kernel executed by a different number of SMX processors are between 0.5 and 0.6 (closer to 0.5). Figure 2 shows details about these efficiencies.

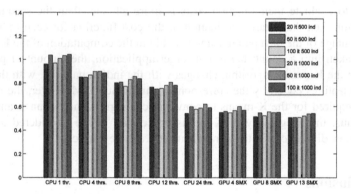

**Fig. 2.** Mean efficiencies for different CPU threads (1-24 threads) and mean highest efficiencies for GPU SMX processors (4, 8, 13 SMX) with populations of 500 and 1000 individuals and 20, 50, and 100 generations.

Figure 3 provides the execution times of the *OpenCL* GPU kernels with different numbers of SMX processors and *WILocal* work-items. As the SMX processors and/or *WILocal* work-items increase, the corresponding execution time decreases. Of course, as the number of individuals in the population or iterations increases, the execution times are larger. Moreover, if we compare the configurations of SMX processors and *WILocal* work-items with the same number of *WIGlobal* work-items, for example 4 SMX processors and *WILocal* = 256 work-items with 8 SMX processors and *WILocal* = 128 work-items (i.e. *WIGlobal* = 1024 work-items), we observe that the configurations with more SMX processors are better. This means that the level of parallelization corresponding to the distribution of the individuals among the SMX

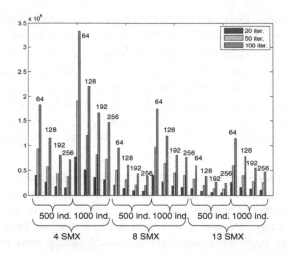

**Fig. 3.** Execution times (in miliseconds) for OpenCL GPU kernel with 4, 8, and 13 SMX processors and *WILocal* equal to 64, 128, 192, and 256 work-items in the GPU (populations of 500 and 1000 individuals and 20, 50, and 100 generations).

processors to evaluate their function costs is more efficient than the second level of parallelization, i.e., the parallel evaluation of the cost function for each individual by taking advantage of the data parallelism available in the computation of the K-means. It has to be taken into account that in our target application, the amount of parallelism available in the K-means algorithm changes with the individuals (i.e. with the number of selected features coded by the corresponding individual). Moreover, the number of iterations required for the K-means to converge can significantly change among individuals. Thus, the irregularity of K-means processing in the considered application implies higher differences in the parallelism that can be extracted.

# 5  Conclusions

Although many works in the literature have shown important speedups achieved by different parallel evolutionary algorithms implemented on GPUs, less details have been reported about the benefits of such many-core architectures in such data mining applications where the evaluation of the fitness for the individuals of the population in the involved evolutionary algorithm implies irregularities in the codes or in the data accesses, along with high dimensional patterns and high volume data. This paper compares parallel implementations for heterogeneous platforms including multicore CPU and GPU architectures of a multi-objective approach to a high-dimensional feature selection problem related with EEG classification on a BCI task. Thus, *OpenCL* CPU and GPU kernels have been developed to take advantage of these heterogeneous architectures, and their respective performances have been analyzed.

More specifically, the application to be parallelized is based on a multi-objective optimization evolutionary algorithm with two cost functions to be evaluated for the individuals in the population. The evaluation of the individuals in the population implies the computation of two validation indices for the clustering obtained through a K-means algorithm applied to the patterns of the dataset. Therefore, two parallelization approaches have been implemented in the *OpenCL* kernels. The first one corresponds to a master-worker parallel implementation of the multi-objective evolutionary algorithm that distributes the evaluation of the individuals among the available threads. The second approach to extract parallelism is applied to the K-means algorithm, and thus, it entails the parallelization of the cost function evaluation. The *OpenCL* CPU kernel only implements the master-worker parallelization while the *OpenCL* GPU kernel implements both approaches. This way, the individuals are distributed among the SMX processors of the GPU and the K-means computation are parallelized among the *WILocal* work-items in the SMX processors. Despite the optimized memory transferences between CPU and GPU implemented, the required memory copies still affect the execution time and the synchronization requirements in the SIMD parallel implementation of the K-means, and the irregularity in the available parallelism for the K-means in our application seems to be the responsible of the lower GPU performances. Thus, although appreciable speedups are also provided by the GPU architecture, more approaches to cope with the irregularities and data profile of the application have to be taken into account in future works.

**Acknowledgements.** This work has been funded by projects TIN2012-32039 and TIN2015-67020-P (Spanish "Ministerio de Econ. y Compet." and FEDER funds).

# References

1. Mukhopadhyay, A., et al.: A survey of multiobjective evolutionary algorithms for data mining: part I. IEEE Trans. Evol. Comput. **18**(1), 4–19 (2014)
2. Mukhopadhyay, A., et al.: A survey of multiobjective evolutionary algorithms for data mining: part II. IEEE Trans. Evol. Comput. **8**(1), 20–35 (2014)
3. Rupp, R., Kleih, S.C., Leeb, R., Millán, J.R., Kübler, A., Müller-Putz, G.R.: Brain-computer interfaces and assistive technology. In: Grübler, G., Hildt, E. (eds.) Brain-Computer Interfaces in Their Ethical, Social and Cultural Contexts. The International Library of Ethics, Law and Technology, vol. 12, pp. 7–38. Springer Science-Business Media, Dordrecht (2014). doi:10.1007/978-94-017-8996-7_2
4. Collet, P.: Why GPGPUs for evolutionary computation? In: Tsutsui, S., Collet, P. (eds.) Massively Parallel Evolutionary Computation on GPGPUs. Natural Computing Series, pp. 3–14. Springer, Heidelberg (2013)
5. Teodoro, G., Kurc, T., Andrade, G., Kong, J., Ferreira, R., Saltz, J.: Application performance analysis and efficient execution on systems with multi-core CPUs, GPUs, and MICs: a case study with microscopy image analysis. Intl. J. of High Perform. Comput. Appl. 1–20 (2015). doi:10.1177/1094342015594519
6. Greg, C.: Hazelwood K. Where is the data? Why you cannot debate CPU vs. GPU performances without the answer. In: IEEE International Symposium on Performance Analysis of Systems and Softwareernational(ISPASS), pp.134–144 (2011)
7. Bellman, G.A.: Adaptive Control Processes: A Guided Tour. Princeton University Press, Princeton (1961)

8. Marinaki, M., Marinakis, Y.: An Island memetic differential evolution algorithm for the feature selection problem. In: Terrazas, G., Otero, F.E., Masegosa, A.D. (eds.) NICSO 2013. SCI, vol. 512, pp. 33–47. Springer, Heidelberg (2014)
9. Emmanouilidis, C., Hunter, A., MacIntyre, J.: A multiobjective evolutionary setting for feature selection and a commonality-based crossover operator. In: Proceedings of the 2000 Congress on Evolutionary Computation, pp. 309–316. IEEE Press, New York (2000). doi:10.1109/CEC.2000.870311
10. Handl, J., Knowles, J.: Feature subset selection in unsupervised learning via multiobjective optimization. Int. J. Comput. Intell. Res. 2(3), 217–238 (2006). doi:10.5019/j.ijcir.2006.64
11. Oliveira, L.S., Sabourin, R., Bortolozzi, F., Suen, C.Y.: A methodology for feature selection using multiobjective genetic algorithms for handwritten digit string recognition. Int. J. Pattern Recognit. Artif. Intell. 17(6), 903–929 (2003). doi:10.1142/S021800140300271X
12. Kim, Y., Street, W.N., Menczer, F.: Evolutionary model selection in unsupervised learning. Intell. Data Anal. 6(6), 531–556 (2002). doi:10.1145/347090.347169
13. Morita, M., Sabourin, R., Bortolozzi, F., Suen, C.Y.: Unsupervised feature selection using multi-objective genetic algorithms for handwritten word recognition. In: Proceedings of the 7th International Conference on Document Analysis and Recognition, 666–670. IEEE Press, New York (2003). doi:10.1109/ICDAR.2003.1227746
14. Mierswa, I., Wurst, M.: Information preserving multi-objective feature selection for unsupervised learning. In: Proceedings of the 8th Annual Conference on Genetic and Evolutionary Computation, GECCO 2006, pp. 1545–1552. ACM, New York (2006). doi:10.1145/1143997.1144248
15. de Souza, T., Matwin, J., Japkowitz, N.: Parallelizing feature selection. Algorithmica 45(3), 433–456 (2006). doi:10.1007/s00453-006-1220-3
16. Zao, Z., Zhang, R., Cox, J., Duling, D., Sarle, W.: Massively parallel feature selection: an approach based on variance preservation. Mach. Learn. 92(1), 195–220 (2013). doi:10.1007/s10994-013-5373-4
17. Kimovski, D., Ortega, J., Ortiz, A., Baños, R.: Parallel alternatives for evolutionary multi-objective optimization in unsupervised feature selection. Expert Syst. Appl. 42(9), 4239–4252 (2015). doi:10.1016/j.eswa.2015.01.061
18. Kimovski, D., Ortega, J., Ortiz, A., Baños, R.: Leveraging cooperation for parallel multi-objective feature selection in high-dimensional EEG data. Concurrency: Pract. Experience 27, 5476–5499 (2015)
19. Arbelaitz, O., Gurrutxaga, I., Muguerza, J., Pérez, J.M., Perona, I.: An externsive comparative study of cluster validity indices. Pattern Recogn. 46(1), 243–256 (2013)
20. Deb, K., Agrawal, S., Pratab, A., Meyarivan, T.: A fast elitist Non-dominated sorting genetic algorithms for multi-objective optimisation: NSGA-II. In: Deb, K., Rudolph, G., Lutton, E., Merelo, J.J., Schoenauer, M., Schwefel, H.-P., Yao, X. (eds.) PPSN 2000. LNCS, vol. 1917, pp. 849–858. Springer, Heidelberg (2000)
21. OpenCL registry. www.khronos.org/registry/cl/
22. Forgy, E.: Cluster analysis of multivariate data: efficiency vs interpretability of classification. Biometrics 21, 768 (1965)
23. Fonseca, C.M., López-Ibáñez, M., Paquete, L., Guerreiro, A.P.: Computation of the Hypervolume indicator (2014). http://iridia.ulb.ac.be/ ∼ manuel/hypervolume
24. Luong, T.V., Melab, N., Talbi, E.-G.: GPU-based island model for evolutionary algorithms. In: GECCO 2010 Proceedings of the 12th Annual Conference on Genetic and Evolutionary Computation, pp. 1089–1096 (2010)
25. Alba, E., Luque, G., Nesmachnow, S.: Parrallel Metaheuristics: recent advances and new trends. Intl. Trans. Op. Res. 20, 1–48 (2013)

26. Pospichal, P., Jaros, J., Schwarz, J.: Parallel genetic algorithm on the CUDA architecture. In: Di Chio, C., Cagnoni, S., Cotta, C., Ebner, M., Ekárt, A., Esparcia-Alcazar, A.I., Goh, C.-K., Merelo, J.J., Neri, F., Preuß, M., Togelius, J., Yannakakis, G.N. (eds.) EvoApplicatons 2010, Part I. LNCS, vol. 6024, pp. 442–451. Springer, Heidelberg (2010)
27. Sharma, D., Collet, P.: Implementation techniques for massively parallel Multi-objective optimization. In: Tsutsui, S., Collet, P. (eds.) Massively Parallel Evolutionary Computation on GPGPUs, pp. 267–286. Springer, Heidelberg (2013)
28. Wong, M.L., Cui, G.: Data mining using parallel multi-objective evolutionary algorithms on graphics processing units. In: Tsutsui, S., Collet, P. (eds.) Massively Parallel Evolutionary Computation on GPGPUs, pp. 287–307. Springer, Heidelberg (2013)
29. Baramkar, P.P., Kulkarni, D.B.: Review for K-means on graphics processing units (GPU). Intl. J. Research & Technology (IJERT) 3(6), 1911–1914 (2014)
30. Kijsipongse, E., U-ruekolan, S.: Dynamic load balancing on GPU clusters for large-scale K-means clustering. In: Proceedings of Ninth International Joint Conference on Computer Science and Software Engineering (JCSSE), pp. 346–350 (2012)
31. Farivar, R., Rebolledo, D., Chan, E., Campbell, R.: A Parallel implementation of K-means clustering on GPUs. In: Proceedings of the International Conference on Parallel and Distributed Processing Techniques and Applications (PDPTA), 14–17 July 2008
32. Wu, R., Zhang, B., Hsu, M.: Clustering billions of data points using GPUs. In: Proceedings of UCHPC-MAW 2009 (2009). doi:10.1145/1531666.1531668
33. Zechner, M., Granitzer, M.: Accelerating K-Means on the Graphics Processor via CUDA. In: Proceedings of First International Conference on Intensive Applications and Services, pp. 7–15 (2009)
34. Fazendeiro, P., Padole, C., Sequeira, P., Prata, P.: OpenCL implementations of a genetic algorithm for feature selection in periocular biometric recognition. In: Panigrahi, B.K., Das, S., Suganthan, P.N., Nanda, P.K. (eds.) SEMCCO 2012. LNCS, vol. 7677, pp. 729–737. Springer, Heidelberg (2012)
35. Asensio-Cubero, J., Gan, J.Q., Palaniappan, R.: Multiresolution analysis over simple graphs for brain computer interfaces. J. Neural Eng. 10(4), 046014 (2013). doi:10.1088/1741-2560/10/4/046014

# Computational Systems for Modelling Biological Processes

# Prediction of Proinflammatory Potentials of Engine Exhausts by Integrating Chemical and Biological Features

Chia-Chi Wang[1,2,3,4], Ying-Chi Lin[1,2], Yuan-Chung Lin[2,3], Syu-Ruei Jhang[3], and Chun-Wei Tung[1,2,4(✉)]

[1] School of Pharmacy, Kaohsiung Medical University, Kaohsiung, Taiwan
cwtung@kmu.edu.tw
[2] Ph.D. Program in Toxicology, Kaohsiung Medical University, Kaohsiung, Taiwan
[3] Institute of Environmental Engineering, National Sun Yat-sen University, Kaohsiung, Taiwan
[4] National Institute of Environmental Health Sciences,
National Health Research Institutes, Miaoli County, Taiwan

**Abstract.** The increasing prevalence of immune-related diseases has raised concerns about immunotoxicity of engine exhausts. The evaluation of immunotoxicity associated with engine exhausts has relied on expensive and time-consuming experiments. In this study, a computational method named CBM was developed for predicting proinflammatory potentials of engine exhausts using chemical and biological data which are routinely analyzed for toxicity evaluation. The CBM model, based on a principal component regression algorithm, performs well with high correlation coefficient values of 0.972 and 0.849 obtained from training and independent test sets, respectively. In contrast, chemical or biological features alone showed poor correlation with the toxicity. The model indicates the importance of the utilization of both chemical and biological features for developing an effective model. The proposed method could be further developed and applied to predict bioactivities of mixtures.

**Keywords:** Engine exhaust · Genotoxicity · Immunotoxicity · Principal component regression · Proinflammatory potential

## 1  Introduction

The toxicity of engine exhausts is of high concern to human health [1]. In addition to the well-known toxicities such as carcinogenic and mutagenic effects, engine exhausts have recently been associated with immune-related diseases such as asthma, chronic obstructive pulmonary disease and acute respiratory distress syndrome, which may cause chronic influences to human health [2, 3]. Due to the negative health impact of engine exhausts, the toxicity assessment of engine exhausts has been suggested to be a required test for developing less toxic alternative fuels [4].

Currently, polycyclic aromatic hydrocarbons (PAHs) and nitro-PAHs from engine exhausts are substances of major concern due to their known toxic effects [4]. PAHs are byproducts of incomplete combustion of organic materials, e.g., coal, diesel, petrol and wood. These airborne pollutants can distribute both in gas and particle phases of ambient air.

© Springer International Publishing Switzerland 2016
F. Ortuño and I. Rojas (Eds.): IWBBIO 2016, LNBI 9656, pp. 293–303, 2016.
DOI: 10.1007/978-3-319-31744-1_26

Many PAHs exhibit genotoxicity and are suspected carcinogens in humans [5]. The chemical structures of PAHs have been linked to their toxicity potentials. PAHs consisting of 2 to 4 benzene rings are of less concern to human health, whereas PAHs with more than or equal to five benzene rings which are considered to be more toxic with higher bioactivation potentials. Nonetheless, PAHs typically appear as mixtures, the estimation of the overall risk of PAHs exposure is difficult.

To give an estimation of overall toxicity posed by PAH mixtures, a toxicity equivalence factor (TEF) method was therefore proposed by U.S. Environmental Protection Agency (EPA). Each PAH has been assigned a TEF value representing its equivalent concentration of the most toxic or carcinogenic congener, i.e. the ratio of the toxicity of a PAH congener to that of Benzo(a)pyrene (BaP), a PAH generally considered as the most toxic and carcinogenic PAH. Based on the TEF values, concentrations of PAHs can be converted to equipotent concentrations of BaP. TEF values in combination with measured concentrations can therefore be applied to calculate the overall toxicity (TEQ value) of the PAH mixtures. The TEF/TEQ method has been applied to calculate the carcinogenicity of PAH mixtures in environmental samples [6–10].

While the TEF method provides an intuitive way for estimating the risk of a mixture, there are major limitations of applying such a method to estimate toxicity of engine exhausts. First, the calculation only bases on routinely tested PAHs. The ignorance of the effects from less abundant PAHs and other non-PAH chemical components in the samples may bias the estimation. Second, the method presumes the effects of the components are additive. In fact, the biological interactions of chemical components can also be antagonistic or synergistic. To fill the data gaps, Ames test [11] are routinely utilized to experimentally determine the genotoxicity of engine exhausts providing an overall genotoxicity of mixtures [12–14].

The aforementioned method only focuses on genotoxicity and carcinogenicity of engine exhausts. However, for the immunotoxicity of engine exhausts that have drawn increasing attentions [15], there is currently no well-established method for the evaluation of immunotoxicity potentials. Instead of conducting experiments for immunotoxicity assessment that are both labor-intensive and expensive, it is desirable to develop a computational method for predicting immunotoxicity of engine exhausts.

In this study, a proinflammatory cytokine of tumor necrosis factor alpha (TNF-alpha) was utilized as an indicator of immunotoxicity. TNF-alpha plays important roles both in immune cell homeostasis and pathophysiology, and has been shown to differentially modulate various toxic endpoints associated with carcinogenesis of different PAHs [16]. The release of TNF-alpha from stimulated monocyte/macrophages is crucial in the initial stage of inflammation in tissue injury. The increased levels of TNF-alpha also lead to the development of inflammatory pathological responses which are hallmarks of many pulmonary disorders, including asthma, chronic bronchitis, chronic obstructive pulmonary disease, acute lung injury and acute respiratory distress syndrome [17–20].

Since genotoxicity has been reported to be associated with immunotoxicity of PAHs [21], the data from routine evaluations on engine exhausts including chemical features derived from PAH concentrations and biological features of Ames test could be useful for predicting immunotoxicity. This study proposes to develop a computational method using chemical and biological features of engine exhausts for predicting proinflammatory potentials of engine exhausts. Principal component regression (PCR) technique, combining principal component analysis and linear regression, was applied to develop the prediction model named CBM. Based on chemical and biological features of 11 training samples, the developed CBM model successfully predicted the inflammatory potential of 5 test samples with a correlation coefficient of 0.847. By contrast, the models built on individual feature of chemical or biological data fail to predict proinflammatory potentials. Altogether, the utilization of both features is required to develop an effective model. The proposed method could be further developed and applied to the evaluation of mixture for other toxicities of concern to human health such as cardiovascular toxicity.

# 2    Materials and Methods

## 2.1    Samples of Engine Exhaust

A total of 16 engine exhaust samples were collected from a six-cylinder engine of Cummins B5.9-160 using various blends of diesel-hydrogen fuels in the Refining and Manufacturing Research Center for heavy-duty diesel engine operation at the Chinese Petroleum Corporation. Engine tests were completed using a Schenck GS-350 dynamometer under several loading conditions. Each collected sample was extracted in a Soxhlet extractor with a mixed solvent (n-hexane and dichloromethane 1:1 (v/v), 750 mL each) for 24 h. The extracts were then poured up into silica gel positioned under a layer of anhydrous $Na_2SO_4$ (about 1 cm high) and above a glass fiber support. The purified solution was concentrated to 1.0 mL by purging with ultra-pure nitrogen for GC/MSD analysis. GC/MSD equipped with a capillary column (HP Ultra 2; 50 m × 0.32 mm × 0.17 μm) was calibrated with a diluted standard solution of 16 PAH compounds (PAH mixture-610 M from Supelco, USA) plus five additional individual PAHs (Merck, Germany). The concentrations of 21 PAH compounds, including naphthalene (Nap), acenaphthylene (AcPy), acenaphthene (Acp), fluorine (Flu), phenanthrene (PA), anthracene (Ant), fluoranthene (FL), pyrene (Pyr), benzo(a)anthracene (BaA), chrysene (CHR), cyclopenta(c,d)pyrene (CYC), benzo(b)fluoranthene (BbF), benzo(k)fluoranthene (BkF), benzo(e)pyrene (BeP), benzo(a)pyrene (BaP), perylene (PER), dibenzo(a,h)anthracene (DBA), benzo(b)chrycene (BbC), indeno(1,2,3,-cd)pyrene (IND), benzo(ghi)perylene (Bghip), and coronene (COR), were then determined as described in our previous studies [6–10].

## 2.2    Fluctuation Ames Test

*Salmonella typhimurium* TA98 and TA100 were grown overnight in nutrient growth supplemented with ampicillin 25 μg/mL under constant shaking at 37°C. The resulting cultures used directly (TA98) or diluted 1:4 (TA100) with exposure medium (Moltox). The test samples, positive and negative controls were prepared in tripliates in 24-well plates. Ten μl of the tester was mixed with the bacterial overnight culture (50 μl) and exposure medium to 250 μl/well and cultured for 90 min at 37°C with 250 rpm constant shaking. After this pre-incubation, 2.5 ml of the histidine-deficient reversion indicator medium were added to each well. The mixtures were then transferred to 384-well plates (48 aliquots per test) and incubated for 48 h at 37°C without agitation. A reversion due to mutation events can be detected by the color shift of the reversion indicator medium from purple to yellow, caused by the pH change of the medium due to the metabolic activity of the revertants. The number of wells containing revertants was determined by an absorption measurement at 590 nm. If a sample is mutagenic, the number of revertant wells is significantly higher compared with the number of revertant wells in the negative control. In this study, the genotoxicity is represented by two proportions of the number of wells with revertants to the number of all tested wells for TA98 and TA100.

### 2.3 Cell Culture and TNF-alpha Detection

Human monocyte THP-1 cells were cultured in RPMI 1640 cell culture medium supplemented with 10 % heat-inactivated fetal bovine serum (Hyclone) and 1 % penicillin/streptomycin (Gibco) at 37°C in 5 % $CO_2$. THP-1 cells ($5 \times 10^5$ cells/mL) were treated with different samples in 48-well plates in the presence of lipopolysaccharide (0.5 μg/mL) for 48 h. The supernatants were collected and quantified for TNF-alpha, as an indicator of cell proinflammatory responses, by standard sandwich enzyme-linked immunosorbent assay (ELISA) as previously described [22]. The level of cytokines in LPS group was designated as 100 %, and the percentage of TNF-alpha level in each group was calculated according to the following formula: Proinflammatory potential = TNF-alpha$_{sample}$/TNF-alpha$_{LPS}$ × 100%, where a value over 100 % means induction and a value smaller than 100 % means inhibition.

### 2.4 Principal Component Regression (PCR)

Principal component regression (PCR) has been extensively used for the development of various predictive regression models [23–25]. The development of PCR model is based on a two-step method. First, principal component analysis (PCA) is applied to extract informative principal components accounting a given number of proportion of variance of data. Subsequently, the regression model is built from selected principal components using linear regression algorithms. The PCA procedure is based on an orthogonal transformation converting potentially correlated variables to linearly uncorrelated principal components. In this study, 95 % of variance is utilized to select informative principal components from correlation matrix for developing regression. M5 algorithm is then utilized for selection of principal components for linear regression. The implementation of the PCR methods is based on WEKA package [26] that has been widely used in various prediction works such as hepatotoxicity [27], antibody amyloidogenesis [28] and esophageal squamous cell carcinoma [29].

## 3  Results and Discussion

### 3.1 Chemical and Biological Features

The preparation of dataset and features for developing prediction models of proinflammatory potentials for engine exhaust is described as follows. First, a total of 16 samples of engine exhausts were collected from the use of fuels consisting of different diesel/hydrogen ratios. Second, the concentrations of 21 PAHs were analyzed by GC-MS as chemical features. Third, the toxic equivalency quotient (TEQ) representing the overall toxicity for each sample was calculated by the summation of multiplication of PAH concentrations and corresponding toxic equivalency factors (TEFs) as shown in the following: $TEQ = \sum_{i=1}^{21} PAHconc_i \times TEF_i$, where the $PAHconc_i$ and $TEF_i$ are the concentration of $i$-th PAH and its corresponding TEF. Details of TEQ calculation and TEF values are provided in our previous papers [6–10]. Although TEQ represents the biological toxicity of a sample, the TEQ value calculated by simply adding up toxicity from each PAH is considered a chemical feature rather than direct biological feature from experiments. The two genotoxicity features were measured by Ames test using strains of TA98 and TA100. The two biological features of TA98 and TA100 combined with 22 chemical features of PAHs and TEQ were utilized to develop models for predicting proinflammatory potentials of engine exhausts. Figure 1 showed the characteristics of dataset for all features. Note that bioactivity data of TA98, TA100 and TNF-alpha (the indicator of proinflammatory responses) were represented as percentage values. Nap was the PAH with highest concentration among these samples. AcPy, Flu, PA and FL constituted the second largest group of PAHs that was higher than the other PAHs. The percentage value of TNF-alpha was ranged from 68.45 to 114.8 %.

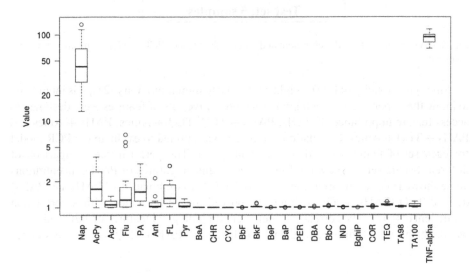

**Fig. 1.** Dataset characteristics

## 3.2 Development of Principal Component Regression Model

To develop and evaluate prediction models for proinflammatory potentials, the dataset of 16 samples were divided into training and independent test sets consisting of 11 and 5 samples, respectively. As the system flow shown in the Fig. 2, the principal component regression (PCR) was applied to develop prediction models using the training set with 21 chemical and 3 biological features and the final model was applied to predict inflammatory potentials of 5 samples in the independent set.

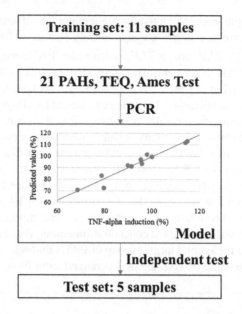

**Fig. 2.** System flow of the development and independent test of PCR model for proinflammatory potentials

Since genotoxicity of PAHs might contribute to immunotoxicity [21], it is interesting to know the importance of each kind of features. Five sets of features were designed to access feature importance of PAHs, PAHs + TEQ, TEQ + Ames, PAHs + Ames and PAHs + TEQ + Ames. Correlation coefficient was applied to evaluate the PCR model for each set of features based on the training set. The performance comparison of different feature sets is shown in Fig. 3. The chemical features of PAH concentrations alone showed poor correlation coefficient of 0.67. The combination of PAHs and PAHs-derived TEQ (PAHs + TEQ) slightly improved the performance with a correlation coefficient of 0.671. The correlation coefficient for biological features of TEQ and Ames was 0.689 that was slightly better than that of PAHs and PAH + TEQ. The results showed

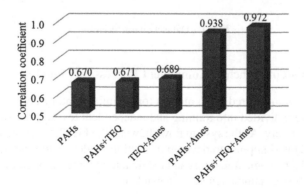

**Fig. 3.** Comparison of regression performances on training set using different feature sets

that individual feature was insufficient for model development. The combinations of chemical and biological features showed great correlation coefficients of 0.938 and 0.972 for PAHs + Ames and PAHs + TEQ + Ames, respectively. Altogether, both chemical and biological features were required for developing a good PCR model.

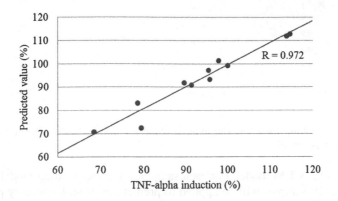

**Fig. 4.** Fitting results on training set based on PAHs, TEQ and Ames

## 3.3   CBM Model and Independent Test

According to the aforementioned results, the final PCR model (named CBM) was developed based on all chemical and biological features of PAHs, TEQ and Ames for best performance and was utilized for subsequent analysis. First, 8 principal components (PCs) were identified from PCA analysis covering 95 % variance. Subsequently, M5 method selected 4 PCs for building the CBM model. The mean absolute error (MAE) and root mean square error (RMSE) of CBM model on training set are 2.630 and 3.172, respectively. Figure 4 showed the fitting results of the developed CBM model. The CBM model was shown in the following Eq. 1. The large value of constant in the Eq. 1 represents the property of TNF-alpha induction values where a value of 100 % indicates a normal condition without any induction or inhibition.

$$TNF\text{-alpha} = -1.902 * PC_2 - 4.4843 * PC_5 + 3.8054 * PC_6 - 27.5738 * PC_8 + 96.9629 \qquad (1)$$

To provide better understanding of the CBM model, training samples were plotted according to PC5 and PC8, the top two PCs with highest absolute values of coefficients. As shown in Fig. 5, the proinflammatory potential was higher in the lower left where the values of PC5 and PC8 were low. Proinflammatory potential tended to decrease with increasing values of PC5 and PC8. It demonstrated that the PCR method was able to identity important PCs correlated with proinflammatory potentials.

Due to the small number of the training set, it is important to avoid overfitting problem in the model construction process. The regression model can be regularized by an extensively utilized ridge parameter. The ridge parameter was set to $10^{-8}$ by default. The leave-one-out cross-validation procedure was applied to evaluate the effect of ridge

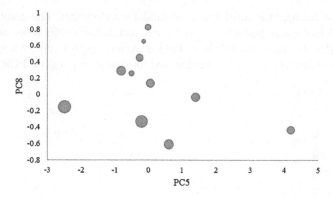

**Fig. 5.** Plot of training samples. The area for each sample represents the squared value of proinflammatory potential.

parameters on the CBM model. For each sample $s$, a corresponding model trained on the remaining 10 samples will be applied to predict the proinflammatory potential of sample $s$. Figure 6 shows the performance of training and leave-one-out cross-validation performances for ridge value $r = 2^1, 2^0, \ldots, 2^{-15}$. As shown in Fig. 6, the ridge parameter has only slight effects on the training performance. However, the best leave-one-out cross-validation performance with a correlation coefficient of 0.839 was achieved for $r = 2^{-3}$. The best ridge parameter of $2^{-3}$ was utilized to construct the final CBM model.

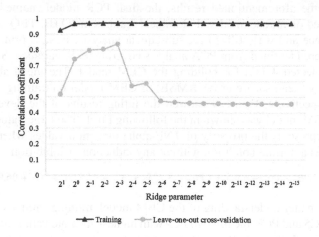

**Fig. 6.** Performance of the CBM model using various ridge parameters

To evaluate the prediction performance of the CBM model, TNF-alpha induction potentials were predicted for the 5 samples in the independent test set. As shown in Fig. 7, a good prediction performance was achieved with a correlation coefficient of 0.849. The MAE and RMSE of independent test were 18.165 and 29.588, respectively. The developed CBM model was able to prioritize the proinflammatory potentials induced by engine exhausts.

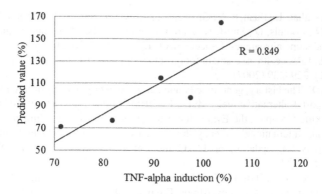

**Fig. 7.** Independent test result

# 4   Conclusions

The immunotoxicity induced by engine exhausts is of increasing concern to human health, however, it has not been included in routinely conducted evaluations. Computational prediction of immunotoxicity could provide an efficient and economic method for identifying fuels with less toxic potentials for further development. In this study, a novel PCR-based method utilizing both chemical and biological features was developed for predicting proinflammatory potentials of engine exhausts. The results showed that neither chemical nor biological features alone are sufficient for predicting the proinflammatory potentials. By using both chemical and biological features, a high correlation coefficient obtained from both training and independent test sets indicated the effectiveness of the developed method. Since the metabolic activation of PAHs could induce higher genotoxicity [30], additional biological features such as Ames tests with the addition of microsomal S9 fraction for metabolic activations of PAHs might serve as useful features for improving the prediction performance. Advanced machine learning methods combined with feature selection algorithms such as support vector machine and genetic algorithm could be further applied to further improve the prediction performance [31, 32]. This study demonstrated a potential application of such models for predicting bioactivity of mixtures by combining chemical and biological features and could be further developed for other bioactivities.

**Acknowledgement.**  The authors would like to acknowledge the financial support from Ministry of Science and Technology of Taiwan (MOST104-2221-E-037-001-MY3) and Kaohsiung Medical University Research Foundation (KMU-M104010 and NSYSUKMU104-I01-2).

# References

1. Krivoshto, I.N., Richards, J.R., Albertson, T.E., Derlet, R.W.: The toxicity of diesel exhaust: implications for primary care. J. Am. Board Family Med. **21**, 55–62 (2008)

2. Benbrahim-Tallaa, L., Baan, R.A., Grosse, Y., Lauby-Secretan, B., El Ghissassi, F., Bouvard, V., Guha, N., Loomis, D., Straif, K.: Carcinogenicity of diesel-engine and gasoline-engine exhausts and some nitroarenes. Lancet Oncol. **13**, 663–664 (2012)
3. Ris, C.: U.S. EPA health assessment for diesel engine exhaust: a review. Inhalation Toxicol. **19**(Suppl. 1), 229–239 (2007)
4. Claxton, L.D.: The history, genotoxicity and carcinogenicity of carbon-based fuels and their emissions: part 4-alternative fuels. Mutat. Res./Rev. Mutat. Res. **763**, 86–102 (2015)
5. IARC Working Group on the Evaluation of Carcinogenic Risks to Humans, World Health Organization, International Agency for Research on Cancer: Overall evaluations of carcinogenicity: an updating of IARC monographs volumes 1 to 42. World Health Organization (1987)
6. Lin, Y.-C., Lee, W.-J., Hou, H.-C.: PAH emissions and energy efficiency of palm-biodiesel blends fueled on diesel generator. Atmos. Environ. **40**, 3930–3940 (2006)
7. Lin, Y.-C., Lee, W.-J., Wu, T.-S., Wang, C.-T.: Comparison of PAH and regulated harmful matter emissions from biodiesel blends and paraffinic fuel blends on engine accumulated mileage test. Fuel **85**, 2516–2523 (2006)
8. Lin, Y.-C., Lee, W.-J., Li, H.-W., Chen, C.-B., Fang, G.-C., Tsai, P.-J.: Impact of using fishing boat fuel with high poly aromatic content on the emission of polycyclic aromatic hydrocarbons from the diesel engine. Atmos. Environ. **40**, 1601–1609 (2006)
9. Lin, Y.-C., Lee, W.-J., Chen, C.-C., Chen, C.-B.: Saving energy and reducing emissions of both polycyclic aromatic hydrocarbons and particulate matter by adding bio-solution to emulsified diesel. Environ. Sci. Technol. **40**, 5553–5559 (2006)
10. Lin, Y.-C., Lee, W.-J., Chen, C.-B.: Characterization of Polycyclic Aromatic Hydrocarbons from the. J. Air Waste Manag. Assoc. **56**, 752–758 (2006)
11. Ames, B.N., Durston, W.E., Yamasaki, E., Lee, F.D.: Carcinogens are mutagens: a simple test system combining liver homogenates for activation and bacteria for detection. Proc. Nat. Acad. Sci. USA **70**, 2281–2285 (1973)
12. Fall, M., Haddouk, H., Loriot, S., Diouf, A., Dionnet, F., Forster, R., Morin, J.-P.: Mutagenicity of diesel engine exhaust in the Ames/ Salmonella assay using a direct exposure method. Toxicol. Environ. Chem. **93**, 1971–1981 (2011)
13. Bunger, J., Bunger, J.F., Krahl, J., Munack, A., Schroder, O., Bruning, T., Hallier, E., Westphal, G.A.: Combusting vegetable oils in diesel engines: the impact of unsaturated fatty acids on particle emissions and mutagenic effects of the exhaust. Archives of Toxicology (2015)
14. Bisig, C., Steiner, S., Comte, P., Czerwinski, J., Mayer, A., Petri-Fink, A., Rothen-Rutishauser, B.: Biological Effects in Lung Cells In Vitro of Exhaust Aerosols from a Gasoline Passenger Car With and Without Particle Filter. Emission Control Sci. Technol. **1**, 237–246 (2015)
15. Che, W., Liu, G., Qiu, H., Zhang, H., Ran, Y., Zeng, X., Wen, W., Shu, Y.: Comparison of immunotoxic effects induced by the extracts from methanol and gasoline engine exhausts in vitro. Toxicol. Vitro **24**, 1119–1125 (2010)
16. Kabatkova, M., Svobodova, J., Pencikova, K., Mohatad, D.S., Smerdova, L., Kozubik, A., Machala, M., Vondracek, J.: Interactive effects of inflammatory cytokine and abundant low-molecular-weight PAHs on inhibition of gap junctional intercellular communication, disruption of cell proliferation control, and the AhR-dependent transcription. Toxicol. Lett. **232**, 113–121 (2014)
17. Lundblad, L.K., Thompson-Figueroa, J., Leclair, T., Sullivan, M.J., Poynter, M.E., Irvin, C.G., Bates, J.H.: Tumor necrosis factor-alpha overexpression in lung disease: a single cause behind a complex phenotype. Am. J. Respir. Crit. Care Med. **171**, 1363–1370 (2005)

18. Lee, W.L., Downey, G.P.: Neutrophil activation and acute lung injury. Curr. Opin. Crit. Care **7**, 1–7 (2001)
19. Mukhopadhyay, S., Hoidal, J.R., Mukherjee, T.K.: Role of TNF alpha in pulmonary pathophysiology. Respir. Res. **7**, 125 (2006)
20. Marcho, Z., White, J.E., Higgins, P.J., Tsan, M.F.: Tumor necrosis factor enhances endothelial cell susceptibility to oxygen toxicity: role of glutathione. Am. J. Respir. Cell Mol. Biol. **5**, 556–562 (1991)
21. Gao, J., Burchiel, S.W.: Genotoxic mechanisms of PAH-induced immunotoxicity. In: Molecular Immunotoxicology, pp. 245–262. Wiley-VCH Verlag GmbH & Co. KGaA (2014)
22. Wang, C.C., Lin, H.L., Wey, S.P., Jan, T.R.: Areca-nut extract modulates antigen-specific immunity and augments inflammation in ovalbumin-sensitized mice. Immunopharmacol. Immunotoxicol. **33**, 315–322 (2011)
23. Adusumilli, S., Bhatt, D., Wang, H., Devabhaktuni, V., Bhattacharya, P.: A novel hybrid approach utilizing principal component regression and random forest regression to bridge the period of GPS outages. Neurocomputing **166**, 185–192 (2015)
24. Dadousis, C., Veerkamp, R., Heringstad, B., Pszczola, M., Calus, M.: A comparison of principal component regression and genomic REML for genomic prediction across populations. Genet. Sel. Evol. **46**, 60 (2014)
25. Mahesh, S., Jayas, D.S., Paliwal, J., White, N.D.G.: Comparison of Partial Least Squares Regression (PLSR) and Principal Components Regression (PCR) Methods for Protein and Hardness Predictions using the Near-Infrared (NIR) Hyperspectral Images of Bulk Samples of Canadian Wheat. Food Bioprocess Technol. **8**, 31–40 (2015)
26. Hall, M., Frank, E., Holmes, G., Pfahringer, B., Reutemann, P., Witten, I.H.: The WEKA data mining software: an update. ACM SIGKDD Explor. Newsl. **11**, 10–18 (2009)
27. Huang, S.H., Tung, C.W., Fulop, F., Li, J.H.: Developing a QSAR model for hepatotoxicity screening of the active compounds in traditional Chinese medicines. Food Chem. Toxicol. **78**, 71–77 (2015)
28. Liaw, C., Tung, C.W., Ho, S.Y.: Prediction and analysis of antibody amyloidogenesis from sequences. PLoS ONE **8**, e53235 (2013)
29. Tung, C.W., Wu, M.T., Chen, Y.K., Wu, C.C., Chen, W.C., Li, H.P., Chou, S.H., Wu, D.C., Wu, I.C.: Identification of biomarkers for esophageal squamous cell carcinoma using feature selection and decision tree methods. Sci. World J. **2013**, 782031 (2013)
30. Topinka, J., Milcova, A., Schmuczerova, J., Mazac, M., Pechout, M., Vojtisek-Lom, M.: Genotoxic potential of organic extracts from particle emissions of diesel and rapeseed oil powered engines. Toxicol. Lett. **212**, 11–17 (2012)
31. Tung, C.W., Ho, S.Y.: POPI: predicting immunogenicity of MHC class I binding peptides by mining informative physicochemical properties. Bioinformatics **23**, 942–949 (2007)
32. Tung, C.W., Ho, S.Y.: Computational identification of ubiquitylation sites from protein sequences. BMC Bioinform. **9**, 310 (2008)

# Calculating Elementary Flux Modes
# with Variable Neighbourhood Search

Jose A. Egea[1]([✉]) and José M. García[2]

[1] Department of Applied Mathematics and Statistics,
Technical University of Cartagena, C/ Dr. Fleming s/n, 30202 Cartagena, Spain
josea.egea@upct.es
[2] Parallel Computer Architecture Group, Facultad de Informática,
University of Murcia, Campus Universitario de Espinardo, 30100 Murcia, Spain
jmgarcia@ditec.um.es

**Abstract.** In this work, we calculate Elementary Flux Modes (EFMs) from metabolic networks using a trajectory-based metaheuristic, *Variable Neighbourhood Search* (VNS). This method is based on the local exploration around an incumbent solution and the subsequent visits to "neighbourhoods" (i.e., other areas of the search space) when the exploration is not successful on improving an objective function. This strategy ensures a suitable balance between exploration and exploitation, which is the key point in metaheuristic-based optimization. Making use of linear programming and the *Simplex* method, a VNS-based metaheuristic has been designed and implemented. This algorithm iteratively solves the linear programs resulting from the formulation of different hypotheses about the metabolic network. These solutions are, when feasible, EFMs. The application of the proposed method on a benchmark problem corroborates its efficacy.

**Keywords:** Elementary Flux Modes · Metaheuristics · Metabolic networks · Variable neighbourhood search

## 1 Introduction

Metabolic network analysis [25] is relevant area of systems biology. Metabolic networks are sets of physic and metabolic processes which determine the biochemical and physiological properties in cells. These networks are usually represented as metabolites linked by biochemical reactions (which can be reversible in some cases). Metabolic networks begin with one or more metabolites named substrates and end with one or more metabolites named products. Discovering new pathways linking substrates and products in metabolic networks, both at experimental and theoretical level, could have an enormous impact in fields like bioengineering, biomedicine, food technology, energy or environmental sciences among others.

Realistic metabolic networks can be extremely complex and their study has been traditionally carried out by dividing them into smaller pathways. However, this approach can limit the discovering of unknown metabolic capabilities

© Springer International Publishing Switzerland 2016
F. Ortuño and I. Rojas (Eds.): IWBBIO 2016, LNBI 9656, pp. 304–314, 2016.
DOI: 10.1007/978-3-319-31744-1_27

in microorganisms [11]. Thus, building genome-scale metabolic networks which model the complete microorganisms metabolism is of great importance [7]. A methodology exhaustively used in the study of metabolic networks is the elementary flux mode analysis [23]. Elementary Flux Modes (EFMs) can be considered as biologically meaningful pathways under steady-state conditions, and they constitute an unique set representing any possible physiological situation in a cellular system. EFM allows to visualize a large amount of unobserved potential pathways, which are the result of combining different reactions in the metabolic network. Discovering those potential pathways may lead to the production of certain metabolites (or degradation of certain substrates) if the conditions to activate such pathways are found. This fact would represent a great advance in metabolic engineering and in the society.

EFM analysis has been used to study relevant metabolic networks [4,8], to analyze some network properties such as fragility or robustness [1,24], or to optimize the production of certain metabolites through microorganisms [28]. The most popular tool to calculate all the EFMs of a metabolic network is *efmtool* [27]. Despite the interesting applications than can arise from EFM analysis, its study has usually been limited to small or medium-scale networks, due to the exponential growth of the number of EFMs with the network size [13]. Yeung, Thiele and Palsson [29] estimated that the number of extreme pathways (which is a subset of EFMs) is around $10^{29}$ for a genome-scale model in humans. Different approaches have been proposed to overcome this limitation. Some of them are based on graph theory [3,5,10,22]. This approach works well from the mathematical point of view but has some drawbacks as pointed out by Planes and Beasley [19]. The most commonly used methods to calculate EFMs are based on a linear programming formulation resulting from analyzing the network subject to a set of stoichiometric, thermodynamic and steady-state constraints. Some recent works using this approach can be found in the literature (see e.g., [6,17,18,20]). Among the different approaches to calculate EFMs in metabolic pathways using linear programming, one of the most promising is based on metaheuristic optimization (in particular, a genetic algorithm [12]) to iteratively solve linear programs formulated to retrieve the EFMs. The advantage of using metaheuristics for solving this problem is that, instead of evaluating all the possible combinations of activated/deactivated reactions in the network (which would involve an unaffordable computation time for networks of medium-large size), the method computes selected solutions by means of *members of the population* which evolve with time based on an objective function. This avoids, on the one hand, to explore all the search space and, on the other hand, to find many duplicated solutions.

In this work, we propose a similar approach using a trajectory-based metaheuristic, *Variable Neighborhood Search* (VNS). Our approach explores different neighborhoods from an incumbent solution to find different EFMs using linear programming. We have shown the steps of our algorithm and the frequentist analysis in which it is based. The algorithm has been implemented in the R language and uses the *lpSolve* library for solving the linear programs. Our algorithm has been tested over the network presented in [26] and has found all the existing EFMs in a shor time. This paper is structured as follows: In Sect. 2 we provide a brief description of VNS; Sect. 3 explains the proposed methodology for calculating EFMs based on VNS; the methodology is applied to a case study in Sect. 4. The paper ends with a set of conclusions.

## 2   Variable Neighborhood Search

Variable Neighborhood Search (VNS) is a metaheuristic method developed by Mladenović and Hansen [15]. It has successfully been applied to different optimization problems in both continuous and integer domains [9]. In the context of function optimization, VNS explores the search space defined by a neighborhood until a local optimum is found, and then moves towards a different neighborhood, repeating the process iteratively for a given number of function evaluations or computation time. One of the key points of the method is the strategy used to visit new neighborhoods. VNS usually defines a new neighborhood by perturbing a set of decision (i.e., independent) variables, using a distance criterion. When the new neighborhood is visited (i.e. a new solution in this neighborhood is evaluated), a local search is applied over it. The typical scheme consists of visiting neighborhoods close to the current solution (for example, perturbing only one variable) until no further improvement is achieved. Then, the method starts exploring other neighborhoods, repeating the process and updating the best found solution.

Unlike other metaheuristics, such as evolutionary algorithms, in which the search is more global at the beginning and becomes more local at the end of the search, VNS starts with a local search over the incumbent solution. This can accelerate the convergence towards optimal solutions. To avoid the risk of getting trapped in local optima, the method starts the global phase later by visiting new neighborhoods. The basic VNS scheme (BVNS) is given in Algorithm 1.

---

**Algorithm 1.** $BVNS(\mathbf{x}, k_{max}, t_{max})$

---
1: **repeat**
2:     $k \leftarrow 1$
3:     **repeat**
4:         $\mathbf{x}' \leftarrow pertub(\mathbf{x}, k)$
5:         $\mathbf{x}'' \leftarrow local\_search(\mathbf{x}')$
6:         $\mathbf{x} \leftarrow update(\mathbf{x}, \mathbf{x}'', k)$
7:     **until** $k = k_{max}$
8:     $t \leftarrow Time()$
9: **until** $t > t_{max}$
10: **end** $BVNS$

---

Where $x, x'$ and $x''$ are, respectively, the current solution, the new solution found in the explored neighborhood, and the solution obtained after applying the local search to the latter. $k$ is the number of perturbed variables (as a measure of distance of the explored neighborhood and the current solution), and $k_{max}$ is the maximum number of variables that can be perturbed at the same time. Finally $t$ represents a parameter for the computation time or number of function evaluations used as stop criterion when it reaches a threshold, $t_{max}$.

# 3  Methodology

## 3.1  EFM Calculation Through Linear Programming

A metabolic network consisting of $m$ metabolites and $n$ reactions is defined by a stoichiometric matrix, $\mathbf{S}$, $m \times n$ dimensional. Only internal metabolites are considered in the matrix. The element $x_{ij}$ is positive when the metabolite $i$ is a product of reaction $j$, and negative in the case of being a substrate. Those elements, in absolute value, are the stoichiometric coefficients in the considered reactions. A flux $\mathbf{v} \in \mathbb{R}^n$ through a reaction must comply with three conditions to be an EFM:

1. Steady-state condition: all the internal metabolites must be balanced.
2. Irreversible reactions must have positive fluxes.
3. An EFM is a minimal, unique set of flux-carrying reactions operating in steady state, i.e., the non-zero indexes of one EFM cannot be a subset of the non-zero indexes of another EFM.

Most of the computational methods to compute EFMs divide a reversible reactions into two irreversible ones with opposite direction. Doing so, all the fluxes must be positive.

Given a set of $K$ knocked-out reactions in a network and an index $\mu$ corresponding to a target reaction from which certain metabolite is produced, the optimization problem to calculate an EFM can be formulated as a linear program as follows:

$$\text{Minimize} \quad \sum_{r=1}^{n} v_r \tag{1}$$

Subject to

$$\mathbf{S} \cdot \mathbf{v} = 0 \tag{2}$$

$$\mathbf{v} \geq 0 \tag{3}$$

$$v_\mu \geq 1 \tag{4}$$

$$\forall i \in K : v_i = 0 \tag{5}$$

Equation 2 defines the steady-state. Equation 3 ensures non-negative fluxes. Equation 4 enforces a positive flux trough a target reaction, $r_\mu$. Finally, Eq. 5 guarantees that no flux can be found using the $K$ knocked out reactions.

This problem can be efficiently solved using the Simplex algorithm [16]. The set of fluxes $\mathbf{v}$ which solves the problem is an EFM. Changing the reactions in the knocked out set, $K$, different linear programming problems can be formulated and thus different EFMs can be found. In fact, this procedure could find all the EFMs in a network as long as all the possible combinations of knocked out reactions are tried. For medium or large scale networks there is a combinatorial explosion which makes impossible to solve the linear programs associated to all the possible combinations in $K$ from the practical point of view. Metaheuristics are a good option for the selection and evolution of different $K$ sets in order to enumerate a significant number of EFMs, avoiding redundant or infeasible solutions. Zanghellini et al. [30] pointed out that a metabolic network can be sufficiently characterized without calculating all the EFMs. Therefore, a method capable to find a number of EFMs diverse enough could be useful to characterize metabolic networks.

## 3.2   Applying the VNS Metaheuristic

In this section we present a VNS-based heuristic method for EFM calculation in metabolic networks to overcome the combinatorial explosion arising in medium and large-scale networks. The method has been implemented in R [21] and it uses one of its libraries for solving linear programming problems, *lpsolve* [2]. To our knowledge, this is the first R implementation for calculating EFMs using metaheuristics.

The different steps followed by the proposed algorithm are enumerated below (please note that a "solution" is a set of knocked-out reactions).

- **Step 0**: User declaration of stoichiometric matrix, target reaction (i.e., Eq. 4) and maximum number of linear programs to be solved.
- **Step 1**: Random solution generation until the first EFM is found.
- **Step 2**: Calculation and update of success/failure frequencies and EFM list. A "success" means that the calculation led to a new EFM. A "failure" means that no feasible solution or a redundant EFM was found.
- **Step 3**: New solution generation based on observed success/failure frequencies.
- **Step 4**: Feasibility and/or redundancy check.
- **Step 5**: Neighborhood update based on success or failure.
- **Step 6**: Stop if the maximum number of iterations is achieved or go to Step 2 otherwise.

Using this procedure, different solutions within the search space are visited. These solutions provide EFMs or infeasible results after solving the associated linear programming problem (Eqs. 1 to 5). As outlined in Step 4, when an EFM is found, we need to check if it was previously calculated to consider a success or a failure.

A key parameter of the method, $kmax$, defines the furthest neighborhood that will be explored with respect to the current solution. In this work, a neighborhood $k$ is composed of any solution which perturbs $k$ variables from the current solution. A solution is a sequence of $n$ binary numbers representing whether a reaction is considered knocked-out (0) or active (1). Perturbing $k$ variables means that $k$ reactions will change their status (from active to knocked-out or vice-versa). Figure 1 illustrates this for a network with 10 reactions. If $k = 1$, only one dimension is perturbed. If $k = 2$ then two dimensions are perturbed and so on. Note that the perturbed dimensions when $k = i + 1$ do not necessarily include the same perturbed dimensions when $k = i$. For large-scale problems $kmax$ is usually limited to a low number to save computational time. The current value of $k$ is automatically calculated by VNS as it will be explained later. However, the actual $k$ dimensions perturbed are given using a success/failure frequentist analysis, as explained below.

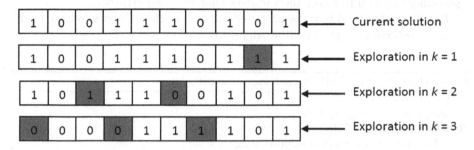

**Fig. 1.** Different neighborhoods from the current solution

**Frequentist Analysis to Select the $k$ Perturbed Dimensions.** After Step 1 of the procedure, the $k$ dimensions to be perturbed are chosen using a frequentist analysis. In particular, if a problem has $n$ reaction, a $n \times 4$ dimensional frequency matrix, $F$, is created. Each of the 4 columns represent the success or failure frequency corresponding to each reaction (in rows). The first column in $F$, called $\mathbf{e}^1$ contains, for each row (i.e., reaction) a counter which is increased when a new EFM is found. Thus, if a new EFM has been found in the current iteration and the value for e.g., variables $v_i$ and $v_j$ is 1, then $e_i^1$ and $e_j^1$ will be increased one unit. Similarly, another column in $F$ called $\mathbf{e}^0$ will contain frequencies that will be increasing when a new EFM is found and the corresponding reactions have values equal to 0. The other two columns in $F$, named $\mathbf{f}^1$ and $\mathbf{f}^0$, are built following the same principle, but their values are increased when an infeasible result or a redundant EFM is found in the current iteration.

Success/failure rates, $\mathbf{c}$, can then be defined for each variable $j$ in two cases: when their value is 1 or when it is 0.

$$c_j^1 = e_j^1 / f_j^1 \quad \forall j \in [1, \ldots, n] \tag{6}$$

$$c_j^0 = e_j^0 / f_j^0 \quad \forall j \in [1, \ldots, n] \tag{7}$$

A new vector, $\mathbf{q}$, is created from the $\mathbf{c}$ rates. This new vector contains, for each variable the opposite to its current value (i.e., 0 if the value is 1 for this variable and vice-versa)

$$q_i = \begin{cases} c_i^1 \text{ if } v_i = 0 \\ c_i^0 \text{ if } v_i = 1 \end{cases} \forall i \in 1, \ldots, n \tag{8}$$

We can then calculate a probability of success when a variable is perturbed. This is the selection criterion to decide which $k$ dimensions will be perturbed during the VNS process. The probability of success when perturbing a variable $j$, $P_j$, is then defined as:

$$P_j = \frac{q_j}{\sum\limits_{i=1}^{n} q_i} \tag{9}$$

To avoid a systematic selection of the $k$ variables with the highest probability, which could result in cycles along the search, a simple random sampling without replacement is used to select the variables that will be perturbed.

Tu fully understand how the proposed method works, the pseudo-code of the complete procedure is presented in Algorithm 2.

---

**Algorithm 2.** $vns\_efm(\mathbf{S}, target, k_{max}, maxiter)$

---

1: $k \leftarrow 1, iter \leftarrow 0$
2: **freq_lists** $\leftarrow init\_freq\_lists()$
3: **while** efm_list $= \emptyset$ **do**
4:     Generate a new random solution $\mathbf{x_{new}}$
5:     sol $\leftarrow lin\_prog(\mathbf{x_{new}}, \mathbf{S}, target)$
6:     $iter = iter + 1$
7:     **if** sol $\equiv EFM$
8:         $update\_efm\_list(\mathbf{efm\_list}, sol)$
9:         $\mathbf{x_{current}} \leftarrow \mathbf{x_{new}}$
10:     **end if**
11:     **freq_lists** $\leftarrow update\_freq\_lists(\mathbf{freq\_lists})$
12: **end while**
13: # Main Loop #
14: **repeat**
15:     $\mathbf{P} \leftarrow calculate\_probs(\mathbf{freq\_lists})$
16:     $\mathbf{y} \leftarrow sample(x_{current}, \mathbf{P}, k), \quad \mathbf{y} \in \mathbf{x_{current}}, \quad \mathbf{y} = \{y_1, \ldots, y_k\}$
17:     $\mathbf{x_{new}} \leftarrow perturb\_vars(\mathbf{x_{current}}, \mathbf{y})$
18:     sol $\leftarrow lin\_prog(\mathbf{x_{new}}, \mathbf{S}, target)$
19:     $iter = iter + 1$
20:     **if** sol $\equiv EFM$ **and** sol $\notin$ **efm_list**
21:         # NEW EFM FOUND #
22:         $update\_efm\_list(\mathbf{efm\_list}, sol)$
23:         $\mathbf{x_{current}} \leftarrow \mathbf{x_{new}}$
24:         $k \leftarrow 1$
25:     **else**
26:         # INFEASIBLE SOLUTION OR REDUNDANT EFM #
27:         $k = k + 1$
28:         **if** $k > k_{max}$
29:             $k \leftarrow 1$
30:         **end if**
31:     **end if**
32:     **freq_lists** $\leftarrow update\_freq\_lists(\mathbf{freq\_lists})$
33: **until** $iter = maxiter$
34: $Output \leftarrow$ **efm_list**

---

## 4   Application Example

In this Section we apply the proposed method to the network presented by Szallasi et al. [26] and represented in Fig. 2. There are two reactions entering the system, labeled as $r_1$ and $r_2$. The latter, as well as $r_8$, is reversible. Both of them are therefore subdivided into two irreversible reactions in our analysis. It is known that this system has 8 EFMs. Its corresponding stoichiometric matrix is:

$$S = \begin{pmatrix} & r_1 & r_2 & r_3 & r_4 & r_5 & r_6 & r_7 & r_8 & r_9 & r_{10} & r_{2,rev} & r_{8,rev} \\ \hline A & 1 & 0 & 0 & 0 & -1 & -1 & -1 & 0 & 0 & 0 & 0 & 0 \\ B & 0 & 1 & 0 & 0 & 1 & 0 & 0 & -1 & -1 & 0 & -1 & 1 \\ C & 0 & 0 & 0 & 0 & 0 & 1 & 0 & 1 & 0 & -1 & 0 & -1 \\ D & 0 & 0 & 0 & 0 & 0 & 0 & 1 & 0 & 0 & -1 & 0 & 0 \\ E & 0 & 0 & 0 & -1 & 0 & 0 & 0 & 0 & 0 & 1 & 0 & 0 \\ P & 0 & 0 & -1 & 0 & 0 & 0 & 0 & 0 & 1 & 1 & 0 & 0 \end{pmatrix}$$

We performed 25 independent runs with our proposed method, with a budget of 1000 iterations ($< 1$ second in a standard workstation). The value of $kmax$ was fixed to 7 (no further investigation about this parameter was done in this work). The target reactions were $r_1$ in a first set of runs and $r_2$ in a second one. In the first set, 7 EFMs were consistently found, obtaining the last one in the iterations 266–267 on average. In the second set, 3 EFMs were found (one complementing the other 7, a repeated one from the first set of runs and a cycle with no biological meaning), obtaining the last one in the iteration 254–255 on average. The solutions are presented in the following matrices.

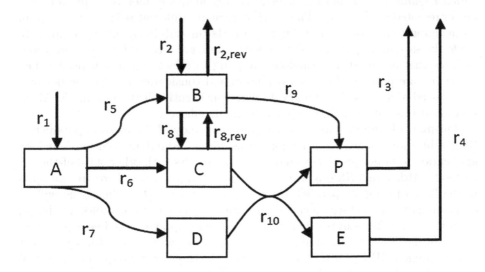

**Fig. 2.** Network from [26]

$$
\begin{array}{c|cccccccccccc}
 & r_1 & r_2 & r_3 & r_4 & r_5 & r_6 & r_7 & r_8 & r_9 & r_{10} & r_{2,rev} & r_{8,rev} \\
\hline
EFM1 & 1 & 1 & 1 & 1 & 0 & 0 & 1 & 1 & 0 & 1 & 0 & 0 \\
EFM2 & 1 & 0 & 1 & 1 & 0 & 1 & 1 & 0 & 0 & 1 & 0 & 0 \\
EFM3 & 1 & 0 & 0 & 0 & 0 & 1 & 0 & 0 & 0 & 0 & 1 & 1 \\
EFM4 & 1 & 0 & 0 & 0 & 1 & 0 & 0 & 0 & 0 & 0 & 1 & 0 \\
EFM5 & 1 & 0 & 1 & 0 & 1 & 0 & 0 & 0 & 1 & 0 & 0 & 0 \\
EFM6 & 1 & 0 & 1 & 0 & 0 & 1 & 0 & 0 & 1 & 0 & 0 & 1 \\
EFM7 & 1 & 0 & 1 & 1 & 1 & 0 & 1 & 1 & 0 & 1 & 0 & 0 \\
\end{array}
$$

$$
\begin{array}{c|cccccccccccc}
 & r_1 & r_2 & r_3 & r_4 & r_5 & r_6 & r_7 & r_8 & r_9 & r_{10} & r_{2,rev} & r_{8,rev} \\
\hline
ART & 0 & 1 & 0 & 0 & 0 & 0 & 0 & 0 & 0 & 0 & 1 & 0 \\
EFM8 & 0 & 1 & 1 & 0 & 0 & 0 & 0 & 0 & 1 & 0 & 0 & 0 \\
EFM1\_REP & 1 & 1 & 1 & 1 & 0 & 0 & 1 & 1 & 0 & 1 & 0 & 0 \\
\end{array}
$$

As it can be observed, all the EFMs from this network can be retrieved using the proposed procedure.

## 5   Conclusions and Future Work

Elementary Flux Mode analysis is a fundamental tool to characterize metabolic network and to discover unknown pathways, which could result in useful applications in biomedicine or bioengineering. Calculating EFMs in medium and large-scale network is a complex task which requires the use of efficient methods to overcome the combinatorial explosion in the field of solutions. Metaheuristic-based optimization seems to be a promising approach to solve this problem due to its nature. A trajectory-based metaheuristic inspired on *variable neighborhood search* for calculating elementary flux modes in metabolic network has been presented in this work. This method explores different neighborhoods from an incumbent solution to find different EFMs through linear programming. To guide the search, frequencies of success or failure are recorded to calculate probabilities that are used in a random sampling to select which dimensions (i.e., reactions) are perturbed in each neighborhood considered, $k$. The method has been tested over a test benchmark, finding consistently all the existing EFMs in very short computation times, thus proving its efficacy.

Future work will include a more exhaustive study of the method parameters (i.e., values for $kmax$, penalties for infeasible or redundant solutions) and its application to larger problems, comparing the results with other state-of-the-art methods. Although VNS is a trajectory based metaheuristic, there are examples in the literature for its parallelization and cooperation [14]. This issue will be considered too. A local search can also be implemented by, for instance, analyzing the adjacent solutions of the incumbent one before applying the linear programming routine. The presented approach is compatible with a mixed-integer linear programming (MILP) formulation, in which the target reaction is an additional degree of freedom in the calculation of new EFMs.

**Acknowledgments.** Author Jose A. Egea acknowledges the funding received from the Spanish Ministry of Economy and Competitiveness through the project "Multi-Scales" (DPI2011-28112-C04-04). This work was jointly supported by the Fundación Séneca (Agencia Regional de Ciencia y Tecnología, Región de Murcia) under grant 15290/PI/2010 and the Spanish MEC and European Commission FEDER under grant TIN2012-31345.

# References

1. Behre, J., Wilhelm, T., von Kamp, A., Ruppin, E., Schuster, S.: Structural robustness of metabolic networks with respect to multiple knockouts. J. Theor. Biol. **252**(3), 433–441 (2008)
2. Berkelaar, M., et al.: lpSolve: Interface to Lp_solve v. 5.5 to Solve Linear/Integer Programs (2015), rpackageversion5.6.11. http://CRAN.R-project.org/package=lpSolve
3. Blum, T., Kohlbacher, O.: Using atom mapping rules for an improved detection of relevant routes in weighted metabolic networks. J. Comput. Biol. **15**(6), 565–576 (2008)
4. Carlson, R., Srienc, F.: Fundamental Escherichia coli biochemical pathways for biomass and energy production: identification of reactions. Biotechnol. Bioeng. **85**(1), 1–19 (2004)
5. Croes, D., Couche, F., Wodak, S.J., van Helden, J.: Inferring meaningful pathways in weighted metabolic networks. J. Mol. Biol. **356**(1), 222–236 (2006)
6. Dávid, L., Bockmayr, A.: Computing elementary flux modes involving a set of target reactions. IEEE/ACM Trans. Comput. Biol. Bioinform. **11**(6), 1099–1107 (2014)
7. Feist, A.M., Palsson, B.Ø.: The growing scope of applications of genome-scale metabolic reconstructions using Escherichia coli. Nat. Biotech. **26**(6), 659–667 (2008)
8. de Figueiredo, L.F., Schuster, S., Kaleta, C., Fell, D.A.: Can sugars be produced from fatty acids? a test case for pathway analysis tools. Bioinformatics **25**(1), 152–158 (2009)
9. Hansen, P., Mladenović, N., Moreno-Pérez, J.: Variable neighborhood search: methods and applications. Ann. Oper. Res. **175**(1), 367–407 (2010)
10. Céspedes, J.F.H., De Asís Guil Asensio, F., Carrasco, J.M.G.: A new approach to obtain EFMs using graph methods based on the shortest path between end nodes. In: Ortuño, F., Rojas, I. (eds.) IWBBIO 2015, Part I. LNCS, vol. 9043, pp. 641–649. Springer, Heidelberg (2015)
11. Kaleta, C., de Figueiredo, L.F., Schuster, S.: Can the whole be less than the sum of its parts? pathway analysis in genome-scale metabolic networks using elementary flux patterns. Genome Res. **19**(10), 1872–1883 (2009)
12. Kaleta, C., de Figueiredo, L.F., Schuster, S.: EFMEvolver: computing elementary flux modes in genome-scale metabolic networks. LNCS **157**, 179–189 (2009)
13. Klamt, S., Stelling, J.: Combinatorial complexity of pathway analysis in metabolic networks. Mol. Biol. Rep. **29**(1–2), 233–236 (2002)
14. Luque, G., Alba, E.: Enhancing parallel cooperative trajectory based metaheuristics with path relinking. In: Proceedings of the 16th Annual Conference on Genetic and Evolutionary Computation (GECCO14), pp. 1039–1046 (2014)
15. Mladenović, N., Hansen, P.: Variable neighborhood search. Comput. Oper. Res. **24**(11), 1097–1100 (1997)

16. Nash, J.C.: The (Dantzig) simplex method for linear programming. Comput. Sci. Eng. **2**(1), 29–31 (2000)
17. Pey, J., Planes, F.J.: Direct calculation of elementary flux modes satisfying several biological constraints in genome-scale metabolic networks. Bioinform. Accepted (2015). doi:10.1093/bioinformatics/btu193
18. Pey, J., Villar, J.A., Tobalina, L., Rezola, A., García, J.M., Beasley, J.E., Planes, F.J.: TreeEFM: calculating elementary flux modes using linear optimization in a tree-based algorithm. Bioinformatics **31**(6), 897–904 (2015)
19. Planes, F.J., Beasley, J.E.: A critical examination of stoichiometric and path-finding approaches to metabolic pathways. Briefings Bioinform. **9**(5), 422–436 (2008)
20. Quek, L.E., Nielsen, L.K.: A depth-first search algorithm to compute elementary flux modes by linear programming. BMC Syst. Biol. **8**(1), 94 (2014)
21. R Development Core Team: R: A Language and Environment for Statistical Computing. R Foundation for Statistical Computing, Vienna, Austria (2008). http://www.R-project.org, ISBN 3-900051-07-0
22. Rahman, S.A., Advani, P., Schunk, R., Schrader, R., Schomburg, D.: Metabolic pathway analysis web service (pathway hunter tool at cubic). Bioinformatics **21**(7), 1189–1193 (2005)
23. Schuster, S., Dandekar, T., Fell, D.A.: Detection of elementary flux modes in bio-chemical networks: a promising tool for pathway analysis and metabolic engineering. Trends Biotechnol. **17**(2), 53–60 (1999)
24. Stelling, J., Klamt, S., Bettenbrock, K., Schuster, S., Gilles, E.D.: Metabolic network structure determines key aspects of functionality and regulation. Nature **420**(6912), 190–193 (2002)
25. Stephanopoulos, G.N., Aristidou, A.A., Nielsen, J.H.: Metabolic engineering : principles and methodologies. California Academic Press, San Diego (1998)
26. Szallasi, Z., Stelling, J., Periwal, V. (eds.): System Modeling in Cellular Biology: From Concepts to Nuts and Bolts, 1st edn. The MIT Press, Canada (2006)
27. Terzer, M., Stelling, J.: Large-scale computation of elementary flux modes with bit pattern trees. Bioinformatics **24**(19), 2229–2235 (2008)
28. Trinh, C.T., Unrean, P., Srienc, F.: Minimal Escherichia coli cell for the most efficient production of ethanol from hexoses and pentoses. Appl. Environ. Microbiol. **74**(12), 3634–3643 (2008)
29. Yeung, M., Thiele, I., Palsson, B.Ø.: Estimation of the number of extreme pathways for metabolic networks. BMC Bioinformatics **8**(1), 363 (2007)
30. Zanghellini, J., Ruckerbauer, D.E., Hanscho, M., Jungreuthmayer, C.: Elementary flux modes in a nutshell: properties, calculation and applications. Biotechnol. J. **8**(9), 1009–1016 (2013)

# Using Nets-Within-Nets for Modeling Differentiating Cells in the Epigenetic Landscape

Roberta Bardini, Alfredo Benso, Stefano Di Carlo[✉], Gianfranco Politano,
and Alessandro Savino

Department of Control and Computer Engineering,
Politecnico Di Torino, Corso Duca Degli Abruzzi 24, Torino, Italy
{roberta.bardini,alfredo.benso,stefano.carlo,
gianfranco.politano,alessandro.savino}@polito.it

## 1 Introduction

In this work the authors propose the use of a high-level Petri net formalism for modeling developmental processes at the cell level, taking explicitly into account the role of epigenetic regulation. The term "epigenetic" can refer to all possible mechanisms "acting on the genomic information between genotype and phenotype" [1], changing the actual condition of a system without changing the underlying DNA sequence. "For example, even though the vast majority of cells in a multicellular organism share an identical genotype, organismal development generates a diversity of cell types with disparate, yet stable, profiles of gene expression and distinct cellular functions. Thus, cellular differentiation may be considered an epigenetic phenomenon, largely governed by changes in what is described as the "epigenetic landscape"" [1, 2]. The epigenetic landscape was initially proposed as a visual metaphor [2] for describing the scenario where developmental processes take place. Cell type differentiation is represented as increasingly irreversible, as ridges rise between the valleys where the different cells are traveling [3]. These two metaphoric terms refer to a quasi-potential scenario, as proposed in [6].

In computational terms, an epigenetic landscape can be interpreted as the ensemble of all possible regulation configurations a complex system can be in. Unfortunately, the huge number of variables involved in epigenetic regulation make this task extremely challenging.

The proposed model, based on Petri Nets, is designed to simulate and conceptually characterize the contributions of the epigenetic regulation from those of other mechanisms within the cells, including transcriptional and post-transcriptional regulation, and metabolism.

In this work we present the idea and the main properties of the model. Experimental results are not yet available but will be presented at the conference.

## 2 Motivation

The complexity of regulation mechanisms within the cell can be addressed choosing different formalisms [4]. Mathematical models, such as those based on differential equations [20], are potentially able to model several activities but suffer from high complexity

© Springer International Publishing Switzerland 2016
F. Ortuño and I. Rojas (Eds.): IWBBIO 2016, LNBI 9656, pp. 315–321, 2016.
DOI: 10.1007/978-3-319-31744-1_28

in terms of modeling effort and computational requirements. Boolean Networks have been also efficiently exploited in modeling different regulation mechanisms.

Previously, we modeled enhanced gene regulatory networks (taking into account the contributions of post-transcriptional actors, such as miRNAs) by means of Boolean Networks [5]. However, Boolean models introduce a strong simplification of the modeled reality posing several limitations:

- the states of a node in the network are described by a Boolean value, while resources in a biological system are better described by continuous quantities;
- they do not support the composition of larger models from smaller ones, while information in the real system often has an encapsulated, modular and hierarchical organization;
- links between nodes are represented with weighted edges carrying either activation or inhibition; the interactions occurring in a biological network require more information than that to be properly described.
- they are deterministic, that is, the outcome of the execution is unique, whereas natural phenomena are intrinsically stochastic.

From these considerations the need emerges for a more suitable formalism for modeling a biological phenomenon and its dynamic evolution.

Developmental processes into the epigenetic landscape have already been addressed with mathematical models applied to gene regulatory networks [6]. Sharing the quantitative interpretation of the epigenetic landscape metaphor proposed there, we aim to further challenge the so-called central dogma of biology, proposing a more powerful computational model whose higher specificity and organization allows for a better representation and subsequent analyses.

## 3    Methods

Epigenetic mechanisms of regulation, which are able to affect the whole network within a cell by controlling the availability of genomic information, overimposes a hierarchically separate level of complexity on systems dynamics. That raises the need of a two-folded approach which allows for taking into account some prerogatives of one layer of regulation above the others. Moreover, the position of the cell into its epigenetic landscape is co-determined by the state of its regulatory network.

Since both these facets of dynamic regulation must be taken into account, the Nets-Within-Net formalism is proposed, which is a high-level Petri net model consisting of a system network whose tokens are, in turn, Petri nets.

### 3.1    Petri Nets

Petri nets can be used as a tool for describing distributed, concurrent asynchronous systems using a low degree of abstraction. The reason for their success as net models lies in their graphical representation, together with a well-defined semantics allowing formal analysis [7]. They've been assessed as valuable tools for modeling biological systems, and examples of their application can be find in the literature [8–10]. Petri nets are bipartite, weighted and directed graphs. Each node belongs to one out of two separate groups: places or transitions. Edges link nodes of different kind only, i.e., a place and a transition or vice-versa, and not two places or two transitions. Input places are those from which edges go to transitions, output places, conversely, are reached by edges from transitions. Each place can contain entities called tokens, which can move along edges and transitions to the next places of the net. The tokens move through a transition if it is enabled and it fires. Enabling occurs when in its input places the number of tokens exceeds a threshold associated with a transition. Each transition fires by activating its specific function, which regulates the way tokens move through it. In fact, each transition can be programmed for functioning in a specific way, under different requirements also in terms of token availability. The enabling of a transition does not directly imply its activation, but it is a necessary condition for it: a transition may fire only once it is enabled, but it will only fire after an interval of time which can be non-deterministic. This peculiar dynamical feature allows for the representation of concurrency in distributed systems, satisfying one of the requirements for a suitable representation of complex biological problems, where interactions among the elements of the same network tend to be parallel and asynchronous.

Some topological structures in the net are related to specific dynamical features. A group of places and transitions is defined as a trap if tokens entering such structures won't exit them. If tokens enter a net structure only to exit it and are not able to come back in, it is a structural deadlock. Sets of places and transitions involving a uniform number of tokens over time are called invariants and have a great importance for model analysis. Each of these can be related to specific dynamical features of biological systems: for example, traps can model cyclic structures in biological regulation that are activated by an input, structural deadlocks can refer to cyclic structures that might produce molecules by consuming themselves, while invariants can implement the assumption of mass conservation in biochemical reactions [11].

Given a topological structure, a Petri net is specified also by its initial marking, that is, the initial distribution of tokens in the places. As the network evolves, subsequent markings are formed, corresponding to some of the states the system can assume. The state space of a Petri net is defined as its reachability graph, i.e., the set of possible markings the initial marking can evolve into.

A transition is described as dead if it is not supposed to fire anymore, in the subsequent evolution of the network. Depending on the capability of a transition for firing, it can be described with different degrees of liveness, the higher corresponding to a transition, which always fires as the net evolves. The whole Petri net is described in terms of liveness depending on that of all its transitions.

A place can be bounded, in the sense that the number of tokens it can contain is limited. It is described as safe if such number is 1, and as $k$-bounded if it is $k$ [12].

Tokens can represent quantities of resources, discrete or continuous. Their transitions can fire once enabled, following deterministic, but also stochastic rules. This allows for the representation of real quantities in biological systems, taking into account the contribution of stochastic events to their evolution in time.

The model described so far is called also Place-Transition network, or low-level Petri net, and it is a suitable formalism for modeling biological networks better than, say, Boolean Networks. Still, it has only one kind of token. The capability of this model can be exploited for overcoming such limitation, building more suitable formalisms. In high-level Petri nets, each token contains complex information or data, providing much more potential for addressing real-world problems. A compact representation of this additional layer of information is achieved in colored Petri nets: to each token can be attached a different, arbitrarily complex data value, which is graphically represented as a color. Different color sets refer to the possible values of the tokens they contain. In this formalism, each place is described by a marking, which is a multi-set over the color set attached to the place [7].

Petri nets can be used for building nets-within-nets, i.e., structures consisting of a system Petri net whose tokens are provided with a structure, which is based on Petri nets too [13]. Hence, a net can contain net items, being able to move in the system net and fire themselves trough its transitions. This allows for the description of mobility, hierarchy and encapsulation, since the net tokens move in a frame whose structure is independent from them, sets up functional constraints on their mobility, and can be nested. Net tokens can be considered as references to net items, if the system is described in terms of reference semantics. They can also be described in terms of their existence in different places and with different internal states, according to value semantics instead. In this frame, copies of a net token can be created to model concurrent execution. In nets-within-nets, net tokens can communicate via predefined interfaces, allowing for insights to the interplay of locality and concurrency [14–17].

The choice of the Petri net formalism allows to overcome most limitations of Boolean Networks:

- the possible states of a place can be described with discrete but also continuous quantities;
- nested Petri nets properly represent hierarchically organized networks; the potential for describing encapsulation and motility in biological systems is expressed at its full in nets-within-nets;
- links between places are described by edges and transitions, whose functioning can be set up independently from that of the others, allowing for a suitable representation for the diversity of interactions described in biological systems;
- choosing to set a non-deterministic delay between enabling and firing for transitions makes the description of stochastic contributions to biological mechanisms possible.

Addressing the particular problem of a differentiating cell going through the epigenetic landscape, the nets-within-nets formalism is potentially able to satisfy the requirement for a representation of intrinsically intertwined, yet hierarchically separable levels of regulation.

In our model, the system Petri net represents the epigenetic landscape, each place referring to a specific epigenetic regulation state. Transitions represent the stages of a developmental process, which is a path undertaken by tokens. Each token is in turn a Petri net, representing the enhanced regulatory network of a specific cell, where the places are the molecules involved, and transitions are the interactions between them. Each cell adapts to the epigenetic constraints linked to the place it is in, and undergoes subsequent network evolution. An abstract and static representation of the model is

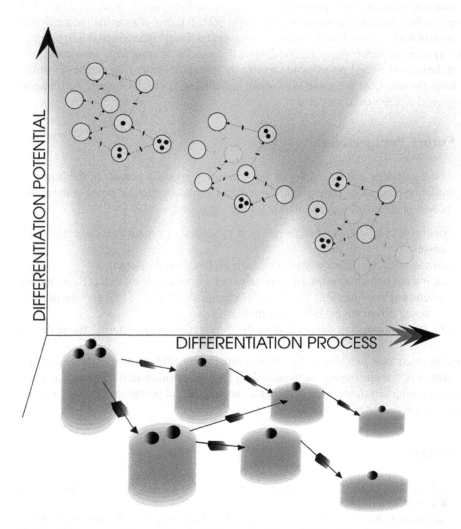

**Fig. 1.** This representation of the model shows the hierarchical relation between the system net and the net tokens. Each place of the system net imposes de facto topological modifications on the net tokens it contains, contextually de-activating some of their components. Traveling through the system net, each net token evolves indipendently, as shown by their different markings evolving also respect to the contraints of each system place. In the context of a developmental process, the vertical position of system places on the axis of differentiation potential refers to the *quasi-potential* landscape [6] they are described in, ranging from maximal stemness to complete differentiation of the net tokens it contains, along the differentiation process. For some net tokens (black spheres) a representation of the Petri net it refers to is provided.

provided in Fig. 1. Transitions in the outer net can be enabled by system regulations as well as by specific conformations of the net tokens in the respective input places. In this way, the transition of a cell from a state to another is co-regulated by both epigenetic and non-epigenetic factors, providing for a more realistic mimicry of the biological system. In fact, differentiation can be explained both in terms of external guidance and by means of intrinsic regulation, since the two are extensively intertwined. The dynamics of the system net, referring to a developmental scenario, or epigenetic landscape, and that of the individual cell, referring to specific differentiation processes, can be discerned one from the other. At the same time, their functioning is strictly intertwined, like in biological developmental processes.

## 4 Future Perspectives

The model proposed can be employed for addressing specific matters in the frame of differentiation processes in both physiological and artificial contexts. Some of the applications Nets-within-nets can be suitable for are suggested:

- modeling the physiological differentiation processes in organismal developmental processes;
- modeling stem cell differentiation into their niches in adult tissues;
- modeling the induction of pluripotent stem cells from differentiated cells by means of genetic engineering (iPSC, [18]), or by chemicals (CiPSC, [19]) and the process of obtaining differentiated cells from them;
- studying the population distribution of the differentiating cells during tissue development;
- testing particular net tokens for their compatibility with particular places of the system net, for classifying them according to their position in the epigenetic landscape, during physiologic od pathogenic developmental processes of development, in a natural or artificial context.

## References

1. Goldberg, A.D., Allis, C.D., Bernstein, E.: Epigenetics: A Landscape Takes Shape. Cell **128**(4), 635–638 (2007)
2. Waddington, C.A.: Organisers & Genes. Cambridge University Press, Cambridge (1940)
3. Wikipedia contributors, "C. H. Waddington– Epigenetic landscape" Wikipedia, The Free Encyclopedia, https://en.wikipedia.org/wiki/C._H._Waddington#Epigenetic_landscape. Accessed 5 December 2015
4. Fisher, J., Henzinger, T.A.: Executable Cell Biology. Nat. Biotechnol. **25**, 1239–1249 (2007)
5. Benso, A., Di Carlo, S., Politano, G., Savino, A., Vasciaveo, A.: An extended gene protein/ products Boolean network model including post-transcriptional regulation. Theor. Biol. Med. Model. **11** (2014)
6. Huang, S.: The molecular and mathematical basis of Waddington's epigenetic landscape: a framework for post-Darwinian biology? Bio Essays **34**, 149–157 (2012)
7. Jensen, K.: Colored Petri Nets: Basic Concepts, Analysis Methods and Practical Use, vol. 1. Springer, Heidelberg (2013)

8. Heiner, M., Gilbert, D., Donaldson, R.: Petri nets for systems and synthetic biology. In: Bernardo, M., Degano, P., Zavattaro, G. (eds.) SFM 2008. LNCS, vol. 5016, pp. 215–264. Springer, Heidelberg (2008)

9. Yang, J., Lian, J., Pu, H., Gao, R.: Modeling the genetic information transmission based on colored petri nets. In: Proceeding of the IEEE International Conference on Information and Automation Hailar, China (2014)

10. Chaouiya, C.: Petri net modelling of biological networks. Oxford J. Brief Bioinform **8**, 210–219 (2007)

11. Koch, I.: Petri nets in systems biology. Softw. Syst. Model. **14**, 703–710 (2015)

12. Wikipedia contributors, "Petri Net" Wikipedia, The Free Encyclopedia. https://en.wikipedia.org/wiki/Petri_net. Accessed 05 December 2015

13. Valk, R.: Petri nets as token objects: an introduction to elementary object nets. In: Silva, M., Desel, J. (eds.) ICATPN 1998. LNCS, vol. 1420, pp. 1–24. Springer, Heidelberg (2000)

14. Wikipedia contributors, "Nets within nets" Wikipedia, The Free Encyclopedia. https://en.wikipedia.org/wiki/Nets_within_Nets. Accessed 05 December 2015

15. Cabac, L., Duvigneau, M., Moldt, D., Rölke, H.: Modeling dynamic architectures using nets-within-nets. In: Ciardo, G., Darondeau, P. (eds.) ICATPN 2005. LNCS, vol. 3536, pp. 148–167. Springer, Heidelberg (2005)

16. Köhler, M.: Mobile Object Net Systems: Petri Nets as Active Tokens, Technical report 320. Universitat Hamburg (2002)

17. Valk, R.: Concurrency in Communicating Object Petri Nets, Concurrent OOP and PN, pp. 164–195. Springer-Verlag, Heidelberg (2001)

18. Takahashi, K., Yamanaka, S., Induction of pluripotent stem cells from mouse embryonic and adult fibroblast cultures by defined factors, Cell, 2006

19. Lin, T., Ambasudhan, R., Yuan, X., et al.: A chemical platform for improved induction of human iPSCs. Nat. Methods **6**, 805–808 (2009)

20. Chen, K., Novak, B.: Network dynamics and cell physiology. Nat. Rev. Mol. Cell Biol. **2**(12), 908–916 (2001). doi:10.1038/35103078

# Simulations of Cardiac Electrophysiology Combining GPU and Adaptive Mesh Refinement Algorithms

Rafael S. Oliveira[1]([✉]), Bernardo M. Rocha[2], Denise Burgarelli[4],
Wagner Meira Jr.[3], and Rodrigo W. dos Santos[2]

[1] Departamento de Ciência da Computação,
Universidade Federal de São João de Rei, São João del Rei, Brazil
Sachetto@ufsj.edu.br
[2] Departamento de Ciência da Computação e Programa em Modelagem
Computacional, Universidade Federal de Juiz de Fora, Juiz de Fora, Brazil
[3] Departamento de Ciência da Computação, Universidade Federal de Minas Gerais,
Belo Horizonte, Brazil
[4] Departamento de Matemática, Universidade Federal de Minas Gerais,
Belo Horizonte, Brazil

**Abstract.** Computer models have become valuable tools for the study
and comprehension of the complex phenomena of cardiac electrophysiology. However, the high complexity of the biophysical processes translates
into complex mathematical and computational models. In this paper we
evaluate a hybrid multicore and graphics processing unit numerical algorithm based on mesh adaptivity and on the finite volume method to cope
with the complexity and to accelerate these simulations. This is a very
attractive approach since the electrical wavefront corresponds to only a
small fraction of the cardiac tissue. Usually, the numerical solution of
the partial differential equations that model the phenomenon requires
very fine spatial discretization to follow the wavefront, which is approximately 0.2 mm. The use of uniform meshes leads to high computational
cost as it requires a large number of mesh points. In this sense, the tests
reported in this work show that simulations of three-dimensional models
of cardiac tissue have been accelerated by more than 626 times using
the adaptive mesh algorithm together with its parallelization, with no
significant loss in accuracy.

## 1 Introduction

The phenomenon of electric propagation in the heart comprises a set of complex
non-linear biophysical processes. Its multi-scale nature spans from nanometre
processes such as ionic movements and protein dynamic conformation, to cen-
timetre phenomena such as whole heart structure and contraction.

Computer models [4] have become valuable tools for the study and comprehension of such complex phenomena, as they allow different information acquired
from different physical scales and experiments to be combined to generate

© Springer International Publishing Switzerland 2016
F. Ortuño and I. Rojas (Eds.): IWBBIO 2016, LNBI 9656, pp. 322–334, 2016.
DOI: 10.1007/978-3-319-31744-1_29

a better picture of the whole system functionality. Not surprisingly, the high complexity of the biophysical processes translates into complex mathematical and computational models. The modern cardiac models are described by non-linear systems of partial differential equations (PDE) coupled to a non-linear set of ordinary differential equations (ODE) resulting in a problem with millions of variables and hundreds of parameters.

The bidomain model [14] is considered to be the most complete description of the electrical activity in cardiac tissue. Under suitable assumptions the bidomain equations (a non-linear system of PDEs) may be reduced to a simpler model, called monodomain, which is less computationally demanding. Unfortunately, large scale simulations, such as those resulting from the discretization of an entire heart, still a computational challenge. In spite of the difficulties and the complexity associated with the implementation and use of these models, the benefits and applications justify their use. Computer models have been used during the tests of new drugs, development of new medical devices, new techniques of non-invasive diagnosis for several cardiac disease, cardiac arrhythmia, reentry, fibrillation or defibrillation and have been the research topic of many studies [2,7,11].

In the simulations of cardiac electrophysiology, the electrical wave front that travels through the heart is very sharp. Due to this sharp spatial variation the numerical methods need fine spatial discretizations to follow the wavefront, which is approximately 0.2 mm [5], to ensure sufficiently accurate results. The execution of cardiac simulations on meshes with a large number of nodes is computationally expensive as it requires repeated solutions of linear systems with millions degrees of freedom. In addition, the memory requirements of such simulations become increasingly large. The use of adaptive mesh methods provides a solution to these problems. By maintaining the extremely fine resolution only where it is needed (i.e., near the wavefront) the number of degrees of freedom is significantly reduced, resulting in faster computations, lower memory usage and reducing the need for disk space for the recording of the output files.

In this paper we extend the parallel accelerated adaptive mesh algorithm, presented in [9] by making the following improvements: 1. extension of the mathematical formulation to be able to make simulations using three-dimensional meshes; 2. inclusion of a graphics processing unit (GPU) parallel implementation to solve the system of ODEs; 3. inclusion of a pre-conditioner for solving the linear system associated to the PDE.

## 2 Monodomain Model

The wave of excitation propagates through the cardiac tissue because the cardiac cells are electrically coupled via special proteins called gap junctions. This phenomenon can be described mathematically by a reaction-diffusion equation called monodomain model, given by

$$\beta C_m \frac{\partial V(x,y,t)}{\partial t} + \beta I_{ion}(V(x,y,t), \boldsymbol{\eta}(x,y,t)) = \nabla \cdot (\boldsymbol{\sigma}(x,y)\nabla V(x,y,t)) + I_{stim}(x,y,t)$$

$$(1)$$

$$\frac{\partial \boldsymbol{\eta}(x,y,t)}{\partial t} = \boldsymbol{f}(V(x,y,t), \boldsymbol{\eta}(x,y,t)), \tag{2}$$

where $V$ is the variable of interest and represents the transmembrane potential, i.e. the difference between intracellular to extracellular potential; $\boldsymbol{\eta}$ is a vector of state variables that also influence the generation and propagation of the electric wave, and usually includes the intracellular concentration of different ions ($K^+$, $Na^+$, $Ca^{2+}$) and the permeability of different membrane ion channels; $\beta$ is the surface-volume ratio of heart cells; $C_m$ is the membrane capacitance, $I_{ion}$ the total ionic current, which is a function of $V$ and $\boldsymbol{\eta}$, $I_{stim}$ is the current due to an external stimulus and $\boldsymbol{\sigma}$ is the monodomain conductivity tensor. We assume that the boundary of the tissue is isolated, i.e., no-flux boundary conditions ($\mathbf{n} \cdot \sigma \nabla V = 0$ on $\partial\Omega$) are imposed.

In this work, two cell models from distinct species with different levels of complexity were considered to simulate the kinetics of the reaction term $I_{ion}$ in Eq. 2. The Bondarenko et al. model [3] that describes the electrical activity of left ventricular cells of mice and the ten Tusscher-Panfilov model for human ventricular tissue [16]. The Bondarenko et al. model (BDK) model consists of the sum of 15 transmembrane currents. In short, Bondarenko's model is based on a system of ODEs with 41 differential variables that control ionic currents and cellular homeostasis. In this model most of the ion channels are represented by Markov chains (MCs). The ten Tusscher-Panfilov (TT2) model has 19 state variables which are described by ODEs, and it also describes the intracellular calcium concentration and conductances of the ionic channels. A complete description of the currents, the equations and parameters of the BDK and TT2 can be found in [3,16] respectively.

## 2.1    Finite Volume Model Applied to Monodomain

In this section we will make a brief description of the Finite Volume Method (FVM) applied to the monodomain equations. Details about the FVM applied to monodomain for two-dimensional problems can be found in [9,10].

The reaction and diffusion part of the monodomain equations can be split by employing the Godunov operator splitting. Each time step involves the solution of two different problems: a nonlinear system of ODEs

$$\frac{\partial V}{\partial t} = \frac{1}{C_m}[-I_{ion}(V, \eta_i) + I_{stim}] \tag{3}$$

$$\frac{\partial \eta_i}{\partial t} = f(V, \eta_i) \tag{4}$$

and a parabolic linear PDE

$$\frac{\partial V}{\partial t} = \frac{1}{\beta C_m}[\nabla \cdot (\boldsymbol{\sigma} \nabla V)] \tag{5}$$

The spatial discretization of the parabolic PDE results in a linear system of equations that has to be solved at each time step.

**Time Discretization.** The time derivative present in Eq. (5), which operates on $V$ is approximated by an implicit first-order Euler scheme:

$$\frac{\partial V}{\partial t} = \frac{V^{n+1} - V^n}{\Delta t}, \tag{6}$$

where $V^n$ represents the transmembrane potential at time $t_n$ and $\Delta t$ the time step.

**Space Discretization.** The diffusion term of Eq. (5) needs to be spatially discretized. To do this we will consider the following relations:

$$J = -\sigma \nabla V \tag{7}$$

where $J$ ($\mu A/cm^2$) represents the density of the intracellular current flow and

$$\nabla \cdot J = -I_v. \tag{8}$$

In this expression, $I_v(\mu A/cm^3)$ is a volumetric current and corresponds to the left side of Eq. (5).

For simplicity, we will consider a tri-dimensional uniform mesh, consisting of cubes (called "Volumes"). Situated in the center of each volume is a node and the transmembrane potential $V$ is associated with each node of the mesh.

After defining the geometry of the mesh and the partitioning of the domain in control volumes, the FVM-specific equations can be presented. Equation (8) can be integrated spatially over a specific cube, leading to:

$$\int_\Omega \nabla \cdot J da = -\int_\Omega I_v \, da. \tag{9}$$

applying the divergence theorem, we find that

$$\int_\Omega \nabla \cdot J da = \int_{\partial\Omega} J \cdot \mathbf{n}, \tag{10}$$

where $\mathbf{n}$ is the vector normal to the surface.

Finally, assuming that $I_v$ represents an average value in each particular cube, and using Eq. (5), we have the following relationship:

$$\beta \left( C_m \frac{\partial V}{\partial t} \right) \Big|_{(i,j,k)} = \frac{-\int_{\partial\Omega} J \cdot \mathbf{n}}{h^3}, \tag{11}$$

where $h^3$ is the volume of the control cell and $\mathbf{n}$ represents the vector normal to the surface.

For the three-dimensional problem, formed by a uniform grid of cubes with face area $h^2$, the calculation of $J$ can be split as the sum of the flows on the six faces:

$$\int_{\partial\Omega} J \cdot \mathbf{n} = h^2 \cdot \sum_{l=1}^{6} J_l \tag{12}$$

where,

$$\sum_{l=1}^{6} J_l = J_{x_{i+1/2,j,k}} - J_{x_{i-1/2,j,k}} + J_{y_{i,j+1/2,k}} - J_{y_{i,j-1/2,k}}$$

$$+ J_{z_{i,j,k+1/2}} - J_{z_{i,j,k-1/2}},$$

The tensor $\sigma = \text{diag}[\sigma_x, \sigma_y, \sigma_z]$ must be determined at the interfaces of the volume. For this, we use the harmonic mean:

$$\sigma_{x_{i+1/2,j,k}} = \frac{2\sigma_{x_{i,j,k}}\sigma_{x_{i+1,j,k}}}{\sigma_{x_{i+1,j,k}} + \sigma_{x_{i,j,k}}} \tag{13}$$

A similar reasoning can be used to calculate $\sigma_{x_{i-1/2,j,k}}$, $\sigma_{y_{i,j+1/2,k}}$, $\sigma_{y_{i,j-1/2}}$, $\sigma_{z_{i,j,k+1/2}}$ and $\sigma_{z_{i,j,k-1/2}}$.

The flows $J_{x_{m,n,o}}$, $J_{y_{m,n,o}}$ and $J_{y_{m,n,o}}$ are calculated at the faces $((m,n,o) = (i+1/2,j,k)$, $(i-1/2,j,k)$, $(i,j+1/2,k)$, $(i,j-1/2,k)$, $(i,j,k+1/2)$ or $(i,j,k-1/2))$ as follows:

$$J_{x_{m,n,o}} = \sigma_x(m,n,o)\frac{\partial V}{\partial x}\Big|_{(m,n,o)}, \tag{14}$$

$$J_{y_{m,n,o}} = \sigma_y(m,n,o)\frac{\partial V}{\partial y}\Big|_{(m,n,o)}, \tag{15}$$

$$J_{z_{m,n,o}} = \sigma_z(m,n,o)\frac{\partial V}{\partial z}\Big|_{(m,n,o)}. \tag{16}$$

**Adaptive Non-uniform Mesh (ALG).** When the electrical wave is propagating through the heart, only a fraction of the excitable medium is occupied by wavefronts. In these regions, the solution or its derivatives change rapidly. Therefore, the numerical solution of the differential equations in these regions requires the use of an extremely fine mesh. Thus, the use of uniform meshes leads to high computational costs. Therefore, adaptive procedures that take into account the scale differences in the phenomena present reliable and efficient solutions.

Recently, the use of adaptive refinement to obtain meshes suitable for the representation of the cardiac electrophysiology equations has been investigated, see, for example [1,15]. The application of ALG in the simulations of cardiac electrophysiology for 2-dimensional meshes can be found in [9,10]. In this work we are proposing the application of ALG in 3-dimensional cardiac meshes.

In order to apply FVM in ALG, we will approximate the partial derivatives of $V$ on the interfaces using the following finite difference scheme, considering uniform discretizations in space ($\Delta x = \Delta y = \Delta z = h$). For sake of simplicity, we only show the equations for direction $x$ since the equations for $y$ and $z$ can be obtained similarly.

$$\frac{\partial V}{\partial x}\Big|_{(i+1/2,j,k)} = \sum_{c=1}^{m_1} \frac{V_{r,c} - V_{i,j,k}}{h_1}, \tag{17}$$

$$\left.\frac{\partial V}{\partial x}\right|_{(i-1/2,j,k)} = \sum_{c=1}^{m_2} \frac{V_{i,j,k} - V_{l,c}}{h_2},\tag{18}$$

where $m_1$ is the number of neighbors at right of the cell centered at $(i, j, k)$ and $m_2$ is the number of neighbors at left; $V_{r,k}$ are neighbors at right, and $V_{l,k}$ are the neighbors at left. The discretizations are defined by:

$$h_1 = h_{i,j} \text{ if } \mathcal{L}_{i,j} > \mathcal{L}_{r,k} \text{ and } h_1 = h_{r,k} \text{ otherwise,}$$
$$h_2 = h_{i,j} \text{ if } \mathcal{L}_{i,j} > \mathcal{L}_{l,k} \text{ and } h_1 = h_{l,k} \text{ otherwise,}\tag{19}$$

where $\mathcal{L}$ is the refinement level of the cell. Rearranging and substituting the discretizations in (11) and decomposing the operators as described by Eqs. (3), (4) and (5) yields:

$$C_m \frac{V_{i,j,k}^* - V_{i,j,k}^n}{\Delta t} =$$
$$-\frac{(S_1 J^*_{x_{i+1/2,j,k}} - S_2 J^*_{x_{i-1/2,j,k}} + S_3 J^*_{y_{i,j+1/2,k}} - S_4 J^*_{y_{i,j-1/2,k}} + S_5 J^*_{z_{i,j,k+1/2}} - S_6 J^*_{z_{i,j,k-1/2}})}{\beta h_{i,j,k}^3}\tag{20}$$

$$C_m \frac{V_{i,j,k}^{n+1} - V_{i,j,k}^*}{\Delta t} = -I_{ion}(V_{i,j,k}^*, \boldsymbol{\eta}^n)\tag{21}$$

$$\frac{\partial \boldsymbol{\eta}^{n+1}}{\partial t} = f(\boldsymbol{\eta}^n, V^*, t)\tag{22}$$

where:

$$S_1 J_{x_{i+1/2,j,k}} = -\sigma_{x_{i+1/2,j,k}} \sum_{c=1}^{m_1} \frac{V_{r,c} - V_{i,j,k}}{h_1} S_1\tag{23}$$

$$S_2 J_{x_{i-1/2,j,k}} = -\sigma_{x_{i-1/2,j,k}} \sum_{c=1}^{m_2} \frac{V_{i,j,k} - V_{l,c}}{h_2} S_2\tag{24}$$

For a regular grid we have $S_1 = h_1^2$ and $S_2 = h_2^2$, i.e., the area of the volume face. Therefore, we can simplify the above equations, obtaining:

$$J_{x_{i+1/2,j,k}} = -\sum_{c=1}^{m_1} \sigma_{x_{r',c}}(V_{r,c} - V_{i,j,k})h_1\tag{25}$$

$$J_{x_{i-1/2,j,k}} = -\sum_{c=1}^{m_2} \sigma_{x_{l',c}}(V_{i,j,k} - V_{l,c})h_2\tag{26}$$

where $\sigma_{x_{r',c}}, \sigma_{x_{l',c}} \sigma_{y_{b',k}}$ are the conductivity values calculated using Eq. 13.

Developing all the equations, we can now define the formula for each volume:

$$\alpha V_{i,j,k}^{*} - \sum_{c=1}^{m_1} \sigma_{x_{r'},c}(V_{r,c} - V_{i,j,k}) + \sum_{c=1}^{m_2} \sigma_{x_{l'},c}(V_{i,j,k} - V_{l,c})$$

$$- \sum_{c=1}^{m_3} \sigma_{y_{t'},c}(V_{t,c} - V_{i,j,k}) + \sum_{c=1}^{m_4} \sigma_{y_{b'},c}(V_{i,j,k} - V_{b,c}) \qquad (27)$$

$$- \sum_{c=1}^{m_5} \sigma_{z_{f'},c}(V_{f,c} - V_{i,j,k}) + \sum_{c=1}^{m_6} \sigma_{z_{bk'},c}(V_{i,j,k} - V_{bk,c})$$

$$= V_{i,j,k}^{n} \alpha$$

where $\alpha = (\beta C_m h_{i,j}^3)/\Delta t$.

# 3  Methods

In this section, we discuss the parallel numerical implementation and experimental setup of the various cardiac simulations we performed.

## 3.1  Parallel Numerical Implementations

Algorithm 1 describes the steps used for the numerical resolution of monodomain model. As can be seen, we have to reassemble the monodomain matrix at each time step if a refinement or derefinement operation has been performed in that step. In this paper, the criteria used for refinement and derefinement are based on the flux across the interface of neighboring cells, as described in [9].

---

**Algorithm 1.** Steps used for the numerical resolution of monodomain model

---
1: set cell model initial conditions;
2: assemble the monodomain matrix (Linear system form PDE);
3: while $t < final\_t$ do
4:     update cell Model state vector;
5:     solve cell model;
6:     solve linear system (PDE) via conjugate gradient method;
7:     refine-derefine
8:     reassemble the monodomain matrix if needed;
9:     $t = t + dt$
10: end while

---

Computer simulations, such as those resulting from fine spatial discretization of a tissue, are computationally expensive. For example, when a $100\,\mu m$ discretization is used in a $5\,cm \times 5\,cm \times 5\,cm$ 3D tissue, and the Bondarenko model, which has 41 differential equations, is used as cardiac cell model a total of $500 \times 500 \times 500 \times 41 = 5,125,000,000$ unknowns must be computed at each time step. In addition, to simulate $150\,ms$ of cardiac electrical activity 5 billions of unknowns of the nonlinear systems of ODEs and the PDE with 125,000,000 of unknowns must be computed 15,000 times (with $\Delta t = 0.01$ ms). To deal with this high computational cost we parallelized, using OpenMP, the functions described in line 8 (assembly of the monodomin matrix) and 6 (conjugate gradient method); and using CUDA the function in line 5 (solution of odes) of Algorithm 1. The full description of the OpenMP implementation can be found in [10].

Differently from [10], in this work we use a Jacobi preconditioner [14] to accelerated the convergence of the conjugate gradient method. Despite being simple, the Jacobi preconditioner suites very well for the adaptive mesh algorithm, as we do not need to rebuild the preconditioning matrix every refine/derefine step, as this method uses only the diagonal of the linear system matrix as the preconditioning matrix.

To solve the non-linear systems of ODEs present in the BDK model, the explicit Euler (EE) method was used. Although it is well known that explicit numerical methods have strong limitations because of stability and accuracy restrictions, they are widely used due to their simplicity of implementation [6].

For the TT2 model, the numerical solution of ODEs at each volume was performed using the Rush-Larsen (RL) method [12]. The RL method is an explicit method, easy to implement and has better stability properties than the explicit Euler method. Thus, it allows the use of larger time steps resulting in an efficient method for the numerical solution of cell models of cardiac electrophysiology. Unfortunately, this method is not suitable for every model, like BDK due to the use of Markov chains [6].

The solution of these ODEs is a embarrassingly parallel problem regardless of the numerical method. No dependency exists between the solutions of the different systems of ODEs at each finite volume $Vol_{i,j,k}$. Therefore, it is quite simple to implement a parallel version of the code: each thread is responsible to solve a fraction of the non-linear systems of ODEs.

In order to accelerate even further our simulations, we also used GPU implementations, using CUDA, for the solutions of ODEs using both numerical methods (RL and EE). A description of these implementations can be found in [13].

## 3.2  Computational Simulations

In this section we present the numerical experiments and the computing environment used to perform them. We also report the results of the experiments and compare the performance of the parallel adaptive mesh approach with the fixed mesh implementation.

**Computing Environment.** All the numerical experiments were performed using a GNU/Linux 4.1.13 machine, with 32 GB of memory and a Intel Core i7-4930K 3.40 GHz processor with 6 cores. Our monodomain solver was implemented in C++ and compiled using GNU GCC 5.2.0 with $-O3$ optimization flag enabled. For the GPU tests we used a GeForce GTX 760, with 1152 CUDA cores organized in 6 multiprocessors. This GPU has a total of 4 GB of memory. The CUDA code was compiled with NVIDIA nvcc compiler version 7.5.17 with the same optimization flags of the CPU code.

**Test Problems.** In order to evaluate the acceleration and to validate our adaptive mesh implementation we used 2 different test problems, using simplified geometries. A brief description of these problems is presented here.

*Benchmark Problem.* In [8], a benchmark problem was proposed to help the validation of implementations of the monodomain model. In this problem, the domain is a rectangular region of size $2 \times 0.7 \times 0.3 \, \text{cm}^3$ and the fibers are parallel to the longitudinal direction and the conductivity tensor is considered transversely isotropic. The conductivity in the fiber direction is $1.334 \, \text{mS/cm}$ and the conductivity in the cross-fiber direction is $0.176 \, \text{mS/cm}$. Other parameters are defined as: $\chi = 1400 \, \text{cm}^{-1}$ and $\text{Cm} = 1 \, \mu\text{F/cm}^2$. The TT2 is used in this test case. An external stimulus of $-50\,000 \, \mu\text{A/cm}^3$ is applied during $2 \, \text{ms}$ in a small cubic region of $0.15 \, \text{cm}^3$ at one corner of the tissue to trigger the electrical activity. This test problem will be referenced as *Test 1*.

*Modified Benchmark.* In order to better evaluate the performance of our implementation, we modified the Benchmark problem described above, by using the BDK model as the cellular model. This test case was develop as we wanted to evaluate the performance of our approach when using a more complex and complicated model to solve. This test problem will be referenced as *Test 2*.

As we interested in the impact of spatial discretisation and the parallel implementation on the execution times of the simulations, we solved the test problems using the following configurations:

– Test 1:
  • 150 ms of cardiac activity simulation.
  • $\Delta t = 0.05$ for both EDOs and EDP.
  • Fixed mesh with $\Delta x = 100 \, \mu\text{m}$. Adaptive mesh with minimum $\Delta x = 100 \, \mu\text{m}$ and maximum $\Delta x = 400 \, \mu\text{m}$ (referred to as 100–400). Adaptive mesh with minimum $\Delta x = 125 \, \mu\text{m}$ and maximum $\Delta x = 500 \, \mu\text{m}$ (referred to as 125–500).
– Test 2:
  • 150 ms of cardiac activity simulation.
  • $\Delta t = 0.05$ for EDP and $\Delta t = 0.0001$ for the EDOs.
  • Fixed mesh with $\Delta x = 100 \, \mu\text{m}$. Adaptive mesh with minimum $\Delta x = 100 \, \mu\text{m}$ and maximum $\Delta x = 400 \, \mu\text{m}$. Adaptive mesh with minimum $\Delta x = 125 \, \mu\text{m}$ and maximum $\Delta x = 500 \, \mu\text{m}$.

Figure 1(a)–(d) shows the propagation of the electrical wave in the benchmark problem mesh from the initial stimulus until its complete activation.

**Fig. 1.** Propagation of the electrical wave in the benchmark problem mesh.

# 4    Results

In this section, we present the results with respect to execution time and speedup associated with the solution of our test problems using our parallel adaptive mesh implementation.

## 4.1    Test 1

The benchmark was solved using the configurations described in Sect. 3.2. All the parallel executions were performed using 6 threads. To measure the accuracy of our implementation we considered the activation time of a node, which was defined as the time at which the transmembrane potential $v$ reaches the value of $0\,mV$. We used same metric that [8] used to compare the results of several codes.

Figure 2a shows the activation times of CARP [17] compared with our implementations using different mesh configurations. The activation times resulting for ALG simulations are almost identical to the one presented by CARP, either using fixed or adaptive meshes. This result shows that our implementation can be used to perform simulations of the electrical activity of cardiac tissue. Is worth noting that the OpenMP and the GPU code resulted in the same activation times.

Table 1 shows the achieved speedups for different mesh and code configurations when compared to the execution with a serial code using $100\,\mu m$ fixed mesh ($\approx 175.2\,min$). By only using an adaptive mesh our implementation became $50.02\times$ faster for the 125–500 configuration and $16.4\times$ faster for the 100–400.

We also tested our implementations using the hybrid OpenMP and GPU version of the code and compared the resolution time with a serial code using $100\,\mu m$ fixed mesh. As can be seen in Table 1, our code using multicore and GPU was $118.17\times$ faster for the 125–500 configuration and $51.02\times$ faster for the 100–400.

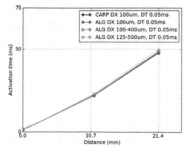

(a) Activation times of CARP, as reported in [8] compared to our implementation using three configurations.

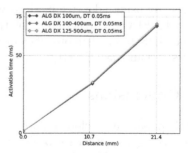

(b) Comparison of the activation times for the modified benchmark.

**Fig. 2.** Comparison of the activation times for Test 1 and Test 2.

**Table 1.** Speedups over a 100 μm fixed mesh for Test 1.

| Mesh | Code | Speedup |
|------|------|---------|
| 125–500 | serial | 50.02 |
| 125–500 | OpenMP+GPU | 118.17 |
| 100–400 | serial | 16.4 |
| 100–400 | OpenMP+GPU | 51.02 |

**Table 2.** Speedups over a 100 μm fixed mesh for Test 2.

| Mesh | Code | Speedup |
|------|------|---------|
| 125–500 | serial | 255.65 |
| 125–500 | OpenMP+GPU | 626.75 |
| 100–400 | serial | 120.28 |
| 100–400 | OpenMP+GPU | 292.70 |

### 4.2 Test 2

After the validation of our implementation using the benchmark problem, we solved the modified benchmark in order to investigate the behaviour of our code when using a more complex cellular model as the BDK model. In Fig. 2b we have the activation times for the three configurations of Test problem 2. As can be seen, the activation time are very close for all mesh configurations.

In Table 2 we show the speedups by using parallel computing and adaptive meshes over the serial fixed mesh code ($\approx$ 392.5 h). For this problem, our parallel implementation achieved speedups of 626.75× and 292.70× for the 125–500 and the 100–400 configurations.

The difference of the achieved speedups between the two test problems can be explained by one main reason: the BDK model used in Test 2 is more complex to solve than the TT2, used in Test 1. Because of this, if we solve less BDK EDOs (by using adaptive meshes) we are saving more time than with we solve less TT2 EDOs. Furthermore, for the same complexity reason, the GPU solver of the BDK model is more efficient as we have more computation over memory accesses than the TT2 GPU solver.

## 5 Conclusions

In this paper we developed, implemented, parallelized and validated an adaptive mesh strategy in order to speed up cardiac electrophysiology simulations for 3D domains. The achieved results are very promising, indicating that the use of ALG and parallel computing is able to reduce the execution time of a simulation by more than 626× (from 16 days to less then 38 min) for a complex cellular model, compared to the use of fixed meshes and serial executions using only a single node with 6 cores and a inexpensive GPU.

**Acknowledgments.** This work was partially funded by CNPq, Capes, Fapemig, UFJF and Finep.

# References

1. Bendahmane, M., Bürger, R., Ruiz-Baier, R.: A multiresolution space-time adaptive scheme for the bidomain model in electrocardiology. Numer. Met. Partial. Diff. Equ. **26**(6), 1377–1404 (2010)
2. Biktashev, V., Holden, A.: Re-entrant activity and its control in a model of mammalian ventricular tissue. Proc. Royal Soc. of London. Ser. B Biol. Sci. **263**(1375), 1373–1382 (1996)
3. Bondarenko, V.E., Szigeti, G.P., Bett, G.C., Kim, S.J., Rasmusson, R.L.: Computer model of action potential of mouse ventricular myocytes. Am. J. Physiol. Heart Circulatory Physiol. **287**(3), H1378–H1403 (2004)
4. Hodgkin, A., Huxley, A.: A quantitative description of membrane current and its application to conduction and excitation in nerve. J. Physiol. **117**, 500–544 (1952)
5. Hunter, P., Borg, T.: Integration from proteins to organs: the physiome project. Nat. Rev. Mol. Cell Biol. **4**(3), 237–243 (2003)
6. Moreira Gomes, J., Alvarenga, A., Silva Campos, R., Rocha, B., Couto da Silva, A., Weber dos Santos, R.: Uniformization method for solving cardiac electrophysiology models based on the markov-chain formulation. IEEE Trans. Biomed. Eng. **62**(2), 600–608 (2015)
7. Morgan, S., Plank, G., Biktasheva, I., Biktashev, V.: Low energy defibrillation in human cardiac tissue: a simulation study. Biophys. J. **96**(4), 1364–1373 (2009)
8. Niederer, S.A., Kerfoot, E., Benson, A.P., Bernabeu, M.O., Bernus, O., Bradley, C., Cherry, E.M., Clayton, R., Fenton, F.H., Garny, A., Heidenreich, E., Land, S., Maleckar, M., Pathmanathan, P., Plank, G., Rodríguez, J.F., Roy, I., Sachse, F.B., Seemann, G., Skavhaug, O., Smith, N.P.: Verification of cardiac tissue electrophysiology simulators using an n-version benchmark. Philos. Trans. R. Soc. A: Math. Phys. Eng. Sci. **369**(1954), 4331–4351 (2011)
9. Oliveira, R.S., Rocha, B.M., Burgarelli, D., Meira Jr., W., dos Santos, R.W.: An adaptive mesh algorithm for the numerical solution of electrical models of the heart. In: Murgante, B., Gervasi, O., Misra, S., Nedjah, N., Rocha, A.M.A.C., Taniar, D., Apduhan, B.O. (eds.) ICCSA 2012, Part I. LNCS, vol. 7333, pp. 649–664. Springer, Heidelberg (2012)
10. Oliveira, R.S., Rocha, B.M., Burgarelli, D., Meira Jr., W., Santos, R.W.D.: A parallel accelerated adaptive mesh algorithm for the solution of electrical models of the heart. Inter. J. High Perform. Syst. Archit. **4**(2), 89–100 (2012)
11. Panfilov, A., Müller, S., Zykov, V., Keener, J.: Elimination of spiral waves in cardiac tissue by multiple electrical shocks. Phys. Rev. E **61**(4), 4644 (2000)
12. Rush, S., Larsen, H.: A practical algorithm for solving dynamic membrane equations. IEEE Trans. Biomed. Eng. BME **25**(4), 389–392 (1978)
13. Sachetto Oliveira, R., Rocha, B.M., Amorim, R.M., Campos, F.O., Meira Jr., W., Toledo, E.M., dos Santos, R.W.: Comparing CUDA, OpenCL and OpenGL implementations of the cardiac monodomain equations. In: Wyrzykowski, R., Dongarra, J., Karczewski, K., Waśniewski, J. (eds.) PPAM 2011, Part II. LNCS, vol. 7204, pp. 111–120. Springer, Heidelberg (2012)
14. Weber dos Santos, R., Plank, G., Bauer, S., Vigmond, E.J.: Preconditioning techniques for the bidomain equations. Lecture Notes in Computational Science and Engineering, vol. 40, pp. 571–580. Springer, Heidelberg (2004)

15. Southern, J., Gorman, G., Piggott, M., Farrell, P.: Parallel anisotropic mesh adaptivity with dynamic load balancing for cardiac electrophysiology. J. Comp. Sci. **3**, 8–16 (2012)

16. ten Tusscher, K.H.W.J., Panfilov, A.V.: Alternans and spiral breakup in a human ventricular tissue model. Am. J. Physiol.: Heart Circulatory Physiol. **291**(3), H1088–H1100 (2006)

17. Vigmond, E.J., Hughes, M., Plank, G., Leon, L.J.: Computational tools for modeling electrical activity in cardiac tissue. J. Electrocardiol. **36**, 69–74 (2003)

# A Plasma Flow Model in the Interstitial Tissue Due to Bacterial Infection

Ruy Freitas Reis[✉], Rodrigo Weber dos Santos, and Marcelo Lobosco

Pós-Graduação em Modelagem Computacional,
Universidade Federal de Juiz de Fora, Juiz de Fora, Brazil
ruyfreis@gmail.com, rodrigo.weber@ufjf.edu.br, marcelo.lobosco@ice.ufjf.br

**Abstract.** Diseases due to infections might lead to death. Fever is often the first sign of an infection; other signs are skin hot to touch, shivering, aching muscles, pain, redness, swelling and so on, depending on the kind of infection. This study is a first attempt to model one of the infection symptoms, the edema. Briefly, edema may be caused by increased blood vessel wall permeability which lead to a swollen, red area. Neutrophil-bacteria iterations trigger a chain of cytokine reactions which in turn change the vessel wall permeability leading to an increase of interstitial fluid pressure. All the iterations are modeled using a n-phase partial differential equation system based on porous media assumptions. Model solutions are obtained using finite-volume method and the upwind scheme. Finally, the numerical results are qualitatively compared with experimental data available from the literature, presenting a good agreement.

**Keywords:** Computational immunology · Edema modeling · Porous media · Partial differential equations

## 1 Introduction

Infectious diseases are responsible for a significant number of deaths [8]. Symptoms of some infectious diseases may include an edema, which is a red, swollen area that feels hot and tender to the touch [15]. Edema is caused by increased blood vessel wall permeability due to the presence of a pathogenic agent. The Human Immune System is responsible for protecting the body against pathogenic agents, however some of the mechanisms used in this task are not completely understood [3]. In this scenario, mathematical and computational models (stochastic or deterministic) can be useful to predict the behavior of the immune system according to several different scenarios.

This paper presents an initial plasma flow model in the interstitial tissue caused by bacterial infection. The model is based on bacteria-neutrophil interactions which trigger some immune system reactions leading to an increase of plasma flow to the interstitial tissue. All interactions are mathematically described as a n-phase porous media model resulting in a system of partial differential equations (PDE) [4,13]. In addition to diffusion-reaction, the neutrophil dynamics is also affected by chemotaxis [18].

F. Ortuño and I. Rojas (Eds.): IWBBIO 2016, LNBI 9656, pp. 335–345, 2016.
DOI: 10.1007/978-3-319-31744-1_30

Finite volume method (FVM)[19] is the numerical method used to solve the resulting PDE system in an one-dimensional domain. In addition, the chemotaxis operator is discretized using a first order upwind scheme (FOU) in order to ensure a stable solution [9].

There are some previous studies which model the interstitial fluid pressure (IFP) dynamics due to cancer growing [7,10]. Also, there are studies about bacteria-neutrophil iterations [11,12]. However, to the best of our knowledge, there are no previous studies about coupling bacterial infection dynamics with the interstitial fluid pressure.

This paper is divided into 6 sections, starting with this introductory section, followed by a brief overview of the immune system, the proposed mathematical model, the numerical strategy used to solve it, the results of our simulations, and finally, the conclusions and future works.

## 2    Immunology Background

Inflammation is an important reaction of the organism due to an injury, *i.e.* any process that cause tissue or cellular lesion. It is also an immunological process, so the cells involved in this process may be distinct, depending on the site of lesion [16]. The main cells involved in an immune response are the leukocytes. There are several kinds of leukocytes, and they can be classified according to their nuclear morphology: mononuclear (T lymphocytes, B lymphocytes, natural killer, monocytes, macrophages and dendritic cells) or polymorphonuclear (neutrophils, eosinophils, basophils and mast cells). The neutrophils represent 70 % of blood leukocytes [1].

When pathogens enter into the body, they encounter cells and molecules of the immune system, such as proteins of the complement system and macrophages, that immediately develop a response to them. The macrophages phagocyte the pathogens and produce proteins called cytokines that signal to other innate cells that their help are needed. Interstitial mobility, *i.e.* the ability to move easily across the space among tissue cells, is an essential aspect of immune cells as they must be able to move to that specific infected site upon demand [14]. Some cytokines trigger a cascade of events which increases the endothelium permeability and helps other immune systems cells to leave the circulatory system, enter into the tissue and migrate to the site of infection [16]. The increased endothelium permeability allows not only immune cells, but also fluids, to enter into the tissue [5]. When the interstice is excessively filled with liquid (called interstitial fluid [5]) occurs an extracellular edema formation.

## 3    Mathematical Modeling

### 3.1    Interstitial Fluid Pressure

The continuity equation for steady-state incompressible flow is given by:

$$\nabla \cdot \boldsymbol{v}_f = q_f, \tag{1}$$

where $\boldsymbol{v}_f$ is the interstitial fluid velocity, and $q_f$ is the fluid source/sink term.

To approach the interstitial fluid pressure $P$ in a porous media, and considering an isotropic and homogeneous tissue, Darcy's law is used as follows:

$$\boldsymbol{v}_f = -\frac{\mathbf{K}}{\mu}\nabla P, \tag{2}$$

where $\mathbf{K}$ is the hydraulic conductivity of the interstice, and $\mu$ is the viscosity of the fluid phase.

It is assumed that a continuous distributed source/sink throughout the tissue is given by Starling's law [17]:

$$q_f = k_f(P_c - P - \sigma(\pi_c - \pi_i)), \tag{3}$$

where $k_f = L_p(S/V)$ is the filter coefficient, $L_p$ is the hydraulic permeability of the microvascular wall, and $(S/V)$ is the surface area of the vessel wall per unit of volume; $P$ and $P_c$ are the interstitial fluid pressure and capillary fluid pressure, respectively; $\pi_c$ and $\pi_i$ are the interstitial osmotic pressures of the capillary and interstice, respectively; $\sigma$ is the osmotic reflection coefficient for plasma proteins.

The presence of bacteria triggers a chain of cytokine reactions which increase the capillary permeability. Thus, it also increases the filter coefficient which in turn increases the plasma flux, from the capillaries to the interstice. So, in order to model filter coefficient dynamics the following equation is used:

$$L_p = \gamma_p(1 + c_{bp}\rho_b s_b), \tag{4}$$

where $\gamma_p$ is the hydraulic permeability of the microvascular wall in a normal tissue, $s_b$ is the bacteria saturation term in the interstice, $\rho_b$ is the bacteria density, and $c_{bp}$ is the influence of bacteria to the hydraulic permeability of the microvascular.

Therefore, the interstitial fluid pressure (IFP) can be mathematically modeled by the resulting equation:

$$\begin{cases} \nabla \cdot \dfrac{\mathbf{K}}{\mu}\nabla P = -k_f(P_c - P - \sigma(\pi_c - \pi_i)) & \text{in } \Omega \\ \alpha P + \beta \nabla P \cdot \mathbf{n} = f_p & \text{on } \partial\Omega, \end{cases} \tag{5}$$

where $\Omega \subset \mathbb{R}$ and $P : \Omega \to \mathbb{R}$.

## 3.2 Bacteria Dynamics

The bacteria growing in the interstitial tissue is described by the following equation:

$$\begin{cases} \dfrac{\partial \phi \rho_b s_b}{\partial t} = \nabla \cdot (D_b \nabla(\rho_b s_b)) - r_b + q_b & \text{in } \Omega \times I \\ \alpha s_b + \beta \nabla s_b \cdot \mathbf{n} = f_b & \text{on } \partial\Omega \times I \\ s_b(\cdot, 0) = s_{b0} & \text{in } \Omega, \end{cases} \tag{6}$$

where $\Omega \subset \mathbb{R}$ and $I = (0, t_f] \subset \mathbb{R}^+$ is the time interval, $s_b : \Omega \times I \to \mathbb{R}^+$ is the bacteria saturation term in the interstice, $\rho_b$ is the bacteria density, $\phi$ is the media porosity, $D_b$ is the bacteria diffusion coefficient in the interstice, $q_b$ is the source term which represents the bacteria growing dynamics, and $r_b$ the sink term which represents the bacteria death dynamics.

Bacteria growing dynamics is based on a previous model [12] and represented by the following equation:

$$q_b = c_b \rho_b s_b, \tag{7}$$

where $c_b$ is the bacteria growing rate in the interstice.

Finally, bacteria death dynamics $r_b$ represents the interaction between neutrophil, a representative of the phagocytic cells, and bacteria. It is also based on a previous model [12] by the following equation:

$$r_n = \lambda_{nb} \rho_n s_n \rho_b s_b, \tag{8}$$

where $\lambda_{nb}$ is the bacteria death rate due to neutrophil-bacteria interaction, *i.e.* rate of phagocytosis.

### 3.3   Neutrophil Dynamics

Neutrophil dynamics in the interstitial tissue is described by the following equation:

$$\begin{cases} \dfrac{\partial \phi \rho_n s_n}{\partial t} = \nabla \cdot (D_n \nabla(\rho_n s_n) - \chi_{nb} \rho_n s_n \nabla(\rho_b s_b)) - r_n + q_n & \text{in } \Omega \times I, \\ \alpha s_n + \beta \nabla s_n \cdot \mathbf{n} = f_n & \text{on } \partial\Omega \times I \\ s_n(\cdot, 0) = s_{n0} & \text{in } \Omega, \end{cases} \tag{9}$$

where $\Omega \subset \mathbb{R}$ and $I = (0, t_f] \subset \mathbb{R}^+$ is the time interval, $s_n : \Omega \times I \to \mathbb{R}^+$ is the neutrophil saturation term in the interstice, $\rho_n$ is the neutrophil density, $D_n$ is the neutrophil diffusion coefficient in the interstice, $\chi_{nb}$ is the influence of chemotaxis in the tissue, $q_n$ is the source term which represents the neutrophil transport across the capillary vessel, and $r_n$ is the sink term which represents the neutrophil death dynamics.

Neutrophil growing dynamics is modeled as a transport through the vessels, which can be described as follows [12]:

$$q_n = \gamma_n \rho_b s_b \rho_n (s_{n,max} - s_n), \tag{10}$$

where $s_{n,max}$ is the blood neutrophil saturation and $\gamma_n$ is the capillary blood vessel permeability to neutrophil.

Finally, neutrophil death dynamics is affected by neutrophil-bacteria interaction and by neutrophil natural death. This can be mathematically described as follows:

$$r_n = \lambda_{bn} \rho_n s_n \rho_b s_b + \mu_n \rho_n s_n, \tag{11}$$

where $\lambda_{bn}$ is the neutrophil death rate due to neutrophil-bacteria interaction, *i.e.* rate of induced apoptosis, and $\mu_n$ represents the neutrophil natural decay.

## 4   Numerical Strategies

Finite volume method (FVM) was applied to Eqs. (5), (6) and (9) to obtain the
so-called discretized equations to solve the system of partial differential equations (PDE). This method is grounded in the evaluation of influx and outflux of
a control volume at each node of the mesh. Let the closed domain $\Omega \cup \Gamma \subset \mathbb{R}$ be
discretized into a set of regular nodal points defined by $S = \{(x_i) ; i = 0, \dots, I_x\}$
with $I_x$ being the number of nodal points spaced with length $\Delta x$. Each nodal
point is surrounded by a control volume. The boundaries (or faces) of control
volumes are positioned mid-way between adjacent nodes. In addition, the physical boundaries are coincided with the control volumes boundaries, as shown
in Fig. 1. In this figure, $A$ and $B$ are the boundaries of the domain. Two distinct numerical strategies must be used to solve the set of equations due to the
presence of steady-state and transient equations.

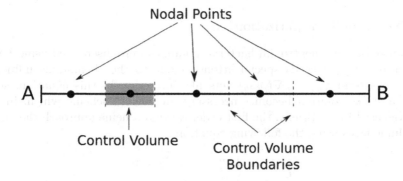

**Fig. 1.** A sample of discretized domain using the FVM.

### 4.1   Pressure Discretization

Pressure equation given by Eq. (5) is discretized using FVM resulting in a linear
system. This linear system was solved using Jacobi method, with the following
resulting series:

$$P_i^{k+1} = \frac{T_{i-\frac{1}{2}} P_{i-1}^k + T_{i+\frac{1}{2}} P_{i+1}^k + q_f \Delta x}{T_{i-\frac{1}{2}} + T_{i+\frac{1}{2}} + k_f \Delta x}, \tag{12}$$

where

$$T_{i-\frac{1}{2}} = T_{i+\frac{1}{2}} = \frac{K}{\mu \Delta x} \tag{13}$$

are the transmissibilities in the control volume faces.

Equation (12) is iterated until a convergence criteria is obtained which, in this
case, is an error smaller then $10^{-8}$. The error is measured using $||P_i^{k+1} - P_i^k||_\infty$.

## 4.2  Bacteria Discretization

Dynamics between neutrophil and bacteria given by Eqs. (6) and (9) are discretized using FVM and the temporal derivatives are discretized using forwarding finite difference scheme so-called Euler method. In addition to spatial discretization the time domain $I$ is partitioned into $N$ equal time intervals of length $\Delta t$, i.e., $(0, t_f] = \cup_{n=0}^{N-1} [t_n, t_{n+1}]$. Thus, Eq. (6) solution is obtained using the following equation:

$$B_i^{n+1} = \frac{\Delta t}{\phi \rho_b} (T_{i+\frac{1}{2}}(B_{i+1}^n - B_i^n) - T_{i-\frac{1}{2}}(B_i^n - B_{i-1}^n) + q_b - r_b) + B_i^n; \qquad (14)$$

where

$$T_{i-\frac{1}{2}} = T_{i+\frac{1}{2}} = \frac{D_b}{\Delta x} \qquad (15)$$

are the transmissibilities in the control volume faces.

## 4.3  Neutrophil Discretization

As discussed before, neutrophil-bacteria dynamics were discretized using FVM. In addition, Eq. (9) needs special attention due to the chemotaxis influence. Chemotaxis term $\chi_{nb}\rho_n s_n \nabla(\rho_b s_b)$ acts like advection in this equation, so to ensure a stable solution becomes necessary an upwind scheme which, in this case, was used FOU. Briefly, the first order upwind scheme approach the flux in the volume faces using the following equation:

$$C_{i+\frac{1}{2}}^{n+1} = \begin{cases} C_i^n & v_{f,i+\frac{1}{2}} < 0 \\ C_{i+1}^n & v_{f,i+\frac{1}{2}} > 0, \end{cases} \qquad (16)$$

where $C$ is the flux approach and $v_f$ is the face velocity. Flux $C_{i-\frac{1}{2}}^{n+1}$ is analogous.

$$N_i^{n+1} = \frac{\Delta t}{\phi \rho_n} ((T_{i+\frac{1}{2}}(N_{i+1}^n - N_i^n) - T_{i-\frac{1}{2}}(N_i^n - N_{i-1}^n)$$

$$-(T_{nb,i+\frac{1}{2}} C_{i+\frac{1}{2}}^n v_{f,i+\frac{1}{2}} - T_{nb,i-\frac{1}{2}} C_{i-\frac{1}{2}}^{n+1} v_{f,i-\frac{1}{2}}) + q_n - r_n) + N_i^n \qquad (17)$$

where

$$T_{i-\frac{1}{2}} = T_{i+\frac{1}{2}} = \frac{\chi_{nb}\rho_n s_n}{\Delta x}, \qquad (18)$$

are the transmissibilities in the control volume faces due to diffusion and

$$T_{q,i-\frac{1}{2}} = T_{q,i+\frac{1}{2}} = \frac{\chi_{nb}\rho_n s_n}{\Delta x} \qquad (19)$$

are the transmissibilities in the control volume faces due to chemotaxis.

## 5    Numerical Experiments

This section presents the numerical results of the simulations using the numerical methods described previously. The simulator was implemented using the C++ programming language. A numerical library, such as NAG, could be used to solve the PDEs. However, if external libraries were used it would not be possible or easy to parallelize the source code to deal with complex scenarios and three-dimensional domains, as we plan to cope with in the near future. Also, few numerical libraries offer functions that are suitable to deal with this type of problem.

GNU GCC 4.8.2 was used to compile the source code. All simulations were performed in a SMP Linux (3.12.11-201) computer consisting of one Intel Core i7-3632QM CPU running at 2.20 GHz and 6 GB of main memory.

### 5.1    Numerical Results

The model's initial conditions and parameters are given in Tables 1 and 2, respectively. In our simulations, we assumed an one-dimensional domain of 1 $cm$ length and a simulation time of 30$s$. In fact, this one-dimensional model is a simplification of a 3D block model in that we have assumed that the lengths associated with $y$ and $z$ are much smaller than the length associated with $x$. All simulations were performed using $\Delta x = 0.02\,cm$ and $\Delta t = 5 \times 10^{-5}s$.

**Table 1.** Initial and boundary conditions

| Variable | Initial condition | Boundary condition |
|---|---|---|
| $s_n$ | $s_n = 0 \; \forall \, x \in \Omega$ | $\nabla s_n \cdot \boldsymbol{n} = 0 \; \forall \, x \in \partial\Omega$ |
| $s_b$ | $s_b = \begin{cases} 0.001 & \text{for } x \in [0.092, 1.0] \\ 0 & \text{otherwise} \end{cases}$ | $\nabla s_b \cdot \boldsymbol{n} = 0 \; \forall \, x \in \partial\Omega$ |
| $P$ | - | $\begin{cases} P = 10.9 & \text{for } x = 0 \\ \nabla P \cdot \boldsymbol{n} = 0 & \text{for } x = 1 \end{cases}$ |

In Table 2, "Estimated" parameters were qualitatively adjusted to reproduce the results. All "Calculated" parameters are variables in the PDE system. In addition, $\mu$ and $\mathbf{K}$ parameters were marked with "*" because it was not possible to obtain their individual values, only the relationship $\mathbf{K}/\mu = 2.5\text{x}10^{-7} cm^2/s/mmHg$ was found.

Model results are shown in Fig. 2. Figures 2(a) and (b) represents the bacteria-neutrophil dynamics modeled by Eqs. (6) and (9). It can be seen that when bacteria start to grow, neutrophils are attracted from blood vessels to the interstice, as it is shown in both figures. In addition, Fig. 2(c) shows the influence of a bacterial infection in the IFP dynamics as a result of Eq. (5). A higher concentration of bacteria induces an increase of the hydraulic permeability of

**Table 2.** Description of the parameters of the model.

| Name | Symbol | Unit | Value | Reference |
|---|---|---|---|---|
| Fluid velocity | $v_f$ | $\dfrac{cm}{s}$ | - | Calculated |
| Pressure | $p$ | $mmHg$ | - | Calculated |
| Viscosity | $\mu$ | $\dfrac{g}{cms}$ | * | [10] |
| Permeability | $\mathbf{K}$ | $cm^2$ | * | [10] |
| Saturation | $s$ | − | - | Calculated |
| Porosity | $\phi$ | − | - | [2] |
| Density | $\rho$ | $\dfrac{g}{cm^3}$ | 1.0 | Estimated |
| Filtering coefficient | $k_f$ | $\dfrac{1}{smmHg}$ | 626.4 | [10] |
| Neutrophil diffusion coefficient | $D_n$ | $\dfrac{cm^2}{s}$ | 0.001 | Estimated |
| Bacteria diffusion coefficient | $D_b$ | $\dfrac{cm^2}{s}$ | 0.001 | Estimated |
| Chemotaxis rate | $\chi_{nb}$ | $\dfrac{cm^5}{sg}$ | 0.001 | Estimated |
| Bacteria reproduction rate | $c_b$ | $\dfrac{1}{s}$ | 0.154 | Estimated |
| Induced apoptosis rate | $\lambda_{bn}$ | $\dfrac{cm^3}{sg}$ | 0.1 | Estimated |
| Phagocytosis rate | $\lambda_{nb}$ | $\dfrac{cm^3}{sg}$ | 1.8 | Estimated |
| Neutrophil source | $\gamma_n$ | $\dfrac{cm^3}{sg}$ | 0.1 | Estimated |
| Apoptosis rate | $\mu_n$ | $\dfrac{1}{s}$ | 0.2 | Estimated |
| Hydraulic permeability | $\gamma_p$ | $\dfrac{cm}{smmHg}$ | $3.6 \times 10^{-8}$ | [10] |
| Bacterial influence in hydraulic permeability | $c_{bp}$ | $\dfrac{cm^3}{g}$ | 1.0 | Estimated |

the microvascular wall which also increases the filtering coefficient. As a result, the IFP increases at the infection site. The IFP increase is the result of plasma accumulation in the interstice. This accumulation can lead to an edema. The boundary condition used in $x = 0$ represents the capillary, and the boundary condition used in $x = 1$ represents the continuum.

As an attempt to validate qualitatively the results obtained numerically, another experiment was performed. Its objective is to measure how pressure changes over time in a specific point of the domain. The chosen point was $x = 0.9\,cm$. Figure 3 presents the results. Comparing them with the experimental data founded in the literature [6], it is possible to qualitatively observe the simulations are similar in their shapes which gives credibility to the model results.

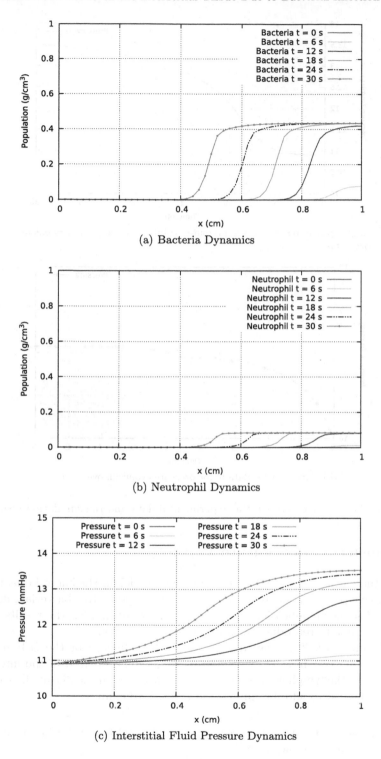

(a) Bacteria Dynamics

(b) Neutrophil Dynamics

(c) Interstitial Fluid Pressure Dynamics

**Fig. 2.** Simulation results

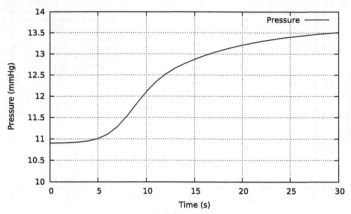

(a) Numerical results for the pressure evolution at the point $x = 0.9cm$ over time.

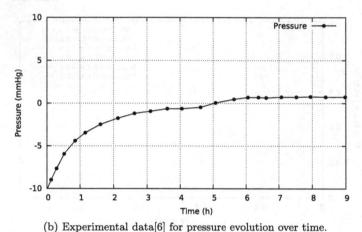

(b) Experimental data[6] for pressure evolution over time.

**Fig. 3.** Comparative between the experimental data and the simulation results.

# 6  Conclusion and Future Works

In this study, it was presented a computational model for the bacteria-neutrophil dynamics coupled with a plasma flow model as a response to a bacterial infection. Despite of some simplifications and limitations of this model, the results shown that it was able to reproduce an initial edema formation.

As future work it is planed to analyze the influence of the IFP increase in the immune system cells due the pressure gradient created in the interstice. Furthermore, the lymphatic system was not considered in the plasma flow model; it is also planned to add its influence in the IFP dynamics.

**Acknowledgements.** The financial support by CNPq, CAPES, UFJF and FAPEMIG is greatly acknowledged.

# References

1. Abbas, A.K., Lichtman, A.H., Pillai, S.: Imunologia Celular e Molecular. Elsevier, Belo Horizonte (2012)
2. Peter, J.: Basser.: Interstitial pressure, volume, and flow during infusion into brain tissue. Microvasc. Res. **44**(2), 143–165 (1992)
3. Bauernfeind, F., Ablasser, A., Bartok, E., Kim, S., Schmid-Burgk, J., Cavlar, T., Hornung, V.: Inflammasomes: current understanding and open questions. Cell. Mol. Life Sci. **68**(5), 765–783 (2011)
4. Chen, Z., Ewing, R.E.: Comparison of various formulations of three-phase flow in porous media. J. Comput. Phys. **132**(2), 362–373 (1997)
5. Guyton, A.C., Hall, J.E.: Textbook of Medical Physiology. Guyton Physiology Series. Elsevier Saunders, Philadelphia (2006)
6. Arthur, C.: Guyton.: Interstitial fluid pressure: Il. pressure-volume curves of interstitial space. Circ. Res. **16**(5), 452–460 (1965)
7. Jain, R.K., Martin, J.D., Stylianopoulos, T.: The role of mechanical forces in tumor growth and therapy. Annu. Rev. Biomed. Eng. **16**, 321 (2014)
8. Koneman, E., Win Jr., W., Allen, S., Janda, W., Procop, G., Scheckenberger, P., Woods, G.: Color Atlas and Textbook of Diagnostic Microbiology, 6th edn. LWW, Philadelphia (2005)
9. McDonald, B.E., Ambrosiano, J.: High-order upwind flux correction methods for hyperbolic conservation laws. J. Comput. Phys. **56**(3), 448–460 (1984)
10. Phipps, C., Kohandel, M.: Mathematical model of the effect of interstitial fluid pressure on angiogenic behavior in solid tumors. Comput. Math. Methods Med. **2011**, 9 (2011)
11. Pigozzo, A.B., Macedo, G.C., dos Santos, R.W., Lobosco, M.: On the computational modeling of the innate immune system. BMC Bioinform. **14**(Suppl 6), S7 (2013)
12. Pigozzo, A.B., Macedo, G.C., dos Santos, R.W., Lobosco, M.: Computational modeling of microabscess formation. Comput. Math. Methods Med. **2012**, 16 (2012)
13. Pinder, G.F., Gray, W.G.: Essentials of Multiphase Flow in Porous Media, 1st edn. Wiley, Hoboken (2008)
14. Pixley, F.J.: Macrophage migration and its regulation by csf-1. Inter. J. Cell Biol. **1–12**, 2012 (2012)
15. Scallan, J., Huxley, V.H., Korthuis, R.J.: Capillary Fluid Exchange: Regulation, Functions, and Pathology, 2nd edn. Morgan & Claypool Publishers, San Rafael (2010)
16. Sompayrac, L.: How the Immune System Works. Wiley-Blackwell, Hoboken (2012)
17. Starling, E.H.: On the absorption of fluids from the connective tissue spaces. J. Physiol. **19**(4), 312–326 (1896)
18. Tepole, A.B., Kuhl, E.: Computational modeling of chemo-bio-mechanical coupling: a systems-biology approach toward wound healing. Comput. Method Biomech. Biomed. Eng. **19**, 1–18 (2014)
19. Versteeg, H., Malalasekra, W.: An Introduction to Computational Fluid Dynamics: The Finite, vol. Method, 2nd edn. Prentice Hall, Upper Saddle River (2007)

# Reactive Interstitial and Reparative Fibrosis as Substrates for Cardiac Ectopic Pacemakers and Reentries

Rafael Sachetto Oliveira[1], Bruno Gouvêa de Barros[2], Johnny Moreira Gomes[2], Marcelo Lobosco[2], Sergio Alonso[3], Markus Bär[4], and Rodrigo Weber dos Santos[2]([✉])

[1] Departamento de Ciência da Computação,
Universidade Federal de São João del Rei, São João del Rei, Brazil
[2] Departamento de Ciência da Computação e Programa em Modelagem
Computacional, Universidade Federal de Juiz de Fora, Juiz de Fora, Brazil
rodrigo.weber@ufjf.edu.br
[3] Departament de Física, Universitat Politècnica de Catalunya, Barcelona, Spain
[4] Physikalisch-Technische Bundesanstalt, Braunschweig, Berlin, Germany

**Abstract.** Dangerous cardiac arrhythmias have been frequently associated with focal sources of fast pulses, i.e. ectopic pacemakers. However, there is a lack of experimental evidences that could explain how ectopic pacemakers could be formed in cardiac tissue. In recent studies, we have proposed a new theory for the genesis of ectopic pacemakers in pathological cardiac tissues: reentry inside microfibrosis, i.e., a small region where excitable myocytes and non-conductive material coexist. In this work, we continue this investigation by comparing different types of fibrosis, reparative and reactive interstitial fibrosis. We use detailed and modern models of cardiac electrophysiology that account for the micro-structure of cardiac tissue. In addition, for the solution of our models we use, for the first time, a new numerical algorithm based on the Uniformization method. Our simulation results suggest that both types of fibrosis can support reentries, and therefore can generate in-silico ectopic pacemakers. However, the probability of reentries differs quantitatively for the different types of fibrosis. In addition, the new Uniformization method yields 20-fold increase in cardiac tissue simulation speed and, therefore, was an essential technique that allowed the execution of over a thousand of simulations.

## 1 Introduction

The function of the heart, in physiological conditions or pathological ones, is a non-linear, multi-scale and multi-physics phenomenon, and as such, poses incredible challenges to science and medicine. Computational and mathematical models of the heart have supported and complemented experimental techniques proposing answers to both basic and applied questions regarding cardiac physiology. In particular, in the area of cardiac electrophysiology, for decades, models have

© Springer International Publishing Switzerland 2016
F. Ortuño and I. Rojas (Eds.): IWBBIO 2016, LNBI 9656, pp. 346–357, 2016.
DOI: 10.1007/978-3-319-31744-1_31

been used to propose and test new theories for basic mechanisms [2,4], during the tests of drugs [20] and development of new medical devices [8].

Dangerous cardiac arrhythmias, such as atrial and ventricular fibrillation have been frequently associated to fibrosis [9,14,24,26]. For instance, during atrial fibrillation the duration of fibrillation episodes was highly correlated to the degree of fibrosis, i.e. the amount of fibrous tissue in the extracellular space between myocytes and/or fibrous tissue that replaced dead myocytes [16]. Nevertheless, in the aforementioned works, whether in animal or computational experiments, an artificial trigger is usually used to induce the arrhythmia, a focal stimulation site that induces rapid pacing. This focal trigger is often called ectopic pacemaker.

Unfortunately, the mechanisms and conditions for the formation of a region that becomes an ectopic pacemaker are until now poorly understood. Some new evidences suggest that two mechanisms (combined or not) may play a very important role in the generation of such dangerous triggers of arrhythmia: abnormal intracellular calcium (Ca2+) signaling [25] and microreentry (reentry inside a fibrotic region).

In our previous works [1,6] we were able to generate an ectopic pacemaker in computer simulations of cardiac electrophysiology by assuming a single hypothesis for substrate: the existence of a region of microfibrosis, a mixture of excitable (healthy myocytes) and non-conducting (fibrosis) areas. The mechanism of this ectopic pacemaker is simple. The two different phases (excitable and non-conducting) form a maze for wave propagation. Propagation fractionates and follows zig-zag pathways. Macroscopic propagation is considerably slowed down. Microscopically, the topology of the maze allows the electric wave to reexcite the fibrotic region before the wave leaves it. Therefore, reentry occurs inside the fibrotic region. This spiral-like wave will continuously try to generate ectopic beats when it touches the border of the microfibrosis region [12].

Here, we continue this investigation by comparing different types of fibrosis, reparative and reactive interstitial fibrosis. The excessive deposition of collagen in the extracellular matrix may physically and electrically separate neighboring myocytes, a process called reactive interstitial fibrosis, or completely replace a dead myocyte, a process called reparative fibrosis [21].

Our computational models are based on the microscopic tissue model that uses subcellular discretization of 8 $\mu$m and detailed and realistic gap junction distribution [5]. From this basic microscopic model we generate an equivalent discrete model [6]. With the discrete or network-like model we simulated reactive interstitial fibrosis by randomly removing a certain percentage of links between neighboring cells; and simulate reparative fibrosis by randomly removing a certain percentage of myocytes from the network.

Since our models account for realistic description of single myocyte electrophysiology and for the microstructure of cardiac tissue such as gap junction distribution and tissue anisotropy, they are computationally very expensive to solve. To speed up the simulations, we use in this work the new Uniformization method, first proposed and tested for single cell simulations [15]. This is the first time this new numerical method is tested in the framework of simulations of cardiac tissue.

## 2   Methods

### 2.1   Modeling Fibrosis

We use a microscopic tissue model with subcellular discretization of $8\,\mu$m to represent detailed shapes of myocytes as well as realistic gap junction distribution [5] and presented in the top of Fig. 1. From this basic microscopic model we generate an equivalent discrete model [6] and presented in the bottom of Fig. 1. With the discrete or network-like model we simulated reactive interstitial fibrosis by randomly removing a certain percentage of links between neighboring cells, $\phi_l$ (see Fig. 2); and to simulate reparative fibrosis we randomly removed a certain percentage of cells, $\phi_c$, from the network (see Fig. 1).

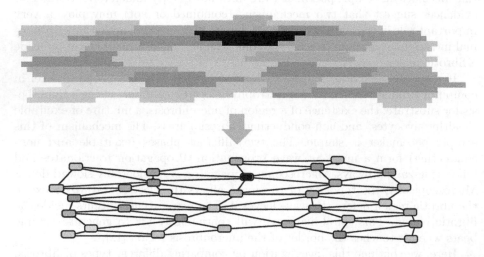

**Fig. 1.** Basic microscopic model with subcellular discretization to represent detailed shapes of myocytes as well as realistic gap junction distribution (Top) and the equivalent detailed discrete model (Bottom) where each myocyte is represented by one cell in the network. A black shaded myocyte represents how reparative fibrosis is modeled in the microscopic model (Top) and in the equivalent detailed discrete model (Bottom).

The purpose of our work is to assess the probability of a certain fibrotic tissue to become an ectopic pacemaker and to study how it depends on the fraction and type of fibrosis, $\phi_c$ and $\phi_l$. To achieve this goal it is necessary to perform thousands of simulations of different microfibrosis models.

For instance, to simulated reparative fibrosis, for each given value of $\phi_c$ we randomly generate a microfibrosis tissue. Fifty realizations of different microfibrosis but with the same value for $\phi_c$ were simulated. This process is represented by Fig. 3. The value of $\phi_c$ was varied from 44 % to 64 %, in steps of 2 %. Therefore a total of $11 \times 50 = 550$ cardiac tissues with different characteristics of reparative fibrosis were simulated. The same was performed for the case of reactive interstitial fibrosis, i.e. the value of $\phi_l$ was also varied from 44 % to 64 %, in steps of 2 %. Again a total of $11 \times 50 = 550$ cardiac tissues with different characteristics of reactive interstitial fibrosis were simulated. For all the results presented in the next section the modeled cardiac tissues had dimensions of $1\,\mathrm{cm} \times 1\,\mathrm{cm}$.

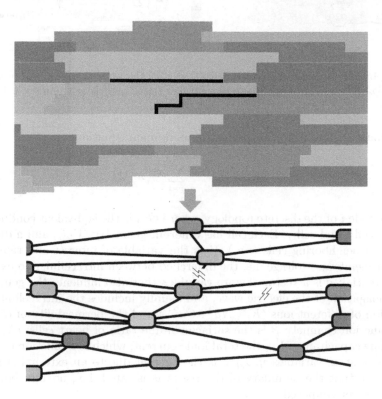

**Fig. 2.** Black shaded areas represent how reactive interstitial fibrosis is modeled in the microscopic model (Top) and in the equivalent detailed discrete model (Bottom).

## 2.2 Numerical Methods

The detailed and discrete model of cardiac tissue was numerically solved [6] using the finite volume method. The reaction and diffusion parts of the discrete monodomain equations were split by employing the Godunov operator splitting [22]. Therefore, each time step involves the solution of two different problems: a nonlinear system of ODEs

$$\frac{\partial V_i}{\partial t} = \frac{1}{C_m} \left[ -I_{ion}(V_i, \boldsymbol{\eta}_i) + I_{stim} \right], \tag{1}$$

$$\frac{\partial \boldsymbol{\eta}_i}{\partial t} = f(V_i, \boldsymbol{\eta}_i); \tag{2}$$

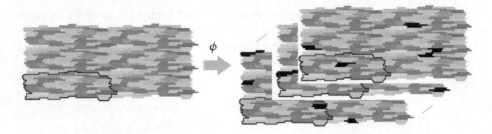

**Fig. 3.** Multiple instances of models generated with the same percentage of reparative fibrosis, $\phi_c$. The size of these illustrative cardiac tissue models is 1.1 mm × 0.4 mm.

and a linear system from the discrete monodomain equation

$$Vol_i\beta(C_m\frac{\partial V_i}{\partial t} + I_{ion}(V_i, \boldsymbol{\eta}_i)) = (\sum_{j=1}^{nn_i}(G_{i,j} * (V_j - V_i)) + I_i^{stim}, \tag{3}$$

for each node $i$ of the discrete topology, where $G_{i,j}$ is the equivalent conductance between cell $i$ and cell $j$. Each cell has different geometry ($Vol_i$) and a different number of neighboring cells ($nn_i$). $V$ is the variable of interest and represents the transmembrane voltage, i.e. the difference between intracellular to extracellular potential; $\boldsymbol{\eta}$ is a vector of state variables that also influence the generation and propagation of the electric wave, and usually includes the intracellular concentration of different ions ($K^+$, $Na^+$, $Ca^{2+}$) and the permeability of different membrane ion channels; $\beta$ is the surface-volume ratio of heart cells; $C_m$ is the membrane capacitance, $I_{ion}$ the total ionic current, which is a function of $V$ and a vector of state variables $\boldsymbol{\eta}$, $I_{stim}$ is the current due to an external stimulus. We assume that the boundary of the tissue is isolated, i.e., no–flux boundary conditions were imposed.

For the discretization of the nonlinear system of ODEs we note that its stiffness demands very small time steps. For simple models based on Hodgkin-Huxley formulation this problem is normally overcome by using the Rush-Larsen (RL) method [18]. However, for the most modern and complex models that are highly based on Markov Chains, the RL method seems to be ineffective in terms of allowing larger time steps during the numerical integration.

In this work we use the modern Markov-based model of Bondarenko *et al.* [7] for myocyte electrophysiology. For the case of the Bondarenko *et al.* model, we tested both methods, Euler and RL, and both demanded the same time step, $\Delta t_o = 0.0001$ ms due to stability issues. Since the RL method is more expensive per time step than the Euler method, in this work, we used either the simple explicit Euler method for the discretization of the nonlinear ODEs or the new Uniformization and SAST1 method [15].

For the discrete monodomain equation we used the unconditionally stable implicit Euler scheme. This allowed us to use longer time steps for the numerical solution of it than those used for the solution of ODEs ($\Delta t_p > \Delta t_o$). This linear system was solved in parallel with the PETSc library [3]. The nonlinear system of ODEs were also solved in parallel using the MPI library [10].

In the next sections we give more details about the new numerical methods of SAST1 and Uniformization.

**The Extended Rush-Larsen Method (SAST1).** The method proposed by Rush-Larsen (RL) [19] is very popular in the community of cardiac electrophysiology and focus on the gating variables from Hodgkin and Huxley's formulation [11]. This method is based on a local linearization of the equations in the form

$$\frac{dy}{dt} = \alpha(V)(1 - y) - \beta(V)y, \tag{4}$$

so the equations associated with the gating variables are numerically solved by:

$$y_{n+1} = (y_n - \frac{\alpha}{\alpha + \beta})e^{-(\alpha+\beta)h} + \frac{\alpha}{\alpha + \beta}, \tag{5}$$

$$k = |\alpha + \beta|, \tag{6}$$

where $h$ is the time step. The remaining equations of the model are evaluated by the Euler method. Therefore, this is a first order method. If the value of $k$, defined by Eq. (6), is close to zero we use the Euler method instead of Eq. (5) for the corresponding gating variable at that point of the simulation.

The first order version of Sundnes *et al.* method (SAST1) [23] extends the RL method for each and every differential equation of the model. This is achieved after the following linearization around $y^j(t_n)$, where $t_n = nh$ and $h$ is the time step:

$$\frac{dy^j(t)}{dt} = f^j(\overrightarrow{Y}(t_n), t_n) + k(y^j(t) - y^j(t_n)), \tag{7}$$

$$k = \frac{\partial f^j(\overrightarrow{Y}(t_n), t_n)}{\partial y^j}, \tag{8}$$

whereas $\overrightarrow{Y}$ is the vector containing the variables of the model $y^j$ as components and $f^j$ is the right-hand side function for variable $y^j$. This can now also be considered a quasi-linear equation with an analytic solution similar to the RL one.

The SAST1 method is presented in the next equations, where the partial derivative is approximated via finite differentiation:

$$y_{n+1}^j = y_n^j + \frac{f^j(\overrightarrow{Y_n}, t_n)}{k}(e^{hk} - 1), \tag{9}$$

$$k = \frac{\partial f^j(\overrightarrow{Y_n}, t_n)}{\partial y^j}. \tag{10}$$

**Uniformization Method.** The Uniformization (or randomization) method [13] is an efficient technique for the transient analysis of Markov models. The method is traditionally used for reliability, dependability and performance analysis of computer systems modeled by Markov models.

Let $\pi(t)$ be the probability line vector of a continuous time Markov Chain (CTMC) and $\mathbf{P}(t)$ the associated transition matrix, whereas $[\mathbf{P}]_{ij}$ is the transition rate from state $i$ to state $j$. So the ODEs system associated with the CTMC is given by $\frac{d\pi}{dt} = \pi\mathbf{P}$. If we consider the terms $[\mathbf{P}]_{ij}$ as constant in a small time interval, the ODEs system will be locally linear. Therefore, the solution of the system within this small time interval is given by $\pi(t + h) = \pi(t)e^{h\mathbf{P}}$, whereas the matrix exponential is defined by the Taylor series as $e^{t\mathbf{P}} = \sum_{i=0}^{\infty} \frac{(t\mathbf{P})^i}{i!}$.

The direct usage of matrix $\mathbf{P}$ leads to numerical errors and slow convergence in the evaluation of the series. The main reasons for this are: (1) diagonal terms of $\mathbf{P}$ are negative and the remaining terms are non-negative and; (2) the existence of terms with magnitude greater than 1. The Uniformization method defines $\mathbf{P}^* = \mathbf{P}/q + I$, where $q \geq max_{1 \leq i \leq N_{mk}} |[\mathbf{P}]_{ii}|$ and $N_{mk}$ is the number of states in the CTMC. Since $\mathbf{P} = q(\mathbf{P}^* - I)$, we have

$$\pi(t + h) = \pi(t)e^{qt(\mathbf{P}^* - I)} = e^{-qt}\pi(t)e^{qt\mathbf{P}^*},$$

so the resulting solution using the Taylor series is

$$\pi(t + h) = e^{-qt}\pi(t) \sum_{i=0}^{\infty} \frac{(qt\mathbf{P}^*)^i}{i!}. \tag{11}$$

In order to truncate the series in Eq. (11) we can use the following relation [17] to choose the number of evaluated terms $N$, given an error tolerance $\lambda$:

$$\lambda \leq 1 - e^{-q(t)h} \sum_{i=0}^{N} \frac{(q(t)h)^i}{i!}. \tag{12}$$

In this paper, the implementation called $Uni + SAST1$ uses the Uniformization method for all the equations modeled by Markov Chains and uses the SAST1 method for the remaining equations in the Bondarenko et al. cardiac electrophysiology model.

## 2.3 Computing Environment

All the numerical experiments were performed using a GNU/Linux 2.6.32 machine, with 12 GB of memory and two Intel Xeon E5620 2.40 GHz processors each with 4 cores and 12 MB of cache memory. Our monodomain solver was implemented in C++ using PETSc 3.5.1 [3] and MPI, and compiled using GNU GCC 4.4.7 with $-O3$ optimization flag enabled.

## 2.4   Computational Simulations

In order to evaluate the performance of the Uniformization+SAST1 method compared to the classical Euler method we executed five simulations using different values of $\phi_c$ to randomly remove a certain percentage of cells and generate reentry (values between 0.44 and 0.46). To analyze if the generated reentries would impact on the performance the Uni+SAST1 method, we fixed seeds in our random generation routine to ensure four simulations with reentry and one without, see Sect. 3.2 for details about the reentries.

The simulations were performed using the detailed discrete model and a cardiac tissue of $1.0\,cm \times 1.0\,cm$ size and were executed for $500\,ms$. The values used for $\beta$ and $C_m$ were set to $0.14\,cm^{-1}$ and $1.0\,\mu F/cm^2$, respectively. The time step used to solve the linear system associated to the discrete monodomain equation (PDE) was set to $\Delta t_p = 0.02\,ms$ and to solve the nonlinear system of ODEs associated was set to $\Delta t_o = 0.0001\,ms$ for the Euler method and 0.02 for the Uni+SAST1.

The same parameters were used to assess the probability of a certain fibrotic tissue to become an ectopic pacemaker and to study how this depends on the fraction and type of fibrosis, $\phi_c$ and $\phi_l$. The value of $\phi_c$, for the case of reparative fibrosis, was varied from $44\,\%$ to $64\,\%$, in steps of $2\,\%$. The value of $\phi_l$, for the case of reactive interstitial fibrosis, was also varied from $44\,\%$ to $64\,\%$, in steps of $2\,\%$. Cardiac tissues of $1.0\,cm \times 1.0\,cm$ size were executed for $200\,ms$.

## 3   Results

### 3.1   Uniformization vs. Euler

Table 1 presents the execution times for the resolution of the PDE, ODEs and full simulation for both numerical methods. The time is an average of all different simulations using Euler or Uniformization and SAST1 (Uni+SAST1). The coefficients of variation (CV) for the Uniformization simulations were $3.71\,\%$ for the PDE time, $0.9\,\%$ for the ODEs and $2.3\,\%$ for the Total time.

**Table 1.** Execution times (average) for the resolution of the PDE, ODEs and full simulation for both numerical methods.

| Method | PDE (s) | ODE (s) | Total (s) |
|---|---|---|---|
| Euler | 3731.32 | 109659.13 | 129682.20 |
| Uni+SAST1 | 3594.73 | 2321.17 | 6839.28 |

Analyzing the results in Table 1, we can see that the ODE time dominates the execution time, being $85\,\%$ when using the Euler method. We can also see that, by using the Uni+SAST1 method, we were able to solve the ODEs $47\times$ faster than using the Euler method. For the simulation total time, the speedup achieved was $19\times$.

## 3.2  Reactive Interstitial Fibrosis vs. Reparative Fibrosis

As in our previous works [1,6] we were able to generate reentry inside fibrotic tissues. Figure 4 shows color maps of the transmembrane voltage at four different instants along one of the simulations. The tissue was stimulated in the left border, the traveling wave fractionates and its zig-zag pattern of propagation supports a sustained reentry pattern of excitation.

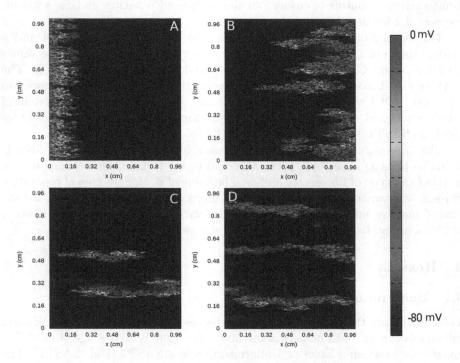

**Fig. 4.** Sustained reentry inside a fibrotic tissue. Four snapshots showing the transmembrane voltage at times $10\,ms$ (A), $80\,ms$ (B), $200\,ms$ (C), and $360\,ms$ (D).

Figure 5 presents the results in terms of probability of reentry for the two different types of fibrosis modeled in this paper: reparative vs. reactive interstitial. For both cases the probability of reentry depends on the fraction of removed cells ($\phi_c$, for the case of reparative fibrosis) or of removed connections or links ($\phi_l$, for the case of reactive interstitial fibrosis). For both types of fibrosis reentry can appear with probabilities higher than 30 %. However, the probability distribution for the case of reactive interstitial fibrosis is shifted to the right, in terms of $\phi$. Therefore, to become an ectopic pacemaker a microfibrosis region with pure reactive interstitial needs higher values of disconnection between cells ($\phi_l$ around 60 %) than in a pure reparative fibrosis case ($\phi_c$ around 45 %). These values are qualitative near the percolation values computed before in [1] for the case of regular and isotropic tissue where each cell connects to six neighboring cells.

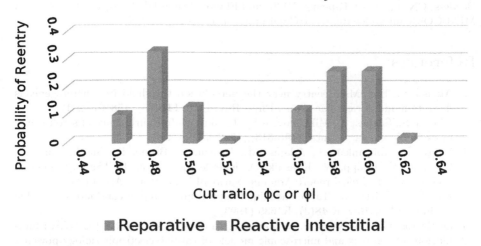

**Fig. 5.** Probability of reentry for the two cases of fibrosis: reparative vs. reactive interstitial fibrosis. For reactive interstitial fibrosis, $\phi_l$ is the percentage of disconnection between cells. For reparative fibrosis, $\phi_c$ is the percentage of removed cells.

## 4   Conclusions

In this paper we have extended our previous works to evaluate how two different types of fibrosis, reactive interstitial and reparative fibrosis, may support the generation of reentry inside a microfibrosis tissue. We have shown that in both types of fibrosis after the traveling wave enters the microfibrosis region propagation is fractionated and its zig-zag patterns supports sustained reentries. In turn, such reentries will continuously try to generate ectopic beats whenever it touches the border of the microfibrosis region. Therefore, fibrosis is shown to be a sufficient substrate for the generation of ectopic pacemakers in diseased cardiac tissues. Although our simulation results supported the thesis that both types of fibrosis can lead to reentries, we have observed that the probability of reentries differs quantitatively for the two different types of fibrosis. To become an ectopic pacemaker a microfibrosis region with pure reactive interstitial needed higher values of disconnection between cells ($\phi_l$ around 60 %) than in a pure reparative fibrosis case ($\phi_c$ around 45 %).

In addition, we have also implemented and tested for the first time the new Uniformization numerical method for simulations of cardiac tissue. With the new Uniformization numerical method cardiac tissue simulations were near 20-fold faster than when using classical methods such as Euler and Rush-Larsen. Therefore, this was an essential technique that allowed fast execution of over a thousand of simulations.

**Acknowledgments.** This work was partially funded by Brazilian Science without Borders, CNPq, Capes, Fapemig, UFJF and Finep; Geman DFG project SFB 910; and MINECO Spain under Ramon y Cajal program RYC-2012-11265.

# References

1. Alonso, S., Bär, M.: Reentry near the percolation threshold in a heterogeneous discrete model for cardiac tissue. Phys. Rev. Lett. **110**(15), 158101 (2013)
2. Alonso, S., Sagués, F., Mikhailov, A.S.: Taming winfree turbulence of scroll waves in excitable media. Science **299**(5613), 1722–1725 (2003)
3. Balay, S., Abhyankar, S., Adams, M., Brown, J., Brune, P., Buschelman, K., Eijkhout, V., Gropp, W., Kaushik, D., Knepley, M., et al.: PETSc users manual revision 3.5. Technical report Argonne National Laboratory (ANL) (2014)
4. Bär, M., Eiswirth, M.: Turbulence due to spiral breakup in a continuous excitable medium. Phys. Rev. E **48**(3), R1635 (1993)
5. de Barros, G.B., Oliveira, S.R., Meira, W., Lobosco, M., dos Santos, W.R.: Simulations of complex and microscopic models of cardiac electrophysiology powered by multi-gpu platforms. Computational and Mathematical Methods in Medicine 2012 (2012)
6. de Barros, B.G., dos Santos, R.W., Lobosco, M., Alonso, S.: Simulation of ectopic pacemakers in the heart: multiple ectopic beats generated by reentry inside fibrotic regions. BioMed Research International 2015 (2015)
7. Bondarenko, V., Szigeti, G., Bett, G., Kim, S., Rasmusson, R.: Computer model of action potential of mouse ventricular myocytes. Am. J. Physiol. Heart Circulatory Physiol. **287**, H1378–H1403 (2004)
8. Dos Santos, R.W., Kosch, O., Steinhoff, U., Bauer, S., Trahms, L., Koch, H.: MCG to ECG source differences: measurements and a two-dimensional computer model study. J. Electrocardiol. **37**, 123–127 (2004)
9. Finet, J.E., Rosenbaum, D.S., Donahue, J.K.: Information learned from animal models of atrial fibrillation. Cardiol. Clin. **27**(1), 45–54 (2009)
10. Groop, W., Lusk, E.: User's guide for mpich, a portable implementation of MPI. Technical report Argonne National Laboratory (1994)
11. Hodgkin, A.L., Huxley, A.F.: A quantitative description of membrane current and its application to conduction in nerve. J. Phisiol. **117**, 500–544 (1952)
12. Hubbard, M.L., Henriquez, C.S.: A microstructural model of reentry arising from focal breakthrough at sites of source-load mismatch in a central region of slow conduction. Am. J. Physiol. Heart Circulatory Physiol. **306**(9), H1341–H1352 (2014)
13. Jensen, A.: Markoff chains as aid in the study of markoff processes. Skandinavisk Aktuarietidskrift **36**, 87–91 (1953)
14. Laurent, G., Moe, G., Hu, X., Leong-Poi, H., Connelly, K.A., So, P.P.S., Ramadeen, A., Doumanovskaia, L., Konig, A., Trogadis, J., Courtman, D., Strauss, B., Dorian, P.: Experimental studies of atrial fibrillation: a comparison of two pacing models. Am. J. Physiol. Heart Circulatory Physiol. **294**(3), H1206–H1215 (2008)
15. Gomes, M.J., Alvarenga, A., Campos, S.R., Rocha, B., da Silva, C.A., dos Santos, W.R.: Uniformization method for solving cardiac electrophysiology models based on the markov-chain formulation. IEEE Trans. Biomed. Eng. **62**(2), 600–608 (2015)
16. Platonov, P.G., Mitrofanova, L.B., Orshanskaya, V., Ho, S.Y.: Structural abnormalities in atrial walls are associated with presence and persistency of atrial fibrillation but not with age. J. Am. Coll. Cardiol. **58**(21), 2225–2232 (2011)

17. Reibman, A., Trivedi, K.: Numerical transient analysis of markov models. Comput. Oper. Res. **15**(1), 19–36 (1988)
18. Rush, S., Larsen, H.: A practical algorithm for solving dynamic membrane equations. IEEE Trans. Biomed. Eng. **4**, 389–392 (1978)
19. Rush, S., Larsen, H.: A practical algorithm for solving dynamic membrane equations. IEEE Trans. Biomed. Eng. **25**(4), 389–392 (1978)
20. dos Santos, R.W., Campos, F., Neumann, L., Nygren, A., Giles, W., Koch, H.: ATX-II effects on the apparent location of M cells in a computational model of a human left ventricular wedge. J. Cardiovasc. Electrophysiol. **17**, S86–S95 (2006)
21. Silver, M.A., Pick, R., Brilla, C.G., Jalil, J.E., Janicki, J.S., Weber, K.T.: Reactive and reparative fibrillar collagen remodelling in the hypertrophied rat left ventricle: Two experimental models of myocardial fibrosis. Cardiovasc. Res. **24**(9), 741–747 (1990)
22. Sundnes, J.: Computing the Electrical Activity in the Heart. Springer, Heidelberg (2006)
23. Sundnes, J., Artebrant, R., Skavhaug, O., Tveito, A.: A second-order algorithm for solving dynamic cell membrane equations. IEEE Trans. Biomed. Eng. **56**, 2546–2548 (2009)
24. Tobon, C., Ruiz-Villa, C.A., Heidenreich, E., Romero, L., Hornero, F., Saiz, J.: A three-dimensional human atrial model with fiber orientation. electrograms and arrhythmic activation patterns relationship. PLoS ONE **8**(2), e50883 (2013)
25. Voigt, N., Dobrev, D.: Cellular and molecular correlates of ectopic activity in patients with atrial fibrillation. Europace **14**(suppl 5), v97–v105 (2012)
26. Weiss, J.N., Karma, A., Shiferaw, Y., Chen, P.S., Garfinkel, A., Qu, Z.: From pulsus to pulseless the saga of cardiac alternans. Circ. Res. **98**(10), 1244–1253 (2006)

# Miyazawa-Jernigan Contact Potentials and Carter-Wolfenden Vapor-to-Cyclohexane and Water-to-Cyclohexane Scales as Parameters for Calculating Amino Acid Pair Distances

Nikola Štambuk[1(✉)], Paško Konjevoda[1], and Zoran Manojlović[2]

[1] Ruđer Bošković Institute, Zagreb, Croatia
{stambuk,pkonjev}@irb.hr
[2] Croatian Institute for Toxicology and Anti-Doping, Zagreb, Croatia
zoran.manojlovic@antidoping-hzta.hr

**Abstract.** The difference between amino acid chemical properties that correlate to the exchangeability of protein sequence residues is often analysed using approach proposed by Grantham (1974). His difference formula, i.e., matrix, for calculating the distances between amino acid pairs of the protein consists of three essential amino acid physicochemical properties – composition, polarity and volume, that are significantly correlated to the substitution frequencies of the protein residues. Miyata et al. (1979) re-evaluated this concept, and showed that the degree of amino acid difference is just as adequately explained by only two physicochemical factors, volume and polarity. Miyazawa-Jernigan relative partition/hydrophobic energies ($\varepsilon = \Delta e_{ir}$), and Carter-Wolfenden vapor-to-cyclohexane scale ($G_{v>c} = \Delta G_{v>c}$) are two alternative amino acid physicochemical parameters that are strongly correlated to their polarity and volume/mass, respectively. We show that the Miyazawa-Jernigan residue contact potential could be used instead of the Grantham polarity and composition parameters to derive an updated Miyata matrix. This substitution permits Miyata matrix correction for the amino acid parameters of: contact energies, repulsive packing energies, secondary structure energies, and Grantham's composition property. Distance values calculated between both (classic and updated) Miyata matrices exhibit a strong correlation of $r = 0.91$. The possibility of analyzing residue distances based on Carter-Wolfenden water-to-cyclohexane (w > c) and vapor-to-cyclohexane (v > c) scales instead of the amino acid polarity and volume parameters is also discussed, and a new distance matrix is derived.

**Keywords:** Contact · Potential · Amino acid · Distance matrix · Protein sequence

## 1 Introduction

In 1974 Richard Grantham developed a difference formula, i.e., Grantham's distance matrix, for calculating the differences between amino acid pairs of the protein sequences [1]. He showed that three amino acid physicochemical properties – composition, polarity

© Springer International Publishing Switzerland 2016
F. Ortuño and I. Rojas (Eds.): IWBBIO 2016, LNBI 9656, pp. 358–365, 2016.
DOI: 10.1007/978-3-319-31744-1_32

and volume, correlate significantly with the protein residue substitution frequencies [1]. Later, in 1979, Takashi Miyata, Sanzo Miyazawa and Teruo Yasunaga re-evaluated this concept [2]. They concluded that the degree of amino acid difference, with respect to the three-dimensional protein conformation and related weak residue interactions, is just as adequately explained by two physicochemical factors – volume and polarity, i.e., by Miyata's distance matrix [2]. In this study we investigate the classic Miyata matrix with respect to three alternative amino acid physicochemical properties that are strongly correlated with their polarity and volume: Miyazawa-Jernigan relative partition energies ($\varepsilon = \Delta e_{ir}$), Carter-Wolfenden water-to-cyclohexane scale ($G_{w>c} = \Delta G_{w>c}$) and vapor-to-cyclohexane scale ($G_{v>c} = \Delta G_{v>c}$) [2–6].

## 2  Results and Discussion

The Miyazawa-Jernigan residue contact potential was calculated using a large number of residue contacts on a real dataset of defined sequence-structure [3–5]. Relative partition energy values ($\Delta e_{ir}$) by Miyazawa-Jernigan correlate strongly with the Grantham amino acid polarity values (Fig. 1).

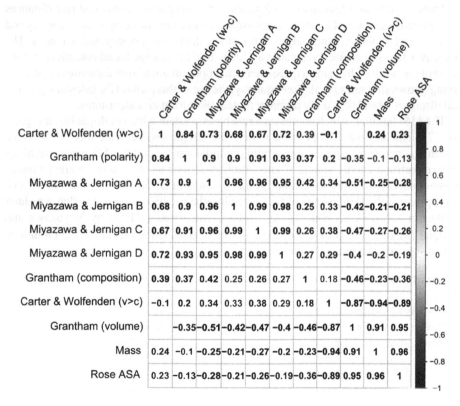

**Fig. 1.** Correlation matrix of different amino acid physicochemical properties

The 1999 update (Methods A-D) by Miyazawa and Jernigan of their original 1985 predictions of the equilibrium distributions of contacts in protein structures (with the Bethe approximation) shows four different characteristics [3, 4].

These updated Methods A-D are characterized by different relative partition energy values $\varepsilon = \Delta e_{ir}$ [4]. For Method A, interaction energies among residues consist of contact energies only, while Method B corresponds to the case in which interactions among residues consist of pairwise contact energies only [4]. In Method C repulsive packing energies are taken into account, and in Method D the total interaction energies consist of contact energies, repulsive packing energies, and secondary structure energies [4].

Depending on the characteristics of the equilibrium distributions of contacts in observed protein structures, the correlation between Miyazawa-Jernigan relative partition energies ($\varepsilon = \Delta e_{ir}$) and Grantham amino acid polarities ranges from $r = 0.90$ to $r = 0.93$ (Fig. 1).

For relative partition/hydrophobic energies ($\varepsilon = \Delta e_{ir}$), calculated by methods A and B, the correlation with Grantham's polarity is $r = 0.90$, and for Methods C and D $r$ values are 0.91 and 0.93, respectively (Fig. 1). This indicates that it is possible to use Miyazawa-Jernigan relative partition energy values instead of the Grantham polarity for deriving new versions of the amino acid substitution matrices.

There are two well-known distance matrices for comparing amino acid pair distances of the protein sequences: Grantham's consisting of three factors (composition, polarity and volume), and Miyata's, with only a two-factor calculation (polarity and volume). The Miyazawa-Jernigan contact potential is a widely used knowledge-based potential for globular proteins, derived by using a quasi-chemical approximation from databases of proteins having known structures [4, 5]. Consequently, the use of this method for calculating amino acid distances would help unify different protein bioinformatics algorithms.

The M5 classifier, which combines a rule based system with a multiple linear regression [7], indicates that it is possible to derive a simple and accurate link between the Grantham chemical factor of amino acid composition (c) and the Miyazawa-Jernigan relative partition energy values. Composition is defined as *"the atomic weight ratio of hetero (noncarbon) elements in end groups of rings to carbons in the side chain"* [1], and represents the third chemical factor of Grantham's distance matrix, that was later omitted in Miyata's version [2]. $\Delta e_{ir}$ values, calculated in 1999 by Miyazawa and Jernigan (Method A), correct the original 1985 values (Method O) for the Grantham chemical factor of amino acid composition (*c*):

```
Method A = -0.002 × composition + 0.1636 × Method O + 0.0018


=== Evaluation on training set ===
Correlation coefficient              0.9999
Mean absolute error                  0.0019
Root mean squared error              0.0021
Relative absolute error              1.2337 %
Root relative squared error          1.2268 %
```

Consequently, it is not surprising that classic Miyazawa-Jernigan residue contact potentials, i.e., relative partition energies $\varepsilon = \Delta e_{ir}$ [4], could be used instead of the Grantham polarity and composition parameters to derive an updated Miyata matrix.

Table 1 presents a newly-derived matrix based on a two-parameter amino acid distance model (Method D). The distance ($d_{ij}$) of amino acid pairs between residue $a_i$ and residue $a_j$ was calculated according to Miyata et al. [2]:

$$d_{ij} = \sqrt{\left(\Delta \varepsilon_{ij}/\sigma_\varepsilon\right)^2 + \left(\Delta v_{ij}/\sigma_v\right)^2} \tag{1}$$

where $\Delta \varepsilon_{ij}$ and $\Delta v_{ij}$ represent relative partition (hydrophobic) energy and the Grantham volume difference induced by amino acid substitution, $\sigma_\varepsilon$ and $\sigma_v$ are standard deviations of $\Delta \varepsilon_{ij}$ and $\Delta v_{ij}$, respectively.

The different distance values of the equilibrium distribution of contacts in protein structures produced according to the method used (Methods A-D) enable a Miyata matrix correction for the amino acid parameters of contact energies and Grantham's composition (Method A), the production of pairwise contact energies (Method B), repulsive packing energies (Method C), and secondary structure energies (Method D) (Table 1). The distances presented in Table 1 also reconstruct the columns of the standard genetic code table (Fig. 2; PAST software version 2.17c, Paired group algorithm, constrained, Euclidean similarity measure [8]).

**Table 1.** Amino acid pair distances calculated from Miyazawa-Jernigan relative partition/hydrophobic energies $\Delta e_{ir}$ (Method D) and Grantham's amino acid volumes

| L | I | M | V | S | P | T | A | Y | H | Q | N | K | D | E | C | W | R | G | |
|---|---|---|---|---|---|---|---|---|---|---|---|---|---|---|---|---|---|---|---|
| 0.60 | 0.66 | 0.89 | 1.48 | 3.81 | 3.80 | 2.98 | 3.38 | 0.59 | 2.06 | 3.08 | 3.29 | 3.66 | 3.73 | 3.42 | 2.18 | 1.13 | 2.38 | 4.07 | F |
| | 0.23 | 0.44 | 0.92 | 3.34 | 3.33 | 2.57 | 2.85 | 0.88 | 1.79 | 2.81 | 2.88 | 3.59 | 3.36 | 3.16 | 1.59 | 1.70 | 2.34 | 3.52 | L |
| | | 0.24 | 0.82 | 3.18 | 3.17 | 2.38 | 2.72 | 0.76 | 1.56 | 2.59 | 2.69 | 3.36 | 3.16 | 2.94 | 1.60 | 1.67 | 2.11 | 3.41 | I |
| | | | 0.61 | 2.93 | 2.92 | 2.14 | 2.48 | 0.89 | 1.35 | 2.38 | 2.45 | 3.20 | 2.92 | 2.73 | 1.47 | 1.84 | 1.98 | 3.18 | M |
| | | | | 2.46 | 2.45 | 1.74 | 1.93 | 1.47 | 1.26 | 2.20 | 2.07 | 3.22 | 2.57 | 2.54 | 0.97 | 2.45 | 2.12 | 2.60 | V |
| | | | | | 0.01 | 0.89 | 0.75 | 3.54 | 1.96 | 1.52 | 0.68 | 2.70 | 0.78 | 1.56 | 2.57 | 4.48 | 2.61 | 1.11 | S |
| | | | | | | 0.88 | 0.75 | 3.53 | 1.95 | 1.50 | 0.67 | 2.68 | 0.76 | 1.54 | 2.57 | 4.47 | 2.60 | 1.12 | P |
| | | | | | | | 0.94 | 2.67 | 1.07 | 0.89 | 0.32 | 2.19 | 0.83 | 1.12 | 2.15 | 3.60 | 1.79 | 1.69 | T |
| | | | | | | | | 3.21 | 1.84 | 1.82 | 0.99 | 3.10 | 1.38 | 1.98 | 1.86 | 4.19 | 2.70 | 0.79 | A |
| | | | | | | | | | 1.66 | 2.63 | 2.96 | 3.10 | 3.36 | 2.95 | 2.35 | 0.99 | 1.82 | 3.96 | Y |
| | | | | | | | | | | 1.03 | 1.33 | 1.96 | 1.70 | 1.38 | 2.09 | 2.55 | 0.98 | 2.63 | H |
| | | | | | | | | | | | 0.87 | 1.30 | 0.91 | 0.35 | 2.85 | 3.42 | 1.18 | 2.52 | Q |
| | | | | | | | | | | | | 2.13 | 0.52 | 0.99 | 2.43 | 3.87 | 1.93 | 1.65 | N |
| | | | | | | | | | | | | | 1.95 | 1.14 | 4.02 | 3.60 | 1.28 | 3.77 | K |
| | | | | | | | | | | | | | | 0.83 | 2.95 | 4.23 | 2.08 | 1.89 | D |
| | | | | | | | | | | | | | | | 3.16 | 3.71 | 1.38 | 2.63 | E |
| | | | | | | | | | | | | | | | | 3.27 | 3.03 | 2.28 | C |
| | | | | | | | | | | | | | | | | | 2.41 | 4.95 | W |
| | | | | | | | | | | | | | | | | | | 3.48 | R |
| | | | | | | | | | | | | | | | | | | | G |

A strong correlation is observed between distances calculated by Miyata et al. in 1979 [2], and the new distance values corrected for statistical averages of the numbers of contacts in proteins of defined structure (Methods A-D) [4]:

1. Miyata et al. (1979) – Method A, $r = 0.88$
2. Miyata et al. (1979) – Method B, $r = 0.89$
3. Miyata et al. (1979) – Method C, $r = 0.90$
4. Miyata et al. (1979) – Method D, $r = 0.91$

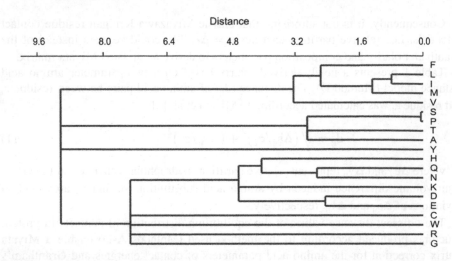

**Fig. 2.** The clustering of the residue distances calculated using Miyazawa-Jernigan relative partition/hydrophobic energies $\Delta e_{ir}$ (Method D) and Grantham's amino acid volumes reconstructs the columns of standard genetic code table

Recently, Charles Carter and Richard Wolfenden developed two new amino acid scales based on the measurement of water-to-cyclohexane (w > c) and vapor-to-cyclohexane (v > c) transfer equilibria and free energy determination [6]. The water-to-cyclohexane (w > c) amino acid scale is based on free energy values ($G_{w>c} = \Delta G_{w>c}$). It measures the "hydrophobic character" of the residue, i.e., amino acid's tendency to leave water and enter a nonpolar condensed phase [6]. Free energy values $G_{w>c}$ are strongly correlated with the Grantham polarity ($r = 0.84$, Fig. 1), and Kyte and Doolittle hydrophobicity scale ($r = -0.86$) [9, 10]. Significant correlation is also observed with Miyazawa-Jernigan relative partition (hydrophobic) energies, especially for Method A and Method D (Fig. 1).

The vapor-to-cyclohexane (v > c) scale reflects the so-called "lipophilic character" of amino acid, and describes "the transfer of an isolated solute molecule from the vapor phase to cyclohexane, at infinite dilution" [6]. Vapor-to-cyclohexane (v > c) transfer equilibria values measure "the van der Waals forces that attract a solute from the vapor phase to the walls of a nonpolar solvent cavity minus the cost of making that cavity" [6]. The free energy values $G_{v>c} = \Delta G_{v>c}$ of this scale are closely related to the amino acid's size or its accessible surface area (ASA), i.e., "the area over which the center of a water molecule can retain van der Waals contacts with the side-chain in a Gly-X-Gly tripeptide without penetrating other atoms" [6], [11–14]. The strong correlation values presented in Fig. 1 indicate that Carter and Wolfenden's vapor-to-cyclohexane amino acid free energy parameter ($G_{v>c}$) is a possible alternative to the factors of amino acid volume, molecular mass and solvent accessible surface area (ASA, standard state – $A^0$ [11]), in situations when amino acid distances are used.

The Carter and Wolfenden results redefine the parameters of the classic Miyata matrix and enable the use of two novel, amino acid, free energy factors based on "hydrophobic" and "lipophilic" residue characteristics. The first "hydrophobic character" corresponds to the Grantham's amino acid polarity and the relative partition (hydrophobic) energies according to Miyazawa and Jernigan (Fig. 1). The second "lipophilic character" of the residue is an alternative to the factors of Grantham's amino acid volume, molecular mass (weight) and solvent accessible surface area (Fig. 1). Table 2 presents a newly-derived two parameter amino acid distance matrix based on Carter and Wolfenden models.

**Table 2.** Amino acid pair distances calculated from Carter-Wolfenden water-to-cyclohexane (w > c) and vapor-to-cyclohexane (v > c) scales

| | L | I | M | V | S | P | T | A | Y | H | Q | N | K | D | E | C | W | R | G |
|---|---|---|---|---|---|---|---|---|---|---|---|---|---|---|---|---|---|---|---|
| F | 0.62 | 0.58 | 0.24 | 0.58 | 1.98 | 1.17 | 1.48 | 1.77 | 1.06 | 1.85 | 1.92 | 2.22 | 2.11 | 2.33 | 1.99 | 0.61 | 2.08 | 3.84 | 3.41 |
| L | | 0.07 | 0.70 | 0.05 | 2.16 | 0.69 | 1.66 | 1.46 | 1.69 | 2.41 | 2.36 | 2.58 | 2.48 | 2.74 | 2.46 | 0.77 | 2.60 | 4.34 | 3.12 |
| I | | | 0.68 | 0.06 | 2.18 | 0.76 | 1.69 | 1.52 | 1.65 | 2.39 | 2.35 | 2.58 | 2.49 | 2.74 | 2.45 | 0.78 | 2.53 | 4.33 | 3.18 |
| M | | | | 0.67 | 1.74 | 1.11 | 1.23 | 1.63 | 1.02 | 1.69 | 1.70 | 1.98 | 1.87 | 2.10 | 1.78 | 0.40 | 2.21 | 3.65 | 3.24 |
| V | | | | | 2.13 | 0.73 | 1.64 | 1.48 | 1.63 | 2.36 | 2.31 | 2.54 | 2.44 | 2.69 | 2.41 | 0.74 | 2.55 | 4.29 | 3.14 |
| S | | | | | | 1.87 | 0.50 | 1.49 | 2.03 | 1.66 | 1.05 | 0.80 | 0.82 | 1.07 | 1.23 | 1.42 | 3.56 | 2.78 | 2.26 |
| P | | | | | | | 1.45 | 0.78 | 2.14 | 2.62 | 2.38 | 2.47 | 2.41 | 2.68 | 2.53 | 0.88 | 3.26 | 4.39 | 2.42 |
| T | | | | | | | | 1.27 | 1.67 | 1.53 | 1.05 | 1.02 | 0.97 | 1.25 | 1.22 | 0.93 | 3.17 | 2.99 | 2.40 |
| A | | | | | | | | | 2.52 | 2.72 | 2.31 | 2.24 | 2.22 | 2.49 | 2.48 | 1.25 | 3.83 | 4.23 | 1.67 |
| Y | | | | | | | | | | 1.02 | 1.42 | 1.88 | 1.74 | 1.86 | 1.39 | 1.31 | 1.51 | 3.01 | 4.02 |
| H | | | | | | | | | | | 0.64 | 1.11 | 0.98 | 0.97 | 0.50 | 1.75 | 2.31 | 2.01 | 3.88 |
| Q | | | | | | | | | | | | 0.47 | 0.34 | 0.44 | 0.20 | 1.59 | 2.87 | 2.03 | 3.29 |
| N | | | | | | | | | | | | | 0.14 | 0.27 | 0.61 | 1.81 | 3.34 | 1.99 | 3.00 |
| K | | | | | | | | | | | | | | 0.26 | 0.48 | 1.72 | 3.19 | 2.00 | 3.07 |
| D | | | | | | | | | | | | | | | 0.51 | 1.99 | 3.30 | 1.73 | 3.26 |
| E | | | | | | | | | | | | | | | | 1.73 | 2.79 | 1.88 | 3.49 |
| C | | | | | | | | | | | | | | | | | 2.60 | 3.62 | 2.83 |
| W | | | | | | | | | | | | | | | | | | 3.99 | 5.43 |
| R | | | | | | | | | | | | | | | | | | | 4.82 |
| G | | | | | | | | | | | | | | | | | | | |

The distance ($d_{ij}$) of amino acid pairs between residue $a_i$ and residue $a_j$ was calculated according to Miyata et al. [2]:

$$d_{ij} = \sqrt{(\Delta G_{w>cij}/\sigma_{Gw>c})^2 + (\Delta G_{v>cij}/\sigma_{Gv>c})^2} \qquad (2)$$

$\Delta G_{w>cij}$ are amino acid free energies from water-to-cyclohexane (w > c), and $\Delta G_{v>cij}$ are amino acid free energies from vapor-to-cyclohexane (v > c). $\sigma_{Gw>c}$ and $\sigma_{Gv>c}$ are standard deviations of $\Delta G_{w>cij}$ and $\Delta G_{v>cij}$, respectively. The residue distances calculated according to Carter-Wolfenden and Miyata matrices exhibit significant correlation ($r = 0.71$). The Carter-Wolfenden distances of Table 2 also reconstruct the columns of the standard genetic code table (Fig. 3, $r = 0.84$ [8]), in line with Masami Hasegawa and Takashi Miyata's observations regarding the physicochemical problem of codon-anticodon interaction energy and Bhyravabhotla Jayaram's "rule of conjugates" [15, 16].

This result of the clustering of amino acids according to the genetic code table confirms Carter and Wolfenden's recent observation that tRNA acceptor stem and anticodon bases form independent codes [6]. "Hydrophobic" residue properties measured in water-to-cyclohexane (w > c) environments are related to the tRNA anticodon base information, while "lipophilic" residue properties measured using vapor-to-cyclohexane (v > c) environments are related to tRNA acceptor stem coding [6]. In this context the results we present in Table 2 and Fig. 3 confirm Carter and Wolfenden's assertions [6] *"that acceptor stems and anticodons, which are at opposite ends of the tRNA molecule,*

364     N. Štambuk et al.

*code, respectively, for size and polarity"* and *"that genetic coding of 3D protein structures evolved in distinct stages, based initially on the size of the amino acid and later
on its compatibility with globular folding in water"*. Consequently, amino acid distances
based on Carter and Wolfenden models could be used to analyse the influence of tRNA
coding on the residue polarity, size, and location in the proteins of different structural
and functional characteristics [6].

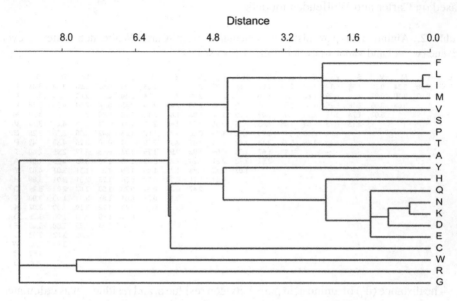

**Fig. 3.** The clustering of the residue distances calculated using Carter-Wolfenden water-to-
cyclohexane (w > c) and vapor-to-cyclohexane (v > c) scales reconstructs the columns of standard
genetic code table

The amino acid pair distance matrices presented could be of importance for the
analyses of amino acid sequences, including the exchangeability of protein sequence
residues during evolution. Substitution matrices, like PAM and BLOSUM, show the
likelihood of substitution in an evolutionary process for each pair of amino acids [17].
However, a researcher must select an appropriate matrix from the series of matrices,
considering evolutionary distance between sequences in analysis [17].

The novel matrix presented in this paper is based on distances between physicochemical properties of amino acids, and does not include evolutionary distance or the
parsimony principle [18]. This approach could be used as a supplement to PAM and
BLOSUM matrices in an analysis of closely-related protein sequences. It could also be
used in QSAR studies to quantify the substitution of one or more amino acid within a
peptide chain [19].

**Acknowledgments.** The support of the Croatian Institute for Toxicology and Anti-Doping, and
the Croatian Ministry of Science, Education and Sports is gratefully acknowledged (grant No.
098-0982929-2524).

# References

1. Grantham, R.: Amino acid difference formula to help explain protein evolution. Science **185**, 862–864 (1974)
2. Miyata, T., Miyazawa, S., Yasunaga, T.: Two types of amino acid substitutions in protein evolution. J. Mol. Evol. **12**, 219–236 (1979)
3. Miyazawa, S., Jernigan, R.L.: Estimation of effective interresidue contact energies from protein crystal structures: quasi-chemical approximation. Macromolecules **18**, 534–552 (1985)
4. Miyazawa, S., Jernigan, R.L.: Self-consistent estimation of inter-residue protein contact energies based on an equilibrium mixture approximation of residues. Proteins **34**, 49–68 (1999)
5. Leelananda, S.P., Feng, Y., Gniewek, P., Kloczkowski, A., Jernigan, R.L.: Statistical contact potentials in protein coarse-grained modeling: from pair to multi-body Potentials. In: Kolinski, A. (ed.) Multiscale Approaches to Protein Modeling, pp. 127–157. Springer, New York (2011)
6. Carter, C.W., Wolfenden, R.: tRNA acceptor stem and anticodon bases form independent codes related to protein folding. Proc. Natl. Acad. Sci. U.S.A. **112**, 7484–7488 (2015)
7. Witten, I.H., Frank, E., Hall, M.A.: Data Mining: Practical Machine Learning Tools and Techniques. Morgan Kaufmann, Amsterdam (2011)
8. Hammer, O., Harper, D.A.T., Ryan, P.D.: PAST: Paleontological Statistics Software Package for Education and Data Analysis. Palaeontologia Electronica. 49, 9 (2001), http://palaeo-electronica.org/2001_1/past/issue1_01.htm
9. Štambuk, N., Konjevoda, P.: Prediction of secondary protein structure with binary coding patterns of amino acid and nucleotide physicochemical properties. Int. J. Quant. Chem. **92**, 123–134 (2003)
10. Kyte, J., Doolittle, R.F.: A simple method for displaying the hydropathic character of a protein. J. Mol. Biol. **157**, 105–132 (1982)
11. Rose, G.D., Geselowitz, A.R., Lesser, G.J., Lee, R.H., Zehfus, M.H.: Hydrophobicity of amino acid residues in globular proteins. Science **229**, 834–838 (1985)
12. Radzicka, A., Wolfenden, R.: Comparing the polarities of the amino acids: side-chain distribution coefficients between the vapor phase, cyclohexane, 1-octanol, and neutral aqueous solution. Biochemistry **27**, 1664–1670 (1988)
13. Lee, B., Richards, F.M.: The interpretation of protein structures: estimation of static accessibility. J. Mol. Biol. **55**, 379–400 (1971)
14. Chothia, C.: The nature of the accessible and buried surfaces in proteins. J. Mol. Biol. **105**, 1–12 (1976)
15. Hasegawa, M., Miyata, T.: On the antisymmetry of the amino acid code table. Orig. Life. **10**, 265–270 (1980)
16. Jayaram, B.: Beyond the wobble: the rule of conjugates. J. Mol. Evol. **45**, 704–705 (1997)
17. Hodgman, T.C., French, A., Westhead. D.R.: Bioinformatics. Taylor & Francis, Abingdon (2009)
18. Pevsner, J.: Bioinformatics and Functional Genomics. Wiley Blackwell, Chichester (2015)
19. Green, D.V.S., Segall, M.: Chemoinformatics in lead optimization. In: Bajorath, J. (ed.) Chemoinformatics for Drug Discovery, pp. 149–168. Wiley, Hoboken (2014)

# eHealth

# Inter-observer Reliability and Agreement Study on Early Diagnosis of Diabetic Retinopathy and Diabetic Macular Edema Risk

Manuel Emilio Gegundez-Arias[1], Carlos Ortega[2], Javier Garrido[2],
Beatriz Ponte[3], Fatima Alvarez[4], and Diego Marin[5(✉)]

[1] Department of Mathematics, University of Huelva, Huelva, Spain
gegundez@dmat.uhu.es
[2] North Area of Sanitary Management of Cordoba,
Andalusian Health Service, Seville, Spain
comillan@gmail.com,
javier.garrido.sspa@juntadeandalucia.es
[3] Virgen de Macarena University Hospital of Seville,
Andalusian Health Service, Seville, Spain
bepontezu@hotmail.com
[4] Juan Ramon Jimenez Hospital of Huelva,
Andalusian Health Service, Seville, Spain
fatimaalvarezgil@yahoo.es
[5] Department of Electronic, Computer Science and Automatic Engineering,
University of Huelva, Huelva, Spain
diego.marin@diesia.uhu.es

**Abstract.** The degree of inter-observer agreement on early diagnosis of diabetic retinopathy (DR) and diabetic macular edema (DME) risk has been assessed in this paper. Three sets of DR and DME risk ratings on 529 diabetic patients were independently built by ophthalmologists of the Andalusian (Spain) Health Service through observation of two macula-centered retinographies from these patients (one image per eye, 1058 images). DR was graded on a 0–3 scale from DR-unrelated to severe DR, while DME risk was graded on a 0–2 scale from no risk to moderate-severe risk. Inter-rater reliability (IRR) assessment was performed by the intra-class correlation (ICC) and two kappa-like statistical variants —Light's kappa and Fleiss' kappa. ICC-computed IRR showed excellent agreement between our three coders: values were 0.844 (95 % CI, 0.822–0.865) and 0.833 (95 % CI, 0.805–0.853) for DR and DME ratings, respectively. Kappa index-quantified assessment resulted in substantial agreement, as both kappa indexes rendered values around 0.60 for DR and 0.75 for DME ratings. All computed IRR metrics proved high inter-observer agreement and consistency among DR degree and DME risk diagnoses. Reliable diagnosis provided by human experts supports the generation of reference standards that can be used in the development of automatic DR diagnosis systems.

**Keywords:** Automated diagnoses · Diabetic retinopathy · Gold standard diagnoses · Inter-observer agreement · Inter-rater reliability · Macular edema

© Springer International Publishing Switzerland 2016
F. Ortuño and I. Rojas (Eds.): IWBBIO 2016, LNBI 9656, pp. 369–379, 2016.
DOI: 10.1007/978-3-319-31744-1_33

# 1   Introduction

Diabetes mellitus (DM) is a metabolic disorder characterized by the presence of hyperglycemia due to defective insulin secretion, defective insulin action or both. DM is a serious and increasing global health burden and it is estimated that 382 million people suffered from diabetes in 2013, and this figure is expected to rise to 592 million by 2035 [1].

Diabetic retinopathy (DR) is the most common microvascular complication in both type 1 and type 2 DM and a leading cause of new cases of legal blindness in working-age population around the world [2]. DR is clinically defined, diagnosed and treated exclusively according to the extent of retinal vascular disease. Three distinct forms of DR are described: (1) macular edema, which includes diffuse or focal vascular leakage at the macula; (2) progressive accumulation of blood vessel change that includes microaneurysms, intraretinal hemorrhages, vascular tortuosity and vascular malformation (all them being known as nonproliferative diabetic retinopathy) that ultimately leads to abnormal vessel growth (proliferative diabetic retinopathy); and (3) retinal capillary closure, a form of vascular change detected on fluorescein angiography that is also well recognized as a potentially blinding complication in diabetes [3].

DR remains an important cause of blindness in Spain. Blindness prevalence among diabetic persons ranges from 4 to 11 %, thus being higher than in other countries (between 1 and 5 %). Overall DR prevalence in Spain varies in the different studies, but it is approximately 40 %, while proliferative DR prevalence is between 4 and 6 %, and macular edema prevalence is between 1.4 and 7.9 % [4]. Although there is some evidence that the incidence of DR is decreasing in Europe and the USA, only a few studies report the same trend in Spain [5]. On the other hand, DR severity is well known to have a close association with DM duration. However, most diabetic patients have no symptoms in early retinopathy stages, when treatments to prevent vision loss (laser therapy and/or vitrectomy and/or intraocular pharmacological intervention) are effective. For all these reasons, a systematic screening program for DR in the diabetes community is needed for early detection, and to reduce blindness in diabetic patients.

DR screening with a non-mydriatic camera has been supported by several guidelines, since this procedure renders the best cost-effectiveness ratio and is easy to use. Screening by retinal photography is not a substitute for a comprehensive ophthalmic examination, but there is evidence that it can serve as a screening tool for DR to identify patients with retinopathy for referral for ophthalmic evaluation and management [6, 7]. Telehealth programs based on digital retinal pictures have the potential to allow increased DR diagnoses, improving access to diagnosis, and thus resulting in timely treatment and preservation of vision.

The high prevalence of diabetes may lead the number of diabetic patients whose screening and monitoring should be continued to exceed the capabilities of health institutions. If the number of people with diabetes is expected to rise up to 592 million in 2035, one single annual evaluation of every patient will mean that 3.2 million eyes are to be examined every day of the year. In this framework, the employment of digital images for eye-disease diagnosis could be exploited for computerized early detection of DR. The benefits of a system for automatic detection of early signs of this disease (i.e., microaneurysms, hemorrhages or exudates) by means of digital retinal image analysis have been widely studied and positively assessed by experts [8, 9]. This system could be used by non-experts to filtrate cases of patients not affected by the disease. Thus, the specialists' workload would be reduced and the effectiveness of preventive protocols and early therapeutic treatments would be increased. Furthermore, it would also result in economic benefits for public health systems, since cost-effective treatments associated to early illness detection have been observed to lead to remarkable cost savings [10, 11].

In recent years, within this context of designing DR diagnosis systems, much effort has been spent on developing algorithms for automatic segmentation of the main anatomical retina components and early DR-related lesions (i.e., [12–16]). Some prototypes of such systems have been already presented [17–20], but telemedicine is complex, requiring the services of expert teams working collaboratively to achieve quality clinical outcomes. System testing and validation are usually carried out by measuring sensitivity and specificity of automated diagnoses relative to the corresponding gold standard diagnoses provided by ophthalmologic specialists. These sets of ground-truth references may be directly built by a single ophthalmologist or —with the purpose of increasing reliability— be produced from several sets of experts' observations. In this framework, inter-expert agreement studies are required to prove consistency between observational diagnoses provided by multiple human experts. In this paper, a validation study of experts at reading fundus to generate reference standards in DR diagnosis —based on assessing inter-rater reliability and agreement— is carried out.

This work is also specifically aimed at contributing to the development of the system for automated detection of DR that is being implemented by the Health Ministry of the Andalusian Regional Government (Andalusia, Spain) with the purpose of enhancing the effectiveness of its screening program for DR in the next years. The DR diagnoses generated by three independent medical experts showed high reliability, which will allow building a reference standard that will be used to carry out initial system testing.

## 2 Materials and Methods

This paper presents a study of interexpert agreement of early diabetic retinopathy (DR) diagnosis and diabetic macular edema (DME) risk. The following design considerations were followed:

### 2.1 Coders

Ophthalmologic specialists of the Andalusian Health Service (Spain) from the North Area of Sanitary Management of Cordoba (NASMC), Virgen de Macarena Universitary Hospital of Seville (VMUHS) and Juan Ramon Jimenez Hospital of Huelva (JRJHH), have taken part assigning ratings to DR diagnosis and DME risk[1].

---

[1] The collaboration of medical experts was formally developed through the Project "Expert System for Early Automated Detection of Diabetic Retinopathy by Analysis of Digital Retinal Images", supported and funded by the Health Ministry of the Andalusian Regional Government (Spain).

## 2.2   Subjects and Images

The subjects under consideration were 529 patients with diabetes at risk for DR. Our image collection consists of 529 image pairs of two macula-centered retinographies of the posterior pole acquired from these patients (one image per eye: 1058 images).

These fundus images are publicly available for noncommercial use through the so-called MESSIDOR program [21]. The MESSIDOR dataset contains 1200 fundus images acquired from diabetic patients by the Paris Hôpital Lariboisière, the Faculté de Médecine St. Etienne, and the LaTIM at Brest (France). Some of these images come in pairs (one image per eye from a patient), some others are single. All available image pairs (i.e., 529 examinations:1058 images) were used in the present study.

The whole set of images were acquired using a Topcon TRC NW6 non-mydriatic retinograph with 45° field-of-view (two-thirds of them with pharmacological pupil dilation and the remaining with no dilation). The images were digitalized to 1440 × 960, 2240 × 1488 or 2304 × 1536 pixel images, corresponding to approximately 910, 1380 and 1455 pixel retina diameters, respectively, being 8 bits per color plane. These images are provided in TIFF format. To ensure utmost protection of patient privacy, information that might allow identifying a patient to be reconstructed was removed, and there is no actual knowledge that the images could be used alone or in combination to identify any subject.

## 2.3   Clinical Diagnosis Generation

Each of the images contained in our image collection was used for making clinical diagnoses based on specifying a DR grade and DME risk per image. For this purpose, the simplified classification described in Table 1 was attended.

This classification has been designed within the framework of the Messidor/Techno-Vision Project [22] based on the recommendations provided by: ALFEDIAM (French association for the study of diabetes and metabolic diseases) [23, 24], ETDRS (Early Treatment Diabetic Retinopathy Study) [25], and Diabetic Retinopathy Screening Services in Scotland [26].

With the help of software, clinicians from VMUHS, NASMC and JRJHH manually and independently marked the red (microaneurysms and hemorrhages) and bright (hard exudates) lesions present in each image, which are the basis of the selected diagnosis criteria. These marks allowed automatic building of three sets of DR and DME risk ratings for each image, each set corresponding to each participant medical center, following strictly the classification criteria described in Table 1.

The final DR and DME risk grading of each patient was established from the highest rate assigned to both available fundus images of the patient under consideration. The distribution of the study cases resulting from the diagnoses —grouped by the different rating categories of DR grade and DME risk— is shown in Table 2.

**Table 1.**  Diabetic retinopathy (DR) and diabetic macular edema (DME) severity scale applied to build disease clinical diagnoses.

| Nonproliferative DR grade | |
|---|---|
| 0 (No apparent retinopathy) | ($\mu A^* = 0$) and ($H^* = 0$) |
| 1 (Mild) | ($0 < \mu A \leq 5$) and ($H = 0$) |
| 2 (Moderate) | (($5 < \mu A < 15$) or ($0 < H < 5$)) |
| 3 (Severe) | ($\mu A \geq 15$) or ($H \geq 5$) |

*\* $\mu A$ and $H$ denotes the number of microaneurysms and hemorrhages, respectively.*

| DME risk | |
|---|---|
| 0 (Apparently absent: no risk) | No visible hard exudates |
| 1 (Apparently present: mild) | Visible hard exudates: shortest distance between macula and hard exudates is higher than one papilla. diameter |
| 2 (Apparently present: moderate-severe) | Visible hard exudates: shortest distance between macula and hard exudates is not higher than one papilla diameter |

**Table 2.**  Number of diabetic patients diagnosed in the different rating categories —DR grade and DME risk— by each of the participating medical centers.

DR rating category

| Medical Center | 0 | 1 | 2 | 3 |
|---|---|---|---|---|
| JRJHH | 198 | 174 | 146 | 11 |
| VMUHS | 276 | 54 | 143 | 56 |
| NASMC | 281 | 91 | 128 | 29 |

DME rating category

| Medical Center | 0 | 1 | 2 |
|---|---|---|---|
| JRJHH | 372 | 54 | 103 |
| VMUHS | 393 | 36 | 100 |
| NASMC | 411 | 35 | 83 |

## 2.4    Reliability and Agreement Metrics

In this study, the assessment of inter-rater reliability (IRR) is mainly performed by the intra-class correlation (ICC) [27]. ICC is one of the most commonly-used statistics for IRR assessment by ordinal variables, being also suitable when —as in our study case— all subjects are rated by multiple coders or by fully-crossed designs.

In addition, in order to complement agreement conclusions provided by ICC results, we have also assessed IRR with kappa statistics. Since the mathematical foundations of kappa provided by Cohen [28] are only suitable for two coders, two kappa-like statistical variants were computed for nominal data designs with three or more coders: Light's kappa [29] and Fleiss' kappa [30]. While the latter formalizes IRR statistic as an extension of Scott's Pi statistic [31], the former assesses IRR by computing kappa for all coder pairs and uses the arithmetic mean of these estimates to provide an overall index of agreement.

All these mentioned IRR metrics were computed using the R irr package (see the irr reference manual for further information [32]).

## 3 Inter-Rater Reliability Results

This section presents the IRR results evaluated through ICC, Light's kappa and Fleiss' kappa for our study case: three coders (VMUHS, NASMC and JRJHH Medical Centers) who independently rated DR grading and DME risk on 529 diabetic patients. DR was graded into four stages on a 0-3 scale (from DR-unrelated to severe DR retinas), while DME risk was evaluated into three grades on a 0-2 scale (from no risk to moderate-severe risk).

### 3.1 ICC Results

The results of the ICC-rendered IRR assessment corresponding to DR-grading and DME-risk agreements are displayed in Table 3. The F-statistic significance tests, p-value and confidence intervals, CI, for 95 % confidence are also listed.

ICC was computed using a two-way mixed model, agreement type, to assess the degree of agreement between the diagnoses by the 3 coders across the 529 subjects. Attending to the obtained p and F values, interrater variations of both DR-grading and ME-risk diagnosis are not significant.

Regarding ICC values, according to the cutoffs for qualitative rating of agreement provided in [33] (see first row in Table 4), ICC-computed IRR shows excellent agreement between the three medical centers in the diagnosis of both DR grading and DME risk.

**Table 3.** ICC-quantified IRR assessment. Study case: three coders (VMUHS, NASMC and JRJHH) rating DR grading on a 0–3 scale and DME risk on a 0–2 scale on 529

| DR | | | |
|---|---|---|---|
| ICC | F-test | p-value | 95 % CI |
| 0.844 | F(528,654) = 17.8 | 1.01e–221 | (0.820, 0.865) |
| DME | | | |
| ICC | F-test | p-value | 95 % CI |
| 0.833 | F(528,768) = 16.1 | 9.23e–238 | (0.805, 0.853) |

**Table 4.** ICC and Kappa Qualitative ratings of agreements (proposed by Cicchetti in 1994 [33] for ICC and by Landis and Koch in 1977 [34] for kappa values).

| Qualitative ratings | | | | | |
|---|---|---|---|---|---|
| Intervals | [0.00, 0.20] | [0.20, 0.40] | [0.40, 0.60] | [0.60, 0.80] | [0.80, 1.00] |
| ICC | poor | poor | fair | good | excellent |
| Kappa | slight | fair | moderate | substantial | almost perfect |

## 3.2 Light's Kappa and Fleiss' Kappa Results

The results of kappa-like statistical variant-estimated IRR assessment are provided in Tables 5 (Light's kappa) and 6 (Fleiss' kappa). As previously, both tables show DR-grading and DME-risk agreements.

Table 5 also includes kappa, z-statistic tests and p-values for all coder pairs. On the other hand, apart from overall results, Table 6 lists kappa values for the different rating categories ($k_j$, kappa for the j-th category) and their corresponding z statistics and p-values.

According to the guidelines provided in [34] for interpreting kappa values (see second row in Table 4), the kappa results in Tables 5 and 6 indicate substantial agreement between the 3 coders in both DR-grading and DME-risk diagnoses. Anyway, this same qualitative agreement is observed to be reached with a higher kappa value in the case of DME-risk than in the case of DR-grading. This difference in kappa magnitudes can be explained by the fact that it is more difficult to detect red lesions than bright ones, mainly due to the small size of microaneurysms.

**Table 5.** IRR assessment quantified by Light's kappa. Study case: three coders, JRJHH, VMUHS and NASMC (1, 2 and 3 respectively) rating DR grading on a 0–3 scale (top panel) and DME risk on 0–2 a scale (bottom panel) on 529 diabetic patients.

| Light's kappa results - DR | | |
|---|---|---|
| Raters | kappa | z-test | p-value |
| 1,2 | 0.502 | 17.4 | 0* |
| 1,3 | 0.617 | 20.5 | 0* |
| 2,3 | 0.714 | 21.6 | 0* |
| 1,2,3 | **0.611** | 40.8 | 0* |
| Light's kappa results - DME | | |
| Raters | kappa | z-test | p-value |
| 1,2 | 0.739 | 14.9 | 0* |
| 1,3 | 0.732 | 14.2 | 0* |
| 2,3 | 0.801 | 14.7 | 0* |
| 1,2,3 | **0.757** | 8.99 | 0* |

*\* The p-value calculated is lower than the smallest number possible in R.*

**Table 6.** IRR assessment quantified by Fleiss' kappa. Study case: three coders (VMUHS, NASMC and JRJHH) rating DR grading on a 0–3 scale (top panel) and DME risk on a 0–2 scale (bottom panel) on 529 diabetic patients.

Fleiss' kappa results - DR

|  | Category 0 | Category 1 | Category 2 | Category 3 |
|---|---|---|---|---|
| kj | 0.7347 | 0.3997 | 0.6357 | 0.4900 |
| kj standard error | 0.0251 | 0.0251 | 0.0251 | 0.0251 |
| kj z-test | 29.2694 | 15.9235 | 25.3240 | 19.5195 |
| p-value | 0[a] | 0[a] | 0[a] | 0[a] |

| overall kappa | z-test | Standard error | p-value | C.I. (95 %) |
|---|---|---|---|---|
| 0.6032 | 36.7705 | 0.0164 | 0[a] | (0.5948, 0.6115) |

Fleiss' kappa results - DME

|  | Category 0 | Category 1 | Category 2 |
|---|---|---|---|
| kj | 0.7866 | 0.5484 | 0.8166 |
| kj standard error | 0.0251 | 0.0251 | 0.0251 |
| kj z-test | 31.3350 | 21.8480 | 32.5310 |
| p-value | 0[a] | 0[a] | 0[a] |

| Overall kappa | z-test | Standard error | p-value | C.I. (95 %) |
|---|---|---|---|---|
| 0.7554 | 37.9574 | 0.0199 | 0[a] | (0.7453, 0.7656) |

[a]The p-value calculated is lower than the smallest number possible in R

It is important to point out that these conclusions are drawn with both formulations of kappa (Light's kappa and Fleiss' kappa). Besides, they confirm the high level of inter-rater reliability obtained for DR grading and DME risk diagnoses from ICC results.

# 4   Conclusions

Automated diabetic retinopathy (DR) screening or pre-screening carried out by systems focused on the detection of the earliest ophthalmic signs of illness can have a major impact in the near future. Since the number of potential patients is very high, development of automatic DR diagnosis systems based on retinal image computer analysis may provide remarkably quicker screening programs for early detection of these disorders. The development, improvement and validation of such systems require reliable diagnosis provided by human experts that can be used as a reference standard. With the purpose of increasing reliability, these ground truths may be produced from several sets of experts' observations. In this framework, the assessment of interrater reliability for observational fundus images is needed to prove consistency among observational ratings.

This work is aimed at quantifying the degree of agreement between ophthalmologic specialists from the North Area of Sanitary Management of Cordoba, Virgen de Macarena Universitary Hospital of Seville and Juan Ramon Jimenez Hospital of Huelva (Andalusia, Spain). These three coders independently rated DR grading and diabetic macular edema (DME) risk on 529 image pairs of macula-centered retinographies of the posterior pole taken from diabetic patients (one image per eye: 1058 images). DR was graded into four stages on a 0–3 scale (from DR-unrelated to severe DR retinas), while DME risk was evaluated into three grades on a 0–2 scale (from no risk to moderate-severe risk). On the other hand, it should be pointed out that this collection of images is publicly available through the so-called MESSIDOR program.

Thus, three sets of DR and DME risk ratings on 529 patients —each set corresponding to each participant medical center— were built and constitute the input data for computing inter-rater reliability (IRR).

In this paper, IRR assessment was mainly performed by the intra-class correlation (ICC), which is commonly used for ordinal variables and ratings provided by multiple coders. ICC confirmed excellent agreement between the three medical centers in the diagnosis of both DR grading and DME risk. ICC values were 0.844 (95 % CI, 0.822–0.865) and 0.833 (95 % CI, 0.805-0.853) for DR and DME-risk ratings, respectively, thus reaching the maximum level of qualitative ratings of agreements.

This high reliability was supported by the results rendered by two kappa-like statistical variants for nominal data and multiple coder designs, Light's kappa and Fleiss' kappa. IRR assessment quantified by these kappa indexes resulted in substantial agreement between the 3 coders in both DR-grading and DME-risk diagnoses (both kappa indexes showing values around 0.60 and 0.75 for DR and DME ratings, respectively).

It is worth mentioning the results obtained by Fleiss' kappa for the rating category "zero" (i.e., kappa statistic used to measure the observed level of agreement between our three coders when only two categorical ratings are considered: No Disease vs. Present Disease). In this case, kappa agreement values increase for both DR and DME diagnoses in comparison to overall kappa results considering all different rating categories. This increase is especially significant for the sets of experts' DR diagnoses (from 0.603 to 0.735). It should be taken into account that computer programs for DR and DME early diagnosis will have to determine —through fundus images— whether a patient presents disease signs with very high sensitivity. This can be done by evaluating their automated "disease/no disease" outputs relative to specialist-provided reference standards that can be built in the same way (disease /no present disease). On the other hand, for a system to be safely implemented into DR screening design, failures in automated pathological diagnoses must be guaranteed to reach no clinical importance. For this reason, reference standard DR diagnoses must be detailed in the different ratings.

Finally, it should be pointed out that this work specifically contributes to the development of the system for automated DR detection implemented by the Health Ministry of the Andalusian Regional Government (Andalusia, Spain) with the purpose of enhancing effectiveness in its DR screening program. Since the sets of DR ratings generated by our three independent medical experts have shown high reliability, a reference standard can be built to carry out initial system testing on the patients considered in this work.

**Acknowledgments.** The authors would like to thank the Messidor program partners for facilitating their database.

This work was carried out as part of the Project "Expert System for Early Automated Detection of Diabetic Retinopathy by Analysis of Digital Retinal Images", supported and funded by the Health Ministry of the Andalusian Regional Government (Spain).

# References

1. Guariguata, L., Whiting, D.R., Hambleton, I., Beagley, J., Linnenkamp, U., Shaw, J.E.: Global estimates of diabetes prevalence for 2013 and projections for 2035. Diabetes Res. Clin. Pract. **103**(2), 137–149 (2014)
2. Klein, B.E.K.: Overview of epidemiologic studies of diabetic retinopathy. Ophthalmic Epidemiol. **14**(4), 179–183 (2007)
3. Boyd, S., Advani, A., Altomare, F., Stockl, F.: Clinical practice guidelines for the prevention and management of diabetes in Canada: Retinopathy. Can. J. Diab. **37**(Suppl. 1), S137–S141 (2013)
4. Vila, L., Viguera, J., Aleman, R.: Diabetic retinopathy and blindness in Spain: epidemiology and prevention. Endocrinol. Nutr. **55**(10), 459–475 (2008)
5. Gibelalde, A., et al.: Prevalence of diabetic retinopathy using non-mydriaticretinography. An. SistSanit. Navar. **33**(3), 271–276 (2010)
6. Sender, M.J., Bagur, S.M., Badia, X., Maseras, M., de la Puente, M.L., Foz, M.: Cámara de retina no midríatica: estudio de coste-efectividad en la detección temprana de la retinopatía diabética. Med. Clín. **121**(12), 446–452 (2003)
7. Aptel, F., Denis, P., Rouberol, F., Thivolet, C.: Screening of diabetic retinopathy: effect of field number and mydriasis on sensitivity and specificity of digital fundus photography. Diab. Metab. **34**(3), 290–293 (2008)
8. Patton, N., Aslam, T.M., MacGillivray, T., Deary, I.J., Dhillon, B., Eikelboom, R.H., Yogesan, K., Constable, I.J.: Retinal image analysis: conc epts, applications and potential. Prog. Retin. Eye Res. **25**, 99–127 (2006)
9. Singalavanija, A., Supokavej, J., Bamroongsuk, P., Sinthanayothin, C., Phoojaruenchanachai, S., Kongbunkiat, V.: Feasibility study on computer- aided screening for diabetic retinopathy. Jpn. J. Ophthalmol. **50**, 361–366 (2006)
10. American Academy of Ophthalmology. Diabetic retinopathy. Preferred practice pattern guidelines (2008). http://www.aao.org/ppp
11. Lairson, D.R., Pugh, J.A., Kapadia, A.S., Lorimor, R.J., Jacobson, J., Velez, R.: Cost effectiveness of alternative methods for diabetic retinopathy screening. Diab. Care **15**, 1369–1377 (1992)
12. Aquino, A., Gegúndez-Arias, M.E., Marín, D.: Detecting the optic disc boundary in digital fundus images using morphological, edge detection, and feature extraction techniques. IEEE Trans. Med. Imag. **29**(11), 1860–1869 (2010)
13. Marín, D., Aquino, A., Gegúndez-Arias, M.E., Bravo, J.M.: A new supervised method for blood vessel segmentation in retinal images by using gray-level and moment invariants-based features. IEEE Trans. Med. Imag. **30**(1), 146–158 (2011)
14. Gegúndez-Arias, M.E., Marin, D., Bravo, J.M., Suero, A.: Locating the fovea center position in digital fundus images using thresholding and feature extraction techniques. Comput. Med. Imaging Graph. **37**, 386–393 (2013)
15. Akram, M.U., Tariq, A., Anjum, M.A., Javed, M.Y.: Automated detection of exudates in colored retinal images for diagnosis of diabetic retinopathy. Appl. Opt. **51**(20), 4858–4866 (2010)
16. Quellec, G., Lamard, M., Josselin, P.M., Cazuguel, G., Cochener, B., Roux, C.: Optimal wavelet transform for the detection of microaneurysms in retina photographs. IEEE Trans. Med. Imag. **27**(9), 1230–1241 (2008)
17. Acharya, U.R., Lim, C.M., Ng, E.Y.K., Chee, C., Tamura, T.: Computer-based detection of diabetes retinopathy stages using digital fundus images. Proc. Inst. Mech. Eng. H **223**(5), 545–553 (2009)

18. Niemeijer, M., Abràmoff, M.D., van Ginneken, B.: Information fusion for diabetic retinopathy CAD in digital color fundus photographs. IEEE Trans. Med. Imag. **28**(5), 775–785 (2009)
19. Philip, S., Fleming, A.D., Goatman, K.A., Fonseca, S., Mcnamee, P., Scotland, G.S., Prescott, G.J., Sharp, P.F., Olson, J.A.: The efficacy of automated "disease/no disease" grading for diabetic retinopathy in a systematic screening programme. Br. J. Ophthalmol. **91**, 1512–1517 (2007)
20. Abràmoff, M.D., Folk, J.C., Han, D.P., Walker, J.D., Williams, D.F., Russell, S.R., Massin, P., Cochener, B., Gain, P., Tang, L., Lamard, M., Moga, D.C., Quellec, G., Niemeijer, M.: Automated analysis of retinal images for detection of referable diabetic retinopathy. JAMA Ophthalmol. **131**(3), 351–357 (2013)
21. MESSIDOR TECHNO-VISION Project, France, MESSIDOR: Digital Retinal Images (Download images section). http://messidor.crihan.fr/download-en.php
22. MESSIDOR TECHNO-VISION Project, France, Methods to evaluate segmentation and indexing techniques in the field of retinal ophthalmology. http://messidor.crihan.fr/index-en.php
23. Massin, P., Angioi-Duprez, K., Bacin, F., Cathelineau, B., Cathelineau, G., Chaine, G., Coscas, G., Flament, J., Sahel, J., Turut, P., Guillausseau, P.J., Gaudric, A.: Recommandations de l'ALFEDIAM pour le d'epistage, et la surveillance de la r'etinopathiediab'etique. Diab. Metab. **22**, 203–209 (1996)
24. Massin, P., Angioi-Duprez, K., Bacin, F., Cathelineau, B., Cathelineau, G., Chaine, G., Coscas, G., Flament, J., Sahel, J., Turut, P., Guillausseau, P.J., Gaudric, A.: Recommandations de lALFEDIAMpour le d'epistage et la surveillance de la r'etinopathiediab'etique. J. Fr. Ophtalmol. **20**, 302–310 (1997)
25. Early Treatment Diabetic Retinopathy Study Research Group: Grading diabetic retinopathy from stereoscopic color fundus photographs an extension of the Modified Airlie House classification: ETDRS report number 10". Ophthalmol. **98**, 786–806 (1991)
26. Diabetic Retinopathy Screening Services in Scotland. Diabetic retinopathy screening: Annex E. Scottish diabetic retinopathy grading scheme. The Scottish Government Publications. http://www.scotland.gov.uk/Publications/2003/07/17638/23088
27. Shrout, P.E., Fleiss, J.L.: Intraclass correlations: uses in assessing rater reliability. Psychol. Bull. **86**(2), 420–428 (1979)
28. Cohen, J.: A coefficient of agreement for nominal scales. Educ. Psychol. Measur. **20**(1), 37–46 (1960)
29. Light, R.J.: Measures of response agreement for qualitative data: Some generalizations and alternatives. Psychol. Bull. **76**(5), 365–377 (1971)
30. Fleiss, J.L.: Measuring nominal scale agreement among many raters. Psychol. Bull. **76**(5), 378–382 (1971)
31. Scott, W.A.: Reliability of content analysis: The case of nominal scale coding. Public Opin. Q. **19**(3), 321–325 (1955)
32. Gamer, M., Lemon, J., Fellows, I., Singh, P.: Various coefficients of interrater reliability and agreement. R package version 0.83 (2010). http://CRAN.R-project.org/package=irr
33. Cicchetti, D.V.: Guidelines, criteria, and rules of thumb for evaluating normed and standardized assessment instruments in psychology. Psychol. Assess. **6**(4), 284–290 (1994)
34. Landis, J.R., Koch, G.G.: The measurement of observer agreement for categorical data. Biometrics **33**(1), 159–174 (1977)

# Automated Detection of Diabetic Macular Edema Risk in Fundus Images

Diego Marin[1(✉)], Manuel Emilio Gegundez-Arias[2], Carlos Ortega[3], Javier Garrido[3], Beatriz Ponte[4], and Fatima Alvarez[5]

[1] Department of Electronic, Computer Science and Automatic Engineering,
University of Huelva, Huelva, Spain
diego.marin@diesia.uhu.es
[2] Department of Mathematics, University of Huelva, Huelva, Spain
gegundez@dmat.uhu.es
[3] North Area of Sanitary Management of Cordoba, Andalusian Health Service, Cordoba, Spain
comillan@gmail.com, javier.garrido.sspa@juntadeandalucia.es
[4] Virgen de Macarena University Hospital of Seville, Andalusian Health Service, Seville, Spain
bepontezu@hotmail.com
[5] Juan Ramon Jimenez Hospital of Huelva, Andalusian Health Service, Huelva, Spain
fatimaalvarezgil@yahoo.es

**Abstract.** This paper is aimed at assessing the initial performance of a computer-based system to detect the risk of diabetic macular edema (DME). The development of this tool was funded by the Health Ministry of the Andalusian Regional Government (Spain) with the purpose of being integrated into a complete system for early diagnosis of diabetic retinopathy (DR).

The algorithmic methods are based on the detection of retinal exudates (early ophthalmic signs of DME) by fundus image processing. It has been tested on a set of 1058 macula-centred retinographies from people with diabetes at risk for retinal diseases. Each of the images was rated on a 0–2 scale (from no DME risk to moderate-severe risk) created from the observations of ophthalmologic specialists of three Andalusian Health Service Medical Centres. Since these three sets of DME expert ratings showed a high agreement and consistency, a consensus diagnosis was built and used as a ground truth. System evaluation was carried out by measuring the sensitivity and specificity of automated DME risk detection regarding this clinical reference diagnosis. In addition, system failures in real cases of DME risk (false negatives) and its clinical importance were also measured.

The system showed several promising operation points, being able to work at a sensitivity level comparable to human experts, with no clinically-important failures, and enough specificity from a hypothetical practical implementation point of view. Thus, it demonstrated 0.9039 sensitivity per image (against 0.7948, 0.9345 and 0.8690 of specialists), with all false negatives graded as mild DME risk, and 0.7696 specificity. This last value indicates that over 75 % of the images with no apparent DME risk under consideration are correctly identified by the system.

Initial performance assessment shows that the presented system for the detection of DME risk is a suitable tool to be integrated into a complete DR pre-screening tool for the automated management of patients within a screening

© Springer International Publishing Switzerland 2016
F. Ortuño and I. Rojas (Eds.): IWBBIO 2016, LNBI 9656, pp. 380–390, 2016.
DOI: 10.1007/978-3-319-31744-1_34

programme. Progress in this integration is definitely associated with the need to carry out a comprehensive system evaluation.

**Keywords:** Diabetic retinopathy · Diabetic macular edema · Early detection system · Automated screening · Retinal image processing · Computer-aided diagnosis

# 1 Introduction

Diabetic Retinopathy (DR) is a retinal disease derived from complications caused by the abnormally high levels of glucosein blood produced by diabetes mellitus (DM). It constitutes the main cause of blindness in diabetic patients and is a leading cause of new cases of legal blindness in working-age population around the world [1]. Although DM does not necessarily involve vision impairment, about 2 % of patients are blind, and 10 % undergo vision degradation after 15 years of diabetes as a consequence of these microvascular complications [2, 3].

The main problem involved in DR diagnosis is that this disorder is usually asymptomatic in its early stages, when medical treatment is more effective and disease progression can be prevented. Therefore, early DR detection is required to ensure treatment is received on time and periodical eye-fundus examinations need to be carried out through systematic screening programmes in the diabetes community [4]. DR screening with a non-mydriatic camera renders the best cost-effectiveness ratio and is easy to use. Screening by retinal photography is not a substitute for a comprehensive ophthalmic examination, but there is evidence that it can serve as a screening tool for DR to identify patients with retinopathy for referral for ophthalmic evaluation and management [5, 6]. However, this preventive action means a huge challenge for health services, since the number of potential patients, whose screening and monitoring should be continued, is very high – it is estimated that 382 million people suffered from diabetes in 2013, and this figure is expected to rise to 592 million by 2035 [7].

In this framework, development of automatic DR diagnosis systems may provide remarkably quicker screening programs for early detection of this disorder. These systems are focused on the detection of early ophthalmic signs of illness (i.e., microaneurysms, hemorrhages or exudates) by retinal image computer analysis. This way, in recent years, much effort has been spent on the development of algorithms for automatic segmentation of these early DR-related lesions (a review can be found in [8]); in addition, some prototypes of such systems have already been presented [9–15].

In Andalusia (a Spanish Autonomous Region of ≈ 8.5 million people and DM prevalence of ≈ 16 %), the Health Ministry of the Regional Government supported and funded the development of a DR/No DR diagnosis system based on microaneurysm and hemorrhage detection [16]. Its initial performance assessment on 529 diabetic patients showed that the system fulfills requirements that would demand its possible implementation as a DR pre-screening tool for the automated management of patients within a screening programme. For instance, it could be implemented to work at sensitivity per patient of 0.9380 (near those provided by other ophthalmologic specialists' diagnoses – 0.9416), with all false negatives being graded as mild DR, and specificity of 0.5098. This last value indicates that over half of the patients with no apparent DR were correctly identified by the system.

Closely associated with DR is the diabetic macular edema (DME), a swelling of the retina due to fluid leaking from blood vessels within the macula. A new computer-based system to diagnose the risk of this retinal disease is introduced in this paper. The system is based on the automated detection of exudates in fundus images. Retinal exudates are lipid and lipoprotein deposits that appear as white or yellowish regions on a retinography. They are considered as one of the earliest possible ophthalmic signs of DME. Although their presence is not always a strong surrogate for DME (they are present in roughly 90 % of DME patients), their detection – specifically their number and position on the retina compared to the macula – can be used for an automatic graduation of the risk of future disease development [17, 18]. Therefore, the integration of this computing tool into the above-mentioned DR diagnosis system will contribute towards obtaining a comprehensive system to detect diabetes-associated retinal disease: in addition to detecting DR signs by the presence of microaneurysms and haemorrhages, it will also diagnose the risk of DME through the detection of exudates in the retina.

The material used for evaluating the DME risk detection system as well as the procedure applied for its generation, are described in the next Section. The comparison results between system and human expert-provided diagnoses are presented in Sect. 3, while the authors' conclusions in Sect. 4 put an end to this paper.

## 2    Materials and Methods

The system has been evaluated by comparing its automated diagnoses with clinical diagnoses provided by ophthalmologic specialists. In this paper and with the purpose of increasing reliability, the diagnosis considered as ground truth has been produced from different sets of ophthalmologists' observations. In this case, the assessment of inter-expert reliability and agreement metrics is required to prove consistency between single observational diagnoses provided by experts. In this section, the images used to evaluate the system, as well as the procedures applied to generate both ground truth and automated diagnoses, are described.

### 2.1    Images

Our image collection consists of 1058 macula-centred retinographies of the posterior pole acquired from diabetic patients by the Paris Hôpital Lariboisière, the Faculté de Médecine St. Etienne, and the LaTIM at Brest (France). These fundus images are publicly available for non-commercial use through the so-called MESSIDOR programme [19]. The whole set of images were acquired using a Topcon TRC NW6 non-mydriatic retinograph with a 45° field-of-view. To ensure utmost protection of patient privacy, information that might allow identifying a patient to be reconstructed was removed and there is no actual knowledge that the images could be used alone or in combination to identify any subject.

## 2.2 Generation of Ophthalmologists' Diagnoses

Each of the images contained in our image collection was used for making clinical diagnoses based on specifying a grade of DME risk per image. For this purpose, the simplified classification described in Table 1 was attended.

**Table 1.** Diabetic macular edema (DME) severity scale applied to build clinical diagnosesof the disease.

| DME risk rating categories | |
| --- | --- |
| 0 (Apparently absent: no risk) | No visible exudates |
| 1 (Apparently present: mild) | Visible exudates: shortest distance between macula and exudates is higher than one papilla diameter |
| 2 (Apparently present: moderate-severe) | Visible exudates: shortest distance between macula and exudates is not higher than one papilla diameter |

This classification has been designed within the framework of the Messidor/Techno-Vision Project [19] based on the recommendations provided by: ALFEDIAM (the French association for the study of diabetes and metabolic diseases) [20, 21], ETDRS (Early Treatment Diabetic Retinopathy Study) [22], and Diabetic Retinopathy Screening Services in Scotland [23].

Ophthalmologic specialists of the Andalusian Health Service (Spain) from the North Area of Sanitary Management of Cordoba (NASMC), Virgen de Macarena University Hospital of Seville (VMUHS) and Juan Ramon Jimenez from the Hospital of Huelva (JRJHH), with the help of software, manually and independently marked the exudates present in each image, which are the basis of the selected diagnosis criteria. These marks allowed the automatic building of three sets of DME risk ratings for each image strictly following the classification criteria described in Table 1, each set corresponding to each participating medical center.

## 2.3 Inter-observer Reliability and Agreement Study on Ophthalmologists' Diagnoses

The assessment of inter-rater reliability (IRR) has been performed by the intra-class correlation (ICC) [24]. ICC is one of the most commonly used statistics for IRR assessment by ordinal variables, being also suitable when —as in our study case— all subjects are rated by multiple coders. In addition, in order to complement agreement conclusions provided by ICC results, we have also assessed IRR with kappa statistics. Since the mathematical foundations of kappa provided by Cohen [25] are only suitable for two coders, a kappa-like statistical variant was computed for nominal data designs with three or more coders: Fleiss' kappa [26].

Table 2 presents IRR results evaluated through ICC and Fleiss' kappa for our study case: three coders (VMUHS, NASMC and JRJHH Medical Centers) who independently rated DME risk into three categories on a 0-2 scale (no DME risk, mild and moderate-severe risk). According to the cut-offs for qualitative rating of agreement provided in [27] (see the first row of Table 3), ICC-computed IRR shows excellent agreement between the three medical centers in the DME risk grading diagnoses. On the other hand, according to the guidelines provided in [28] to interpret kappa values (see the second row of Table 3), Fleiss' kappa result indicates substantial agreement between the 3 coders, thus confirming the high level of inter-rater reliability obtained from ICC.

**Table 2.** IRR assessment quantified by ICC and Fleiss' kappa.

| ICC | 95 % Confidence interval | Fleiss' kappa | 95 % Confidence interval |
|-----|--------------------------|---------------|--------------------------|
| 0.830 | (0.805, 0.853) | **0.7554** | (0.7453, 0.7656) |

**Table 3.** ICC and Kappa Qualitative ratings of agreements (proposed by Cicchetti in 1994 [27] for ICC and by Landis and Koch in 1977 [28] for kappa values).

Qualitative ratings

| Intervals | [0.00, 0.20) | [0.20, 0.40) | [0.40, 0.60) | [0.60, 0.80) | [0.80, 1.00] |
|-----------|--------------|--------------|--------------|--------------|--------------|
| ICC | poor | poor | fair | good | excellent |
| Kappa | slight | fair | moderate | substantial | almost perfect |

## 2.4   Generation of Ground-Truth Clinical Diagnoses

All computed IRR metrics proved high inter-observer agreement and consistency among the diagnoses of DME risk grading provided by the three participating medical centers, NASMC, VMUHS and JRJHH. This fact supports the generation of reliable diagnoses that can be used as reference standards in the evaluation of the DME risk detection system presented in this paper. Thus, a consensus diagnosis for each of the 1058 eye-fundus color images was produced from the sets of our three experts' observations of exudates. This diagnosis was built following the same procedure described above for the generation of ophthalmologists' diagnoses, considering that a region of retinal exudate is present in a retinography if it has been observed by at least two of our three coders. The distribution of the study cases resulting from this diagnosis —grouped by the different rating categories of DME risk grade— is shown in Table 4.

**Table 4.** Number of images diagnosed in the different DME risk rating categories by all generated diagnoses.

| Total: 1058 images | DME risk rating category | | |
|---|---|---|---|
| Medical center | 0 | 1 | 2 |
| JRJHH | 872 | 57 | 129 |
| VMUHS | 856 | 57 | 145 |
| NASMC | 872 | 57 | 129 |
| Consensus Diagnosis (Ground truth) | 829 | 102 | 127 |

### 2.5 Generation of Automated Diagnoses

The implemented DME risk diagnostic system processes a retinal photography and applies digital image treatment algorithms with the aim of detecting exudates (*Ex*), first ophthalmologic sign of the disease. The following general stages can be distinguished:

- **Detection of *Ex* candidates:** In this stage, the system first generates the most suitable images to extract bright pixel regions. In addition, it also produces a segmentation of the optic disk – like exudates, the optic disk usually appears in eye fundus images as a yellowish region and its segmentation is particularly important to reduce false positives. The output of this stage is a binary image in which regions susceptible of being exudates (*Ex* candidates) have been identified from the retinal background.
- **Detection of *Ex* lesions:** A supervised classification scheme is applied to produce *Ex* candidate probability maps indicating the probability for particular candidates to be a lesion or not. Then, each candidate region is classified as *Ex* lesion or non-lesion depending on if its associated probability is greater than a certain threshold. The setting values for thresholding the *Ex* candidate probability maps will determine the final detection of exudates and thus the operating point of the system.

The system analyzed each of the 1058 retinographies of our testing database and diagnosed them as DME risk-affected or not according to the presence or absence of exudates. A sweep of *Ex* thresholds was carried out to analyze the system behavior at its different operation points.

## 3    Experimental Results

System performance was assessed in terms of sensitivity (Se) and specificity (Sp) by comparing the resulting automated diagnoses on each fundus image with the corresponding ground-truth diagnosis. Taking Table 5 into account, these metrics are defined as:

$$Se = \frac{TP}{TP + FN}; \quad Sp = \frac{TN}{FP + TN} \tag{1}$$

**Table 5.** Contingency DME risk detection

|  | DME risk present | DME risk absent |
|---|---|---|
| DME risk detected | True positive (TP) | False positive (FP) |
| DME risk not detected | False negative (FN) | True negative (TN) |

Se and Sp are the ratios of well-detected pathological (DME risk-affected) and non-pathological (no DME signs) cases, respectively. On the other hand, false negatives (FN), this is, failures in automated diagnoses of real pathological cases, also play an important role in evaluating this kind of diagnosis computer programmes. If this system had to be safely implemented into DR screening designs, it would have to determine whether fundus images present early signs of DME risk with high Se, guaranteeing that FN cases have no clinical importance. In addition, it should also diagnose images who do not present signs of the disease with enough Sp to result in economic benefits for public health systems.

Table 6 presents the *Se*, *Sp* and FN results of system diagnoses on the images considered in this work. The distribution of FN cases within the considered DME risk grades (listed in Table 1), which has been generated according to our ground-truth diagnosis, is also specified. The table shows several system operating points, which were obtained by thresholding the exudate probability maps at different threshold values. These points are marked in the receiver operating characteristic curve (ROC curve), showed in Fig. 1, where sensibility is plotted versus false positive fractions by varying this threshold.

**Table 6.** DME risk-detection system performance in terms of sensibility (*Se*) and specificity (*Sp*) on the 1058 fundus images considered in this work. Detection failures of pathological cases (*FN*) are presented within different categories of DME risk grade (attending to our reference clinical diagnoses). A possible system operating point is marked in bold and italic. The evaluation of the clinical diagnoses provided by the participating medical centres (NASMC, JRJHH and VMUHS) is also offered.

| Diagnoses | Se | Sp | FN – mild | FN – Moderate-severe |
|---|---|---|---|---|
| System | 0,9738 | 0,4608 | 6 | 0 |
|  | 0,9520 | 0,5718 | 11 | 0 |
|  | 0,9301 | 0,6876 | 16 | 0 |
|  | *0,9039* | *0,7696* | 22 | 0 |
|  | 0,8515 | 0,8444 | 32 | 2 |
| NASMC | 0,7948 | 0,9952 | 47 | 0 |
| JRJHH | 0,9345 | 0,9638 | 15 | 0 |
| VMUHS | 0,8690 | 0,9964 | 30 | 0 |

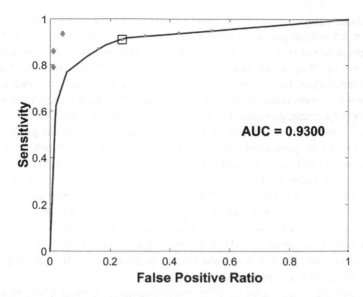

**Fig. 1.** Receiver operating characteristic curve. System performance (filled circles and solid line) is illustrated together with that provided by human observers' diagnosis (filled diamonds). A practical operating point is indicated by the open square.

The representation includes the value of area under the curve (AUC), metric that is usually used as a single measure to quantify global system performance.

In addition, in order to contextualize the results of system diagnoses with those provided by ophthalmologic specialists, the evaluation of the diagnostic data sets built from the observations of the three medical centers participating in this work has been included in Table 6 and Fig. 1.

An overview of the results shows that the system could be set to work fulfilling the requirements that would demand its hypothetical practical integration into a DR pre-screening tool. Table 6 and Fig. 1 show that system could operate at a sensitivity level comparable to human experts, with no clinically-important failures, and high specificity. For instance, a possible operating point may be that resulting in a sensitivity-specificity given by 0.9039-0.7696 (point indicated in bold and italic in Table 6 and by the open square in Fig. 1). At this setting point, the system correctly diagnosed 207 of the 229 pathological images, 22 failures being graded as mild DME risk in our ground-truth diagnosis – clinical observations by NASMC, JRJHH and VMUHS showed 47, 15 and 30 mild-graded false negatives, respectively. On the other hand, the system succeeded in detecting 638 of the 829 non-pathological images – the success percentage of NASMC, JRJHH and VMUHS was higher: 99.52 %, 96.38 % and 99.64 %, respectively.

## 4   Conclusions

Diabetic retinopathy (DR) is a general term for all disorders of the retina caused by diabetes. DR remains an important cause of blindness in Spain – blindness prevalence among diabetic people ranges from 4 to 11 %. In Andalusia (a Spanish Autonomous Region of ≈ 8.5 million people), DR prevalence was 6.92 % in December 2013, with a diabetes prevalence of 16.3 % (6 % unknown) attending to the data collected in its Digital Retinal Screening Programme since it was established in 2005 within the Andalusian Integral Diabetes Plan. This means that 4713 eye tests would have to be made every day of the year within systematic screening programmes. Automated diabetic retinopathy screening or pre-screening carried out by systems focused on the detection of the earliest ophthalmic signs of the illness can have a major impact on such DR programmes in shortening times with greater patient safety. This way, the Health Ministry of the Andalusian Regional Government funded the development of a DR diagnosis system with the purpose of enhancing the effectiveness of its DR screening programme. This system is based on the detection of the earliest DR-related lesions, microaneurysms and haemorrhages. In addition, in order to obtain a comprehensive system, it has also supported the design of a new computer-based tool to detect exudates in the retina and thus, diagnose risks of diabetic macular edema (DME) – exudates are prime markers of this retinal disease. The first results of this system for the detection of DME risk, implemented within this framework, are presented in this paper.

The system was run on a set of 1058 macula-centered retinographies from people with diabetes at risk for DR to diagnose the presence or absence of DME risk signs. Performance evaluation was carried out by measuring sensitivity and specificity of automated DME risk detection relative to a clinical reference diagnosis, which was generated from the diagnoses provided by ophthalmologic specialists of three Andalusian medical centers (NASMC, JRJHH and VMUHS). Since these three sets of DME risk expert ratings showed high agreement and consistency, a consensus diagnosis could be built with the purpose of increasing reliability in the clinical diagnosis used as ground truth. This reference standard plays an important role for system testing since it allows evaluating automated "yes/no detected DME risk signs" outputs relative to a ground truth built in the same way ("yes/no observed DME risk signs"). In addition, it allows categorizing clinical importance of system failures in real pathological cases (false negatives).

System performance results showed several promising operation points from a hypothetical practical implementation point of view. For instance, the system could be implemented to work at sensitivity of 0.9039, with all false negatives being graded as mild DME risk, and specificity of 0.7696. On one hand, attending to safety concerns, this system sensitivity is near or even higher than those provided by other ophthalmologic specialists' diagnoses (0.7948, 0.9345 and 0.8690 for NASMC, JRJHH and VMUHS diagnoses, respectively), thus, the system works at a similar level of false negatives (22 of 229 patients with DME risk signs were diagnosed by the system as non DME risk-affected against 47, 15 and 30 for NASMC, JRJHH and VMUHS, respectively – all of them being mild-graded DME risk cases). On the other hand, attending to specialists' workload reduction concerns, although system specificity is lower than those reached by medical experts (0.7696 against 0.9952, 0.9638 and 0.9964 for NASMC, JRJHH and VMUHS, respectively), its value indicates that over 75 % of the images under consideration with no apparent DME risk signs have been correctly identified.

Therefore, if the system were integrated into a complete DR pre-screening tool for the automated management of patients within a screening programme, these results would mean that the system would work diagnosing images acquired from DME risk-affected people with sensitivity comparable to human specialists, without failing in clinical relevant cases. In addition, the number of images with no DME risk signs that are usually examined by human specialists would be significantly reduced by the system. In any case, it should be assumed that these conclusions have been drawn from results obtained in our collection of 1058 fundus images corresponding to diabetic people. Progress in the definitive integration of this system into retinal DR screening programmes for clinical practice is definitely associated with the necessity of carrying out a comprehensive system evaluation.

**Acknowledgements.** This work was carried out as part of the Project "Automatic System for Early Diabetic Retinopathy Detection by Retinal Digital Images Analysis", supported and funded by the Health Ministry of the Andalusian Regional Government (Spain).

# References

1. Goldenberg, R., Punthakee, Z.: Definition, classification and diagnosis of diabetes, prediabetes and metabolic syndrome. Can. J. Diabetes 37(Suppl. 1), S8–S11 (2013)
2. Klein, R., Meuer, S.M., Moss, S.E., Klein, B.E.: Retinal microaneurysm counts and 10-year progression of diabetic retinopathy. Arch. Ophthalmol. **113**, 1386–1391 (1995)
3. Massin, P., Erginay, A., Gaudric, A.: Rétinopathie Diabétique. Elsevier, New York (2000)
4. Fong, D.S., Aiello, L., Gardner, T.W., King, G.L., Blankenship, G., Cavallerano, J.D., Ferris, F.L., Klein, R.: Diabetic retinopathy. Diab. Care **26**, 226–229 (2003)
5. Sender, M.J., Bagur, S.M., Badia, X., Maseras, M., de la Puente, M.L., Foz, M.: Cámara de retina no midríatica: estudio de coste-efectividad en la detección temprana de la retinopatía diabética. Med. Clín. **121**(12), 446–452 (2003)
6. Aptel, F., Denis, P., Rouberol, F., Thivolet, C.: Screening of diabetic retinopathy: effect of field number and mydriasis on sensitivity and specificity of digital fundus photography. Diab. Metab. **34**(3), 290–293 (2008)
7. Guariguata, L., Whiting, D.R., Hambleton, I., Beagley, J., Linnenkamp, U., Shaw, J.E.: Global estimates of diabetes prevalence for 2013 and projections for 2035. Diab. Res. Clin. Pract. **103**(2), 137–149 (2014)
8. Mookiah, M.R.K., Acharya, U.R., Chua, C.K., Lim, C.M., Ng, E.Y.K., Laude, A.: Computer-aided diagnosis of diabetic retinopathy: A review. Comput. Biol. Med. **43**, 2136–2155 (2013)
9. Usher, D., Dumskyj, M., Himaga, M., Williamson, T., Nussey, S., Boyce, J.: Automated detection of diabetic retinopathy in digital retinal images: a tool for diabetic retinopathy screening. Diabet. Med. **21**(1), 84–90 (2004)
10. Philip, S., Fleming, A.D., Goatman, K.A., Fonseca, S., Mcnamee, P., Scotland, G.S., Prescott, G.J., Sharp, P.F., Olson, J.A.: The efficacy of automated "disease/no disease" grading for diabetic retinopathy in a systematic screening programme. Br. J. Ophthalmol. **91**, 1512–1517 (2007)
11. Abramoff, M.D., Niemeijer, M., Suttorp-Schulten, M.S.A., Viergever, M.A., Russell, S.R., van Ginneken, B.: Evaluation of a system for automatic detection of diabetic retinopathy from color fundus photographs in a large population of patients with diabetes. Diab. Care **31**(2), 193–198 (2008)

12. Niemeijer, M., Abràmoff, M.D., van Ginneken, B.: Information fusion for diabetic retinopathy CAD in digital color fundus photographs. IEEE Trans. Med. Imag. **28**(5), 775–785 (2009)
13. Dupas, B., Walter, T., Erginay, A., Ordonez, R., Deb-Joardar, N., Gain, P., Klein, J.C., Massin, P.: Evaluation of automated fundus photograph analysis algorithms for detecting microaneurysms, haemorrhages and exudates, and of a computer-assisted diagnostic system for grading diabetic retinopathy. Diab. Metab. **36**, 213–220 (2010)
14. Abràmoff, M.D., Folk, J.C., Han, D.P., Walker, J.D., Williams, D.F., Russell, S.R., Massin, P., Cochener, B., Gain, P., Tang, L., Lamard, M., Moga, D.C., Quellec, G., Niemeijer, M.: Automated analysis of retinal images for detection of referable diabetic retinopathy. JAMA Ophthalmol. **131**(3), 351–357 (2013)
15. Abràmoff, M.D., Suttorp-Schulten, M.S.A.: Web-based screening for diabetic retinopathy in a primary care population: The eye check project. Telemed. J. E Health **11**(6), 668–674 (2005)
16. M.E. Gegundez-Arias, D. Marin, C. Ortega, J. Garrido, B. Ponte, F. Alvarez, J. M. Bravo, M.J. Vasallo, A tool for automated diabetic retinopathy pre-screening based on retinal image computer analysis, Submitted to Computers in Biology and Medicine
17. Nayak, J., Bhat, P.S., Acharya, U.R.: Automatic identification of diabetic maculopathy stages using fundus images. J. Med. Eng. Technol. **33**(2), 119–129 (2009)
18. Giancardo, L., Meriaudeau, F., Karnowski, T.P., Li, Y., Garg, S., Tobin, K.W., Chaum, E.: Exudate-based diabetic macular detection in fundus images using publicly available datasets. Med. Image Anal. **16**(1), 216–226 (2012)
19. MESSIDOR TECHNO-VISION Project, France, MESSIDOR: Digital Retinal Images, Available: (Download images section). http://messidor.crihan.fr/download-en.php
20. Massin, P., Angioi-Duprez, K., Bacin, F., Cathelineau, B., Cathelineau, G., Chaine, G., Coscas, G., Flament, J., Sahel, J., Turut, P., Guillausseau, P.J., Gaudric, A.: Recommandations de l'ALFEDIAM pour le d´epistage et la surveillance de la r´etinopathie diab´etique. Diab. Metab. **22**, 203–209 (1996)
21. Massin, P., Angioi-Duprez, K., Bacin, F., Cathelineau, B., Cathelineau, G., Chaine, G., Coscas, G., Flament, J., Sahel, J., Turut, P., Guillausseau, P.J., Gaudric, A.: Recommandations de lALFEDIA Mpour le d´epistage et la surveillance de la r´etinopathiediab´etique. J. Fr. Ophtalmol. **20**, 302–310 (1997)
22. Early Treatment Diabetic Retinopathy Study Research Group: Grading diabetic retinopathy from stereoscopic color fundus photographs an extension of the Modified Airlie House classification: ETDRS report number 10. Ophthalmology **98**, 786–806 (1991)
23. Diabetic Retinopathy Screening Services in Scotland, 2003. Diabetic retinopathy screening:Annex E. Scottish diabetic retinopathy grading scheme. The Scottish Government Publications. http://www.scotland.gov.uk/Publications/2003/07/17638/23088
24. Shrout, P.E., Fleiss, J.L.: Intraclass correlations: Uses in assessing rater reliability. Psychol. Bull. **86**(2), 420–428 (1979)
25. Cohen, J.: A coefficient of agreement for nominal scales. Educ. Psychol. Measur. **20**(1), 37–46 (1960)
26. Fleiss, J.L.: Measuring nominal scale agreement among many raters. Psychol. Bull. **76**(5), 378–382 (1971)
27. Cicchetti, D.V.: Guidelines, criteria, and rules of thumb for evaluating normed and standardized assessment instruments in psychology. Psychol. Assess. **6**(4), 284–290 (1994)
28. Landis, J.R., Koch, G.G.: The measurement of observer agreement for categorical data. Biometrics **33**(1), 159–174 (1977)

# Use of Mobile Application for Nutrition Health Education

Hsiao-Hui Li[1], Mei-Hua Luo[2], and Yuan-Hsun Liao[3(✉)]

[1] Department of Mobile Technology, Toko University, Chiayi, Taiwan
Xiasohui@gmail.com
[2] Division of Nutrition, Chiayi Branch, Taichung Veterans General Hospital, Chiayi, Taiwan
mhlu@vghtc.gov.tw
[3] College of Information Science and Engineering,
Fujian University of Technology, Fuzhou, China,
yuanhsunliao@gmail.com

**Abstract.** People haven't enough activity and bad eating habits in a busy life. And, people aren't understanding nutrition health knowledge for foods and drink. People suffering from obesity and chronic disease risk are relatively high. People how to learn nutrition health knowledge are very important for reducing those risks. But, the applying of nutrition health education general is speech communication by nurse at the hospital. That learning situated is short time. People cannot enough understanding information. So, this paper proposed the use of mobile application to perform nutrition health education. When people or patients need health knowledge, they can use application system on the mobile device to get relation information anytime anywhere. Because the contents of nutrition health education are very large and complex, it's impossible what people or patients want understanding at a short time. Therefore, this application can to apply people focusing on self-requirement to get relation information.

**Keywords:** Nutrition health education · Mobile application · Health management

## 1 Introduction

More and more people pay attention to nutrition and health. The relation of nutrition and health can get more information from internet [1]. But, that is difficult how to search and analyze suitable information from a large nutrition database for customer requirement. Recently, people almost got the nutrition and health information from the hospital. Chronic conditions such as diabetes, hypertension and obesity are effectively controlled. In addition to coordinate drug treatment, diet control is a very important. In research results, teenagers cannot do a healthy diet because they in order to save money, eating, not habit, and not feel the need of the reason [2]. Eating too much of the behavior of adults are eat sweets drink the most, followed by snack intake [3]. So, we propose the APP to assist people early detection and correction of adverse health behaviors. People can convenient and easy to get the screening services of behavioral risk factor. On the other hand, it should assist the known behavioral risk factors to receive adequate guidance. These unhealthy behaviors or called high-risk behavior, such as excessive

© Springer International Publishing Switzerland 2016
F. Ortuño and I. Rojas (Eds.): IWBBIO 2016, LNBI 9656, pp. 391–399, 2016.
DOI: 10.1007/978-3-319-31744-1_35

drinking, eating breakfast irregular, regular intake of high fat or high cholesterol foods, have affect the health, shortened life expectancy, and a close relationship with morbidity and mortality [4].

Smart phone has become an important partner of modern life, social and carry companionship. We hope that using the phone APP's health education to increase consumer acceptance of health education of intent. The main purpose of this study is to enhance consumer acceptance of mobile technology through health education messages, promote health knowledge, correct eating behaviors and appropriate exercise habits, so that consumers can perform autonomous health management, healthy weight loss, thereby reducing the risk of chronic diseases such as diabetes, the risk of hypertension.

## 2 Mobile Technology Application

The development of smart phones already has a functional PC with smart phones and interactive interface development and popularization. We can use the Internet and applications on a smart phone [5]. Smart phone provides voice control and recognition of the diversity of functions and data processing capability.

On the mobile technology have many applications. Kousaridas scholar use the smart phone features to integrate the cash flow design of bank and corporate and effective integrate the business practices, to promote the business in action more effective application [6]. Ramlakhan et al. [7] and Wadhawan et al. [8, 9] scholars propose that the mold of a melanoma skin cancer detection system made into a set on the smart phone application. Melanin musty is a rare skin disease. It happens in every detail and the appearance easily be mistaken for a mole. So, it is more difficult to distinguish not easy to find. Once it is detected the onset of the high mortality rate. Therefore, this condition can be detected to design an application of this disease. Grimaldi et al. [10] scholars use the emit light camera function of smart phone. The blood flow after the bright light will produce different gloss to reach the heart rate detection. Oresko et al. [11] and Lin et al. [12] scholars use a sensor to capture the heartbeat of the signal to analysis and judgment of some pathological information on this unit. Banitsas et al. [13] and Doukas et al. [14] scholars ported the medical imaging to smart phone, let the doctor outside the hospital can advance access to this image, in order to enhance the quality of medical care. Further, they have about video compression technology and some additional features research, for example that use the smart phone to establish inspection reports. Consolvo et al. [15] design a healthy sport system. The system records the user's movement and rest each day by a simple display interface. The interface is presented in a flower garden. When user movement and regular exercise every day, this flower garden flower more diversity and more open the more beautiful. With this fun interface interact with the user to encourage more regular movement of the user to achieve the movement concept of health. Nakauchi [16] proposes diet and message recording system for wearable. The system in addition to mobile phones is still a sphygmomanometer and calories composed monitors to record a healthy diet and measured value. Matsumoto [17] establish the alimentary therapy assist system because the prevalence of the metabolic syndrome of the Japanese rose sharply. This system records the user's dining history, and on the basis of medical information, physical condition, diet records the user's personal preferences and the market price of cooking materials to provide health menu to automatically calculate for user. That system effectively prevents the disease. Antoniou [18] design the intelligent web-based system for chronic illnesses and have adapted to individual needs of a week personalized diet plan and provide medical personnel to monitor the user of the distal end of the measured value and Food History. When the measured value exceeds the equivalent of setting standards, the system may issue a warning or reminder messages through phone or email.

## 3   Nutrition Health Education APP Design

We use mobile technology combined with health education programs and diet. According to the recommendations of the needs of users and nutritionists to develop design a nutrition health education APP for people, patients and dietitians. Through the provision of health education information, users can pay attention to diet anytime and understand the status of own diet and precautions.

### 3.1   Design Architecture

Nutrition health education APP mainly uses the mobile device. Therefore, the use of smart phone features to be designed. Mainly operated by a finger touch screen pong and designed by the handheld device information display moderate interface that information display section nor for excessive. The system is following those concepts to design. This system architecture is following (Fig. 1):

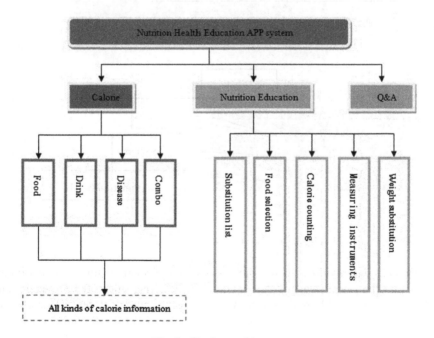

**Fig. 1.** Design architecture

## 3.2   Function Design

In the user interface, this system has been divided by three features which are calorie, nutrition education, and Q&A as Fig. 2. In calorie part, the main service is to provide the nutrient content of foods for users to fully understand the contents of the selected food. In nutrition education part, the main service is to provide relevant health education knowledge for users to understand the health education content at the eating foods. In Q&A part, the main content is to descript the system development and all of the relevant operating description and Q&A. The following sections illustrate the relevant content.

**Fig. 2.** Function design

**Calorie.**   In this function, the main service is to provide the calorie information of diets. Users can got the detail calorie information of the selected food by tapping screen. The items of foods are very much. So, this system divided classification into three parts as general foods, wines, disease type and combination meals, as shown Fig. 3.

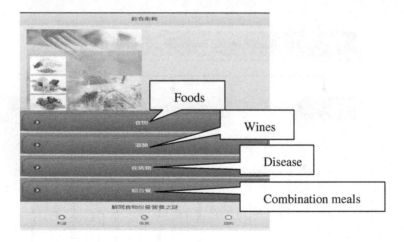

**Fig. 3.** The foods classification of calorie

*General foods information.* In the general foods information, this system divided features into the seven items as dairy, grains, fishes, vegetables, fruits, sugars, and snacks. Users can quickly find the information to get requirement item in the classification.

*Disease type information.* In the disease type information, this system divided disease type into diabetes, kidney disease and other items. Users can quickly find the information to get requirement item in the classification. Patients can know what foods to eat more appropriate for the patient's diet more helpful. After clicking its desire that the disease diet, appears the recommended dietary content of the disease, such as clicking kidney disease. As shown in Fig. 4, it shows recommend eating calorific value. And then, users can click the recommended calorie nutritional analysis to show the recommended diet content such as click the 1500 calories nutrition analysis, as shown in Fig. 5. And, users can pat the nutrition analysis if users want to know the detail nutrition content. The nutrition analysis divided into breakfast, lunch, and dinner. Users can pat what want the information on time such as patting nutrition analysis, as shown in Fig. 6.

**Nutrition Education Information.** In this function, the main service is to provide the nutrition education information of all diets. Users can pat what want to know the detail information of nutrition education. That nutrition education information have apply "what is food substitution table", "how to choices food", "how to calculate the calories by myself needs each day", and "how to use measuring instruments" etc. information, as shown in Fig. 7.

User can pat what want to need the nutrition education item to show the relevant nutrition education information such as patting what is food substitution table, as shown in Fig. 8.

**Fig. 4.** The recommended dietary content of kidney disease.

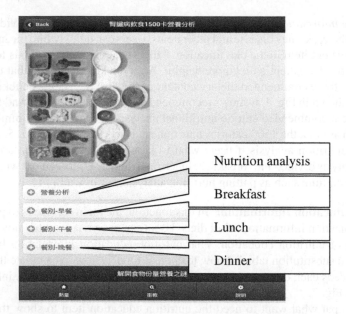

**Fig. 5.** The 1500 calories nutrition analysis of kidney disease diet.

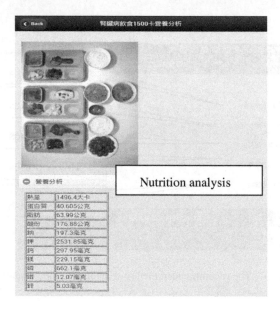

**Fig. 6.** The detail nutrition analysis of kidney disease diet.

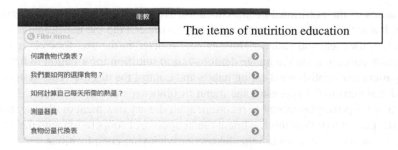

**Fig. 7.** Nutrition education information.

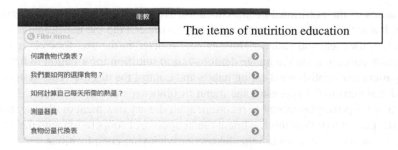

**Fig. 8.** What is food substitution table.

## 4    Conclusion

The goal in this paper is to investigate consumer acceptance intention of nutrition health education through mobile technology. Through the results, Nutrition Health Education APP can help people to know nutrition and health education knowledge helpful.

The features of the nutritional health education APP system are using the pictures of food with a nutritional analysis entity and easy to understand the nutritional education information for users and patient. This system involved in the user food choices skills from health education via the image demonstration nutrition information to help users or the patient to establish good eating habits and control the disease. This system has a professional nutritionist suggested the menu of diabetes and kidney disease. Not only conform to the principle amount of recommended dietary treatment of disease, but also allows the patient to follow this treatment menu uptake. In cooperation with the doctor's treatment, the patient can control the disease and delaying complications.

## References

1. Scaioli, G., Bert, F., Galis, V., Brusaferro, S., De Vito, E., La Torre, G., Manzoli, L., Messina, G., Torregrossa, M.V., Ricciardi, W., Gualano, M.R., Siliquin, R.: Pregnancy and internet: sociodemographic and geographic differences in e-health practice. Results from an Italian multicenter study. Public Health 129(9), 1258–1266 (2015)
2. Rosen, L.D., Lim, A.F., Felt, J., Carrier, L.M., Cheever, N.A., Lara-Ruiz, J.M., Mendoza, J.S., Rokkum, J.: Media and technology use predicts ill-being among children, preteens and teenagers independent of the negative health impacts of exercise and eating habits. Comput. Hum. Behav. 35, 364–375 (2014)
3. Tani, Y., Kondo, N., Takagi, D., Saito, M., Hikichi, H., Ojima, T., Kondo, K.: Combined effects of eating alone and living alone on unhealthy dietary behaviors, obesity and underweight in older Japanese adults: Results of the JAGESO. Appetite 95, 1–8 (2015)
4. Mayer-Brown, S., Lawless, C., Fedele, D., Dumont-Driscoll, M., Janicke, D.M.: The effects of media, self-esteem, and BMI on youth's unhealthy weight control behaviors. Eating Behaviors, (in press) Available online 12 November 2015
5. Punja, S.G., Mislan, R.P.: Mobile device analysis. Small Scale Digit. Device Forensics J. 2(1), 1–16 (2008)
6. Kousaridas, A., Parissis, G., Apostolopoulos, T.: An open financial services architecture based on the use of intelligent mobile devices. Electron. Commer. Res. Appl. 7(2), 232–246 (2008)
7. Ramlakhan, K., Shang, Y.: A mobile automated skin lesion classification system. In: 2011 23rd IEEE International Conference on Tools with Artificial Intelligence (ICTAI), pp. 138–141 (2011)
8. Wadhawan, T., Situ, N., Rui, H., Lancaster, K., Yuan, X., Zouridakis, G.: Implementation of the 7-point checklist for melanoma detection on smart handheld devices. In: 2011 Annual International Conference of the IEEE Engineering in Medicine and Biology Society, EMBC, pp. 3180–3183 (2011)
9. Wadhawan, T., Situ, N., Lancaster, K., Yuan, X., Zouridakis, G.: SkinScan©: a portable library for melanoma detection on handheld devices. In: 2011 IEEE International Symposium on Biomedical Imaging: From Nano to Macro, pp. 133–136 (2011)
10. Grimaldi, D., Kurylyak, Y., Lamonaca, F., Nastro, A.: Photoplethysmography detection by smartphone's videocamera. In: 2011 IEEE 6th International Conference on Intelligent Data Acquisition and Advanced Computing Systems (IDAACS), vol. 1, pp. 488–491 (2011)
11. Oresko, J.J., Jin, Z., Cheng, J., Huang, S., Sun, Y., Duschl, H., Cheng, A.C.: A wearable smartphone-based platform for real-time cardiovascular disease detection via electrocardiogram processing. IEEE Trans. Inf. Technol. Biomed. 14(3), 734–740 (2010)

12. Lin, C.T., Chang, K.C., Lin, C.L., Chiang, C.C., Lu, S.W., Chang, S.S., Ko, L.W.: An intelligent telecardiology system using a wearable and wireless ECG to detect atrial fibrillation. IEEE Trans. Inf. Technol. Biomed. **1,4**(3), 726–733 (2010)
13. Banitsas, K.A., Georgiadis, P., Tachakra, S., Cavouras, D.: Using handheld devices for real-time wireless teleconsultation. In: 26th Annual International Conference of the IEEE Engineering in Medicine and Biology Society, IEMBS 20104, vol. 2, pp. 3105–3108 (2004)
14. Doukas, C., Pliakas, T., Maglogiannis, I.: Mobile healthcare information management utilizing Cloud Computing and Android OS. In: 2010 Annual International Conference of the IEEE Engineering in Medicine and Biology Society (EMBC), pp. 1037–1040 (2010)
15. Consolvo, S., Landay, J.A., McDonald, D.W.: Designing for behavior change in everyday life, pp. 405–414. Focus, San Francisco (2009)
16. Nakauchi, Y., Kozakai, K., Taniguchi, S., Fukuda, T.: Dietary and health information logging system for home health care services. In: IEEE Symposium on Foundations of Computational Intelligence, FOCI 2007, pp. 275–280 (2007)
17. Matsumoto, T., Shimada, Y., Teo, T., Kawaji, S.: A design of information system improving dietary habit based on individual clinical data and life style. In: 21st IEEE International Symposium on Computer-Based Medical Systems, CBMS 2008, pp. 173–175, June 2008
18. Antoniou, I., Nanou, T., Fotiadis, D., Likas, A.: An intelligent system for the provision of personalized dietary plans and health monitoring. In: 4th International IEEE EMBS Special Topic Conference on Information Technology Applications in Biomedicine, pp. 70–73 (2003)

# Tools for Next Generation Sequencing Data Analysis

# Automatic Workflow for the Identification of Constitutively-Expressed Genes Based on Mapped NGS Reads

Rosario Carmona[1], Pedro Seoane[2], Adoración Zafra[1], María José Jiménez-Quesada[1], Juan de Dios Alché[1], and M. Gonzalo Claros[2(✉)]

[1] Plant Reproductive Biology Laboratory, Department of Biochemistry, Cell and Molecular Biology of Plants, Estación Experimental Del Zaidín, CSIC, Granada, Spain
{rosario.carmona,dori.zafra,mariajose.jimenez, juandedios.alche}@eez.csic.es
[2] Departamento de Biología Molecular y Bioquímica, Universidad de Málaga, Málaga, Spain
{seoanezonjic,claros}@uma.es

**Abstract.** Expression analyses such as quantitative and/or real-time PCR require the use of reference genes for normalization in order to obtain reliable assessments. The expression levels of these reference genes must remain constant in all different experimental conditions and/or tissues under study. Traditionally, housekeeping genes have been used for this purpose, but most of them have been reported to vary their expression levels under some experimental conditions. Consequently, the election of the best reference genes should be tested and validated in every experimental scenario. Microarray data are not always available for the search of appropriate reference genes, but NGS experiments are increasingly common. For this reason, an automatic workflow based on mapped NGS reads is presented with the aim of obtaining putative reference genes for a giving species in the experimental conditions of interest. The calculation of the coefficient of variation (CV) and a simple, normalized expression value such as RPKM per transcript allows for filtering and selecting those transcripts expressed homogeneously and consistently in all analyzed conditions. This workflow has been tested with Roche/454 reads obtained from olive (*Olea europaea* L.) pollen and pistil at different developmental stages, as well as with Illumina paired-end reads from two different accessions of *Arabidopsis thaliana*. Some of the putative candidate reference genes have been experimentally validated.

**Keywords:** Housekeeping genes · Normalization · Real-time PCR · Olive (*Olea europaea* L.)

## 1 Introduction

For a long time, gene expression studies have been carried out by non-quantitative or semi-quantitative RNA gel blots and reverse transcription-polymerase chain reaction (RT-PCR) analyses. Later, real-time PCR revolutionized the field, because of its higher sensitivity, specificity and broad quantification range, in comparison to classical RT-PCR.

© Springer International Publishing Switzerland 2016
F. Ortuño and I. Rojas (Eds.): IWBBIO 2016, LNBI 9656, pp. 403–414, 2016.
DOI: 10.1007/978-3-319-31744-1_36

The selection of an appropriate normalization method is crucial for reliable quantitative gene expression assessments in real-time PCR [1] in order to correct for non-specific variation, such as differences in RNA quality and quantity. The most widely spread method for normalization consists of relative quantification: gene expression level is normalized for an internal reference gene. The expression of an ideal reference gene should remain practically constant in all tissues and under different experimental conditions being studied. This appears to be the case of housekeeping genes, which encode proteins involved in basic cell metabolism and basal functions. Their expression levels are relatively constant in most tissues, making them useful as references when quantifying gene expression. However, it has been reported that several of the most commonly used housekeeping genes exhibit considerable variability in microarray data sets or under different experimental conditions [2, 3]. Therefore, for studies comparing gene expression among different tissues, cell types, environmental conditions or developmental stages, the choice of these reference genes should be based on previous experimental evidence, and must also be carefully tested and validated [4].

In some species, a data mining strategy based on publicly available microarray data repositories for genes displaying stable expression throughout different conditions is possible for the selection of candidate reference genes [5]. However, this kind of data is still unavailable for non-model organisms such as olive tree (*Olea europaea* L.). Although its genome is not available yet, transcriptomic sequencing analyses are becoming more a more frequent. Regarding to the identification of putative reference genes in olive, several attempts have been carried out, evaluating olive genes orthologous to the best-ranked reference genes from other crops. Reference genes were chosen based on their stability in olive tissues, as occurs in other plants, and throughout different experimental conditions: fruit developmental/ripening stages and leaves subjected to wounding [6, 7], and at different developmental stages of the olive mesocarp tissue across different cultivars [8]. The peculiarity of plant reproductive tissues makes the search of such reference genes particularly tricky, as some well known housekeeping genes display differential expression in pollen, pistil and other floral organs [5]. However, other analyses indicate that a large proportion of constitutive transcripts are shared by most somatic, reproductive, and haploid tissues [9].

Since RNA-seq experiments usually require validation and real-time PCR has become the necessary routine, here it is presented an automatic workflow constructed with AutoFlow [10], wherein data from next-generation sequencing (NGS) experiments are exploited for obtaining putative reference genes which fit better to the particular experimental conditions of each case. NGS reads belonging to different conditions (i.e., tissues, developmental stages or environmental states) in a given species are mapped to the transcriptome of such species. Mapped reads are counted and the coefficient of variation (CV) is calculated along with normalized expression values by RPKM. It is possible to filter results by a minimum number of mapped reads per transcript and condition (by default 10), as well as a cut-off CV (by default 10 %). Once identified those transcripts expressed considerably and homogeneously in all conditions, they are considered as specific reference genes for real-time PCR experiments for this species in these particular experimental conditions. This workflow has been developed with

Roche/454 reads for olive tree and then confirmed with Illumina reads for *Arabidopsis thaliana* in a recent study about the effect of the temperature in the flowering time [11].

## 2    Materials and Methods

### 2.1    Olive Tree Reads and Transcriptome

As detailed in the recently published database ReprOlive (http://reprolive.eez.csic.es) [12], RNAs were obtained from pollen and pistil at three different developmental stages for cultivar 'Picual'. cDNA libraries were generated using the cDNA Synthesis System Kit (Roche) and sequenced with a Roche GS-FLX Titanium + at the research facilities of the University of Malaga. Raw reads are available in the SRA database with BioProject ID PRJNA287107. Reads were mapped against the transcriptome (63,965 transcripts) described in ReprOlive [12].

### 2.2    *Arabidopsis* Reads and Transcriptome

Two *Arabidopsis thaliana* accessions, Killean-0 (Kil-0), and late flowering, Columbia-0 (Col-0) [11], were compared to determine genes involved in early flowering. Three biological replicates from ten day-old seedlings were paired-end sequenced (100 bp) on a HiSeq 1000. Raw reads were downloaded from the SRA database with BioProject ID PRJEB9470. The *Arabidopsis* transcriptome (35,386 transcripts) used was downloaded from Phytozome (https://phytozome.jgi.doe.gov) and refers to TAIR10 [13].

### 2.3    Detection of Reference Genes

An automatic workflow for the whole process has been constructed using AutoFlow [10], a workflow builder developed in Ruby in our laboratory. The workflow input consists of raw reads and transcriptome files in Fasta format (Fig. 1). First of all, raw reads are pre-processed using SeqTrimNext [14] in order to go forward only with reliable reads. Then, reads of every experimental condition are mapped to complete transcripts using Bowtie2 [15], allowing each read to map in every complementary transcript. Mapped reads are counted with Bio-samtools from BioRuby [16]. Coefficient of variation (CV, ratio of the standard deviation to the mean, expressed as a percentage) is calculated per transcript based on its normalized expression value by RPKM. A table containing all this information is generated and then filtered by two parameters: CV and minimum number of mapped reads per transcript and condition. Default values are 10 % and 10, but both are customizable.

It is often the case that several transcripts share the same closest ortholog. Taking this into account, a final summary of the table is done by selecting just one of these orthologs, the one with the greatest RPKM value. Finally a Venn diagram showing the number of specific and common orthologs is generated to visualize the results and check the suitability of parameters.

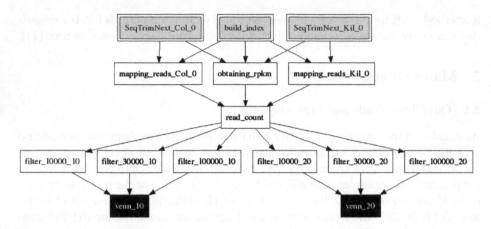

**Fig. 1. Flow diagram for the detection of reference genes in *Arabidopsis*.** Values of filtering parameters tested: 10 % and 20 % for the coefficient of variation (CV), and 10,000, 30,000 and 100,000 reads for the minimum number of mapped reads per transcript and condition. One Venn diagram by each CV cut-off is obtained.

## 2.4    Experimental Validation

For a preliminary validation of some of the candidates, a semi-quantitative RT-PCR analysis was carried out. For 18S analysis, next primers were used: primers Oe 18SF: 5'-TTT GAT GGT ACC TGC TAC TCG GAT AAC C and Oe 18SR: 5'-CTC TCC GGA ATC GAA CCC TAA TTC TCC. RT-PCR analysis of ubiquitin monomer to polyubiquitin pentamer according to Alché et al. [17]. The primers used for actin analysis were: Oe AF: 5'-TTG CTC TCG ACT ATG AAC AGG and Oe AR: 5'-CTC TCG GCC CCA ATA GTA ATA. Transcripts present in different olive plant material were subjected to 25 cycles of amplification in order to ensure an exponential amplification rate. 5 µl of RT-PCR reaction were loaded per lane and separated on 2 % agarose gels in Tris-borate-EDTA (TBE). Equal loading of the RT mixture used for PCR was ensured by using Bioanalyzer (Agylent Technologies) accurate quantitation.

## 3    Results and Discussion

### 3.1    Read Pre-processing

Tables 1 and 2 show statistics for each species about the total of raw and useful reads (after pre-processing with SeqTrimNext) at each library or replicate employed along the workflow, including mean length of useful reads.

**Table 1.** Statistics about libraries used in olive tree workflows.

| Olive Roche/454 sequences | | | | | |
|---|---|---|---|---|---|
| Gene library | Tissue | Developmental stage | Raw reads | Useful reads | Mean length (nt) |
| PM | Pollen | Mature | 217,163 | 111,760 | 385 |
| PG1 | Pollen | 1 h germination | 258,167 | 141,232 | 380 |
| PG5 | Pollen | 5 h germination | 233,921 | 120,276 | 384 |
| S2 | Pistil | Stage2 | 257,813 | 138,077 | 379 |
| S3 | Pistil | Stage3 | 247,401 | 141,903 | 394 |
| S4 | Pistil | Stage4 | 262,749 | 150,185 | 384 |

**Table 2.** Statistics concerning sequences employed in *A. thaliana* workflows.

| *Arabidopsis thaliana* Illumina paired-end sequences (100 bp) | | | | |
|---|---|---|---|---|
| Accession | Replicate | Raw reads ($\times$ 2) | Useful reads ($\times$ 2) | Mean length (nt) |
| Columbia-0 | Col-0-rep1 | 12,251,921 | 11,861,826 | 96 |
| Columbia-0 | Col-0-rep2 | 12,297,628 | 11,912,459 | 96 |
| Columbia-0 | Col-0-rep3 | 13,076,233 | 12,678,437 | 97 |
| Columbia-0 | Col-0(rep1 + rep2 + rep3) | 37,625,782 | 36,452,722 | 96 |
| Killean-0 | Kil-0-rep1 | 11,210,913 | 10,914,810 | 97 |
| Killean-0 | Kil-0-rep2 | 11,523,570 | 11,198,260 | 97 |
| Killean-0 | Kil-0-rep3 | 9,107,610 | 8,859,088 | 97 |
| Killean-0 | Kil-0(rep1 + rep2 + rep3) | 31,842,093 | 30,972,158 | 97 |

## 3.2 Putative Reference Genes in Olive Tree

Since some transcripts are better suited references for expression experiments within tissues, whereas others may be more appropriate for experiments between tissues [18], three different executions of the workflow were carried out in olive: pollen, pistil and both together. Regarding the filtering, a reference gene should have a CV < 10 % (default) or 20 % (non stringent); since reads come from Roche/454 platform, the minimum number of mapped reads per gene should be 10, 50 or 100. A comparative summary of results is shown on Fig. 2. While a significant number of candidates are obtained for pollen and pistil with the most stringent conditions (> 100 reads, CV < 10 %), only one candidate is obtained for both pollen and pistil. As an example, the list of the best reference genes for olive pistil is presented in Table 3.

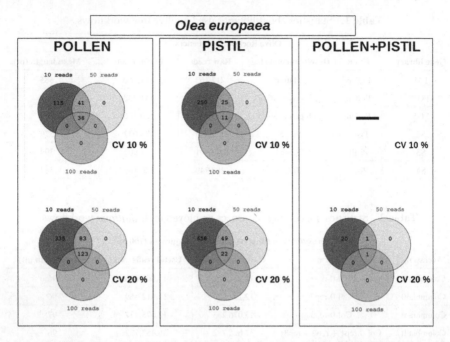

**Fig. 2.   Venn diagrams summarizing the number of putative reference genes obtained for olive tree.** Two cut-off values were used for CV and three different counts for the mapped reads.

Many of the candidate reference genes obtained in olive pistil were already suggested and/or used as reference genes in others species. This is the case of:

– Polyubiquitin 10 (UBQ10), which shows highly stable expression in *Arabidopsis* [5], and has been validated as reference gene in blueberry, cotton and poplar, and used for normalizing in a work regarding olive fruit development and ripening [19];
– Glyceraldehyde-3-phosphate dehydrogenase (GADPH), which has been identified as one of the best reference genes for olive fruit development and ripening [6] and used as normalizer for the analysis of cDNAs associated with alternate bearing in olive [20];
– Elongation factor 1-α (ELNFa), which was evaluated as candidate for reference gene in potato (*Solanum tuberosum*), resulting in the most stable among the group tested during biotic and abiotic stresses [21]; it has also been validated as best reference gene in plenty of species.

Other candidates obtained in olive pistil have never been used as reference genes, but they show outstanding RPKM values. This is the case of methylesterase 1 (MES1) and salicylic acid-binding protein 2 (SABP2) (Table 3). Their use should be carefully considered and evaluated in the near future.

**Table 3.**  **The best proposed reference genes in olive tree pistil** obtained for three stages (S2, S3 and S4) of development in olive pistil, with CV < 10 % and minimum number of mapped reads of 100. *Transcript_id*: transcript identifiers in the ReprOlive transcriptome.

| PISTIL | No. MAPPED READS | | | CV (%) | RPKM | Best hit | Description |
|---|---|---|---|---|---|---|---|
| transcript_id | S_2 | S_3 | S_4 | | | | |
| rp11_olive_045557 | 107 | 105 | 106 | 0.77 | 0.614 | K7US22 | Ubiquitin2 *Zea mays* |
| rp11_olive_006473 | 260 | 272 | 266 | 1.84 | 0.624 | Q8H159 | Polyubiquitin 10 *Arabidopsis thaliana* (Mouse-ear cress) |
| rp11_olive_003751 | 143 | 139 | 135 | 2.35 | 0.276 | P26520 | Glyceraldehyde-3-phosphate dehydrogenase, cytosolic *Petunia hybrida* |
| rp11_olive_000305 | 117 | 108 | 109 | 3.62 | 0.123 | Q84JM4 | Topless-related protein 3 *Arabidopsis thaliana* (Mouse-ear cress) |
| rp11_olive_002595 | 116 | 110 | 106 | 3.71 | 0.198 | O24661 | Asparagine synthetase [glutamine-hydrolyzing] *Triphysaria versicolor* |
| rp11_olive_006479 | 130 | 144 | 147 | 5.28 | 0.329 | P69310 | Ubiquitin *Avena sativa* (Oat) |
| rp11_olive_000229 | 102 | 111 | 116 | 5.28 | 0.115 | A0A022R633 | Uncharacterized protein *Erythranthe guttata* |
| rp11_olive_031243 | 170 | 156 | 188 | 7.64 | 0.789 | P69313 | Ubiquitin *Helianthus annuus* (Common sunflower) |
| rp11_olive_019507 | 810 | 910 | 1,011 | 9.01 | 3.307 | Q6RYA0 | Salicylic acid-binding protein 2 *Nicotiana tabacum* (Common tobacco) |
| rp11_olive_015883 | 810 | 912 | 1,011 | 9.01 | 3.008 | Q8S8S9 | Methylesterase 1 *Arabidopsis thaliana* (Mouse-ear cress) |
| rp11_olive_018099 | 166 | 181 | 209 | 9.62 | 0.649 | Q39196 | Probable aquaporin PIP1-4 *Arabidopsis thaliana* (Mouse-ear cress) |

In olive pollen, α-tubulin (TUA3) and β-tubulin (TUBb) also emerge as candidates. Although they have been extensively used as reference genes, controversial data have been reported on its reliability, being both considered the best in certain species and the worst ones in others [6]. It seems that they can be used in our work.

Actin 7 (ACT7) was also suggested as a good candidate in olive pollen. Actin in general is an extensively reported reference gene. It has been described as one the most stable in chicory [22], berry [23] and pea [24]. It was also considered as a potential candidate in olive fruits [6], however, it was not among the best in our analysis.

For both together, olive tree pollen and pistil, the best reference gene suggested by the workflow is S-adenosylmethionine decarboxylase (SAMDC), which was previously pointed out as one of the most abundant sequences in expressed sequence tag (ESTs) libraries of potato (*Solanum tuberosum*) [25].

A preliminary RT-PCR validation of polyubiquitin and actin in olive tissues in comparison to 18S, a widely used housekeeping gene, is shown in Fig. 3, showing both thick and similar expression levels in reproductive tissues (mature pollen and pistil).

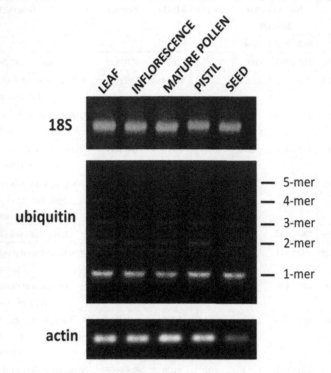

**Fig. 3. Preliminary RT-PCR validation** of polyubiquitin and actin in olive tissues in comparison to18S.

### 3.3 Putative Reference Genes in *Arabidopsis*

In *Arabidopsis*, four executions were made: one execution with each (three) sample replicates and the fourth joining all replicates (Table 2). CVs < 10 and 20 % were considered, but since these reads were obtained with the Illumina platform, a greater minimum number of mapped reads is required: 10,000, 30,000 and 100,000 were selected for detached replicates and 100,000, 300,000 and 700,000 for joined replicates. Additionally, only paired-end reads with both endings aligning were taken into account. A comparative summary of results is shown on Fig. 4, showing that the number of candidates is very small. The list of the best reference genes in this experiment for *Arabidopsis* is given in Table 4. Interestingly, the same best candidates have been obtained for separated replicates than joined. It suggests that the average number of reads per replicate ($\approx$ 11,000,000 reads) results enough for the aim of the workflow, as well as the possible variability between replicates appears not to be affecting the estimation.

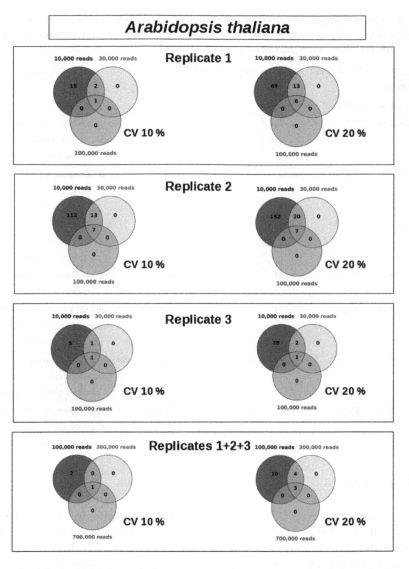

**Fig. 4.** Venn diagrams for candidate reference genes in *Arabidopsis*.

Ribulose-1,5-bisphosphate carboxylase/oxygenase (RuBisCO) is one of the putative reference genes, which has been previously used for such purpose in tea leaf tissues [26]. However, it is not a good candidate for non-green tissues like those of the anther, neither for pollen [27]. Other candidates have never been used before as reference genes in any other species. Their use should be carefully considered and evaluated. For example, although not previously reported, extensin-3 (EXT3) is by far the best candidate.

**Table 4.**  **The best proposed reference genes in** *Arabidopsis* obtained for one of the replicates from *Arabidopsis accessions* Kil-0 and Col-0, with CV < 10 % and a minimum number of mapped reads of 100,000. T*ranscript_id*: transcript identifiers in the TAIR10 transcriptome.

| *ARABI-DOPSIS* transcript_id | No. MAPPED READS | | CV (%) | RPKM | Best hit | Description |
|---|---|---|---|---|---|---|
| | Col_0 rep2 | Kil_0 rep2 | | | | |
| AT5G38410.1 | 277,446 | 274,674 | 0.50 | 8.307 | P10798 | Ribulose bisphosphate carboxylase small chain 3B, chloroplastic *Arabidopsis thaliana* |
| AT5G38420.1 | 261,642 | 265,320 | 0.70 | 9.061 | P10797 | Ribulose bisphosphate carboxylase small chain 2B, chloroplastic *Arabidopsis thaliana* |
| AT5G38430.1 | 250,158 | 254,916 | 0.94 | 9.254 | P10796 | Ribulose bisphosphate carboxylase small chain 1B, chloroplastic *Arabidopsis thaliana* |
| AT1G21310.1 | 276,110 | 259,674 | 3.07 | 4.770 | Q9FS16 | Extensin-3 *Arabidopsis thaliana* (Mouse-ear cress) |
| AT1G67090.1 | 335,538 | 313,654 | 3.37 | 9.377 | P10795 | Ribulose bisphosphate carboxylase small chain 1A, chloroplastic *Arabidopsis thaliana* |
| AT2G39730.2 | 153,776 | 140,718 | 4.43 | 2.364 | P10896-2 | Short of Ribulose bisphosphate carboxylase/oxygenase activase, chloroplastic *Arabidopsis thaliana* |
| AT2G39730.1 | 153,782 | 140,708 | 4.44 | 2.379 | P10896 | Ribulose bisphosphate carboxylase/oxygenase activase, chloroplastic *Arabidopsis thaliana* |

# 4    Conclusion

The workflow and underlying algorithm described in this manuscript seems to be independent of the number of mapped reads, and their length since it seems to work equally well with a few long reads (Roche/454) and with many short reads (Illumina). Furthermore, the fact of that some of the suggested candidates have been previously described as reference genes supports this experimental approach. There is no doubt that the expression levels and stability of the best control genes hinted at the described workflow must be validated by highly sensitive methods like real-time or digital PCR under each specific experimental condition prior to its utilization for normalization. The use of statistical algorithms for the evaluation of best-suited reference genes is highly recommended. However, the utilization of the described workflow where possible can be quite helpful as a preliminary approximation about the best candidates.

# 5   Conflict of Interest

The authors confirm that this article content has no conflicts of interest.

**Acknowledgments.** This work has been supported by co-funding from the ERDF (European Regional Development Fund) and (i) MINECO (grants BFU2011-22779 and RECUPERA2020-3.1.4.), (ii) INIA (grant RTA2013-00068-C03-02), and (iii) PAI (grants P10-CVI-6075, P10-AGR-6274 and P11-CVI-7487). The authors also thankfully acknowledge the computer resources and the technical support provided by the Plataforma Andaluza de Bioinformática of the University of Málaga.

# References

1. Freeman, W.M., Walker, S.J., Vrana, K.E.: Quantitative RT-PCR: pitfalls and potential. Biotechniques **26**, 112–125 (1999)
2. Lee, P.D., Sladek, R., Greenwood, C.M.T., Hudson, T.J.: Control genes and variability: absence of ubiquitous reference transcripts in diverse mammalian expression studies. Genome Res. **12**, 292–297 (2002)
3. Suzuki, T., Higgins, P.J., Crawford, D.R.: Control selection for RNA quantitation. Biotechniques **29**, 332–337 (2000)
4. Brunner, A.M., Yakovlev, I.A., Strauss, S.H.: Validating internal controls for quantitative plant gene expression studies. BMC Plant Biol. **4**, 14 (2004)
5. Czechowski, T., Stitt, M., Altmann, T., Udvardi, M.K.: Genome-wide identification and testing of superior reference genes for transcript normalization. Society **139**, 5–17 (2005)
6. Nonis, A., Vezzaro, A., Ruperti, B.: Evaluation of RNA extraction methods and identification of putative reference genes for real-time quantitative polymerase chain reaction expression studies on olive (olea europaea L.) fruits. J. Agric. Food Chem. **60**, 6855–6865 (2012)
7. Resetic, T., Stajner, N., Bandelj, D., Javornik, B., Jakse, J.: Validation of candidate reference genes in RT-qPCR studies of developing olive fruit and expression analysis of four genes involved in fatty acids metabolism. Mol. Breed. **32**, 211–222 (2013)
8. Ray, D.L., Johnson, J.C.: Validation of reference genes for gene expression analysis in olive (Olea europaea) mesocarp tissue by quantitative real-time RT-PCR. BMC Res. Notes. **7**, 304 (2014)
9. Ma, J., Skibbe, D.S., Fernandes, J., Walbot, V.: Male reproductive development: gene expression profiling of maize anther and pollen ontogeny. Genome Biol. **9**, R181 (2008)
10. Seoane, P., Carmona, R., Bautista, R., Guerrero-Fernández, D., Claros, G.: AutoFlow: an easy way to build workflows. In: Proceedings IWBBIO, pp. 342–349 (2014)
11. Lutz, U., Posé, D., Pfeifer, M., Gundlach, H., Hagmann, J., Wang, C., Weigel, D., Mayer, K.F.X., Schmid, M., Schwechheimer, C.: Modulation of Ambient Temperature-Dependent Flowering in Arabidopsis thaliana by Natural Variation of FLOWERING LOCUS M. PLoS Genet. **11**, e1005588 (2015)
12. Carmona, R., Zafra, A., Seoane, P., Castro, A.J., Guerrero-Fernández, D., Castillo-Castillo, T., Medina-García, A., Cánovas, F.M., Aldana-Montes, J.F., Navas-Delgado, I., de Alché, J.D., Claros, M.G.: ReprOlive: a database with linked data for the olive tree (Olea europaea L.) reproductive transcriptome. Front. Plant Sci. **6**, 625 (2015)

13. Lamesch, P., Berardini, T.Z., Li, D., Swarbreck, D., Wilks, C., Sasidharan, R., Muller, R., Dreher, K., Alexander, D.L., Garcia-Hernandez, M., Karthikeyan, A.S., Lee, C.H., Nelson, W.D., Ploetz, L., Singh, S., Wensel, A., Huala, E.: The Arabidopsis Information Resource (TAIR): Improved gene annotation and new tools. Nucleic Acids Res. **40**, 1202–1210 (2012)
14. Falgueras, J., Lara, A.J., Fernández-Pozo, N., Cantón, F.R., Pérez-Trabado, G., Claros, M.G.: SeqTrim: a high-throughput pipeline for pre-processing any type of sequence read. BMC Bioinformatics **11**, 1–12 (2010)
15. Langmead, B., Salzberg, S.L.: Fast gapped-read alignment with Bowtie 2. Nat. Methods **9**, 357–359 (2012)
16. Goto, N., Prins, P., Nakao, M., Bonnal, R., Aerts, J., Katayama, T.: BioRuby: Bioinformatics software for the Ruby programming language. Bioinformatics **26**, 2617–2619 (2010)
17. de Dios, A.J., Castro, A.J., Olmedilla, A., Fernández, M.C., Rodríguez, R., Villalba, M., Rodríguez-García, M.I.: The major olive pollen allergen (Ole e I) shows both gametophytic and sporophytic expression during anther development, and its synthesis and storage takes place in the RER. J. Cell Sci. **112**, 2501–2509 (1999)
18. Coker, J.S., Davies, E.: Selection of candidate housekeeping controls in tomato plants using EST data. Biotechniques **35**, 740–748 (2003)
19. Vezzaro, A., Krause, S.T., Nonis, A., Ramina, A., Degenhardt, J., Ruperti, B.: Isolation and characterization of terpene synthases potentially involved in flavor development of ripening olive (Olea europaea) fruits. J. Plant Physiol. **169**, 908–914 (2012)
20. Yanik, H., Turktas, M., Dundar, E., Hernandez, P., Dorado, G., Unver, T.: Genome-wide identification of alternate bearing-associated microRNAs (miRNAs) in olive (Olea europaea L.). BMC Plant Biol. **13**, 10 (2013)
21. Nicot, N.: Housekeeping gene selection for real-time RT-PCR normalization in potato during biotic and abiotic stress. J. Exp. Bot. **56**, 2907–2914 (2005)
22. Maroufi, A., Van Bockstaele, E., De Loose, M.: Validation of reference genes for gene expression analysis in chicory (Cichorium intybus) using quantitative real-time PCR. BMC Mol. Biol. **11**, 15 (2010)
23. Reid, K.E., Olsson, N., Schlosser, J., Peng, F., Lund, S.T.: An optimized grapevine RNA isolation procedure and statistical determination of reference genes for real-time RT-PCR during berry development. BMC Plant Biol. **6**, 27 (2006)
24. Die, J.V., Román, B., Nadal, S., González-Verdejo, C.I.: Evaluation of candidate reference genes for expression studies in Pisum sativum under different experimental conditions. Planta **232**, 145–153 (2010)
25. Ronning, C.M., Stegalkina, S.S., Ascenzi, R.A., Bougri, O., Hart, A.L., Utterbach, T.R., Vanaken, S.E., Riedmuller, S.B., White, J.A., Cho, J., Pertea, G.M., Lee, Y., Karamycheva, S., Sultana, R., Tsai, J., Quackenbush, J., Griffiths, H.M., Restrepo, S., Smart, C.D., Fry, W.E., Van Der Hoeven, R., Tanksley, S., Zhang, P., Jin, H., Yamamoto, M.L., Baker, B.J., Buell, C.R.: Comparative analyses of potato expressed sequence tag libraries. Plant Physiol. **131**, 419–429 (2003)
26. Gohain, B.: Rubisco-bis-phosphate oxygenase (RuBP)- A potential housekeeping gene forqPCR assays in tea. African J. Biotechnol. **11**, 11193–11199 (2012)
27. Hoedemaekers, K., Derksen, J., Hoogstrate, S.W., Wolters-Arts, M., Oh, S.-A., Twell, D., Mariani, C., Rieu, I.: BURSTING POLLEN is required to organize the pollen germination plaque and pollen tube tip in Arabidopsis thaliana. New Phytol. **206**, 255–267 (2015)

# Influence of Normalization on the Analysis of Electroanatomical Maps with Manifold Harmonics

Margarita Sanromán-Junquera[1]([✉]), Inmaculada Mora-Jiménez[1],
Arcadio García-Alberola[2], Antonio Caamaño-Fernández[1],
and José Luis Rojo-Álvarez[1]

[1] Department of Signal Theory and Communications, Telematics and Computing,
Rey Juan Carlos University, Madrid, Spain
`margarita.sanroman@urjc.es`
[2] Arrhythmia Unit, Hospital Universitario Virgen de la Arrixaca, Murcia, Spain

**Abstract.** Electrical and anatomical maps (EAM) are built by cardiac navigation systems (CNS) and by Electrocardiographic Imaging systems for supporting arrhythmia ablation during electrophysiological procedures. Manifold Harmonics Analysis (MHA) has been proposed for analyzing the spectral properties of EAM of voltages and times in CNS by using a representation of the EAM supported by the anatomical mesh. MHA decomposes the EAM in a set of basis functions and coefficients which allow to conveniently reconstruct the EAM. In this work, we addressed the effect of normalization of the mesh spatial coordinates and the bioelectrical feature on the EAM decomposition for identifying regions with strong variation on the feature. For this purpose, a simulated EAM with three foci in a ventricular and in an atrial tachycardia was used. These foci were located at different distances amongst themselves, and different voltages were also considered. Our experiments show that it is possible to identify the foci origin by considering the first 3–5 projections only when normalization was considered, both for atrial and ventricular EAM. In this case, better quality in the EAM reconstruction was also obtained when using less basis functions. Hence, we conclude that normalization can help to identify regions with strong feature variation in the first stages of the EAM reconstruction.

**Keywords:** Cardiac navigation system · Electroanatomical maps · Manifold harmonics · Normalization · Reconstruction · Regionalization · Focal tachycardia location · Electrophysiological study

## 1 Introduction

Cardiac arrhythmia mechanism knowledge is crucial for the application of successful therapy. Given that there is no specific medical image modality for visualizing cardiac bioelectricity, several technologies have been proposed for supporting

© Springer International Publishing Switzerland 2016
F. Ortuño and I. Rojas (Eds.): IWBBIO 2016, LNBI 9656, pp. 415–425, 2016.
DOI: 10.1007/978-3-319-31744-1_37

and guiding cardiac ablation, such as cardiac navigation systems (CNS) or Electrocardiographic Imaging (ECGI). Cardiac ablation in electrophysiological studies (EPS) is one of the most effective treatments for a number of arrhythmias, in which several catheters are introduced inside the heart for sensing the electrical field, and one of them is used to sear the arrhythmogenic diseased tissue by means of radiofrequency or intense cold [1,2]. Current invasive and noninvasive systems for cardiac ablation support are capable of first building a mesh for visualizing the anatomical structure of the cardiac chamber under study, and then providing with three-dimensional (3D) electroanatomical maps (EAM) of a bioelectrical feature, including activation time, or voltage amplitude.

Anatomical meshes are composed by vertices joined by triangular faces, and bioelectrical features in EAM are only measured at some of the mesh vertices. In [3], a simple methodology for spectral analysis processing was introduced to provide useful quantitative and qualitative magnitudes such as bandwidth, spectral content, or frequency bands, from EAM and anatomical meshes usually obtained in current systems for EPS during arrhythmia ablation.

However, preceding works on Manifold Harmonics Analysis (MHA) did not pay special attention to the issue of different orders of magnitude among geometrical units (often mm) and physiological features units (mV for voltage maps, or ms for activation maps). As far as the matrix operations involved in MHA are related to eigendecomposition operators working on mixed magnitudes, it should be expected that different orders of magnitude can have some impact on the spectral magnitudes estimated from this technique. Hence, in this work we analyzed with detail, in terms of normalization, the impact of the different orders of magnitudes on the first set of projections (sorted by their eigenvalues magnitude) for the EAM reconstruction, in order to identify the larger variations of the bioelectrical feature.

The rest of the paper is structured as follows. Next section summarizes the theoretical MH framework for 2-manifold surfaces and its extension for considering features on the surface. The effect of normalization on projections and first stages of reconstruction is also presented in Sect. 2. A detailed proof of concept of the EAM reconstruction and the projections direction is presented in Sect. 3, with emphasis on the advantage provided by normalization. Finally, discussion and conclusions are summarized in Sect. 4.

## 2   Manifold Harmonics and Normalization

MHA has often been used in the Computer Graphics field as a tool for 2-manifold analysis (3D shapes) in clustering, mesh parametrization, mesh compression, mesh segmentation, remeshing, surfaces reconstruction, texture mapping, or watermarking applications [4,5]. The MH approach allows the representation of a triangulated mesh (defined as a set of vertices, edges, and triangles) as a combination of elements in an orthogonal basis by generalizing the Fourier Transform to 2-manifold surfaces [6]. In the same way that the Fourier Transform decomposes a function as the sum of basis functions (sines and cosines)

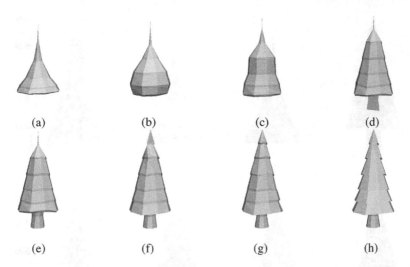

**Fig. 1.** Reconstruction of a tree mesh with 144 vertices by using the first $m = 6$ (a), $m = 10$ (b), $m = 20$ (c), $m = 31$ (d), $m = 40$ (e), $m = 50$ (f), $m = 60$ (g), and $m = 144$ (h) basis.

of increasing frequencies, MH define a Fourier-like function basis to perform spectral analysis on manifolds by considering the eigenfunctions of the Laplace Operator (LO).

For the analysis of shapes in a 3D geometrical space, the LO $\boldsymbol{\Delta}$ is defined on a 2-dimensional Riemannian manifold as the divergence of the gradient of function $\boldsymbol{f} : \mathbb{R}^3 \rightarrow \mathbb{R}$. That is, $\boldsymbol{\Delta f} = \mathrm{div}(\mathrm{grad}\ \boldsymbol{f}) = \sum_i \delta_i^2 \boldsymbol{f}$, where $\delta_i$ denotes differentiation with respect to the $i$-th coordinate function [11]. The LO formulation can be simplified by defining the Laplace-Beltrami Operator (LBO):

$$\boldsymbol{\Delta} = \sum_i \frac{1}{\sqrt{|G|}} \delta_i \left( \sqrt{|G|} \sum_j G^{ij} \delta_j f \right) \tag{1}$$

where $\boldsymbol{G}$ is the metric tensor, $|\boldsymbol{G}| = det(\boldsymbol{G})$ , and $\boldsymbol{G}^{ij}$ is the $(i,j)$ component of the inverse of $\boldsymbol{G}$ [7]. Metric tensor $\boldsymbol{G}$ is used to determine geometric values of the manifold, such as length of curves or angles. The spectrum of the LBO contains information about the geometry of the manifold.

Among the different ways for discretizing LO on a mesh, this work follows the proposal by Vallet and Lévy [6], hence using the extension of LO to manifolds called Laplace-de Rham operator (LRO), defined by

$$\boldsymbol{\Delta}_{ij} = -\frac{cotan(\beta_{ij}) + cotan(\alpha_{ij})}{\sqrt{|\boldsymbol{v}_i||\boldsymbol{v}_j|}}; \boldsymbol{\Delta}_{ii} = -\sum_j \boldsymbol{\Delta}_{ij}, \quad \forall i,j = 1, \cdots, n \tag{2}$$

where $\boldsymbol{v}_i = (x_i, y_i, z_i)$ and $\boldsymbol{v}_j = (x_j, y_j, z_j)$ are vertices of $\boldsymbol{f}$ linked by an edge; $\beta_{ij}$ and $\alpha_{ij}$ are the opposite angles to the edge between $\boldsymbol{v}_i$ and $\boldsymbol{v}_j$; and $|\boldsymbol{v}_i|$ is the area of the Voronoi region of the vertex $\boldsymbol{v}_i$ in its 1-ring neighborhood [8].

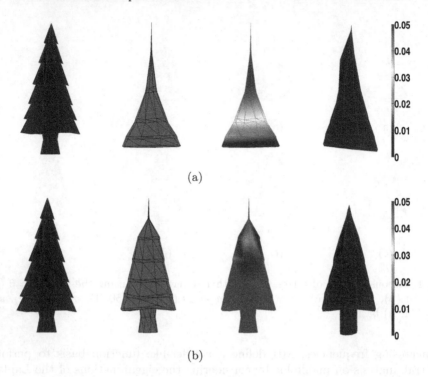

**Fig. 2.** From left to right: original tree with feature variation and reconstructions considering just geometry (gray shapes), geometry and feature altogether, and normalized geometry and feature. Reconstructions have considered $m = 6$ (a) and $m = 35$ (b).

The LRO uses a cotangent scheme which is symmetrized by using the area of the Voronoi region. This symmetrization is required in order to get a positive semidefinite $\boldsymbol{\Delta}$ matrix with positive eigenvalues and orthogonal eigenvectors [9]. Eigenvectors and eigenvalues $\{\boldsymbol{H}^k, \lambda_k\}_{k=1}^n$ of symmetric matrix $\boldsymbol{\Delta}$ satisfy

$$- \boldsymbol{\Delta} \boldsymbol{H}^k = \lambda_k \boldsymbol{H}^k \tag{3}$$

where $n$ is the number of vertices in the mesh, and the set $\{\boldsymbol{H}^k\}_{k=1}^n$ are the so-called *Manifold Harmonic Basis* (MHB). The MHB compose an orthonormal basis $\boldsymbol{H}^k \in \mathbb{R}^n$ which is invariant to scale and surface rotation [10]. The $k$-th element of the MHB is a piecewise linear function given by its values $\boldsymbol{H}_i^k$ at vertices $i$ of the surface [6].

Eigenvalues $\lambda_k$ correspond to the spectrum of the manifold, which is an isometric invariant that only depends on the Riemannian structure of the manifold. Eigenvalues are often sorted from lower to higher values, which correspond to lower and higher eigenvalues of $\boldsymbol{\Delta}$, respectively. Smaller (larger) eigenvalues of the spectrum are related to coarser (finer) structures of the manifold.

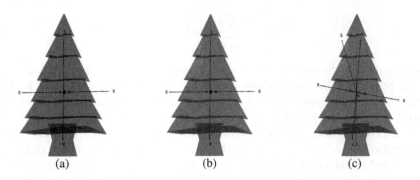

(a)                    (b)                    (c)

**Fig. 3.** Original meshes and first six projections (red lines, labeled with a number) from the inverse MHT: (a) geometrical mesh; (b) geometrical mesh with feature variation; and (c) geometrical mesh with feature variation and normalization (Color figure online).

The Manifold Harmonic Transform (MHT) maps the coordinates of each vertex in the spatial domain into the frequency space, by projecting them onto the MHB. Projections of vertices onto the MHB are named *coefficients* $\hat{a}_k$, which are obtained as

$$\hat{a}_k = \sum_{i=1}^{n} v_i H_i^k \tag{4}$$

Application of the inverse MHT reconstructs the mesh back into the geometric space by using the first $m$ frequencies, given by

$$\hat{v}_i = \sum_{k=1}^{m} \hat{a}_k H_i^k \tag{5}$$

where $m$ is the number of basis used for reconstruction. If $m = n$, then $\hat{v}_i = v_i$ and the original surface is obtained. When $m << n$, a low-pass filtered version of the shape is obtained, since details (high spatial frequencies) are discarded [6]. Figure 1 shows the reconstruction of a tree shape of 144-vertices by using the first $m = 6, 10, 20, 31, 40, 50, 60, 144$. In this case, a highly similar version of the original shape is obtained with $m = 50$.

The above MHT can be extended to consider the variation of a feature $h$ upon the mesh. This extension is specially useful to represent similar 3D shapes with different appearance (e.g. due to lighting conditions or color), and also more in line with the current application of this paper, when dealing with 3D cardiac EAM. In any of these cases, couples $\{H^{ke}, \lambda_{ke}\}$ are obtained by computing the eigenvectors and eigenvalues of $\Delta$ when vertices are defined as $u_i = (x_i, y_i, z_i, h_i)$, where $h_i$ is the feature to represent on the mesh and $(x_i, y_i, z_i)$ are the geometrical coordinates for mesh vertex $u_i$. *Coefficients* are now obtained as $\hat{a}_{ke} = \sum_{i=1}^{n} u_i H_i^{kc}$, and the inverse MHT allows to reconstruct the feature variation on the geometric shape by

$$\hat{\boldsymbol{u}}_i = \sum_{ke=1}^{m} \hat{a}_{ke} \boldsymbol{H}_i^{ke}. \tag{6}$$

Note here that the spatial vertex coordinates and the associated feature have different physical units, so they are quite likely to have different ranges for their values. Though the application of a scaling parameter $\gamma$ to the manifold is known to result in eigenvalues scaled by a factor $1/\gamma^2$ [11], we empirically explore in this work how the normalization of every dimension in vector $\boldsymbol{u}$ affects to the reconstruction when $m \ll n$ is considered. For our approach, every dimension in $\boldsymbol{u}$ is scaled to zero mean and unit standard deviation, so that both geometry and feature variation are balanced in terms of energy. Left panel in Fig. 2 shows the original geometric tree shape (144-vertex) and the variation of a feature on the surface (see an intense focus in the left upper part of the tree). Second panel in Fig. 2 represent the geometric tree shape reconstruction when $m = 6$ and $m = 35$ elements are considered. Next panels correspond to reconstruction of the geometrical shape and the feature with the original values in $\boldsymbol{u}$ (third panel) and when they are normalized (fourth panel). When dimensions in $\boldsymbol{u}$ are normalized, the dual process (de-normalization) is performed on $\hat{\boldsymbol{u}}$ to represent the tree shape in the original scale. Comparison of reconstructions in Fig. 2 with the same number of elements $m$ ($m < n$) reveals that both geometrical shape and feature are better represented when normalization is performed. Remark that this last comment becomes less true as $m \to n$, since normalization does not affect to the reconstruction when all $m = n$ is considered.

When the inverse MHT is computed, each part of the sum in (5) and (6) corresponds to a vector (named *projection* in this paper) pointing to a specific direction in the original geometric space. To analyze how directions of projections change when an additional feature is considered on the surface, Fig. 3 represents the first six projections as red lines labeled from 1 to 6. These projections are superimposed on the 3D shape they come from: (a) just the geometrical shape; (b) the geometrical shape and the variation of a feature on the surface with original values; and (c) the same as (b) but with normalized values. Comparison of Fig. 3(a) and (b) shows that projections are quite similar and geometrical characteristics are predominant to $h$ in the first reconstruction stages. This is the reason why geometrical shapes in the reconstruction with $m = 6$ in Fig. 2(a) are so similar when considering $\boldsymbol{v}$ and $\boldsymbol{u}$ without normalization. However, note the change in the first projections when $\boldsymbol{u}$ is normalized, with the projection labeled with number 3 pointing to the region of the tree with the highest value in feature $h$. These differences are also seen in the reconstruction with $m = 6$ (right panel in Fig. 2(a)), where the shape is more similar to the original tree and the region with the highest value of $h$ is also better localized.

## 3  Experiments

A detailed proof of concept is presented in this section in order to evaluate both the EAM reconstruction and the direction of the first projections when reconstructing with the MH-based methodology. For this purpose, a set of simulated focal tachycardia voltage (FTV) EAM was used by projecting 3 simultaneously activated foci onto a real ventricular and atrial mesh, respectively. Foci with the same and different voltage amplitude were simulated, hence mimicking the same or different instants of initial activation. Ventricular and atrial meshes were obtained from the segmentation of real left ventricular and atrial computed tomography images [3]. Since these meshes had hundred of thousands of vertices, and CNS-EAM had few hundreds of vertices, original meshes were decimated to 500-vertices by using the Quadrid Edge Collapse Decimation

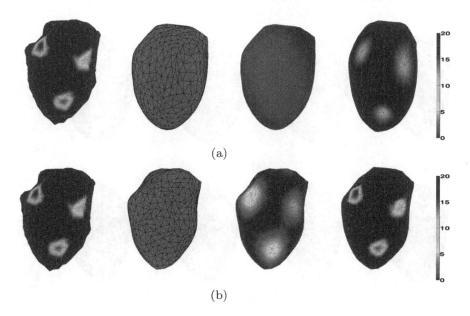

**Fig. 4.** From left to right: original simulated ventricular FTV-EAM with 3 foci of different voltages (mV) and reconstructions considering just anatomy, non-normalized, and normalized EAM. Reconstructions have considered $m = 10$ (a) and $m = 55$ (b).

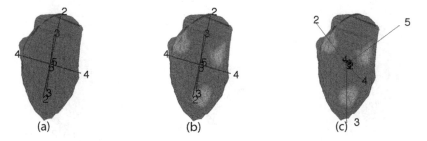

**Fig. 5.** First 5 projections (red lines, labeled with a number) from the inverse MHT onto the ventricular: (a) anatomical mesh, (b) non-normalized, and (c) normalized simulated FTV-EAM with 3 foci of different voltages (Color figure online).

method [12]. This method reduces the number of vertices in mesh surfaces by using iterative contraction of edges in terms of quadratic forms.

Figure 4 shows a simulated ventricular FTV-EAM, with 3 foci activated at different time instants (first left panel). Next panels show the reconstruction for the anatomical mesh, and for the non-normalized and normalized EAM using $m = 10$ (a) and $m = 55$ (b). Note that the 3 foci were defined in the normalized version with just $m = 10$, while this number was not enough to determine the foci in the non-normalized one. Apart from slight difference in the anatomical meshes, the EAM reconstruction with $m = 55$ in the normalized version was almost the same as the original one, while the non-normalized EAM identified the foci but

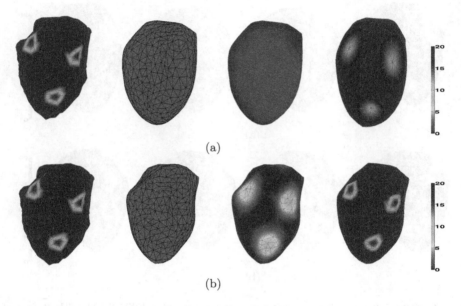

Fig. 6. From left to right: original simulated ventricular FTV-EAM with 3 foci of equal voltage (mV) and reconstructions considering just anatomy, non-normalized, and normalized EAM. Reconstructions have considered $m = 10$ (a) and $m = 57$ (b).

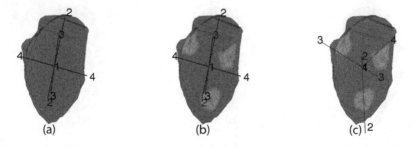

Fig. 7. First 4 projections (red lines, labeled with a number) from the inverse MHT onto the ventricular: (a) anatomical mesh, (b) non-normalized, and (c) normalized simulated FTV-EAM with 3 foci of equal voltages (Color figure online).

with not enough definition. The anatomical reconstruction was more similar to the original mesh for the non-normalized than for the normalized EAM. Figure 5 shows the first 5 projections onto the original anatomical mesh and EAM (in left panel of Fig. 4) considering: (a) the anatomical mesh, (b) the non-normalized EAM, and (c) the normalized EAM. While similar projections were obtained for both the anatomical mesh and the non-normalized EAM, the second, third, and fifth projections pointed to the origin of the 3 foci just for normalized-EAM.

Figure 6 shows a simulated ventricular FTV-EAM, with 3 foci activated at the same time instant (first left panel). Next panels show the reconstruction for the anatomical mesh, and the non-normalized and normalized EAM using $m = 10$ (a) and $m = 57$ (b). As previous EAM, the 3 foci were defined in the normalized version with just $m = 10$, and the EAM reconstruction with $m = 57$

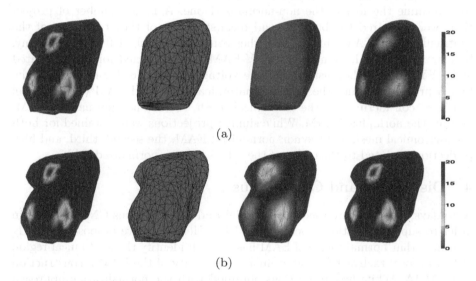

**Fig. 8.** From left to right: original simulated atrial FTV-EAM with 3 foci of different voltages (mV) and reconstructions considering just anatomy, non-normalized, and normalized EAM. Reconstructions have considered $m = 10$ (a) and $m = 88$ (b).

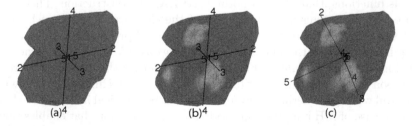

**Fig. 9.** First 5 projections (red lines, labeled with a number) from the inverse MHT onto the atrial: (a) anatomical mesh, (b) non-normalized, and (c) normalized simulated FTV-EAM with 3 foci of different voltages (Color figure online).

in the normalized version was almost the same as the original one. Figure 7 shows the first 4 projections onto the original anatomical mesh and EAM (in left panel of Fig. 6) considering: (a) the anatomical mesh, (b) the non-normalized EAM, and (c) the normalized EAM. While similar projections were obtained for both the anatomical mesh and the non-normalized EAM, the second, third, and fourth projections pointed to the origin of the 3 foci just for normalized-EAM. Comparison of Figs. 4 and 6 shows that less number of projections was required when the foci were activated at the same time instant.

Figure 8 shows a simulated atrial FTV-EAM, with 3 foci activated at different time instants (first left panel). Next panels show the reconstruction for the anatomical mesh, and the non-normalized and normalized EAM using $m = 10$ (a) and $m = 88$ (b). As previous ventricular EAM, the 3 foci were defined in the normalized version with just $m = 10$, while this number was not enough to determine the foci in the non-normalized one. A larger number of projections were required to obtain a good reconstruction of the atrial than of the ventricular FVT-EAM, due to the higher complexity of the left atrial anatomy. Similar results to the ventricular FTV-EAM were obtained for the simulated atrial FTV-EAM when using the same voltage amplitude. Figure 9 shows the first 5 projections onto the original anatomical mesh and EAM (in left panel of Fig. 8) considering: (a) the anatomical mesh, (b) the non-normalized EAM, and (c) the normalized EAM. While similar projections were obtained for both the anatomical mesh and the non-normalized EAM, the second, third, and fifth projections pointed to the origin of the 3 foci just for normalized-EAM.

## 4    Discussion and Conclusions

Nowadays, 3D EAM are used in sequential and instantaneous CNS or ECGI in order to support ablation procedures in EPS. Here, we have proposed to study the effect that normalization of EAM has to both identify the anatomical region of foci in focal tachycardia and evaluate the quality of the EAM reconstruction using MHA. While first projections obtained with the normalization approach pointed to tachycardia foci, the projections with non-normalized EAM were similar to those obtained using only the anatomical information. The focal identification in the normalized EAM reconstruction is also obtained earlier, i.e. uses less basis functions, than the non-normalized EAM reconstruction. The remarkable progress in systems for visualization of the electrical activity, such as CNS and ECGI, is bringing a new landscape for the signal processing techniques that are required to work with multivariate recordings as EAM. The old paradigm of processing a signal at a time and then putting together all the individual results will be soon needing from new digital signal processing techniques, which will be more useful when considering simultaneously the geometrical information. In this setting, the use of MH is a promising and emerging direction that enables revisiting the spectral analysis by accounting for the geometrical information involved on a natural way. It is not surprising that we will probably have to revisit and reconsider the new signal processing tools from their very foundations, and fundamental tools as normalization need to be clearly established in this new scenario.

**Acknowledgments.** This work was supported in part by the Spanish Government with Research Projects TEC2013-48439-C4-1-R and TSI-020100-2010-0469.

# References

1. Feld, G.K., Fleck, R.P., Chen, P.S., Boyce, K., Bahnson, T.D., Stein, J.B., Calisi, C.M., Ibarra, M.: Radiofrequency catheter ablation for the treatment of human type 1 atrial flutter. Identification of a critical zone in the reentrant circuit by endocardial mapping techniques. Circulation **86**(4), 1233–1240 (1992)
2. Klein, L.S., Shih, H.T., Hackett, F.K., Zipes, D.P., Miles, W.M.: Radiofrequency catheter ablation of ventricular tachycardia in patients without structural heart disease. Circulation **85**(5), 1666–1674 (1992)
3. Sanromán-Junquera, M., Mora-Jiménez, I., Saiz, J., Tobón, C., García-Alberola, A., Rojo-Álvarez, J.L.: Quantitative spectral criteria for cardiac navigation sampling rate using manifold harmonics analysis. Comput. Cardiol. **39**, 357–360 (2012)
4. Zhang, H., van Kaick, O., Dyer, R.: Spectral methods for mesh processing and analysis. In: Proceedings of the Eurographics State-of-the-art Report, pp. 1–22 (2007)
5. Zhang, H., van Kaick, O., Dyer, R.: Spectral mesh processing. Comput. Graph. Forum **29**(6), 1865–1894 (2010)
6. Vallet, B., Lévy, B.: Spectral geometry processing with manifold harmonics. Comput. Graph. Forum **27**(2), 251–260 (2008)
7. Mahadevan, S.: Representation Discovery using Harmonic Analysis. Morgan and Claypool Publishers, San Rafael (2008)
8. Meyer, M., Desbrun, M., Schrder, P., Barr, A.: Discrete Differential-Geometry Operators for Triangulated 2-Manifolds. Visualization and Mathematics III (Proceedings of VisMath), pp. 35–54. Springer, Heidelberg (2003)
9. Petronetto, F., Paiva, A., Helou, E.S., Stewart, D.E., Nonato, L.G.: Mesh-free discrete laplace-beltrami operator. Comput. Graph. Forum **32**(6), 214–226 (2013)
10. Botsch, M., Kobbelt, L., Pauly, M., Alliez, P., Lévy, B.: Polygon Mesh Processing. AK Peters Ltd., Natick (2010)
11. Reuter, M., Wolter, F.E., Peinecke, N.: Laplace-beltrami spectra as 'Shape-DNA' of surfaces and solids. Comput. Aided Des. **38**(4), 342–366 (2006)
12. Garland, M., Heckbert, P.: Simplifying surfaces with color and textures using quadric error metrics. In: Proceedings of IEEE Visualization, pp. 263–270 (1998)

# AC-DIAMOND: Accelerating Protein Alignment via Better SIMD Parallelization and Space-Efficient Indexing

Huijun Mai[1], Dinghua Li[1], Yifan Zhang[1], Henry Chi-Ming Leung[1], Ruibang Luo[1,2,3], Hing-Fung Ting[1(✉)], and Tak-Wah Lam[1,2]

[1] HKU-BGI Bioinformatics Algorithms and Core Technology Laboratory, Department of Computer Science, University of Hong Kong, Hong Kong, China
{hjmai,dhli,yfzhang2,cmleung2,rbluo,hfting,twlam}@cs.hku.hk
[2] L3 Bioinformatics Limited, Hong Kong, China
[3] United Electronics Co., Ltd, Beijing, China

**Abstract.** To speed up the alignment of DNA reads or assembled contigs against a protein database has been a challenge up to now. The recent tool DIAMOND has significantly improved the speed of BLASTX and RAPSearch, while giving similar degree of sensitivity. Yet for applications like metagenomics, where large amount of data is involved, DIAMOND still takes a lot of time. This paper introduces an even faster protein alignment tool, called AC-DIAMOND, which attempts to speed up DIAMOND via better SIMD parallelization and more space-efficient indexing of the reference database; the latter allows more queries to be loaded into the memory and processed together. Experimental results show that AC-DIAMOND is about 4 times faster than DIAMOND on aligning DNA reads or contigs, while retaining the same sensitivity as DIAMOND.For example, the latest assembly of the Iowa praire soil metagenomic dataset generates over 9 milllion of contigs, with a total size about 7 Gbp; when aligning these contigs to the protein database NCBI-nr, DIAMOND takes 4 to 5 days, and AC-DIAMOND takes about 1 day. AC-DIAMOND is available for testing at http://ac-diamond.sourceforge.net.

**Keywords:** DNA-protein alignment · SIMD · Dynamic programming · Compressed indexing

## 1 Introduction

In recent years, the rapid advancement of next-generation sequencing (NGS) technologies has made it feasible to generate massive sequencing data for studying individual microbes and microbial communities. In such studies, it often involves aligning a large amount of reads and/or assembled contigs to existing databases of protein sequences so as to determine which existing species or genes can be found in the sample. On one hand, complex metagenomic data, like soil data, use reads with hundreds of Gbp, and the resulting contigs can have a total size of several

© Springer International Publishing Switzerland 2016
F. Ortuño and I. Rojas (Eds.): IWBBIO 2016, LNBI 9656, pp. 426–433, 2016.
DOI: 10.1007/978-3-319-31744-1_38

Gbp to tens of Gbp. On the other hand, the protein databases are also growing rapidly; at present the most popular reference protein database is the NCBI-nr database[1], which has a total size of 23.8G amino acids as of May 2015.

Existing alignment tools are mainly designed for DNA-DNA alignment. There are only a few DNA-protein alignment tools and their speed is still a concern when handling large volume of data. BLASTX [6], in view of its superior sensitivity, has been the golden standard for DNA-protein alignment for over 30 years. Yet BLASTX is prohibitively slow and can only handle small datasets. To ease our discussion, let us consider the alignment of 10 Gbp of contigs (or reads) to NCBI-nr, BLASTX using a computer with 12 CPU cores would take over 60 years. RAPSearch [2,3] improved the speed of BLASTX drastically by using suffix array and reduced amino acid alphabet. For 10 Gbp of contigs or reads, RAPSearch takes about 40 days. On the other hand, by trading sensitivity for speed, PAUDA [4] and Ghostz [5] can further decrease the alignment time to about 10 days. Early this year, the release of DIAMOND [1] finally showed that speed can be improved drastically with scarifying sensitivity; for 10 Gbp of contigs or reads, DIAMOND takes about 6 days and gives similar sensitivity as BLASTX. Nevertheless, it is still disappointing that aligning DNA reads to NCBI-nr indeed takes much longer time than generating the reads (e.g., Illumina MiSeq and HiSeq X can generate 7.5 and 600 Gbp of reads per day, respectively).

In this paper we present AC-DIAMOND, an alignment tool that is based on DIAMOND. AC-DIAMOND resolves the computation bottlenecks of DIAMOND and is a few times faster than DIAMOND when aligning reads or contigs to proteins. With respect to the above example of aligning 10 Gbp of contigs to NCBI-nr, AC-DIAMOND takes only 1.3 days and gives the same sensitivity as DIAMOND.

## 2   Methodologies

Following DIAMOND, AC-DIAMOND uses double indexing to locate all spaced seeds (in a lexicographical order) that appear in both queries and references. The two lists of seeds are traversed together linearly to report all matching seeds. All matching seeds are then extended to a certain length ($\leq 80\,$aa) by using either ungapped extension or Smith-Watherman alignment [9]. Finally, seeds with sufficient extension score are kept to produce alignments using dynamic programming (DP). AC-DIAMOND aims at solving the following three main bottlenecks in DIAMOND.

1. Although DIAMOND has applied the SIMD (Single instruction multiple data) technology to parallelize the Smith-Watherman alignment, it does not apply SIMD technology to parallelize the dynamic programming step (this is primarily because the DP table size can vary among different pairs of queries and references). However, over 90 % of DIAMOND's total running time is on the dynamic programming step and DIAMOND fails to speed up this step.

---

[1] ftp://ftp.ncbi.nih.gov/blast/db/FASTA/nr.gz.

2. As the lists of sorted seeds for the indexes are large, the queries and references must be loaded into the memory bucket by bucket such that the same references should be loaded multiple times for different buckets of queries and the indexes for the same reference should be reconstructed multiple times. Note that we should not store and reuse the constructed indexes in the hard disk because constructing "on the fly" is much faster than loading it from hard disk.

3. DIAMOND uses SIMD for Smith-Watherman alignment of the same seed in queries against multiple matched seeds in references. However, since the Smith-Watherman alignment is performed only if the ungapped extension fails, around 75 % seeds in queries need to perform exactly once Smith-Watherman alignment. Thus, SIMD cannot speed up the Smith-Watherman alignment processes by packing multiple processes but increases overheads.

To solve the first problem, we study the distribution of matching seeds. Our main observation is that over 90 % of seeds in a query have a few matching seeds from different references. Since seeds from the same position of the same query are identical, the DP table sizes for these alignments are similar because there are restrictions on the number of insertions and deletions for DNA-protein alignment. Thus, we can apply the SIMD technique in the dynamic programming step. If the corresponding seeds are the same in query, up to 8 such DP tables are packed together. The DP tables are filled together using the same set of SIMD instructions. Theoretically, we can get 8 times speed-up in this step. However, because of the number of matched seeds is not a multiple of 8 and there are some overheads in packing the DP tables, we achieve a speed-up of 6 to 7 times in the alignment of contigs and reads, respectively.

For the second problem, as the indexes are sorted in lexicographical order, the distance between two adjacent seeds is usually small. We find that 83.6 % of the seed distances are smaller than $2^8$ for NCBI-nr. Thus, we can compress the indexes for protein references such that more queries can be loaded and be aligned in each round. By compressing the reference indexes, we can have a reduction of 50 % of time for loading and construction of indexes using a similar amount of memory as DIAMOND. Note that we can speed-up DIAMOND by allocating more memory and increasing its bucket size. However, we can achieve the same speed-up ratio by compressing the indexes for references and loading more queries in each round.

For the third problem, we redesigned the packing strategy for SIMD acceleration. Since the Smith-Watherman alignment is performed for a predefined length (80 aa), the size of all DP tables are the same ($80 \times 80$) for all seeds. For seeds in queries with less than 4 matched seeds in references, AC-DIAMOND packs up to 16 tables together to perform DP. Note that these DP instances are different from each other in both query and reference sequences, there is no efficient way to load the 16 alignment scores from a scoring-matrix (BLOSUM-62 by default) in parallel, as limited by current instruction sets of SIMD. But the SIMD acceleration still works for the other computational part of DP, and AC-DIAMOND gives a 4-fold speed-up to DIAMOND in this step.

We have implemented the above three approaches and refined the programs in some parts of DIAMOND, the resulting software, i.e., AC-DIAMOND, can achieve at least 4 times of speed-up when compared with DIAMOND on aligning different sets of contigs or reads. Since the programming workflow of these two alignment tools are basically the same, the contigs or reads that can be aligned often have over 99.9 % in common, and also over 99 % of alignment results are the same. The exceptions are due to two special cases.

– DIAMOND's application of a heuristic method in filling the DP table can speed up the dynamic programming step while AC-DIAMOND does not.
– DIAMOND allows a long insertion or deletion at the end of an alignment whose size is larger than the pre-defined restriction by the users.

Although DIAMOND and AC-DIAMOND, for a very small percentage of reads or contigs, might have little difference in the alignments reported, we found that it often only affects the ordering of their alignment results, e.g., the best alignment found by DIAMOND becomes the second best alignment of AC-DIAMOND (in our experiment, for contigs, they are 99.4 % consistent in best alignments; and for reads, over 99.9 % consistent best alignments), the sensitivity of AC-DIAMOND and DIAMOND are indeed very similar.

## 3    Experimental Results

To compare the speed and sensitivity of DIAMOND and AC-DIAMOND, we benchmarked the two alignment tools with large datasets of queries (either contigs or reads) on a computer with 12 CPU cores. The first dataset contains assembled contigs constructed by MEGAHIT v1.0.1 [7] for the Iowa Native Prairie soil metagenome sequencing data [8] (minimum length $\geq$ 300 bp; average length 727 bp). The size of the dataset is about 7 Gbp. The second dataset contains MiSeq sequencing data from a single bacteria Chromohalobacter salexigens DSM-3043 (downloaded from NCBI with SRA number SRP057274)[2]. There are 40.6 million reads with average length 300 bp and the total query size is about 12 Gbp. The contigs and reads were aligned to the NCBI-nr database (as of 2015/05/09) which contains 23.8G amino acids.

First, we compared the performance of DIAMOND and AC-DIAMOND on aligning contigs to the NCBI-nr database. Six sets of query contigs were extracted from the first dataset, containing 1 Gbp, 2 Gbp, 3 Gbp, 4 Gbp, 5 Gbp and 6 Gbp of contigs, respectively. These six query sets, after translated into protein sequences, contained 2G, 4G, 6G, 8G, 10G and 12G amino acids, respectively. They are each aligned to the NCBI-nr database using DIAMOND and AC-DIAMOND with their default memory settings. Table 1 shows the running time. DIAMOND and AC-DIAMOND, running with own default settings, had similar peak memory consumption, namely 37 GB and 40 GB, respectively (note that DIAMOND loaded at most 2G aa of queries and reference sequences in each round, while AC-DIAMOND,

---

[2] http://www.ncbi.nlm.nih.gov/sra/SRX1000158[accn].

**Table 1.** Comparison of DIAMOND and AC-DIAMOND on aligning contigs (average length 727 bp) to NCBI-nr. Their running times (both DP only & overall) were recorded for six query sets of contigs with increasing total size (measured by the number of amino acids after translation). AC-DIAMOND was consistently faster, as reflected by the speed-up ratios below. DIAMOND and AC-DIAMOND were running with their default settings and had similar peak memory usage, which was recorded to be 37 GB and 40 GB in all six experiments, respectively.

| Total size (aa) of query contigs | Dynamic programming time (hours) | | | Total running time (hours) | | |
|---|---|---|---|---|---|---|
| | DIAMOND | AC-DIAMOND | Speed-up | DIAMOND | AC-DIAMOND | Speed-up |
| 2G | 15.54 | 2.37 | 6.56 | 17.39 | 4.05 | 4.29 |
| 4G | 30.84 | 4.68 | 6.59 | 34.52 | 7.57 | 4.56 |
| 6G | 45.62 | 6.92 | 6.59 | 51.14 | 11.50 | 4.45 |
| 8G | 60.02 | 9.08 | 6.61 | 67.35 | 14.83 | 4.54 |
| 10G | 74.22 | 11.23 | 6.61 | 83.37 | 18.61 | 4.48 |
| 12G | 87.05 | 13.16 | 6.61 | 97.85 | 21.60 | 4.53 |

being more memory-efficient, loaded 4G aa of queries and 2G aa of references in each round). We will consider the performance with more memory later.

Table 1 shows that AC-DIAMOND achieved a speed-up ratio of about 4.5 for the total running time and 6.6 for the DP (dynamic programming) step only. Note that the DP step is the bottleneck. Making the DP step faster gives significant improvement in the overall performance. The speed-up ratios for different query sets are consistent. The speed-up for the 4G, 8G and 12G query sets is a bit higher than the other three query sets because the former are multiples of 4G, i.e., the batch query batch size used by AC-DIAMOND. We also compared the alignment results produced by the two tools. We find that 99.99 % of contigs reported by these two tools are the same. On one hand, there are 397 contigs reported by DIAMOND but not by AC-DIAMOND. On the other hand, there are 804 contigs reported by AC-DIAMOND but not by DIAMOND.

Next, we compared DIAMOND and AC-DIAMOND on aligning reads (Table 2). We extracted the first 1 Gbp, 2 Gbp, 3 Gbp, 4 Gbp, 5 Gbp and 6 Gbp of reads from the second dataset and constructed six sets of queries of size 2G, 4G, 6G, 8G, 10G and 12G aa respectively. The reads were aligned to NCBI-nr database using DIAMOND and AC-DIAMOND with default setting. Note that the reads were much shorter than contigs of the previous experiment, yet AC-DIAMOND still achieved a speed-up of over 4.1 on total running time. Similar to the previous experiment, the speed-up ratio for query sets with size multiple of 4G were a bit higher than the others. But the speed-up ratio for the DP step was higher than in the experiment on contigs (7.3 against 6.6), this was because for the read dataset, a seed in a query read often matched more seeds in reference sequences (precisely, the average number increased from 12.2 to 14.2), and hence there was a higher chance for AC-diamond to pack more DP steps together for SIMD processing.

Similar to the contig experiment, we find that over 99.99 % of reads reported by these two tools are the same. More specifically, there are 55 reads reported

**Table 2.** Comparison of DIAMOND and AC-DIAMOND on aligning reads (average length 300 bp) to NCBI-nr. Six input files of reads with increasing total sizes were used. DIAMOND and AC-DIAMOND were running with their default settings. Their peak memory usages were 33 GB and 35 GB in all six experiments, respectively.

| Total size (aa) of query reads | Dynamic programming time (hours) | | | Total running time (hours) | | |
|---|---|---|---|---|---|---|
| | DIAMOND | AC-DIAMOND | Speed-up | DIAMOND | AC-DIAMOND | Speed-up |
| 2G | 14.12 | 1.93 | 7.32 | 16.39 | 4.05 | 4.05 |
| 4G | 28.40 | 3.88 | 7.32 | 32.96 | 7.71 | 4.27 |
| 6G | 42.26 | 5.77 | 7.32 | 49.05 | 11.72 | 4.19 |
| 8G | 53.40 | 7.29 | 7.33 | 62.16 | 14.60 | 4.26 |
| 10G | 64.53 | 8.81 | 7.32 | 75.25 | 18.00 | 4.18 |
| 12G | 72.60 | 9.91 | 7.33 | 84.83 | 20.03 | 4.24 |

**Table 3.** Comparison of DIAMOND and AC-DIAMOND in six different memory settings on aligning contigs to NCBI-nr. All six experiments used the same query set of contigs with 12G aa. When more memory was used, both software run faster. Note that the speed-up ratio of AC-DIAMOND increased with more memory. In each setting, we adjusted the maximum size of contig queries and reference sequences DIA-MOND and AC-DIAMOND loaded into the memory in each round, and then we measured the peak memory usage. Details of each setting: **(a)** DIAMOND: default (contigs 2G aa; references 2G aa); AC-DIAMOND: (contigs 2G aa; references 2G aa); **(b)** DIA-MOND: default; AC-DIAMOND: default (contigs 4G aa; references 2G aa); **(c)** Both DIAMOND and AC-DIAMOND: (contigs 3G aa; references 3G aa); **(d)** DIAMOND: (contigs 3G aa; references 3G aa); AC-DIAMOND: (contigs 6G aa; references 3G aa) **(e)** Both DIAMOND and AC-DIAMOND: (contigs 4G aa; references 4G aa); **(f)** DIA-MOND: (contigs 4G aa; references 4G aa); AC-DIAMOND: (contigs 8G aa; references 4G aa)

| Memory usage Setting | Peak memory (GB) | | Total running time (hours) | | |
|---|---|---|---|---|---|
| | DIAMOND | AC-DIAMOND | DIAMOND | AC-DIAMOND | Speed-up |
| (a) | 37 | 34 | 97.85 | 22.99 | 4.26 |
| (b) | 37 | 40 | 97.85 | 21.60 | 4.53 |
| (c) | 49 | 44 | 93.99 | 20.35 | 4.62 |
| (d) | 49 | 57 | 93.99 | 19.56 | 4.81 |
| (e) | 62 | 56 | 93.39 | 19.36 | 4.82 |
| (f) | 62 | 73 | 93.39 | 18.97 | 4.92 |

by DIAMOND but not by AC-DIAMOND, and 288 reads reported by AC-DIAMOND but not by DIAMOND.

Finally, we evaluated the performance of DIAMOND and AC-DIAMOND when more queries or reference sequences were loaded into the memory in each round (i.e., more memory was required). All experiments were based on aligning the same query set (12G aa of contigs, the largest query set in Table 1) to NCBI-nr. Results are shown in Table 3 and Fig. 1.

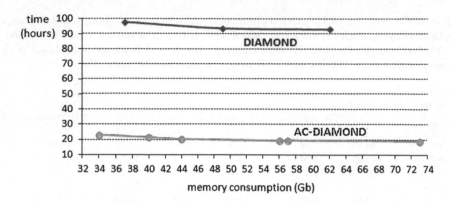

**Fig. 1.** The running time of DIAMOND and AC-DIAMOND, when using different memory settings, on aligning contigs (total size 12G aa) to NCBI-nr

By default DIAMOND loaded 2G aa queries and 2G aa references in each round and used 37 GB of memory. For this setting, AC-DIAMOND used 34 GB of memory (see Table 3, setting (a)). AC-DIAMOND with its own default loaded 4G aa queries and 2G aa references, and used 40 GB of memory. Note that DIA-MOND does not allow the sizes of queries and references to be independently configured.

With more memory, both software run faster. Percentage-wise, Table 3 shows that AC-DIAMOND benefited more than DIAMOND. For example, let us compare Settings (a) and (e), in which respectively 2G and 4G aa queries and references were loaded in each round. DIAMOND's running time decreased by 4.56 %, and AC-DIAMOND has an improvement of 15.79 %. Note that the speed-up ratio of AC-DIAMOND increased from Setting (a) to Setting (f). Intuitively, more memory favors AC-DIAMOND because it allows a query seed to match more reference seeds and more DP steps to pack together for SIMD processing.

## 4    Conclusion

AC-DIAMOND takes advantage of the SIMD technology and compressed index-ing to resolve the computational bottleneck of DIAMOND and obtain a few-fold speed up for the DNA-protein alignment. It is worth-mentioning that the speed-up is not a the sacrifice the alignment sensitivity of DIAMOND. To obtain further speed-up, we plan to investigate how to exploit GPUs to share the workload.

## References

1. Buchfink, B., Xie, C., Huson, D.H.: Fast and sensitive protein alignment using DIA-MOND. Nat. Methods **12**(1), 59–60 (2015)
2. Ye, Y., Choi, J.H., Tang, H.: RAPSearch: a fast protein similarity search tool for short reads. BMC Bioinform. **12**(1), 159 (2011)

3. Zhao, Y., Tang, H., Ye, Y.: RAPSearch2: a fast and memory-efficient protein similarity search tool for next-generation sequencing data. Bioinformatics **28**(1), 125–126 (2012)
4. Huson, D.H., Xie, C.: A poor man's BLASTX-high-throughput metagenomic protein database search using PAUDA. Bioinformatics, btt254 (2013)
5. Suzuki, S., Kakuta, M., Ishida, T., Akiyama, Y.: Faster sequence homology searches by clustering subsequences. Bioinformatics, btu780 (2015)
6. Altschul, S.F., Gish, W., Miller, W., Myers, E.W., Lipman, D.J.: Basic local alignment search tool. J. Mol. Biol. **215**(3), 403–410 (1990)
7. Li, D., Liu, C.M., Luo, R., Sadakane, K., Lam, T.W.: MEGAHIT: an ultra-fast single-node solution for large and complex metagenomics assembly via succinct de Bruijn graph. Bioinformatics, btv033 (2015)
8. Howe, A.C., Jansson, J.K., Malfatti, S.A., Tringe, S.G., Tiedje, J.M., Brown, C.T.: Tackling soil diversity with the assembly of large, complex metagenomes. Proc. Nat. Acad. Sci. **111**(13), 4904–4909 (2014)
9. Smith, T.F., Waterman, M.S.: Identification of common molecular subsequences. J. Mol. Biol. **147**(1), 195–197 (1981)

# Prioritization of Schizophrenia Risk Genes by a Network-Regularized Logistic Regression Method

Wen Zhang, Jhin-Rong Lin, Rubén Nogales-Cadenas,
Quanwei Zhang, Ying Cai, and Zhengdong D. Zhang[✉]

Department of Genetics, Albert Einstein College of Medicine,
Bronx, NY 10461, USA
zhengdong.zhang@einstein.yu.edu

**Abstract.** Schizophrenia (SCZ) is a severe mental disorder with a large genetic component. While recent large-scale microarray- and sequencing-based genome wide association studies have made significant progress toward finding SCZ risk variants and genes of subtle effect, the interactions among them were not considered in those studies. Using a protein-protein interaction network both in our regression model and to generate a SCZ gene subnetwork, we developed an analytical framework with Logit-Lapnet, the graphical Laplacian-regularized logistic regression, for whole exome sequencing (WES) data analysis to detect SCZ gene subnetworks. Using simulated data from sequencing-based association study, we compared the performances of Logit-Lapnet with other logistic regression (LR)-based models. We use Logit-Lapnet to prioritize genes according to their coefficients and select top-ranked genes as seeds to generate the gene sub-network that is associated to SCZ. The comparison demonstrated not only the applicability but also better performance of Logit-Lapnet to score disease risk genes using sequencing-based association data. We applied our method to SCZ whole exome sequencing data and selected top-ranked risk genes, the majority of which are either known SCZ genes or genes potentially associated with SCZ. We then used the seed genes to construct SCZ gene subnetworks. This result demonstrates that by ranking gene according to their disease contributions our method scores and thus prioritizes disease risk genes for further investigation. An implementation of our approach in MATLAB is freely available for download at: http://zdzlab.einstein.yu.edu/1/publications/LapNet-MATLAB.zip.

## 1 Introduction

SCZ is a common and severe lifelong brain disorder. It is a major cause of disability and reduces life expectancy by ~25 years on average. With its substantial mortality and morbidity, SCZ causes enormous personal and community burdens (Darves-Bornoz et al. 1995). In the United States, about 1 % of the general population, or 3 million

---

**Electronic supplementary material** The online version of this article (doi:10.1007/978-3-319-31744-1_39) contains supplementary material, which is available to authorized users.

F. Ortuño and I. Rojas (Eds.): IWBBIO 2016, LNBI 9656, pp. 434–445, 2016.
DOI: 10.1007/978-3-319-31744-1_39

Americans, suffer from this lifelong disabling illness (Regier et al. 1993). Thus, elucidating the etiology of the disease and developing effective treatment are of great medical
urgency. The heritability of SCZ is well established. Recent studies have revealed a
complex genetic architecture of the disease, involving multiple and heterogeneous
genetic factors. Risk variants range in frequency from common to extremely rare and
size from single nucleotide variants (SNVs) to large copy number variants (CNVs).
Since 2009, GWASs have identified around 50 SCZ-associated loci with genome-wide
statistical significance ($P < 5 \times 10^{-8}$) (Regier et al. 1993). Recently, a meta-analysis of
SCZ discovered 108 risk loci, providing a significant source for identifying causal
variants and causal genes of SCZ (Schizophrenia Working Group of the Psychiatric
Genomics 2014). Rare congenital disorders associated with structural variants at
22q11.2, 15q13.3, 1q21.1, and several other genomic locations count for relatively
small proportion of cases with SCZ (Bergen et al. 2012; Betcheva et al. 2013; Huang
et al. 2010; Irish Schizophrenia Genomics and the Wellcome Trust Case Control 2012;
Jeffrey A. Lieberman 2006; Shi et al. 2009; Shi et al. 2011; Wong et al. 2014).
Increasing number of structural variation burden has been also observed in SCZ cases
(Walsh et al. 2008).

Next-generation sequencing (NGS) has made it possible to evaluate the role of de
novo or rare SNVs, both previously essentially inaccessible, in SCZ with DNA samples
from parent-child trios or case-control cohorts. Using WES, instead of SNP microarray,
as the genotyping tool to obtain a complete picture of genetic variants in coding
sequences, a recent study assayed rare coding SNVs and small insertions and deletions
(indels) in 2,536 SCZ cases and 2,543 normal controls and demonstrated a polygenic
burden primarily arising from rare disruptive mutations distributed across many genes
(Purcell et al. 2014). Recently, a number of statistical tests have been designed for
WES-based variant analysis (Asimit and Zeggini 2010; Bansal et al. 2010; Basu and
Pan 2011; Stitziel et al. 2011). Most of these methods first aggregate variants in each
gene and then consider the association of each gene with the disease/phenotype separately. Hoffman et al. have developed a framework for applying a family of penalized
regression methods that simultaneously consider multiple susceptibility loci in the same
statistical model (Hoffman et al. 2013). In a more recent work, Larson and Schaid drew
on penalized regression in combination with variant collapsing measures to identify
rare variant enrichment in exome sequencing data (Larson and Schaid 2014).

Here we present a penalized regression method with graphical Laplacian network
regularization and variant aggregation measures for case-control WES data analysis to
assess gene contributions to the disease phenotypes. We first compared the performance of our regression method with other existing similar approaches using simulation under different scenarios. We then applied our method to the SCZ case-control
WES data to prioritize SCZ risk genes. We discuss how the genes and pathways that
we identified to make high contributions to SCZ may shed new light on genetic
structure behind the SCZ in general.

## 2  Results and Discussion

**Analysis of Simulated Phenotype and Genotype Data.** We first simulated WES data sets with phenotypes under four different scenarios (Supplementary Table 1 and Supplementary Methods) and then used them to evaluate the performance of our network-regularized regression method (Logit-Lapnet) and three existing ones (Logit, Lasso, and Enet). Each simulation was replicated 50 times. After computing the sensitivity and specificity on the cutoff paths, we plotted the receiver-operating characteristic curve (ROC) and calculate the area under curve (AUC) of each method (Fig. 1). As its AUC is the largest under all four different simulation scenarios, the Logit-Lapnet method outperforms all other three. A similar performance assessment can also be made on the regularization path (Wan et al. 2013; Zhang et al. 2013) (Supplementary Fig. 1). From all the results, we could conclude that given available alternatives Logit-Lapnet is the best choice for prioritizing candidate genes among this class of algorithms.

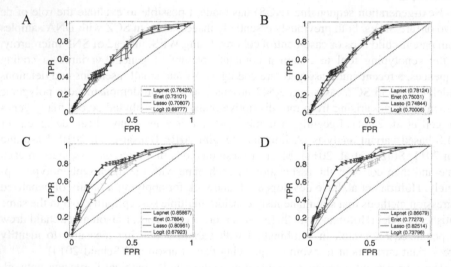

**Fig. 1.** ROCs with simulated samples. (A)–(D). Simulation under scenarios 1–4. All simulations were done with 100 samples. Average ROCs of four logistic regression methods are shown in different colors. The error bar indicates the standard deviation of all replicates for each sample. Their corresponding AUCs are given in parentheses in the figure legend.

To assess how the sample size affects the performance, we also simulated data sets with different sample sizes ranging from 10 to 200 under each scenario. For each sample size, we simulated every model 40 times and calculated the average AUCs (Supplementary Figs. 2 and 3). Over all, the LR and its extensions gave good analysis results for simulated case-control phenotype-genotype data– the AUC ranged from 0.6 to 0.9. The assessment of the Lasso and the Enet methods is less clear: one outperforms the other in each of two simulation scenarios. We also calculated and compare the F1 scores of regression methods. F1 score is the harmonic mean of precisions and recalls and acts as the integration of both of these two evaluations. This property makes it an informative and efficient measurement of performance of different methods. Here, the plot shows that under all four simulation conditions the F1 scores of the Logit-Lapnet method are the highest among four methods being studied (Supplementary Fig. 4). The F1 score comparison indicates that the Logit-Lapnet algorithm is more accurate than Logit, Lasso, and Enet methods.

Due to feasibility and efficiency of the Logit-Lapnet algorithm in prioritizing risk genes, we could further apply the method to real WES data set and get prioritizations of the genes so as to identify important genes relating to the disease that under consideration. Our motivation to develop the method for association studies is based on the hypothesis that integration of interaction networks improves prediction precision of logistic models. The network-constrained algorithm has been proved to out-perform alternative options such as lasso and elastic net analyses that are implemented separately from biological input (Li and Li 2008; Wan et al. 2013; Zhang et al. 2013). Enlightened by the application of Logit-Lapnet to efficiently identify molecular pathways and cancer biomarkers, we adapted this class of methods to analyze a set of WES data for SCZ. We use the ratio matrix of damaging allele counts over neutral allele counts to represent normalized population genotype information. The Logit-Lapnet approach is more sensitive for identifying disease genes because the relevant network modules are considered by using the regularization based on the network. Laplacian graphs are derived from gene networks, for which we used High-quality INTeractomes (HINT) network in our study. HINT is a database of high-quality protein-protein interactions in different organisms, which have been compiled from different data sources and then filtered both systematically and manually to remove erroneous and low-quality interactions (Das and Yu 2012).

Combining information of gene interactions, the Laplacian graphs form the penalized term with regard to contribution coefficient of each gene. The $L_2$-normalized item incorporates network information into the estimation procedure of the regression model and encourages smoothness in the estimate of contributions of candidate genes. Incorporation of a gene network contributes to the advantages of Logit-Lapnet over the other methods since in this way the method integrates into its calculation a vast amount of *a priori* biological information from the network, which is ignored in either lasso or elastic net methods. In summary, our method takes advantage of the information obtained about genotype relationships beyond the scope of other single regression study.

**Analysis of SCZ WES Data.** The simulation results clearly indicate that the LR and its extensions can be effectively applied to case-control genotype data to identify genes related to the phenotype or disease under consideration and the Logit-Lapnet method gives the best performance. Here, we applied this method to the SCZ WES data to estimate the corresponding coefficients as phenotypic contributions of SCZ target genes. First, we derived from the WES data the phenotype vector **y**, the gene evaluation data matrix **X**, and the normalized graph Laplacian matrix **L**. Corresponding to the SCZ patient cohort, **y** is a binary column vector with the elements: 1's for patients with SCZ and 0's for ones without SCZ. **X** and **L** are $n \times 844$ and $844 \times 844$ matrices, respectively, where $n$ is the number of individuals (cases and controls) and 844 is the number of candidate genes, some of which could be responsible for SCZ in the original WES study cohort. In the SCZ WES data set from dbGaP, there are 2,545 cases and 2,545 controls in total. Maximizing the Logit-Lapnet function (Eq. 3) is a computationally intensive process (Supplementary Fig. 5), which necessitates parallel processing of large data sets. We randomly divided the SCZ WES data set into 25 subsets, each with $\sim 100$ cases and $\sim 100$ controls. We analyzed them in parallel and then integrated gene scores. Using the Logit-Lapnet method, we estimated from these subsets in parallel the coefficients of candidate genes as their contributions to SCZ. We arranged the genes in each list in descending order of their coefficients and integrated the ranked gene lists using a robust rank aggregation method (Kolde et al. 2012). We randomized the SCZ WES data set to evaluate this ranked gene list and to remove possible false positives. In each iteration, we randomized the disease labels among the samples and processed the random data set in the same manner as the real one. After many iterations of randomization, for each gene we calculated the probability that the rank of this gene based on a random data set is the same as or even better than that based on the real data set. After removing genes with probability ≥0.05, we then considered the top 20 genes as the most promising candidates for SCZ in the WES study cohort (Supplementary Table 2). 10 of them including CARD10, TIMP2, PPP2CA, and PTPRB were also identified as SCZ risk genes by the original exome sequencing study (Purcell et al. 2014).

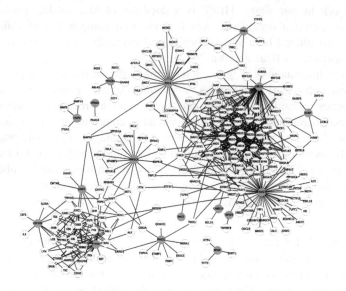

**Fig. 2.** SCZ gene subnetwork. After filtering by randomization, 20 genes with largest regression coefficients were used as seeds to form the subnetwork in HINT. Gray and white nodes represent the seed genes and their direct neighbors, respectively, in the network. Three seed genes are not included in HINT and thus are omitted from the subnetwork.

Because most cellular components exert their functions through interactions with other cellular components, such inter- and intracellular inter-connectivity implies that the impact of a specific genetic variation is not restricted to the activity of an SCZ-related gene product that carries it, but can spread along the links of the network and alter the activity of other SCZ-related gene products that otherwise carry no changes. Therefore, an understanding of SCZ genes' network context is essential to understand the genetics of this disease. Using the aforementioned top 20 genes as seeds and the 'extraction by shell' method implemented in SubNet (Lemetre et al. 2013), we extracted a GsN (Fig. 2) from HINT (Das and Yu 2012), a high-quality protein-protein interaction network. This SCZ GsN contains 223 proteins and 546 interactions among them. The majority of the proteins (207, 92.83 %) form a connected component. Despite its small size, this subnetwork clearly shows a power law distribution for its node degrees: a prominent characteristic of complex biological networks. Unlike the majority of proteins in the subnetwork, a few of them have a large number of inter-action partners and are functionally more important. Many seed genes are such network hubs in the SCZ GsN. PHC2, one network hub, was found to be affected by mutations in SCZ patients (Purcell et al. 2014). Known SCZ loci 5q31.1, 6p22, and 12q22 contain, respectively, PPP2CA, TRIM27, and UBE2 N, which are among top 20 risk genes that we identified for SCZ in the WES study cohort. In addition, these three genes contain nonsynonymous coding variants with minor allele frequencies (MAF) less than 0.1 %. Located in 12q15-q21, PTPRB, another seed gene with high contribution, contain nonsynonymous coding variants with MAF < 0.5 %.

As a proof of principle, our SCZ study demonstrates that our data analysis work-flow and methods can be successfully applied to WES data sets to identify disease risk genes and subnetworks. Although they are designed to be applicable to large GWAS and WES data sets, such as those provided by dbGaP, the implementation in main text of this paper processes data sets with moderate sample sizes. Because individual genotypes have larger effect on transcript abundance than on disease risk, a small sample size can still be powerful to detect disease variants and genes (Gibson 2014). To analyze larger ones, the optimization problem – the rate limiting step – needs to be solved by more efficient optimizers or parallel computing or both. This difficulty is still an open challenge and an active research area. For comparison of running time with larger different sample sizes, please see Supplementary Fig. 5.

# 3   Methods

**Input data for WES Regression Analysis.** After SNVs and indels in the sequenced subjects are identified and their genotypes called, while assuming the reference alleles (RAs) of these variants to be neutral, we predicted the functional consequence (i.e., neutral/tolerable/benign or damaging/deleterious) of the alternative alleles (AAs) using computational programs. Let $n$ and $p$ be the numbers of genes and sequenced subjects (samples), respectively. To carry out the regression analysis of the case-control WES data, we first summarized the genotypes and the allelic functional annotations on gene level in two $n$-by-$p$ matrices, $\mathbf{D}$ and $\mathbf{N}$, which hold all of the numbers regarding the neutral and damaging allele counts of each gene $i$ respectively in each sample $j$. As longer genes tend to have more variants, to prevent the gene length from skewing up the analysis, we normalized the damaging allele counts through the way of getting them divided by neutral allele counts. Since each gene may contain multiple variants (SNPs and indels), we counted the damaging and the neutral alleles of all variants within a gene. In this way, multiple variants mapped to one gene are combined to obtain the allele counts on the gene level. Most of the sequencing errors are filtered out at the quality control steps as part of the variant calling process. For any remaining ones, because they occur randomly, their effect will be cancelled out in a case-control study. Thus, we define the input ratio matrix $X$ as in Eq. 1.

$$X := [X_{ij}], D := [D_{ij}], N := [N_{ij}], X_{ij} = \frac{D_{ij}}{N_{ij} + \tau}, \tau = \begin{cases} 0 & \text{if } N_{ij} \neq 0 \\ 1 & \text{otherwise} \end{cases} \quad (1)$$

Due to computational intensity, instead of scoring the whole gene set, the current implementation of our method scores and thus prioritizes a set of genes preselected based on prior knowledge given the genotypic data. This set of genes can be obtained in two steps. First, a core group of genes can be collected from various disease gene databases (Goh et al. 2007; Pinero et al. 2015; Pletscher-Frankild et al. 2015; Rappaport et al. 2014). Then, adding their neighboring genes in a gene network can augment this core group of genes. Even for less commonly studied diseases, this approach can procure a set of relevant genes for scoring. Genes selected for scoring and prioritization are included in matrices $X$ and $L$ for analysis. It is the disease, not the sample of the data set, that determines what genes to be selected. Therefore, the same set of genes included in $X$ and $L$ will be used for two different samples of the same disease. If the disease has different genetic risk factors in these two samples, then the ranking of selected genes will be different for these two samples in the results. For flowchart to further illustrate the process of input data integration, please refer to Supplementary Fig. 6. Pre-processing of the WES data by generating the ratio data matrix is a novelty of the method. Current studies do not provide this kind of data matrix generation in analyzing WES data.

**Graphically Laplacian Network-Regularized LR Method.** In a recent study a network-regularized linear regression method has been proposed for variable selections (Li and Li 2008). Compared with Lasso and elastic net (Enet) methods, they show advantages of using the network-regularized process, which include higher precisions and comparable or even better sensitivity and specificity. In other variable selection problems where the output vectors only contain binary values (i.e., either 0 or 1), the LR with network regularization outperforms previous alternatives (Zhang et al. 2013). The Laplacian graphical network-regularized LR methods (Logit-Lapnet) have been used to identify molecular pathways of breast cancers (Zhang et al. 2013). Its application to simulated gene expression data showed excellent sensitivity and specificity and higher accuracy than Lasso and Enet methods. Inspired by its original application, we adapted the Logit-Lapnet method with significant redesign to analyze the case-control genotype data, focusing on NGS-generated data. Different from previous studies which use Logit-Lapnet to analyze the gene expression data (Zhang et al. 2013), we utilize the method to WES data analysis.

Let $\mathbf{y} = [y_1, ..., y_n]^T$ be the phenotype vector, where $y_i = 0$ for control and 1 for case ($i = 1, ...., n$). Given matrix $\mathbf{X}$, the aforementioned gene evaluation data matrix $\mathbf{Y}$ is modeled by logistic function:

$$Y = \Pr(y = 1|\mathbf{X}; \boldsymbol{\theta}) = \frac{e^{\mathbf{X}\theta}}{1 + e^{\mathbf{X}\theta}} \tag{2}$$

The regression coefficients in $\boldsymbol{\theta}$ quantify the effect of the genotypes of genes on the 'odds ratio' of having the disease and thus represent the importance of genes. Genes with larger coefficients will be ranked higher than others (Li and Li, 2008). The coefficient vector $\boldsymbol{\theta}$ is estimated by minimizing the negative log-likelihood function, or equivalently, maximizing the positive function, of logistic model combined with penalized terms. The mitigated formula of logistic graph Laplacian net criteria is

$$\mathbb{C}(\boldsymbol{\theta}, \lambda, \alpha) = \sum_{i=1}^{n} \left[ -y_i X_i \boldsymbol{\theta} + \ln\left(1 + e^{X_i \theta}\right) \right] + \lambda \alpha |\boldsymbol{\theta}|_1 + \lambda(1 - \alpha)\boldsymbol{\theta}^T \mathbf{L}\theta \tag{3}$$

where $X_i$ is the $i$-th row-vector of data matrix $\mathbf{X}$. $\mathbf{L}$ is the normalized graph Laplacian matrix, $|\boldsymbol{\theta}|_1$ is the $L_1$ norm of $\boldsymbol{\theta}$, i.e., $|\boldsymbol{\theta}|_1 = \sum_{j=1}^{p} \theta_j$, with $\theta_j$ corresponding contribution coefficient of each gene. Suppose $\mathbf{A}$ and $\mathbf{E}$ are adjacency matrix and degree matrix of the network, respectively, then $\mathbf{L}$ is given as:

$$\mathbf{L} = \mathbf{I} - \mathbf{E}^{-\frac{1}{2}}\mathbf{A}\mathbf{E}^{-\frac{1}{2}} \tag{4}$$

where $\mathbf{I}$ is the identity matrix with the same dimension to that of $\mathbf{A}$ or $\mathbf{E}$.

Equation 3 contains three terms: the negative log-likelihood function; the L1 normalized penalty term, which L1 penalizes the norm of $\theta$; and the graph Laplacian term, which is formulated as the inner product of $\theta$ regarding to Laplacian matrix L in (Zhang et al. 2013). The last term can be treated as the L2 normalized item as well. Equation 3 also makes it clear that the Laplacian network-regularized LR is the ordinary form of general logistic regressions, which include several more special types of logistic models. In case when L = I, the algorithm becomes an Enet module. When $\alpha = 1$ and $\lambda \neq 0$, the method is regressed to Lasso. If $\lambda = 0$, the method is further simplified as a standard LR model without penalties (referred to as the Logit model hereafter). It is clear that Logit-Lapnet, Enet, Lasso, and Logit are LR methods with different levels of constraints on regression coefficients, known as contributions in our methods. Given the data matrix X, the optimal values for the model parameters $\alpha$ and $\lambda$ are determined by a leave-one-out cross validation (CV) procedure, and the optimal coefficients $\theta$ are estimated by minimizing the criteria $\mathbb{C}(\theta, \lambda, \alpha)$ given optimal parameters $\lambda_{opt}$ and $\alpha_{opt}$:

$$\theta^* = \underset{\theta}{\arg\min}\, \mathbb{C}\big(\theta, \lambda_{\mathrm{opt}}, \alpha_{\mathrm{opt}}\big) \tag{5}$$

Convexity of Eq. 3 guarantees minimal index performance $\mathbb{C}\big(\theta^*, \lambda_{\mathrm{opt}}, \alpha_{\mathrm{opt}}\big)$, and the corresponding $\theta^*$ could be worked out by a standard optimizer such as the Matlab-based software CVX (Grant and Boyd 2014). For theoretical properties of Logit-Lapnet, please refer to Lemma 1, 2, 3 and Theorem 1 in (Zhang et al. 2013) and the references therein. The Lemma 3 and Theorem 1 in (Zhang et al. 2013) provide grouping effects for the Logit-Lapnet procedure. Mathematical formulation in (Zhang et al. 2013) supports our novel application of Logit-Lapnet to risk gene prioritization.

**Processing SCZ WES Genotype Data.** The Swedish SCZ population-based case and control WES data set (study accession ID number: phs000473.v1.p1) was downloaded from dbGaP. Excluding the variants in non-coding regions, we extracted two sets of variants – SNVs and indels – in exons of the sequenced subjects from the WES data set. For SNVs, the functional roles of their AAs were predicted by SIFT (Hu and Ng 2013), PolyPhen2 (Sachdev and Keshavan 2010), and Blosum62 scoring matrices. We combined these three scores using a so-called 'damaging-dominant' rule – the AA was considered as damaging as long as one method predicts so. We used this policy since the variant annotations are usually predicted by and selected among high impacts and damaging, in this sense, is the higher impact compared with neutral/tolerant. For indels, we used SIFT to predict the functional roles of their AAs. RAs were considered neutral in this regard. We derived the **D** and **N** matrices (Eq. 1) for SNVs and indels separately from their functional annotations. After combining **D** and **N** of SNVs and indels separately, we calculated **X** and use it as the input data matrix. The ratio matrix reflects the relative genetic influences of damaging alleles in each gene on the disease status.

**Compiling SCZ-related Genes.** Derived from the SCZ WES genotype data, the input matrix X holds the damage load for 13,899 sequenced genes in 200 samples. Given the large number of genes and the modest sample size, to keep the statistical analysis tractable we focused our analysis on genes likely to be related to SCZ according to prior knowledge. Our strategy for gene selection was to include both genes most relevant to SCZ and ones with potential but unknown associations with SCZ. We used a two-tier approach. First, we compiled a list of 308 genes that have been shown to be SCZ-related:

- 217 genes with prior evidence for association with SCZ, which are prioritized in the data source SZGR (Jia et al. 2010).
- 91 genes collected from published literatures (Supplementary Table 3).

Next, based on the 'guilt-by-association' principle, we collected 536 direct neighbors of these 308 genes in the HINT. Together, we selected 844 candidate genes.

**Networks used in SCZ WES Data Analysis.** We used HINT with the Logit-Lapnet method for our SCZ WES data analysis. From HINT, the Laplacian matrix of the aforementioned target genes was generated and then used as the graphical Laplacian normalized term in Eq. 2.

# 4    Conclusion

We developed a computational framework for WES data analysis that combines both prioritization of disease risk genes by graphical Laplacian regularized LR and extraction of disease-related GsN by SubNet. Although the Logit-Lapnet method has been used before to analyze gene expression data and somatic mutation profiles (Betcheva et al. 2013; Hoffman et al. 2013; Shi et al. 2009; Stitziel et al. 2011), our study demonstrates here that after data transformation it can also be efficiently applied to exome sequencing-based GWAS genotype data. Method assessment by simulation shows that Logit-Lapnet is more sensitive for identifying seed genes with higher priorities than other related methods. We applied our method to SCZ WES data. Top-ranked genes are either known SCZ risk genes or closely related to SCZ. Using them as seeds, we extracted the SCZ GsN from known protein interaction network. It provides a valuable subnetwork for pathway and gene module detection of SCZ.

**Acknowledgements.** This work was supported by the NIH Pathway to Independence Award from National Library of Medicine (5R00LM009770-06) and the American Heart Association Grant-in-Aid (13GRNT16850016) to Z.D.Z.

# References

Asimit, J., Zeggini, E.: Rare variant association analysis methods for complex traits. Ann. Rev. Genet. **44**, 293–308 (2010)

Bansal, V., Libiger, O., Torkamani, A., Schork, N.J.: Statistical analysis strategies for association studies involving rare variants. Nature Rev. Genet. **11**, 773–785 (2010)

Basu, S., Pan, W.: Comparison of statistical tests for disease association with rare variants. Genet. Epidemiol. **35**, 606–619 (2011)

Bergen, S.E., O'Dushlaine, C.T., Ripke, S., Lee, P.H., Ruderfer, D.M., Akterin, S., Moran, J.L., Chambert, K.D., Handsaker, R.E., Backlund, L., et al.: Genome-wide association study in a Swedish population yields support for greater CNV and MHC involvement in schizophrenia compared with bipolar disorder. Mol. Psychiatry **17**, 880–886 (2012)

Betcheva, E.T., Yosifova, A.G., Mushiroda, T., Kubo, M., Takahashi, A., Karachanak, S.K., Zaharieva, I.T., Hadjidekova, S.P., Dimova, I.I., Vazharova, R.V., et al.: Whole-genome-wide association study in the Bulgarian population reveals HHAT as schizophrenia susceptibility gene. Psychiatr. Genet. **23**, 11–19 (2013)

Danecek, P., Auton, A., Abecasis, G., Albers, C.A., Banks, E., DePristo, M.A., Handsaker, R.E., Lunter, G., Marth, G.T., Sherry, S.T., et al.: The variant call format and VCFtools. Bioinformatics **27**, 2156–2158 (2011)

Darves-Bornoz, J.M., Lemperiere, T., Degiovanni, A., Gaillard, P.: Sexual victimization in women with schizophrenia and bipolar disorder. Soc. Psychiatry Psychiat. Epidemiol. **30**, 78–84 (1995)

Das, J., Yu, H.: HINT: High-quality protein interactomes and their applications in understanding human disease. BMC Syst. Biol. **6**, 92 (2012)

Gibson, G.: A primer of human genetics (Sinauer Associates, Inc.) (2014)

Goh, K.I., Cusick, M.E., Valle, D., Childs, B., Vidal, M., Barabasi, A.L.: The human disease network. Proc. National Acad. Sci. US Am. **104**, 8685–8690 (2007)

Grant, M., Boyd, S.: CVX: Matlab software for disciplined convex programming, version 2.1 (2014)

Hoffman, G.E., Logsdon, B.A., Mezey, J.G.: PUMA: A unified framework for penalized multiple regression analysis of GWAS data. PLoS Comput. Biol. **9**, e1003101 (2013)

Hu, J., Ng, P.C.: SIFT Indel: Predictions for the functional effects of amino acid insertions/deletions in proteins. PLoS ONE **8**, e77940 (2013)

Huang, J., Perlis, R.H., Lee, P.H., Rush, A.J., Fava, M., Sachs, G.S., Lieberman, J., Hamilton, S.P., Sullivan, P., Sklar, P., et al.: Cross-disorder genomewide analysis of schizophrenia, bipolar disorder, and depression. Am. J. Psychiatry **167**, 1254–1263 (2010)

Irish Schizophrenia Genomics, C., and the Wellcome Trust Case Control, C. Genome-wide association study implicates HLA-C*01:02 as a risk factor at the major histocompatibility complex locus in schizophrenia. Biological psychiatry **72**, 620–628 (2012)

Jeffrey, A., Lieberman, T.S.S., Perkins, D.O.: Textbook of Schizophrenia, The American Psychiatric Publishing, Arlington, Virginia, USA (2006)

Jia, P., Sun, J., Guo, A.Y., Zhao, Z.: SZGR: A comprehensive schizophrenia gene resource. Mol. Psychiatry **15**, 453–462 (2010)

Kim, S., Jeong, K., Bafna, V.: Wessim: A whole-exome sequencing simulator based on in silico exome capture. Bioinformatics **29**, 1076–1077 (2013)

Kolde, R., Laur, S., Adler, P., Vilo, J.: Robust rank aggregation for gene list integration and meta-analysis. Bioinformatics **28**, 573–580 (2012)

Larson, N.B., Schaid, D.J.: Regularized rare variant enrichment analysis for case-control exome sequencing data. Genet. Epidemiol. **38**, 104–113 (2014)

Lemetre, C., Zhang, Q., Zhang, Z.D.: SubNet: A Java application for subnetwork extraction. Bioinformatics **29**, 2509–2511 (2013)

Li, C., Li, H.: Network-constrained regularization and variable selection for analysis of genomic data. Bioinformatics **24**, 1175–1182 (2008)

Ng, S.B., Turner, E.H., Robertson, P.D., Flygare, S.D., Bigham, A.W., Lee, C., Shaffer, T., Wong, M., Bhattacharjee, A., Eichler, E.E., et al.: Targeted capture and massively parallel sequencing of 12 human exomes. Nature **461**, 272–276 (2009)

Pinero, J., Queralt-Rosinach, N., Bravo, A., Deu-Pons, J., Bauer-Mehren, A., Baron, M., Sanz, F., Furlong, L.I.: DisGeNET: A discovery platform for the dynamical exploration of human diseases and their genes. Database J. Biol. Databases Curation **2015**, 28 (2015)

Pletscher-Frankild, S., Palleja, A., Tsafou, K., Binder, J.X., Jensen, L.J.: DISEASES: Text mining and data integration of disease-gene associations. Methods **74**, 83–89 (2015)

Purcell, S.M., Moran, J.L., Fromer, M., Ruderfer, D., Solovieff, N., Roussos, P., O'Dushlaine, C., Chambert, K., Bergen, S.E., Kahler, A., et al.: A polygenic burden of rare disruptive mutations in schizophrenia. Nature **506**, 185–190 (2014)

Rappaport, N., Twik, M., Nativ, N., Stelzer, G., Bahir, I., Stein, T.I., Safran, M., Lancet, D.: Malacards: A comprehensive automatically-mined database of human diseases. Current protocols in bioinformatics/editoral board, Andreas D Baxevanis [et al.] 47, 1 24 21–21 24 19 (2014)

Regier, D.A., Narrow, W.E., Rae, D.S., Manderscheid, R.W., Locke, B.Z., Goodwin, F.K.: The de facto US mental and addictive disorders service system. Epidemiologic catchment area prospective 1-year prevalence rates of disorders and services. Arch. Gen. Psychiatry **50**, 85–94 (1993)

Sachdev, P.S., Keshavan, M.S.: Secondary Schizophrenia. United Kingdom at the University Press, Cambridge (2010)

Schadt, E.E., Linderman, M.D., Sorenson, J., Lee, L., Nolan, G.P.: Computational solutions to large-scale data management and analysis. Nature Rev. Genet. **11**, 647–657 (2010)

Schizophrenia Working Group of the Psychiatric Genomics Consortium: Biological insights from 108 schizophrenia-associated genetic loci. Nature **511**, 421–427 (2014)

Shi, J., Levinson, D.F., Duan, J., Sanders, A.R., Zheng, Y., Pe'er, I., Dudbridge, F., Holmans, P. A., Whittemore, A.S., Mowry, B.J., et al.: Common variants on chromosome 6p22.1 are associated with schizophrenia. Nature **460**, 753–757 (2009)

Shi, Y., Li, Z., Xu, Q., Wang, T., Li, T., Shen, J., Zhang, F., Chen, J., Zhou, G., Ji, W., et al.: Common variants on 8p12 and 1q24.2 confer risk of schizophrenia. Nat. Genet. **43**, 1224–1227 (2011)

Sim, N.L., Kumar, P., Hu, J., Henikoff, S., Schneider, G., Ng, P.C.: SIFT web server: predicting effects of amino acid substitutions on proteins. Nucleic Acids Res. **40**, W452–W457 (2012)

Stitziel, N.O., Kiezun, A., Sunyaev, S.: Computational and statistical approaches to analyzing variants identified by exome sequencing. Genome Biol. **12**, 227 (2011)

Walsh, T., McClellan, J.M., McCarthy, S.E., Addington, A.M., Pierce, S.B., Cooper, G.M., Nord, A.S., Kusenda, M., Malhotra, D., Bhandari, A., et al.: Rare structural variants disrupt multiple genes in neurodevelopmental pathways in schizophrenia. Science **320**, 539–543 (2008)

Wan, Y.W., Nagorski, J., Allen, G.I., Li, Z.H., Liu, Z.D.: Identifying cancer biomarkers through a network regularized Cox model. Genomic Signal Processing and Statistics (GENSIPS), 2013 IEEE International Workshop on (Houston, pp. 36–39. IEEE, TX (2013)

Wang, K., Li, M., Hakonarson, H.: ANNOVAR: Functional annotation of genetic variants from high-throughput sequencing data. Nucleic Acids Res. **38**, e164 (2010)

Wong, E.H., So, H.C., Li, M., Wang, Q., Butler, A.W., Paul, B., Wu, H.M., Hui, T.C., Choi, S.C., So, M.T., et al.: Common variants on Xq28 conferring risk of schizophrenia in Han Chinese. Schizophr. Bull. **40**, 777–786 (2014)

Zhang, W., Wan, Y.W., Allen, G.I., Pang, K., Anderson, M.L., Liu, Z.: Molecular pathway identification using biological network-regularized logistic models. BMC Genom. **14**(Suppl 8), S7 (2013)

# GNATY: Optimized NGS Variant Calling and Coverage Analysis

Beat Wolf[1,2(✉)], Pierre Kuonen[1], and Thomas Dandekar[2]

[1] Insitute of Complex Systems, University of Applied Sciences Western Switzerland,
1700 Fribourg, Switzerland
**beat.wolf@hefr.ch**
[2] University of Würzburg, Biozentrum Universität Würzburg,
97074 Würzburg, Germany

**Abstract.** Next generation sequencing produces an ever increasing amount of data, requiring increasingly fast computing infrastructures to keep up. We present GNATY, a collection of tools for NGS data analysis, aimed at optimizing parts of the sequence analysis process to reduce the hardware requirements. The tools are developed with efficiency in mind, using multithreading and other techniques to speed up the analysis. The architecture has been verified by implementing a variant caller based on the Varscan 2 variant calling model, achieving a speedup of nearly 18 times. Additionally, the flexibility of the algorithm is also demonstrated by applying it to coverage analysis. Compared to BEDtools 2 the same analysis results were found but in only half the time by GNATY. The speed increase allows for a faster data analysis and more flexibility to analyse the same sample using multiple settings. The software is freely available for non-commercial usage at http://gnaty.phenosystems.com/.

**Keywords:** Next generation sequencing · Variant calling · Algorithmics

## 1 Background

During recent years, next generation sequencing (NGS) technologies became faster and produced increasing amounts of data. While in the beginning of DNA sequence analysis the majority of the time and money was spent on sequencing, the NGS speed increases started to shift the focus towards the analysis of the produced data. Time intensive parts of the NGS data analysis, such as sequence alignment, received many improvements over the years. Various sequence aligners were created, outperforming each other in various domains such as precision and speed [1]. But not all analysis steps received the same amount of attention, such as variant calling and coverage analysis.

In this paper we want to show that there is a big potential for speed increases in those areas, without lowering the analysis quality. We present GNATY, which stands for GensearchNGS Analysis Tools librarY. It is a stand alone, freely available collection of tools, based on the code of GensearchNGS [2], a commercial

© Springer International Publishing Switzerland 2016
F. Ortuño and I. Rojas (Eds.): IWBBIO 2016, LNBI 9656, pp. 446–454, 2016.
DOI: 10.1007/978-3-319-31744-1_40

NGS data analysis software. The goal is to achieve those performance increases with the usage of modern development techniques like stream processing, multithreading and the usage of a modular architecture which allows to share code between the different types of analyses. This with the objective to be able to handle more data on already existing infrastructure.

Current NGS data pipelines commonly use a series of processing steps to transform the raw output from next generation sequencers into a BAM [3] file. This file contains the aligned sequences which can be used to perform the final analysis of the sequenced sample. In this paper we focus on two aspects of the post alignment sequence analysis, variant calling and coverage analysis.

Various tools exist for variant calling. They can be split into two main categories: those that determine the presence of a variant through probabilistic methods and those using a heuristic/statistical approach. The tools in the first category, which contains popular ones such as samtools [3] or GATK [4], are mostly based on the use of Bayes theorem to determine the probability of a genotype at a certain position in the alignment. The second group of variant callers, containing notably Varscan 2 [5], relies on heuristic/statistical methods to determine if a variant is present. Working with user defined threshold values, a variant is called based on rules such as minimum frequency or coverage.

While the first group of variant callers remains more popular, it has recently been shown [6] that by using appropriate parameters, the results between both approaches are comparable. GNATYs variant calling model replicates the one used in Varscan 2, with the goal of recreating the results of Varscan 2 as close as possible. The design of GNATY, which will be elaborated further in the Sect. 2 is based on a modular stream based approach.

The proposed architecture can be used for other types of analysis that access the aligned genome in a sequential way. This is why to further validate the approach taken by GNATY we also implemented a tool to perform coverage analysis. Coverage analysis is in the context of this paper is to produce a BED file containing coverage information for all regions in the alignment. A few tools exist to perform this task, the most popular one being BEDtools [7]. The GNATY reimplements the algorithm used in BEDtools, with the goal to get the same analysis results in a shorter amount of analysis time.

In this paper we first present the software architecture used by GNATY to implement those two tools. We then compare the heuristic variant calling included in GNATY with Varscan 2 as well as the coverage analysis tool of GNATY with BEDtools.

## 2   Implementation

To achieve the goal of improving the speed of variant calling and coverage analysis for NGS data, a modular architecture was created. The advantage of this architecture is to be able to reuse parts of the processing for both proposed use cases. The main goal of the modular architecture is to separate all I/O operations (input/output) in their own modules, which can work as separate threads.

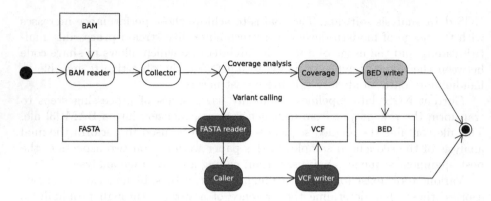

**Fig. 1.** UML (unified modeling language) activity diagram GNATY shared architecture between variant caller and coverage analysis.

This decoupling of I/O operations from the calculations of variant statistics allows for both tasks to progress independently.

Figure 1 shows the general architecture of GNATY, with the the modules shared between variant calling and coverage analysis colored in white, the variant calling modules in dark grey and the coverage analysis ones in light grey. The workflow has been separated into different steps, chained through queues using a stream based approach. Splitting the workflow into independent processes also allowed to share much of the code between the variant calling and coverage analysis tool. Every module in the architecture runs in its own thread.

The first modules used in the processing chain are the BAM reader and Collector, which are in charge of parsing the BAM file and creating a *mpileup* [3] like datastructure. To create an internal mpileup like structure, GNATY uses an expanding circular buffer, allowing for a stream based processing of the aligned reads. After this step, variant calling and coverage analysis split up with specific modules. For the coverage analysis, the Coverage module detects the individual regions that have the same coverage, and sends them to the BED writer which writes the calculated coverage information into a BED file. For variant calling, the FASTA reader adds the reference information to the alignment data created by the Collector. Finally, the Caller module performs the actual variant calling and sends the detected variants to the VCF writer module which writes them into a VCF file. The architecture of this system carefully separates the computationally heavy operations from all I/O operations, thus reducing the CPU downtime during I/O operations.

GNATY is implemented in Java, using the HTSjdk 1.122 library to access BAM files. HTSjdk is an open source library written in Java to read and write BAM files, used by many Java based NGS tools like the popular genome browser IGV [8]. The variant calling uses a heuristic approach like Varscan 2, allowing the user to set specific filters such as minimum frequency, coverage or p-value. For low coverage positions (below 200 reads) Fishers exact test is used to calculate the p-value, for higher coverage positions the Chi-Square test is used. The reason a different statistical test is used for low and high coverage regions is because for low coverage regions, Fishers exact test gives more accurate results than Chi-square, but has performance issues on high coverage regions. Those high coverage regions can be analysed with the Chi-square test which performs much better. This allows GNATY, unlike other variant callers, to not downsample the reads on high coverage regions and work with arbitrarily high coverages.

The default variant calling model of GNATY and Varscan 2 is slightly different, but for a maximum compatibility, a Varscan 2 compatibility mode was implemented and used for the rest of this paper. Nonetheless we will quickly take a look at the differences of the default variant calling model in GNATY and Varscan 2. By default, GNATY and Varscan 2 handle indels slightly differently. One example is that GNATY considers the start of a deletion to be the first base which got deleted, whereas Varscan 2 considers the first base to be the last base covered before the deletion. This leads to differences in the coverage and frequency information used to filter the variants. GNATY also allows to call multiple variants at the same position, if they all pass the defined filters. Varscan 2 does not call multiple variants at the same position, it only reports the most likely variant (the one with the highest frequency). This can cause issues in diploid organisms where two heterozygous variants share a position. Other differences include the calling of variants at positions where the reference sequence has the nucleotide N, which is used as a wildcard and signifies that all nucleotides are possible. Varscan 2 does call variants on those positions, whereas GNATY does not by default. An option has been developed for GNATY which replicates the exact behaviour of Varscan 2. This option was introduced to make it easier to compare the performance of Varscan 2 and GNATY, but also to make it easier for users of Varscan 2 to migrate to GNATY without getting different results.

# 3    Results and Discussion

The implementation of GNATY has been packaged as a self-containing jar file, which can be used as a command line tool. The software can be downloaded at http://gnaty.phenosystems.com/ and can be used for free for non commercial usages.

To compare the speed of our variant caller and coverage calculator, two different datasets from the genome comparison & analytic testing project (GCAT) [9] project were used. The GCAT project provides standardized datasets that can be used to compare different variant calling pipelines. The first dataset from GCAT is ion-torrent-215bp-se-exome-123x, which is an ion torrent exome dataset with a mean coverage of 123x and read lengths of 215 base pairs. The second one is the illumina-100bp-pe-exome-150x dataset, a paired end Illumina HiSeq exome dataset with a mean coverage of 150x and read lengths of 100bp. Both datasets are available on the GCAT website (http://www.bioplanet.com/gcat/). Variant calling and coverage analysis require sorted BAM files. To produce a sorted

BAM file of both datasets, the raw sequencing data has been aligned against the human reference genome HG19 using BWA-MEM 0.7.8-r1039 [10]. The resulting SAM files have been converted to sorted BAM files using samtools 1.0.

The alignment took 23 min for the dataset one and 41 min for dataset two. Converting the SAM files to sorted BAM files took 37 min for the dataset one and 63 min for the second one.

The two resulting BAM files have a size of 5.2 GB for dataset 1 and 6.8 GB for dataset 2. The BAM files for both datasets can be downloaded at ftp://bwolf-1. tic.hefr.ch/ to recreate the results of this paper. All tests have been performed using a dual processor setup with two quad core Intel Xeon E5-2609 2.5 GHz and 32 GB of RAM. The hard disk used has a 128 MB/s read and 94 MB/s write speed. For the variant calling tests, GNATY is compared to Varscan 2.3.7, which uses as its input the mpileup files generated by samtools 1.0. During the performed benchmarks the output of samtools was directly input into Varscan 2 through a standard unix shell pipe, to avoid the overhead of creating a temporary file. The coverage analysis is done using BEDtools 2.21.0.

### 3.1 Variant Calling

The GNATY variant caller is compared to Varscan 2 on both previously described datasets. The run times and variant counts for both tools are compared on both datasets, using the same filters. The filters used are based on the conservative filter values recommended by [6], which are 20 % minimum frequency, 10 coverage and 4 allele reads. Varscan 2 takes as its input the mpileup data generated by samtools, which is the tool responsible to parse the BAM files. GNATY directly reads the BAM files using the HTSjdk library. For samtools, the default options have been used, except for two options: The option -B to disable probabilistic realignment for the computation of base alignment quality has been set, as this is the recommended setting by Varscan 2. The option -x, which disables the reduction of the base quality of overlapping read-pairs, has also been used, as this functionality is not yet present in GNATY. GNATY used the equivalent options, with the additional options -one and -vs2. The option -one limits the amount of reported variants per position to one and -vs2 enables a special mode which reproduces the way Varscan 2 operates. Both of those options have been enabled to guarantee that the speed comparison is fair. As both tools implement the same variant calling method and use the same options, we expect to get the same results.

All tests have been repeated 3 times for every dataset using both tools, averaging the run times for comparison. The comparison of the VCF files created by GNATY and Varscan 2 shows that both called a nearly identical set of variants on both datasets. Both called the same 425'866 variants over both datasets combined. GNATY called 1 additional variant which was not found by Varscan 2. Manual inspection of that variant showed that it is a SNP at a position where the reference sequence as the base N. GNATY only calls variants at locations with the reference N when using the Varscan 2 compatibility mode. As all downstream variant analysis tools are expected to ignore such variants, there is no practical

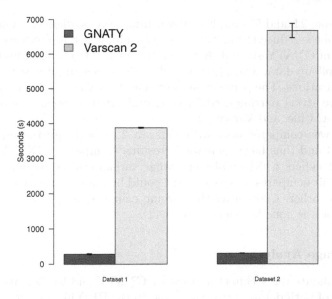

**Fig. 2.** Comparison of variant calling times for GNATY and Varscan 2 on the two tested datasets

difference in the variants called. But to further understand the nature of this difference and to achieve a complete compatibility with Varscan 2, the authors of Varscan 2 have been contacted.

Whereas the results of both tools are practically identical, there is a big difference in terms of run times to perform the analysis. When combining the times of both datasets, GNATY is about 18 times faster than Varscan 2. As shown in Fig. 2, GNATY performed the analysis in 589 s (9 min and 49 s), 283 s ($\sigma$ 6.9 s) for dataset 1 and 306 s ($\sigma$ 1.7 s) for dataset 2. Varscan 2 required 10'559 s (2 h 55 min), 3'876 s ($\sigma$ 5.2 s) (1 h 4 min) for dataset 1 and 6'683 s ($\sigma$ 103 s) (1 h 51 min) for dataset 2.

The runtime reduction surpasses what was initially hoped for and shows the efficiency of the chosen architecture. A time reduction of this magnitude increases the flexibility the user has for the analysis, as it makes it easier to rerun the analysis multiple times with different parameters. Going from almost 3 h to less than 10 min to perform the variant calling does indeed change the way the workflow can be approached. Instead of running the variant calling over night, it can be performed during the time it takes to perform a small miscellaneous task. As Varscan 2 is dependent on the performance of a single CPU core, a CPU would be required that is over 15 times faster on a single core than a currently available CPU to achieve a similar speed increase. This speed increase is also interesting when looking at the complete NGS data analysis chain. In our testcase, aligning both datasets took 64 min and transforming the resulting SAM file into a sorted BAM file took another 100 min. Whereas previously variant calling took a similar amount of time, 2 h and 55 min, it is now comparably negligible. Using GNATY the overall time is reduced from 339 min to 174 min, which is a reduction of 48 %.

Profiling of GNATY showed that the performance bottleneck lies in the creation of the pileup data, which is expected, and creates an interesting target for future optimizations. The performance bottleneck in Varscan 2 is located in the time intensive string parsing used for communication between samtools, which parses the BAM file, and Varscan 2.

GNATY was compared only with Varscan 2 as both use the same variant calling model and thus have comparable results. Comparing GNATY with any other variant callers which implement other variant calling models would not have been a fair comparison, as one model could be more computationally intensive than the other. Comparing the variant calling quality itself between the different models is done by studies like [6,11].

## 3.2 Coverage Analysis

To further validate the architecture used by GNATY and to show its flexibility, the coverage of both datasets was calculated in the BED file format. To perform this analysis, BEDtools 2 was used as the reference implementation. GNATY implemented the coverage analysis in a completely compatible way, with the goal to produce the exact same result. To verify that the results of both tools are identical, the MD5 hashes of the resulting files have been calculated. Both tools did indeed create identical output files, which allows them to be used interchangeably in an analysis pipeline.

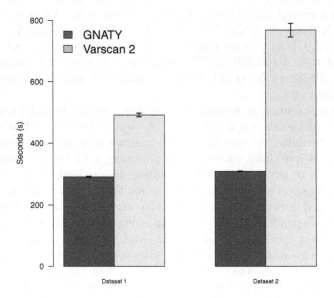

**Fig. 3.** Comparison of coverage analysis times for GNATY and Bedtools 2 on the two tested datasets

Producing exactly the same result, using the default BEDtools settings, GNATY was 52 % (599 s ($\sigma$ 0.6 s) vs 1260 s ($\sigma$ 12 s)) faster over both datasets. Figure 3 nicely shows how this is halving the time needed to analyse the data.

The time required to perform the coverage analysis is similar to the variant calling, but slightly slower. This is interesting, as algorithmically, calculating the coverage of the sample is much easier than detecting variant. The difference can be attributed to the fact that the output file of the analysis, the BED file, is much bigger than the VCF file created during variant calling. This difference (2600 MB vs 32 MB) increases the workload on the hard disk, reducing the overall analysis speed.

# 4   Conclusion

We were able to show that there is still a lot of performance potential in the existing methods for next generation sequence data analysis. Through the decoupling of I/O operations from the calculation heavy parts we were able to drastically decrease the time needed for the analysis, without changing the results. The speed increases in GNATY range from 52 % for coverage analysis compared to BEDtools 2, up to a 18 fold speed increase for variant calling compared to Varscan 2. The speed increase of GNATY compared to Varscan 2 in variant calling transforms the critical step of variant calling from a time intensive processing step to one that no longer takes a critical amount of time in the complete workflow. This shows that there is still a lot of potential for speed increases in NGS analysis pipelines, without having to invest in faster hardware. Future work on GNATY includes, the introduction of a probabilistic variant calling method, similar to the one used in samtools and GATK, variant calling on multiple samples simultaneously and additional performance improvements. Having demonstrated that there is still a lot of optimization potential in default NGS data analysis methods, we will also investigate the possibility of optimizing other tools. In the context of this paper, the conversion of unsorted SAM files to sorted BAM files appears to be an interesting target for future optimizations.

**Conflicts of Interest.** The authors have no conflict of interest to declare.

**Acknowledgements.** The authors thank Phenosystems SA for the opportunity to release part of their software for free.

# References

1. Li, H., Homer, N.: A survey of sequence alignment algorithms for next-generation sequencing. Briefings Bioinform. **11**(5), 473–483 (2010)
2. Wolf, B., et al.: DNAseq workflow in a diagnostic context and an example of a user friendly implementation. BioMed Res. Int. **2015**, 11 (2015). Article ID 403497
3. Li, H., et al.: The sequence alignment/map format and SAMtools. Bioinform. (Oxford, England) **25**(16), 2078–2079 (2009)

4. DePristo, M.A., et al.: A framework for variation discovery and genotyping using next-generation DNA sequencing data. Nat. Genet. **43**(5), 491–498 (2011)
5. Koboldt, D.C., et al.: Somatic mutation and copy number alteration discovery in cancer by exome sequencing. Genome Res. **22**(3), 568–576 (2012)
6. Warden, C.D., et al.: Detailed comparison of two popular variant calling packages for exome and targeted exon studies. PeerJ **2**, 600 (2014)
7. Quinlan, A.R., et al.: A flexible suite of utilities for comparing genomic features. Bioinformatics **26**(6), 841–842 (2010)
8. Thorvaldsdottir, H., et al.: Integrative Genomics Viewer (IGV): high-performance genomics data visualization and exploration. Briefings Bioinform. **14**(2), 178–192 (2013)
9. Highnam, G., et al.: An analytical framework for optimizing variant discovery from personal genomes. Nat. Commun. **6**, 16 (2015)
10. Li, H., et al.: Fast and accurate long-read alignment with burrows-wheeler transform. Bioinform. (Oxford, England) **26**(5), 589–595 (2010)
11. O'Rawe, J., et al.: Low concordance of multiple variant-calling pipelines: practical implications for exome and genome sequencing. Genome Med. **5**(3), 28 (2013)

# De Novo Assembly and Cluster Analysis of Siberian Larch Transcriptome and Genome

Michael Sadovsky$^{(\boxtimes)}$, Yulia Putintseva, Vladislav Birukov,
Serafima Novikova, and Konstantin Krutovsky

Institute of Computational Modelling of SB RAS, Akademgorodok,
660036 Krasnoyarsk, Russia
msad@icm.krasn.ru, yuliya-putintseva@rambler.ru,
vladbir2010@gmail.com, kempfana@mail.ru
http://icm.krasn.ru

**Abstract.** We studied Siberian Larch (*Larix Sibirica*) transcriptome making *de novo* assembly and cluster analysis of contigs frequency dictionaries. Also, some preliminary results of similar study of the larch genome are present. It was found that the larch transcriptome yields a number of unexpected symmetries in the statistical and combinatorial properties of the entities.

**Keywords:** Frequency · Triplet · Order · Cluster · Elastic map · Evolution

## 1   Introduction

A search for an order and structuredness in bulky sets of objects (considerably homogeneous) is one of the key scientific problems. Modern genomics (as well as linguistics) is the second to none in that respect. Indeed, amount of raw nucleotide sequence data grows daily for billions of megabytes. Those sequences are symbol sequences based mainly on the four-letter alphabet $\aleph = \{A, C, G, T\}$. Further we also assumed that neither other symbols, nor blank spaces are supposed to be found in a sequence; a sequence under consideration is also supposed to be coherent (i.e. consisting of a single piece).

We studied an order and structuredness over a set of sequences from finite alphabet $\aleph$; all these sequences represented the transcriptome of Siberian larch (*Larix sibirica* Ledeb.). Transcriptome represents sequences of expressed genes and corresponds to the mRNAs molecules isolated from biological cells or tissues.

Usually, a distance between sequences is used to find out irregularities or similarities among them. Distances between sequences are most often based on sequence alignments (see [6] for the first computer algorithm for aligning two sequences and [9] for a recent review). However, there could be serious problems and constraints in generating reasonable alignments for highly divergent sequences. Meanwhile, there are other concepts and methods that could be much more powerful than those that are based on alignments (e.g., [10,11]), although

© Springer International Publishing Switzerland 2016
F. Ortuño and I. Rojas (Eds.): IWBBIO 2016, LNBI 9656, pp. 455–464, 2016.
DOI: 10.1007/978-3-319-31744-1_41

they still need further investigation of their applicability for addressing biological problems.

Conifers play the key role in climate formation, in Northern hemisphere; besides, they are the most abundant (in biomass) plants growing there. In such capacity, conifers become a matter of study, since one may hope to affect some climatic aspects in some areas, through a controlled plantation of these latter. Such plantation may not be realized without a study of genetic background, at least, in due time and with proper efficiency. Siberian larch is the core species in the boreal forests of Siberia, so the study of that latter is of great importance; so, this paper present the *de novo* assembly results of larch transcriptome, since it is the firstly done.

Key idea in our search for a structure and order in a set of symbol sequences (transcriptome nucleotide sequences) is to translate sequences into their frequency dictionary [1–3,5]. There could be a number of various definitions of a frequency dictionary, but we will use the basic one that is a list of all the strings of a given length accompanied with a frequency of each string (a detailed description is given below). It is crucial that the transformation of a symbol sequence into a frequency dictionary allows us to map a set of sequences into a metric space. The latter provided us with powerful and extended tools for analysis.

We will briefly outline the concept of our study and then demonstrate the main results obtained. First, we changed each symbol sequence (that is a nucleotide sequence in the Siberian larch transcriptome set) into a frequency dictionary. Then, we studied distribution of those dictionaries in a multidimensional space trying to infer any regularities and clusters. Second, for each clustering we checked for stability of clustering. This clustering was carried out using the $K$-means technique. Third, we compared the statistical properties of the clusters identified by $K$-means and found that these clusters demonstrated a very strong symmetry in terms of the statistical properties. In brief, the clusters showed extremely low level of discrepancy in the Chargaff's second parity rule. This low discrepancy is the most intriguing fact concerning the properties of the studied transcriptome sequence set.

## 2 Materials and Methods

### 2.1 Transcriptome Nucleotide Sequence Data

The transcriptome of Siberian larch was originally sequenced in the project on the whole genome sequencing of Siberian larch [12]. The sequence data of *L. sibirica* were obtained using Illumina MiSeq sequencer at the Laboratory of forest genomics of the Siberain Federal University. The RNA was isolated from buds [7]. Total number of sequences in the transcriptome set was 25 748. The shortest sequence had 128 nucleotide base pairs or bp (symbols), while the longest one had 8 512 bp. An average length of the sequences in the transcriptome was 656 bp, with the standard deviation of 565 bp. The histograms of the distribution of the transcriptome sequences entries over their length are presented in Fig. 1. Evidently, the distribution resembles quite strongly Poisson distribution.

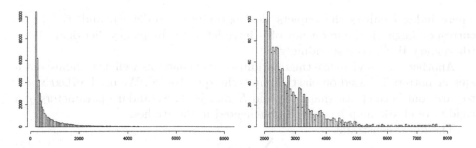

**Fig. 1.** Distribution of the *L. sibirica* transcriptome sequence lengths for all sequences (left) and for sequences $\geq 2000$ bp (right).

We excluded from the further consideration all the sequences entries shorter than 2 000 bp to avoid a degeneracy of frequency dictionaries developed from shorter sequences. Surely, this part of the transcriptome requires special studies. Similar distribution is observed for the contigs of the nuclear genome of *L. sibirica*. Of course, the threshold in 2 000 bp is rather arbitrary: one may choose another length to cut off the set of contigs. We used this figure for two reasons: first, it is approximately twice longer an average gene length, and second, this threshold cuts off a proper number of contigs (1 436 entries) to be checked with BLAST (further, we suppose to implement BLAST to check a contig location in leading vs. lagging strand).

### 2.2 Frequency Dictionary

Previously [1–3,5], a frequency dictionary was proposed to be a fundamental structure of a symbol sequence. The word $\omega = \nu_1\nu_2\ldots\nu_{q-1}\nu_q$ of the length $q$ is a string occurred in the text $\mathfrak{T}$. Here $\nu_j$ is a symbol occupying the $j$-th position at the word; $\nu_j \in \aleph$. Frequency dictionary $W(q)$ is the set of all the words of the length $q$ counted within the text $\mathfrak{T}$ so that each word is accompanied with its frequency. Hereinafter we shall consider the dictionaries $W(3)$, only. $W(3)$ unambiguously maps a text $\mathfrak{T}$ into a 64-dimensional space, where the triplets are the axes, and frequencies are the coordinate figures.

### 2.3 Clusterization Techniques

We used $K$-means technique [19] to analyze the transcriptome; Euclidian distance as a metrics has been used in our study. That latter is defined traditionally: a distance between two frequency dictionaries is

$$\rho\left(W^{(1)}, W^{(2)}\right) = \sqrt{\sum_{i=1}^{64}\left(f_i^{(1)} - f_i^{(2)}\right)^2}, \tag{1}$$

where index $i$ enlists the triplets (lexicographically ordered), and $f_i^{(1)}$ ($f_i^{(2)}$, correspondingly) is the frequency of $i$-th triplet at the frequency dictionary $W^{(1)}$ (dictionary $W^{(2)}$, correspondingly).

Another way to visualize the multidimensional data, as well as to figure out a cluster pattern is based on elastic map technique [16–18]. We used *ViDaExpert*[1] to carry out both elastic mapping, and $K$-means. The standard parameters of a rigidity of elastic map has been implemented in the studies.

The linear constraint

$$\sum_\omega f_\omega = 1 \tag{2}$$

yields an additional parasitic signal. Hence, a triplet must be eliminated from the set. Formally, any one could be eliminated. Practically, the choice affects seriously the results of the treatment. We eliminated the triplet CGC with the minimal standard deviation over the entire dataset (the ensemble of contigs), since it provides the least discrimination of the contigs.

# 3   Results

## 3.1   *K*-means Classification

We developed consequently four classifications using $K$-means technique with number of classes varied from two to five. For each number of classes 350 runs of $K$-means were executed to study the stability of classification. The stable subsets of sequences comprising the transcriptome were determined for each classification. We assumed a classification to be stable, if not less than 95 % of all runs yielded the same distribution of sequences. Moreover, the radii of classes were calculated for each classification, as well as the distances between them. All four classifications were very stable for 350 runs. Same has been done for nuclear genome.

It is a well-known fact that any symbol sequence from a four-letter alphabet ℵ follows Chargaff's (generalized) second parity rule [8]. The rule stipulates that frequencies of any two words composing a complimentary palindrome should be very close. Complimentary palindrome is a couple of words (strings) that read equally in opposite directions, with respect to the symbol substitution according to the first Chargaff's rule. In any frequency dictionary of a thickness $q$ there always exists $(1/2)4^q$ couples of complimentary palindrome; for $W(3)$ there are 32 couples of such triplets. Thus, we have checked the pattern of the second Chargaff's parity rule feasibility for three cases:

(1) within the first class identified through $K$-means;
(2) within the second class identified through $K$-means, and
(3) between these two classes.

The most surprising thing was that the second (generalized) Chargaff's parity had significantly less discrepancy in the third case (the comparison of the

---

[1] http://bioinfo-out.curie.fr/projects/vidaexpert/.

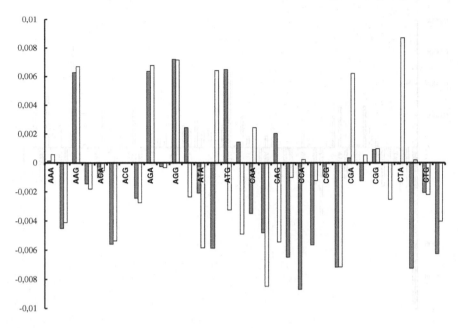

**Fig. 2.** The discrepancy in distribution of two classes obtained through $K$-means based on formula (3). Figure shows the differences between the frequencies of the triplets comprising the complementary palindromes: gray bars show class 1, white bars show class 2.

centers of two classes of the sequences comprising the transcriptome). Indeed, the discrepancy $\mu$ defined according to [4] as

$$\mu = \frac{1}{\|\Omega\|} \sqrt{\sum_{\omega \in \Omega} (f_\omega - f_{\overline{\omega}})^2} \tag{3}$$

showed the drastically different distributions for those three cases mentioned above. Here $\Omega$ means the set of complementary palindromic triplets (consisting of 32 entries for the cases one and two, and of 64 entries for the third case), while $f_\omega$ and $f_{\overline{\omega}}$ are the words making the complementary palindrome; $\| \cdot \|$ means the capacity of a set. The figure $\mu$ defined according to (3) looks like a distance, meanwhile it is not. It does not provide a measure between two points, but presents a discrepancy of a given dictionary. Figures 2 and 4 illustrate this fact.

The most surprising thing is that similar results for the nuclear genome show drastically opposite result: discrepancy (3) determined within two classes is an order less in comparison to that latter determined between the classes. Figure 4 illustrates this fact. Take a note that an average in-class discrepancy (3) is almost five times greater than that latter determined between the classes.

Significantly less discrepancy (3) observed for transcriptome forced the idea that the contigs occupying to the different classes belong the opposite strands of DNA molecule, indeed. This point has been checked through the alignment of

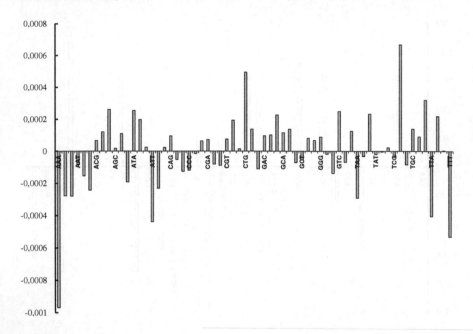

**Fig. 3.** The discrepancy between two classes obtained through $K$-means based on formula (3). For each couple of triplets the difference of the relevant frequencies is shown.

the contigs with BLAST; we aligned our sequences on the overall GenBank non-redundant nucleotide collection. The idea holds true: Fig. 6 illustrates this fact. The contigs belonging to the opposite classes are mainly labeled as located to the opposite strands. In fact, this observation allows to decrease significantly the time for the processing of any bulk genetic data against BLAST: it took about five days to align the contigs taken into consideration (it should be reminded that is was just one eighth of the entire set of contigs), while $K$-mean analysis takes about few seconds, even for a moderate PC. In other words, a contig belonging to a specific class is expected with very high probability to belong to the specific strand (Fig. 3).

Besides, we have developed specific frequency dictionaries for the contigs under consideration. These are the dictionaries where the triplets are counted with a three-nucleotide shift. In other words, the triplets to comprise a dictionary do not overlap, but have no gaps between each other. Obviously, one can develop three different dictionaries differed with the start position of the first nucleotide.

For such dictionaries, famous seven cluster structure in genomes has been found [13–15]. Similar pattern we have observed, also. We have checked all three versions of the modified (as described above) frequency dictionaries developed for the contigs. For all of them a six-cluster pattern has been found: the vertices arrange a kind of hexagonal, while no central cluster takes place. Careful examination of the composition of the vertices in the pattern shows that the vertices actually go through a circular permutation: a change of a set of dictionaries

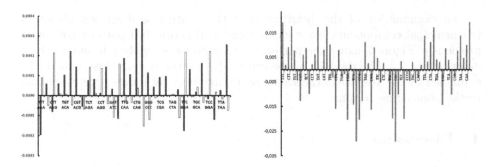

**Fig. 4.** Discrepancy (3) within two classes (left), and between the classes (right) observed for nuclear larch genome.

**Fig. 5.** Octahedral structure of the contigs distribution in 63-dimensional space provided by the specific frequency dictionaries. Colored triangles correspond to permutation cycles.

(determined by the reminder of the division by 3 of the distance from the starting nucleotide in a contig) just resulted in a rotation of the group of contigs (frequency dictionaries, to be exact) clockwise over the image provided by the elastic map.

An examination of the distribution of the contigs as described above, in the principal components space yields wonderful octahedral pattern not found previously. Figure 5 shows the pattern. This pattern is highly relevant to those described in [13–15]; the lack of the central (the seventh one) cluster might be explained by the absence of the sequence fragments in the dataset corresponding to the non-coding regions.

## 4  Discussion

We found out that the key issue of statistical structure of the *L. sibirica* transcriptome was a high symmetry (resembling two subsets) of the sequences comprising the transcriptome set. The symmetry results from the occupation of the contigs within two opposite strands. Yet nothing is clear concerning the similar while opposite behaviour of the contigs assembled from a nuclear genome: an inversion of the discrepancy measure observed for these entities still awaits an explanation.

It should be also stressed out that the very low discrepancy based on formula (3) observed for the centers of two classes mentioned above requires that a number of opposite genes located in complementary strands must also overlap. Such condition is unlikely, but it does not mean

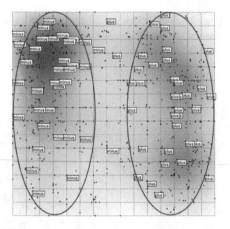

**Fig. 6.** Distribution of the contigs over two classes (left and right clusters, labeled with the marks provided by BLAST).

that such overlapping is not possible. Evidently, the first step to verify it may consist in checking the sequences belonging to various classes (obtained due to $K$-means technique) by BLAST (see Fig. 6). It yields a list of sequences homological to the main strand, and to the complementary one, correspondingly. Such verification may also bring another advantage: it is a common fact that BLAST is a very time-consuming procedure. A combination of $K$-means technique to figure out the tentative specific strand strings with BLAST may seriously decrease the resource demand due to the specific pre-treatment of a set to be checked.

The unusual symmetry manifesting in statistical properties of triplet frequency dictionaries was found in the *L. sibirica* transcriptome. Namely, the nucleotide sequences that were transformed into the frequency dictionaries demonstrated unusually high level of the coincidence of the frequencies of (rather long) oligonucleotides with their counterparts in the opposite strand of genomic DNA. Such high coincidence may confirm an occurrence of a rather significant number of genes occupying the opposite strands and possibly also overlapping each other. It is not observed for the protein coding genes in eukaryotes, but could hypothetically occur for other gene types.

The uncertainty level may be decreased through the comprehensive analysis of all the sequences comprising the transcriptome by BLAST so that the strings belonging to opposite strands would be identified. Further, we plan to carry out similar studies on the same matter with competing clusterization technique, to avoid a possible dependence of the results shown above from the specific type of the clusterization method.

This study was supported by a research grant No. 14.Y26.31.0004 from the Government of the Russian Federation.

# References

1. Bugaenko, N.N., Gorban, A.N., Sadovsky, M.G.: Towards the definition of information content of nucleotide sequences. Mol. Biol. **30**(5), 529–541 (1996)
2. Bugaenko, N.N., Gorban, A.N., Sadovsky, M.G.: The information capacity of nucleotide sequences and their fragments. Biophysics **5**, 1063–1069 (1997)
3. Bugaenko, N.N., Gorban, A.N., Sadovsky, M.G.: Maximum entropy method in analysis of genetic text and measurement of its information content. Open Syst. Inf. Dyn. **5**(2), 265–278 (1998)
4. Grebnev Ya, V., Sadovsky, M.G.: Chargaff's second rule and symmetry in genomes. Fundam. Stud. **12**(5), 958–965 (2014)
5. Hu, R., Wang, B.: Statistically significant strings are related to regulatory elements in the promoter regions of Saccharomyces cerevisiae. Physica A **290**, 464–474 (2001)
6. Needleman, S.B., Wunsch, C.D.: A general method applicable to the search for similarities in the amino acid sequence of two proteins. J. Mol. Biol. **48**(3), 443–453 (1970)
7. Oreshkova, N.V., Putintseva, Yu.A., Kuzmin, D.A., Sharov, V.V., Biryukov, V.V., Makolov, S.V., Deych, K.O., Ibe, A.A., Shilkina, E.A., Krutovsky, K.V.: Genome sequencing and assembly of Siberian larch (Larix sibirica Ledeb.) and Siberian pine (Pinus sibirica Du Tour) and prelimenary transcriptome data. In: Proceedings of the 4th International Conference on Conservation of Forest Genetic Resources in Siberia, Barnaul, Russia, 24–29 August 2015, pp. 127–128 (2015)
8. Qu, H., Wu, H., Zhang, T., Zhang, Z., Hu, S., Yu, J.: Nucleotide compositional asymmetry between the leading and lagging strands of eubacterial genomes. Res. Microbiol. **161**, 838–846 (2010)
9. Tsiligaridis, J.: Multiple sequence alignment, clustering with dot matrices, entropy, genetic algorithms. In: Li, K.-C., Jiang, H., Yang, L.T., Cuzzocreapp, A. (eds.) Big Data: Algorithms, Analytics, and Application, Chap. 4, pp. 71–88. CRC Press (2015)
10. Znamenskij, S.V.: Modeling of the optimal sequence alignment problem. Program Syst. Theor. Appl. **4**(22), 257–267 (2014). In Russian
11. Znamenskij, S.V.: A model and algorithm for sequence alignment. Program Syst. Theor. Appl. **1**(24), 189–197 (2015)
12. Krutovsky, K.V., Oreshkova, N.V., Putintseva, Y., Ibe, A.A., Deutsch, K.O., Shilkina, E.A.: Some preliminary results of a full genome de novo sequencing of Larix sibirica Ledeb., Pinus sibirica Du Tour. Siberian For. J. **1**(4), 79–83 (2014). (in Russian, English abstract)
13. Gorban, A.N., Zinovyev, A.Y., Popova, T.G.: Seven clusters in genomic triplet distributions. Silico Biol. **3**, 39–45 (2003)

14. Gorban, A.N., Zinovyev, A.Y., Popova, T.G.: Four basic symmetry types in the universal 7-cluster structure of microbial genomic sequences. Silico Biol. **5**, 25–37 (2005)
15. Gorban, A.N., Zinovyev, A., Popova, T.G.: Universal seven-cluster structure of genome fragment distribution: basic symmetry in triplet frequencies. In: Kolchanov, N., Hofestaedt, R. (eds.) Bioinformatics of Genome Regulation and Structure II, pp. 153–163. Springer Science+Business Media Inc., New York (2005)
16. Gorban, A.N., Zinovyev, A.Y.: Principal manifolds and graphs in practice: from molecular biology to dynamical systems. Int. J. of Neural Syst. **20**, 219 (2010)
17. Gorban, A.N., Kögl, B., Wünsch, D.C., Zinovyev, A.Yu. (eds.): Principal Manifolds for Data Visualisation and Dimension Reduction. Lecture Notes in Computational Science and Engineering, vol. 58, 332 p. Springer, Heidelberg (2007)
18. Gorban, A.N., Zinovyev, A.Yu.: Principal graphs and manifolds. In: Olivas, E.S., et al. (eds.) Handbook of Research on Machine Learning Applications and Trends: Algorithms, Methods and Techniques, pp. 28–59. Information Science Reference, IGI Global, Hershey (2009)
19. Fukunaga, K.: Introduction to Statistical Pattern Recognition, 2 edn., 591 p. Academic Press, London (1990)

# Assistive Technology for People with Neuromotor Disorders

# Sainet: An Image Processing App for Assistance of Visually Impaired People in Social Interaction Scenarios

Jesus Salido, Oscar Deniz, and Gloria Bueno[(✉)]

Department of IEEAC, Castilla-La Mancha University, Ciudad Real, Spain
{jesus.salido,oscar.deniz,gloria.bueno}@uclm.es
http://visilab.etsii.uclm.es/

**Abstract.** This work describes a mobile application (Sainet) for image processing as an assistive technology devoted to visually impaired users. The app is targeted to the Android platform and usually executed in a mobile device equipped with a back camera for image acquisition. Moreover, a wireless bluetooth headphone provides the audio feedback to the user. Sainet has been conceived as an assistance tool to the user in a social interaction scenario. It is capable of providing audible information about the number and position (distance and orientation) of the interlocutors in the user frontal scenario. For validation purposes the app has been tested by a blind user who has provided valuable insights about its strengths and weaknesses.

**Keywords:** Image processing · Mobile computing · Visually impaired assistance

## 1 Introduction

In the last few years mobile devices have become a pervasive technology in our world. Together with the increasing use of mobile devices the images taken by them are reaching one trillion during 2015. Nowadays the pictures and videos captured by smartphones and tablets surpass those taken by digital cameras while the number of mobile apps available in the online stores is almost 4 millions (www.statista.com). In such scenario apps for image and video processing are reaching a significant role.

Usually the computer vision software for real time processing requires an expensive and high performance hardware. Today the hardware required for astonishing mobile apps capable of video and image real time processing is available in our hands thanks to multi core processing units and high resolution cameras present in the current commercial smartphones.

In addition to hardware, software also facilitates development of computer vision software on mobile devices. Since OpenCV [1,2,4] was released, this open source library has become the standard in the software development community and hopefully available for the main mobile platforms. Besides being a toolbox

© Springer International Publishing Switzerland 2016
F. Ortuño and I. Rojas (Eds.): IWBBIO 2016, LNBI 9656, pp. 467–477, 2016.
DOI: 10.1007/978-3-319-31744-1_42

for computer vision, OpenCV constitutes an standard benchmark to compare different solutions to the same problem.

In this work the VISILAB Group[1] has been exploring the capabilities of mobile computer vision as an assistive technology for the visually impaired (VI). The technologies for assistance of VI people have been around for decades [5–7,12] and the solutions are classified into three categories:

1. Vision sense enhancement. They are devoted to image capture to process it and to display it in such a way that be more easily perceived (as augmented or sharpened on screen).
2. Vision sense recovery. They are systems devoted to driving visual signals directly to the neural cortex.
3. Vision sense substitution. Similar to the first one, however in this case visual information is converted to auditory or tactile information.

Multiple integrated solutions have been provided that might be considered into the category of Vision sense substitution:

- Electronics Travel Aids (ETAs). They are systems to support navigation, usually as substitutive of the white can. The information about the environment can be gathered by cameras, laser scanners or sonars.
- Electronic Orientation Aids (EOAs). They are devices that provide orientation prior to, or during the movement. They can be external to the VI user and/or can be carried by the VI user (e.g., infrared light transmitters and handheld receivers).
- Position Locator Devices (PLDs). When used outdoors they rely on technologies like GPS. Localization in indoor environments still remains a challenge.

In all the mentioned systems the information supplied by human vision is substituted by hearing and touch inputs. There is a general agreement about the requisites that solutions should meet depending on the viewpoint. From the VI user perspective: fast feedback to the user allowing real time operation without interfering other sensory inputs, unobtrusive and lightweight so that they can be carried during long distances and times, reliable even under unexpected circumstances, affordable for most users, easy to use and having valuable functionality. From the developer side the requirement should be: simplicity, robustness, connectivity, performance, originality and improvement capability.

The previous classification on three categories (ETAs, EOAs and PLDs) for vision substitutive systems does not include systems aiming to facilitate daily tasks (e.g. buying in a marketplace) where object and people recongnition is desirable for goal achievement and fulfil the social interactions required [3,14–18]. In particular this work addresses the assistance to blind people in social interaction scenarios which had been subject for a short extent of previous efforts [9,10].

In social sitiuations VI people do not have visual cues and they would value information about their interlocutors such as: appearance, position, facial expressions, and so on. For example a typical situation is the inference of the distance to our interlocutor and his/her degree of interest on the conversation which might be a cue to change or stop the conversation. At times, the scarceness of such information may derive on social exclusion and alienation [9,10] assuming that at least 65 % of a two-person conversation is non-verbal.

The majority of works focused, until now, to VI users consist on very specific "portable" devices. Only recently the smartphone has been considered as a feasible assistive technology [5–7,12]. In fact, it has become not only an outstanding contender for mobile assistive technologies but one of the principal *mHealth* platforms [13]. Specially for VI assistance, the smartphone is having a main role derived from its ability to embed heavy processing tasks as required by computer vision applications.

## 2    Objective

This work describes the development of Sainet, a mobile application for image processing with assistive purposes for VI people in social interaction scenarios [9,10,12]. The app includes vision algorithms to allow people detection (number of persons) and estimation of distance and position of individuals in front of the VI user. The scene in front of the VI user is captured by the back camera of the VI user's smartphone while the information is converted to audible feedback supplied to the VI user by a portable earphone (wirelessly or not).

The main innovative aspects of the proposed solution with Sainet are:

1. Portability. The application is embedded into a commercial smartphone with VI user feedback by an earphone. In that way the solution becomes unobtrusive and easy to use.
2. Devoted to provide visual information related to the social interactions. As social interactions deal with people on the surrounding environment, people detection is one of the key points.
3. Without interference to other sensory inputs. The visual information must be fed back to the VI user as a sound input, however it must be done carefully to cope with possible overload of the auditory channel.
4. Real time operation. The system developed should drive feedback at almost the same pace as social interactions take place, although some task are more demanding than others (i.e. face detection versus gesture detection). For a computer vision app this means being able to process several frames per second (fps).
5. Scalability. Although the objective is very concrete, the app has been conceived as a platform to easily incorporate new funcionality (e.g. smile detection, gesture detection, face recognition, etc.).

## 3    Requirements and Dependencies

The app requirements and dependencies include the operating system (OS) and auxiliary software that have to be installed in the VI user smartphone:

- OS Android 4.0+. This choice relies on the principal adoption of this OS in the Spanish market (about 92 %). OS version dependency is important because the native API (Application programming Interface) is used for face detection.
- Auxiliary software.
  • Text-to-Speeech (TTS) synthesis engine (for instance PicoTTS). It is responsible of the auditory feedback to the VI user through voice messages. This engine is included with the OS and it can be also downloaded from Google Play.
  • OpenCV Manager [1,2,4]. It is the OpenCV binary core library for Android downloaded from the Google App Store the first time the app Sainet is executed.

**Fig. 1.** The VI user testing the Sainet app

The hardware requirements for the mobile phone are truly basic and fulfilled by a wide range of available devices at the present market:

- Back camera. It captures the scene in front of the VI user.
- Bluetooth. It allows wireless connectivity between the smartphone and a wireless earphone for auditory messaging feedback to the user.
- Earphone. It is recommended a wireless earphone for a more unobtrusive set up, however a wire plug earphone is also possible.

In the validation test for Sainet (see Fig. 1) a mobile phone with the following characteristics has been used: model HTC One S (Z520e) 1.5 GHz dual core with 1 GB of RAM and a rear camera with 8 Mega pixels (f2.0, 28 mm, AF).

## 4   Functional Description

Sainet has been developed with the Java language programming using the Android native APIs for TTS and face detection on the images captured. Moreover the OpenCV library is used for torso detection allowing a more robust detection of people on the scene.

Once the application is initiated, a continuous loop with the following four stages is executed:

1. image capture,
2. person localization by face detection,
3. position and distance estimation, and
4. auditory feedback if the user requests it.

## 4.1 Image Capture

The image is captured by the smartphone back camera. As soon as one frame is processed another is captured trying to get the highest rate. Each camera provides different frame resolutions (i.e. number of pixels) that affect the amount of memory needed and the processing time. By default the frame resolution is chosen at mid-range, however this value can be modified (e.g. in the tested device resolutions can be selected among $1280 \times 720$, $480 \times 20$ and $176 \times 144$ pixels).

## 4.2 Person Localization by Face Detection

Person localization is a key problem in social interaction. This is a common problem in other computer vision scenarios (e.g. surveillance). The solution to the problem relies on detecting each person on the image and tracking them while they are in the scene. Many works have been devoted to cope with this problems [9]. The simple approach adopted in the Sainet system derives from two assumptions:

1. Close face-to-face social interactions are the most important. In this context 'close' means 3 meters or less.
2. The VI user needs assistance only when visual information is relevant and unreachable by other sensory inputs.

Human body detection on images is a tough task because the non rigid shape of human body adds a new challenge to the object detection problem. However under the first assumption listed above, person detection can be solved by face detection. Then 'person localization by face detection' becomes an easier problem with available implementations in both the Android native and OpenCV APIs.

Face detection with OpenCV is achieved using Viola-Jones algorithm [8] that uses Haar cascade of classifiers. This algorithm applies naive classifiers to detect single features (i.e. Haar-like features) on the image building an 'stronger' classifier for face detection by training an AdaBoost system. The Viola-Jones algorithm has a very high rate of accuracy, offering about 1 % or less for false negatives (i.e. missing a face that is present) and under 40 % for false positives (detecting a face when there is none).

The main advantages of Viola-Jones algorithm are:

– Availability of open source implementation (e.g. in OpenCV) being freely to use.

- Short processing time. Around 3 fps (aprox. 300 ms between consecutive captures) in the smartphone used for testing. This time depends on frame resolution and the performance of the processing unit.
- Far from 'close' detection. Depending on frame resolution the algorithm can detect faces up to 8 m for a 1920 × 1080 frame resolution under testing conditions.

Android also provides a native algorithm for face detection. The algorithm was patented by Neven Vision, a company acquired by Google in 2006. This algorithm can be only used as part of the Android API and his performance (5 fps) surpass the Viola-Jones implementation in OpenCV for Android in the testing device. The Neven algorithm offers a quite independent accuracy from frame resolution selected under test, however its detection range degrades beyond 3 m.

The main drawback observed for the two aforementioned algorithms arises from the moderately high false positive detection rates. In testing conditions false positives produce too frequent 'phantoms' that generate erroneous useless feedback to the VI user. To decrease the false positive rate (i.e. increasing detection robustness) the Sainet system adds a confirmation stage based on torso detection implemented with a Haar filter cascade implemented with the OpenCV library. The combination of face and torso detection offers a more robust detection with a total processing time around 2 fps.

### 4.3   Position and Distance Estimation

Once the persons are detected on the image the step is to estimate their distance and position relative to the observer. An accurate estimation is difficult to reach from a monocular sequence of frames. However, metric accuracy on distance and position estimation is meaningless from the viewpoint of a VI person. From the VI user perspective a rough estimation is in fact more significant. Furthermore this kind of estimation is quite fast to obtain.

Because the social interaction of interest covers a range of 3 m from the VI user, distance and position may be quantized onto three values respectively ('next', 'close' and 'away' for distance, and 'left', 'central' and 'right' for position). Thereby the scene is divided in a simple 3 × 3 shaped space. To estimate position the image is evenly divided in three vertical regions (see Fig. 2) while distance measure is obtained by setting two thresholds for the area of face detection (i.e. more area for closer face detection).

### 4.4   Auditory Feedback to the VI User

When the required information is extracted from the image in the previous step, it is time to feedback this information to the VI user by a different communication channel rather than visual. The alternative interfaces reported in the specialized literature [5–7, 11, 12, 16] can be classified in two categories: tactile (or haptic) interfaces and sound interfaces. None of these categories is optimal because they overlap the substitute senses of vision for VI users.

**Fig. 2.** Detection visualization and discrete scales of position a distance (just for testing purposes)

In Sainet, speech synthesis was chosen as interface with the VI user. Rather than a continuous feedback to the user, the information is only sent when the user requests it. Whenever a request is initiated by the user, the system converts the information gathered from the scene into a voice message. Wearing an earpiece properly connected to the smartphone, the user listens the voice message played by the system (e.g. *"Next to two people at the right. One more to the far left"*).[2]

The main design decisions around this interface can be explained as follows:

- Sound vs. Haptic. Depending on the familiarity of sound this kind of interface requires short time for training (that may be untrue for an odd tone codification). Moreover sounds interfaces are easier to implement because they do not require an specific device rather than the smartphone and an earpiece.
- Voice vs. tone codification. During previous stages of the work a sound codification, rather than synthetic voice, was tested. In this interface a sound localization technique provides the spatial awareness about distance and position. This solution was rejected during validation with the VI user because the complexity for system calibration and the saturation obtained for the auditory sense. On the contrary, synthetic voice is inherently easy to understand without training nor calibration stages. In the final interface a module generates a text message sent to the Android TTS (Text-to-Speech) engine. The inputs to the module are the number of individuals in scene, their distance and position.
- Feedback under request vs. continuous. One of the principal requisites for interfaces devoted to VI users is that they do not interfere with their remain senses, specially touch and hearing. A continuous artificial feedback to the user usually tends to saturate the communication channel and cause stress on VI users (as reported in validation stages with Sainet prototype). Then, a more natural approach results from an strategy of *'feedback under request'* (i.e. *"get information only when needed"*).

---

[2] This message is a translation for the actual Spanish implementation.

**Fig. 3.** User validation of detection under request

In the *'feedback under request'* approach the unsolved question is: *how to notify the request to the system?* Several options were tested with the VI user: gestures on terminal, pressing of smartphone's volume buttons (see Fig. 3), use of NFC technology (Near Field Communication) and miniature bluetooth switches. The next section explains the validation process and discuss the main lessons learned.

## 5    Validation and Discussion

From the beginning, Sainet was focused on the end-user. Therefore, a VI user (blind) was included as a team member for co-developing the SAINET project. His role was to participate in each validation step of the development moreover his experience becomes a source of endless insight to direct the team's efforts. The system validation was organized in two stages directed to gather the experience from the user. First stage was *'practice with the system'* and second, *'explaining the experience'*. Before the experience began, the user was informed about the purposes of the experience and questions are answered about how to *'play with'* the system. After the experience, the team had a meeting session driven by a questionnaire to the user for recording his answers, comments and suggestions.

The main aspects validated by the user were:

- Robustness of person detection. The strategy combining face and torso detection is helpful under not extreme illumination conditions even in outdoors scenarios, although detection rates are affected by illumination conditions and relative movement between camera and individuals.
- Auditory feedback of scene information. The mobile app (i.e. Sainet system) provides a valuable information about position and distance of people in the scene. Rather than continuous, this info is fed back to the VI user in a *'request*

*under demand'* mode so that the user can decide when to be informed by the system. After that, the information extracted from the scene was converted into a single voice message.

– Application accessibility. Although Sainet does not aim to provide additional accessibility mechanisms to the available ones from the Android OS (e.g. Talk-Back, quick widget launcher, etc.), it is compatible with such as mechanisms.

Among all the lessons learned by the Sainet team with the help of our mate (i.e. the VI user), several statements can be established as principles of assistive technologies for the VI:

1. The substitutive innate senses of a VI user must be free as much as possible. The assistive technology should never saturate nor interfere with the natural senses of the user. Under this principle a *'request under demand'* mode was adopted in the Sainet system and a alternatives based on audio 3D and continuous feedback were discarded.
2. The assistive system must be easy to use. To promote the use of technology it has to be user friendly. In Sainet, voice messages generated by a TTS engine proved to be quite simple and intuitive as the way for communicating the relevant information to the user.
3. The assistive technology should be unobtrusive to interlocutors. To make VI users feel self-confident the assistive technology needs to be easy to wear and unobtrusive. Smartphone use is common even in social interaction scenarios, however Sainet only meets this principle partially. In the user opinion the system should be improved in two related aspects:
   (a) Image capture. The image capture in Sainet is done with the smartphone located on the user torso, to use the back camera as capture device. However, this configuration raises some objections derived from this unnatural setting. In fact, during validation the VI user decided by himself just to take the smartphone in his hands holding it at his eyes height (see Fig. 3). The favourite alternative from the user's perspective would be a capture device included in the protective glasses they usually wear. This option requires a miniature camera with wireless connection to the smartphone where the image would be processed. In this solution the glasses could also include the earphone.
   (b) Feedback request. Several options were tested to provide a request mechanism in Sainet system: pressing of volume buttons, terminal gestures, triggering with NFC tags, and bluetooth clickers. Although the bluetooth clickers provide an unobtrusive mechanism for triggering a request, the VI user felt unwilling to include additional devices to the system. In this sense an smartwatch could play the role of triggering requester without adding auxiliary devices for this exclusive purpose.

**Fig. 4.** Smile detection in the image captured

# 6    Conclusion

After VI user validation it was concluded that Sainet brings an easy usage without demanding special requirements from present mid-range marketed smartphones. The application provides valuable information about the scenario captured by a mobile camera, being also scalable to incorporate new functionality. The additional functionalities considered for futures versions are:

1. Recognition of facial expressions, hands and head gestures (including gaze direction). To investigate the feasibility of new feature addition in the system an experimental smile detection feature was included in the final version (see Fig. 4).
2. Automatic text detection and reading (i.e. posters, bills, product labels, etc.). These functionalities are very common in daily tasks with valuables outcomes for the VI users.
3. Re-recognition of previously tagged people and objects.

**Acknowledgments.** This work describes the results for the project SAINET funded by a grant from the Indra-UCLM university Chair and the Adecco Foundation. The authors want to acknowledge the received collaboration from the VISILAB Research Group and specially to Sergio Vera, Francisco Torres and Jesús Manzano.

# References

1. Baggio, D.L., Emami, S., et al.: Mastering OpenCV with Practical Computer Vision Projects. Packt Publishing, Birmingham (2012)
2. Deniz, O., Salido, J., Bueno, G.: Programación de Apps de Visión Artificial. Bubok Publishing S.L. (2013). http://visilab.etsii.uclm.es
3. Deniz, O., et al.: A vision-based localization algorithm for an indoor navigation app. In: 8th International Conference on Next Generation Mobile Applications, Services and Technologies, pp. 7–12. IEEE (2014)

4. Bueno, G., Deniz, O., et al.: Learning Image Processing with OpenCV. Packt Publishing, Birmingham (2015)
5. Manduchi, R., Coughlan, J.: (Computer) vision without sight. Commun. ACM **55**, 96–104 (2012)
6. Dakopoulos, D., Bourbakis, N.G.: Wearable obstacle avoidance electronic travel aids for blind: a survey. Trans. Syst. Man Cybern. Part C **40**(1), 25–35 (2010)
7. Velázquez, R.: Wearable assistive devices for the blind. In: Lay-Ekuakille, A., Mukhopadhyay, S.C. (eds.) Wearable and Autonomous Biomedical Devices and Systems for Smart Environment. LNEE, vol. 75, pp. 331–349. Springer, Heidelberg (2010)
8. Viola, P., Jones, M.J.: Robust real-time face detection. Int. J. Comput. Vis. (IJCV) **57**(2), 137–154 (2004)
9. Gade, L., Krishna, S., Panchanathan, S.: Person localization in a wearable camera platform towards assistive technology for social interactions. Special Issue on Media Solutions that Improving Accessibility to Disabled Users, Ubiquitous Computing and Communication Journal (2010)
10. Krishna, S., Colbry, D., et al.: A systematic requirements analysis and development of an assistive device to enhance the social interaction of people who are blind or visually impaired. In: Workshop on Computer Vision Applications for the Visually Impaired Conducted Along with European Computer Vision Conference (ECCV), Marseille, France (2008)
11. Krishna, S., Panchanathan, S.: Assistive technologies as effective mediators in inter-personal social interactions for persons with visual disability. In: Miesenberger, K., Klaus, J., Zagler, W., Karshmer, A. (eds.) ICCHP 2010, Part II. LNCS, vol. 6180, pp. 316–323. Springer, Heidelberg (2010)
12. Terven, J.R., Salas, J., Raducanu, B.: New opportunities for computer vision-based assistive technology systems for the visually impaired. J. Comput. **4**, 52–58 (2014). IEEE Computer Society
13. Becker, S., Miron-Shatz, T., et al.: mHealth 2.0: experiences, possibilities, and perspectives. JMIR mHealth uHealth **2**(2), 1–12 (2014)
14. Zhu, X., Ramanan, D.: Face detection, pose estimation and landmark localization in the wild. In: Computer Vision and Pattern Recognition (CVPR) (2012)
15. Yi, C., et al.: Finding objects for assisting blind people. Netw. Model. Anal. Health. Inform. Bioinf. **2**, 71–79 (2013). Springer
16. Ivanchenko, V., Coughlan, J.M., Shen, H.: Crosswatch: a camera phone system for orienting visually impaired pedestrians at traffic intersections. In: Miesenberger, K., Klaus, J., Zagler, W.L., Karshmer, A.I. (eds.) ICCHP 2008. LNCS, vol. 5105, pp. 1122–1128. Springer, Heidelberg (2008)
17. Yang, X., Tian, Y.: Robust door detection in unfamiliar environments by combining edge and corner features. In: Proceedings of IEEE Conference on Computer Vision and Pattern Recognition (CVPR), pp. 57–64 (2010)
18. Winlock, T., Christiansen, E., Belongie, S.: Toward real-time Grocery detection for the visually impaired. In: Proceedings of IEEE Conference on Computer Vision and Pattern Recognition (CVPR), pp. 49–56 (2010)

# The Effect of Transcranial Direct Current Stimulation (tDCS) Over Human Motor Function

Cristian Pérez-Fernández[1], Ana Sánchez-Kuhn[1,2(✉)], Rosa Cánovas[2],
Pilar Flores[1,2], and Fernando Sánchez-Santed[1,2]

[1] Department of Psychology, University of Almeria, Almeria, Spain
{cpf603,pflores,fsanchez}@ual.es
[2] Instituto de Neurorrehabilitación Infantil InPaula, Almería, Spain
asanchezkuhn@ual.es, neuro.rosa@inpaula.com

**Abstract.** Transcranial Direct Current Stimulation (tDCS) is a non-invasive, weak cortical neurostimulation technique which implements direct currents through two electrodes with opposite polarization when both are placed over a conductive surface (e.g. the scalp). It has demonstrated positive effects in a wide range of psychopathologies and neurological disorders in the last 15 years, being its neurophysiological modulatory effect on neuro-motor impairments one of the most important targets in tDCS researching. Thus, different motor-related pathologies have been improved by tDCS, such motor alterations after stroke, Parkinson's disease, cerebral palsy in childhood, multiple sclerosis, etc. The positive effects of tDCS on motor abilities, both pathological condition or in healthy population, define it as an interesting option to induce neurophysiological changes complementing the traditional rehabilitation procedures. The comprehension of its neurophysiological and biochemical effects, the development of more ideographic procedures, and its integration with pharmacological treatments are mandatory in order to further improve its usage in rehabilitation approaches.

## 1 Introduction

From ancient times, human being has tried to modulate the behavior and neural function by using electrical impulses directly on the Central Nervous System (CNS) [1]. The main goal was to "cure" or palliate neurological and psychiatric pathologies [2]. Instead, it was used for a better understanding of the brain physiology too [3]. Thus, the technological revolution and its integration in medical sciences in this past Century has let us going from lesser-controlled and dangerous tools to more efficient and safety stimulation devices, from paradigms of invasive deep stimulation (Deep Brain Stimulation), to non-invasive approaches as the Electro-convulsive therapy and transcranial electro-stimulation techniques.

This last group represents the present and future of the main rehabilitation models, in both neurological and psychiatric alterations [4]. Thus, Transcranial Magnetic Stimulation (TMS) is the more-developed and researched methodology last decades.

---

C. Pérez-Fernández1, A. Sánchez-Kuhn contributed equally to the present manuscript.

© Springer International Publishing Switzerland 2016
F. Ortuño and I. Rojas (Eds.): IWBBIO 2016, LNBI 9656, pp. 478–494, 2016.
DOI: 10.1007/978-3-319-31744-1_43

This tool generates rapid changes in the magnetic field in order to induce electrical currents through the brain, letting the specific modulation of the cortical excitability, both single-program and repetitive stimulation [5]. The use of TMS has demonstrated positive effects in multitude of pathologies [5, 6].

Meanwhile, another device has recently taken the world by storm in the neuroscience researching field, the transcranial Direct Current Stimulation (tDCS). This tool has made a revolution in the last 15 years in research, due to its better side effects, lesser cost, easier application and better management of control condition in experimental procedures than the methodologies described above [4]. Thus, the present chapter has the main goal of exploring the features and possibilities which tDCS offers as a present and future model for the rehabilitation of multitude of alterations which take place in the CNS, especially in pathologies with motor affectation.

## 2    What is tDCS?

tDCS is a non-invasive, weak cortical neurostimulation technique which implements a direct current though two electrodes with opposite polarization [7]. It is composed by one anode (positive pole) and one cathode (negative pole), both connected to a 9 volt battery, and covered by two conductive sponges soaked in saline [8]. When both electrodes are placed over a conductive surface (e.g. the scalp), a direct current from the anode to the cathode is generated, and triggers specific changes in cortical excitability in the tissue which is under the anode and the cathode. In research, the current intensity varies from 0.5 milliamps (mA) to 2 mA, with a variable time of application from 5 to 30 min directly over the scalp. Electrodes size is variable too, with a $4 \times 4$ (16 cm$^2$), $5 \times 5$ (25 cm$^2$), $5 \times 7$ (35 cm$^2$) and $6 \times 6$ (36 cm$^2$).

Both current intensity and electrode sizes are very important in order to know the current density, which is defined as the current intensity (expressed in amps -A-, mA and/or microamps -$\mu$A-), divided by the total surface volume which the electrode occupies, being expressed in research as $\mu A/cm^2$, $mA/cm^2$ and $A/cm^2$ [9], being the range from 0.028 to 0.06 mA/cm$^2$ the commonly used in research [10]. The electrode which is placed on the area of interest is called "active electrode", meanwhile the other one is called as the "reference electrode". In most of studies, the reference electrode has been placed over the contralateral orbit (just above the contralateral eye) [7], albeit it is usual its placement on neck, arms, chin, etc. Some recent studies have demonstrated that it is better considered to use a smaller active electrode [11–13].

### 2.1    Handicaps in the Current Density Distribution

One of the main handicaps that we find in the development of the stimulation is the high current density which is concentrated in the edges of the electrode. This is undesirable for safety considerations and it can raise the aversive sensation during the transcutaneous stimulation. Indeed, various authors have defined and tested the current technique protocols by the development of circular electrodes instead of square-shaped ones [14]. This has to be added to the importance of the control of the saline quantity which is incorporated to both sponges, which main function is to facilitate the electric current transmission [14], and the consideration of the "Shunting effect", which is defined as the important amount of electric charge which is lost by the effect of the skin (maximum skin current density), the skull and the cerebrospinal liquid [15]. Current research is focused too in the spatial distribution of the electrodes and how it can facilitate the "Shunting effect" in order to avoid it [13].

## 2.2  Place of Stimulation and Focality

The emplacement of both electrodes is essential in order to develop an efficient stimulation. Neuroimaging and neurophysiological studies confirmed that polarizing effects of tDCS under the electrodes [16, 17]. The specific stimulation of primary motor cortex (M1), primary visual cortex (V1), somatosensory cortex and areas of frontal lobe has reaffirmed this conclusion, defending the focality and specificity of the stimulation with tDCS over the scalp [24], above all when electrodes size is reduced and the inter-electrode distance is controlled [13].

Nevertheless, recent studies with functional Magnetic Resonance (fMRI) and positron emission tomography (PET) have demonstrated that, despite of the fact that tDCS has its largest effects under the electrodes [18], the stimulation generates sustained and generalized effects in other areas of the CNS [19]. These data can be added to studies with electroencephalography (EEG), which have shown that through the modulation of certain areas by cathodic stimulation, diffused effects throughout the CNS can be observed [20]. Thus, these evidences suggest that the effects of stimulation with tDCS are temporal and spatially specific, but not limited to a single place (site specific but not site limited [10]). That is to say, the stimulation of one area will probably affect others throughout neural circuit networks [21]. This is not surprising due to the complex anatomy and function of the human CNS, but it forces to make an exhaustive analysis of the stimulated area and its relationship with closer structures, it does not matter cortical or subcortical in nature [22, 23].

# 3   Neurophisiology of tDCS: General Concepts

The specific effects of tDCS into the CNS are still unknown. Nevertheless, it is widely confirmed that its general effect is a modulation of the resting membrane potentials of the cortical neurons which are placed under the electrodes. An increment or decrease of the cortical excitability can be observed depending on the direction and the current intensity applied [24]. After the stimulation, the main effects duration of a single 10 min session can go further 90 min [25]. These changes depend on the polarity of the stimulation, differentiating between anodal and cathodic stimulation.

The area which is directly stimulated by the anode shows a depolarization of the resting membrane potential, which raises the neuronal excitability up and lets a faster and more spontaneous response of the neurons under the electrode. When the area is stimulated by the cathode, the direct current causes a hyperpolarization of the resting potential of the membrane, dropping down the excitability of the neurons under the electrode [7].

## 3.1 Neurophysiology of tDCS: What Do We Know?

Ardolino et al. [20] observed that short-term tDCS effects are due to local changes in ionic concentrations, specific alterations in transmembrane proteins and to electrolysis related with subtle changes in the Hydrogen protons concentration. As we mentioned above, the mechanisms are not clear, but the tDCS action seems to be related with a hyperpolarization and depolarization combination in the membranes of the neurons, as well as certain alterations in the synaptic efficiency [26, 27]. Following this idea, recent studies have observed a large accumulation of myoinositol into the phospholipid membrane after atDCS on the right frontal lobe by a proton magnetic resonance spectroscopy [28] and a high increase of oxyhemoglobin concentrations after 1 mA of atDCS by a near-infrared spectography [29].

Despite the large number of works which have elucidated many physiological aspects underlying the tDCS effects, it still does not exist a potent theoretical model which explains all its features. Anyway, some authors as Molaee-Ardekani et al., [30] have recently developed a mesoscopic model (the study of population of neurons) over the somatosensory cortex in order to generate a basic structure to explain tDCS effects. Thus, they proposed that the transformation of the presynaptic pulse density of the afferent action potentials into modulated potentials in postsynaptic membrane drives to sustained changes in the local field potentials, and this changes the neural firing rates at different neural sub-population levels (Pyramidal cells, types I and I' Interneurons cells).

As these authors pointed out, the main effects of tDCS seem to be focused in pyramidal cells in the cortex. At the same time, different neuronal groups play the role of facilitating/inhibiting the electric impulse propagation and, thus, are critical in the generation of the evoked potentials (EPs). This neuronal set is composed by Interneurons, whose ortodromic direction (soma/dendrite axe) will facilitate (alignment with stimulation direction) or difficult (non-alignment with stimulation direction) the change in transmembrane potential, as well as happens with pyramidal cells (more information in Molanee-Ardekani et al., [30]). Thus, atDCS seems to be implicated in the selective depolarization of Pyramidal cells and type I interneurons, meanwhile ctDCS hyperpolarizes such cells, showing the opposite effect. Such model could explain short-term plasticity following tDCS [31].

This pyramidal cells/interneurons interaction is regulated by the excitatory mechanisms, guided by the action of the glutamatergic system, and the inhibitory action is controlled, eminently, by gamma-Aminobutyric acid (GABA) tone (Neural mass type model). Therefore, different factors like intensity and density of stimulation, and soma-dendrite axe orientation are essential for the final EPs generation after tDCS intervention. These, finally, will have repercussions in the selective modulation of the spatial and functional conformation of the dendritic ramifications at cortex level, altering their shape/orientation, as well as the total number of functional synapses which they can develop. This phenomenon of dendritic modulation, added to the later phosphorylation of proteins and transcription factors in the cellular nucleus, represents a process very similar to long-term potentiaton [32] and depression, bases of long-term cerebral plasticity [24].

## 3.2  Biochemistry of tDCS

Many studies have demonstrated substantial neurochemical differences between both types of stimulation (anodal and cathodic). Researchers have agreed that the excitatory effects of atDCS are mediated, at least in part, by an important reduction of the GABAergic activity and a facilitation of the glutamatergic N-methyl-D-aspartate receptors (NMDAr) [16, 33].

Otherwise, the inhibitory effects of the cathodic tDCS (ctDCS) seem to be mediated by an important reduction of the glutamatergic excitatory system [34]. It has been also observed a correlation between the increase of the monoaminergic tone and a strongest facilitation of the neuroplastic changes induced by the tDCS [35]. Such neuroplastic phenomena can be facilitated by certain pharmacological compounds like Citalopram (selective serotonin reuptake inhibitor) [36]. More recently, tDCS has been linked with adenosine A1 receptor [37, 38].

Finally, pharmacological studies have demonstrated that the tDCS effects, both immediate and long-term ones, can be eliminated by the selective blockade of Sodium (carbamazepine) and Calcium (flunarizine) channels, as well those long-term effects can be disturbed by the blockade of NMDAr (dextromethorphane) [16]. Added to this, the administration of L-Dopa [39] can turn into inhibition those neuroplastic effects after atDCS application and enhance inhibitory influence of ctDCS.

## 4  Safety Considerations for tDCS

Many studies corroborate that tDCS is a safety and suitable-for-use methodology in humans, and that it is linked to adverse effects only in rare occasions [40]. As well as in its clinical/experimental usage, variables, which we have to consider for safety concerns, are: current intensity, electrode size, current density, total stimulation time and number of sessions [41–43].

Iyer et al. [42] did not observed any sort of side effects, neither in cognitive nor psychomotor measures after 20 min of stimulation, both anodal and cathodic, and both current intensity of 1 mA and 2 mA in the prefrontal cortex. Added to this, indirect biomarkers of brain damage have demonstrated that tDCS application does increase neither the serum levels of the molecular markers of neuronal lesion (N-acetil-aspartate), nor the specific neuronal Enolasa [28]. Thus, no pathological changes by fMRI and EEG have been related to the application of tDCS [17, 42]. Outer the CNS, researchers have focused in the tDCS effects on the heart function, being minimal and non-harmful at any grade in this organ [44].

An exhaustive analysis conducted by Poreisz et al. [40] demonstrated that, after 567 tDCS sessions (both cathodic and anodal) in 102 subjects (both healthy subjects and with diverse neuropathologies patients), no significant negative effects were observed. The stimulation process, with a mean of 12 min, a current intensity of 1 mA, and a reference electrode size of 35 cm$^2$, showed better side effects than previous researches which used TMS.

Despite of the fact that tDCS is a widely known safety tool, is not free of side effects. The more common observed side effects are: subtle tingle, moderate fatigue, soft itching sensations, mild burning and slight pain just under the electrodes [40]. Less common, head ache, difficulty to concentration, nausea and sleep disturbances [40]. In rare occasions, skin lesions by burns after the stimulation have been produced [45]. Anyway, this last phenomenon is related to the electrode shapes used at early stages of the use of tDCS in science, and some researchers argue that it is very important to discern the independence between side skin effects and its null repercussion in the cerebral tissue [15]. No convulsive effects have been linked to tDCS application by now [46].

# 5 tDCS on Motor Function in Patients with Motor-Related Pathologies

## 5.1 Stroke and tDCS

Motor recovery after stroke is linked to the maintenance of ipsilesional motor networks and interactions between ipsilateral and contralesional hemispheres [47], and the tDCS application seems to be capable to modulate them in post-stroke patients. Schlaug et al. [48] defended a hypothetic inter-hemispheric inhibition model by comparing the use of atDCS on the lessoned hemispheric and the cathode in the preserved one, demonstrating that larger effects can be observed by the cathodic stimulation of non-affected hemisphere, presumably due to the current density distribution is not disturbed by a lesion, with intact intracortical networks.

Following this line, Lidenberg et al. [47] showed that, by combining the stimulation on both hemispheres (atDCS on ipsilateral M1 and ctDCS on contralesional M1), 1.5 mA $(0.09 \text{ mA/cm}^2)$ during 30 min, added to orthodox occupancy and physical therapy (OT/PT), motor functions of post-stroke patients were improved stronger than those which receive OT/PT and Sham stimulation. Nevertheless, not all the studies have found this superiority of bilateral approach over single anodal/cathodal interventions [49], with 20 min of tDCS by 1 mA $(0.029 \text{ mA/cm}^2)$, showing strongest behavioral and physiological (Motor evoked potentials -MEPs-) changes at single interventions.

Otherwise, a significant improvement assessed in a specific hand function task for daily living activities (Jebsen-Taylor hand function test) was observed after the application of 1 mA $(0.04 \text{ mA/cm}^2)$ of atDCS during 20 min on post-stroke patients' M1 [50], but not with sham stimulation, resulting in functional gains in motor function of the paretic hand. This ipsilesional stimulation has been successfully confirmed in acute interventions too, when stroke is recent [51], showing significant improvements further to 15 days after the stimulation.

Such unilateral tDCS positive effects have been demonstrated further over hand function after stroke. Indeed, post-stroke aphasia has been selected as a key target for tDCS intervention. Rosso et al. [52] demonstrated that, after the application of 1 mA $(0.028 \text{ mA/cm}^2)$ of ctDCS during 15 min over the right Broca's area, improved language performance was developed, defending the idea that ctDCS can suppress inhibitory inter-hemispheric influences from the right Broca's area to the affected one, and that inter-individual differences are key in order to design accurately stimulation procedures. Those effects could be mediated by GABAergic intracortical and inter-hemispheric function [53].

## 5.2  Dysphagia and tDCS

Dysphagia is a high disrupting possible post-stroke effect. It consists on the impossibility of the patient to start and accomplish the voluntary or involuntary behavior of swallowing. This issue causes serious consequences as nutritional problems; complications of pulmonary aspiration and daily life deficits for the impossibility of swallow even saliva [54, 55].

It is well known that the primary motor cortex plays the principal role on the voluntary activation of the swallowing behavior, appearing both hemispheres to be responsible for this process [56]. However, neuroimaging techniques showed that in the majority of individuals, the projection during swallowing was larger in one hemisphere than in other, being this fact independent of handedness and showing to be different between a pair of identical right-handed twins. Besides the primary motor cortex, it is important to highlight that also the insula, predominantly on the right side, and the cerebellum, mainly on the left side, are also recruited by the swallowing mechanism [57]. Therefore, various areas are involved on the development of this complex mechanism but, at any rate, the primarily motor cortex is a key target area for the use of tDCS stimulation as a treatment for dysphagia.

Studies related to this possibility have been performed across the last years with positive effects over post-stroke dysphagia patients (for review, see Sandrini & Cohen [58]). For instance, in the study of Kumar et al. [59] tDCS or sham on the motor cortex was applied in conjunction with standardized swallowing maneuvers. Anodal or sham tDCS with 2 mA for 30 min were administered over the undamaged hemisphere during 5 consecutive days providing a significant improvement of the scores obtained in the Dysphagia Outcome and Severity Scale (DOSS). Also Yang et al. [60] showed a positive effect of anodal tDCS, 1 mA, for 20 min during 10 consecutive days during swallowing training over the pharyngeal motor cortex of the affected hemisphere. Measures were taken by the Functional dysphagia scale (FDS) using the video fluoroscopic swallowing (VFSS) immediately after the intervention and three months later. Both, the anodal and the sham condition group improved their scores immediately after the intervention equally. However, the differences between the groups emerged on the second evaluation, scoring the anodal stimulated patients significantly higher three months later.

## 5.3  Parkinson's Disease and tDCS

Broeder et al. [61] have recently made an exhaustive review of the main behavioral and physiological effects of tDCS application in patients with Parkinson's disease. Despite of the importance of cognitive alterations, we will only focus on motor-type ones. Thus, Fregni et al. [62] demonstrated that, after only one session of atDCS on left M1 in patients in OFF-phase, with a current density of 0.029 $mA/cm^2$ (1 mA, 35 $cm^2$) for 20 min, a significant increase in MEPs amplitude were produced compared to sham condition, with a clear decrease of MEPs amplitude after ctDCS. All these results correlated with motor improvements as well (tremor, bradykinesia, rigidity, postural instability, gait, etc.). Other studies, which implemented 5 consecutive days of atDCS, 2 mA (0.057 $mA/cm^2$) have supported these results in patients in ON-phase [63]. Specific motor positive effects have been observed on gait and upper limb performance too [64].

Following this motor improvement, an original single-case study by Kaksi et al. [65] demonstrated that, after the application of atDCS, with a current intensity of 2 mA ($0.05$ mA/cm$^2$), over both superior M1 and preMotor cortices (Cz position in 10-20 EEG System), a significant improvement during a tango dancing in trunk and lower extremity movements was produced. This complements the general improvements in upper limbs widely demonstrated by other authors.

Finally, a study conducted by Pereira et al. [66] showed specific linguistic and physiological modulations in the CNS after atDCS, with a current intensity of 2 mA ($0.057$ mA/cm$^2$), during 20 min on both the left dorsolateral prefrontal cortex (DLPFC) and left temporo-parietal cortex (TPC). Phonemic fluency enhancements were observed in patients with Parkinson's disease after the stimulation of left DLPFC, but not after the stimulation on TPC, by increasing the connectivity of verbal fluency networks involving frontal, parietal and fusiform areas, changes quite larger in phonemic fluency than semantic, representing a clear proof of the widely known dissociation between both functions into the CNS.

## 5.4 Spinal Cord Injury and tDCS

After the widely known (but limited) effects of tDCS and TMS on the management of neuropathic pain after spinal cord injury [67], motor-related effects have been converted as a key target in researching. Thus, Silva et al. [68] demonstrated that, after the application of atDCS on both M1 (Cz position in EEG 10-20 system) with a current intensity of 2 mA ($0.057$ mA/cm$^2$) for 12 min in a male with total chronic spinal cord injury, a general improvement in exercise tolerance was observed by measuring specific changes in exercise time and power, perceived exertion, glucose levels, and variability of the time needed to reach the threshold of heart rate.

Otherwise, Murray et al. [69] demonstrated that, after the stimulation by atDCS on left M1 (extensor carpi radialis muscle representation) of 9 patients with chronic spinal cord injury, with a current intensity of 2 mA ($0.64$ mA/cm$^2$) for 20 min, three sessions in 3 weeks, an important increase (up to 40 %) in corticospinal excitability (MEPs) amplitude was observed. This did not happen with 1 mA stimulation procedure. Instead of the high current density implemented by the authors, no significant side effects were observed.

Nevertheless, the strongest positive tDCS effects in this pathologic population have not been reached by traditional cortex stimulation approach. Transcutaneous spinal direct current stimulation (tsDCS) represents a novel practice where the active electrode is placed over the back. Thus, Hubli et al. [70] placed the active electrode longitudinally between the spinous processes of T11 and T12, with a current intensity of 2.5 mA ($0.056$ mA/cm$^2$) for 20 min, both cathodal and anodal stimulation and both patients with spinal cord injury and healthy subjects. That procedure showed specific differences in spinal reflex behavior between pathologic and healthy subjects, where patients showed higher changes in spinal reflex amplitude after atsDCS, being even better modulation than a single session of assisted walking in the driven gait orthosis "Lokomat".

Also, changes in conduction along lemniscal pathway (specific somatosensory evoked potentials amplitude P30) in healthy subjects after the application of atsDCS over the spinous process of the T10 have been observed, with a current intensity of 2.5 mA (0.071 mA/cm$^2$) during 15 min [71].

### 5.5    Cerebral Palsy in Children and tDCS

Cerebral palsy refers to permanent, mutable motor development disorders stemming from a primary brain lesion, causing secondary musculoskeletal problems and limitations in activities of daily living [72]. This disease is the most common motor disorder in children [73].

As we will see, there are several studies of tDCS involving children suggesting the safety of this method. However, it is showed that lesser intensity is required than that used normally on adults to produce cortex stimulation on children, as the peak electrical fields for a given stimulus intensity in the adolescent brain were twice as high as in the adult brain for conventional tDCS [74].

Recently, some studies have found positive effect of tDCS on cerebral palsy when combined with training and rehabilitation. On a study with 24 children with cerebral palsy, positive effects of tDCS combined with treadmill training were observed on balance and functional performance. 1 mA anodal tDCS was administered over 5 weekly sessions of 20 min during 2 weeks over the primary motor cortex of the non-dominant hemisphere during the performance of the treadmill training. The evaluation was carried out with the Pediatric Balance Scale (PBS) and the Pediatric Evaluation of Disability Inventory (PEDI) finding positive effects on the experimental group on balance one week and one month after the treatment, however, no positive effects were found on the self-care and mobility PBS subscales [72].

Spasticity, which is one of the most common symptoms of cerebral palsy [75], is an upper motor neuron syndrome characterized by a velocity-dependent increase in the tonic stretch reflexes with exaggerated tendon jerks resulting from hyperexcitability of this reflex [76]. In a recent study carried out by Aree-uea et al. [77] 46 children between 8 and 18 years with cerebral palsy were tested on spasticity before and after a 1-mA anodal tDCS treatment over the left primary motor cortex during 5 consecutive sessions combined with physical stretching exercises. Results showed a reduction of finger spasticity immediately after the treatment, a reduction of the elbow spasticity immediately and 24 after the treatment and a reduction of wrist spasticity immediately, 24 and 48 h after the treatment. However, no effects of tDCS were found regarding elbow spasticity.

tDCS combined with virtual reality training has also shown to produce positive effects on the body sway velocity [78], as well as regarding spatiotemporal gait variables (velocity and cadence), gross motor function and mobility [79]. Furthermore, in this last study anodal tDCS led to a significant change in motor cortex plasticity, as evidenced by the increase in the amplitude of the motor evoked potential.

Therefore, accompanied with motor training and using specific parameters, tDCS might be a promising technique to improve the symptoms of cerebral palsy in children.

## 5.6  Multiple Sclerosis/Amyotrophic Lateral Sclerosis and tDCS

Few studies from 2010 have focused on different motor and cognitive effects of the application of tDCS in patients with multiple sclerosis [80]. Thus, Cuypers et al. [81] demonstrated that, after the application of atDCS with a current intensity of 1 mA (0.04 mA/cm$^2$) for 20 min on M1 (First Dorsal Interosseous), contralateral to the more impaired hand, a significant corticospinal excitability increase was observed evaluated by MEP variations, effect non-observed after sham stimulation. This cortical modulation triggered to a recruitment-curve plateau increase, something which could be explained by distal effects mediated by large-diameter myelinated axons. Nevertheless, no functional effects were studied. However, no motor improvement facilitation were observed by Meesen et al. [82] after atDCS with a current intensity of 1 mA (0.04 mA/cm$^2$) for 20 min on contralateral to impaired hand M1, compared to sham condition. Further researches in motor function are needed.

Otherwise, and due to the lack of motor assessment studies, positive sensory modulations have been observed after atDCS application in patients with multiple sclerosis [80]. Mori et al. [83] demonstrated that, by applying atDCS with a current intensity of 2 mA (0.057 mA/cm$^2$), during 20 min/5 consecutive daily stimulation on somatosensory cortex (S1), temporally ameliorated sensory deficits (spatial discrimination thresholds on the hypoesthetic hand) further to 2 weeks after treatment were observed in patients with multiple sclerosis. These sorts of positive sensory modulations have been observed too in pain self-sensation in patients with multiple sclerosis [84], after the application of atDCS with a current intensity of 2 mA (0.057 mA/cm$^2$), during 20 min/5 consecutive daily stimulation on contralateral to somatic painful area M1, with a clear decreasement of values in standarized pain scales.

On the other hand, little use of tDCS has been developed on Amyotrophic lateral sclerosis (ALS). As Di Lazzaro et al. [85] pointed out, after the variable results of the application of TMS in patients with ALS, tDCS could be considered as a better intervention tool due to its longer-lasting effects on cortical excitability. Thus, and as a preliminary study, these authors showed no significant effects after ctDCS on M1 (the cortical representation of the first dorsal interosseous muscle), with a current intensity of 1 mA (0.029 mA/cm$^2$) for 20 min on both hemispheres in two different patients. Related to this, Munneke et al. [86] demonstrated no significant cortical excitability variations after 1 mA ctDCS during 7, 11 and 15 min. However, an important effect was observed in healthy subjects, indicating that patients with ALS could have less responsive corticospinal pathways to the inhibitory ctDCS effects. Such results were early demonstrated by Quartarone et al. [87] in both anodal and cathodic stimulation types. Nevertheless, it has been demonstrated that continuous theta burst stimulation (cTBS) by TMS can induce an inhibitory effect on corticospinal excitability in patients with ALS only after 5 daily sessions [88], so it is not unreasonable to postulate that repetitive ctDCS training could generate similar effects on patients with ALS.

## 6    Conclusions

TDCS is a stimulation device whose neuromodulation properties have been widely demonstrated last decade. At the same time, its soft side effects, easy administration, good sham condition control and low price are essential keys in order to understand the reason why hundred of researches are focusing on its study [26].

The large ranges of variants for usage of tDCS are vast, due to the cortical stimulation affects to many behaviors and neurological process. This fact is observable both healthy subjects and patients with diverse neuropathologies, being the neuronal plasticity effect of tDCS its main encouraging aspect for future therapeutic programs. Specifically, the widely known positive effects of tDCS over motor functions, it does not matter cortical (M1), cerebellar or extra-encephalic disposition, and in pathological (Parkinson's disease, Restless syndrome, etc.) or healthy population (functioning of upper/lower limbs, etc.) administration, show that this device represents a fantastic opportunity to improve the rehabilitation effects of orthodoxical training-procedures. For this, the achievement of international protocols and safety guides which ensure a proper use of this technique are mandatory [89].

Basic research of neuroanatomical/functional structures are essential, with the main goal of generate the needed knowledge in order to work with reliably data for the development of more ideographic and specific programs of stimulation. On this sense, the improvement of neuroimaging techniques and scanners with a better neuroanatomical resolution is fundamental [89].

At a microscopic level, it is primary to still improving the researching on the biochemical base of tDCS in the human cortex, due to the current information do not let us to generate a complete theory about its action over the plastic mechanisms inherent to the human brain [90]. Added to this, the development of more sophisticated methodologies for the analysis of the axon orientation, the dendritic arborization, the role of astrocytes and the electrical field threshold of cortical cells is very important too [91].

Otherwise, its use in patients who are refractory to certain pharmacological treatments, as well as in those people in whose state (e.g. pregnancy) the pharmacological intervention is discouraged [92] is another guideline for non-invasive technologies in the future. Added to this last issue, tDCS could play a key role as a useful modulator of pharmacological treatments [93, 94].

The high impact in children with cerebral palsy of tDCS might be due to the general higher impact of its application in immature brains, as the electric peaks in children and adolescents are twice as higher than in adults by using the same parameters [74]. In addition, brain plasticity which characterizes these early ages could contribute to its potentiated effect. For this reason, in our point of view, the effect of tDCS on children is still needed to be further studied as well as the development stages of the CNS to ensure its safety.

The tDCS technique is already being used in the private as well as in the public clinic context. However, there are still some handicaps which should be taken into consideration on its application. Namely, the exact parameters which are effective for each particular pathology have to be defined. In addition, individual differences regarding the effect of tDCS are also a factor which is needed to be studied. Finally, mechanism of action of tDCS must still be explained in order to exactly define the neurophysiology that underlies the present technique, which is a cue factor for its specific and safety use.

**Acknowledgements.** This study was funded by the grants from the Ministerio de Economía y Competitividad, Spanish Government (PSI2012-31660 and PSI2014-55785-C2-1-R) and counted with the participation of the Instituto de Neurorehabilitación InPaula.

# References

1. Largus, S.: De Compositionibus Medicamentorum. Wechel, Paris (1529)
2. Kellaway, P.: The part played by the electric fish in the early history of bioelectricity and electrotherapy. Bull. Hist. Med. **20**, 112–137 (1946)
3. Zago, S., Ferrucci, R., Fregni, F., Priori, A.: Bartholow, Sciamanna, Alberti: pioneers in the electrical stimulation of the exposed human cerebral cortex. Neurosci. **14**(5), 521–528 (2008)
4. Priori, A., Hallett, M., Rothwell, J.C.: Repetitive transcranial magnetic stimulation or transcranial direct current stimulation? Brain Stimul. **2**(4), 241–245 (2009)
5. Pascual-Leone, A., Tormos-Muñoz, J.M.: Estimulación magnética transcraneal: fundamentos y potencial de la modulación de redes neurales específicas. Rev Neurol **46**(S1), S3–S10 (2008)
6. Kobayashi, M., Pascual-Leone, A.: Transcranial magnetic stimulation in neurology. Lancet Neurol. **2**(3), 145–156 (2003)
7. Nitsche, M.A., Cohen, L.G., Wassermann, E.M., Priori, A., Lang, N., Antal, A., et al.: Transcranial direct current stimulation: State of the art 2008. Brain Stimul. **1**(3), 206–223 (2008)
8. George, M.S., Aston-Jones, G.: Noninvasive techniques for probing neurocircuitry and treating illness: vagus nerve stimulation (VNS), transcranial magnetic stimulation (TMS) and transcranial direct current stimulation (tDCS). Neuropsychopharmacol. **35**(1), 301–316 (2010)
9. Ruffini, G., Wendling, F., Merlet, I., Molaee-Ardekani, B., Mekonnen, A., Salvador, R., et al.: Transcranial current brain stimulation (tCS): models and technologies. IEEE Trans Neural Syst Rehab Eng. **21**(2), 333–345 (2013)
10. Costa, T., Lapenta, O., Boggio, P., Ventura, D.: Transcranial direct current stimulation as a tool in the study of sensory-perceptual processing. Atten., Percept., Psychophys. **77**(6), 1813–1840 (2015)
11. Nitsche, M.A., Doemkes, S., Karaköse, T., Antal, A., Liebetanz, D., Lang, N., et al.: Shaping the effects of transcranial direct current stimulation of the human motor cortex. J. Neurophysiol. **97**(4), 3109–3117 (2007)
12. Bastani, A., Jaberzadeh, S.: a-tDCS differential modulation of corticospinal excitability: The effects of electrode size. Brain Stimul. **6**(6), 932–937 (2013)
13. Faria, P., Hallett, M., Miranda, P.C.: A finite element analysis of the effect of electrode area and inter-electrode distance on the spatial distribution of the current density in tDCS. J. Neural Eng. **8**(6), 1–24 (2011)
14. Minhas, P., Datta, A., Bikson, M.: Cutaneous perception during tDCS: Role of electrode shape and sponge salinity. Clin. Neurophysiol. **122**(4), 637–638 (2011)
15. Bikson, M., Datta, A., Elwassif, M.: Establishing safety limits for transcranial direct current stimulation. Clin. Neurophysiol. **120**(6), 1033–1034 (2009)
16. Nitsche, M.A., Fricke, K., Henschke, U., Schlitterlau, A., Liebetanz, D., Lang, N., et al.: Pharmacological modulation of cortical excitability shifts induced by transcranial direct current stimulation in humans. J. Physiol. **553**(pt1), 293–301 (2003)

17. Nitsche, M.A., Niehaus, L., Hoffmann, K.T., Hengst, S., Liebetanz, D., Paulus, W., et al.: MRI study of human brain exposed to weak direct current stimulation of the frontal cortex. Clin. Neurophysiol. **115**(10), 2419–2423 (2004)

18. Kwon, Y.H., Ko, M.H., Ahn, S.H., Kim, Y.H., Song, J.C., Lee, C.H., et al.: Primary motor cortex activation by transcranial direct current stimulation in the human brain. Neurosci Lett. **435**(1), 56–59 (2008)

19. Lang, N., Siebner, H.R., Ward, N.S., Lee, L., Nitsche, M.A., Paulus, W., et al.: How does transcranial DC stimulation of the primary motor cortex alter regional neuronal activity in the human brain? Euro J. Neurosci. **22**(2), 495–504 (2005)

20. Ardolino, G., Bossi, B., Barbieri, S., Priori, A.: Non-synaptic mechanism sunderlie the after-effects of cathodal transcutaneous direct current stimulation of the human brain. J. Physiol. **568**, 653–663 (2005)

21. Lefaucheur, J.P.: Principles of therapeutic use of transcranial and epidural cortical stimulation. Clin. Neurophysi. **119**(10), 2179–2184 (2008)

22. Boggio, P.S., Zaghi, S., Fregni, F.: Modulation of emotions associated with images of human pain using anodal transcranial direct current stimulation (tDCS). Neuropsychol. **47**(1), 212–217 (2009)

23. Boggio, P.S., Zaghi, S., Lopes, M., Fregni, F.: Modulatory effects of anodal transcranial direct current stimulation on perception and pain thresholds in healthy volunteers. Eur. J. Neurol. **15**(10), 1124–1130 (2008)

24. Miranda, P.C., Lomarev, M., Hallett, M.: Modeling the current distribution during transcranial direct current stimulation. Clin. Neurophysiol. **117**(7), 1623–1629 (2006)

25. Nitsche, M.A., Paulus, W.: Sustained excitability elevations induced by transcranial DC motor cortex stimulation in humans. Neurol. **57**(10), 1899–1901 (2001)

26. Been, G., Ngo, T.T., Miller, S.M., Fitzgerald, P.B.: The use of tDCS and CVS as methods of non-invasive brain stimulation. Brain Res. Rev. **56**(2), 346–361 (2007)

27. Parasuraman, R., McKinley, R.A.: Using noninvasive brain stimulation to accelerate learning and enhance human performance. Hum. Factors J. Hum. Factor Ergon. Soc. **56**(5), 816–824 (2014)

28. Rango, M., Cogiamanian, F., Marceglia, S., Barberis, B., Arighi, A., Biondetti, P., et al.: Myoinositol content in the human brain is modified by transcranial direct current stimulation in a matter of minutes: a 1H-MRS study. Magn. Reson. Med. **60**(4), 782–789 (2008)

29. Merzagora, A.C., Foffani, G., Panyavin, I., Mordillo-Mateos, L., Aguilar, J., Onaral, B., et al.: Prefrontal hemodynamic changes produced by anodal direct current stimulation. Neuroimage. **49**(3), 2304–2310 (2010)

30. Molaee-Ardekani, B., Márquez-Ruiz, J., Leal-Campanario, R., Gruart, A., Merlet, I., Birot, G., et al.: Effects of transcranial direct current stimulation (tDCS) on sensory evoked potentials: a computational modeling study. Brain Stimul. **6**(1), 25–39 (2013)

31. Chaieb, L., Antal, A., Paulus, W.: Gender-specific modulation of short-term neuroplasticity in the visual cortex induced by transcranial direct current stimulation. Visual Neurosci. **25**(1), 77–81 (2008)

32. Fritsch, B., Reis, J., Martinowich, K., Schambra, H.M., Ji, Y., Cohen, L.G., et al.: Direct current stimulation promotes BDNF-dependent synaptic plasticity: potential implications for motor learning. Neuron **66**(2), 198–204 (2010)

33. Liebetanz, D., Nitsche, M.A., Tergau, F., Paulus, W.: Pharmacological approach to the mechanisms of transcranial DC-stimulation-induced after-effects of human motor cortex excitability. Brain. **125**(pt10), 2238–2247 (2002)

34. Charlotte, J.S., Jonathan, G.B., Stephenson, M.C., O'Shea, J., Wylezinska, M., Kincses, Z.T., et al.: Polarity-sensitive modulation of cortical neurotransmitters by transcranial stimulation. J. Neurosc. **29**(16), 5202–5206 (2009). doi:10.1523/JNEUROSCI.4432-08.2009
35. Nitsche, M.A., Grundey, J., Liebetanz, D., Lang, N., Tergau, F., Paulus, W.: Catecholaminergic consolidation of motor cortical neuroplasticity in humans. Cereb. Cortex **14**(11), 1240–1245 (2004)
36. Nitsche, M.A., Kuo, M.F., Karrasch, R., Wachter, B., Liebetanz, D., Paulus, W.: Serotonin affects transcranial direct current-induced neuroplasticity in humans. Biol. Psychiatry **66**(5), 503–508 (2009)
37. Izumi, Y., Zorumski, C.F.: Direct cortical inputs erase long-term potentiation at Schaffer collateral synapses. J. Neurosci. **28**(38), 9557–9563 (2008)
38. Márquez-Ruiz, J., Leal-Campanario, R., Sánchez-Campusano, R., Molaee-Ardekani, B., Wendling, F., Miranda, P., et al.: Transcranial direct-current stimulation modulates synaptic mechanisms involved in associative learning in behaving rabbits. PNAS **109**(17), 6710–6715 (2012)
39. Kuo, M.F., Paulus, W., Nitsche, M.A.: Boosting focally-induced brain plasticity by dopamine. Cereb. Cortex **18**(3), 648–651 (2008)
40. Poreisz, C., Boros, K., Antal, A., Paulus, W.: Safety aspects of transcranial direct current stimulation concerning healthy subjects and patients. Brain Res. Bull. **72**(4–6), 208–214 (2007)
41. Nitsche, M.A., Paulus, W.: Excitability changes induced in the human motor cortex by weak transcranial direct current stimulation. J. Physiol. **527**, 633–639 (2000)
42. Iyer, M.B., Mattu, U., Grafman, J., Lomarev, M., Sato, S., Wassermann, E.M.: Safety and cognitive effect of frontal DC brain polarization in healthy individuals. Neurology. **64**(5), 872–875 (2005)
43. Wassermann, E.M., Grafman, J.: Recharging cognition with DC brain polarization. Trends Cogn. Sci. **9**(11), 503–505 (2005)
44. Vandermeeren, Y., Jamart, J., Ossemann, M.: Effect of tDCS with an extracephalic reference electrode on cardio-respiratory and autonomic functions. BMC Neurosci. **11**, 38 (2010)
45. Palm, U., Keeser, D., Schiller, C., Fintescu, Z., Reisinger, E., Padberg, F., et al.: Skin lesions after treatment with trancranial direct current stimulation (tDCS). Brain Stimul. **1**(4), 386–387 (2008)
46. Tanaka, S., Watanabe, K.: Transcranial direct current stimulation: a new tool for human cognitive neuroscience. Brain Nerve **61**, 53–64 (2009)
47. Lindenberg, R., Renga, V., Zhu, L.L., Nair, D., Schlaug, G.: Bihemispheric brain stimulation facilitates motor recovery in chronic stroke patients. Neurology. **75**(24), 2176–2184 (2010)
48. Schlaug, G., Renga, V., Nair, D.: Transcranial direct current stimulation in stroke recovery. Arch. Neurol. **65**(12), 1571–1576 (2008)
49. O'Shea, J., Boudrias, M.H., Stagg, C.J., Bachtiar, V., Kischka, U., Blicher, J.U., et al.: Predicting behavioral response to TDCS in chronic motor stroke. NeuroImage. **85**, 924–933 (2014)
50. Hummel, F., Celnik, P., Giraux, P., Floel, A., Wu, W.H., Gerloff, C., et al.: Effects of non-invasive cortical stimulation on skilled motor function in chronic stroke. Brain. **128**(3), 490–499 (2005)
51. Sattler, V., Acket, B., Gerdelat-Mas, A., Raposo, N., Albucher, J.F., Thalamas, C., et al.: Effect of repeated sessions of combined anodal tDCS and peripheral nerve stimulation on motor performance in acute stroke: A behavioral and electrophysiological study. Ann Phys Rehab Med. **55**(S1), S5–S6 (2012)

52. Rosso, C., Perlbarg, V., Valabregue, R., Arbizu, C., Ferrieux, S., Alshawan, B., et al.: Broca's area damage is necessary but not sufficient to induce after-effects of cathodal tDCS on the unaffected hemisphere in post-stroke aphasia. Brain Stimul. **7**(5), 627–635 (2014)

53. Mälly, J.: Non-invasive brain stimulation (rTMS and tDCS) in patients with aphasia: Mode of action at the cellular level. Brain Res. Bull. **98**, 30–35 (2013)

54. Gordon, C., Hewer, R.L., Wade, D.T.: Dysphagia in Acute Stroke. Br. Med. J. (Clin. Res. Ed). **295**, 411–414 (1987)

55. Barer, D.H.: The Natural History and Functional Consequences of Dysphagia after Hemispheric Stroke. J. Neurol. Neurosurg. Psychiatry **52**(2), 236–241 (1989)

56. Martin, R.E., Sessle, B.J.: The Role of the Cerebral Cortex in Swallowing. Dysphagia **8**(3), 195–202 (1993)

57. Hamdy, S., Rothwell, J.C., Aziz, Q., Thompson, D.G.: Organization and reorganization of human swallowing motor cortex: implications for recovery after stroke. Clin. Sci. **99**(2), 151–157 (2000)

58. Sandrini, M., Cohen, L.G.: Noninvasive brain stimulation in neurorehabilitation. Handb Clin. Neurol. **116**, 499–524 (2013)

59. Kumar, S., Wagner, C.W., Frayne, C., Zhu, L., Selim, M., Feng, W., et al.: Noninvasive brain stimulation may improve stroke-related dysphagia: a pilot study. Stroke **42**(4), 1035–1040 (2011)

60. Yang, E.J., Baek, S.R., Shin, J., Lim, J.Y., Jang, H.J., Kim, Y.K., et al.: Effects of transcranial direct current stimulation (tDCS) on post-stroke dysphagia. Restor Neurol Neurosci. **30**(4), 303–311 (2012)

61. Broeder, S., Nackaerts, E., Heremans, E., Vervoort, G., Meesen, R., Verheyden, G., et al.: Transcranial direct current stimulation in Parkinson's disease: Neurophysiological mechanisms and behavioral effects. Neurosci. Biobehav. Rev. **57**, 105–117 (2015)

62. Fregni, F., Boggio, P.S., Santos, M.C., Lima, M., Vieira, A.L., Rigonatti, S.P., et al.: Noninvasive cortical stimulation with transcranial direct current stimulation in Parkinson's disease. Mov. Disord. **21**, 1693–1702 (2006)

63. Valentino, F., Cosentino, G., Brighina, F., Pozzi, N.G., Sandrini, G., Fierro, B., et al.: Transcranial direct currentstimulation for treatment of freezing of gait: a cross-over study. Mov. Disord. **29**, 1064–1069 (2014)

64. Benninger, D.H., Lomarev, M., Lopez, G., Wassermann, E.M., Li, X., Considine, E., et al.: Transcranial direct current stimulation for the treatment of Parkinson's disease. J. Neurol. Neurosurg. Psychiatry **81**, 1105–1111 (2010)

65. Kaski, D., Allum, J.H., Bronstein, A.M., Dominguez, R.O.: Applying anodal tDCS during tango dancing in a patient with Parkinson's disease. Neurosci. Lett. **568**, 39–43 (2014)

66. Pereira, J.B., Junqué, C., Bartrés-Faz, D., Martí, M.J., Sala-Llonch, R., Compta, Y.: Modulation of verbal fluency networks by transcranial direct current stimulation (tDCS) in Parkinson's disease. Brain Stimul. **6**, 16–24 (2013)

67. Mehta, S., McIntyre, A., Guy, S., Teasell, R.W., Loh, E.: Effectiveness of transcranial direct current stimulation for the management of neuropathic pain after spinal cord injury: a meta-analysis. Spinal Cord. **53**(11), 780–785 (2015)

68. Silva, F.T.G., Rêgo, J.T.P., Raulino, F.R., Silva, M.R., Reynaud, F., Egito, E.S.T., et al.: Transcranial direct current stimulation on the autonomic modulation and exercise time in individuals with spinal cord injury: A case report. Autonomic Neurosci : Basic Clin. **193**, 152–155 (2015)

69. Murray, L.M., Edwards, D.J., Ruffini, G., Labar, D., Stampas, A., Pascual-Leone, A., et al.: Intensity dependent effects of transcranial direct current stimulation on corticospinal excitability in chronic spinal cord injury. Arch. Phys. Med. Rehab. **96**(4 Suppl), S114–S121 (2015)
70. Hubli, M., Dietz, V., Schrafl-Altermatt, M., Bolliger, M.: Modulation of spinal neuronal excitability by spinal direct currents and locomotion after spinal cord injury. Clin Neurophysiol: Off. J. Inter. Fed. Clin. Neurophysiol. **124**(6), 1187–1195 (2013)
71. Cogiamanian, F., Vergari, M., Pulecchi, F., Marceglia, S., Priori, A.: Effect of spinal transcutaneous direct current stimulation on somatosensory evoked potentials in humans. Clin Neurophysiol: Off. J. Inter. Fed. Clin. Neurophysiol. **119**(11), 2636–2640 (2008)
72. de Almeida, J.R., Guyatt, G.H., Sud, S., Dorion, J., Hill, M.D., Kolber, M.R., et al.: Management of bell palsy: clinical practice guideline. CMAJ **186**, 917–922 (2014)
73. Swaiman, K.F.: Cerebral palsy. In: Swaiman, K.F., Ashwal, S., Ferriero, D.M., Schor, N.F. (eds.) Swaiman's Pediatric Neurology: Principles and Practice, 5th edn, pp. 999–1008. St. Louis: Elsevier, New York (2012)
74. Minhas, P., Bikson, M., Woods, A.J., Rosen, A.R., Kessler, S.K.: Transcranial direct current stimulation in pediatric brain: a computational modeling study. In: Annual International Conference of the IEEE Engineering in Medicine and Biology Society (EMBC) (2012)
75. Bax, M., Tydeman, C., Flodmark, O.: Clinical and MRI correlates of cerebral palsy: the european cerebral palsy study. JAMA **296**, 1602–1608 (2006)
76. Lance, J.W.: Pathophysiology of Spasticity and Clinical Experience with Baclofen. In: Feldman, R.G., Young R.R., Koella, W.P., (eds.) Spasticity, Disordered Motor Control. Miami, FL: Symposia Specialists, pp. 185–203 (1980)
77. Aree-uea, B., Auvichayapat, N., Janyacharoen, T., Siritaratiwat, W., Amatachaya, A., Prasertnoo, J., et al.: Reduction of spasticity in cerebral palsy by anodal transcranial direct current stimulation. J. Med. Assoc. Thai. **97**(9), 954–962 (2014)
78. Lazzari, R.D., Politti, F., Santos, C.A., Dumont, A.J.L., Rezende, F.L., Grecco, L.A.C., et al.: Effect of a single session of transcranial direct-current stimulation combined with virtual reality training on the balance of children with cerebral palsy: a randomized, controlled, double-blind trial. J. Phys Ther. Sci. **27**(3), 763–768 (2015)
79. Grecco, L.A.C., de Almeida, N., Mendonça, M.E., Galli, M., Fregni, F., Santos, C.: Effects of anodal transcranial direct current stimulation combined with virtual reality for improving gait in children with spastic diparetic cerebral palsy: a pilot, randomized, controlled, double-blind, clinical trial. Clin Rehabil. **29**(12), 1212–1223 (2015)
80. Palm, U., Ayache, S.S., Padberg, F., Lefaucheur, J.-P.: Non-invasive brain stimulation therapy in multiple sclerosis: a review of tDCS, rTMS and ECT results. Brain Stimul. **7**(6), 849–854 (2014)
81. Cuypers, K., Leenus, D.J.F., Van Wijmeersch, B., Thijs, H., Levin, O., Swinnen, S.P., et al.: Anodal tDCS increases corticospinal output and projection strength in multiple sclerosis. Neurosci. Lett. **554**, 151–155 (2013)
82. Meesen, R.L.J., Thijs, H., Daphnie, J.F., Leenus, D.J.F., Cuyper, K.: A single session of 1 mA anodal tDCS-supported motor training does not improve motor performance in patients with multiple sclerosis. Restore Neurol Neurosci. **32**(2), 293–300 (2014)
83. Mori, F., Nicoletti, C.G., Kusayanagi, H., Foti, C., Restivo, D.A., Marciani, M.G., et al.: transcranial direct current stimulation ameliorates tactile sensory deficit in multiple sclerosis. Brain Stimul. **6**(4), 654–659 (2013)
84. Mori, F., Codeca, C., Kusayanagi, H., Monteleone, F., Buttari, F., Fiore, S., et al.: Effects of anodal transcranial direct current stimulation on chronic neuropathic pain in patients with multiple sclerosis. J. Pain. **11**(5), 436–442 (2010)

85. Di Lazzaro, V., Ranieri, F., Capone, F., Musumeci, G., Dileone, M.: Direct current motor cortex stimulation for amyotrophic lateral sclerosis: a proof of principle study. Brain Stimul: Basic, Transl., Clin. Res. Neuromodul. 6(6), 969–970 (2015)
86. Munneke, M.A., Stegeman, D.F., Hengeveld, Y.A., Rongen, J.J., Schelhaas, H.J., Zwarts, M.J.: Transcranial direct current stimulation does not modulate motor cortex excitability in patients with amyotrophic lateral sclerosis. Muscle Nerve 44(1), 109–114 (2011)
87. Quartarone, A., Lang, N., Rizzo, V., Bagnato, S., Morgante, F., Sant'angelo, A., et al.: Motor cortex abnormalities in amyotrophic lateral sclerosis with transcranial direct-current stimulation. Muscle Nerve. 35(5), 620–624 (2007)
88. Munneke, M.A.M., Rongen, J.J., Overeem, S., Schelhaas, H.J., Zwarts, M.J., Stegeman, D.F.: Cumulative effect of 5 daily sessions of θ burst stimulation on corticospinal excitability in amyotrophic lateral sclerosis. Muscle Nerve 48(5), 733–738 (2013)
89. Brunoni, A.R., Nitsche, M.A., Bolognini, N., Bikson, M., Wagner, T., Merabet, L., et al.: Clinical research with transcranial direct current stimulation (tDCS): Challenges and future directions. Brain Stimul. 5(3), 175–195 (2012)
90. Javadi, A.H., Cheng, P.: Transcranial direct current stimulation (tDCS) enhances reconsolidation of long-term memory. Brain Stimul. 6(4), 668–674 (2012)
91. Radman, T., Ramos, R.L., Brumberg, J., Bikson, M.: Role of cortical cell type and morphology in subthreshold and suprathreshold uniform electric field stimulation. Brain Stimul. 2(4), 215–228 (2009)
92. Zhang, X., Liu, K., Sun, J., Zheng, Z.: Safety and feasibility of repetitive transcranial magnetic stimulation (rTMS) as a treatment for major depression during pregnancy. Arch Womens Ment Health. 13(4), 369–370 (2010)
93. Brunoni, A.R., Ferrucci, R., Bortolomasi, M., Scelzo, E., Boggio, P.S., Fregni, F., et al.: Interactions between transcranial direct current stimulation (tDCS) and pharmacological interventions in the Major Depressive Episode: Findings from a naturalistic study. Eur Psychiatry. 28(6), 356–361 (2013)
94. Brunoni, A.R., Valiengo, L., Baccaro, A., Zanão, T.A., de Oliveira, J.F., Goulart, A., et al.: The Sertraline vs. electrical current therapy for treating depression clinical study: Results from a factorial, randomized, controlled trial. JAMA Psychiatry. 32(1), 90–98 (2013)

# Evaluation of Cervical Posture Improvement of Children with Cerebral Palsy After Physical Therapy with a HCI Based on Head Movements and Serious Videogames

Miguel A. Velasco[1(✉)], Rafael Raya[1,2(✉)], Luca Muzzioli[3], Daniela Morelli[3], Marco Iosa[3], Febo Cincotti[3], and Eduardo Rocon[1]

[1] Centro de Automática Y Robótica CAR, UPM-CSIC,
Ctra. Campo Real Km 0.2, Arganda del Rey, 28500 Madrid, Spain
miguel.velasco@csic.es, rayalopez.rafa@gmail.com
[2] Department of Information Technology, Universidad CEU San Pablo, Ctra. Boadilla del Monte,
Km. 5.300, 28925 Alcorcón Madrid, Spain
[3] Fondazione Santa Lucia FSL, Via Ardeatina, 306, 00142 Rome, Italy

**Abstract.** This paper presents the preliminary results of a novel rehabilitation therapy for cervical and trunk control of children with cerebral palsy (CP). The therapy is based on the use of an inertial sensor that will be used to control a set of serious videogames with movements of the head. Ten users with CP participated in the study, in the experimental and control groups. Ten sessions of therapy provided improvements in head and trunk control that were higher in the experimental group for Visual Analogue Scale (VAS), Goal Attainment Scaling (GAS) and Trunk Control Measurement Scale (TCMS). Significant differences (27 % vs. 2 % of percentage improvement) were found between the experimental and control groups for TCMS ($p < 0.05$). The kinematic assessment shows that there are some improvements in active and passive range of motion, but no significant differences were found pre- and after-therapy. This new strategy, together with traditional rehabilitation therapies, could allow the child to reach maximum levels of function in the trunk and cervical regions.

**Keywords:** Cerebral palsy · Cervical posture · Inertial sensor · Serious games

## 1 Introduction

Cerebral palsy (CP) is a disorder of posture and movement due to a defect or lesion in the immature brain [1]. CP affects between 2 to 3 per 1000 live-births, reported for the European registers by the Surveillance of Cerebral Palsy European Network (SCPE) [2], and there is a prevalence of three to four per 1000 among school-age children in the USA [3]. CP is the most common cause of permanent serious physical disability in childhood, and the prospect of survival in children with severe level of impairment has increased in recent years, [4]. CP is often associated to sensory deficits, cognition impairments, communication and motor disabilities, behavior issues, seizure disorder, pain and secondary musculoskeletal problems. CP can be classified according to different criteria:

© Springer International Publishing Switzerland 2016
F. Ortuño and I. Rojas (Eds.): IWBBIO 2016, LNBI 9656, pp. 495–504, 2016.
DOI: 10.1007/978-3-319-31744-1_44

the distribution of the deficits, the gross motor function, the predominant abnormality, and other systems. The "Surveillance of Cerebral Palsy in Europe (SCPE): a collaboration of cerebral palsy surveys and registers" presented a consensus on the definition, classification and description of CP [5, 6].

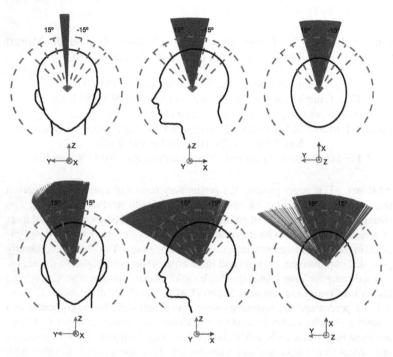

**Fig. 1.** Angular orientations in frontal, sagittal and transverse planes measured with the ENLAZA interface. Above, an individual without motor disorders; below, one with CP.

Poor head and trunk control in CP produce limitations beyond function. In hypotonic CP infants (as in Fig. 1) they can be so severe that the infant may experience difficulty breathing or swallowing since effective oral functioning for feeding begins with attaining better head stability to improve jaw control [7]. Moreover, the head is responsible for the directional orientation of the special senses and its movements are influenced by the information these provide. It is not unexpected that certain disorders of the special senses may lead to unusual head movements and that disorders of head movement may force unusual conditions upon the special senses [8].

Posture improvement is important not only because of functional reasons, but also to improve some secondary conditions related with health and social interaction. Kramer et al. [9] compared the semi-prone (SP) and sitting (SIT) training positions with respect to head control in children with CP, before and after 5 weeks biofeedback training using a head position trainer (HPT). They concluded that biofeedback training with a HPT can be effective in either the SIT or the SP positions, with improvement lasting at least 16 weeks after training is discontinued.

A program was recently developed to promote adaptive responses and upright head position in students with multiple disabilities (cerebropathy and spastic tetraparesis and their head tended to be tilted forward) through the use of micro-switch clusters (i.e., combinations of two micro-switches) during a stimulation period [10]. The study consisting in measuring their actual level of head control, that is, the length of time they kept the upright head position. The five students exposed to the program showed a significant increase in adaptive responses performed with head upright. A recent research investigated the head position correction using a commercial inertial product. Two subjects with severe multiple disabilities with spastic quadriplegic cerebral palsy with limited trunk and head control participated. Results showed significantly increased their time duration of maintaining upright head position to obtain the desired environmental stimulation [11].

This study is presented as a proof of concept of a rehabilitation therapy for the improvement of head and trunk posture in children with CP based on active head exercises performed through serious videogames that will be accessed with an interface based on inertial technology. We aim to develop evidence-based criteria for the integration of these exercises into the traditional therapies and to determine their role in maximizing head control in children with CP. We hypothesize that the user can improve his/her head posture by using the ENLAZA interface based on the neuroplasticity and the capacity to learn new motor skills.

## 2 Background

### 2.1 Assessment of Cerebral Palsy

When assessing people with CP, many factors and the symptoms need to be monitored. The WHO International Classification of Functioning, Disability and Health (ICF) along with several other recent publications such as the SCPE have sensitized health professionals to the importance of evaluating the functional consequences of different health states. For ambulation, the Gross Motor Function Classification System (GMFCS) [12] has been widely employed internationally to group individuals with CP into one of five levels based on functional mobility or activity limitation. So has the bimanual fine motor function system BFMF [13], or, in prospective studies, the Manual Ability Classification System MACS [14].

Most methods are subjective measures that classify the motor involvement on the basis of functional abilities. In milder cases, the assessment and conclusions may vary by the subjective examinations of various professionals. Therefore, a combination of significant motor developmental delay and abnormalities in the neurologic examination is required to make the diagnosis. A promising approach is the use of normal and abnormal general movement patterns. This method appears to have high sensitivity and specificity for the diagnosis of CP [15, 16]. Motion sensing, by means of MEMS inertial sensors a real scientific breakthrough in the medical field, where there is a need for small ambulatory sensor systems for measuring the kinematics of body segments [17]. As a result, inertial sensors have been chosen for different applications focused on people with motor disorders, such as the evaluation of

**Fig. 2.** The ENLAZA interface (Werium Solutions S.L., Spain)

clinical spasticity assessment (by measuring the range of motion) [18] and the quantification of standing balance by assessing displacement of the center of mass in CP [19] and clinical assessment of tremor in Parkinson [20].

## 2.2 Therapies and Treatments

Treatments for CP patients depend on the patient's presentation and range from physical therapy to medication and surgery. They follow these basic principles [21]: (1) emphasis on normalization of the quality of movement; (2) emphasis on functional activities, which focuses on the development of skills necessary for the performance of activities of daily living (ADL). Superior clinical results have been observed in children participating in this intervention [22].

The priorities in the management of CP are currently moving from traditional strategies focusing on promoting compensation towards new strategies aiming on restore motor function with the increasing evidence of neuroplasticity and motor learning theories. These strategies have a higher impact during early ages because of neuroplasticity, i.e. the ability of the neurons and other human brain cells to reorganize their structure and function after an injury, in response to different external and internal factors, including physical training [23]. Task-oriented therapies aim to improve the movement and the posture of the user with CP by the repetitive training performing a certain functional task. People with CP frequently show impaired limb, trunk and head control, which affect performances of ADLs. The majority of research in children with CP focuses on assessment and treatment of upper and lower extremities. In contrast, literature on trunk and head control in children with CP is scarce [24].

## 3   The ENLAZA Human-Computer Interface (HCI)

Raya et al. proposed the ENLAZA interface [25], an adapted input device for users with severe motor disorders (especially CP) that cannot use traditional solutions such as mice, joysticks or trackballs to access the computer. ENLAZA (Fig. 2) allows users to control the cursor of the computer with movements of their heads and consists of a headset with a cap and an inertial measurement unit, IMU, (Werium S.L., Spain) that integrates a tridimensional (3D) accelerometer, a 3D gyroscope and a 3D magnetometer mounted on a commercial helmet. A calibrated IMU measures 3D acceleration (caused by motion and gravity), 3D angular velocity and 3D earth magnetic field. A data fusion algorithm presented in [26] estimates the IMU orientation and enables measurement of inclination changes less than $1.0°$ and $1°\text{-}2°$ heading accuracy.

The user wears the ENLAZA device at the beginning of the experiment. After wearing the device, the calibration procedure starts. This procedure consisted of maintaining the head in front on the computer screen (zero position) and calibrate angular rotation required to reach the horizontal and vertical bounds of the screen. Once calibrated for the first session, it is not necessary to repeat it for the next sessions.

For the purpose of this study, the mouse pointer is controlled with an absolute control, meaning that there is a unique mapping between head orientation and location of the pointer. After a calibration process, all pixels in the screen are reachable for the user's Cervical Range of Motion, CROM. During the calibration, a therapist adjusts the gain of the transfer function that translates the orientation of the head into a location of the pointer on the screen. A Robust Kalman Filter (RKF) [27] was developed to facilitate fine motor control based on the characterization of involuntary movements found in users with cerebral palsy. In addition, the designed software captures data used to assess performance in the task. In particular, the application captures the positions of the mouse pointer and target during the session.

## 4  Methodology

Five users with CP from the "Fondazione Santa Lucia" (FSL, Rome, Italy), a center specialized in the treatment of CP participated on this study. The inclusion/exclusion criteria can be found in Table 1. Participants wore ENLAZA and played serious videogames. Six videogames have been specially designed and developed in C# and the framework.NET 4.0 to be played with the ENLAZA interface, and another set of six commercial off-the-shelf videogames have been adapted to be played with this system. These videogames gather the following characteristics:

**Table 1.** Inclusion/exclusion criteria

| Inclusion criteria | Exclusion criteria |
| --- | --- |
| -Males and females, aged 4-21 years old | -Aggressive or self-injure behavior. |
| -Diagnosed CP and cervical hypotonia or difficulties on head control | -Involuntary movements on the head. |
| -Cognitive capacity and behavior appropriate to understand the tasks and follow simple instruction and active participation in the study, | -Cervical surgery within the previous 6 months. |
| -Signed written informed consent by parents or legal guardian. | -Inability to control the ENLAZA system during the first testing session. |
| -Medically stable. | -Severe visual limitations |

- Fun and systematic.
- Visual and auditory feedback
- Clear objective for the user: Task and duration.
- Different levels of difficulty.
- To use colours or images to represent abstract concepts as time.

Another 5 users with CP were recruited for the control group. They followed the traditional physical and occupational therapy.

### 4.1 Assessment of Head Posture

We propose a kinematic and functional analysis of the improvement hypothesized.

**Kinematic Assessment.** We will use ENLAZA to measure (1) the CROM, i.e. flexion-extension, rotations and lateral flexion, during active movements directly performed by the child and (2) the CROM during passive mobilization of the therapist.

**Functional Assessment.** The outcome measures to measure the head posture improvement will be:

- Gross Motor Function Measure-88 (GMFM-88). We will use the items 21 and 22 depending on the child's skills [28]. They assess whether a child can lift and maintain his/her head in a vertical position with trunk support by a therapist while sitting.
- Visual Analogue Scale (VAS). VAS is a valid scale that consists of a line of 100 mm separating two labels: 0 = "No head control" and 10 = "Perfect head control". We asked parents, children and therapist to put a cross on the line of 100 mm separating these two labels to indicate on the VAS the level of head control of the children.
- Goal attainment scaling (GAS). GAS allows the therapist to program a desired improvement and to judge if the child achieved it (the fact that the goal is chosen by therapist made this scale very sensitive). Goal 1 will be related to head movement, and goal 2 will be optional and related to choking/swallowing, if the child has daily trouble swallowing.

### 4.2 Data Analysis

Mean and standard deviation have been used for the description of clinical scale scores. Percentage improvement has been evaluated as (post- and pre-value) / pre-value*100. Non parametric inferential statistical tests have been used: in particular the Wilcoxon signed ranks test was used for within group analyses in order to compare clinical scores at T0 and T1, whereas the Mann-Whitney u-test was used for the between group comparisons at T0 and at T1, separately. The alpha level of statistical significant threshold was set at 0.05 for all the analyses.

# 5 Results

## 5.1 Functional Assessment

Tables 2 and 3 show the mean clinical scores pre and post intervention. The changes resulted statistically significant in terms of control of the head (p = 0.034), visuomotor control assessed by GAS-score (p = 0.043) and TCMS (p = 0.042). The gross motor functioning slightly improved but the value of p did not achieve the statistical significant threshold (p = 0.048). The items 21 and 22 of GMFM-88 remained unaltered both in experimental as well as in control group.

For the control group, statistically significant improvements occurred in terms of head control (p = 0.046) and visuomotor control (p = 0.042), but neither in terms of trunk control (p = 0.317) nor of gross motor functioning (p = 0.317). It implied that the improvement of trunk control was significantly higher in Experimental Group with respect to Control Group (about +27 % vs. +2 %, respectively, as reported in Table 3 and Fig. 3). The other percentage changes, despite quite higher in experimental group, were not statistically different between the two groups.

**Table 2.** Clinical scores for experimental group (p-value refers to Wilcoxon signed rank test)

| Experimental Group | Pre | Post | p-value |
|---|---|---|---|
| VAS | 6.4 ± 1.1 | 7.6 ± 1.3 | **0.034** |
| GAS | 22.8 ± 0.4 | 64.3 ± 3.6 | **0.043** |
| TCMS | 19.4 ± 14.1 | 24.2 ± 17.9 | **0.042** |
| GMFM-88 | 44.4 ± 23.2 | 50.2 ± 27.8 | 0.068 |
| GMFM-88 Item 21 | 3 ± 0 | 3 ± 0 | 0.999 |
| GMFM-88 Item 22 | 3 ± 0 | 3 ± 0 | 0.999 |

**Table 3.** Percentage improvements in clinical scores for experimental vs- control group (p-value refers to Mann Whitney u test)

| Scale | Experimental group | Control group | p-value |
|---|---|---|---|
| VAS | 18.9 ± 6.0 % | 15.2 ± 9.4 % | 0.690 |
| GAS | 181.3 ± 17.0 % | 155.6- ± 27.9 % | 0.222 |
| TCMS | 27.2 ± 11.5 % | 1.8 ± 4.1 % | **0.008** |
| GMFM-88 | 11.5 ± 18.7 % | 0.8 ± 1.77 % | 0.151 |

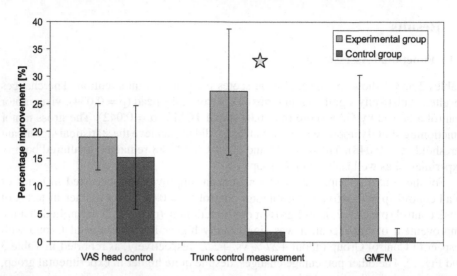

**Fig. 3.** Mean and standard deviations of percentage improvements in the groups. Stars indicate a statistically significant difference between groups assessed by Man Whitney u test (p < 0.05)

### 5.2 Kinematic Assessment

The ranges of motion measured before and after the therapy showed a rise of AROM in all three movements. The percentages of increment were +20 %, +38 % and +85 % in order to achieve 93°, 90° and 145° for active flexion-extension, lateral flexion and rotation, respectively. Unfortunately, no statistically significant differences were found (p = 0.18, p = 0.62, p = 0.43).

The passive range of motion, PROM, presented smaller changes: an increment of +5 % and +57 % to achieve 77° and 140° of passive lateral flexion and rotation. Passive range of motion during flexion experienced a small reduction (-6 %) and decremented from 86° to 81°. Once again, no statistical significance was found (p = 0.43, p = 0.62, p = 0.43).

## 6   Discussion and Conclusion

We found improvements in four metrics (VAM, GAS, TCMS and GMFM-88) in the experimental group, although they were not significant for GMFM-88. The improvements were generally better for the experimental group and significant differences were found in TCMS between the groups. The percentage of improvement in trunk control is indeed remarkable and shows the potential of this kind biofeedback in rehabilitation therapies.

Despite the lack of statistical significance in the improvements, the values AROM and APROM for flexion, lateral-flexion and extension after 10 work sessions are closer to the physiologically normal ROMs, i.e. 90°, 160° and 90° for flexion-extension, lateral flexion and rotation.

We proved a therapy for the rehabilitation of head and trunk motor control with inertial sensors and serious games as a complement to traditional therapies is possible, and that the improvements of this novel therapy are better that those achieved with traditional therapies alone. In future experiments, we will recruit a larger group in a multi-center study in order to look for greater significance in the functional and kinematic evaluation.

**Acknowledgements.** Authors would like to thank the members and staff in FSL. This work was possible thank to ABC EU Project, CPWALKER Project, INTERPLAY Project and IVANPACE Project, which is funded by Obra Social de Caja Cantabria.

# References

1. Bax, M., Goldstein, M., Rosenbaum, P., Leviton, A., Paneth, N., Dan, B., Jacobsson, B., Damiano, D.: Proposed definition and classification of cerebral palsy. Dev. Med. Child Neurol. **47**(8), 571–576 (2005)
2. Johnson, A., Perinatal, N., Unit, E., Road, O.: Prevalence and characteristics of children with cerebral palsy in Europe. Dev. Med. Child Neurol. **44**(9), 633–640 (2002)
3. Winter, S., Autry, A., Boyle, C., Yeargin-Allsopp, M.: Trends in the prevalence of cerebral palsy in a population-based study. Pediatrics **110**(6), 1220–1225 (2002)
4. Blair, E.: Epidemiology of the cerebral palsies. Orthop. Clin. North Am. **41**(4), 441–455 (2010)
5. Cans, C.: Surveillance of cerebral palsy in Europe: a collaboration of cerebral palsy surveys and registers. Dev. Med. Child Neurol. **42**(12), 816–824 (2007)
6. Krägeloh-Mann, I., Cans, C.: Cerebral palsy update. Brain Dev. **31**(7), 537–544 (2009)
7. Redstone, F., West, J.F.: The importance of postural control for feeding. Pediatr. Nurs. **30**(2), 97–100 (2004)
8. Gresty, M.A., Halmagyi, G.M.: Abnormal head movements. J. Neurol. Neurosurg. Psychiatry **42**(8), 705–714 (1979)
9. Kramer, J.F., Ashton, B., Brander, R.: Training of head control in the sitting and semi-prone positions. Child Care Health Dev. **18**(6), 365–376 (1992)
10. Lancioni, G.E., Singh, N.N., O'Reilly, M.F., Sigafoos, J., Didden, R., Oliva, D., Severini, L.: Fostering adaptive responses and head control in students with multiple disabilities through a microswitch-based program: follow-up assessment and program revision. Res. Dev. Disabil. **28**(2), 187–196 (2007)
11. Shih, C.-H.: Assisting people with disabilities improves their collaborative pointing efficiency through the use of the mouse scroll wheel. Res. Dev. Disabil. **34**(1), 1–10 (2013)
12. Palisano, R.R., Rosenbaum, P., Bartlett, D., Livingstone, M.: Gross Motor Function Classification System - Extended & Revised. CanChild Centre for Childhood Disability Research, McMaster University (Reference: Dev. Med. Child. Neurol. **39**, 214–223 (1997))
13. Beckung, E., Hagberg, G.: Correlation between ICIDH handicap code and gross motor function classification system in children with cerebral palsy. Dev. Med. Child Neurol. **42**, 669–673 (2000)
14. Eliasson, A.C., Krumlinde-Sundholm, L., Rösblad, B., Beckung, E., Arner, M., Öhrvall, A.M., Rosenbaum, P.: The manual ability classification system (MACS) for children with cerebral palsy: scale development and evidence of validity and reliability. Dev. Med. Child Neurol. **48**(7), 549–554 (2006)

15. Mac, R., Lecture, K.: The 1999 Ronnie Mac Keith Lecture, pp. 836–842 (2001)
16. van der Heide, J., Paolicelli, P.B., Boldrini, A., Cioni, G.: Kinematic and qualitative analysis of lower-extremity movements in preterm infants with brain lesions. Phys. Ther. **79**(6), 546–557 (1999)
17. Kim, M., Kim, B.H., Jo, S.: Quantitative evaluation of a low-cost noninvasive hybrid interface based on EEG and eye movement. IEEE Trans. Neural Syst. Rehabil. Eng. **23**(2), 159–168 (2015)
18. Cutti, A., Cappello, A., Davalli, A.: A new technique for compensating the soft tissue artefact at the upper-arm: in vitro validation. J. Mech. **5**(2), 1–15 (2005)
19. Ghasemzadeh, H., Jafari, R., Prabhakaran, B.: A body sensor network with electromyogram and inertial sensors: Multimodal interpretation of muscular activities. IEEE Trans. Inf. Technol. Biomed. **14**(2), 198–206 (2010)
20. Powell, H.C., Hanson, M., Lach, J.: A wearable inertial sensing technology for clinical assessment of termor. In: Circuits Systems Conference BIOCAS 2007, pp. 9–12 (2007)
21. Bower, E.: The multiply handicapped child. In: Wilson, B.A., McLellan, L.D. (eds.) Rehabilitation Studies Handbook, pp. 315–354. Cambridge University Press (1997)
22. Ketelaar, M., Vermeer, A., Hart, H., van Petegem-van Beek, E., Helders, P.J.: Effects of a functional therapy program on motor abilities of children with cerebral palsy. Phys. Ther. **81**(9), 1534–1545 (2001)
23. Aisen, M.L., Kerkovich, D., Mast, J., Mulroy, S., Wren, T.A.L., Kay, R.M., Rethlefsen, S.A.: Cerebral palsy: clinical care and neurological rehabilitation. Lancet Neurol. **10**(9), 844–852 (2011)
24. Heyrman, L., Molenaers, G., Desloovere, K., Verheyden, G., De Cat, J., Monbaliu, E., Feys, H.: A clinical tool to measure trunk control in children with cerebral palsy: the trunk control measurement scale. Res. Dev. Disabil. **32**(6), 2624–2635 (2011)
25. Raya, R., Ceres, R., Rocon, E., González, T., Ruiz, A.: Desarrollo de un interfaz inercial orientado a terapias alternativas en la parálisis cerebral. In: Actas de las IV Jornadas Iberoamericanas de Tecnologías de Apoyo a la Discapacidad, pp. 57–62 (2009)
26. Raya, R., Roa, J.O., Rocon, E., Ceres, R., Pons, J.L.: Wearable inertial mouse for children with physical and cognitive impairments. Sens. Actuators A Phys. **162**(2), 248–259 (2010)
27. Raya, R., Rocon, E., Gallego, J.A., Ceres, R., Pons, J.L.: A robust kalman algorithm to facilitate human-computer interaction for people with cerebral palsy, using a new interface based on inertial sensors. Sens. (Basel) **12**(3), 3049–3067 (2012)
28. Weis, R.: Gross motor function measure (GMFM-66 and GMFM-88) user's manual. Eur. J. Paediat. Neurol. **8**, 111–112 (2004)

# Enriched Environment Affects Positively a Progression of Neurodegeneration: Elastic Maps-Based Analysis

Michael Sadovsky[1,2]([⊠]), Andrey Morgun[1,2], Alla Salmina[1,2],
Natalia Kuvacheva[1,2], Elena Khilazheva[1,2], and Elena Pozhilenkova[1,2]

[1] Institute of Computational Modelling of SB RAS, Akademgorodok,
Krasnoyarsk 660036, Russia
msad@icm.krasn.ru, {441682,allasalmina,elena.hilazheva}@mail.ru,
sadovsky.mikhail@gmail.com, natalya.kuvacheva@gmail.com
[2] Krasnoyarsk State Medical University, p. Zheleznyaka str., 1,
660022 Krasnoyarsk, Russia
http://icm.krasn.ru

**Abstract.** We studied the model to figure out the factors that may affect and retard the development of Alzheimer's disease. The experimental rats have been kept in two kinds of environment: standard one vs. enriched one, and amiloid protein has been injected to both groups of rats to simulate Alzheimer's disease. It is found the enriched environment is the key factor to retard the development of neurodegenerative disorder.

**Keywords:** Cluster · Morphology · Behaviour · Model · Alzheimer's Disease

## 1 Introduction

Alzheimer's Disease (AD) is a neurodegenerative pathology with no efficient therapy. AD is characterized by prominent behavioral alterations in the social sphere (social memory loss, disruption of social connections, aggression) and non-social sphere (alterations of spatial memory, difficulties in memory, anxiety and depression) that are resulted in aberrant socialization and communication [4]. The main cause of Alzheimer's type of neurodegeneration is intracerebral deposition of amyloid-beta protein ($A\beta$). This is associated with the development of oxidative stress, ion homeostasis disruption, metabolic alterations, synaptic dysfunction, and neuronal cells loss. Etiology and pathogenesis of AD are relatively well-studied, but still there are no effective pharmacological strategies for AD treatment.

Complex etiology and pathogenesis of AD suggests that treatment protocols should be complex as well. Correction of molecular alterations seen in AD should be combined with some rehabilitative protocols. The main hypothesis of the present work stands on the positive influence of enriched environment (EE)

© Springer International Publishing Switzerland 2016
F. Ortuño and I. Rojas (Eds.): IWBBIO 2016, LNBI 9656, pp. 505–514, 2016.
DOI: 10.1007/978-3-319-31744-1_45

on AD progression, in comparison to standard conditions (SC), at the level of brain progenitor cells proliferation/adult neurogenesis that might be further used for AD prevention and/or treatment. This hypothesis is based on numerous experimental and clinical data on the effects of enrichment of the environment with various stimuli on neurodegenerative changes resulting in slowing down the progression of neurodegeneration. In other words, living in the environment enriched with sensor and cognitive factors would be beneficial for the people with AD. Out study reveals some new aspects of EE positive influence on AD development and suggest novel approaches to their analysis.

## 2    Materials and Methods

Experiments have been performed on Wistar rats, 300–350 g, according to the principles of the European Regulations (86/609/). The study was approved by the Local Ethic Committee of the Prof. V.F. Voino-Yasenetsky Krasnoyarsk State Medical University (No. 35/2011).The following age groups have been used:

(i) 7–9 months old (young rats, $n = 12$) that corresponds to the period of social maturation and completion of brain development;
(ii) 23–25 months old (aging rats, $n = 12$) that corresponds to the initiation of the period of senescence in rats.

The animals have been kept under the conditions of vivarium at $21 \pm 1°C$ and 12 h/12 h day/night regimen with free access to food and water.

Young and aging rats have been kept under different conditions. For producing EE, we used the standard protocol of long-lasting (60 days) keeping the animals in the multistimuli environment. EE was reproduced in the cage ($78 \times 48 \times 39$ cm) divided into two floors connected with stairs. The cage was filled with plastic tunnels, houses, stairs, boxed, and wheels. Each cage contained 12 animals for intensive social interactions.

Human A$\beta$1-42 (Sigma-Aldrich, USA) was prepared as stock solutions at the concentration of 1 mg/ml in sterile 0.1 mol/L PBS ($p$H 7.4), and aliquots were stored at $-20°C$. A$\beta$1-42 solutions were aggregated by incubation at $37°C$ for 4 days before use as described previously. Rats were anesthetized with Chloral hydrate (0.35 mg/kg) and mounted in a stereotaxic frame (Narishige Scientific Instrument Lab, Japan). Bilateral injections of A$\beta$1-40 (5 $\mu$l, 2$\mu$g/$\mu$l) into the CA1 region of the hippocampus were done in the following coordinates: anterior/posterior (A/P) = 3.0 mm, medio/lateral (M/L) = $\pm$2.2 mm, dorsoventral (D/V) = 2.8 mm. Further work with rats was as long as up to 10 days after the surgery [5]. Analysis of A$\beta$ accumulation was done starting from the 10th day after A$\beta$ injection with thioflavin S staining [6].

The following groups have been used for further analysis:

(1) young animals kept in SC;
(2) young animals kept in EE;
(3) aging animals kept in SC;

(4) aging animals kept in EE;
(5) aging animals with the AD model kept in SC;
(6) aging animals with the AD model kept in EE. Sham-operated animals have
    been used as a control group.

## 2.1  Cell Culture

Isolation and culture of neural progenitor cells in all the experimental groups
of animals were carried our simultaneously according to the standard protocol.
Briefly, rats were anesthetized (Chloral hydrate, 120 mg/kg) and decapitated on
the ice. The brains were isolated and transferred into 35 mm culture dish with
2 % glucose in PBS. The hippocampus was isolated and excised to size of 1 mm$^3$.
After dissection, the hippocampal tissue was places in fresh 2 % glucose in PBS
and centrifuged twice (150 g×5 min). The pellet was resuspended in 1 ml medium
NeuroCult' NS-A (Sigma). The numbers of cells were calculated and live cells
were determined with Trypan Blue. $(1.2 \div 1.5) \times 10^5$ viable cells were cultured
in the T-25 culture flacons with 10 ml NeuroCult' NS-A Proliferation (Sigma)
medium in the incubator (37° C, 5 % $CO_2$). On the next day, the formation of
neurospheres has been observed.

By phase contrast, viable neurospheres were identified as translucent accumulation plurality of cells bearing on their surface microspikes. In 2 to 4 days of
cultivation (depending on the density and size of the spheres), neurospheres have
been subcultured for the measurements. Real-time assessment of neurospheres
proliferation (cell index) was obtained with XCelligence methodology (Roche,
Switzerland). We identified several phases of neurospheres growth in the culture: **A** — deposition of cells under the influence of gravity; **B** — phase of
relative stability; **C** — phase of active spheroid cell colonies formation; **D** and **E**
were two consecutive phases characterized by slow proliferation or steady-state
period of the growth of neurospheres.

## 2.2  Structure of Experimental Data

For each experimental group we used 4 wells to produce the data array. Measurements of cell index have been done every 5 min from zero point of the experiment
in vitro. We measures cell index (CI) for all the samples. CI represents the value
corresponding to the sample status. CI was analyzed every 15 min for 24 h. Original data base was the table with the 32 experimental samples (lines) including
control, and 132 experimental points (for each the object). Formally, the experimental data could be viewed as time series, and we will discuss this in detail
below. There was important problem in the above-mentioned experimental conditions such as significant excess of data space dimensions over the number of
objects (wells) available for the analysis.

## 2.3  Methods

The main goal of this study was to retrieve the data from the above-mentioned
structure. Data processing has been done in several steps. In the first steps,

we were solving the following task: all the experimental objects have been assigned as indicators of dependent variables whereas time moments when CI measurements have been done have been assigned as the objects. The first question was on clustering the time moments for making the data similar of hardly distinguished in the "mice space". In other words, this procedure might be considered as segmenting the time series.

Two methods of clusterization has been used: $K$-means and elastic maps technique; we used ViDaExpert[1] for this. $K$-means is a "classic" method of unsupervised classification [1] when entire set of objects is randomly divided into $K$ classes (groups) followed by identification of the kernel as an average for all the points within the class. After that, for all the points regardless their attribution to a class the distance between the point and the kernel is calculated. If the point appears to be closer to the kernel of the same class, it remains to stay the same class. If the point from $l$-class locates closer to the kernel of $m$-class, it is redirected to the latter one. Obviously, $1 \leq l, m \leq K$. Upon the redistribution completion of all the points, the procedure runs again for newly determined dynamic kernels followed by the procedure of redistribution. The procedure runs up to the moment of a complete stop of all the points; this former converges always. Finally, the obtained classes should be checked for a discernibility; however, we did not do it in the study.

$K$-means has several features that make an implementation rather intricate. At first, the number of classes could not be pre-determined. Secondly, there is no guarantee for stability of a convergence of redistribution of the points within the classes, particularly, for various types of realization, since classifying starts from random distribution of points in $K$ classes. Thirdly, quite often it is necessary to ensure the discernibility by combining two (or more) indistinguishable.

Visualization and clustering with elastic maps were used as a second method of data analysis [2]. The main idea of the method is in the approximation of multidimensional data with the small dimensional manifold. In our case with the 2-dimensional manifold (a plane). As the first step, we identify two main components in the distribution of points in the space. Then, the plane is put on these two lines as over two axes; each point, then, is projected to the plane and connected to the projections point with the "spring" (link). After that, the plane is allowed to stretch and warp, but its topology must keep the same. A researcher can use two fitting parameters: rigidity of the spring and elasticity of the platform. The system is released to allow accepting the conformation with the minimal energy of stretching along all the directions.

After all, the image of each original point is re-located over the jammed surface so that it is defined as the point on the surface located at the shortest distance from the object. Finally, the obtained elastic map is straightened with dual non-linear transformation, for further visualization of multidimensional data.

---

[1] http://bioinfo-out.curie.fr/projects/vidaexpert/.

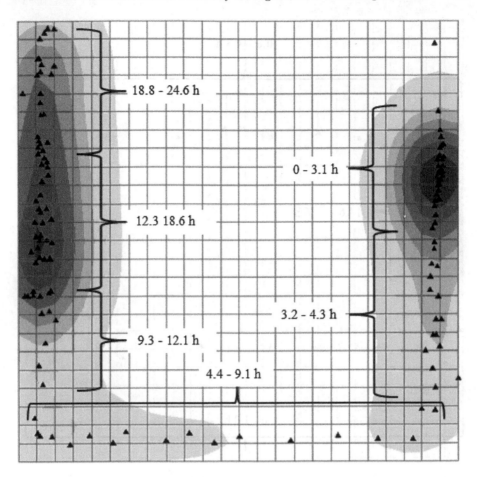

**Fig. 1.** Elastic map of time moments distribution drawn in the coordinates of standard conditions (SC) (12 coordinates: progenitor brain cells obtained from all young animals in SC, all aging animals in SC, or all A$\beta$-treated animals kept in SC).

## 3    Results

A problem in data analysis links to the excessive dimensionality of the data. Here we have got the dimensionality (as the number of observation moments in data stipulated to be the time series) exceeding the number of the objects available for observations. The problem comes from the methodology of experiment or its protocol, and may not be rejected. Therefore, one has to choose the method providing a retrieval the knowledge from the available experimental material. Standard approach force to segment the time series furthered with a study of the fragments [3].

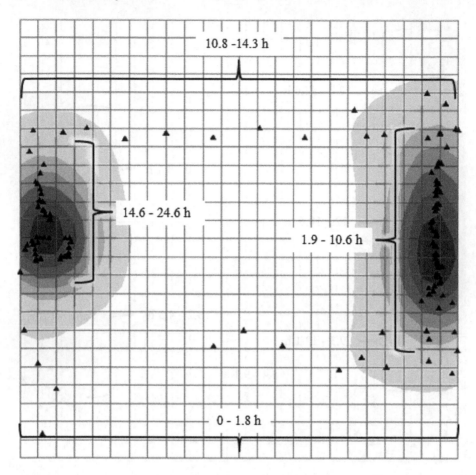

**Fig. 2.** Elastic map of time moments distribution drawn in the coordinates of enriched environment (EE) (12 coordinates: progenitor brain cells obtained from all young, aging, or Aβ-treated animals kept in EE).

We have used two approaches to such segmenting: first uses conditionally naïve bases for segmentation whereas the second approach utilizes elastic map construction.

Indeed, a segmentation of time series means the identification of rather long fragments appropriate for further approximation of the series with the intuitively clear and simple relationship. As an example, it is possible to allocate the fragments with the stationary behavior of the time series or well-described with an exponent (attenuation of a process). Implementation of such approach requires description of methods to identify such fragments and critical points. The latter ones are the time moments when the time series changes its behavior. As a rule, the second task is solved in a relatively simple way: these are the points where a curve corresponding to the time series demonstrates something like a break.

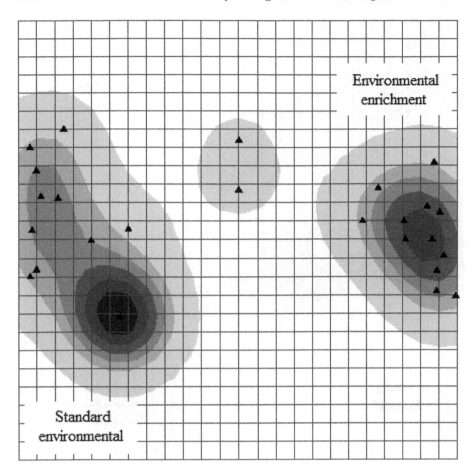

**Fig. 3.** Elastic map of wells distribution shown in the coordinates of time moments (10 coordinates: the data obtained at 4,8; 5,8; 6,8; 7,8; 8,8; 9,8; 10,8; 11,8; 12,8; 13,8 h).

One can fix some level of changes of a derivative and to identify the points. Exactly this approach has been used in the present study for the primary analysis of the data obtained.

Another possible approach is based on the analysis of less or more strict procedures of time series segmenting. We will not discuss here the whole spectrum of all the methods and approaches (for details see [3]), but we will describe the procedure which was used in the present study. This procedure is based on the clustering of the time series points with the non-linear statistics, so, we have used the method of elastic maps [2]. The method is described above, therefore we will emphasize its application to the problem of segmenting the time series.

It should be noted whether all the tested time series were homogenous: they have similar length, and the data are obtained at the same time moments. Since the number of series points (130) is bigger that the number of the experimental

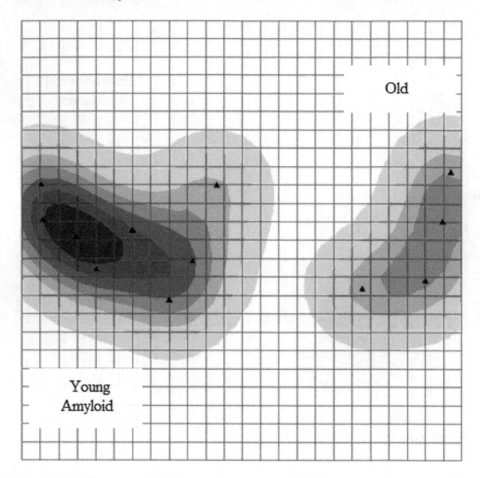

**Fig. 4.** Elastic map showing the distribution of wells in the coordinates of seven time moments (4,5; 5,3; 6,0; 6,8; 7,5; 8,3 and 9,0 h) for standard conditions.

samples (32 wells), any possible direct analysis of wells distribution in the space of time points would have no sense: all the experimental points could be described with some hyperplane with high level of accuracy (dimension of 32). It is clear that this hyperplane may reflect internal data structure or may not.

We identified the fragments of the series whose length allowed direct analysis of the experimental data. For that, we solved a dual task: construction of an elastic map with time moments but not wells considered as independent variables (objects). Experimental wells were considered as coordinates for studying the distribution of time moments. As a result, we have obtained clusters of time moments that were very close within the coordinates consisted of the wells. Then, we analyzed the distribution of wells at the points that have been selected according to the results of above-mentioned clusterization.

Figure 1 shows the distribution of time moments in the coordinates of experimental wells (dual task). It is clear that all the time moments are divided into three zones: two of them are the clusters (at the boarders of the elastic maps) while third one represents a transition from one cluster to another one (point at the bottom of the map). Exactly these points that were not included into any cluster, were further used for the direct analysis.

We should mention that in fact, the time moments used for direct analysis almost completely coincide with the phase C selected in "naïve" segmentation in the primary data processing (4.4 to 9.1 h of the experiment). This coincidence indirectly confirms that phase C designated on physiological and biological reasons appears to be relevant to the structure of the data that are used for the analysis. In other words, phase C is determined not only by physiological behavior of the cell population but corresponds to the real proliferation pattern of cells grown in wells. Therefore, content analysis of multidimensional data with the space dimensionality exceeding the number of experimental points is rather possible.

Results of clustering for the dual task determined within the coordinates of EE allows identifying not only phase C, but phase A as well which is not visible at the elastic map within the coordinates of SC (10.8 to 14.3 hrs for phase C and 0.0 to 1.8 hrs for phase A). This is shown on Fig. 2. It is very important that time periods corresponding to phase C for SC and EE found with the method of elastic maps (Figs. 1 and 2) do not coincide. The absence of coincidence comes from the influence of various environments for animals affecting processes of neurogenesis and development of progenitor cells in vivo and in vitro.

## 4    Discussion

Differences in cluster structure within the coordinates of wells for SC and EE allow to conclude that influence of such difference on the progenitor cells development is confirmed.

Let us come to the analysis of experimental data made in "direct" coordinates (coordinates of various time moments). For that, we selected time moments corresponding to phase C found in "naïve" segmentation. However, selection of the moments was done according to statistic characteristics of the primary data but not on the biological basis. Total number of the points (time moments) falling within the required range still is very high for reliable direct data analysis, thus, we faced the problem of reducing the number of experimental points. This problem was solved with the simple method: we selected a subset of points (at 8 time moments) that have been chosen evenly along whole set of points out of all the appropriate points.

Duration and mutual location of time intervals corresponding to the phase C in SC and EE do not coincide (Fig. 3). Party, this situation leads to significant elevation of the number of experimental points suitable for direct analysis. Since all the points within the SC and EE data sets could not be subjected to the analysis, we have reduced this number by shifting to a subset. This subset was constructed to include (less or more in proportion) the points corresponding to the phase C of the cells obtained from the rats kept in SC or EE.

Figure 4 shows the age-dependence of proliferative activity of brain progenitor cells determined for phase C. Here the data for the animals kept in SC only (progenitor cells isolated from young rats, aging rats, or rats subjected to intrahipocampal injection of A$\beta$) are shown. Obviously, the age becomes the key factor of differentiation for this group of animals: cells obtained from young, control or amyloid-treated rats form one cluster whereas the cells obtained from aging rats form another one.

Thus, we have shown that the results and conclusions made on traditional methods of discrimination between various experimental groups fit well the results of analysis based on the approaches of non-linear statistics. Such coincidence of the conclusions allows validating the data obtained with traditional methods of data processing that is important, particularly, because of widely-spread criticism about the reliability of conclusions made with conventional statistics in experimental biology.

Another important finding of our study is confirmed influence of EE at the proliferative activity of neural progenitor cells in (patho)physiological conditions. Shortly speaking, enriched environment may prevent progression of neurodegenerative changes acting at adult neurogenesis, thus suggesting prospects for further development of preventive and rehabilitation strategies based on this phenomenon.

**Acknowledgement.** The study was supported by the grant of the President of the Russian Federation for the Leading Scientific Teams (project No. 1172.2014.7).

# References

1. Gorban, A.N., Rossiev, D.A.: Neuronal networks at PC. Novosibirsk: Science. 276 p. (1996) (in Russian)
2. Zinoviev, A.Y., Pitenko, A.A., Popova, T.G.: Application of elastic maps. Neurocomputers, no. 4, pp. 31–39 (2002) (in Russian)
3. Murphy J.J. Technical analysis of the financial markets. A comprehensive guide to trading methods and applications, New York Institute of Finance, 530 p. (1999)
4. Crews, L., Masliah, E.: Molecular mechanisms of neurodegeneration in Alzheimer's disease. Hum. Mol. Genet. **19**(R1), 12–20 (2010)
5. Li, X., Yuan, H.F., Quan, Q.K., Wang, J.J., Wang, N.N., Li, M.: Scavenging effect of Naoerkang on amyloid beta-peptide deposition in the hippocampus in a rat model of Alzheimer's disease. Chin. J. Integr. Med. **17**(11), 847–853 (2011)
6. Sipos, E., Kurunczi, A., Kasza, A., Horváth, J., Felszeghy, K., Laroche, S., Toldi, J., Párducz, A., Penke, B., Penke, Z.: Beta-amyloid pathology in the entorhinal cortex of rats induces memory deficits: implications for Alzheimer's disease. Neuroscience **147**(1), 28–36 (2007)

# Fundamentals of Biological Dynamics and Maximization of the Information Extraction from the Experiments in the Biological Systems

# Clustering of Multi-image Sets Using Rényi Information Entropy

Renata Rychtáriková[✉]

Faculty of Fisheries and Protection of Waters, South Bohemian Research Center
of Aquaculture and Biodiversity of Hydrocenoses, Institute of Complex Systems,
University of South Bohemia in České Budějovice, Zámek 136,
373 33 Nové Hrady, Czech Republic
rrychtarikova@frov.jcu.cz
http://www.frov.jcu.cz/cs/ustav-komplexnich-systemu-uks

**Abstract.** We propose a clustering method based on the calculation of
variables derived from the $\alpha$-dependent Rényi information entropy – a
point information gain entropy $(H_\alpha)$ and point information gain entropy
density $(\Xi_\alpha)$, which measure an information-entropic distance between
two multidimensional distributions. The matrices of $H_\alpha/\Xi_\alpha$ values as
functions of the parameter $\alpha$ and a label of a multidimensional set's
object are classified into groups using a standard k-means algorithm.
The method is presented on two multi-image series which in the origin,
the number of images in the sets, the number of image color channels,
and the pixel resolution differ.

**Keywords:** Rényi entropy · Image series' clustering · Multidimensional
data analysis

## 1 Introduction

In our previous works, we presented an effective method for the measurement of
an information content of multidimensional objects [1–3]. The method is primarily based on the calculation of an information-entropic variable point information
gain

$$\Gamma_{\alpha,i} = \frac{1}{1-\alpha} \log_2 \frac{\sum_{j=1}^{k} p_{i,j}^{\alpha}}{\sum_{j=1}^{k} p_i^{\alpha}}, \tag{1}$$

which is in fact a subtraction of two Rényi entropies – with and without an examined element of the $j$-property in the probability distributions $p_i$ and $p_{i,j}$ with
$k$ phenomena (properties), respectively. The variable $\alpha$ denotes a dimensionless
Rényi coefficient.

In comparison to generally known information measures like Kullback-Leibler
divergence [4], the advantages of $\Gamma_{\alpha,i}$ lies in these aspects:

- Due to the parametrization by $\alpha$, the $\Gamma_{\alpha,i}$ describes the multifractal character
  of the distribution. At low values of $\alpha$, the elements of rare occurrences are

© Springer International Publishing Switzerland 2016
F. Ortuño and I. Rojas (Eds.): IWBBIO 2016, LNBI 9656, pp. 517–526, 2016.
DOI: 10.1007/978-3-319-31744-1_46

separated and detected, whereas the elements of the high occurrences are merged. With increasing $\alpha$, the $\Gamma_{\alpha,i}$-distribution becomes uniform.

- At neglecting computer averaging, the $\Gamma_{\alpha,i}$-distribution is a monotonous function, which changes from a convex course at low $\alpha$, over a linear dependence at $\alpha = 2$, to a concave course.
- Apart from the global information context, the $\Gamma_{\alpha,i}$-calculation enables to specify the local information context of the element in the distribution, which is evaluated from the local distribution around the examined point.
- Summations of $\Gamma_{\alpha,i}$ give macroscopic variables a point information gain entropy

$$H_\alpha = \sum_{j=1}^{k} n_j \Gamma_{i,\alpha} \tag{2}$$

and a point information gain entropy density

$$\Xi_\alpha = \sum_{j=1}^{k} \Gamma_{i,\alpha}, \tag{3}$$

whose $\alpha$-dependent spectra characterize the distribution sufficiently ($n_j$ is a number of the occurrence of the $j$-th property in the distribution).

Earlier, we briefly presented the usage of $H_\alpha$- and $\Xi_\alpha$-spectra for recognition of pattern origin in the image. Whereas a unifractal image showed a monotonous increasing dependency of $H_\alpha$ and $\Xi_\alpha$ on $\alpha$, a multifractal image exhibited the spectral course with one dominant peak [3]. Current growing interest in big data clustering and categorization led us to the idea that these $H_\alpha$- and $\Xi_\alpha$-spectra (or, in case may be, $H_\alpha$ and $\Xi_\alpha$ at suitably chosen $\alpha$) as inputs into a standard clustering algorithm could split multi-object data sets into groups with objects of similar properties.

In this contribution, we show and discuss the usage of $H_\alpha$- and $\Xi_\alpha$-spectra for clustering two different sets of multifractal images. The first set is a *in silico* simulation of self-organizing Belousov-Zhabotinsky chemical reaction [6], where the clustering method describes its course. The second series is a sequence of color digital camera images – live cell's microscopic images obtained *via* focusing along the z-axis, where the clustering is used for finding the position of cell in its spread function.

## 2    Methods

All information-entropic values was computed using the Image Info Extractor Professional software (ICS FFPW USB) and stored in double precision floating point format in *.mat files (Matlab®, Mathworks, USA). Details of the computation are described in [3].

The variables $H_{\alpha,Wh}$ and $\Xi_{\alpha,Wh}$ for the 12-bit image series simulating self-organizing reaction were calculated for $\alpha = \{0.1, 0.3, 0.99, 1.3, 1.5, 1.7, 2.0, 2.5,$

3.0, 3.5, 4.0} from probability intensity histogram obtained from whole image (global information entropy).

In case of 12-bit RAW bright-field microscopic images, the clustering was extended about values of $H_{\alpha,Cr}$ and $\Xi_{\alpha,Cr}$ calculated from probability intensity histogram created from pixels which form a cross around an examined pixel (local information entropy) and about $\alpha = \{0.5, 0.7\}$. The red, green, and blue pixels of each RAW file were processed by method which connects the non-interpolating debayeriation algorithm [5] with calculation of $\Gamma_{\alpha,i}$ and give a resulted quater-resolved $\Gamma_{\alpha,i}$-image (matrix): The calculation of $\Gamma_{\alpha,i}$ for the red and blue channel was performed via removing one pixel of the respective color from the respective histogram, whereas, in case of the green channel, two pixels of the intensities which occupy the respective BGGR quadruplet were removed.

Consequently, the matrices of $H_{\alpha,Wh/Cr/Wh+Cr}$ and $\Xi_{\alpha,Wh/Cr/Wh+Cr}$ as a dependence of $\alpha$, order of the image in the series, and, for microscopic images, also a color channel were prepared in the Matlab® software. The matrices were imported into the Unscrambler® X software, where they underwent the z-score standardization over $\alpha$ and clustering via k-means algorithm with squared Euclidian distance into 19 (simulated self-organizing reaction) and 2 groups (microscopic images), respectively.

For better illustration of the course of the simulated reaction, the groups were re-numbered in Microsoft Office Excel (USA) to be consecutive (i.e., img. 1 and the first following image of the different group are classified into gr. 1 and 2, respectively, etc.). The principal component analysis (PCA) of the simulated reaction was also done using Unscrambler® X software after the z-score normalization of the input matrix.

# 3    Results and Discussion

## 3.1    Simulation of the Belousov-Zhabotinsky Reaction

Despite its inorganic character, the Belousov-Zhabotinsky reaction [7–10] as the simplest example of self-organization can help us to understand the development of the embryos and creation of life. Using rules of multilevel cellular automata, we generated a systematic sequence of 14,287 multifractal grayscale images (Fig. 1), which fully simulate the behavior of this reaction that each image corresponds to a reaction step [6].

Due to the smooth, well distinguished, and developing structures which form the simulation steps (Fig. 1), the computation of global information-entropic variables from the intensity histogram of each image at 11 values of $\alpha$ (see Sect. 2) was enough for its characterization and, consequently, for clustering the series. The number of groups (19) was chosen in order to correspond, in the late phase of simulation, to the number of states as observed in the real Belousov-Zhabotinsky reaction [11]. The trajectory of the simulation (Fig. 2) can be approximately separated into four main blocks, where the first two phases are developmental and very heterogenous:

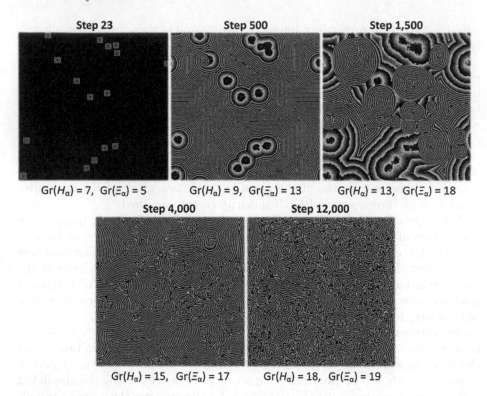

**Fig. 1.** Structures in steps 23, 500, 1500, 4000, and 12,000 of the simulated Belousov-Zhabotinsky reaction. Visualization of original 12-bit images in 8 bits. Values of $Gr(H_\alpha)$ and $Gr(\Xi_\alpha)$ correspond to the group of images in Figs. 2 and 3 computed from $H_\alpha$- and $\Xi_\alpha$-spectra, respectively.

1. The initial phase (img. 1–250) is never observed in the reaction and contains the images of primary octagonal structures (*e.g.*, step 23 in Fig. 1). Due to the development of structures, clustering *via* $H_\alpha$-spectra as well as $\Xi_\alpha$-spectra classified this initial phase into 11 groups (Fig. 2b).
2. The images of the second phase (img. 251–1001) depict a development of waves with diffusive circular centers (step 500 in Fig. 1). Using $\Xi_\alpha$- and $H_\alpha$-spectra, this phase was separated into 8 and 12 groups (Fig. 2b).
3. The next block (img. 1001–4000) is that, where the waves in images broaden and break into ram-horn-like spiral doublets (step 1500 in Fig. 1) which, in later phase, give rise to dense smooth thin circular waves (step 500 in Fig. 1). Clustering of $H_\alpha$-spectra assigned the images into gr. 9 and 11–19, while $\Xi_\alpha$-spectra gave gr. 11, 12, 14–19 (Fig. 2a).
4. Remaining ca. 8000 images exhibit a pseudoperiodic behavior (step 12,000 in Fig. 1), which oscillate between 7 (gr. 9, 11, 12, 14, 15, 18, 19) and 6 groups (gr. 12, 14–17, 19) for computation from $H_\alpha$- and $\Xi_\alpha$-spectra, respectively (Fig. 2a). Here, the wave and spirals interchange and show a strong

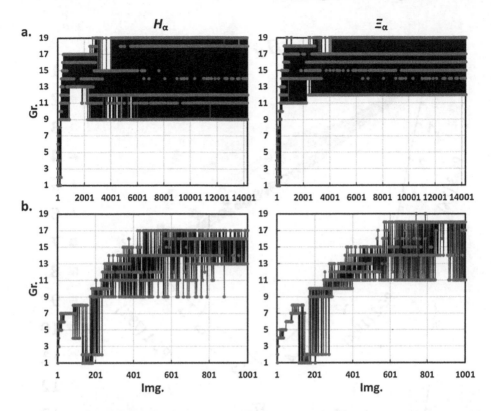

**Fig. 2.** k-Means clustering (squared Euclidian distance) of images of the simulated Belousov-Zhabotinsky reaction using $H_\alpha$- and $\Xi_\alpha$-spectra, respectively. Graphs **b** are expansions of graphs **a** in the range of img. 1–1001.

multifractal character. It is possible that another statistical method which corresponds better to the multifractal character of the image, *e.g.* the method of principal manifolds [12], would not lead to the pseudoperiodic behavior but to one final state.

Clustering of the image set simulating the Belousov-Zhabotinsky reaction using $\Xi_\alpha$-spectra is more homogenous and accurate than that using $H_\alpha$-spectra. The reason is that the first 3 principal components (PCs) obtained *via* PCA, which is part of the clustering algorithm, already completely describe the PC scores – PC-1, PC-2, and PC-3 describes 77, 16, and 7% of the PC scores, respectively (Fig. 3). It is a consequence of the fact that, during the computation of $H_\alpha$-values, $\Gamma_\alpha$-values are not weighted by their frequency of occurrence and, thus, the $\Gamma_\alpha$-values for pixels of low frequency of occurrence are not suppressed by the $\Gamma_\alpha$-values for pixels of high frequency of occurrence. Let us note that the PC scores also represent the information distance/similarity between the image groups and images themselves (Fig. 3).

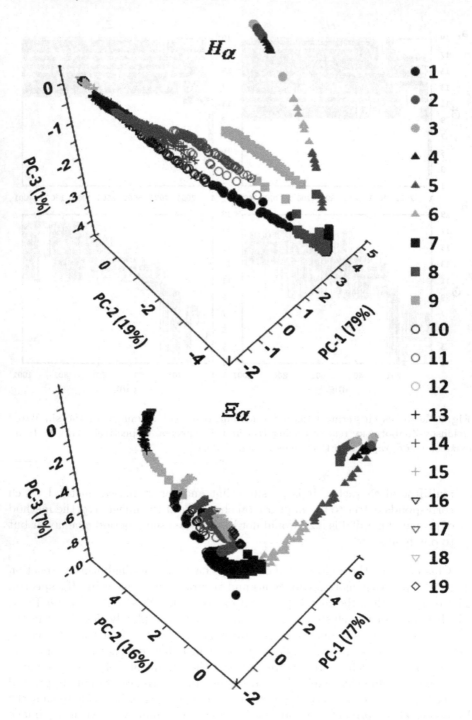

**Fig. 3.** Scores of the first 3 principal component for the images of the simulated Belousov-Zhabotinsky reaction computed from $H_\alpha$- and $\Xi_\alpha$-spectra. The legend shows labels of groups.

## 3.2   Microscopic Images

Microscopic techniques play an important role in the medical diagnostics. A potential for the evaluation of the clinical tests is also hidden in the classical optical bright-field microscopy [5].

In this section, we used the method of clustering using $H_\alpha$- and $\Xi_\alpha$-spectra for a more practical task – clustering of a 93-image z-stack of 12-bit RAW bright-field micrographs of a live mammalian cell into two groups – focused and unfocused – with the aim to find a region of cell's occurrence in the object spread function. For better characterization, the $H_\alpha$- and $\Xi_\alpha$-spectra were calculated for pixels of each color separately. In addition, apart from a global

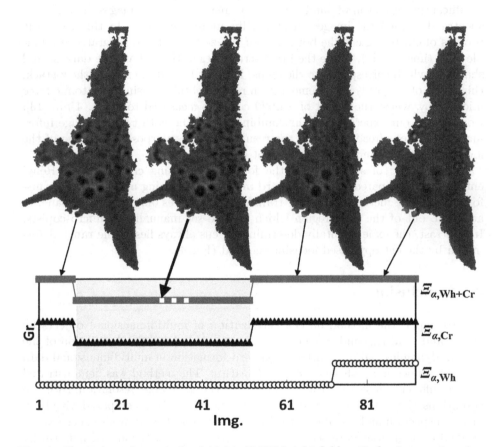

**Fig. 4.** Clustering of a z-stack of 12-bit RAW bright-field optical micrographs of a live L929 cell using $\Xi_\alpha$-spectra into two groups. Calculated from statistics done on all pixels of a respective color ($\Xi_{\alpha,Wh}$), pixels of the same color forming a cross around each examined pixel and combination of the both ($\Xi_{\alpha,Wh+Cr}$). The focal region is highlighted by gray color. The white points in the ($\Xi_{\alpha,Wh+Cr}$ plot correspond to positions of a focal plane which was set by microscopists (3 independent determinations). The original 12-bit RAW files of the cell are visualized in 8-bit/c RGB images.

(whole image) information (indexed $_{Wh}$), the images were further specified by local cross information-entropic kind of surroundings (indexed $_{Cr}$) due to the distribution of camera noise. The number of $\alpha$ was extended to 13 (see Sect. 2).

Figure 4 shows only the results of clustering using $\Xi_\alpha$-spectra, since the results for $H_\alpha$-spectra were identical. This is due to the facts that intensity histograms of the images are tail-free and, also, almost every pixel has its own unique local type of pixels' surroundings. For each image, the first fact led to the similar spectra of global information, whereas the latter gave the almost identical $H_{\alpha,Cr}$- and $\Xi_{\alpha,Cr}$-spectra.

As seen in Fig. 4, the clustering $via$ $\Xi_{\alpha,Wh}$-spectra distinguished only the most blurred part of the z-stack (img. 73–93). These spectra classified the images insufficently by reason of similar intensity distributions of images. The $\Xi_{\alpha,Cr}$-spectra described the images substantially more accurately. In this case, the process of clustering cut the both tails of the z-stack off and separated its middle part (img. 10–51), where the inner structures of the cell are the darkest and sharpest, which corresponds to the focus. Provided a 100-nm step in the z-stack, this range of images is in agreement with results obtained using the atomic force microscopy, where the height of a L929 cell was measured to be ca. 4 $\mu$m [13]. Contrary to our expectations, the combination of the both type of image information did not already improve the results. For this purpose, the usage of the local type of surroundings was sufficient enough.

The method of searching for the focal region using clustering $via$ Rényi entropy was compared with a standard method of focusing using Fourier transformation. The method of Fourier transformation gives only one focal plane, in addition, out of the focal region which can be set manually by microscopists. In contrast, the experimentally determined focus always lies in the range determined by the entropy-based focusing method (Fig. 4).

## 4    Conclusion

In this paper, the method on characterization of multidimensional objects by global and local $H_\alpha$- and $\Xi_\alpha$-spectra was extended about the description of the system dynamics and recognition of pattern formation in multidimensional data sets $via$ connection with a clustering algorithm. The method was demonstrated on two different image sets, where images had the same pixel resolution. At given conditions, the method classified the image sets efficiently as compared with both visual inspection and a real experiment. The method has been also successfully verified on numerous data sets – on the course of a real Belousov-Zhabotinsky reaction [14], on classification of living cells' states during the cell cycle [15], and on numerous determination of focal ranges of biological and non-biological samples in optical microscopy. The method can be further improved by finding a suitable entropic power [16], which would extend it to clustering and comparison of the image sets of different pixel resolution.

**Acknowledgments.** This work was financially supported by CENAKVA (No. CZ.1.05/2.1.00/01.0024), CENAKVA II (No.LO1205 under the NPU I program) and The CENAKVA Centre Development (No. CZ.1.05/2.1.00/19.0380).

# References

1. Stys, D., Urban, J., Vanek, J., Cisar, P.: Analysis of biological time-lapse microscopic experiment from the point of view of the information theory. Micron **42**, 360–365 (2010)
2. Urban, J., Vaněk, J., Štys, D.: Preprocessing of microscopy images via shannons entropy. In: Proceedings of Pattern Recognition and Information Processing, pp. 183–187. Minsk, Belarus (2009)
3. Rychtáriková, R., Korbel, J., Macháček, P., Císař, P., Urban, J., Soloviov, D., Štys, D.: Point information gain, point information gain entropy and point information gain entropy density as measures of semantic and syntactic information of multidimensional discrete phenomena, under the revision. In: EEE Transactions on Information Theory. http://arxiv.org/pdf/1501.02891.pdf
4. Kullback, S., Leibler, R.A.: On Information and sufficiency. Ann. Math. Stat. **22**(1), 79–86 (1951)
5. Rychtáriková, R., Náhlík, T., Smaha, R., Urban, J., Štys Jr., D., Císař, P., Štys, D.: Multifractality in imaging: application of information entropy for observation of inner dynamics inside of an unlabeled living cell in bright-field microscopy. In: Sanayei, A., et al. (eds.) Emergence, Complexity and Computation: ISCS 2014, vol. 14, pp. 261–267. Springer, Switzerland (2015)
6. Štys, D., Náhlík, T., Zhyrova, A., Rychtáriková, R., Papáček, Š., Císař, P.: Model of the Belousov-Zhabotinsky reaction. Submitted to Lecture Notes in Computer Science
7. Belousov, B.P.: A Periodic Reaction and Its Mechanism, Collection of Short Papers on Radiation Medicine. Medical Publishing, Moscow (1959)
8. Zhabotinsky, A.M.: Periodical process of oxidation of malonic acid solution (a study of the Belousov reaction kinetics). Biofizika **9**, 306–311 (1964)
9. Zhabotinsky, A.M.: Periodic liquid phase reactions. Proc. Acad. Sci. USSR **157**, 392–395 (1964)
10. Rovinsky, A.B., Zhabotinsky, A.M.: Mechanism and mathematical model of the oscillating bromate-ferroin-bromomalonic acid reaction. J. Phys. Chem. **88**, 6081–6084 (1984)
11. Štys, D., Jizba, P., Papáček, Š., Náhlík, T., Císař, P.: On measurement of internal variables of complex self-organized systems and their relation to multifractal spectra. In: Kuipers, F.A., Heegaard, P.E. (eds.) IWSOS 2012. LNCS, vol. 7166, pp. 36–47. Springer, Heidelberg (2012)
12. Gorban, A.N., Kégl, B., Wunsch, D.C., Zinovyev, A. (eds.): Principal Manifolds for Data Visualisation and Dimension Reduction. LNCSE, vol. 58. Springer, Heidelberg (2007). ISBN: 978-3-540-73749-0
13. Malakhova, D., Štys, D., Rychtáriková, R.: Adjustment of dynamic high resolution images of living cells by combination of an optical microscopy in transmitting light, atomic force microscopy and image information analysis. Chem. Listy **107**, s402–s404 (2013)

14. Zhyrova, A., Stys, D., Cisar, P.: Macroscopic description of complex self-organizing system: Belousov-Zhabotinsky reaction. In: Sanayei, A., et al. (eds.) ISCS 2013: Emergence, Complexity and Computation, vol. 8, pp. 109–115. Springer, Heidelberg (2014)
15. Romanova, K., Cisar, P., Stys, D.: Living cell state trajectory in time-lapse microscopy. In: Foret, F. et al. (eds.) Conference: 9th International Interdisciplinary Meeting on Bioanalysis, Brno, Czech Republic, 01–02 November 2012, CECE 2012: 9th International Interdisciplinary Meeting on Bioanalysis, pp. 33–35 (2012)
16. Jizba, P., Dunningham, J.A., Joo, J.: Role of information theoretic uncertainty relations in quantum theory. Ann. Phys. **355**, 87–114 (2015)

# Least Information Loss (LIL) Conversion of Digital Images and Lessons Learned for Scientific Image Inspection

Dalibor Štys[✉], Tomáš Náhlík, Petr Macháček, Renata Rychtáriková, and Mohammadmehdi Saberioon

Faculty of Fisheries and Protection of Waters, South Bohemian Research Center of Aquaculture and Biodiversity of Hydrocenoses, Institute of Complex Systems, University of South Bohemia in České Budějovice, Zámek 136, 373 33 Nové Hrady, Czech Republic
stys@jcu.cz
http://www.frov.jcu.cz/cs/ustav-komplexnich-systemu-uks

**Abstract.** Nowadays, most digital images are captured and stored at 16 or 12 bit per pixel integers, however, most personal computers can only display images in 8 bit per pixel integers. Besides, each microarray experiment produces hundreds of images which need larger storage space if images are stored in 16 or 12 bit. This is in most cases done by conversion of single images by an algorithm, which is not apparent to the user. A simple method to avoid the problem is converting 16 or 12-bit images to 8 bit by direct division of the 12-bit intervals into 256 sections and counting the number of points in each of them. Although this approach preserves the proportion of camera signals, it leads to severe loss of information due to losses in intensity depth resolution. The main aim of this article is introducing least information loss (LIL) algorithm as a novel approach to minimize the information loss caused by the transformation the primary camera signals (16 or 12 bit per pixels) to 8 bit per pixel. Least information loss algorithm is based on the omission of unoccupied intensities and transforming remaining points to 8 bit. This approach not only preserve information by storing intervals in the image EXIF file for further analysis, but also it improves object contrast for better visual inspection and object oriented classification. LIL algorithm may be applied also in image series where it enables comparison of primary camera data at scales identical over the whole series. This is particularly important in cases that the coloration is only apparent and reflect various physical processes such as in microscopy imaging.

**Keywords:** Least information loss algorithm · Digital image visual inspection · Image series comparison · Image conversion

## 1 Introduction

Capture and visual inspection of digital images is a very important part of research activities in biology and medicine as well as in numerous other scientific

© Springer International Publishing Switzerland 2016
F. Ortuño and I. Rojas (Eds.): IWBBIO 2016, LNBI 9656, pp. 527–536, 2016.
DOI: 10.1007/978-3-319-31744-1_47

disciplines. Digital images, being integral to microscopic series or behavioral records, are used for obtaining the information about the identity of a research object in the whole (time and spatial) context. Many other spatially resolved datasets are further provided by other various instruments as well. Since human perception is primarily visual, these types of data are also examined in the form of digital images. It is the reason why the primary signal – an electrical response at a detector – is often overlooked. This primary response is digitized into a limited number of levels which lead to an irrecoverably lost of some details of the signal. Indeed, this primary loss is necessary and enables any further data processing. Unfortunately, the world of digital imaging is full of images which are further compressed and transformed in various ways. The visual inspection using a computer screen is no less aberrated. In such transformations, users are not often aware of distortions and consider these artifacts as research findings.

In our previous article, we present the software Image Explorer for the most objective exploration of image details of an 12 bit dataset [1]. Provided that 8-bit color depth resolution is necessary, on some specific examples in this article, we introduce a methodical approach how to avoid imaging artifacts by preservation most possible information in images as well as image series during 12-bit to 8-bit compression. We further demonstrate a software tool, which was developed for this purpose, and compare the results with standard algorithms of intensity depth compression.

## 2    Results

### 2.1    Methodological Approach for Obtaining and Transformation of Images

The method of bit conversion was primary developed to understand a local chemical composition in the course of the Belousov-Zhabotinsky reaction [2] and local optical properties captured in a bright-field microscopic z-stack of a live cell [3]. The digital camera served as a colorimeter or optical densitometer for the capture, in intensity of each pixel, the best possible time and spatial representation of a respective region of wavelengths.

A standard digital color camera is a grayscale camera, where each quadruplet of pixels on the chip, the Bayer mask [4] is covered by two green filters, one blue filter, and one red filter, which transmit light in the regions of middle, short, and long wavelengths, respectively. The camera of industrial standard digitizes a response of the pixel into 12 bit levels which are stored in the computer as a grayscale 16-bit image (called *a raw file*). From these signals, the color is constructed by process of debayerization, where the grayscale matrix is separated into intensities of individual color (red, green, and blue) channels. Image outputs from most standard imaging software provides the same number of pixels as at the camera chip. This is achieved by intensity interpolation.

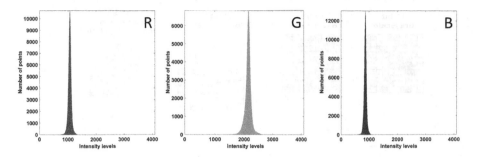

**Fig. 1.** Red, green, and blue intensity histogram of a 12-bit camera record (raw file) of an image from a z-stack of a MG63 cell (Color figure online).

Figure 1 shows a typical range of occupied digital levels in a 12-bit image from a bright-field microscopic z-stack of live cell. The resolution and offset may be only set for the whole camera chip which constrain the possible extent of recorded intensity levels. Therefore, in a carefully designed experiment (Fig. 1), the camera was set to record as much changes of the 12-bit signal over the whole series and for all three color channels as possible. The maxima of peaks were localized in the middle of the intensity scale in order to accommodate all intensities in the record of the z-stack, without any saturation of intensity histograms. The signal obtained by the bright-field microscopy is, in the case of the sample, a result of combination of diffraction and absorption of transmitted light and object autofluorescence and, in the case of image formation, a result of destructive and constructive interferences of light waves, which are further discretized by the camera chip. The resulted colors of the observed digitized objects are then only coincidental and a next, carefully chosen, image transformation do not distort the information in the image so much, yet.

For purposes of visual inspection, a 12-bit image must be always transformed into an image of 8-bit intensity depth. Most imaging software, which open 12-bit images, in fact, do the 8-bit transformation by an unknown and uncontrollable algorithm. The most straightforward transformation is merging of a total number of elements in neighboring 16 intensity levels into one bin to obtain a 256-intensity scale, here called *simple transformation* (Fig. 2, bottom row). Despite this approach preserves the proportions of the original camera signal, it leads to substantial losses in intensity depth resolution and, thus, in information content.

For an unbiased visual inspection of larger image series, we developed a least information loss – LIL – bit compression algorithm as part of the LIL Convertor software (see Sect. 4). An input image is a vice-bit raw file, where the step of debayerization (Fig. 2, upper row) gives a quater-resolved non-interpolated RGB image: Red and blue pixels of the Bayer mask was directly adopted to the respective color channel, two green pixels of the quadruplet of Bayer mask were averaged to create a green channel. Consequently, each color channel undergoes

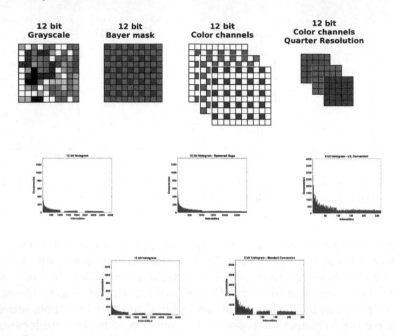

**Fig. 2.** Comparison of the LIL conversion with standard simple 12-bit to 8-bit conversion. *Upper row* – The schematic description of the LIL debayerization. *Middle row* – Mechanism of the 12-bit to 8-bit LIL conversion applied to one color channel. *Bottom row* – The simple standard 12-bit to 8-bit image conversion (Color figure online).

omission of unoccupied intensities and, in case may be, replacing of intensities, whose occurrence is below a threshold, by averages of their neighbors (Fig. 2, middle row). The subsequent reduction of the bit depth is achieved after subtraction of values of the maximal and minimal occupied intensity level, followed by binning of the region of occupied intensity levels into 256 levels. Starting and ends of each intensity interval are stored in the EXIF file of the image and enable any future analysis.

The difference in the outcomes of these two approaches is seen in Fig. 3. The left panel shows the simple 12 bit to 8 bit transformation. The prevalence of green tone in the simply transformed image is a result of the location of the green histogram at higher intensity levels (cf. Fig. 1). Under the close inspection, the LIL converted image (Fig. 3, *middle*) preserves many more image details as quantified in Table 1, where we can see that the intensity depth of red and blue channel after LIL conversion increased more than 3×.

Neither the LIL approach, nor the simple approach is never utilized in any of the transformation or visualisation software. This causes serious misconceptions in data analysis.

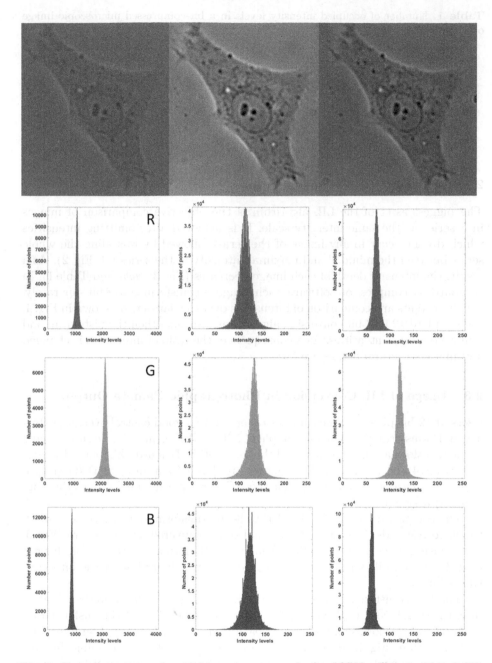

**Fig. 3.** Transformations of a 12-bit camera record of a MG63 cell into 8-bit RGB image by the simple algorithm (*left column*), LIL algorithm applied to individual image (*middle column*), and LIL algorithm applied to the whole z-stack (*right column*). $1^{st}$ row – Transformed images. $2^{nd}$–$4^{th}$ row – red, green, and blue intensity histograms, respectively (Color figure online).

**Table 1.** Number of occupied intensity levels in a bit-compressed microscopic image of a MG63 cell.

| Conversion method | Red channel | Green channel | Blue channel |
|---|---|---|---|
| 12-bit original | 805 | 1735 | 964 |
| 8-bit standard conversion | 84 | 132 | 89 |
| 8-bit LIL conversion | 256 | 256 | 256 |
| 8-bit LIL conversion on the whole series | 179 | 235 | 150 |

## 2.2  LIL Conversion of Image Series

The biggest asset of the LIL algorithm is the objective comparison of images in a series at the same intensity scale. It is achieved after omitting intensities which do not occur in any image of the series followed by re-scaling the whole series between the minimal and maximal intensity in the series (cf. Fig. 2). As a result, the intensity depth of each image decreases (Fig. 3, *right* and Table 1). In standard bit-compression software, each image is transformed to 8 bit separately, which prevents any comparison of intensities over the dataset. As shown in Fig. 4, the standard 12 to 8-bit conversion algorithms also smoothen the histogram and remove outlying intensities. As a consequence, the ratio of intensities is changed and many structural details are lost.

## 2.3  Usage of LIL Conversion in Photographic Camera Output

Users of 12-bit photographic cameras do not have a good control over data storage and transformation. On an example of Nikon NEF format [5], we show results of the transformation of one image of the series of the Belousov-Zhabotinsky reaction (Figs. 4 and 5). The NEF histogram analyzed from an open DNG (Fig. 4, *upper*) format does not indicate that the NEF transformation modifies the original 12-bit signal.

The comparison of the simple NEF to 8-bit transformation (Fig. 4, $2^{nd}$ row), which corresponds the best to the original signal captured by the camera (Fig. 5, *left*), with a back transformation from the LIL format to the 12-bit format (Fig. 4, $3^{rd}$ row) illustrates the extent of details which are lost by the bit-depth transition.

The LIL transformation (Fig. 4, $4^{th}$ row and Fig. 5, *middle*) preserves ratios from the signal. NEF to PNG transformation using DNG to PNG transformation freeware [6] transforms and re-weights each image separately using an interpolation algorithm (Fig. 4, *bottom*). This approach is suitable for obtaining the best possible genre photograph (Fig. 5, *right*), but is misleading for scientific analysis.

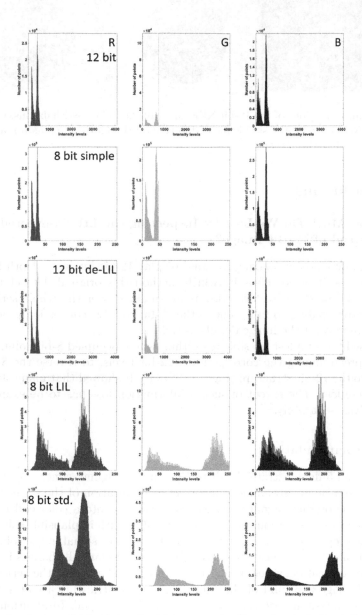

**Fig. 4.** Red, green, and blue intensity histograms for (*from top to bottom*) the original 12-bit NEF image of the Belousov-Zhabotinsky reaction, the respective simply transformed 8-bit image, the 12-bit reconstructed image after de-LILing, the 8-bit LIL-converted image, and the 8-bit image obtained using NEF to PNG freeware (Color figure online).

**Fig. 5.** Transformations of the 12-bit NEF image of the Belousov-Zhabotinsky reaction to 8-bit PNG image by the simple algorithm (*left*), LIL algorithm (*middle*), and using NEF to PNG freeware (*right*).

## 3    Conclusions

### 3.1    How Much Do We Lose by Inspecting the LIL-Converted Image Instead of 12-Bit Image?

Despite many kinds of imaging freeware display 12-bit intensity depth formats, in most cases, it is not clear by which method this original dataset is transformed and how the transformation is loss-making over the whole series. The LIL converter enables to fully control the 12-bit to 8-bit conversion process and to preserve most of the image details as well.

We used the simplest possible algorithm, where occupied 8-bit columns were equally spread among the corresponding 12-bit levels, to reverse the 8-bit LIL image into the 12-bit structure (Fig. 4, $3^{rd}$ row). It showed that many structural details were lost. The extent of such a information loss has to be examined in each analysis separately.

### 3.2    Recommendations

Figures 4 and 5 demonstrate the distortions of an original camera signal, which may lead to misunderstanding or incorrect interpretation of the scientific data.

When the original signal is transformed directly into a color-coded matrix, during the visual inspection, we may be unsatisfied with implausible colors. However, such a representation is equivalent to the primary signal in the colorimetric measurement. If the intensity scale is set to accommodate all changes in concentrations or optical densities during the capture of the series, the series of LIL converted image is a kind of the calibrated colorimetric (in case of a chemical reaction) or densitometric (in case of a cell culture) signal. In addition, a LIL-converted dataset allows to examine most faithfully faint details in the image. The best way, certainly, is to work with the original 12-bit dataset, but according to our knowledge – apart from our Image Explorer [1] – there is no software which would do that without dataset transformation.

For computer processing it is indeed recommended to use the original 12 bit signal. When a kind of signal and level grouping or averaging is necessary to do, it should be done on purpose. It means that we should either know the anticipated statistical distribution in the image, and, thus, the process which generated the image, or search for a given detail, such as object size, etc. An approach to such a targeted transformation of data was developed by us ourselves [7].

Indeed, similar analogies may be drawn for all types of images, let it be x-ray or infrared signals recorded by the space telescope or local magnetic field intensities derived from the NMR signal.

## 4 Methods

### 4.1 LIL Converter

The LIL convertor is available upon request from the authors. The input file into the LIL convertor software is a 12-bit PNG format which is equivalent to digital negative – DNG – file. The GUI enables to select a type of Bayer mask or the monochromatic option for conversion of grayscale images. The software is equipped by three debayerization approaches (Fig. 6):

– Simple, where the value for each pixel is taken as an average of the nearest pixels of a given, the resulted image has the same size as the input raw file.

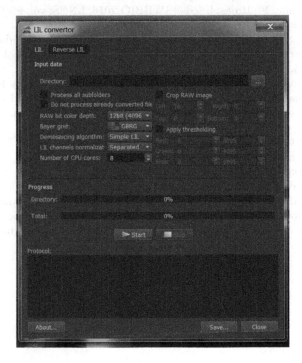

**Fig. 6.** Graphical user interface of the LIL converter software.

- AHL (adaptive homogeneity-directed algorithm), which selects the direction of interpolation in order to maximize a homogeneity metric and minimize color artifacts. The size of the resulted image is the same as the input raw file.
- Simple LIL, which is the approach discussed in this article and which returns an 8-bit color image of 1/4 of the size of the original image.

Normalization (finding the minimal and maximal intensities) may be done for each color channel separately or for the whole image. Image may be cut at borders (for the conversion of objects of interest and removing the columns and rows for calibration of camera). Threshold of rare intensities may be set for each color channel separately. The protocol on the computation is stored and may be repeatedly uploaded.

**Acknowledgments.** This work was financially supported by CENAKVA (No. CZ.1.05/2.1.00/01.0024), CENAKVA II (No. LO1205 under the NPU I program) and The CENAKVA Centre Development (No. CZ.1.05/2.1.00/19.0380).

# References

1. Macháček, P., Císař, P., Náhlík, T., Rychtáriková, R., Štys, D.: Visual exploration of principles of microscopic image intensities formation using image explorer software. In: Ortuño, F., Rojas, I. (eds.) IWBBIO 2016. LNCS, vol. 9656, pp. 537–544. Springer, Heidelberg (2016)
2. Zhyrova, A., Štys, D.: Construction of the phenomenological model of Belousov-Zhabotinsky reaction state trajectory. Int. J. Comput. Math. **91**, 4–13 (2014)
3. Rychtáriková, R., Náhlík, T., Smaha, R., Urban, J., Štys Jr., D., Císař, P., Štys, D.: Multifractality in imaging: application of information entropy for observation of inner dynamics inside of an unlabeled living cell in bright-field microscopy. In: Sanayei, A., et al. (eds.) ISCS 2014, vol. 14, pp. 261–267. Springer, Switzerland (2015)
4. Bayer, B.E.: Color imaging array. U. S. Patent No. 3,971,065 (1976)
5. http://www.nikonusa.com/en/learn-and-explore/article/ftlzi4ri/nikon-electronic-format-nef.html
6. http://www.coolutils.com/online/NEF-to-PNG
7. Rychtáriková, R., Korbel, J., Macháček, P., Císař, P., Urban, J., Soloviov, D., Štys, D.: Point information gain, point information gain entropy and point information gain entropy density as measures of semantic and syntactic information of multi-dimensional discrete phenomena, arxiv:1501.02891 (2015)

# Visual Exploration of Principles of Formation of Microscopic Image Intensities Using Image Explorer Software

Petr Císař, Tomáš Náhlík, Renata Rychtáriková$^{(\boxtimes)}$, and Petr Macháček

Faculty of Fisheries and Protection of Waters, South Bohemian Research Center
of Aquaculture and Biodiversity of Hydrocenoses, Institute of Complex Systems,
University of South Bohemia in Česke Budějovice, Zámek 136,
373 33 Nové Hrady, Czech Republic
rrychtarikova@frov.jcu.cz
http://www.frov.jcu.cz/cs/ustav-komplexnich-systemu-uks

**Abstract.** The article demonstrates the most frequent mistakes made upon the transformation of digital images in biology. An image is formed by a few physical processes which contribute to each color channel in a different way. In the case of microscopic image in transmitting light, these processes are mainly light diffraction and absorption and autofluorescence of objects, which are all followed by distortion of wavefronts by the microscope optics. The final image is then a result of these processes in the plane of the camera chip. The article further reports methods to avoid (i) misconceptions due to apparent coloration after transformation of the original signal on the camera chip into color image and (ii) loss of the resolution after reduction of intensity depth from 12-bit to 8-bit.

**Keywords:** Living cell imaging · Bit depth · Superresolution microscopy

## 1 Introduction

Visual inspection is still an inevitable part of analysis of biological datasets. In the previous article of this volume [1], we demonstrated how the transformation of the image may obscure the primary dataset and ways how such a problem may be avoided. A next step is the analysis in ways by which the image is formed.

In case of a transmitted-light microscope, there are two main contributions to the formation of final image: (a) diffraction/emission and/or absorption of light by the sub-objects of the observed sample and (b) an image build-up along the optical path. In the image, we observe a sum of these effects originating from all observed sub-objects which lie in different distances from a lens and are projected by the lens on different points in the space. All phenomena which lead to the actual image build-up are finally summed into the intensity recorded in each camera pixel, possibly also partly separated into regions of wavelengths by the color camera filter [2]. The goal of the analysis of a microscopy dataset should be

© Springer International Publishing Switzerland 2016
F. Ortuño and I. Rojas (Eds.): IWBBIO 2016, LNBI 9656, pp. 537–544, 2016.
DOI: 10.1007/978-3-319-31744-1_48

to separate these elementary contributions and to identify the positions and optical properties of objects, which give rise to diffraction/absorption/fluorescence. Recently, we have elaborated a method how to come as close as possible to such a localisation [3].

In this paper, we use an example of the microscopy image series to demonstrate some elements of the visual analysis of biological data. For this purpose, we present a novel software tool – Image Explorer – which was designed to facilitate a visual image inspection.

## 2 Results

### 2.1 Image Explorer

Ordinary computer monitors and printers display only images of 8-bpc intensity depth. In order to visualize vice-bpc data, we developed a least information loss (LIL) algorithm [1]. This algorithm utilizes the whole intensity scale of each color channel for creation of a 8-bpc RGB pseudocolor image. In the 8-bit LIL format, all intensity levels are occupied and, thus, the maximum of information is visualized. The color of the LIL-transformed image (e.g., Fig. 1, *left panel*) is similar to colors of images transformed by standard image conversion algorithms.

As seen in Fig. 1, despite the visualization of maximum information, an image of a living cell in transmitting light after LIL transformation (Fig. 1) does not show many structured details any more. Even, the following expansion of a part of this image (Fig. 2) did not lead to the understanding of meaning content of the image. The signal is a result of wavelength dependent light refraction on intracellular objects and a consequent interference of light waves projected on the

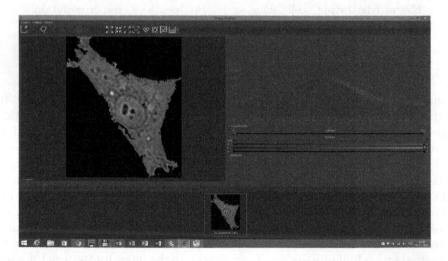

**Fig. 1.** Image of a living cell converted using the least information loss (LIL) algorithm from the 12-bpc into 8-bpc intensity depth.

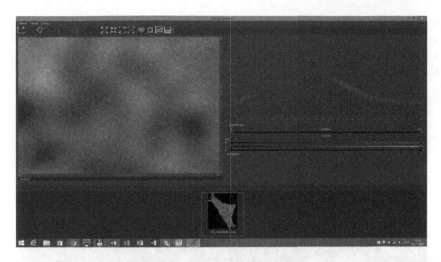

**Fig. 2.** A focused part of the image of a living cell in Fig. 1.

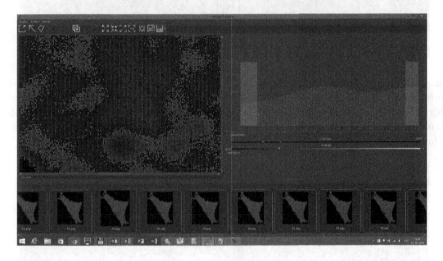

**Fig. 3.** The image of a living cell in Fig. 2 in selected blue intensities with numerous precise details of the cellular interior. The coloration was changed upon the selection of levels of individual color contributions (*right panel*). For comparison, the original image remains in the bottom image row (Color figure online).

camera chip. Therefore, the combination of LIL-transformed RGB intensities still keep a visual impression of a blurry image which does not contain many details. The best possible approximation to the original 12-bpc image is an examination of individual color channels and selected intensities.

The Image Explorer software was developed for examination of the signal from individual color channels and to select individual levels from the vice-bit intensity histograms from focused part of the image. The effect of such selections are shown in Figs. 3 and 4. After the selection of the respective (blue and green) channels, many additional details of cell interior emerged. However, for a correct interpretation of the results, it is necessary to further examine the physical origin of the image, since in most cases a new pseudocolor arises only by coincidence of the rescaled image channels.

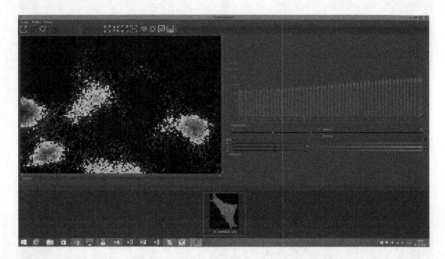

**Fig. 4.** The combination of selected blue and green levels provides insight into the mechanism of the creation of image coloration. Pixels of selected intensities of both color channels are localized in the color map of a magnified part of the image. More distinct points (possibly objects of the cell interior) are distinguished there (Color figure online).

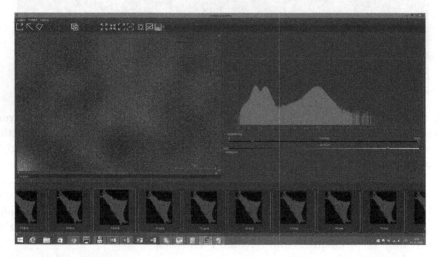

**Fig. 5.** Transformation of an original 12-bit grayscale response of camera chip (raw format) into 8-bit intensity depth. The grayscale intensity histogram (*right panel*) is split into three partly overlapping regions which correspond to the three color channels (Color figure online).

## 2.2    Information Loss upon Reduction of 12-Bit to 8-Bit Intensity Depth

Figure 5 depicts a section of (computer interpretation of) the original 12-bit camera signal. Figure 6 is an expansion of image in Fig. 5, where the elementary pixels are color-coded. Here is possible to examine positions of elementary

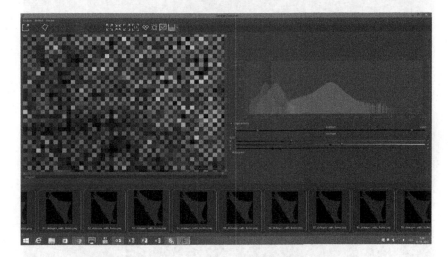

**Fig. 6.** The same focused part of the image of living cell as in Fig. 5 but with individual color-coded channels. The scales are expanded only to intensity levels which are present in the focused window. For visual inspection, the numbers of original levels (more than 150) in each 12-bit color channel are rescaled to 256 levels. After this rescaling, the tiniest details of the camera response are seen (Color figure online).

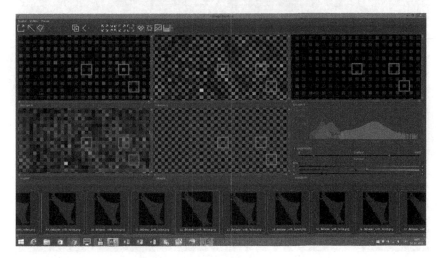

**Fig. 7.** Section of Fig. 5 with the positions of intensity maxima in different color channels. Points with maximal intensity are marked by squares (Color figure online).

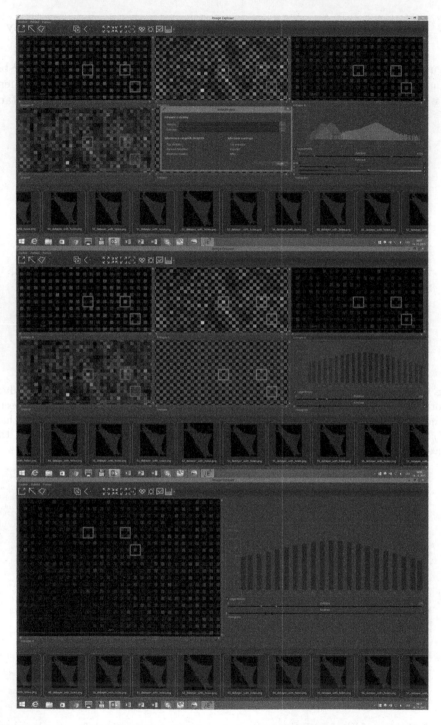

**Fig. 8.** Upper panel: opening of the image series. Overview of images in the series are given in bottom panel of the GUI. Middle panel: examination of individual channels. Lower panel: separated examination of the blue channel (Color figure online).

objects, which build the image, with the precision of one camera pixel, i.e. $34 \times 34\,\mathrm{nm}^2$. Upon a standard 12 to 8-bit depth transformation, all these details are lost by shrinkage to less than 100 intensity levels.

Figure 7 shows differences in the positions of maxima in individual color channels in a raw format. These differences occurred probably due to both different properties of individual channels, since the camera parameters may be set only for the whole chip, and differences in image build-up by the lenses. To visualize this kind of image data, each span of 12-bit intensities, which correspond to a respective color channels, are rescaled separately between 256 intensity levels.

## 3   Conclusion

The Image Explorer software enables to examine systematically, level by level, intensity changes in the image. The analysis presented in this article showed two types of image distortion which are almost always overlooked.

During the examination of color-coded image, it is overlooked that the signal represented by different wavelength is a result of simultaneous action of different physical processes. It is thus not relevant to analyze the color tones on the computer display. A separate analysis of the three color channels is more adequate approximation to spectrally resolved single response detection.

A next distortion of the signal comes from reduction of intensity levels. Even at the best information preserving transformation ([1], this volume), the reduction of the number of levels is in the order of $1/100$. In Figs. 6 and 7 we show that relevant differences may be found at the level of the original 12-bit image intensities.

The software Image Explorer further allows a detailed visual inspection of original data matrix obtained by the camera. We believe that such a visual inspection should precede any data analysis using any other image transformation software and might avoid an overinterpretation of artifacts as well as a loss of relevant information. The algorithms developed for the Image Explorer software are relevant for any $n \times x \times y$ data matrix of any precision.

## 4   Methods

Image Explorer is a constantly evolving software tool for visual inspection and comparison of digital camera images. It allows an inspection of primary images of intensity depths up to 16 bit. The inspection includes a standard combination of channels into color tones at possibility of examination of an individual channel. For comparison of objects, there is an object marker tool which enables to mark objects for further localization in the image. Figure 8 shows some of the features of the Image Explorer interface: opening of an image series, visual inspection of all three color channels and expansion of one individual color channel.

**Acknowledgments.** This work was financially supported by the Ministry of Education, Youth and Sports of the Czech Republic – projects CENAKVA (No. CZ.1.05/2.1.00/01.0024), CENAKVA II (No. LO1205 under the NPU I program) and The CENAKVA Centre Development (No. CZ.1.05/2.1.00/19.0380).

# References

1. Náhlík, T., Macháček, P., Rychtáriková, R., Saberioon, M., Štys, D.: Least information loss (LIL) conversion of digital images and lessons learned for scientific image inspection. In: Ortuño, F., Rojas, I. (eds.) IWBBIO 2016. LNCS, vol. 9656, pp. 527–536. Springer, Heidelberg (2016)
2. Náhlík, T., Štys, D.: Microscope point spread function, focus and calculation of optimal microscope set-up. Int. J. Comput. Math. **91**, 221–232 (2014)
3. Rychtáriková, R., Náhlík, T., Smaha, R., Urban, J., Štys D., Jr., Císař, P., Štys, D.: Multifractality in imaging: application of information entropy for observation of inner dynamics inside of an unlabeled living cell in bright-field microscopy. In: Sanayei, A., et al. (eds.) Emergence, Complexity and Computation: ISCS 2014, vol. 14, pp. 261–267. Springer, Switzerland (2015)

# On Optimization of FRAP Experiments: Model-Based Sensitivity Analysis Approach

Štěpán Papáček[1](✉) and Stefan Kindermann[2]

[1] Faculty of Fisheries and Protection of Waters, South Bohemian Research Center of Aquaculture and Biodiversity of Hydrocenoses, Institute of Complex Systems, University of South Bohemia in České Budějovice, Zámek 136, 373 33 Nové Hrady, Czech Republic
spapacek@frov.jcu.cz
[2] Industrial Mathematics Institute, Johannes Kepler University of Linz, Altenbergerstr. 69, 4040 Linz, Austria
kindermann@indmath.uni-linz.ac.at

**Abstract.** The accuracy in the determination of model parameters from data depends on the experimental setup. The advance in this area is often hindered by lack of communication between experimentalists and mathematical modelers. We aim to point out a potential benefit in parameter inference when the design variables are chosen optimally. Our approach, although case independent, is illustrated on FRAP (Fluorescence Recovery After Photobleaching) experimental technique. The core idea is closely related to the sensitivity analysis, namely to the maximization of a sensitivity measure depending on experimental settings. The proposed modification of the FRAP experimental protocol is simple and the enhancement of the whole parameter estimation process is significant.

**Keywords:** FRAP · Sensitivity analysis · Optimal experimental design · Parameter estimation

## 1 Introduction

Performing experiments within a biological system can be viewed as an activity involving four interconnected parts: (i) measurement method and equipment, (ii) setting of an experimental protocol (e.g., experimental conditions), (iii) the proper measurement and data acquisition, (iv) data processing, usually in accordance to certain mathematical model of a presumably known underlying processes. Close collaboration within research teams dedicated to each one of the above mentioned parts (i–iv) is the key factor for the final success of an experimental research project in biology or biomedicine. However, it seems, the communication between experimentalists and data processing staff is rather complicated, which results in disruption of the link between the aforementioned parts (ii) and (iv).

The common practice of experimental condition 'tuning' resides on trial-error method performed by experimentalists while the subsequent data processing is

© Springer International Publishing Switzerland 2016
F. Ortuño and I. Rojas (Eds.): IWBBIO 2016, LNBI 9656, pp. 545–556, 2016.
DOI: 10.1007/978-3-319-31744-1_49

not always taken into account and frequently left to some 'external forces'. It is not a rare case that large amount of data is routinely generated without a clear idea about further data processing.

Here, we suggest to analyze simultaneously both the data (i.e., the processes hidden in data) and the experimental protocol, aiming to establish the link between experimental conditions and the accuracy of our results. The key role plays a reliable mathematical model of all involved processes. The whole idea is presented in a simplified case study of FRAP data processing. In following Sect. 2 we provide the background information concerning FRAP method. In Sect. 3 we rigorously formulate the problem and in Sect. 4 we provide two practical results: (i) a proposition for an optimal (i.e., assuring smallest confidence intervals of the parameter estimates) setting of the bleached region size, and (ii) relevant data space selection, getting rid of irrelevant data while the confidence interval enlargement is controlled by the factor $1 + \eta$. The novelty and benefits of our approach as well as outlooks for further research are discussed in the final Sect. 5.

## 2    FRAP Measurement Technique and FRAP Data Structure

Fluorescence Recovery After Photobleaching (FRAP) is a well-established measurement technique for determination of the mobility of fluorescent molecules in membranes of the living cells [1]. The FRAP technique is based on measuring the fluorescence intensity (proportional to non-bleached particles concentration) in a region of interest (being usually an Euclidian 2D domain) in response to a high-intensity laser pulse. We suppose the laser pulse (the so-called *bleach*) causes an irreversible loss in fluorescence of some particles (originally in the bleached area) without any damage to intracellular structures. After the bleach, we observe the change in fluorescence intensity in a monitored region,[1] presumably reflecting the diffusive transport of fluorescent compounds from the area outside the bleach.[2] Based on the spatio-temporal FRAP data, the effective diffusion coefficient was estimated using a closed form model [4–6] in past decades. Nowadays, a computationally more expensive approach based on numerical simulation of a corresponding model is preferred, because it does not need some unrealistic conditions to be assumed [7–11].

In the illustrative Fig. 1, we see the time sequence of FRAP images, one pre-bleach image and 5 post-bleach images. The first post-bleach image was taken 8 s after application of high intensity laser pulse for bleaching the strip shaped region reducing the phycobilisome fluorescence in *P. cruentum* to about 40 % of the initial value (in the bleached region) due to the destruction of a portion of the phycobilin pigments.

---

[1] In general we observe both recovery and loss in fluorescence in different regions corresponding to FRAP or FLIP (Fluorescence Loss in Photobleaching), respectively, see [2,3] for more details.

[2] Obviously, there is the transport of bleached particles from the bleached region to originally non-bleached regions.

A FRAP data structure usually consists of a rectangular matrix, where each entry quantifies the fluorescence intensity (proportional to the non-bleached particle concentration) at a particular point in a spatio-temporal domain (e.g., by a number between 0 and 255)

$$u(x_l, t_j)_{l=1}^{N_x}, \quad j = 0...N_t, \tag{1}$$

where $l$ is a spatial index uniquely identifying the position where the signal $u$ is measured and $j$ is the time index. Moreover, the data points are uniformly distributed both in space (on an equidistant 1D or 2D mesh) and time (the time interval between two consecutive measurements is usually constant).

*Remark 1.* In the following, we adopt the simplified notation consisting in using only one index $i$ ($i = 1...N_{data}$) for all data in the space-time domain. This data storing is independent of spatial dimensionality, i.e., the same for both 2D and 1D fluorescence profiles. For the axis-symmetric problem, e.g. strip-shaped bleach, see Fig. 1, the reduction of dimensionality can be done by signal averaging along the axis parallel to bleach axis, i.e., the dynamics perpendicular to bleach is analyzed only.

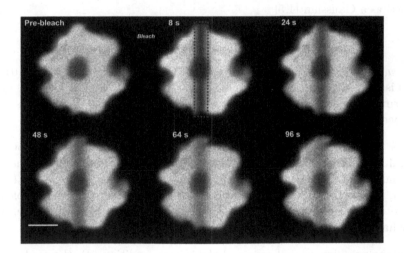

**Fig. 1.** Representative image sequence from FRAP experiments for a single cell of red algae *Porphyridium cruentum* for phycobilisome fluorescence [2]. First, a fluorescence image before bleaching was detected (Pre-bleach), and then the phycobilisome fluorescence was bleached out across the middle of the cell in the vertical direction. The recovery dynamics in bleached region (red dashed rectangle in the first post-bleach image) is clearly seen on sequence of 5 post-bleached images, taken 8, 24, 48, 64 and 96 s after bleach, respectively. The length of the scale bar is 3 μm.

# 3  Mathematical Model

## 3.1  Initial Boundary Value Problem Formulation

Consider the Fickian diffusion equation describing the spatio-temporal dynamics of concentration $u(x,t)$ of one component system

$$\frac{\partial}{\partial t}u(x,t) = p\Delta u(x,t), \quad \text{in } \Omega \times [0,T], \quad u(x,0) = u_0\phi(x), \tag{2}$$

where $x$, $t$, $p$, $\phi(x)$, are space coordinate (of an appropriate dimension), time, the diffusion coefficient, and some given initial shape, respectively.

Boundary conditions could be, e.g.,

$$u(x,t) = 0, \quad \text{or} \quad \frac{\partial}{\partial n}u(x,t) = 0 \quad \text{on } \partial\Omega \times [0,T]. \tag{3}$$

We also consider the simplest case of unbounded domains, $\Omega = \mathbb{R}^n$, in which case we set appropriate decay conditions at $\|x\| \to \infty$, $t \in [0,T]$. Equations (2–3) and variants are the basis for all the further analysis.

In FRAP experiments, the initial condition, i.e., the first post-bleach profile with the background or pre-bleach fluorescent intensity subtracted, is often modeled as a Gaussian [4,5]

$$u_0(x) = u_{0,0}e^{-\frac{2\|x\|^2}{r_0^2}}. \tag{4}$$

where $u_{0,0} > 0$ is the maximum depth at time $t = 0$ (i.e., for $x = 0$), and $r_0 > 0$ is the half-width of the bleach at height (depth) $u_{0,0}e^{-2}$; see e.g., [5]. By this formulation, the spatio-temporal distribution of bleached particles is in fact described.

*Remark 2.* Having the pre-bleach intensity profile $u_{pre}(x)$ we can subtract it from all post-bleach values $u(x,t)$. The quantity $|u(x,t) - u_{pre}(x)|$ is then proportional to the bleached particle concentration $c_b$. Notice, $c_b(x,t) + c_{nb}(x,t) \sim u_{pre}(x)$, hence, the study of $c_b(x,t)$ is complemental to the study of non-bleached particle concentration $c_{nb}(x,t)$. Later on, we opt for the second choice, which somewhat simplifies our calculation.

For the one-dimensional case and the initial condition (4), the explicit solution for $u_b$ (in sake of simplicity the subindex $b$ is omitted) is

$$u(x,t) = u_{0,0}\frac{r_0}{\sqrt{r_0^2 + 8pt}} e^{-\frac{2x^2}{r_0^2 + 8pt}}. \tag{5}$$

The above Eq. (5) results from an oversimplified formulation, nevertheless, it facilitates presentation of the model-based sensitivity approach in Sect. 4. In the next subsection, the inverse problem of parameter estimation is formulated as an ordinary least squares problem (8). However, the detailed analysis is presented elsewhere [9,10].

## 3.2   Model Parameter Estimation and Model-Based Sensitivity Analysis

Let us define the forward map (also called parameter-to-data map)

$$F : (p) \rightarrow u(x_i, t_i)_{i=1}^{N_{\text{data}}}, \ \mathbb{R}^q \rightarrow \mathbb{R}^{N_{\text{data}}} \tag{6}$$

where $p = [p_1, \ldots, p_q]^T$, $q \in \mathbb{N}$ and $u$ is as in (1). Our regression model is now

$$F(p) = \text{data}, \tag{7}$$

where the data are modeled as contaminated with additive white noise

$$\text{data} = F(p_T) + e = u(x_i, t_i)_{i=1}^{N_{\text{data}}} + (e_i)_{i=1}^{N_{\text{data}}}.$$

Here $p_T$ denotes the true coefficient and $e$ is a data error vector which we assume to be normally distributed with variance $\sigma^2$, $(e_i)_{i=1}^{N_{\text{data}}} \in \mathbb{R}^{N_{\text{data}}}$, $e_i = \mathcal{N}(0, \sigma^2)$.

Given some data, the aim of the parameter estimation problem is to find $p$, such that (7) is satisfied in some appropriate sense. Since (7) usually consists of an over-determined system (there are more data points than unknowns), it cannot be expected that (7) holds with equality, but instead an appropriate notion of solution (which we adopt for the rest of the paper) is that of a least-squares solution $p_c$ (with $\|.\|$ denoting the Euclidean norm on $\mathbb{R}^{N_{\text{data}}}$):

$$\|F(p_c) - \text{data}\|^2 = \min_p \|F(p) - \text{data}\|^2. \tag{8}$$

The above defined parameter estimation problem is usually ill-posed for non-constant coefficients, such that regularization has to be employed; see, e.g., [12]. The solution of an practical example based on FRAP data was presented in [9].[3]

For the sensitivity analysis, we require the sensitivity matrix composed from the Fréchet-derivatives $F'[p] \in \mathbb{R}^{N_{\text{data}} \times q}$ of the forward map $F$, that is $F'[p] = [\frac{\partial}{\partial p} u(x_1, t_1, p), \ldots, \frac{\partial}{\partial p} u(x_{N_{\text{data}}}, t_{N_{\text{data}}}, p)]^T$. Based on $F'[p]$, for $p = [p_1, \ldots, p_q]^T$, we construct the Fisher information matrix $M$

$$M[p_1, \ldots, p_q] = F'[p]^T F'[p] \in \mathbb{R}^{q \times q}. \tag{9}$$

Suppose we have computed $p_c$ as a least-squares solution in the sense of (8). Let us define the residual as

$$res^2(p_c) = \|F(p_c) - \text{data}\|^2 = \sum_{i=1}^{N_{\text{data}}} [\text{data}_i - u(x_i, t_i, p_c)]^2, \tag{10}$$

then according to [13], it is possible to quantify the error between the computed parameters $p_c$ and the true parameters $p_T$, see [10] for more details.

---

[3] Papáček et al. in [9] allow to change the diffusion coefficient $p$ in time, i.e., $p = (p_j)_{j=1}^{N_t}$. This approach in fact permits to model the anomalous diffusion as well as normal diffusion.

Further, in Sect. 4, we provide an example of the Fickian isotropic diffusion in one component system, i.e., with one time and space independent scalar parameter $p$ as unknown. The Fisher information matrix $M$ then collapses into the scalar quantity $\sum_{i=1}^{N_{\text{data}}} \left[ \frac{\partial}{\partial p} u(x_i, t_i, p) \mid_{p=p_c} \right]^2$, and the $1 - \alpha$ confidence interval, is described as follows, cf. [13],

$$(p_c - p_T)^2 \sum_{i=1}^{N_{\text{data}}} \left[ \frac{\partial}{\partial p} u(x_i, t_i, p) \mid_{p=p_c} \right]^2 \leq \frac{res^2(p_c)}{N_{\text{data}} - 1} f_{1, N_{\text{data}}-1}(\alpha), \qquad (11)$$

where $f_{1, N_{\text{data}}-1}(\alpha)$ corresponds to the upper $\alpha$ quantile of the Fisher distribution with 1 and $N_{\text{data}} - 1$ degrees of freedom.

*Remark 3.* In (11), several simplifications are possible. Note that according to our noise model, the residual term $\frac{res^2(p_c)}{N_{\text{data}}-1}$ is an estimator of the error variance such that the approximation $\frac{res^2(p_c)}{N_{\text{data}}-1} \sim \sigma^2$ holds if $N_{\text{data}}$ is large. Moreover, the term $f_{1, N_{\text{data}}-1}(\alpha)$ can approximately be viewed as independent of $N_{\text{data}}$ as well and of a moderate size.

## 4  Results on Optimization of FRAP Experimental Design

Many empirical recommendations are related to the design of FRAP experiments, e.g., how to set shape and size of the bleach, location and size of the monitored region, the time span of the measurement, cf. [1,6,11]. However, based on the sensitivity analysis—by maximizing the sensitivity of the measured output (as a whole in the monitored region) on the estimated parameter with respect to some design variable—we can obtain a rigorous tool for the optimal (model-based) design of experiment. I.e., we can propose an optimal choice of some design factors.

### 4.1  Optimizing the Bleaching Spot

In FRAP measurements, the size of the bleaching spot, characterized by the parameter $r_0$, can be varied, which leads to the question if there is an optimal size that can be used. We may based our analysis on relation (11) and look how the quantity $\sum_{i=1}^{N_{\text{data}}} \left[ \frac{\partial}{\partial p} u(x_i, t_i, p) \mid_{p=p_c} \right]^2$ depends on the size of the bleaching spot. We can therefore try to maximize the above quantity corresponding to global sensitivity, since this corresponds to minimal confidence intervals (for comparable experiments).

Let us consider the problem (2) on free space in the one-dimensional case with one unknown scalar parameter $p$ and a Gaussian as in (4) as initial condition. From the solutions (5) we obtain for the derivative

$$\frac{\partial}{\partial p} u(x, t) = -u_{0,0} \frac{r_0}{(r_0^2 + 8pt)^{\frac{5}{2}}} e^{-\frac{2x^2}{r_0^2 + 8pt}} 4t(r_0^2 + 8pt - 4x^2). \qquad (12)$$

A corresponding quantity $S_{loc} = [\frac{\partial}{\partial p}u(x,t)\frac{p}{u_{0,0}}]^2$ is the (local) squared relative sensitivity. By an appropriate scaling based on the experimental time span $T$ and the characteristic diffusion length $\sqrt{2pT}$, we have

$$S_{loc}(\tilde{x}, \tilde{t}) = \frac{16\tilde{t}^2 \frac{r_0^2}{pT}}{(\frac{r_0^2}{pT} + 8\tilde{t})^3} e^{-\frac{4\tilde{x}^2}{\frac{r_0^2}{pT} + 8\tilde{t}}} (1 - \frac{4\tilde{x}^2}{\frac{r_0^2}{pT} + 8\tilde{t}})^2. \tag{13}$$

where the new dimensionless variables for space resp. time are defined $\tilde{x} := \frac{x}{\sqrt{2pT}}$, and $\tilde{t} := \frac{t}{T}$, respectively. The quantity $S_{loc}$ depends on the experimental design parameter $r_0$. Introducing the same spatial scaling for $r_0$ as for $x$, i.e., $\tilde{r}_0 := \frac{r_0}{\sqrt{2pT}}$, we finally obtain

$$S_{loc}(\tilde{x}, \tilde{t}, \tilde{r}_0) = \frac{4\tilde{t}^2 \tilde{r}_0^2 (\tilde{r}_0^2 + 4\tilde{t} - 4\tilde{x}^2)^2}{(\tilde{r}_0^2 + 4\tilde{t})^5} e^{-\frac{4\tilde{x}^2}{\tilde{r}_0^2 + 4\tilde{t}}}. \tag{14}$$

Let us further consider the case of a dense set of observations on a space-time cylinder $Q = [-L, L] \times [0, T]$, where $[-L, L]$ is the space interval of observations and $[0, T]$ the time interval: $(x_i, t_i) \in ([-L, L] \times [0, T]) \cap \{(k\Delta x, l\Delta t) \mid (k, l) \in N\}$. Here $\Delta x$ and $\Delta t$ are the spacings of the grid in space and time direction, respectively. The term corresponding to the Fischer information matrix in relation (11), i.e., the (global) squared relative sensitivity $S_{gl}$, can be approximated by an integral

$$\sum_{i=1}^{N_{data}} (\frac{\partial}{\partial p}u(x_i, t_i)\frac{p}{u_{0,0}})^2 \sim \frac{2}{\Delta\tilde{x}\Delta\tilde{t}} \int_0^{\tilde{L}} \int_0^1 S_{loc}(\tilde{x}, \tilde{t}, \tilde{r}_0)d\tilde{x}d\tilde{t}.$$

Finally, we find

$$S_{gl} = \frac{1}{\Delta\tilde{x}\Delta\tilde{t}} \int_0^{\tilde{L}} \int_0^1 \frac{8\tilde{t}^2 \tilde{r}_0^2 (\tilde{r}_0^2 + 4\tilde{t} - 4\tilde{x}^2)^2}{(\tilde{r}_0^2 + 4\tilde{t})^5} e^{-\frac{4\tilde{x}^2}{\tilde{r}_0^2 + 4\tilde{t}}} d\tilde{x}d\tilde{t} = \frac{1}{\Delta\tilde{x}\Delta\tilde{t}}K(\frac{L}{\sqrt{2pT}}, \frac{r_0}{\sqrt{2pT}}) \tag{15}$$

with

$$K(L, r_0) = \int_0^L \int_0^1 \frac{8t^2 r_0^2 (r_0^2 + 4t - 4x^2)^2}{(r_0^2 + 4t)^5} e^{-\frac{4x^2}{r_0^2 + 4t}} dxdt. \tag{16}$$

The plot of the (global) squared relative sensitivity $S_{gl}$, depending on the size of scaled monitored region $\frac{L}{\sqrt{2pT}} \in [10, 20]$ and bleach size $\frac{r_0}{\sqrt{2pT}} \in [0, 3]$, is shown in Fig. 2, left.

A particular case happens if $L \to \infty$ in the above integral (16). Then

$$K(\infty, r_0) = \frac{\sqrt{\pi}r_0^2 (r_0^4 + 6r_0^2 - 4\sqrt{r_0^2 + 4}r_0 - \sqrt{r_0^2 + 4}r_0^3 + 6)}{8(r_0^2 + 4)^{3/2}}. \tag{17}$$

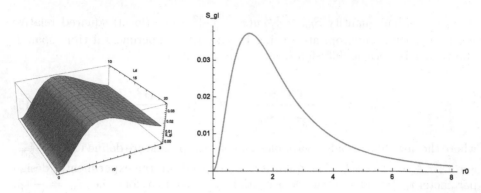

**Fig. 2.** Left: 3D plot of the sensitivity $S_{gl}$, depending on monitored region size ($\frac{L}{\sqrt{2pT}} \in$ [10, 20]) and bleach size ($\frac{r_0}{\sqrt{2pT}} \in [0,3]$). The optimum for $r_0$, independent on the size of monitored region, does exist for the fixed measurement time $T$. Right: For the size of monitored region of $L \to \infty$ and fixed measurement time $T$, the maximal sensitivity $S_{gl}$ is reached at the size of bleach $\frac{r_0}{\sqrt{2pT}} = 1.222$.

Based on the above (17), we try to infer the $r_0$ with maximal sensitivity. Hence, when the grid factor $\frac{1}{\Delta \tilde{x} \Delta t}$ is assumed to be fixed, we try to find out about the maximal value of the function

$$K(\infty, r_{opt}) = \max_{r_0 > 0} K(\infty, r_0).$$

For this special case, i.e., $L = \infty$, it turns out that in this setting the function $K(\infty, r_0))$ has a unique maximum (Fig. 2, right), which location can be calculated numerically as

$$r_{opt} = 1.222\sqrt{2p_cT}. \tag{18}$$

*Remark 4.* Conversely, if a size of a fixed bleaching spot is given, we can find the measurement time $T_{opt}$, which corresponds to an optimal radius $r_0$ set according to (18), as

$$T_{opt} = 0.335\frac{r_0^2}{p_c}.$$

## 4.2 How to Get Rid of Irrelevant Data

Another issue with great practical relevance, mainly when the cost of observations and experimental runs can not be neglected, resides in an adequate choice of the observation region. This issue is closely related with the amount of data to process. Now, we briefly resume our previous results extensively studied in [10] and draw a simple recommendation for data space selection.

The confidence interval estimate (11) gives us useful information on the quality of a least-squares estimate. The sensitivity is the central ingredient in this estimate, being the main factor controlling the error $|p_c - p_T|$. Let us consider the question if we can use less data without increasing the error too much. Notice that our results on data space selection are model independent in this Subsect. 4.2.

Let us assume two cases: (i) the usual situation that *full* data are given on a space-time cylinder $Q = [-L, L] \times [0, T]$: $(x_i, t_i) \in ([-L, L] \times [0, T]) \cap \{(k\Delta x, l\Delta t) \mid (k, l) \in N\}$, and (ii) *reduced* data, i.e., less data points $u(\bar{x}_i, \bar{t}_i)_{i=1}^{\bar{N}_{\text{data}}} \in \mathbb{R}^{\bar{N}_{\text{data}}}$, where $\bar{N}_{\text{data}} < N_{\text{data}}$, and $\{(\bar{x}_i, \bar{t}_i)\} \subset \{(x_i, t_i)\}$, and a corresponding least-squares estimate $\bar{p}_c$ calculated using these (reduced) data. Without loss of generality, let us impose the following ordering of the data $\{(\bar{x}_i, \bar{t}_i)_{i=1}^{\bar{N}_{\text{data}}}\} = \{(x_i, t_i)_{i=1}^{\bar{N}_{\text{data}}}\}$, i.e., that the new data are just the first $\bar{N}_{\text{data}}$ values of the original data.

The corresponding confidence interval estimate (11) still holds with a (usually larger) confidence interval

$$(\bar{p}_c - p_T)^2 \sum_{i=1}^{\bar{N}_{\text{data}}} (\frac{\partial}{\partial p} u(x_i, t_i))^2 \leq \frac{res^2(\bar{p}_c)}{\bar{N}_{\text{data}} - 1} f_{1, \bar{N}_{\text{data}}-1}(\alpha). \tag{19}$$

In [10] we developed a novel approach, which aims to reduce the amount of data without shortening the confidence interval, based on relation

$$(\bar{p}_c - p_T)^2 \sum_{i=1}^{N_{\text{data}}} (\frac{\partial}{\partial p} u(x_i, t_i))^2 \leq (1 + \eta) \left( \frac{res^2(\bar{p}_c, R_c)}{\bar{N}_{\text{data}} - 1} f_{1, N_{\text{data}}-1}(\alpha) \right). \tag{20}$$

We can see almost the same type of estimates as in the full data case (11), but the upper bound is enlarged by the factor $(1 + \eta)$, where $\eta > 0$ is chosen according to

$$\eta \geq \frac{\sum_{i=\bar{N}_{\text{data}}+1}^{N_{\text{data}}} (\frac{\partial}{\partial p} u(x_i, t_i))^2}{\sum_{i=1}^{\bar{N}_{\text{data}}} (\frac{\partial}{\partial p} u(x_i, t_i))^2}. \tag{21}$$

The above relation (21) serves as the criterion for selection of $\bar{N}_{\text{data}}$. If $\eta$ is chosen first as an upper bound, then (21) is relating $\bar{N}_{\text{data}}$ to $\eta$. Choosing $\bar{N}_{\text{data}}$ as small as possible such that (21) holds, we can pick a subset $u(\bar{x}_i, \bar{t}_i)_{i=1}^{\bar{N}_{\text{data}}}$, which contains almost the same information as the full data, out of the original data $u(x_i, t_i)_{i=1}^{N_{\text{data}}}$. If (21) holds with small $\eta$, we can get rid of *irrelevant data* (the complement $\{u(x_i, t_i)_{i=1}^{N_{\text{data}}}\} \setminus \{u(\bar{x}_i, \bar{t}_i)_{i=1}^{\bar{N}_{\text{data}}}\}$) and further process only the reduced amount of *relevant data* $u(\bar{x}_i, \bar{t}_i)_{i=1}^{\bar{N}_{\text{data}}}$.

*Remark 5.* The complement of the relevant data relative to the full data (the *irrelevant data*) is actually defined by our choice of the confidence interval enlargement compared to the full data case, i.e., on $\eta$, see [10] for more details. Hence, in sake of unambiguity, we could define our terms *relevant data* resp. *irrelevant data* more rigorously, e.g., as $\eta$-*relevant data* and $\eta$-*irrelevant data*, respectively.

## 4.3   Selection of Monitored Region for the Fickian Diffusion in Free Space Case

In our previous works [3,10], based on the previous theoretical results presented in Subsect. 4.2 and mainly in (20), we found the relevant data space as lotus-like shaped plot in space-time coordinates. Now, let us study the procedure of data selection from the point of view of practical relevance. I.e., we want to find simple formulas for selecting a *rectangular* spatio-temporal region which yield approximately similar error bounds compared to the case when all the spatial points $x \in \mathbb{R}^n$ are used as data.

Based on our previous findings in Subsect. 4.2, we propose the following procedure for the selection of monitored region. The process of data selection (reduction) involves estimates of the sensitivity and, hence, of $K(y,t)$ as in (15). Suppose we want to find the relevant data on a space-time cylinder $Q = [-L, L] \times [0, T]$ which yields a comparable confidence interval to full data given on the whole spatial domain $[-L', L'] \times [0, T]$ with $L'$ being large and with equal grid spacings $\Delta t, \Delta x$. Assuming $L'$ large, we can approximately view the full data interval as $[-L', L'] \sim \mathbb{R}$ and, in the following, treat this as the case when the full data are given on the whole space interval.

The condition which we impose now is (21) with some chosen small $\eta$. We have the situation that the ideal $N_{\text{data}}$ corresponds to data on $\mathbb{R} \times [0, T]$, while the actual data are given on $[-L, L] \times [0, T]$. Condition (21), assuming that the data points are dense such that the sums can be approximated by integrals, can be stated as

$$\eta \geq \frac{\int_0^T \int_{-\infty}^{\infty} |\frac{\partial}{\partial p} u(x,t)|^2 dx dt - \int_0^T \int_{-L}^{L} |\frac{\partial}{\partial p} u(x,t)|^2 dx dt}{\int_0^T \int_{-L}^{L} |\frac{\partial}{\partial p} u(x,t)|^2 dx dt}. \tag{22}$$

Rearranging terms and with the notation as before, we come to the condition

$$1 + \eta \geq \frac{K(\infty, r_0)}{K(L, r_0)}. \tag{23}$$

This inequality can be solved by numerical means, for instance, if we allow for a 1 % increase of the confidence interval, we take $\eta = 0.01$ and find numerically the following recommendation for the length of data interval $L$:

$$L \geq 3.82 \sqrt{T p_c + \frac{1}{8} r_0^2} \qquad \text{for } \eta \leq 0.01, \tag{24}$$

which leads to the confidence intervals that are almost identical to the case of full spatial data.

## 5   Concluding Remarks

In this paper, we propose the interconnection of two important activities in performing experiments in biology and biomedicine: (i) design of experiment, i.e., optimal or near-optimal setting of experimental design factors, and (ii) data processing based on a complex mathematical model containing the specific experimental conditions as parameters. Despite the fact that some recommendations and findings concerning the FRAP experimental protocol exist, cf. [1,7], their applicability is limited because they are based on specific experimental conditions.

Our approach is more general. We formulate the problem of parameter estimation in precise terms of parameter-to-data map, parameter estimates and their confidence intervals. Then, we introduce the key concept of local and global sensitivity of measured data on estimated parameters. Based on this quantities, we propose a method to get rid of irrelevant data points before they are actually used for the parameter inference, i.e., we propose a novel method how to separate *relevant* and *irrelevant data sets*.

Finally, as a proof-of-concept, we derived two results for the simplified model of FRAP experiments. The first one concerns the optimization of one of the most important experimental design factors in FRAP experiments, the bleach size $r_0$. The second one proposes a formula for the optimal selection of the size of monitored region $L$, taking into account only the relevant data, while the confidence interval for the parameter estimate is controlled by the 'growth factor' $1 + \eta$. Our findings are expected to be incorporated into a FRAP experimental protocol – it is not computationally expensive and the enhancement of the parameter estimation process is substantial, cf. Fig. 2.

Certainly, the more realistic model formulation should be conceived, e.g., taking into account anisotropic diffusion on finite two dimensional domain, binding reaction, bleaching during scanning.

This is the subject of our ongoing research together with an ambitious goal consisting of the computationally effective *on-line* model-based sensitivity analysis. The idea is to suggest the optimal values of experimental design variables *on-line*, i.e., to perform the experimental protocol modification (or tuning) during FRAP measurements. The main drawback is not the mathematical or technical difficulty but the complicated communication between the members of mathematical and biological community.[4]

Last but not least, in order to enhance the quality of the whole process concerning experiments in biology and biomedicine, we propose to pay interest and keep the track of the experimental data and metadata, and the protocol development as well. This is why initiatives like the BioWes system [15] has been developed at the Institute of Complex Systems, FFPW USB.

**Acknowledgement.** This work was supported by the Ministry of Education, Youth and Sport of the Czech Republic - projects 'CENAKVA' (No. CZ.1.05/2.1.00/01.0024) and 'CENAKVA II' (No. LO1205 under the NPU I program), and by the OeAD (Austrian agency for international mobility and cooperation in education, science and research) within the programme "Aktion Oesterreich-Tschechien (AOeCZ-Universitaetslehrerstipendien)". I (SP) would like to thank to my colleagues Jiří

---

[4] Despite the aforementioned problems, some successful stories already exist [14].

Jablonský (for valuable discussions) and Radek Kaňa (for valuable discussions and real FRAP data and pictures), and all members of the Industrial Mathematics Institute, Johannes Kepler University of Linz, for their kind support during my "November-Aktion-stay".

# References

1. Mueller, F., Mazza, D., Stasevich, T.J., McNally, J.G.: FRAP and kinetic modeling in the analysis of nuclear protein dynamics: what do we really know? Curr. Opin. Cell Biol. **22**, 1–9 (2010)
2. Kaňa, R., Kotabová, E., Lukeš, M., Papáček, Š., Matonoha, C., Liu, L.N., Prášil, O., Mullineaux, C.W.: Phycobilisome mobility and its role in the regulation of light harvesting in red algae. Plant Physiol. **165**(4), 1618–1631 (2014)
3. Papáček, Š., Jablonský, J., Matonoha, C., Kaňa, R., Kindermann, S.: FRAP & FLIP: Two sides of the same coin? In: Ortuño, F., Rojas, I. (eds.) IWBBIO 2015. LNCS, vol. 9044, pp. 444–455. Springer, Heidelberg (2015)
4. Axelrod, D., Koppel, D.E., Schlessinger, J., Elson, E., Webb, W.W.: Mobility measurement by analysis of fluorescence photobleaching recovery kinetics. Biophys. J. **16**, 1055–1069 (1976)
5. Mullineaux, C.W., Tobin, M.J., Jones, G.R.: Mobility of photosynthetic complexes in thylakoid membranes. Nature **390**, 421–424 (1997)
6. Ellenberg, J., Siggia, E.D., Moreira, J.E., Smith, C.L., Presley, J.F., Worman, H.J., Lippincott-Schwartz, J.: Nuclear membrane dynamics and reassembly in living cells: targeting of an inner nuclear membrane protein in interphase and mitosis. J. Cell Biol. **138**, 1193–1206 (1997)
7. Travascio, F., Zhao, W., Gu, W.Y.: Characterization of anisotropic diffusion tensor of solute in tissue by video-FRAP imaging technique. Ann. Biomed. Eng. **37**(4), 813–823 (2009)
8. Sbalzarini, I.F.: Analysis, modeling and simulation of diffusion processes in cell biology. VDM Verlag Dr. Muller (2009)
9. Papáček, Š., Kaňa, R., Matonoha, C.: Estimation of diffusivity of phycobilisomes on thylakoid membrane based on spatio-temporal FRAP images. Math. Comput. Model. **57**, 1907–1912 (2013)
10. Kindermann, S., Papáček, Š.: On data space selection and data processing for parameter identification in a reaction-diffusion model based on FRAP experiments. Abstr. Appl. Anal., Article ID 859849, 17 (2015)
11. Blumenthal, D., Goldstien, L., Edidin, M., Gheber, L.A.: Universal approach to FRAP analysis of arbitrary bleaching patterns. Scientific reports 5 (2015). doi:10.1038/srep11655
12. Engl, H., Hanke, M., Neubauer, A.: Regularization of Ill-posed Problems. Kluwer, Dortrecht (1996)
13. Bates, D.M., Watts, D.G.: Nonlinear Regression Analysis: Its Applications. Wiley, New York (1988)
14. Mai, J., Trump, S., Ali, R., Schiltz, R.L., Hager, G., Hanke, T., Lehmann, I., Attinger, S.: Are assumptions about the model type necessary in reaction-diffusion modeling? A FRAP application. Biophys. J. **100**(5), 1178–1188 (2011)
15. http://www.biowes.org/

# A Semantic-Based Metadata Validation for an Automated High-Throughput Screening Workflow: Case Study in CytomicsDB

Enrique Larios Vargas, Zhihan Xia, Joris Slob, and Fons J. Verbeek$^{(\boxtimes)}$

Section Imaging and Bioinformatics, LIACS, Leiden University,
Leiden, The Netherlands
{e.larios.vargas,z.xia,j.slob,f.j.verbeek}@liacs.leidenuniv.nl

**Abstract.** High-Throughput Screening (HTS) techniques are typically used to identify potential drug candidates. These type of experiments require invest in large amount of resources. The appropriate data management of HTS experiments has become a key challenge in order to succeed in the target validation. Current developments in imaging systems has to cope with computational requirements due to the significant increment of volumes of data. However, no special care has been taken to ensure the consistency, integrity and reliability of the data managed in HTS experiments. The appropriate validation of the data used in an HTS experiment has turned to be a key success factor in the target validation, thus a mandatory process to be included in the HTS workflow.

This paper describes our research in the validation process as performed in CytomicsDB. This system is a modern RDBMS-based platform, designed to provide an architecture capable of dealing with the strict validation requirements during each stage of the HTS workflow. Furthermore, CytomicsDB has a flexible architecture which support easy access to external repositories in order to validate experiments data.

## 1 Introduction

High-troughput screening (HTS) assays provide a way for researchers to simultaneously study interactions between large numbers of potential drug candidates with a target of interest. Significant volumes of data are generated as a consequence of conducting HTS experiments, making it cumbersome to store, interact, and throughly mine the data for biological significance [11].

Due to the large amount of resources invested and the complexity of the processes performed during an HTS experiment, it is convenient and necessary to build an automated workflow system which will be in charge of the management, supervision and validation of the data in every stage of the workflow. Our automated HTS workflow described in [9] is divided into 5 stages: (1) Design, (2) Image Acquisition, (3) Analysis, (4) Visualization and (5) Storage. This last stage is performed in parallel with the first 3 stages. It is extremely relevant to give special care to the information stored in the platform in order to have reliable output available to display in the Visualization stage.

© Springer International Publishing Switzerland 2016
F. Ortuño and I. Rojas (Eds.): IWBBIO 2016, LNBI 9656, pp. 557–572, 2016.
DOI: 10.1007/978-3-319-31744-1_50

Appropiate management of the HTS data is one of the key challenges in drug discovery. Our work is focused on the research of how to use validation strategies to ensure data integrity for data analysis in HTS experiments. In order to ensure consistency, integrity and reliability of the data stored in the platform it is compulsory to perform a strict validation process in every stage of the HTS workflow. CytomicsDB facilitates this validation process using web services which will prove each critical entry with an internal or external repository. The metadata that we store become a key parameter for performing further image/data analysis and drill down the results of different experiments datasets.

To sum up, the contributions of this work comprehend:

1. The application of validation strategies with a case study in High-Throughput. CytomicsDB (Sect. 2).
2. The validation workflow developed by CytomicsDB (Sect. 2).
3. The analysis of the strategies performed in CytomicsDB for solving inconsistency conflicts (Sect. 2).

In Sect. 3 we present the results of the implementation of the validation strategies. Additionally, in Sect. 4 we present our conclusions and we also provide an interpretation of our results.

## 2    Materials and Methods

We introduce CytomicsDB approach for metadata validation, the architecture style implemented, the validation strategies developed and finally, the validation workflow is described in two scenarios such as compounds and siRNAs.

### 2.1    CytomicsDB

An initial design of a platform for managing HTS experiments data is presented in [9]. Our platform is based on an automated HTS workflow which integrates all stages from experiment design to data analysis. The platform relies on a modern relational database system called *MonetDB* (www.monetdb.org). Traditional databases systems generally carry too much overhead when processing analytical queries [3]. What we need is a database optimised for data mining applications. MonetDB is a leading open source database system that has been designed specially for such applications [3]. It has been well-known for its performance in processing analytical queries on large scale data [4].

The data model in [8] has been rigorously revised and generalised to manage all kinds of metadata produced by automated HTS systems. Our system is called *CytomicsDB*, a cytomics-based platform for managing, visualizing and querying HTS data.

## 2.2   Platform Architecture for Validation

The architecture is designed to accurately follow the HTS experiment workflow described in [8], considering four key activities: Acquisition, Visualization, Integration and Exploration. The acquisition consists of two steps: image acquisition and metadata formulation. Visualization is a key factor in this architecture due to the large amount of images produced in HTS. All these images should be linked to the plates they originate from. The integration is concerned with the tasks of image and data analysis which are performed by external applications. However, the output related to certain metadata is also included in the experiment. Therefore it is necessary to integrate external APIs to the architecture through web services. Finally, the exploration is associated to querying the data and the results. This task is also a key step in order to verify if the experiment requires a new iteration, possibly making some adjustments to the plate design.

Following the workflow of an HTS experiment described in [8], CytomicsDB uses a four-layered architecture (Fig. 1). The performance, stability, speed and scalability are main concerns for the architecture due to the overhead of connecting to external web services and loading a BLAST+ local sequence database into the RAM which are relatively heavy tasks during the validation process. Besides that, since the workflow relies on external applications (e.g. BLAST+ gets different I/O schema in Windows and Linux), it is needed to concern about the compatibility across different operation systems.

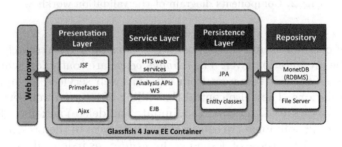

**Fig. 1.** CytomicsDB architecture

The validation process consists of four main activities: retrieving candidates, screening candidates, reporting inconsistences, and updating master tables according to user's decision. The four activities are distributed in several components, which interact with each other. The component diagram in Fig. 2 shows their interaction. These components operate within CytomicsDB using a four-layered architecture [9]. The validation is mainly performed in the *Service layer* and *Persistence layer* which are described hereafter.

**The Service Layer.** The service layer is described in Fig. 2(b), it includes *manage beans* and several *utilities* such as parsers and comparators. They work as the pivot in the validation process to control the generation of candidates, the

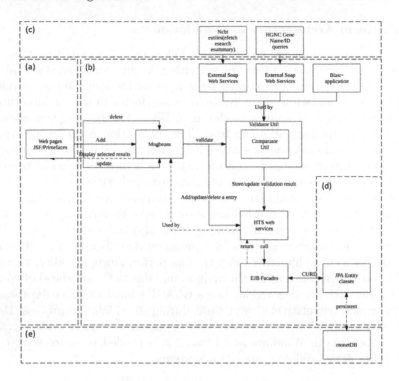

**Fig. 2.** Components diagram of the validation workflow

screening of candidates and the responding actions after the researcher makes a choice on the presentation layer. The *manage beans* are controllers which request to the *utilities* to visit resources from external web services (or applications) and do calculations. They also control the calling of internal web services in the service layer to do CRUD (create, read, update and delete) actions in the master tables. The results obtained are collected and sent back to the presentation layer (c.f. Fig. 2(a)). The purpose of designing the *utilities* as independent components from the *manage beans* is to make the parallel execution of the *utility* instances easier.

The service layer also consists of multiple web services that support every step in the HTS workflow. These web services invoke different APIs which are in charge of the Experiment design, Image Analysis and Data Analysis [9]. This structure allows easy scalability adding more functional modules. For instance, parsers in the *utility module* use web services to access external data sources. The Simple Object Access Protocol (SOAP) messages are selected for invoking the web services and receiving results because of its approved interoperability in web applications and heterogeneous environments. For these web services, one big portion of work is keeping the persistence in the database by using modules from the persistence layer (c.f. Fig. 2(d)).

**The Persistence Layer.** The persistence layer (see Fig. 2(d)) is based on the principle of object-relational mapping (ORM) [5] which involves delegating access to a relational database and, which in turn gives an object-oriented view of the relational data, and vice versa. The Java Persistence API (JPA) framework has been implemented in this layer to keep a bidirectional correspondence between the database and objects. Those Java objects used in this framework are known as Java Entities [7]. The entities are objects that shortly live in memory but persistent in the database. Besides that, they have all the features of a Java class like instantiation, abstraction, inheritance, relationships and so on. The entities used in CytomicsDB follow the same structure as the tables they map to. CRUD operations are registered as named query methods which are written in Java Persistence Query Language (JPQL). These customized queries can be attached to entities as native queries via JPA.

## 2.3 Validation Strategies

The validation process can be abstracted as a strategy. In this strategy, each object which objectively exists in the real world, e.g., a *Compound*, or a *siRNA* is defined as an entity E. For each entity, several attributes are assigned to it, like the name, the ID number and the publisher. The attributes that are used to describe one entity can be defined as a set: $A=\{a_1,a_2,...,a_n\}$, in which $a_i$ *(1 ≤ i ≤n)* means the $i^{th}$ attribute of the entity. In this strategy, multiple data sources are involved as well. They can be categorized as two types. One set is from the lab in which researchers use CytomicsDB to manage their experiment data. In CytomicsDB, this set is uploaded by researchers and stored as *master tables* in the database. Another group of sources are from external databases. They are used to validate the metadata uploaded by researchers. All these data sources can be expressed as a collection $S=\{s_1,s_2,...,s_m\}$ in which $s_i$ $(1 \le i \le m)$ represents the $i^{th}$ data source among the $m$ data sources. Adopted from [15], the data source $s_i$ offers a fact value f for the attribute $a_j$ of an entity E. Different data sources may have different fact values for a same attribute of the entity. For the entity $E$, if $a_j \in [a_1, a_n]$, $f_{(s_i,a_j)} \neq f_{(s_l,a_j)}, i \neq l$, then a conflict or inconsistency is found between data source $s_i$ and $s_l$. In all fact values from all data sources, those who correspond to the attribute value in the real world are referred to as *true value*. So the validation process is in fact a process for identifying *true values* among all conflicts between data sources. The relationships among the entity, its attributes and fact values from different data sources in CytomicsDB are sketched in Fig. 3. In CytomicsDB, the idea is to validate the metadata in $S_2$ by data from data sources $S_1$ who have the same attributes as $S_2$. Conflicts found among data sources during this process indicate potential inconsistency. There are several available strategies to perform the conflict resolution which are described in the following section.

It is common to have an implicit assumption that the content of all the information sources should be mutually consistent [13] in the approaches for solving inconsistency issues between different data sources. A data inconsistency exists

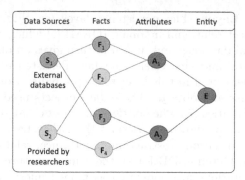

**Fig. 3.** Relationship between the entity, its attributes and fact values from data sources

when two entries (or tuples in the relational data model) coming from different information sources are identified as versions of each other i.e. they represent the same real-world object and some of the values of their corresponding attributes differ [14]. The available conflict resolving strategies are then identified by the ways they elaborate this basic assumption e.g. *Trust your friends* strategy assumes that the tuples from trustful data sources are most likely to be the real-world object, then the data under validation should be consistent with them. Some standard inconsistency resolutions [2] are summarized hereafter.

**No Gossiping Strategy.** The basic idea here is that if multiple objects are retrieved from the excution of queries from different data sources and it is unsure about which object or which value for an attribute in the object matches to the real-world one, then just leave them out and only report on sure facts, or directly report all of them. This is the strategy used by the consistent query answering approaches [6]. This strategy leaves the decision force to users if all query results are reported, which makes it simple to be implemented.

**Trust your Friends.** In the *Trust your Friends* strategy [1] it is required to trust a third party who will provide the correct entries. An assumption is considered that a fact values from reliable external databases can be treated as true values. Especially when those fact values are identical to the ones given by researchers, the probability that those fact values are true values becomes quite high which can be assumed as 100 %. The key point of *Trust your Friends* is having reliable data sources.

**Cry with the Wolves.** The *Cry with the wolves* strategy pursues a different approach, the entries which correctly describe the real-world object prevail over the incorrect ones, given enough evidence. It reflects the principle of following the decision of the majority, of choosing the most common entries among candidates from all data sources and compare to the one under validation [1]. The more data sources involved in the validation process, the better the strategy works.

**Meet in the Middle.** In contrast with the previous strategies, the *Meet in the middle* strategy follows the principle of compromise and does not prefer one value over the other but instead tries to invent a value that is as close as possible to all present values from all data sources and compare it to the one under validation [2].

**Keep Up to Date.** This strategy uses the most recent entries from external data sources to compare to the one under validation. Some additional time-stamp information about the recentness is required in order to do the comparison [1].

### 2.4 Validation Workflow

Theoretically, data sources can not be 100 % accurate in describing all entities in the real world. Since the researches should be experts in the metadata that they upload, the researcher's decisions are involved as a part of the conflict resolving strategy. The validation result from *"Trust Your Friends"* strategy is "passed on" (*Pass it on* [1]) to other researchers to let them decide how to handle possible conflicts. For an entity, the validation result includes the status of its correctness and some possible solutions when some conflicts are detected in some attributes. For the entity, the *"Highest Quality"* [2] entries (in all attributes) obtained from external databases are given as recommended possible solutions to conflicts. The researchers have the final word on the conflict resolutions. For entities with only one unidentifiable attribute, additional fact values from external data sources should be used in the validation strategy. For entities with several attributes, some multi-objective decision algorithms should be implemented during the selection of the *"Highest Quality"* options. The validation workflow follows *"Trust Your Friends"* and *"Pass It On"* strategies while using *"Levenshtein distance"* and *"Multi-Objective Decision"* algorithms. There are two branches separately focusing on single-attribute and multi-attributes situations in the validation workflow. The two branches follow a common principle of the validation workflow. The principle is first parsing each fact value, i.e. the attribute value of metadata from researchers, into a standard unique identifier value by querying it as a keyword in an external data source, then getting the entries from an external database, considered as reliable, which uses the unique identifier as a primary key. The validation of *Compounds* and the validation of *siRNAs* can be viewed as two scenarios corresponding to the two branches of the workflow, respectively.

**Compound Validation.** The whole workflow can be divided into four stages: (1) *Get candidates*, (2) *Screening and marking duplex* (c.f. Fig. 4), (3) *Updating duplex marks* and (4) *Cleaning up the validation result table* (c.f. Fig. 5). When the user wants to insert or update a compound into the master table, the process starts to check whether the compound name exists in the master table. If so, the name will not be stored. Otherwise, the process will call the components for validation in this order, i.e. *get candidates → screening → update duplex marks*. If the user wants to ignore the validation results and keep the compound in the master table as it is, the user can select "ignore" which will call the *"clean up"* component. If the user wants to delete the compound from the master table,

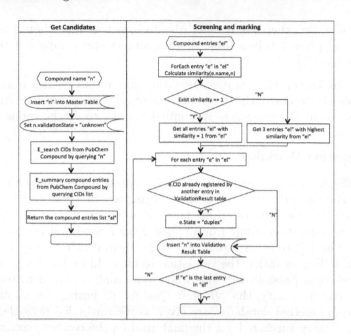

**Fig. 4.** Compound validation: get candidates and screening - marking

then after calling the "clean up" component, the deleting action in the master table will follow.

**siRNAs Validation.** This is the case for validating multi-attributes entities. 5 types of siRNA attributes can be validated with external data sources. They are: the *Gene ID, the Gene Symbol*, the *Accession Number*, the *GI number* and the *Sequence*. Since the *Duplex Number* is only used on the master table as a unique identifier which cannot be validated with external data sources, only a basic duplex validation of siRNAs is adopted, i.e., assuming that the Duplex Number of the siRNA provided by the user is reliable, then only checking if the duplex number exists in the master table is enough. Figure 6 describe the workflow for an siRNA validation. The last step of the workflow uses the multi-objective decision method to show the best candidates to the user for further decision. If a perfect matched candidate is found then the user will get a positive feedback. If no perfect matched candidates are found, then an error message along with the top 3 best matched candidates decided by the multi-objective decision method will be sent to the user. If the perfect matched candidate is targeted, it is still needed to check if a new version of the gene exists in an external database or if the candidate is using a synonym name. If so, a warning message will be sent to the user. The user can choose to ignore solutions from the validation process or accept one as a correction. Figure 7 shows details of the comparison applied to the siRNA attributes.

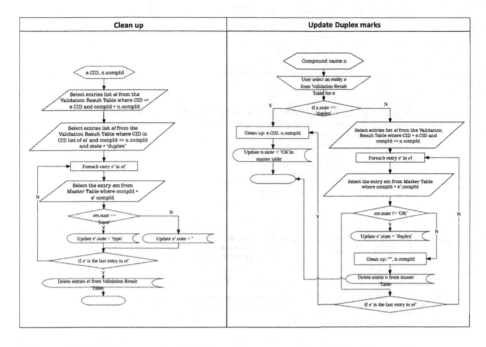

**Fig. 5.** Compound validation: clean up and update duplex marks

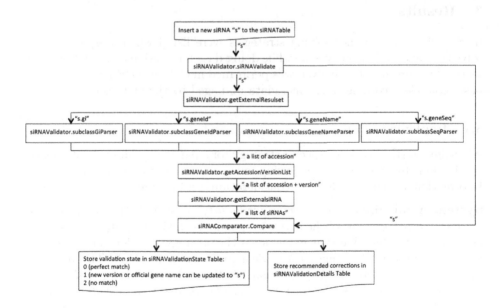

**Fig. 6.** siRNA validation workflow

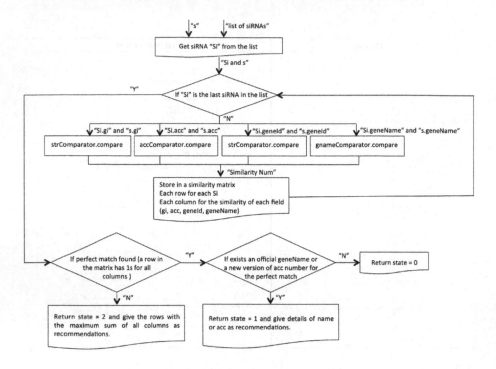

**Fig. 7.** siRNA comparator workflow

# 3   Results

In Sect. 2 several conflict solving strategies were listed which are available to solve the data inconsistency issue while doing data integration or validation. This section explains how the analysis was performed in CytomicsDB for selecting the best strategie according to the metadata managed by the platform.

## 3.1   Implementation

According to [1], choosing a specific strategy for a particular data validation issue can be done by analyzing the following four aspects: (1) System availability, (2) Information availability, (3) Cost considerations and (4) Quality considerations.

**System Availability.** The *"system availability"* constraints mainly come from the data properties. One of the major properties is the input/output value types accepted by the function that implements the strategy. The four most common types are: numerical, strings, categorical and taxonomical [2]. The functions available are: (1) *Vote*, (2) *Average/Sum/Median*, (3) *First/Last*, (4) *Most Complete*, (5) *Most General*, (6) *Most Similar*, (7) *Choose Corresponding*, and (8) *Most Recent*.

The functions that can be used to implement each strategy are listed in Table 1.

**Table 1.** Functions used to implement each strategy

| Strategies | Functions |
|---|---|
| Trust your friends | First/Last, Most similar, Most complete, Choose corresponding |
| Cry with the wolves | Vote |
| Meet in the middle | Average, Median, Most general, Vote |
| Keep up to date | Most recent, first |
| No gossiping | Not applicable |

Each function has some preference on value types and data entry types. These preferences are listed in Table 2.

**Table 2.** Conflict handling functions and their applicable properties

| Functions | Supported input/output types | Supported data entry types |
|---|---|---|
| Vote | All | Single-Column/Multi-Column |
| Average/Sum/Median | Numerical | Single-Column |
| First/Last | All | Single-Column |
| Most complete | All | Single-Column |
| Most general | Taxonomical | Single-Column |
| Most similar | String | Single-Column/Multi-Column |
| Choose corresponding | All | Multi-Column |
| Most recent | All | Single-Column/Multi-Column |

All attributes for siRNA and compound entries are strings, and the functions in each strategy should support both single and multiple-columns situations, thus some functions are not applicable for the test cases any more. Table 3 shows the list of available functions and strategies that are supported under the system constraints.

**Information Availability.** There are three major nucleotide databases: GenBank (National Centre for Biotechnology Information, or also called "*NCBI*",

**Table 3.** Strategies and functions available for siRNAs and compounds

| Strategies | Functions | System availability |
|---|---|---|
| Trust your friends | Most similar | Implementable |
| Cry with the wolves | Vote | Implementable |
| Meet in the middle | Vote | Implementable |
| Keep up to date | Most recent | Implementable |
| No gossiping | Not applicable | Non Implementable |

*EMBL* (European Molecular Biology Laboratory) and *DDBJ* (The DNA Data-bank of Japan). All of them can be queried using SOAP based web services which their integration to the CytomicsDB platform makes more easy. Although these three data sources are available to all strategies, the strategies *Cry with wolves*, *Meet in the middle* and *Keep up to date* may need more data sources in order to be implemented according to the principle of "more data sources the better result".

In case of the siRNAs, the sequences can be queried locally in the web server using the BLAST+ application with its embedded database which is synchronized to the NCBI repository.

The smaller databases may have comparatively less maintenance and data are less reviewed than the other three major data sources. Since there are a lot of them, if they are included as valid data sources for *"Cry with wolves"* or *"Meet in the middle"* strategies under the principle of more data sources are better, then the list of candidates from all data sources will likely includes entries which do not fully correct describe the real-world facts.

**Cost Considerations.** CytomicsDB architecture supports multi-threading with minimized I/O (input/output) operations to all web services. Thus, this makes CytomicsDB comparatively light weighted. The cost considerations are calculated in terms of the overhead needed in order to decide the "correct" result for each strategy.

*Costs for Single-Column Data Attributes.* The *Vote* function or the *Most recent* function does not require some particular complex algorithms and have $O_{(n)}$ as time and space complexity respectively, where $n$ is the number of candidates. The *Most similar* function can be implemented with other similar algorithms. The available algorithms for implementing the *Most similar* function are:

1. *Longest Common Subsequences (LCS) algorithm:* The time complexity is $O_{(m_1 m_2 n)}$ where $m_1$, $m_2$ represent the length of the string and $n$ is the number of candidates. An optimized implementation is built in The *BLAST* application for sequence similarity calculation. The space complexity is expected to be $O_{(m_1 m_2 n)}$ as well.
2. *Levenshtein Distance (LD) algorithm:* Levenshtein [10] proposed the edit distance algorithm which calculates the minimum numbers of operations (insertions, deletions and substitutions) for editing one string. The algorithm is sensitive to local changes in a single word. The time complexity for this algorithm is $O_{(m_1 m_2 n)}$ where $m1$, $m2$ are the length of strings and $n$ is the number of candidates. The space complexity is also $O_{(m_1 m_2 n)}$.
3. *RKR-GST:* The major drawback of this algorithm is the time complexity which is $O_{(m^3 n)}$ (assuming the length of the two strings are almost the same value $m$) in the worst scenario where $n$ is the number of candidates. RKR-GST uses a hash function to calculate the hash number for each division in the two strings. This optimized the complexity of GST to $O_{(m^2 n)}$. The space complexity of this algorithm is $O_{(n m_{max})}$ where $m_{max}$ is the length of the longest string in the two strings.

*Costs for Multi-column Data Attributes.* When the *Most similar* function runs in multi-column mode, it can be attached with different multi-objective decision algorithms as well. A multi-objective decision algorithm is used to model the decision-maker preferences. Algorithms are categorized depending on how the decision-maker articulates these preferences. Considering the overhead and easy-implementing factors, according to [12], in the class of algorithms with no articulation of preferences, the *Objective Sum Method* is one of the most computationally efficient, easy-to-use, and common approaches whose time complexity and space complexity are both $O_{(n)}$. Therefore, this work uses this approach during the accuracy test and implementation.

The calculation costs for different strategies in single and multi-column mode are listed in Table 4.

**Quality Considerations.** The quality of each strategy is based on the accuracy of the functions used to implement the strategy. The quality is measured from two perspectives. First, whether the strategy is able to identify inconsistency in the data under validation. Second, whether the strategy-selected candidate really reflects the real world object. To do these evaluations, 300 Compounds and 300 siRNAs are used as test cases. 150 items in each category are randomly modified, forcing them to be invalid by adding, modifying or deleting one character or digit on each attribute value. To make these manual errors uniformly distributed, for the 150 modified compound name, every 50 names are modified by deletions, insertions or substitutions respectively. Similarly for the 150 modified siRNAs, they are divided into three groups (50 siRNAs in each group) and each group is modified by using one of the three different operations (deletion, insertion or substitution). In each group, the operation is performed on a different attribute (Gene Symbol, Gene Id, Accession Number, GI Number or Sequence) for every 10 entries. The *F-measure* which is a common measure for test accuracy is adopted as an indicator to identify the accuracy in finding inconsistency for each strategy. The F-measure considers both the precision $p$ and the recall $r$ of the test to compute the score.

$$p = \frac{tp}{(tp + fp)} \qquad r = \frac{tp}{(tp + fn)} \qquad F = \frac{(2)(p)(r)}{(p + r)}$$

The unmodified 150 consistent compound names or siRNAs are considered as the positive class, and tp, fp and fn denote the number of true positives, false positives, and false negatives, respectively. For the second criteria, a percentage number named *"True-hit"* is considered as the indicator. In this case, 600 (300 Compound names for the single-column case and 300 siRNA for the multi-column case respectively) proved, unchanged test cases are considered as the representation of real-world objects. Then the measure evaluates if each strategy yielded a "correct" candidate based on the 600 real objects (if not, it means that the "correct" candidate is not consistent with the real-world object). The percentage number shows the percent of matched ones in the 600 real-world objects. In the test, the NCBI database is used as the trustful data source for *Trust your friends* strategy. The NCBI database and EMBL database are used as the data sources for *Cry with the wolves*, *Meet in the middle* and *Keep up to date* strategies. As the *No gossiping* strategy does not work with the candidates retrieved from external data sources, it is not applicable for measuring accuracy in this case.

Table 4 shows the results of measurement on functions and algorithms in both single-column and multi-column mode.

## 3.2   Strategy Selection

The analysis for each strategy is listed in Table 4. In CytomicsDB, special care has been taken for considering the highest quality result with as little cost as possible. From the perspective of the quality, the larger value for *F-measure* and *True-hit* indicates the best quality of the result. According to Table 4:

These results indicate that *Trust your friends* strategy with *Most similar* function gets the highest values among all the strategies. Among all the algorithms in single-column mode, the *LD* algorithm gets 0.983 which is slightly higher than *LCS* algorithm (0.98) and *RKR-GST* algorithm (0.925) in *F-measure*, and it gets 0.573 which is higher than *LCS* algorithm (0.567) and RKR-GST algorithm (0.527) in True-hit measurement.

Among all the algorithms in multi-column mode, the *LD* algorithm scores 0.974 which equals to the score for *LCS* algorithm and is higher than *RKR-GST* algorithm (0.949) in *F-measure*, and it gets 0.933 in True-hit measurement which equals to the True-hit for *LCS* algorithm and better than 0.89 for *RKR-GST* algorithm.

These statistics show that *LD* algorithm has better quality results than the other two algorithms (*LCS* and *RKR-GST*) which are available for implementing the *MOST SIMILAR* function.

The drawback of *Trust your friends* strategy is the cost concern. For the space overhead, the memory use when the implementations are running for all strategies does not have any significant difference in a machine with 8 Gb of RAM. For the time complexity, *Trust your friends* strategy has the highest time among all strategies $(O_{(m^2n)} + O_{(n)} > O_{(n)} > O_{(1)})$. However, the complexity cost could be fetched up a little bit by using the architecture of CytomicsDB. A rough performance test shows, running with un-optimized *RKR-GST* algorithm (which time complexity is $O_{(m^3n)}$) in a multi-threads pathway with minimized I/O (input/output) to all web services, it will take 200ms (not including the time for fetching candidates from external data sources) to validate one single-attribute entry or 600ms (not including the time for fetching candidates from external data sources) to validate a multi-attributes entry in a Linux system with a stable Internet connection. In the same environment, the time for fetching all the candidates from all the external data sources takes less than 1 second in a single-column mode and around 4 seconds in a multi-column mode for *Cry with the wolves* strategy. So the calculation overhead is a comparatively small portion in the whole overhead for the validation process. Therefore, the time overhead for *Trust your friends* strategy is still in an acceptable range compared to its accuracy. As the validation process is performed during the design stage, once the metadata is uploaded and validated, changes on metadata will be very limited

**Table 4.** Strategies and criterias

| Strategies | Functions | System availability | Information availability | Cost considerations | | Quality considerations | | | |
|---|---|---|---|---|---|---|---|---|---|
| | | | | Time complexity | Space complexity | Single column | | Multi column | |
| | | | | | | F-measure | True-hit | F-measure | True-hit |
| Trust your friends | Most similar | Yes | Available | $O_{(m^2 n)} + O_{(n)}$ | $O_{(mn)} + O_{(n)}$ | LCS: 0.98 | LCS: 0.567 | LCS: 0.974 | LCS: 0.93 |
| | | | | | | LD: 0.983 | LD: 0.573 | LD: 0.974 | LD: 0.93 |
| | | | | | | RKR-GST: 0.925 | RKR-GST: 0.527 | RKR-GST: 0.949 | RKR-GST: 0.89 |
| Cry with the wolves | Vote | Yes | Less available | $O_{(n)}$ | $O_{(n)}$ | 0.918 | 0.513 | 0.894 | 0.823 |
| Meet in the middle | Vote | Yes | Less available | $O_{(n)}$ | $O_{(n)}$ | 0.918 | 0.513 | 0.136 | 0.053 |
| Keep up to date | Most recent | Yes | Less available | $O_{(n)}$ | $O_{(n)}$ | 0.902 | 0.507 | 0.355 | 0.193 |
| No gossiping | — | No | Available | $O_{(1)}$ | $O_{(1)}$ | — | — | — | — |

(i.e. "once created, use forever"). So, the overhead of one round validation can be a less important issue than its accuracy.

Another major concern is the information availability criteria. As shown in Table 4, only *Trust your friends* strategy (except *No gossiping* strategy) has full available data sources because theoretically it requires only one data source, it makes *Trust your friends* strategy more implementable than other strategies. The results for system availability are the same for all the strategies except *No gossiping* strategy. Out of above considerations, the strategy "*Trust your friends*" implemented with the *Most similar* function (using Levenshtein distance algorithm) is chosen for the validation process.

## 4    Conclusions and Discussion

In this paper, we have presented a semantic-based metadata validation approach for an automated High-Throughput Screening workflow. Our main goal is to ensure the integrity, consistency and reliability of the data stored in the platform. This is a critical requirement for image and data analysis and further data exploration of the experiment's results. CytomicsDB architecture has been designed to facilitate the integration to external repositories. The use of web services enhance the flexibility to access to external databases and validate the key metadata with this public repositories. Furthermore, aligning the metadata in CytomicsDB to public databases allow the platform to become an ontology based framework capable to handle semantic queries and turning the architecture to a web based interactive semantic platform for cytomics.

We plan to optimize the recomendation results by tuning the multi-objective decision formula including weights associated to the user previous decisions. Currently we are giving the same weight to all attributes during the metadata validation, but including user decision will allow the users to have more accurate candidates to select.

**Acknowledgements.** This work was partially supported by the Erasmus BAPE program (EL) and Cyttron II project (JS).

# References

1. Bleiholder, J., Naumann, F.: Conflict handling strategies in an integrated information system (2006)
2. Bleiholder, J., Naumann, F.: Data fusion. ACM Comput. Surv. **41**(1), 1:1–1:41 (2009)
3. Boncz, P.A.: Monet: a next-generation DBMS kernel for query-intensive applications. Ph.d. thesis, Universiteit van Amsterdam, Amsterdam, The Netherlands, May 2002
4. Boncz, P.A., Manegold, S., Kersten, M.L.: Database architecture evolution: mammals flourished long before dinosaurs became extinct. PVLDB **2**(2), 1648–1653 (2009)
5. Fussell, M.L.: Foundations of object-relational mapping (1997)
6. Fuxman, A., Fazli, E., Miller, R.J., Conquer: efficient management of inconsistent databases. In: Proceedings of the ACM SIGMOD International Conference on Management of Data, SIGMOD 2005, pp. 155–166. ACM, New York, NY, USA (2005)
7. Keith, M., Schincariol, M.: Pro EJB 3: Java Persistence API (Pro). Apress, Berkely (2006)
8. Larios, E., Zhang, Y., Cao, L., Verbeek, F.J.: CytomicsDB: a metadata-based storage and retrieval approach for high-throughput screening experiments. In: Comin, M., Käll, L., Marchiori, E., Ngom, A., Rajapakse, J. (eds.) PRIB 2014. LNCS, vol. 8626, pp. 72–84. Springer, Heidelberg (2014)
9. Larios, E., Zhang, Y., Yan, K., Di, Z., LeDévédec, S., Groffen, F., Verbeek, F.J.: Automation in cytomics: a modern RDBMS based platform for image analysis and management in high-throughput screening experiments. In: He, J., Liu, X., Krupinski, E.A., Xu, G. (eds.) HIS 2012. LNCS, vol. 7231, pp. 76–87. Springer, Heidelberg (2012)
10. Levenshtein, V.I.: Binary codes capable of correcting deletions, insertions, and reversals. Sov. Phys. Dokl. **10**(8), 707–710 (1966)
11. Malik, P., Chan, T., Vandergriff, J., Weisman, J., DeRisi, J., Singh, R.: Information management, interaction in high-throughput screening for drug discovery. In: Ma, Z., Chen, J. (eds.) Database Modeling in Biology: Practices and Challenges. Springer, Heidelberg (2006)
12. Marler, R.T., Arora, J.S.: Survey of multi-objective optimization methods for engineering. Struct. Multi. Optim. **26**(6), 369–395 (2004)
13. Motro, A.: Multiplex: a formal model for multidatabases and its implementation. In: Pinter, R., Tsur, S. (eds.) NGITS 1999. LNCS, vol. 1649, pp. 138–158. Springer, Heidelberg (1999)
14. Motro, A.: Data integration, inconsistency detection and resolution based on source properties. In: Workshop on Foundations of Models for Information Integration (FMII), pp. 429–444 (2001)
15. Yong-Xin, Z., Qing-Zhong, L., Zhao-Hui, P.: A novel method for data conflict resolution using multiple rules. Comput. Sci. Inf. Syst. **10**(1), 215–235 (2013)

# Reachability of the Therapeutic Target in the Systems with Parameters Switch

Magdalena Ochab$^{(\boxtimes)}$, Krzysztof Puszynski, and Andrzej Swierniak

Institute of Automatic Control, Silesian University of Technology,
Akademicka 2A, 44-100 Gliwice, Poland
{Magdalena.Ochab,Krzysztof.Puszynski,Andrzej.Swierniak}@polsl.pl

**Abstract.** Human organism is a complex system whose functioning is still under investigation. The biological models of intercellular interactions are created for better understanding of the complex system behaviour and prediction of the system response to the given stimuli. Medical system such as drug application to organism can be described as a piece-wise non-linear model. In our work we consider the influence of the parameter deviations of the systems with fixed terminal state. There are investigated two types of deviations: small changes of system parameters and changes in particular parameter switching time. We considered three different types of the systems, without self-regulation, with negative feedback loop and with positive feedback loop. We considered differences between these types and influence of the small changes in parameters values to the reachability of therapeutic goal after the drug application.

**Keywords:** Biological model · Switches · Piece-wise non-linear models

## 1 Introduction

Switching control systems have become an object of interest not only in control and system engineering (see [1]) but also in systems biology. Switchings in biological processes are very common phenomenon. The most typical example is the alleles activation or inactivation for a specified gene, that results in significant change in proteins level and finally modification in cell behaviour. Mathematical modelling of biological processes requires some simplifications and assumptions due to enormous quantity of dependencies between variety of organic and non-organic molecules. The constant rate of the reactions in cells similarly to any chemical reactions depends on physical and chemical conditions such as temperature or pH. However the parameters must be estimated or calculated based on the biological experiments. The inaccuracy in establishing values of the parameters can have crucial influence on the computational results. The main goal of this study is to examine effect of the parameters deviation in switching, biological control systems. We consider systems with or without feedback loops with fixed terminal state describing the desired therapeutic effect.

© Springer International Publishing Switzerland 2016
F. Ortuño and I. Rojas (Eds.): IWBBIO 2016, LNBI 9656, pp. 573–584, 2016.
DOI: 10.1007/978-3-319-31744-1_51

## 1.1  Biological Background

**Biological Switches.** Plenty of biological processes proceed in switch-like manner. Changes in number of active alleles in cell result in completely different cell behaviour due to increase or decrease of proteins level. The activation or inactivation of genes should be considered as a switch because it results in different cells decisions, which can be production of substances, like hormones and inflammatory agents, cells division and even programmed cell death [2].

Another group of processes, which can be classified as biological switch, is transmembrane transport. Ion channels are membrane proteins, which are able to form pore to transport molecules across the cell membrane. A wide variety of biological processes involve ions channels, because they enable rapid changes in cells. As en example we can mention cardiac, skeletal and muscle contraction, epithelial transport, T-cell activation and insulin release. Channels are responsible for establishing a resting membrane potential and shaping action potentials by gating the flow of ions. They are a key component of the nervous system cause they enable conduction of nerve impulses. The reaction of open and close the channels can be dependent on voltage or ligands. After the channel opening the conditions are rapidly changed and the cell functioning is significantly altered [3].

Drug application is another process which can be considered as biological switch. Usually drug administration is a sudden step change. In real, biological system, such as humans body, drug level in single cell is dynamically changing and it depends on the type of the medication (for example oral or by a drip-bag). The processes such as transport through the organism, adsorption export and import to the cell and drug degradation result in gradual increase and decrease in active drug level in cell. After drug application the cell functioning is completely altered [4]. The exemplary drug action is protein inactivation, induction of protein degradation or blockade of its transport. As a result the decrease of active protein level is observed resulting in change of the cell behaviour [5].

**Parameters Changes.** Cellular processes base on biochemical reactions between organic macromolecules such as carbs, proteins, lipids and nucleic acids and many different non-organic molecules and ions. Inside cell the chemical and physical conditions change dynamically because of the extracellular and intracellular processes such as cellular export and import, fluctuation of transcription factors, intracellular transport and proteins modifications such as phosphorylation or ubiquitylation [2]. Additional biological process which has got a significant influence on condition is cell growth. Gradual volume and size increase cause changes in all intracellular processes due to changes in substrate concentration and activation or deactivation high energy-consuming processes [6]. Mathematical modelling of biological processes requires determination of the model parameters, which can be estimated or calculated using biological experiment results. Knowing that the established parameters are highly dependent on biological, chemical and physical cell conditions and it is beyond any doubt that the model results can responds to specified cell conditions.

Another impediment in the parameter estimation is difficulty in its person-alisation. The processes rates differ between humans as an innate characteristic feature. Of course the reaction rates cannot be altered in a huge scale, but if the minimal therapy is considered (i.e. therapy which because of drug toxicity is based on minimal drug doses) these small interpersonal variations can result in inefficient therapy.

## 1.2  Proposed Models

To illustrate the problem we consider a simple model of proteins productions. The proposed models refers to cell, where the observed proteins level is estab-lished on the 160 000 molecules, and the therapeutic goal is to decrease proteins number under 70 000 molecules. The analysed models consists of three vari-ables: the gene state, the mRNA molecules and the proteins. We assume that the system is described by the state equations:

$$\frac{dG_A}{dt} = q_a * (N_A - G_A(t)) - q_{d1} * G_A(t). \tag{1}$$

$$\frac{dmRNA}{dt} = t_1 * G_A(t) - d1 * mRNA(t). \tag{2}$$

$$\frac{dA}{dt} = t_2 * mRNA(t) - d_2 * A(t). \tag{3}$$

We introduce a drug to the system which role is to degrade protein A. Because protein A degrades also spontaneously, drug application which increases protein A degradation rate can be considered as the switch of the parameter responsible for degradation rate. We assume that therapeutic effect is achieved when protein A level stays below the given threshold.

So we assume that after the switching time $t_s$ the third model Eq. (3) has the following form:

$$\frac{dA}{dt} = t_2 * mRNA(t) - (d_2 + d_3 * DRUG) * A(t). \tag{4}$$

In the initial consideration we assume that the drugs degradation and spread through the organism is neglected so drug is maintained on a stable level. Because biological processes are often regulated by feedback loops we consider also the models with negative and positive feedback. The second model reflects the sit-uation when the protein acts as their own transcription repressor, so it induces deactivation of its own gene constituting negative feedback. In this model the first Eq. (1) is modified in the following way:

$$\frac{dG_A}{dt} = q_a * (N_A - G_A(t)) - q_{d2} * G_A(t) * A(t). \tag{5}$$

The last analysed case is the model with proteins acting as transcription fac-tor for its own gene. Consequently this model contains a positive feedback loop.

**Table 1.** The values of the model parameters

| Parameter | Description | Value | Unit |
|-----------|-------------|-------|------|
| $q_{a1}$ | Gene activation | $2.78 * 10^{-4}$ | 1/sec |
| $q_{a2}$ | Gene activation induced by protein | $4 * 10^{-9}$ | 1/sec |
| $q_{d1}$ | Gene deactivation | $2.78 * 10^{-4}$ | 1/sec |
| $q_{d2}$ | Gene deactivation (negative feedback loop) | $1.75 * 10^{-9}$ | 1/sec |
| $q_{d3}$ | Gene deactivation (positive feedback loop) | $9.15 * 10^{-4}$ | 1/sec |
| $t_1$ | mRNA transcription | 0.05 | Molecules/sec |
| $t_2$ | Protein translation | 0.1 | 1/sec |
| $d_1$ | mRNA degradation | $1.5 * 10^{-4}$ | 1/sec |
| $d_2$ | Protein degradation | $2.0822 * 10^{-4}$ | 1/sec |
| $d_3$ | Protein degradation dependent on drug | $3 * 10^{-7}$ | 1/sec |
| $DRUG$ | Number of drug molecules | $1.4 * 10^3$ | Molecules |
| $N_A$ | Number of alleles | 2 | Molecules |

In this model gene can also be activated spontaneously and the first equation takes the form:

$$\frac{dG_A}{dt} = (q_a + q_{a2} * A(t)) * (N_A - G_A(t)) - q_{d3} * G_A(t). \tag{6}$$

The basic assumption is that the proteins and mRNA numbers (without drug) are almost the same in all the cases. To ensure that we adjust the rates of the genes activation and deactivation for each model. Parameters values of all models are presented in Table 1.

The differential equations of the systems were transformed to compute the steady state in cases with and without drug application. The equation for the gene steady state differs between model and they have the following form:

$$without\ feedback: \quad G_A = \frac{N_A * q_a}{q_a + q_{d1}};$$

$$negative\ feedback: \quad G_A = \frac{N_A * q_a}{q_a + q_{d2} * A};$$

$$positive\ feedback: \quad G_A = \frac{N_A * q_a + N_A * q_{a2} * A}{q_a + q_{a2} * A + q_{d3}}. \tag{7}$$

The equation for the mRNA and protein level are he same in all models, however the equations for the protein steady state depends on the drug application.

$$mRNA = \frac{t_1 * G_A}{d_1};$$

$$without\ drug: \quad A = \frac{t_2 * mRNA}{d_2};$$

$$with\ drug: \quad A = \frac{t_2 * mRNA}{d_2 + d3 * DRUG}. \tag{8}$$

**Table 2.** The values of the steady states in models

| Model | Drug | $G_A$ | mRNA | A | |
|---|---|---|---|---|---|
| Without feedback | − | 1.00 | 333.33 | 160 | 087 |
| Without feedback | + | 1.00 | 333.33 | 69 | 268 |
| Negative feedback | − | 0.99 | 332.47 | 159 | 675 |
| Negative feedback | + | 1.39 | 464.72 | 69 | 030 |
| Positive feedback | − | 1.03 | 334.27 | 160 | 535 |
| Positive feedback | + | 0.75 | 251.47 | 69 | 042 |

The calculated values of the steady states in each case are presented in Table 2.

## 1.3 Methods

We perform deterministic simulation based on Runge-Kutta 4th order algorithm. In the deterministic simulations the number of molecules or active alleles can be a decimal number, that is explained as the mean response of the whole cell population.

We assume that the control objective - the therapeutic target is to drive the system to terminal state $x(t) < Th$ in time $T$, where $Th$ is an established threshold. We consider the effect of small changes of parameters in two different cases: uncertainty of their estimates or inaccurate estimation of switching time. In the first case we focused on the change in parameter $q_a$ which stands for gene activation. The process of gene activation is tightly coupled with transcription factors, whose level continuously fluctuates so estimation of these parameter is very difficult and thus biased. Additionally it is dependent on the person physical conditions, and it can vary between individuals. Inaccuracy in switching time results from the deviation in time of drug application and differences in time of the drug distribution in the organism.

## 2 Results

### 2.1 Results for Different Models

We have examined three different models, whose parameters were adjusted to reach the stable proteins number (before drug application) on similar level. All the models are exposure to the same drug dose. We assume that therapeutic target is achieved when drug decrease number of protein molecules under 70 000. In the model without feedback loops the target level is reached, however the proteins levels in stable state is much higher comparing to model with positive feedback loop. The model with negative feedback loop is significantly less sensitive for drug activity so the drug level in stable state is higher then threshold and the therapeutic goal is not reached (see Fig. 1).

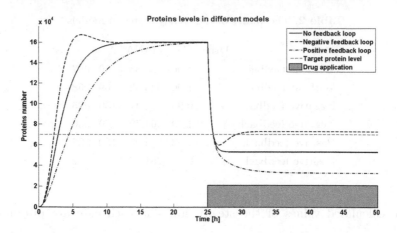

**Fig. 1.** Number of proteins in different models: without feedback loop (*solid line*), with negative feedback loop (*dashed line*) and with positive feedback loop (*dash-dotted line*) after drug application.

## 2.2   Parameter Changes

Please note this, that because of the drug toxicity we consider the minimal therapy and we fit the drug dose to every case. For model without feedback loops number of drug molecules stands at 910, for model with negative feedback it stands at 1550 and for model with positive feedback loop - 520. Because in our work we consider the influence of the parameter deviations from the nominal values on the reachability of the assumed goal and these deviations are given in percentage, the fact that each of the considered model requires different drug dose do not influence the result.

The uncertainty in parameter values can have essential impact on model results. We investigate the effect of the change in parameter $q_a$ which can responds to gene activation rate.

The parameter $q_a$ was altered to 80 %, 90 %, 110 % and 120 % of the original value from the beginning of the simulation. As a result the proteins levels in stable state are significantly altered before drug application as well as after it. In all cases the target protein number was reached in the model with original values and after decrease of the $q_a$ parameters. After increase of the parameter $q_a$ value the proteins number exceeds the target threshold (Fig. 2). In model with positive feedback small increase of the gene activation rate did not indispose achievement of the goal (Fig. 2C).

Variations in gene activation rate result in deviation between proteins number in the steady state. In model without any feedback loop observed deviations are the biggest (Fig. 3 (squares)). The models with auto-regulation are less sensitive for the parameter changes (Fig. 3 (triangles and circles)). The deviations between protein levels in the steady state are the smallest in model with negative feedback loop (Fig. 3 (circles)). In models with feedback loops we can assume

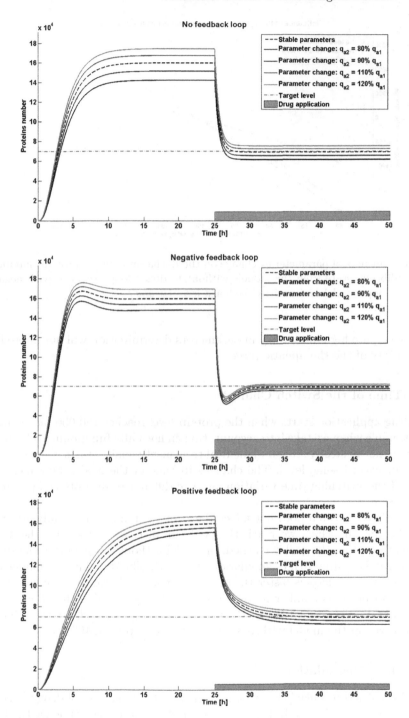

**Fig. 2.** Reachablity of the target protein number with parameters $q_a$ change for models without feedback loop (A), with negative feedback loop (B) and with positive feedback loop (C)

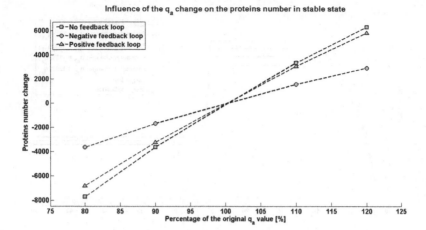

**Fig. 3.** Influences of parameter $q_a$ change on the deviation between proteins number in the equilibrium state from the models without feedback loop (*square*), with negative feedback loop (*circle*) and positive feedback loop (*triangle*).

that even quite high inaccuracy in parameters determination will not altered the reachability of the therapeutic target.

## 2.3 Time of the Switch Change

The drug application starts when the protein level reaches 150 000 molecules. It is reasonable when we take into account, that in normally functioning cell protein levels are not in steady state and medical treatment usually starts when specified indicator exceed some level. The changes in time of the goal achievement is a result of the switching time variation and the differences in proteins level at the time of switch.

The time of switch does not affect the proteins number in steady state but only the time in which we reach the goal. In all cases the sooner the switch happen the sooner the goal is reached. In model without any feedback the steady state is reached after 10 h of simulation (Fig. 4A). The dynamics in model with negative feedback loop is faster then in other models, because the increase and decrease of proteins number is rapidly changed, however due tu the overshoot the steady state is reached after 10 h of simulation (Fig. 4B). The slowest changes are observed in the model with positive feedback loop (Fig. 4C).

## 2.4 Drug Degradation

In biological systems like humans organisms the stable number of drug molecules is impossible to maintain. In live systems variety of processes influence the number of active drug molecules. Firstly the application of drug is relatively short action (even in the case of intravenous drip it takes not longer then 2 h) and

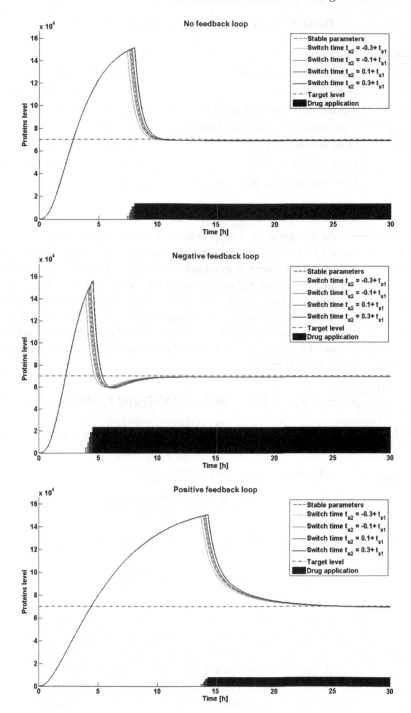

**Fig. 4.** Time of the achievement of target protein number with switch time change for models without feedback loop (A), with negative feedback loop (B) and with positive feedback loop (C)

**Table 3.** The values of the model parameters

| Parameter | Description | Value | Unit |
|---|---|---|---|
| $p_{oral}$ | Dose conversion factor | $1 * 10^{-8}$ | M/(mg/Kg) |
| $U$ | Drug dose rate | 180 | mg/(Kg*sec) |
| $Ka$ | Equilibrium association rate | $8.5 * 10^4$ | 1/M |
| $Bmax$ | Concentration of protein binding sites | $2.86 * 10^{-6}$ | M |
| $delta$ | Drug elimination rate | $1.5 * 10^{-2}$ | 1/sec |
| $i$ | Drug import to cell | $4 * 10^3$ | molec/(sec*M) |
| $e$ | Drug export from cell | $2 * 10^{-3}$ | 1/sec |

active drug levels do not grow in a step-manner because of the transport delay. Moreover the number of active molecules depends on the intracellular import and export. The level of the drug is gradually decreased due to the degradation which finally leads to total drug removal from the human body [7]. We take into account all this dependencies and add two more equations to our model, which describe the total drug number and active drug number (available for cell). This equation for total drug number consists of the part for the oral drug application and its degradation. The process of degradation concerns only molecules not connected to any other particles.

$$\frac{dDRUG_{tot}}{dt} = p_{oral} * U(t) - delta * DRUG_N(t) * DRUG_{tot}(t). \qquad (9)$$

where following [8] $DRUG_N(t)$ is number of free and degradable molecules defined by the equation:

$$DRUG_N(t) = \frac{-(1 + Ka * Bmax - Ka * DRUG_{tot}(t))}{2 * Ka}$$
$$+ \frac{\sqrt{(1 + Ka * Bmax - Ka * DRUG_{tot}(t))^2 + 4 * Ka * DRUG_{tot}(t)}}{2 * Ka}. \qquad (10)$$

The number of active drug molecules depends on the export and import rate as follows:

$$\frac{dDRUG}{dt} = i * DRUG_{tot}(t) - e * DRUG(t). \qquad (11)$$

The values of parameters and its meaning are presented in Table 3.

In the case with drug degradation, drug level is dynamically changed, so the steady state is not reached in analysed time period. In all cases protein number is reduced under therapeutic threshold, but it cannot be maintained on this level. In model with negative feedback loop protein level rapidly exceeded the threshold, so the illness probably cannot be cured. In model with the positive feedback loop the protein level grows very slowly, so there is an opportunity for the success in treatment (Fig. 5).

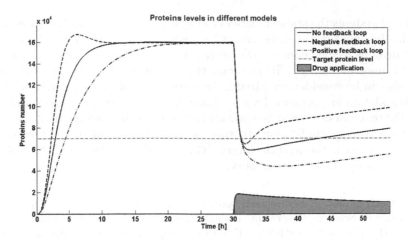

**Fig. 5.** Number of proteins in different models: without feedback loop (*solid line*), with negative feedback loop (*dashed line*) and with positive feedback loop (*dash-dotted line*) after drug application with drug degradation.

## 3   Discussion

Switches in biological models are quite common phenomena, however they have not been fully examined in the literature yet. We were interested how small parameters deviations affect the nominal state of the medical therapy model in the system with switches.

Created models show significant differences in dynamics and sensitivity to the switch. Systems with positive feedback loop are the most sensitive for drug application, so small therapeutic dose is sufficient to reach the goal. However dynamics in this model is the slowest and reaching the therapeutic target takes a long time. In the model with negative feedback loop the opposite situation is observed, the dynamic of the system is rapid but the model is robust so high drug dose is needed to reach the therapeutic goal. Moreover in models with negative feedback loop occurs overregulation which means that even if the threshold is reached, after a time the system can stabilize over the threshold (Fig. 1).

In medical treatment the crucial problem is determining the smallest effective drug level, because of the toxicity of the drugs. Inaccuracy in the parameter estimations affects the reachability of the goal. Treatment with small drug dose might be insufficient in the case of the change in the reactions rates. The reaction constants rate can be easily altered by the physical and chemical conditions. Moreover they are characteristic features for every single person. Taking into account the changes in parameters may be the way to create personalised more efficient and less toxic therapy for a particular person.

Presented results show that the models with feedback loops are less sensitive for parameter deviations than the model without any feedback loop. Changes between the protein number in steady state are much smaller due to the existence of feedback loop (Fig. 3A). The systems with feedback loops are less vulnerable to the parameter deviations so the risk of the underestimation of the drug dose is small.

In real systems the stable number of drug molecules is impossible to maintain and the dynamics of the drug should be taken into account. With the decrease of the drug molecules number, the protein level is increased, so the therapeutic target cannot be reached. To maintain the protein level under threshold a few methods can be considerate. Firstly, there can be applied big drug dose, which enable maintain low proteins level at time long enough to cure, however in some cases the required doses can be toxic and harmful to the patient. In other option is periodical drug application. In this method even small drug doses can be sufficient to reach therapeutic target. This so-called metronomic drug dosege will be the scope of our future work.

**Acknowledgments.** The research presented here was partially supported by the National Science Centre in Poland granted with decision number DEC-2012/05/D/ST7/02072 (for KP), DEC-2014/13/B/ST7/00755 (for AS) and BKM - 514/RAU1/2015, number 17 (MO).

# References

1. Liberzon, D.: Switchings in Systems and Control. University of Illionis at Urbana-Campaign, USA (2003)
2. Alberts, B., Johnson, A., Lewis, J., Raff, M., Roberts, K., Walter, P.: Molecular Biology of the Cell, 4th edn. Garland Science, New York (2002)
3. Sachs, F.: Mechanical transduction by membrane ion channels: a mini review. Mol. Cell. Biochem. **104**, 57–60 (1991)
4. Benet, L.Z., Sheiner, L.B.: Pharmacokinetics: the dynamics of drug absorption, distribution, and elimination. In: The Pharmacological Basis of Therapeutics, pp. 15–20. Macmillian Publishing Co. (1985)
5. Bantscheff, M., Scholten, A., Heck, A.J.R.: Revealing promiscuous drug target interactions by chemical proteomics. Drug Discov. Today **14**(21–22), 1021–1029 (2009)
6. Hartwell, L.H., Weinert, T.A.: Checkpoints: controls that ensure the order of cell cycle events. Science **246**, 629–634 (1989)
7. Buxton, I.L.O., Benet, L.Z.: Pharmacokinetics: the dynamics of drug absorption, distribution, matebolism and elimination. In: Goodman and Gilman's the Pharmacological Basis of therapeutics, 12th edn. McGraw-Hill Education (2011)
8. Puszynski, K., Gandolfi, A., d'Onofrio, A.: The pharmacodynamics of the p53-Mdm2 targeting drug nutlin: the role of gene-switching noise. PLOS Comput. Biol. **10**(12), e1003991 (2014)

# High Performance Computing in Bioinformatics, Computational Biology and Computational Chemistry

# The Case for Docker in Multicloud Enabled Bioinformatics Applications

Ahmed Abdullah Ali[2], Mohamed El-Kalioby[2], and Mohamed Abouelhoda[1,2(✉)]

[1] Faculty of Engineering, Cairo University, Giza, Egypt
mabouelhoda@yahoo.com
[2] Center for Informatics Sciences, Nile University, Sheikh Zaid City, Egypt

**Abstract.** The introduction of next generation sequencing technologies did not bring only huge amounts of biological data but also highly sophisticated and versatile analysis workflows and systems. These new challenges require reliable and fast deployment methods over high performance servers in the local infrastructure or in the cloud. The use of virtualization technology has provided an efficient solution to overcome the complexity of deployment procedures and to provide a safe personalized execution box. However, the performance of applications running in virtual machines is worse than that of those running on the native infrastructure. *Docker* is a light weight alternative to the usual virtualization technology achieving notable better performance. In this paper, we explore the use case scenarios for using *Docker* to deploy and execute sophisticated bioinformatics tools and workflows, with a focus on the sequence analysis domain. We also introduce an efficient implementation of the package *elasticHPC-Docker* to enable creation of a docker-based computer cluster in the private cloud and in commercial clouds like Amazon and Google. We demonstrate by experiments that the use of *elasticHPC-Docker* is efficient and reliable in both private and commercial clouds.

**Keywords:** Bioinformatics · Sequence analysis · High performance computing · Cloud computing · *Docker*

## 1 Introduction

The use of next generation sequencing technology has changed the traditional bioinformatics practice revolving around single-problem single-analysis-tool. Nowadays, sophisticated multi-step workflows have to be used to transform the raw sequence data into meaningful information and knowledge. A recent workflow can include tens of tasks and hundreds of information sources integrated together to achieve the analysis goals. In addition, database systems and tables have to be packaged with the analysis tools to manage the data and results. An example workflow is the medical variant detection workflow [1,2] shown in Fig. 1. This workflow is the most data intensive workflow in sequence analysis and it is highly demanded these days as many mega human genome

© Springer International Publishing Switzerland 2016
F. Ortuño and I. Rojas (Eds.): IWBBIO 2016, LNBI 9656, pp. 587–601, 2016.
DOI: 10.1007/978-3-319-31744-1_52

sequencing projects are running. Examples include the EXAC project (60,000 exomes, see http://exac.broadinstitute.org), and the Genomics England (www. genomicsengland.co.uk), among others. The variant detection workflow starts with assessing the quality of the reads and excluding those (or the regions of them) that are of low quality. Depending on the sequencing platform, this step involves running different quality assurance programs like Fastx [3] or FastQC [4]. It also includes proper formatting of the reads and computing related statistics. Then the reads are aligned (mapped) to a reference human sequence using multiple programs considering manufacture specific parameters. Example mapping programs include BWA [5] and Isaac [6]. To run on high performance computing infrastructure, some extra steps have to be executed to decompose the input into blocks to be processed by different nodes in parallel and to merge the results back in one file. After the mapping, the variant detection sub-workflow is executed. It includes a number of steps for determining (calling) the variants (mutations) compared to the reference genome and discriminating it from errors. Specifically, it includes realignment of the reads around variable positions to increase confidence of it, score recalibration, inclusion of statistical models that take machine-specific errors into account, and removal of redundancy. After the calling step, the variants are annotated with all possible information to identify disease related ones and their effect. The use of an annotation source like Annovar [7] requires integration of about 100 databases and querying them to produce the final annotated variant file. Add to this that the modules (or the programs) and parameters of the system might change

- per user, where each one may require certain modules, annotation databases, and special post-processing;
- per experiment type, where the sequencing may be done for whole genome, whole exome, or RNAseq in a single or multiplexed mode (multiple samples per run); or
- sequencing platforms, where the sequencing technology may be illumina, Ion Torrent, or any other one.

Putting this workflow with all these complexities in a bigger system including sample tracking as well as input/output data management renders the administration process highly complicated, and requires an efficient solution.

Virtualization technology provides a solution to the deployment problem, where the whole system with all modules, databases and the related dependencies are packaged in a virtual machine image. These images can be then used to instantiate a virtual machine running in private or public cloud. Examples of bioinformatics tools using this model of deployment include Crossbow [8] for NGS read alignment and SNP calling, RSD-Cloud [9] for comparative genomics, and CloVR [10] and QIIME [11], among others. Example cloud based tools for structural bioinformatics include those presented in [12–15], among others. Interestingly, the work of Guerrero et al. [12] introduces a GPU-based cloud tool and evaluates its cost/performance in commercial cloud.

The traditional engine for running the virtual machine instances is based either on Oracle Virtual Box [16], KVM [17], Xen Hypervisor [18], or

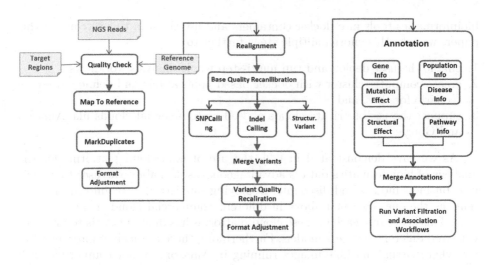

**Fig. 1.** The variant detection workflow. For space limitations, we reported the most important steps. Other steps related to formatting, decomposition of data, statistics, and plotting are not shown.

VMware [19]. Some of these virtualization engines support GPUs: VMware (Vsphere), Xen (Passthrough vGPU), and also KVM (Kernel 3.9 and above). Regarding commercial cloud products, Amazon is based on Xen virtualization engine and it supports GPU through three instance types: The first two are g2.2xlarge and g2.8xlarge and are referred to as G2 instances, as they include NVIDIA (GRID series, 1536 CUDA cores). The second is cg1.4xlarge and is referred to as CG1 (NVIDIA Tesla series). Google is based on KVM virtualization engine and Azure is based on Hyper-V. Both of Google and Azure do not support GPU so far. The use of GPU within *Docker* can be achieved by installing the GPU driver within the *Docker* image. But the image produced this way are not really portable and have to be modified for each GPU card/driver. A better solution has been recently provided by NVIDIA, where a *Docker* plugin (nvidia-docker) has been made available to overcome this problem.

Docker [20] has been introduced to provide a new level of virtualization, in which the computing machine (including the operating system) is not virtualized, but only the application and the related dependencies are encapsulated in a 'virtual' isolated process environment (referred to as a container) running on the host operating system. The advantage is that the application communicates directly with the host operating system with minimum overhead compared to the usual virtual machine setting. Figure 2 shows the software layers for *Docker* compared to that of the usual virtual machine.

Although *Docker* is gaining popularity in the software community, there is little work done to use it in the bioinformatics domain. To the best of our knowledge, only the paper of NGSEasy [21] discussed how to port individual bioinformatics tools into docker containers and use them in this setting. In this paper, we take this effort multiple steps further to

1. enable the user deploy and run multi-step whole analysis workflows,
2. create computer cluster with docker based applications and define a use case scenario for that, and
3. support the use of private clouds as well as commercial clouds like Amazon and Google.

As we have demonstrated in [22], the use of multicloud platforms for the analysis of bioinformatics data is advantageous, as it enables the user to exploit best business models and discounts. Deploying sophisticated bioinformatics systems as the one described above in different commercial clouds is not an easy task. This is because each of these cloud platforms has different APIs to interface with and different business models. Furthermore, there is a lack of interoperability, where virtual machine images running in Amazon are not compatible with those running in Google and vice versa, which directly lead to duplication of work to prepare new images with each deployment. *Docker*, in prinicipal, provides an effective solution to the image portability problem and assures reliable and fast deployment from a single *Docker* image, running on all the clouds. However, extra layers need to be built on top of *Docker* to enable the use of multi-cloud platforms and to enable running data intensive applications using computer clusters.

In this paper, we define use case scanrio for using *Docker* within a computer cluster for bioinformatics workflows and evaluate its performance in comparison to the use of native hardware and usual virtual machines, in private and public cloud. We also present a new version of our multicloud *elasticHPC*, referred to as *elasticHPC-Docker*, to support *Docker*, and create different *Docker* based computer clusters in a local infrastructure or in the commercial Amazon and Google clouds, as mentioned above. In addition to the usual features related to the creation and management of a computer cluster, the new package *elasticHPC-Docker* can have nodes from Amazon and Google based on *Docker*. Our package is easy-to-use and provides a single interface to the different cloud platforms solving the difficult challenges related to the different cloud settings, APIs, and business models, which is not a straightforward task.

This paper is organized as follows: In the following section, we review the basic features of *Docker*, and compare how Amazon and Google support *Docker*. In Sect. 3, we introduce the features of our system *elasticHPC-Docker* and its implementation details. Sections 4 and 5 include experiments and conclusions, respectively.

## 2  *Docker*

*Docker* [20] is a new virtualization scheme based on encapsulating applications and related software dependencies in what is called container images (*Docker*

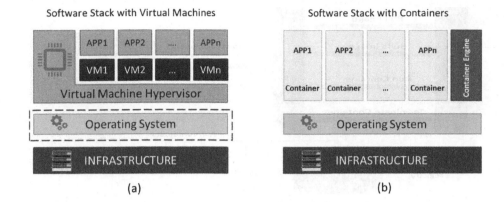

**Fig. 2.** The software layers when using virtual machines (a) and containers (b).

image). In our view, a *Docker* image can be regarded as a lightweight application-centered virtual machine image, and when instantiated it runs as an isolated process within the host operating system. The advantage is that there is minor or even neglected loss of performance when compared to the use of virtual machines. To facilitate the use of *Docker*, the *Docker* team provides a centralized registry to download basic images for different operating systems. The user can then extend these images with own application. It is also possible to publish updated images on the registry website and share it with other users.

As shown in Fig. 3, *Docker* is implemented according to a client-server architecture. The client communicates with the *Docker* server (running daemon) to perform different tasks. It can pull an image from the *Docker* repository to the execution infrastructure. When an image is available, the client can request from the server to run it specifying the number of cores and the size of the RAM. All required TCP and UDP ports are set-up so that data flows from the host machine to the container (e.g., TCP 80 in case of HTTP Server). The user can add more tools and tasks to the running container and can even create new image or update the current one. The APIs for updating the container image is similar to the git clone/commit interface, which can also be done on the image in the repository. In Appendix A, we give an example for creating and defining an image using a docker specification file and we show also how to instantiate it in a container to execute different tasks.

## 2.1  Containers in the Cloud

Currently, Google and AWS are the most prominent commercial cloud service providers who provide container services using *Docker* or equivalent services. (Azure container service is still a beta version.)

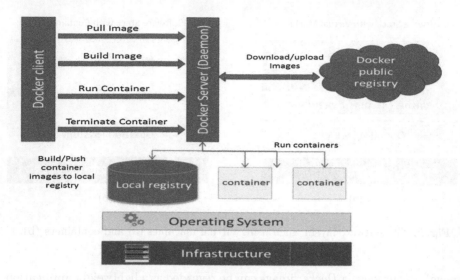

**Fig. 3.** Architecture of *Docker*: Clients communicate with server to execute different tasks related to the retrieval of container images and running the respective containers.

## Google Containers

Google Cloud offers a container service in the form of two products: First, Google provides container-optimized virtual machine images, where such machine image includes programs to run standard *Docker* images, according to a user defined file in YAML format.

Second, Google provides *Google Kubernetes Engine* (GKE) to create a cluster of virtual machines that can run *Docker* images. GKE is based on *pods* [23], where each pod has one or more containers that are dedicated to specific tasks. GKE provides extra interesting features, such as container health monitoring, automatic service restart in case of failure, redundancy of components, and a load balancer. To facilitate the use of container images, Google has established *Google container registry* (GCR) to host the images and users can access from their cloud environment in Google.

The optimized container images and GKE run at no extra cost. That is, the user pays the usual price of the used virtual machines. GKE charges an extra fee of $0.15 per hour per cluster on top of the usual machine price, and this is only if the cluster size is larger than 5 nodes.

GKE has two limitations: First, it does not support *Docker* s private images. Second, the cluster size in GKE cannot exceed 100 nodes.

## Amazon Containers

Amazon provides *Elastic Container Service* (ECS), which enables the deployment of *Docker* containers on Amazon EC2. Amazon uses *docker-compose* [24] to manage docker containers. Docker-compose facilitates the process of setting up a multi-container application by defining the application and all its dependencies in a single file using YAML format. Then, the user chooses the EC2 instance

**Table 1.** Features of supporting containers in both Amazon and Google.

| Feature | Google | Amazon |
|---|---|---|
| **Compute** | | |
| Optimized virtual machines | Yes | Yes |
| Container cluster manager | GKE | Amazon ECS |
| Redundancy | Yes | No |
| Container cluster monitoring | Yes | Yes (If node balancer started) |
| **Storage** | | |
| Root Disk size | 100 GB | 30 GB |
| Attachment of Persistent Disks | Yes | Not yet in ECS |
| Private Image Registry | Yes (GCR) | Yes (ECR) |
| Attachment of Data Volumes | Yes | Yes |
| **Networking** | | |
| Virtual Private Cloud (VPC) | Yes | Yes |
| Mapping Ports from VM to Container | Yes | Yes |
| Load balancing | Yes | Yes |
| **Pricing** | | |
| Optimized VM | Same GCE pricing | Same EC2 pricing |
| Container Cluster | $0.15 per hour per cluster (when above 5 nodes) | NA |
| Load balancer for cluster | included | $0.025 per hour per cluster |
| Image Registry | $0.026 per GB-month | $0.10 per GB-month |

type which hosts the container(s), number of instances and security parameters. Finally, the user starts the application. The Amazon machine instances created to host the container by ECS are optimized virtual machine instances based on RedHat Enterprise Linux (RHEL). The instantiated machines include programs to automatically configure the *Docker* environment.

To use ECS, the user pays for the selected optimized virtual machine types, which is the same as that of the usual instance types. If the load balancing service is selected, the user pays an extra small cost of $0.025 per hour and $0.008 per GB transferred between instances.

ECS of Amazon has one limitation: It does not support attaching EBS volumes to the running containers. One should use only ephemeral disks, which requires copy operations of any needed databases or tools.

In Table 1, we compare the different features of container solutions in Amazon and Google. We also list different products and different pricing options.

# 3   elasticHPC-Docker

## 3.1   elasticHPC-Docker Features

We present the package *elasticHPC-Docker* which has the following set of features:

– Ability to port and run any docker image to either private or commercial clouds.
– Creation and management of a cluster of containers. The cluster can use single or multiple machines.
– The computer cluster can have nodes from different cloud providers; i.e., some nodes can come from Amazon and some can come from Google.
– Ability to create and destroy containers in the run-time. This makes it possible to run multiple containers on the same machine, one at a time. For example, in the variant detection workflow one can run a container for alignment on one virtual machine. By completion of this task, this container cab be killed and a new one can start in the same machine to execute the following task.
– The package supports addition (scaling up) and removal (scaling down) of virtual machines (worker nodes) in a running clusters.
– The package allows mounting of virtual disks and establishment of a shared file system to the containers (Default option is the NFS). In AWS, we use EBS volumes and in Google we use persistent storage disks.
– *elasticHPC-Docker* automatically configures a job scheduler (including security settings among the different providers) among the containers. The default job schedule is PBS Torque, but SGE is also supported.
– The current package includes more than 150 *Docker* specification files (DockerFile) for different bioinformatics tools, including the most important NGS tools, like Fastx [3], BWA [5], GATK [2], among others. Furthermore, it includes a number of structural bioinformatics tools, including Vina AutoDock, AMBER http://ambermd.org, GROMACS (http://www.gromacs.org), and Frodock [25], among others; see the package website for more tools. These tools become ready to use without extra effort, once the computer cluster is created.

## 3.2   Use Case Scenario

The use case scenario for creating and using a *Docker* based cluster in the cloud is as follows: The user installs the client module of *elasticHPC-Docker* on a local machine and uses it to define and start a cluster in either Amazon, Google, or private cloud (OpenStack based). Once the cluster creation command is submitted, the master node of the cluster is first created in the specified cloud environment. The master node will in turn creates all worker nodes in parallel. The master node then distributes the URL of the container image in the registry specified by the user over the worker nodes. The default registries that are supported include (1) Docker Hub, (2) Google Container Registry, and (3) AWS ECS Registry. The master and worker nodes retrieve the container images and start the containers. Once all of them are done, the master node sets up the ports and finalizes the configuration of the cluster in terms of setting up the job scheduler and the shared storage. The cluster is now ready and the user can execute the workflow or application contained in it.

### 3.3   Implementation Details

**Setting Up Compute Nodes**

Fig. 4 shows the architecture of *elasticHPC-Docker*. This architecture supports the above mentioned use case scenario. The client component is a python based program installed in the user's local machine, which can be a simple desktop computer. The main or the base virtual machine image is a prepared image in advance and already available in Google and Amazon clouds. We also provide an OpenStack compatible image for private clouds. These images include the necessary software programs to start *Docker* containers from any *Docker* image or *Docker* specification file. To facilitate the use of images for bioinformatics, we provide a *Docker* image with many NGS tools already pre-installed.

For the topology of the cluster, our default setting is that one container runs on virtual machine (respectively one application) at a time. If the workflow moves from one step to the next, the container which finished the job can be killed and a new container starts to execute the next job. Our *elasticHPC-Docker* allows the user to create some nodes on Amazon and some nodes on Google. This option should be used with caution in case of transferring huge data among remote clouds, as it can be very time consuming and expensive.

The setup of the cluster requires the automatic setup of a job scheduler. *elasticHPC-Docker* achieves this and can install SGE or PBS Torque on the created containers. The master node configures the IPs of the worker nodes and the included cluster containers. TCP port 5000 will be used for communication among virtual instances. Containers can communicate using TCP port 5555. The package web-site includes more implementation and setup details.

**Data Management**

The data is made available to the cluster nodes through data volumes, which allows sharing data between the host machine and a container running in it. We first make the data available for all nodes using a shared file system (the default is NFS). Then we use the data volumes to share data between containers where all containers can access the same shared data.

## 4   Experiments

We conducted two experiments: The first is to measure the time for establishing container clusters over different cloud platforms. The second is to measure the performance of using *Docker* when running the variant detection workflow in Fig. 1. It is worth mentioning that this workflow is the most data intensive workflow based on NGS and the steps in it are analogous to other NGS workflows, like metagenomics, epigenomics, and RNAseq, among others. The selected test data is a typical exome data used in most genome projects.

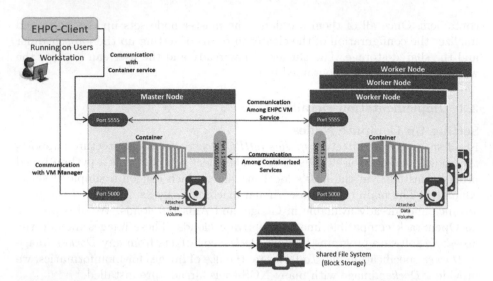

**Fig. 4.** Architecture of *elasticHPC-Docker*: The client of *elasticHPC-Docker* starts a cloud computer cluster and instantiate containers in it using container images. The created cluster automatically adjusts the ports among the machines and among the containers. Shared storage is also created to share data among the containers using data volumes.

**Experiment 1:** Table 2 compares the running times for setting up the computer cluster and the respective *Docker* containers in Google and Amazon using their native solutions against our *elasticHPC-Docker*. As we mentioned before, our *elasticHPC-Docker* builds the cluster from scratch and uses neither Google Kubernetes nor Amazon ECS products to save cost and to have more flexibility to configure images for the purpose of bioinformatics applications.

**Table 2.** Cluster set-up time in minutes for different number of nodes and in different clouds. GKE is Google Kubernetes and ECS is elastic container service of Amazon. *elasticHPC-Docker* (GC) and *elasticHPC-Docker* (AWS) are for our *elasticHPC-Docker* on Google and Amazon, respectively.

| Cluster size | GKE | AWS ECS | *elasticHPC-Docker* (GC) | *elasticHPC-Docker* (AWS) |
|---|---|---|---|---|
| 2 | 02:26 | 03:38 | 01:02 | 03:12 |
| 4 | 02:27 | 03:45 | 01:05 | 03:20 |
| 8 | 02:33 | 04:12 | 01:17 | 03:33 |
| 16 | 02:38 | 04:16 | 01:18 | 04:19 |
| 32 | 03:11 | 04:18 | 01:55 | 04:36 |
| 64 | 03:41 | 04:25 | 01:56 | 04:51 |
| 100 | 03:56 | 04:30 | 02:05 | 05:09 |

**Table 3.** Total running times in minutes and total costs (between brackets) in USD of the whole variant detection pipeline for 4, 8, and 16 nodes.

| Infrastructure | Nodes | | |
|---|---|---|---|
| | 4n | 8n | 16n |
| **Local infrastructure** | | | |
| Physical machine | 174(N.A) | 141(N.A) | 129(N.A) |
| *Docker* | 179 (NA) | 142 (NA) | 130 (NA) |
| OpenStack | 197 (NA) | 161 (NA) | 140 (NA) |
| **Google cloud** | | | |
| GCE VM | 115 ($3.9) | 86 ($5.8) | 74 ($9.9) |
| GCE-containers | 124 ($4.2) | 93 ($6.2) | 81 ($10.9) |
| EHPC-Docker GCE | 120 ($4.0) | 91 ($6.1) | 77 ($10.3) |
| **Amazon cloud** | | | |
| AWS VM | 116 ($4.3) | 87 ($8.5) | 81 ($17.0) |
| Amazon ECS | 119 ($4.3) | 93 ($8.5) | 81 ($17.0) |
| EHPC-Docker EC2 | 120 ($4.3) | 95 ($8.5) | 82 ($17.0) |

From the table, we can observe that the cluster setup time is almost the same for AWS ECS and *elasticHPC-Docker* in Amazon. As for Google Cloud, our *elasticHPC-Docker* was little bit faster than Google Kubernetes. It is also important to note that the creation of cluster worker nodes and container takes place in parallel, which explains the almost constant time for cluster creation.

**Experiment 2:** In Table 3, we compare the performance of the variant detection workflow using (1) native *Docker* installation (wrapped by our *elasticHPC-Docker*), (2) using private cloud software (OpenStack), (3) using virtual machines in Google and Amazon without *Docker*, and (4) using different container solutions in Amazon and Google, and finally using *elasticHPC-Docker* in the Amazon and Google clouds. For this experiment, we used an exome dataset from [2] of size $\approx$9 GB. (The exome is a set of NGS reads sequenced only from the whole coding regions of a genome.) The workflow was executed three times independently on Google, AWS, and private cloud based on OpenStack. In each cloud, the 9 GB input data is divided into blocks to be processed in parallel over the cluster nodes. For fair comparison, we used machines of as similar specifications as possible. We used nodes of type m3.2xlarge (8 Cores, Intel CPU 2.5 GHz, 30 GB RAM, SSD disks, $0.532 h) for Amazon, n1-highmem-8 (8 Cores, Intel CPU 2.5 GHz, 52 GB RAM, SSD disks, $0.504 h) for Google, and an OpenStack virtual machine with 8 Cores, 56 GB RAM on local infrastructure.

For private cloud, we compared *elasticHPC-Docker* against the use of virtual machines created using OpenStack-KVM cloud middleware. *Docker* is container based solution but KVM establishes complete virtual machine. From Table 3, we can see that the use of *Docker* (wrapped by *elasticHPC-Docker*) is more efficient than using whole machine virtualization.

For Google, we compared our solution *elasticHPC-Docker* to the use of the Kubernetes and also compared this to the case of running the workflow directly on the virtual machines without *Docker*. We did also the same for Amazon. As expected, we see that the use of direct installation lead to a little bit faster running time. *elasticHPC-Docker* was little bit faster than solutions based on Kubernetes and ECS. We also note that the running time on local infrastructure is slower compared to that in Amazon or Google. This is because of the slower processors and the slower electromagnetic disks in our local machines.

Table 3 also includes the total costs in \$USD of the whole variant detection pipeline for each cluster size in different public cloud platforms. It shows that Google provides lower prices compared to AWS and this is due to the fact that Google charges per minutes and Amazon charges per hour. Also we noticed that doubling cluster size may reduce the computational time however it increases the cost by factor of 50 % in Google and 100 % in Amazon.

## 5   Conclusions

In this paper, we have introduced *elasticHPC-Docker* based on container technology. Our package enables the creation of a computer cluster with containerized applications and workflows in private and in commercial clouds using single interface. It also includes options to manage the cluster, to deploy and run bioinformatics applications for large datasets, and to interface with image registries.

Through the container technology, *elasticHPC-Docker* provides an efficient solution to the inter-operability issue among commercial clouds, where it enables the user to run the same container image on any of these clouds. Our implementation of the cluster solution in *elasticHPC-Docker* has similar performance compared to Google GKE and Amazon in terms of creating the cluster nodes. But *elasticHPC-Docker* offers more options, including set up of the computer cluster and automatic installation and configuration of the job scheduler to parallelize the jobs. It also has more options for bioinformatics, as the created cluster nodes already include more than 150 bioinformatics tools ready to use. The package also sets-up a shared file system to make the input data available to all the cluster nodes and respective containers during the execution. It also provides commands to import the input data from local environment or cloud storage and export the results to local or remote storage.

It is also worth mentioning that *Docker* based clusters created using Google and Amazon *Docker* solutions are slightly more expensive than the clusters created using *elasticHPC-Docker*, because Amazon and Google incur extra charge for setting up the *Docker* environment.

Regarding the performance of the variant detection workflow using *elasticHPC-Docker* compared to the direct execution on the native hardware, we observed very a neglected runtime overhead. On the other hand, the use of *Docker* was superior to the use of virtual machines in local infrastructure.

(The overhead using *Docker* was 1 %–2 % and the one using OpenStack was 11 %–13 %.) In addition, we tested the performance of the structural bioinformatics tool Frodock [25] using *elasticHPC-Docker* compared to the direct installation and use of usual virtual machines. The use of *elasticHPC-Docker* was very efficient and the overhead of using docker was neglected compared to the use of VM. (The overhead was 2 %–3 % with docker against 20 %–30 % using virtual machine.) Hence, we can conclude that the advantages of using *elasticHPC-Docker* cannot be outweighed by such minor overhead.

In future work, we will integrate *elasticHPC-Docker* into the Tavaxy workflow system [26], where *Docker* images can be used within workflow nodes. *elasticHPC-Docker* is available for academic use at www.elastichpc.org.

# A    Appendix

The following Dockerfile is used to build Docker image for Variant Calling detection:

```
FROM elastichpc:base
MAINTAINER Bob
# Update APT repo
RUN apt-get update -y
# Install Required Packages
RUN apt-get -y install default-jre default-jdk python \
python-pip wget unzip zip vim
# Picard Tools Installation
RUN wget -O \/tmp\/picard-tools-1.129.zip https://github.com/broadinstitute/\
    picard/releases/download/1.129/picard-tools-1.129.zip \
    && mkdir -p /usr/local/picardtools \
    && unzip /tmp/picard-tools-1.129.zip -d /usr/local/picardtools/ \
    && chown -R root:root /usr/local/picardtools
RUN rm -rf /tmp/
# BAM, SAM Tools Installation
RUN apt-get install bwa samtools
# GATK Installation
ADD GenomeAnalysisTK-3.4-46.tar.bz2 /usr/local/
RUN mkdir -p /usr/local/GenomeAnalysisTK-3.4-46
ADD GenomeAnalysisTK-3.4-46.tar.bz2 /usr/local/GenomeAnalysisTK-3.4-46/ \
RUN cd /usr/local/GenomeAnalysisTK-3.4-46 && \
    echo 'export PATH=$PATH:/usr/local/GenomeAnalysisTK-3.4-46' >>  /root/.bashrc && \
    chmod 777 /usr/local/GenomeAnalysisTK-3.4-46/* && \
    cp -rfv /usr/local/GenomeAnalysisTK-3.4-46/* /usr/local/bin/
RUN rm -rf /tmp/*
# Open ports private only
EXPOSE 22
# Start Bash shell at startup
CMD ["/bin/bash"]
```

**Build Docker Image:** To build a docker image you have to install Docker engine on your local host; as explained in http://docs.docker.com/engine/installation/. Once installed, write the code as shown above in a file called Dockerfile on the same directory where you will build your image. Finally run the following command line to build variant calling detection Docker image.

```
$ docker build -t elastihpcc:variant-analysis
```

**Start Docker Container:** To start container using Docker engine, run the following command line:

```
$ docker run -it -p 2200:22 elastihpcc:variant-analysis  bash root@73e44b29d686 \#
```

Now user is ready to call any program for the variant detection workflow.

# References

1. Gonzalez-Garay, M.: The road from next-generation sequencing to personalized medicine. Pers. Med. **11**(5), 523–544 (2014)
2. DePristo, M., Banks, E., et al.: A framework for variation discovery and genotyping using next-generation DNA sequencing data. Nature Genet. **43**(5), 491–498 (2011)
3. FASTX-Toolkit. http://hannonlab.cshl.edu/fastx_toolkit
4. FASTQC. http://www.bioinformatics.babraham.ac.uk/projects/fastqc
5. Li, H., Durbin, R.: Fast and accurate short read alignment with burrows and wheeler transform. Bioinformatics **25**(14), 1754–1760 (2009)
6. Raczy, C., Petrovski, R., Saunders, C.T., et al.: Isaac: ultra-fast whole-genome secondary analysis on Illumina sequencing platforms. Bioinformatics **29**(16), 2041–2043 (2013). (Oxford, England)
7. Wang, K., Li, M., Hakonarson, H.: Annovar: functional annotation of genetic variants from high-throughput sequencing data. Nucleic Acids Res. **38**(16), e164 (2010)
8. Langmead, B., Schatz, M., Lin, J., Pop, M., Salzberg, S.: Searching for SNPs with cloud computing. Genome Biol. **10**, R134 (2009)
9. Wall, D., Kudtarkar, P., Fusaro, V., Pivovarov, R., Patil, P., Tonellato, P.: Cloud computing for comparative genomics. BMC Bioinformatics **11**, 259 (2010)
10. Angiuoli, S., Matalka, M., Gussman, A., et al.: CloVR: a virtual machine for automated and portable sequence analysis from the desktop using cloud computing. BMC Bioinformatics **12**(1), 356+ (2011)
11. Gregory, J., Kuczynski, J., Stombaugh, J., et al.: QIIME allows analysis of high-throughput community sequencing data. Nat. Meth. **7**(5), 335–336 (2010)
12. Guerrero, G., Wallace, R., Vázquez-Poletti, J., et al.: A performance/cost model for a cuda drug discovery application on physical and public cloud infrastructures. Concurrency Comput.: Pract. Experience **26**(10), 1787–1798 (2014)
13. Mrozek, D., Malysiak-Mrozek, B., Klapcinski, A.: Cloud4Psi: cloud computing for 3D protein structure similarity searching. Bioinformatics **30**(19), 2822–2825 (2014)
14. Mrozek, D., Gosk, P., Malysiak-Mrozek, B.: Scaling ab initio predictions of 3D protein structures in Microsoft Azure cloud. J. Grid Comp. **13**(4), 561–585 (2015)
15. Hung, C.-L., Hua, G.-J.: Cloud computing for protein-ligand binding site comparison. Biomed. Res. Int. **2013**, Article ID 170356, 1–7 (2013)
16. Oracle VirtualBox. http://www.virtualbox.org/
17. Kernel Virtual Machine. http://www.linux-kvm.org
18. Xen Project. http://www.xenproject.org/
19. VMware. http://www.vmware.com/
20. Docker. http://docker.com/
21. Folarin, A., Dobson, R., Newhouse, S.: NGSeasy: a next generation sequencing pipeline in Docker containers. F1000Research **4**, 997 (2015)

22. Ali, A.A., El-Kalioby, M., Abouelhoda, M.: Supporting bioinformatics applications with hybrid multi-cloud services. In: Ortuño, F., Rojas, I. (eds.) IWBBIO 2015, Part I. LNCS, vol. 9043, pp. 415–425. Springer, Heidelberg (2015)
23. Pods. http://cloud.google.com/container-engine/docs/pods
24. Docker Compose. https://www.docker.com/docker-compose
25. Garzon, J., Lopéz-Blanco, J., Pons, C., et al.: Frodock: a new approach for fast rotational protein-protein docking. Bioinformatics **25**(19), 2544–2551 (2009)
26. Abouelhoda, M., Issa, S., Ghanem, M.: Tavaxy: integrating Taverna and Galaxy workflows with cloud computing support. BMC Bioinformatics **13**(1), 77 (2012)

# Real Time GPU-Based Segmentation and Tracking of the Left Ventricle on 2D Echocardiography

Sidi Ahmed Mahmoudi[1]([✉]), Mohammed Ammar[2], Guillaume Luque Joris[1], and Amine Abbou[2]

[1] Faculty of Engineering, University of Mons, 9, rue de Houdain, Mons, Belgium
sidi.mahmoudi@umons.ac.be
[2] Faculty of Sciences, Physics Department, University of Boumerdes, Boumerdes, Algeria

**Abstract.** Left ventricle segmentation and tracking in ultrasound images present necessary tasks for cardiac diagnostic. These tasks are difficult due to the inherent problems of ultrasound images (i.e. low contrast, speckle noise, signal dropout, presence of shadows, etc.). In this paper, we propose an accurate and automatic method for left ventricle segmentation and tracking. The method is based on optical flow estimation for detecting the left ventricle center. Then, the contour is defined and tracked using convex hull and spline interpolation algorithms. In order to provide a real time processing of videos, we propose also an effective and adapted exploitation of new parallel and heterogeneous architectures, that consist of both central (CPU) and graphic (GPU) processing units. The latter can exploit both NVIDIA and ATI graphic cards since we propose CUDA and OpenCL implementations. This allowed to improve the performance of our method thanks to the parallel exploitation of the high number of computing units within GPU. Our experiments are conducted using a set of 11 normal and 17 disease hearts ultrasound video sequences. The related results achieved automatic and real-time left ventricle detection and tracking with a rate of 92 % of success.

**Keywords:** Left ventricle detection and tracking · Ultrasound videos · Optical flow · GPU · CUDA · OpenCL

## 1 Introduction

Cardiovascular diseases are the leading cause of death in the world. Therefore, cardiac anomalies diagnosis and detection present a fundamental concern of cardiologists. The left ventricle and particularly the endocardium is a structure of particular interest, since it performs the task of pumping oxygenated blood to the entire body. The left ventricle segmentation presents so a very important task. Several modalities are used to explore the cardiac structures such as:

© Springer International Publishing Switzerland 2016
F. Ortuño and I. Rojas (Eds.): IWBBIO 2016, LNBI 9656, pp. 602–614, 2016.
DOI: 10.1007/978-3-319-31744-1_53

magnetic resonance (MRI), X-ray, ultrasound, or computed tomography (CT) images. Amongst these techniques, ultrasound imaging is a popular tool used in clinical practice since it allows real time visualization of the heart motion.

Actually, the estimation of the left ventricular (LV) volumes and ejection fraction (EF) requires manual tracings of the LV cavity. This manual process can lead to errors and takes a long time. Echocardiography technique provides a real time visualization of the left ventricle motion. It is inexpensive, portable and uses a non-ionizing radiation. For these reasons, many works have been proposed in order to delineate the left ventricle within ultrasound imaging.

Otherwise, this kind of methods consists of several image processing steps that make the computation time so elevated. As result, the constraint of real time processing cannot be satisfied, mainly when treating high definition videos. In this context, one can imagine to exploit parallel-based architectures such as cluster, grid, FPGAs [2,14], graphic processing units, etc. The GPUs present an efficient solution, which is seriously hampered by the high costs of data transfer between CPU and GPU memories. Therefore, we developed a version that exploit in an efficient way, all the available computing units within computers (CPUs or/and GPUs). This implementation can exploit either NVIDIA or ATI graphic cards, based on CUDA[1] or OpenCL[2] respectively. In our case, the high intensive steps are executed in parallel on GPU in order to reduce their computing time. The less intensive ones remain implemented on CPU. The selection of resources (CPU or GPU) is based on estimating the complexity of each step.

The remainder of the paper is organized as follows: Sect. 2 presents the related works, while the Sect. 3 is devoted to describe the proposed approach for left ventricle segmentation and tracking. In Sect. 4, we present the GPU-based implementation that allowed to reduce the computation time and hence offer a real time processing. Experimental results are discussed in the Sect. 5. Finally, conclusion and future works are presented in the Sect. 6.

## 2   Related Work

In literature, several techniques for left ventricle segmentation and tracking have been proposed such as presented in [10,19], where authors developed a recognition algorithm for detecting the contour of left ventricular cavity. The method is based on considering all of the combinations between motion vectors of the ventricular wall in order to extract intersection points. This software was applied on 40 normal and 113 disease hearts where automatic calculation of ejection fraction was agreed with the results of conventional method by a sonographer. Shlinet *et al.* [18] developed a new optical-flow-based method for estimating heart motion from two-dimensional ultrasound sequences. In order to analyze the images locally, authors used a local-affine model for the velocity in space and a linear model in time. To increase the computational efficiency, a wavelet-like algorithm was used for computing B-spline-weighted inner products and

---

[1] CUDA. https://developer.nvidia.com/cuda-zone.

[2] OpenCL. www.khronos.org/opencl/.

moments at dyadic scales. The algorithm was first validated on synthetic data sets that simulate a beating heart with a speckle-like appearance of echocardiograms. Apademetris *et al.* [12] proposed a method for estimating cardiac deformations from 3D echocardiography. A dense motion field is used to calculate the deformation of the heart wall in terms of strain in cardiac specific directions. The strains obtained using this approach in open-chest dogs before and after coronary occlusion, show good agreement with previously published results in the literature. They also exhibit a high correlation with strains produced in the same animals using implanted sonomicrometers. This method provides quantitative regional 3D estimates of heart deformation from ultrasound images. Authors in [4] described a set of domain-specific source transformations on CUDA C that improved performance by 6.7 × on a system of ODEs arising in cardiac electrophysiology.

On the other hand, Image and video processing algorithms present prime candidates for parallelization on GPU since they apply similar or even the same calculations over many pixels. In case of image processing applications, authors in [21] proposed CUDA implementations of several classic image processing algorithms. CUDA-based implementations were also proposed for the median filter using the branchless vectorized median (BVM) filter [1] and a fixed-size kernel median filter [13]. Some GPU works were dedicated to medical imaging in [16] which presents a survey of GPU based medical applications, related to segmentation, registration and visualization methods. Otherwise, in case of video pressing and more particularly motion tracking methods, one can distinguish two categories of optical flow based motion estimation approaches.

The first are called dense optical flow which tracks all frame pixels without selecting any features. In this category, Marzat *et al.* [9] proposed a CUDA implementation of the Lucas-kanade method for optical flow estimation, that allowed to compute dense and accurate velocity field at about 15 frames per second (Fps) with 640 × 480 image resolution. Authors in [11] presented the GPU (CUDA) implementation of the Horn-Shunck optical flow method that offered a real time processing of low resolution videos (316 × 252).

The second category includes software tools tracking selected image features only. Sinha *et al.* [17] developed a GPU implementation of the popular KLT feature tracker [20] and the SIFT feature extraction algorithm [5]. This allowed to extract about 800 features from 640 × 480 video at 10 fps which is approximately 10 times faster than the corresponding CPU implementation. Authors in [15] proposed a GPU-based block matching technique using OpenGL. This implementation offered a real-time processing of 640×480 video. Recently, we developed a Multi-GPU implementation [6] of sparse optical flow computation that allowed real time motion tracking with high definition videos. The latter is exploited within our method of left ventricle tracking.

Our contribution can be summarized in three points:

1. CUDA parallel implementation that allowed to improve performance thanks to the exploitation of NVIDIA GPUs computing units in parallel.

2. OpenCL parallel implementation that allowed to improve performance by exploiting ATI GPUs computing units in parallel.
3. As result of the fast GPU treatment, we refresh the detection of the gravity center at each video frame. In fact, authors in [10,19] apply this step only if the difference between two consecutive gravity centers is bigger than a fixed threshold, which made their method less precise and not fully automatic.

## 3 CPU Implementation

Before presenting the GPU-based implementation, we start by describing the main steps of our left ventricle detection and tracking method. The latter can be summarized within two main steps (Fig. 1): gravity center detection and left ventricle tracking.

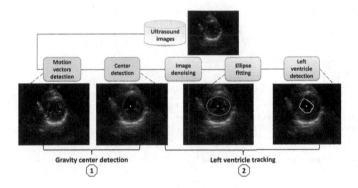

**Fig. 1.** Main steps of our method of left ventricle detection and tracking

### 3.1 Gravity Center Detection

This step consists of detecting the gravity center of the left ventricle. This detection allows to recognize and track the shape of left ventricle within the next step. The gravity center detection is performed with three sub-steps: pre-processing, motion vectors detection and center detection.

### 3.1.1 Pre-processing

First, we acquire the ultrasound videos and apply a noise elimination algorithm. Indeed, we apply morphological operations to remove smaller parts. A median filter is then applied and followed by a binarization (adaptive thresholding) that allowed to eliminate the remaining noise.

### 3.1.2   Motion Vectors Detection

In this step, we apply the sparse optical flow method [6] in order to estimate the optical flow motion vectors related to meaningful image points (corners).

### 3.1.3   Center Detection

Second, we compute the intersection points between the detected motion vectors. This intersection computation is applied within the algorithm described in Listing 1.1. The cavity center of gravity is obtained by computing the mean position between the detected intersections.

**Listing 1.1.** Center detection

```
1  [float=*][language=C++]
2                              // V : list of motion vectors
3  for (i= 0; i<V.size(); i++) { // for each motion vector
4    x1 = Start[i].x;
5    y1 = Start[i].y;           // x1,y1 : begin coordinates of the segment "i"
6    x2 = End[i].x;
7    y2 = End[i].y;             // x2,y2 : end coordinates of the segment "i"
8    for (j=i; j<V;j++) {       // comparison between pairs of motion vectors
9      x3 = Start[j].x;
10     y3 = Start[j].y;
11     x4 = End[j].x;
12     y4 = End[j].y;
13                              // computation of a coeff for each segment
14                              //(exp, for segment D1, y1= a1 * x1 + b1)
15     a1 = (y2 - y1) / (x2 - x1); // a coeff for the first segment
16     a2 = (y4 - y3) / (x4 - x3); // a coeff for the second segment
17
18     if (a1==a2)
19        cout << "parallel segments" << endl;   // No intersection
20     else {
21                              // computation of b coefficients
22        b1 = y1 - (a1 * x1);
23        b2 = y3 - (a2 * x3);
24
25        float inter=(b2-b1)/(a1-a2); // intersection between i and j points
26        inters.push_back(inter);     // include intersection point in an array
27        }
28 }
29 }
```

## 3.2   Left Ventricle Tracking

The second step consists of exploiting the detected gravity center in order to detect and track the shape of left ventricle in real time. This step can be described within two sub-steps: approximation of left ventricle contour and its detection.

### 3.2.1   Approximation of Left Ventricle Contour

Using the gravity center detected previously, we define a radius for researching the maximum brightness point along this radius. This operation is applied in all directions around the gravity center. An ellipse is fitted from the detected points of interest.

### 3.2.2 Left Ventricle Detection

In the last step, the exact contour of the left cavity is detected without the mitral valve by using the convex hull algorithm for which the binary shape of the left ventricle is approximated with a parametric elliptic curve. Fitzgibon *et al.* [3] proposed an ellipse fitting technique which is robust to noise. This algorithm uses a least square fitting to find the best ellipse that describes the extracted contour.

Given $(x_i, y_i)(i = 1, ..., n)$ an $n$ points contour. The objective is to minimize the error between an ellipse $Ax^2 + Bxy + Cy^2 + Dx + Ey + F = 0$ and the contour. The method uses a least square optimization with the algebraic criteria.

$$\begin{cases} \min\limits_{\alpha} \sum_{i=0}^{n}(Ax^2 + Bxy + Cy^2 + Dx + Ey + F)^2 \\ \qquad\qquad B^2 - 4AC = 1 \end{cases} \tag{1}$$

where the constraint $B^2 - 4AC = 1$ ensures that the problem is elliptical. Indeed, solving the system (1) enables to link each stored contour to an ellipse. As a result, the average position of the contour corresponds to the approximated center of gravity calculated at each frame.

## 4  GPU Implementation

Despite the high accuracy of the above-mentioned method, its computing time is so significant, which makes our method not adapted for real time processing of ultrasound images. The high computing time is due to:

1. Several image processing algorithms that are applied for each video frame.
2. The high computational intensity of these steps:
   - The pre-processing step
   - Optical flow based motion vector calculation
   - Computation of the intersection between optical flow vectors.
3. The use of high definition ultrasound images that requires more treatment.

To overcome this constraint, we developed a GPU-based portable implementation that can exploit NVIDIA or ATI GPUs. For a better exploitation of GPUs, we ported and parallelized the high intensive steps (pre-processing, motion vectors and gravity center detection) of our method on GPU, while the less intensive steps (contour approximation and left ventricle detection) remain implemented on CPU. Notice that the high intensive steps are detected within our complexity estimation equation defined in [8]. The less intensive step consists of using the convex hull algorithm which present a high dependency of its tasks and which is not so adapted for GPU programming. Notice that at the end of GPU treatment (pre-processing, motion vectors and gravity center detection), the result data are copied to the CPU memory for completing the next steps. This data transfer is not so consuming in time compared to the total time of the application. This section is presented with three parts. The first one presents our CUDA-based functions that have been used to exploit NVIDIA graphic cards, while the second part is devoted to present the OpenCL-based functions. The third part illustrates the exploitation of these functions within our method of left ventricle tracking.

## 4.1    CUDA-Based Functions

As described above, the high intensive steps are implemented on GPU (CUDA 6.5) in order to reduce the computation time. These steps are implemented as follows:

### 4.1.1    CUDA-Based Pre-processing

This step consists of applying morphological operations and median filter in order to reduce noise. We developed a GPU version that consists of selecting the same number of CUDA threads as the number of image pixels. This allows for each CUDA thread to apply the multiplication of one pixel value with filter (or structural element) values. All the CUDA threads are launched in parallel. More details about this implementation are presented in [7].

### 4.1.2    CUDA-Based Motion Vectors Detection

This step consists of detecting features that are good to track, *i.e.* corners. This detection is implemented on GPU by applying parallel processing on pixels using a GPU grid with a number of threads equal to the number of pixels. Each thread computes if its corresponding pixel represents a corner or not. Once the corners detected, we select a GPU grid that contains the same number of threads as for detected corners. In this way, each thread computes the optical flow (movement) of its corresponding corner. This implementation is well detailed in our previous publication [6].

### 4.1.3    CUDA-Based Gravity Center Detection

The CUDA-based computation of intersection points consists of porting on GPU the algorithm described in Listing 1.1. Indeed, we selected a GPU grid with a matrix of two dimensions. Each dimension consists of N threads, where N represents the number of detected motion vectors. With this representation, each CUDA thread is identified, within the GPU grid, by its x and y coordinates, that allows to compute the intersection point between i and j motion vectors. This implementation allows to replace the two loops presented in Listing 1.1 (lines 2 and 8) by the CUDA parallel treatment applied with the N × N selected threads. As result, each CUDA thread provides one intersection point if it exists. These intersection points are stored in an array called "intersections". The cavity center of gravity is then computed on GPU by selecting one CUDA thread that computes the mean position between the detected intersections stored in "intersections" array. This cavity center is copied to the CPU memory in order to apply the next steps of our algorithm on CPU. These steps are developed with the OpenCV library[3].

## 4.2  OpenCL-Based Functions

The OpenCL-based functions (OpenCL 2.0) are developed for the same steps (pre-processing, motion vectors and gravity center detection) in order to provide a portable solution allowing to exploit ATI graphic cards also.

### 4.2.1  OpenCL-Based Pre-processing and Motion Vectors Detection

This implementation is so similar to the corresponding CUDA version. The main difference between CUDA and OpenCL methods is that with OpenCL, we have to create a context in order to specify the device. In this way, the same code can be used for programming either CPU or GPU. Our OpenCL-based pre-processing and motion vectors detection steps are developed with the OpenCL module of OpenCV[4].

### 4.2.2  OpenCL-Based Gravity Center Detection

The OpenCL-based computation of gravity center is based on the same process described in Sect. 4.1.2. The main difference between CUDA and OpenCL versions are:

– The OpenCL version starts by creating the context in order to select the CPU or GPU
– The OpenCL function is executed by Work-items (equivalent to CUDA thread), which are grouped in work groups (equivalent to blocks in CUDA)
– With OpenCL, the same code can be used for CPU or GPU computing.

## 4.3  GPU-Based Left Ventricle Detection and Tracking

The above-mentioned CUDA and OpenCL implementations are included within our method in order to provide a portable GPU-based left ventricle detection and tracking. The latter is described within five steps: GPU selection, images loading, GPU-based gravity center detection, CPU-based left ventricle tracking and result visualization (Fig. 2).

### 4.3.1  GPU Selection

First, the program detects the type of available GPU. In case of NVIDIA cards, CUDA implementations are called for the next steps. Otherwise, in case of ATI graphic cards, the OpenCL implementations are called. An OpenCL context is also created for specifying the GPU for computation.

---

[3] OpenCV. www.opencv.org.
[4] OpenCV OpenCL. www.opencv.org/modules/ocl.

### 4.3.2   Image Loading

After selecting the available GPU, the input ultrasound images are uploaded and copied to the selected GPU.

### 4.3.3   GPU-Based Gravity Center Detection

This step affects CUDA treatments in case of NVIDIA GPUs, and OpenCL treatments in case of ATI graphic cards.

– CUDA-based gravity center detection: this implementation is described in Sect. 4.1
– OpenCL-based gravity center detection: this implementation is described in Sect. 4.2

### 4.3.4   CPU-Based Left Ventricle Tracking

This step applies the treatment described in Sect. 3.2.

### 4.3.5   Result Visualization

The results are displayed in a window showing the left ventricle contour and its center for each video frame. Thanks to the parallel processing, a real time treatment is achieved.

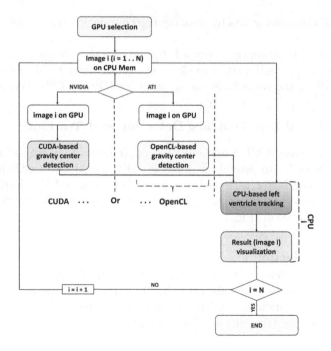

**Fig. 2.** GPU-based left ventricle detection and tracking

Notice that the above-mentioned CUDA and OpenCL functions exploit the texture and shared memories of GPU that offer a fast access to data (image pixels). The CUDA streaming technique was also exploited in order to overlap data transfers (Sect. 4.3.2) by the CUDA kernels execution (Sect. 4.3.3).

## 5    Experimental Results

The tests were run on the following hardware:

- CPU: Intel Core (TM) i5, 2520M CPU@ 2,50 GHz 2,50 GHz, RAM: 4 GB;
- GPU NVIDIA: GeForce GTX 580, RAM: 1.5 GB, 512 CUDA cores.

Experimentations are conducted using a set of 11 normal and 17 disease hearts echographic video sequences. The sequences are provided from the department of cardiology at the Hospital of Tlemcen in collaboration with Dr. Abbou[5]. The ultrasound videos present different resolutions:

- Low resolution videos: $320 \times 240$
- Medium resolution videos: $640 \times 480$
- High resolution videos: $1058 \times 794$

**Table 1.** Comparison of our method with literature algorithms using a video (normal case) of $1058 \times 794$ pixels

|                | Masuda *et al.* | Takahashi *et al.* | Our method |
| --- | --- | --- | --- |
| Success rate   | 85 %            | 88 %               | 92 %       |
| Automatic      | Semi-automatic  | Semi-automatic     | Fully-automatic |
| Computing unit | CPU             | CPU                | GPU        |
| Frame rate     | 4 fps           | 4 fps              | 27.8 fps   |

The image data set was inspected by an experienced cardiologic, from our collaborative hospital, who validated the results obtained within our method. Our above-mentioned method allowed to obtain accurate, automatic and real-time left ventricle detection and tracking with a rate of 92 % of success. Table 1 presents a comparison of our results with Masuda [10] and Takeshima [19] methods that have been published recently. This comparison is performed with the same data set described above. As shown in Table 1, our method outperforms the state of art algorithms in terms of success rate and computation time. The accuracy of our approach is due to the computation and refresh of gravity center at each frame of the video, which allowed to improve the precision. The latter was reduced in [10] since they refresh the gravity center only if the difference

---

[5] Department of cardiology, Tlemcen University Hospital, Tlemcen, Algeria.

between two consecutive gravity centers is bigger than a fixed threshold. This made their method less precise and not fully automatic. Moreover, the quality of our method was improved thanks to the pre-processing step (morphological operations, median filtering and binarization) that allowed to reduce noise, and hence improve the left ventricle detection. Otherwise, our CUDA and OpenCL implementations of the high intensive steps allowed to provide an accelerated method which can exploit both NVIDIA and ATI graphic cards. This acceleration allowed to obtain a real time processing either when treating high definition videos as shown in Table 2. Notice that the use of GPU offers low acceleration in case of processing low resolution videos. This is due to the weak exploitation of graphic processing units. Indeed, GPUs are more adapted for massively parallel applications. We note also that CUDA offers better performance than OpenCL since CUDA presents the most performant GPU programming language. Notice that the OpenCL performance are obtained using a GPU NVIDIA. This allowed to obtain a fair comparison of performance. Our OpenCL implementation was developed in order to offer a portable solution for left ventricle detection and tracking.

**Table 2.** GPU performances of left ventricle detection and tracking

| Resolution | 2 CPU | GPU (CUDA) | | GPU (OpenCL) | |
|---|---|---|---|---|---|
| | | fps | Acc (x) | fps | Acc (x) |
| 320 × 240 | 23 fps | 35.7 | 1.55 × | 31.3 | 1.36 × |
| 640 × 480 | 14 fps | 30.6 | 2.19 × | 27.5 | 1.96 × |
| 1058 × 794 | 4 fps | 27.8 | 6.95 × | 26.1 | 6.52 × |

# 6   Conclusion

In this paper, we proposed an accurate and automatic method for left ventricle segmentation and tracking in real time. The method is based on optical flow estimation for detecting the left ventricle center. The contour is then detected and tracked within the convex hull algorithm. Experimentations showed promising results thanks to the parallel exploitation of GPUs computing units that offered a real time processing even when treating high definition echographic videos. The accelerated GPU treatment (28 fps), which was 7 times faster than the CPU version (4 fps), allowed to refresh the gravity center calculation at each video frame that offered a better success rate of detection (92 % instead of 85 %). Notice also that our application can exploit in an efficient way both NVIDIA and ATI as a result of using CUDA and OpenCL APIs. As future work, we plan to improve our complexity estimation technique in order to have a better distribution of tasks between the available computing units (CPUs or/and GPUs). We plan also to apply our method for detecting and tracking left ventricle with larger data sets of 3D ultrasound videos.

# References

1. Chen, W., Beister, M., Kyriakou, Y., Kachelries, M.: High performance median filtering using commodity graphics hardware. In: Nuclear Science Symposium Conference Record (NSS/MIC), pp. 4142–4147. IEEE (2009)
2. da Cunha Possa, P., Mahmoudi, S., Harb, N., Valderrama, C.: A new self-adapting architecture for feature detection. In: 2012 22nd International Conference on Field Programmable Logic and Applications (FPL), pp. 643–646 (2012)
3. Fitzgibbon, A., Fisher, R. B.: A buyer's guide to conic fitting. In: British Machine Vision Conference, pp. 513–522 (1995)
4. Lionetti, F.V., McCulloch, A.D., Baden, S.B.: Source-to-source optimization of CUDA C for GPU accelerated cardiac cell modeling. In: D'Ambra, P., Guarracino, M., Talia, D. (eds.) Euro-Par 2010, Part I. LNCS, vol. 6271, pp. 38–49. Springer, Heidelberg (2010)
5. Lowe, D.G.: Distinctive image features from scale-invariant keypoints. Int. J. Comput. Vis. (IJCV) **60**(2), 91–110 (2004)
6. Mahmoudi, S.A., Kierzynka, M., Manneback, P., Kurowski, K.: Real-time motion tracking using optical flow on multiple GPUs. Bull. Pol. Acad. Sci. Tech. Sci. **62**, 139–150 (2014)
7. Mahmoudi, S. A., Manneback, P.: Efficient exploitation of heterogeneous platforms for images features extraction. In: 3rd International Conference on Image Processing Theory, Tools and Applications (IPTA), pp. 91–96 (2012)
8. Mahmoudi, S.A., Manneback, P.: Multi-CPU/multi-GPU based framework for multimedia processing. In: Amine, A., Bellatreche, L., Elberrichi, Z., Neuhold, E.J., Wrembel, R. (eds.) Computer Science and Its Applications. IFIP AICT, vol. 456, pp. 54–65. Springer, Heidelberg (2015)
9. Marzat, J., Dumortier, Y., Ducrot, A.: Real-time dense and accurate parallel optical flow using CUDA. In: Proceedings of WSCG, pp. 105–111 (2009)
10. Masuda, K., Takahashi, R., Yoshinaga, T., Uchibori, S.: Elucidation of intersection distribution in motion vectors from successive echocardiograms and its application for heart disease recognition. In: Dössel, O., Schlegel, W.C. (eds.) World Congress on Medical Physics and Biomedical Engineering, pp. 572–574. Springer, Heidelberg (2009)
11. Mizukami, Y., Tadamura, K.: Optical flow computation on compute unified device architecture. In: Proceedings of the 14th International Conference on Image Analysis and Processing, pp. 179–184 (2007)
12. Papademetris, X., Sinusas, A.J., Dione, D.P., Duncan, J.S.: Estimation of 3D left ventricular deformation from echocardiography. Med. Image Anal. **5**(1), 17–28 (2001)
13. Perrot, G., Domas, S., Couturier, R.: Fine-tuned high-speed implementation of a GPU-based median filter. J. Sig. Process. Syst. **75**, 185–190 (2014)
14. Possa, P., Mahmoudi, S., Harb, N., Valderrama, C., Manneback, P.: A multi-resolution FPGA-based architecture for real-time edge and corner detection. IEEE Trans. Comput. **63**(10), 2376–2388 (2014)
15. Ready, J.M., Taylor, C.N.: GPU acceleration of real-time feature based algorithms. In: Proceedings of the IEEE Workshop on Motion and Video Computing, p. 8 (2007)
16. Shi, L., et al.: A survey of GPU-based medical image computing techniques. Quant. Imaging Med. Surg. **2**(3), 188–206 (2012)

17. Sinha, S.N., Fram, J.-M., Pollefeys, M., Genc, Y.: GPU-based video feature tracking and matching. In: EDGE, Workshop on Edge Computing Using New Commodity Architectures (2006)
18. Suhling, M., Arigovindan, M., Jansen, C., Hunziker, P., Unser, M.: Myocardial motion analysis from B-mode echocardiograms. IEEE Trans. Image Process. **14**(4), 525–536 (2005)
19. Takeshima, S., Matsuda, H., Yoshinaga, T., Masuda, K.: Development of automatic recognition software of left ventricle by time series processing echocardiograms and application to disease heart. In: 2011 Biomedical Engineering International Conference (BMEiCON), pp. 165–168, January 2011
20. Tomasi, C., Kanade, T.: Detection and tracking of point features.: Technical report CMU-CS-91-132, CMU, pp. 1–4 (1991)
21. Yang, Z., Zhu, Y., Pu, Y.: Parallel image processing based on CUDA. In: International Conference on Computer Science and Software Engineering, China, pp. 198–201 (2008)

# Parallel Ant Colony Optimization for the HP Protein Folding Problem

Antonio Llanes, Carlos Vélez, Antonia M. Sánchez, Horacio Pérez-Sánchez, and José M. Cecilia[✉]

Bioinformatics and High Performance Computing Research Group (BIO-HPC),
Computer Science Department,
Universidad Católica San Antonio de Murcia (UCAM), Murcia, Spain
{allanes,asanchez,hperez,jmcecilia}@ucam.edu, cvelez@alu.ucam.edu

**Abstract.** Ant Colony Optimisation (ACO) is a bio-inspired population-based metaheuristic which emulates the ant colony's behavior to solve problems computationally. Indeed, it is a *swarm*-based algorithm as it needs the interactions among all ants to provide good solutions to a particular problem. This collective computation is theoretically well-suited for parallelisation as several ants run in parallel looking for solutions, sharing their findings among them. In this paper, we design an ACO metaheuristic to solve the Protein Folding Problem using a simplified model (HP) that identifies amino acids like Hydrophobic (H) or Polar (P), attending to the attraction or the rejection that the amino acid present against water. We also propose a parallel ACO version applied to the HP model on Graphics Processing Units (GPUs) using Compute Unified Device Architecture (CUDA). Our results reveal up to 7× speed-up factor compared to a sequential counterpart version. Results and conclusions about this parallel version suggests a broader area of inquiry, where researchers within the fields of Bioinformatics may learn to adapt similar problems to the tupla of an optimization method and GPU architecture.

**Keywords:** ACO · HP · GPUs · HPC · CUDA

## 1 Introduction

Ant Colony Optimization (ACO) was originally introduced by Dorigo et al. [12]. This is a stochastic algorithm used to solve several computational problems by simulating the behavior of an ant colony. As in the real life, an ant does not have enough intelligence to solve a particular problem by itself, but all ants within the colony can cooperate to solve problems efficiently. This algorithm can be classified as *Swarm Intelligence* [20], or *Metaheuristics* [17], both of them belong to a large number of algorithms within the umbrella of *Soft Computing* [3,27]. ACO has been probed in a variety of problems, including vehicle routing [28], feature selection [6], or autonomous robot [15].

Of particular interest to us is hydrophobic-hydrophilic (HP) model introduced by Dill [8] to reduce the protein folding complexity. This model assumes

© Springer International Publishing Switzerland 2016
F. Ortuño and I. Rojas (Eds.): IWBBIO 2016, LNBI 9656, pp. 615–626, 2016.
DOI: 10.1007/978-3-319-31744-1_54

that the hydrophobic interactions make an important contribution to the free energy of the folding process, so a protein is modeled as an specific sequence of hydrophobic (H for nonpolar) or hydrophilic (P for polar) monomers. The optimal solution to this model is the conformation with more number of adjacencies between H's that originally were not contiguous. This problem is a NP-complete optimization problem according to [2].

Due to the relevance of Protein Folding Problem and the effectiveness of the HP model, intensive research work in this line has been recently developed. Therefore, several approaches based on the application of different optimization methods are described in literature including Monte Carlo methods [25], evolutionary algorithms [16,24], and particle swarm optimization [21], just to name a few.

The choice of model and its associated algorithm is mainly motivated by the required objectives, but it is also constrained by the computer hardware characteristics attainable in the relevant time frame. The role of the software developer is increasingly important as their algorithms are expected to handle a soft balance between performance, power consumption and the quality obtained in the results. About performance, developers have to test different implementations, considering which code fits perfectly with the hardware platform where they run their codes. About power consumption, this issue is increasingly importance specially in large clusters [23]. Finally, developers have to take care about quality due to the inherit stochastic nature of ACO, so the goal of ACO is to reduce drastically the computation time maintaining the quality of the results.

This paper shows the parallelisation of ACO metaheuristic on Graphics Processing Units to solve the Protein Folding Problem using HP model. Our implementation leads to factor gains exceeding 7× as applied to the protein folding when compared to its sequential counterpart version running on a similar single-threaded high-end CPU. Moreover, an extensive discussion focused on different implementation paths on GPUs shows the way to deal with parallel graph connected components. The rest of the paper is structured as follows: First, a description of ACO and HP are introduced, next we describe the process used to implement the parallel version, giving details of our design. Then, preliminary experimental results are shown to finish with some conclusions and directions for future work.

## 2    Methodology

This section describes HP model for the protein folding as well as the way we adapt the ACO algorithm to optimize this problem. Moreover, we briefly review the main characteristics of CUDA for those who are not familiar with this programming model.

### 2.1    Description of HP Model

HP model for the protein folding problem was introduced by Dill et. al [8], and it has been widely used to predict protein structures, like [1,4,22]. In HP model each protein sequence is represented as a string $A = a_1, a_2, ..., a_n$, where $a_i \in H, P$ and $1 \leq i \leq n$. A conformation of $A$ is defined by a sequence of fold directions starting from the lattice site occupied by the first amino acid $a_1$. The different protein conformations are restricted to self-avoiding paths on two or three dimensions. Most protein structure prediction methodologies assume that the native state of the protein is defined by the lowest value of the Gibbs free energy what is estimated by a specific scoring function, which strongly depends on the coarse-grained or all-atom model used for representing the protein structure [9].

In HP model, the energy of a conformation is defined by a scoring function that assumes that the hydrophobic interactions make an important contribution to the free energy of the folding process. The optimal solution to this model will be the conformation with more number of adjacencies between H's (topological contacts) that originally weren't contiguous in the given sequence. Thus, the Protein Folding Problem is translated into an optimization problem as follows: Given an amino acid sequence $A = a_1, a_2, ..., a_n$, find an energy minimizing conformation $\mathcal{C}^o$.

$$EC^o = min\{E(\mathcal{C}) \forall \mathcal{C}\} \tag{1}$$

where $\mathcal{C}$ is a valid conformation of the string $A$ and $E(\mathcal{C})$ is defined by

$$EC = \sum_{i,j} e(a_i, a_j) \tag{2}$$

where

$$e(a_i, a_j) = \begin{cases} -1 & \text{if } a_i, a_j = HH \text{ and they form a topological contact} \\ 0 & \text{otherwise} \end{cases}$$

Therefore, the objective function for the HP protein folding problem is defined by Eq. 2, and its values will be referred as scoring values. The goal of our parallel version of ACO is the computation of the conformations which achieve minimum values of the energy function according to Eq. 2.

## 2.2 Ant Colony Optimization for the Protein Folding Based on HP Model

*Ant Colony Optimization* (ACO) [7,11,13] is based on foraging behavior observed in colonies of real ants. The method generally uses simulated "ants" (i.e., mobile agents), which first construct tours or paths on a network structure (corresponding to solutions for a problem), and then deposit "pheromone" (i.e., signaling chemicals) according to the quality of the solution generated. The algorithm takes advantage of emergent properties of the multi-agent system, where positive feedback (facilitated by pheromone deposition) quickly drives the population to high quality solutions.

The original ACO method (called the *Ant System* [12]) was developed by Dorigo in the 1990s, and this version (or slight variants thereof, such as the MAX-MIN Ant System (MMAS) [26]) is still in regular use [5,14,19]. The Ant System (AS) algorithm is divided into two main stages: *Conformation construction* and *Pheromone update*. Conformation construction is based on $m$ ants building protein conformations in parallel. Those protein conformations are constructed based on a probabilistic action choice rule, called the *random proportional rule* in order to decide which position to place the next amino acid (restricted to H or P). The probability for ant $k$, placed at position $i$, of amino acid $j$ is given by the Eq. 3

$$p_{i,j}^k = \frac{[\tau_{i,j}]^\alpha [\eta_{i,j}]^\beta}{\sum_{l \in N_i^k} [\tau_{i,l}]^\alpha [\eta_{i,l}]^\beta}, \qquad if\ j \in N_i^k, \tag{3}$$

where $\eta_{i,j} = C_{i,j}$ is a heuristic value that represents the number of non- contiguous H-H contacts, $\alpha$ and $\beta$ are two parameters which determine the relative *influences* of the pheromone trail and the heuristic information respectively, and $N_i^k$ is the feasible neighbourhood of ant $k$ when at position $i$. This latter set represents the set of positions that ant $k$ has not yet visited; the probability of choosing a position outside $N_i^k$ is zero (this prevents an ant returning to a position, which is not allowed in the HP model). By this probabilistic rule, the probability of choosing a particular edge $(i,j)$ increases with the value of the associated pheromone trail $\tau_{i,j}$ and of the heuristic information value $\eta_{i,j}$. The random proportional rule ends with a selection procedure, which is done analogously to the *roulette wheel* selection procedure of evolutionary computation (for more detail see [10,18]). Each value $\tau_{i,j}^\alpha \eta_{i,j}^\beta$ of a position j that ant k has not visited yet determines a slice on a circular roulette wheel, the size of the slice being proportional to the weight of the associated choice. Next, the wheel is spun and the position to which the marker points is chosen as the next position for ant k. Furthermore, each ant $k$ maintains a memory, $M^k$, called the *tabu list*, which contains the positions already visited. This memory is used to define the feasible neighbourhood, and also allows an ant to both to compute the conformation's score $T^k$ it generated, and to retrace the path to deposit pheromone.

After all ants have constructed their protein conformations, the pheromone trails are updated. This is achieved by first lowering the pheromone value on all edges by a constant factor, and then adding pheromone on edges that ants have crossed in their conformations. Pheromone evaporation is implemented by

$$\tau_{i,j} \leftarrow (1 - \rho)\tau_{i,j}, \qquad \forall(i,j) \in L, \tag{4}$$

where $0 < \rho \leq 1$ is the pheromone evaporation rate. After evaporation, all ants deposit pheromone on their visited edges:

$$\tau_{i,j} \leftarrow \tau_{i,j} + \sum_{k=1}^m \Delta\tau_{i,j}^k, \qquad \forall(i,j) \in L, \tag{5}$$

where $\Delta\tau_{ij}$ is the amount of pheromone ant $k$ deposits. This is defined as follows:

$$\Delta\tau_{i,j}^k = \begin{cases} 1/C^k if\ e(i,j)^k \text{ belongs to } T^k \\ 0 \qquad \qquad \text{otherwise} \end{cases} \tag{6}$$

where $C^k$, the score of the conformation $T^k$ built by the $k$-th ant, is computed as the sum of number of non-contiguous H-H contacts belonging to $T^k$. According to Eq. 6, the better an ant's conformation, the more pheromone the edges belonging to this conformation receive. In general, edges that are used by many ants, receive more pheromone, and are therefore more likely to be chosen by ants in future iterations of the algorithm.

### 2.3 CUDA Programming Model

Before we discuss our parallel versions, we briefly review the main characteristics of CUDA, for the benefit of readers who are unfamiliar with the programming model. CUDA is based on a hierarchy of abstraction layers; the *thread* is the basic execution unit; threads are grouped into *blocks*, each of which runs on a single multiprocessor, where they can share data on a small but extremely fast memory. A *grid* is composed of blocks, which are equally distributed and scheduled among all multiprocessors. The parallel sections of an application are executed as *kernels* in a SIMD (Single Instruction Multiple Data) fashion, that is, with all threads running the same code. A kernel is therefore executed by a grid of thread blocks, where threads run simultaneously grouped in batches called *warps*, which are the scheduling units.

## 3    Parallel Ant Colony Optimization on GPUs for the HP Protein Folding

This Section summarizes the parallelization process of the Ant Colony Optimization as applied to the HP protein folding using CUDA. Algorithm 1 shows Single Program Multiple Data (SPMD) pseudocode for the AS. Firstly, all AS structures for the HP protein folding problem (Conformation matrix, number of amino acids,...) are initialized. Next, the conformation construction and pheromone update stages are performed until the convergence criterion is reached (number of iterations in our case). Both stages are executed by each ant and it can be repeated several times from the beginning to restart the computation. This restarts are implemented to avoid stalling in local optimum.

---

**Algorithm 1.** Sequential pseudocode of ACO-HP

---
1: **for** each epoch **do**
2:     **for** each ant **do**
3:         conformationConstruction()
4:         updatePheromoneMatrix()
5:     **end for**
6: **end for**

---

In what follows, we describe the parallelization approach for both main kernels of ACO algorithm: Conformation construction and Pheromone Update.

## 3.1    Conformation Construction Parallelization

The "traditional" task parallelism approach to conformation construction is based on the observation that ants run in parallel looking for the best protein conformation they can find. Therefore, any inherent parallelism exists at the level of individual ants. To implement this idea of parallelism using CUDA, each ant is identified as a CUDA thread, and threads are equally distributed among CUDA thread blocks. Each thread deals with the task assigned to each ant; i.e., maintenance of an ant's memory (list of all visited positions, and so on) and movement which is mainly based on the *random proportional rule* previously explained.

## 3.2    Pheromone Update Parallelization

The final stage in the ACO algorithm is pheromone update, which comprises two main tasks: pheromone evaporation and pheromone deposit. The first step is quite straightforward to implement in CUDA, as a single thread can independently calculate the Eq. 4 for each entry of the pheromone matrix, thus lowering the pheromone value on all edges by a constant factor.

Ants then deposit different quantities of pheromone on the edges that they have crossed to create their conformations. As stated previously, the quantity of pheromone deposited by each ant depends on the quality of the protein conformation found by that ant. This kernel allocates a thread per each amino acid, which is placed at different positions. Each ant generates its own private conformation in parallel, and they may place an amino acid into the same position as another ant. This fact forces us to use atomic instructions for accessing the pheromone matrix.

# 4    Experimental Results

This section briefly shows the experimental results obtained with our implementation. First of all we review our experimental set up. Then we proceed with an evaluation of both sequential and parallel versions before we present the experimental results.

## 4.1    Hardware Environment

During our experimental study, we have used the following platforms:

- **On the CPU side:** Four Intel Xeon X7550 processors running at 2 GHz and plugged into a quad-channel motherboard endowed with 128 GB of DDR3 memory.
- **On the GPU side:** GPU NVIDIA Tesla Kepler K40c with 2.880 cores, (15 multiprocessors with 192 cores each), running at 880 MHz, offering a processing power up to 5.068 GFLOPS. It also have 12 GB of GDDR5 RAM with ECC capability and a buswidth of 384 bits, giving a bandwidth of 288 GB per second.

**Table 1.** Hardware Description

| | | Intel CPU | NVIDIA |
|---|---|---|---|
| | Vendor and type | Intel CPU | NVIDIA |
| | Family | Haswell | Kepler |
| | Class | Xeon | Tesla |
| | Model | X7750 | K40c |
| | Year | 2015 | 2014 |
| Processing elements | Cores per multiprocessor | (does not apply) | 192 |
| | Number of multiprocessors | | 15 |
| | Total number of cores | 8 | 2880 |
| | Clock frequency (MHz) | 2000 | 880 |
| Maximum number of GPU threads | Per multiprocessor | (does not apply) | 2048 |
| | Per block | | 1024 |
| | Per warp | | 32 |
| SRAM memory (per multiprocessor in GPU) | Shared (only GPU) | 32 KB L1D | 16 or 48 KB |
| | L1 cache | and | 48 or 16 KB |
| | (Shared + L1) | 32 KB L1IZ | 64 KB |
| L2 cache | (shared by all cores) | 256 KB | 1536 KB |
| L3 cache | | 16 MB | (d.n.a) |
| DRAM memory | Size (MB) | 131072 | 11520 |
| | Speed (MHz) | $2 \times 666$ | $2 \times 3004$ |
| | Width (bits) | 256 | 384 |
| | Bandwidth (GB/s) | 42.66 | 288.34 |
| | Tecnology | DDR3 | GDDR5 |
| CUDA compute capabilities | | (d.n.a.) | 3.5 |

Table 1 shows a detailed descriptions of all these platforms. Moreover, we use gcc 4.8.2 with the -O3 flag to compile on the CPU, and the CUDA compiler/driver/runtime version 6.5 to compile and run on the GPU.

## 4.2 Profiling

This section briefly shows the performance analysis for both GPUs and CPU codes. The profiling is performed with Microsoft Visual Profiler for the sequential code, and with NVIDIA Visual Profiler for the CUDA code. Figure 1 shows the sequential code profiling where the Conformation Construction stage takes more than half of total computation time, this function is the responsible to create a

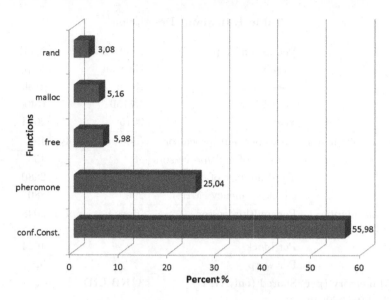

**Fig. 1.** Main functions for our ACO implementation

**Table 2.** Kernel occupancy

| Function name | Duration ($\mu s$) | Occupancy |
|---|---|---|
| conformationConstruction_Cuda | 1.888.756,607 | 62.50 % |
| startPheromoneMatrix_Cuda | 65.182,431 | 100 % |
| pheromoneVaporize_Cuda | 51.420,896 | 100 % |
| stateVectorGenerator | 1.198,016 | 75 % |
| updatePheromoneMatrix_Cuda | 51,904 | 100 % |
| iamax_kernel | 8,479 | 100 % |
| updatePheromoneMatrix_BestSolution_Cuda | 5,216 | 100 % |

valid conformation for each ant. In second position is the pheromone stage which is actually parallelized as well.

The Table 2 summarizes all the kernels implemented in the new parallel code, with theirs occupancy rate. All kernels have a high occupancy rate.

### 4.3   Execution Results

This section shows our experimental results from several points of view. Firstly, we evaluate the scalability of our parallel implementation, varying the number of ants from 256 to 8192 ants (see Table 3). In the conformation construction stage, each ant is identify to a CUDA thread and those threads are equally divided into blocks. Therefore, the number of threads per block depends on the number of ants set. Our algorithm prevents this situation by setting our

**Table 3.** Execution time (in seconds) for both: sequential and CUDA implementations of HP protein folding problem by varying the number of ants.

| | Execution time (s) | |
|---|---|---|
| Number of ants | Sequential version | CUDA version |
| **256** | *16.188* | 53.2158 |
| **512** | *30.114* | 53.4919 |
| **1024** | 59.035 | *53.891* |
| **2048** | 117.434 | *53.2006* |
| **4096** | 234.533 | *57.8629* |
| **8192** | 483.835 | *67.3611* |

empirically demonstrated optimum thread block layout for each case. Whenever the number of ants is large enough, the best configuration is for 256 threads per block.

Table 3 shows a great scalability along with the number of ants. However, it also shows that with a low number of ants, sequential code run even faster than the parallel version. It is due to the extra charge of transferring of data between host and device, in these cases, CPU can support the charge of the complexity of the problem in order to maintain advantage from the parallel implementation. But, in the other hand, as soon as we increment the number of ants, the sequential code experiments an exponential increment of time, unlike the CUDA version. This is exactly what we expect as the data parallelism in this problem is not very high, besides, we have implemented the kernel with more time consumption "conformationConstruction()", and we established the association from one ant to one thread, our benefits become higher when we increase the number of ants.

Table 3 shows the execution times in seconds for both implementations we have developed in this work. These experimental results are obtained by using as a benchmark the "1tuk" protein which contains up to 67 amino acids in a 3-D fashion. Moreover, ACO parameters include the following: 1000-independent ACO runs, $\alpha = 1$ and $\beta = 3$.

Figure 2 shows the execution time fixing the number of ants (up to 2048), and varying proteins in order to test the algorithm with several number of amino acids. It is noteworthy to point out that the parallel version is faster than the sequential counterpart version in all the proteins tested. In this case, when the number of amino acids is increased, the differences between sequential and parallel version get closer, this is for the same reason that previously was presented, since we established the association of one thread to one ant, our benefits become higher when the number of ants is increased, beating the sequential code by a wide margin.

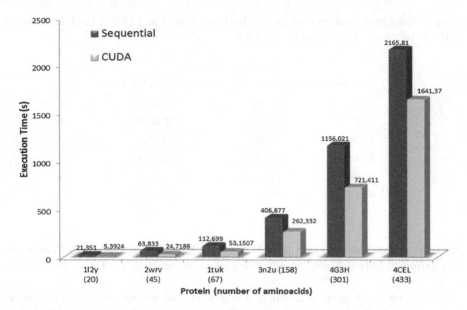

**Fig. 2.** Execution time in seconds for the execution of sequential and CUDA ACO implementations for different proteins that have different number of amino acids (Color figure online).

## 5    Conclusions and Outlook

Ant Colony Optimization (ACO) belongs to the family of population-based metaheuristics that has been successfully applied to many NP-complete problems. In this work, we present a parallel version of ACO algorithm as applied to the protein folding problem on Graphics Processing Units. We use a well-known coarse grained HP model that classifies amino acids into Hydrophobic (H) or Polar(P), attending to the attraction or the rejection that the amino acid present against water. We identify the natural parallelism of ACO; i.e., ants running in parallel to find out a solution with threads in the cuda programming model. Our experimental results leads performance gains up to 7× speedup factor compared to its sequential counterpart version.

The tupla ACO and protein folding on GPUs is still at a relatively early stage, and we acknowledge that we have tested a relatively simple variant of the algorithm and the protein model. But, with many other types of combinations still to be explored, this field seems to offer a promising and potentially fruitful area of research. Especially due to the nature of the problem, this is not a problem with a lot of data to compute, only 4 (in 2 dimensions) or 6 (in 3 dimensions) operations we can do simultaneously. However, some algorithmic improvements may be introduced to enhance performance. Among them, we may highlight to provide a data-parallelism design that takes advantage of vector-fashion execution in current processor architectures.

**Acknowledgements.** This work has been funded by grants from the Fundación Séneca of the Región of Murcia (18946/JLI/13) and by the Nils Coordinated Mobility under grant 012-ABEL-CM-2014A, in part financed by the European Regional Development Fund (ERDF). We also thank Nvidia for the hardware donation under GPU Research and Educational Center Program.

# References

1. Backofen, R., Will, S.: A constraint-based approach to fast and exact structure prediction in three-dimensional protein models. Constraints **11**(1), 5–30 (2006)
2. Berger, B., Leighton, T.: Protein folding in the hydrophobic-hydrophilic (HP) model is NP-complete. J. Comput. Biol. **5**(1), 27–40 (1998)
3. Bonissone, P.P.: Soft computing: the convergence of emerging reasoning technologies. Soft Comput.-Fusion Found. Methodol. Appl. **1**(1), 6–18 (1997)
4. Bui, T.N., Sundarraj, G.: An efficient genetic algorithm for predicting protein tertiary structures in the 2D HP model. In: Proceedings of the 7th Annual Conference on Genetic and Evolutionary Computation, pp. 385–392. ACM (2005)
5. Chang, R.-S., Chang, J.-S., Lin, P.-S.: An ant algorithm for balanced job scheduling in grids. Future Gener. Comput. Syst. **25**(1), 20–27 (2009)
6. Chen, Y., Miao, D., Wang, R.: A rough set approach to feature selection based on ant colony optimization. Pattern Recogn. Lett. **31**(3), 226–233 (2010)
7. Di Caro, G., Dorigo, M.: Ant colony optimization: a new meta-heuristic. In: Proceedings of the Congress on Evolutionary Computation (1999)
8. Dill, K.A., Bromberg, S., Yue, K., Fiebig, K.M., Yee, D.P., Thomas, P.D., Chan, H.S.: Principles of protein folding–a perspective from simple exact models. Protein Sci. **4**(4), 561 (1995)
9. Dill, K.A., MacCallum, J.L.: The protein-folding problem, 50 years on. Science **338**(6110), 1042–1046 (2012)
10. Dorigo, M., Stützle, T.: Ant Colony Optimization. Bradford Company, Scituate (2004)
11. Dorigo, M., Birattari, M., Stützle, T.: Ant colony optimization. IEEE Comput. Intell. Mag. **1**(4), 28–39 (2006)
12. Dorigo, M., Maniezzo, V., Colorni, A.: Ant system: optimization by a colony of cooperating agents. IEEE Trans. Syst. Man Cybern. Part B: Cybern. **26**(1), 29–41 (1996)
13. Dorigo, M., Stützle, T.: Ant colony optimization: overview and recent advances. In: Gendreau, M., Potvin, J.-Y. (eds.) Handbook of Metaheuristics. ISOPMS, vol. 146, pp. 227–263. Springer, Heidelberg (2010)
14. Eberhart, R.C., Kennedy, J.: A new optimizer using particle swarm theory. In: Proceedings of the 6th International Symposium on Micro Machine and Human Science, New York, NY, vol. 1, pp. 39–43 (1995)
15. Garcia, M.A.P., Montiel, O., Castillo, O., Sepúlveda, R., Melin, P.: Path planning for autonomous mobile robot navigation with ant colony optimization and fuzzy cost function evaluation. Appl. Soft Comput. **9**(3), 1102–1110 (2009)
16. García-Martínez, J.M., Garzón, E.M., Cecilia, J.M., Pérez-Sánchez, H., Ortigosa, P.M.: An efficient approach for solving the hp protein folding problem based on UEGO. J. Math. Chem. **53**(3), 794–806 (2015)
17. Glover, F., Kochenberger, G.A.: Handbook of Metaheuristics. Springer, Heidelberg (2003)

18. Golberg, D.E.: Genetic Algorithms in Search, Optimization, and Machine Learning. Addion Wesley, Boston (1989)
19. Ke, B.-R., Chen, M.-C., Lin, C.-L.: Block-layout design using max-min ant system for saving energy on mass rapid transit systems. IEEE Trans. Intell. Transp. Syst. **10**(2), 226–235 (2009)
20. Kennedy, J., Kennedy, J.F., Eberhart, R.C., Shi, Y.: Swarm Intelligence. Morgan Kaufmann, Burlington (2001)
21. Kondov, I.: Protein structure prediction using distributed parallel particle swarm optimization. Nat. Comput. **12**(1), 29–41 (2013)
22. Liu, J., Li, G., Jun, Y., Yao, Y.: Heuristic energy landscape paving for protein folding problem in the three-dimensional HP lattice model. Comput. Biol. Chem. **38**, 17–26 (2012)
23. Pénzes, P.I., Martin, A.J.: Energy-delay efficiency of VLSI computations. In: Proceedings of the 12th ACM Great Lakes Symposium on VLSI, pp. 104–111. ACM (2002)
24. Schug, A., Wenzel, W.: An evolutionary strategy for all-atom folding of the 60-amino-acid bacterial ribosomal protein L20. Biophys. J. **90**(12), 4273–4280 (2006)
25. Strunk, T., Wolf, M., Wenzel, W.: Peptide structure prediction using distributed volunteer computing networks. J. Math. Chem. **50**(2), 421–428 (2012)
26. Stützle, T., Hoos, H.H.: Max-min ant system. Future Gener. Comput. Syst. **16**(8), 889–914 (2000)
27. Verdegay, J.L., Yager, R.R., Bonissone, P.P.: On heuristics as a fundamental constituent of soft computing. Fuzzy Sets Syst. **159**(7), 846–855 (2008)
28. Bin, Y., Yang, Z.-Z., Yao, B.: An improved ant colony optimization for vehicle routing problem. Eur. J. Oper. Res. **196**(1), 171–176 (2009)

# Neuroimaging Registration on GPU: Energy-Aware Acceleration

Francisco Nurudín Álvarez[1], José Antonio Cabrera[1], Juan Francisco Chico[2], Jesús Pérez[1], and Manuel Ujaldón[1(✉)]

[1] Computer Architecture Department, University of Málaga, Malaga, Spain
ujaldon@uma.es
[2] Brain Dynamics, Andalusian Technological Park, Malaga, Spain

**Abstract.** We present a CUDA implementation for Kepler and Maxwell GPU generations of neuroimaging registration based on the NiftyReg open-source library [1]. A wide number of strategies are deployed to accelerate the code, providing insightful guidelines to exploit the massive parallelism and memory hierarchy within emerging GPUs. Our efforts are analyzed from different perspectives: Acceleration, numerical accuracy, power consumption and energy efficiency, to identify potential scenarios where performance per watt can be optimal in large-scale biomedical applications. Experimental results suggest that parallelism and arithmetic intensity represent the most rewarding ways on the road to high performance bioinformatics when power is a major concern.

## 1 Introduction

Neuroimaging provides a crucial perspective for basic and clinical human neuroscience. A variety of neuroimaging technologies allow the structure and function of the intact human brain to be studied with minimal invasion, presenting a tremendous opportunity for a better understanding of healthy states and damaging when clinical surgery is applied.

Bioinformatics tools are key at all stages of neuroimaging, allowing scientists to control highly sophisticated imaging instruments and to make sense of the vast amounts of complex data generated by them. Over the past several years, members of the neuroimaging research community mainly sponsored by the National Institutes of Health have developed a number of reliable, accurate and easy to use tools. Among them, we highlight the Neuroimaging Informatics Technology Initiative (NIfTI) [2], created to provide coordinated and targeted service, training and research to speed the development and enhance the utility of informatics tools related to neuroimaging. NIfTI focuses on tools that are used within functional Magnetic Resonance Imaging (fMRI) [3]. First, because fMRI is rapidly growing, and second, because there are a small number of widely used informatics tools in the fMRI research community.

To facilitate inter-operation of functional MRI data analysis software packages, the NIfTI Data Format Working Group proposed the new analyze-style NIfTI data format. Thereafter, a number of NIfTI-aware toolkits were born

© Springer International Publishing Switzerland 2016
F. Ortuño and I. Rojas (Eds.): IWBBIO 2016, LNBI 9656, pp. 627–638, 2016.
DOI: 10.1007/978-3-319-31744-1_55

(e.g., FSL, AFNI, SPM, Freesurfer), with `dicomnifti` allowing the conversion from DICOM images into the NIfTI format for Linux users [4]. Later, `libnifti` provided a reference implementation of a C library to read, write and manipulate NIfTI images, and similarly, counterpart versions written in Java and Matlab were created too. The source code for those libraries was put into the public domain, and corresponding projects were hosted at SourceForge [5].

Once all those basic pillars consolidated, scientists started to develop libraries based on NIfTI format for many different purposes. Among them, `NiftyReg` [1] emerged to focus on rigid, affine and non-linear registration, and when CUDA was born, they incorporated a GPU-based implementation for Tesla (2008) and Fermi (2010) generations. However, no similar efforts were made on Kepler (2012) and Maxwell (2015) GPUs. This work contributes with new and upgraded kernels to take advantage of those features introduced over the last five years. In addition to GPU performance, our implementation worries about power consumption. And ultimately, we analyze correlation (if any) between speed-up and energy, to identify those potential scenarios where performance per watt can be maximized.

The rest of the paper is organized as follows. Section 2 briefly describes the way we use the `NiftyReg` library. Section 3 outlines the methods we have implemented as CUDA kernels for a GPU acceleration, and the set of optimizations performed on each kernel. Section 4 introduces the experimental setup used during our testbed, followed by Sect. 5 where our measurements are shown and analyzed. We end in Sect. 6 with the conclusions drawn from this work.

## 2    The NiftyReg Library

NiftyReg is a medical image registration library mostly used for brain analysis. It was developed by Marc Modat et al. at UCL (University College London), along with its parent project, NifTK [6].

The library has been fully implemented on CPU using C++, and some methods have a CUDA implementation available for Fermi GPUs [7]. Methods can also be executed from a standalone console application, with a flag to determine whether the CPU or GPU-based version is used.

Based on this software infrastructure, we take as departure point the execution of the `reg_f3d` method to compute the non-rigid registration for a set of consecutive images. In our particular case, this produces as output the corresponding 3D volume of a human brain for the zone filmed using MRI images. Figure 1 shows the sequence of functions that `reg_f3d` requires to be completed, and Table 1 summarizes the functionality for each of these functions, which have been transformed into optimized CUDA kernels as next section outlines.

## 3    Set of Optimizations Performed

The original GPU implementation of NiftyReg extends the CPU classes, which limits the amount of parallelism extracted via CUDA streams. In terms of accuracy, it makes use of single precision floating-point arithmetic, which is fine for

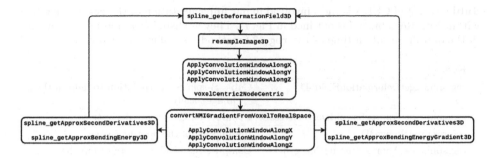

**Fig. 1.** Sequence of CUDA kernels involved during the execution of the `reg_f3d` application within NiftyReg library.

image registration algoritms where interpolations are frequent. Nevertheless, our GPU variants were all validated using an internal tool within NiftyReg to compare results versus those obtained on the CPU.

We identify several scenarios for a performance improvement of NiftyReg code on GPUs. Opportunities can be grouped into two different categories:

- **Parallelism.**
  1. Strategy: *Discontinue the use of conditional statements.* Rationale: Increase throughput in vector processing to benefit from the warp (an effective vector unit of size 32 for threads in most operations).
  2. Strategy: *Remove control and data dependencies* (mostly by reordering instructions whenever feasible). Rationale: Increase ILP (instruction-level parallelism) to take advantage of the graphics pipeline.
  3. Strategy: *Promote loop unrolling.* Rationale: Squeeze performance of vector units.
- **Memory.**
  1. Strategy: *Replace textures memory when accessed via 1D and 3D spaces.* Rationale: Textures are mostly optimized for 2D spaces (those used when rendering).
  2. Strategy: *Replace shared memory (cache) by constant memory.* Rationale: Exploit the memory hierarchy wisely.
  3. Strategy: *Promote the use of unified memory.* Rationale: Take advantage of new hardware features and adapt the code to future memory capabilities.

Optimizations related to parallelism will be tagged as versions 1.x of the library. Those regarding memory are numbered as 2.x for textures, 3 for constant and 4 for unified memory from now on (see Table 3).

## 3.1   Enhancing Parallelism

The best opportunity to improve parallelism in NiftyReg using the GPU was found in convolution kernels. As Fig. 1 outlines, there are three CUDA kernels

**Table 1.** List of CUDA kernels involved in our implementation of the `reg_f3d` method within NiftyReg library (as outlined in Fig. 1). They are listed from more to less weight in the overall execution time of the `reg_f3d` method (as Table 5 later reveals).

| Kernel | Description |
|---|---|
| reg_spline_getDeformationField3D | Cubic B-splines interpolation to deform the image locally |
| reg_spline_getApproxSecondDerivatives3D | Compute all the second derivatives. |
| reg_resampleImage3D | Resampling of the floating image |
| reg_spline_getApproxBendingEnergy3D | Compute the bending energy from the second derivatives |
| reg_ApplyConvolutionWindowAlongZ | NMI gradient field smoothing along Z axis |
| reg_ApplyConvolutionWindowAlongY | NMI gradient field smoothing along Y axis |
| reg_ApplyConvolutionWindowAlongX | NMI gradient field smoothing along X axis |
| reg_spline_getApprox-BendingEnergyGradient3D | Compute the gradient from the second derivatives |
| _reg_convertNMIGradient-FromVoxelToRealSpace | The similarity measure gradient is converted from voxel space to real space |
| reg_voxelCentric2NodeCentric | From the voxel-centric gradient values, it extracts the analytical node-centric derivative of the similarity measure |

making use of this operation (on X, Y and X directions), twice per iteration. That way, a successful optimization on any of these would benefit six kernels at a time. Those kernels receive as input parameters convolution windows with sizes ranging from 20 to 80, resulting in a loop with the number of iterations unkwnown at compile-time. We precalculated loop bounds by estimating minimum and maximum values within data structures for a given thread, and also reduced the number of instructions within loops to minimize branching costs. But preprocessing was costly and performance numbers were disappointing.

Afterwards, we studied the weighted sum of adjacent values in the data, that all convolutions perform using rounding error reduction schemes. This process, known as Kahan summation (or compensated summation) [8], computes the error produced by the addition of each weighted value for a total sum. The summation included conditional statements to make sure that final values remain within data boundaries, and again the loop with length depending on convolution window sizes introduces control and data dependencies to reduce opportunities for loop unrolling. With all these constraints, optimizations were tough.

Our first successful try was conducted using Pichat's summation [8], an error-accumulative version of Kahan's where instruction interleaving was found to be beneficial for instruction-level parallelism on GPUs. This resulted in version 1.1 for the kernel with a 1.50x speed-up, but at the expense of launching a higher number of kernels (more iterations are required for the method to finalize, reducing overall performance), and a slight loss in accuracy.

Then we took a look at the PTX (assembly) code to identify opportunities for window summations using FMAs (Fused Multiply-Add). This is a special machine instruction able to perform two floating-point operations per cycle on Nvidia GPUs. If we are lucky, compiler automatically reorganizes the code to take advantage of this feature, but often it is required for the programmer to rearrange the PTX code manually. This was our case for NiftyReg, and version 1.2 characterizes this effort, with a rewarding 1.32 speed-up factor and less number of kernel launches required for a similar accuracy to the original GPU implementation (but a little accuracy loss versus the CPU counterpart).

We introduce FMAs in Pichat method to produce version 1.3, where number of kernel launches again increase (to compensate accuracy loss), leading to similar performance results. And yet a version 1.4 was created with no error compensation as a way to minimize the number of iterations and maximize overall performance. Finally, unrolling was applied over the loop iterating on the window size. Given that popular window sizes are multiple of 20 plus 1, wise unrolling factors are 5 and 20.

Table 2 summarizes our performance and accuracy findings. Results correspond to the `ApplyConvWindowAlongX` kernel, and can easily be replicated for `ApplyConvWindowAlongY` an `ApplyConvWindowAlongZ`. Table 5 later reveals that these three methods over more than 17 % of the total execution time for the `reg_f3d` method.

Minor optimization chances for enhancing parallelism were found in other kernels with light workload, so we considered it was not worth the programing effort.

## 3.2   Changes in Memory Management

The GPU implementation of NiftyReg makes an extensive use of textures, in particular 1D and 3D textures. However, 2D are the most effective access to textures allocated in GPUs, because they are by far the most commonly used in video-games and the underlying hardware contains specific features to speed-up their handling. Additionally, texture reads in NiftyReg are linearized (accessed as if they were global memory arrays), and kernels using esoteric texture features (like coordinate normalization or wrapping) worked around them to produce the same result as having them disabled.

Our first attempt to verify 1D/3D textures efficiency was to replace them by global memory arrays. However, we found textures memory to outperform global memory by a 2x factor on Maxwell GPUs, and removal of texture-specific features was proven not to be sensitive to performance. This result also discouraged us from using shared memory (caches within each GPU multiprocessor), as it is not that effective for textures. Moreover, the size of the arrays was huge compared to caches holding less than 100 KB, and arrays are often accessed with poor locality.

**Table 2.** Set of GPU optimizations performed on convolution kernels to improve performance and accuracy of NiftyReg. Error is computed using the root mean square difference. "# of calls" is the number of kernel launches that are required for each version to complete the execution (overall speed-up is lower when a higher number of launches is required versus our departure point, that is, the baseline implementation shown as version 1.0).

| Optimization performed | # of calls | GPU error | CPU error | Unroll factor | Kernel time | Speed-up per kernel | Overall speed-up |
|---|---|---|---|---|---|---|---|
| Version 1.0 (no FMA, | 372 | 0.00 | 45.39 | No | 0.45 | 1.00x | 1.00x |
| Kahan method for | | | | 5 | 0.39 | 1.15x | 1.15x |
| error compensation) | | | | 20 | 0.46 | 0.98x | 0.98x |
| Version 1.1 | 444 | 20.86 | 45.40 | No | 0.30 | 1.50x | 1.26x |
| (no FMA and | | | | 5 | 0.27 | 1.67x | 1.40x |
| Pichat method) | | | | 20 | 0.34 | 1.32x | 1.11x |
| Version 1.2 | 360 | 17.97 | 45.55 | No | 0.34 | 1.32x | 1.37x |
| (FMAs enabled and | | | | 5 | 0.30 | 1.50x | 1.55x |
| Kahan method) | | | | 20 | 0.35 | 1.29x | 1.33x |
| Version 1.3 | 444 | 20.86 | 45.40 | No | 0.30 | 1.50x | 1.26x |
| (FMAs enabled and | | | | 5 | 0.27 | 1.67x | 1.40x |
| Pichat method) | | | | 20 | 0.27 | 1.67x | 1.40x |
| Version 1.4 | 312 | 20.89 | 45.21 | No | 0.37 | 1.22x | 1.45x |
| (FMAs enabled, | | | | 5 | 0.33 | 1.36x | 1.63x |
| no error compensation) | | | | 20 | 0.39 | 1.15x | 1.38x |

Negative performance of these efforts are later illustrated in Table 5 (see versions 2.x for each kernel), where we will discuss speed-ups versus the energy required for each execution.

Our second attempt was to use constant memory for those parameters remaining invariable during a kernel execution. Those values were already considered to be wisely stored in shared memory (that is, cached) during the baseline implementation of NiftyReg, and CUDA advises that this cache has similar latency to the register file. The hypothesis was proven in our case, producing similar performance in both cases, but allowing us to study energy concerns as well (see version 3 for each kernel in Sect. 5).

Our final effort was dedicated to unified memory. This is a new feature introduced in Maxwell GPUs to provide a single address space for the CPU and GPU codes which relieves programmers from declaring memory in both sides and explicitly inserting communication functions via PCI-express. Unified memory is supposed to shine with the advent of Pascal GPUs in 2016, where 3D memory is placed on chip using Stacked DRAM. In the meantime, the use of unified memory is premature, since CUDA drivers have to handle a manual migration of memory pages back and forth between main memory (DDR3) on

**Table 3.** List of optimizations performed over the NiftyReg library at a glance.

| CUDA code | Description |
|---|---|
| Baseline (v. 1.0) | Departure point released at sourceforge.net |
| Version 1.1 | Replace Kahan's method in v. 1.0 by Pichat's method |
| Version 1.2 | Apply FMA (Fused Multiply-Add) in v. 1.0 |
| Version 1.3 | Apply FMA in v. 1.1 |
| Version 2.0 | Replace 1D textures in v. 1.0 by 1D arrays in global memory |
| Version 2.1 | Replace 3D textures in v. 1.0 by 3D arrays in global memory |
| Version 3 | Replace in v. 1.0 derivative arrays declared in shared memory by constant memory |
| Version 4 | Enabling unified memory in v. 1.0 for main memory (CPU) and global memory (GPU) |

CPU and video memory (GDDR5) on GPU via a slow PCI-express bus whenever each side touches data. Therefore, we expect the use of unified memory to hurt performance at this point of the evolution of memory technology, but this feature is a solid investment for future GPUs and will soon be rewarded on more matured versions of CUDA drivers, particularly CUDA 8.0 and beyond. Sooner or later, all CUDA programmers will have to adapt their codes to the new memory hierarchy, and we presume to be early adopters with our version 4 of NiftyReg kernels. In the meantime, we can keep an eye on power consumption and study its evolution through GPU generations. Section 5 wil inform about the energy spent on Maxwell in the meantime Pascal emerges.

In general, we expect the benefits of all our memory optimizations to increase in future GPU generations, further diverging from rendering towards general approaches. The way we handle global and unified memory, first transforming textures into global memory and then unifying main and global memory, shows the road for upgrading old CUDA codes which will soon be obsolete given the new memory hierarchy introduced by Pascal GPUs in early 2016.

This way, we believe our software implementation is investing in solid pillars the hardware layer will soon reward, versus the old era of GPU programming where the emphasis was made on first memory layers (registers and caches).

# 4   Monitoring Energy

## 4.1   Hardware Infrastructure

We have used a measurement system based on an Arduino Mega 2560 microcontroller [9], where we have attached four INA219 sensors [10] to measure currents at real-time. Those sensors are based on shunt resistors, where voltage falls are used as a measurement proportional to the current flow [11]. The communication protocol between Arduino and sensors is $I^2C$ [12].

In order to measure currents, we have taken into account all power supply inputs available within the graphics card, that is, power pins in PCI-Express (12 and 3.3 V) and auxiliary power inputs through a couple of twin connectors of 6 pins usually located in the back corner of the printed circuit board of the graphics card [13]. A PCIe 16x adapter and two 6 pins PCIe extenders were required to carry out the measurements.

## 4.2    Software Tool

Our Arduino was connected to the main computer, a regular PC, through the USB port. On a single measurement, a function written in C language sends a start command through the serial port to the Arduino, who starts measuring upon reception. Whenever we want to finalize the measurement, the end command is sent to the Arduino, which generates a file in JSON format with the information and sends it back to the PC through the serial bus. Finally, a visual interface displays detailed information on screen.

## 4.3    Methodology for Measuring Energy

For the total energy to be calculated for each kernel, we are required to obtain the execution time plus the power consumed over that time period. This is accomplished with a sequence of steps upon kernel launching:

1. Call the function for starting the timer.
2. Call the function for starting the power measurement.
3. Launch the CUDA kernel.
4. Call the function for ending the power measurement.
5. Call the function for ending the timer.
6. Save the execution time and power values.

At the end of the process, our tool gathers 28 samples per measurement for each of the four sensors connected to the 12 V. and 3.3 V. pins of the PCIe, plus the two auxiliary power inputs. We then acumulate the values gathered by all these sensors to figure out the total energy spent on each sample point. Finally, we split the execution time evenly among all samples and multiply this result by the sum of all power samples to estimate the total energy consumed by each kernel. Note that this process performs the usual approximation for the integration of a continuous function through aggregation of discrete bins.

In order to ensure numerical stability and filter potential spikes attributed to noise and/or side-effects, we have run 10 times the reg_f3d method, using as final values for execution time and power consumption the average and median of these 10 runs, respectively.

# 5    Experimental Results

Our benchmark for the non-rigid registration of brain images gathers a set of 20 images obtained from a MRI device, producing as output structural images of type 1 in clinical practice. The execution of our `reg_f3d` method within NiftyReg library takes 531,65 ms to complete the non-rigid registration of a single image composed of $512 \times 512 \times 53$ voxels as departure point.

Table 4 characterizes the system where we have run our experiments, and Table 5 collects all results for the time and energy spent by all the optimization efforts we have conducted along this paper. With all those experimental numbers, our main goal is to demonstrate whether the GPU behaves like our car, where the more we push the throttle the more fuel we spend per kilometer traversed.

**Table 4.** Hardware platform used along our experimental results.

| GPU | Video memory | CPU |
|---|---|---|
| Model: GTX 980 (Maxwell) | Family: GDDR5 | Model: Intel Core 2 Duo |
| Number of cores: 2048 | Size: 4 GB | Main memory: 4 GB DDR3 |
| Frequency: 1126 MHz | Frequency: 7 GHz | O. S.: Ubuntu 14.04 LTX 64 bits |

Unfortunately, our optimization effort did not reach impressive speed-ups to analyze a large range of accelerations, but 2.87x for the K4:v2.0 (that is, kernel `GetApprBendingEnergy3D`, version 2.0) suffices to tell us that the maximum speed-up matches the maximum energy spent, 79.25 J/s of execution (note that the last two columns in the table are the most interesting ones). The lowest acceleration, 0.18x, is found in row K10:v2.1, but in this case a power/time of 44.46 mJ/ms does not hold the minimum value (there are many thirties and even few twenties in the last column). As this kernel consumes the lowest fraction of the execution time, we believe power measurements can be more unstable/unreliable here.

In general, we validate the expected behaviour in all versions of most kernels. Good examples are K3, where all versions slow down to relax the power consumption proportionally, and K10, where versions 2.0 and 4 speed-ups also increase energy spent in similar percentages.

But there are occasional exceptions too: K4:v2.1 is the most remarkable one, as performance doubles while reducing power consumption by more than 40 %. And K8:v2.1 behaves exactly the opposite way, to attain a poor 0.12x speed-up factor while consuming more than twice the baseline version. We also find minor anomalies in several versions 4, as anticipated given the immature state of drivers handling unified memory.

And we leave a nice surprise for the final: Optimizations performed via exploiting additional parallelism (see Sect. 3.1) accelerate and reduce energy spent (see K7, versions 1.1, 1.2 and 1.3). A symmetric behaviour was reflected in K5 and K6, but numbers are ommited given the huge size of the table.

**Table 5.** Power measurements, execution time and speed-up versus baseline versions for the kernels we have optimized within NiftyReg library.

| Abbreviated name for the NiftyReg kernel [number of times launched] (% of time used) | Power consumption (median in mJ.) | Execution time (avg. in ms.) | Acceleration versus baseline | Power / Time |
|---|---|---|---|---|
| K1. GetDeformationField3D [80] (35,93%) | 7,1387 | 0,1619 | 1,0000 | 44,0959 |
| Version 2.0 | 9,1410 | 0,2117 | 0,7648 | 43,1893 |
| Version 2.1 | 9,3170 | 0,2138 | 0,7572 | 43,5864 |
| Version 3 | 7,2939 | 0,1544 | 1,0486 | 47,2370 |
| K2. GetApproxSecondDerivative [80] (20,00%) | 2,7671 | 0,0901 | 1,0000 | 30,7117 |
| Version 2.0 | 2,0813 | 0,0402 | 2,2413 | 51,7232 |
| Version 2.1 | 2,3525 | 0,0417 | 2,1607 | 56,2790 |
| Version 3 | 2,4943 | 0,0881 | 1,0227 | 28,2956 |
| K3. ResampleImage3D [80] (15,80%) | 3,1660 | 0,0712 | 1,0000 | 44,4969 |
| Version 2.0 | 3,0299 | 0,0863 | 0,8250 | 35,0966 |
| Version 2.1 | 3,1677 | 0,1004 | 0,7092 | 31,5442 |
| Version 3 | 3,1331 | 0,0719 | 0,9903 | 43,5875 |
| Version 4 | 3,6709 | 0,0967 | 0,7363 | 37,9778 |
| K4. GetApprBendingEnergy3D [72] (7,97%) | 1,9497 | 0,0399 | 1,0000 | 48,8652 |
| Version 2.0 | 1,1033 | 0,0139 | 2,8705 | 79,2566 |
| Version 2.1 | 0,5438 | 0,0191 | 2,0890 | 28,4245 |
| Version 3 | 1,9928 | 0,0415 | 0,9614 | 48,0072 |
| K5. ApplyConvolWindowAlongZ [8] (6,13%) | 9,9851 | 0,2763 | 1,0000 | 36,1439 |
| Version 2.0 | 14,1203 | 0,3874 | 0,7132 | 36,4518 |
| Version 2.1 | 13,7341 | 0,3856 | 0,7165 | 35,6148 |
| Version 3 | 10,8477 | 0,2827 | 0,9774 | 38,3676 |
| Version 4 | 14,3782 | 0,3795 | 0,7281 | 37,8911 |
| K6. ApplyConvolWindowAlongY [8] (5,57%) | 7,3503 | 0,2510 | 1,0000 | 29,2876 |
| Version 2.0 | 14,8894 | 0,3988 | 0,6294 | 37,3326 |
| Version 2.1 | 14,7958 | 0,4054 | 0,6191 | 36,4986 |
| Version 3 | 8,3243 | 0,2336 | 1,0745 | 35,6381 |
| Version 4 | 15,2838 | 0,4076 | 0,6158 | 37,4962 |
| K7. ApplyConvolWindowAlongX [8] (5,26%) | 8,6946 | 0,2369 | 1,0000 | 36,6954 |
| Version 1.1 | 7,8835 | 0,2268 | 1,0445 | 34,7627 |
| Version 1.2 | 7,9481 | 0,2280 | 1,0390 | 34,8556 |
| Version 1.3 | 7,4438 | 0,2233 | 1,0609 | 33,3325 |
| Version 2.0 | 17,2757 | 0,4755 | 0,4982 | 36,3308 |
| Version 2.1 | 17,5457 | 0,4832 | 0,4903 | 36,3093 |
| Version 3 | 8,2099 | 0,2437 | 0,9721 | 33,6843 |
| Version 4 | 17,5009 | 0,5000 | 0,4738 | 35,0018 |
| K8. GetApprBendingEneGrad... [8] (1,53%) | 1,6589 | 0,0691 | 1,0000 | 24,0149 |
| Version 2.0 | 3,3063 | 0,0550 | 1,2564 | 60,1142 |
| Version 2.1 | 3,3313 | 0,5610 | 0,1232 | 59,3818 |
| Version 3 | 2,6021 | 0,0483 | 1,4306 | 53,8405 |
| K9. ConvertNMIGradFromVox... [8] (1,05%) | 2,2256 | 0,0474 | 1,0000 | 46,9540 |
| Version 2.0 | 1,1783 | 0,0219 | 2,1644 | 53,8757 |
| Version 2.1 | 1,3353 | 0,0219 | 2,1644 | 60,9461 |
| Version 3 | 1,0434 | 0,0293 | 1,6177 | 35,6110 |
| Version 4 | 2,0163 | 0,0391 | 1,2123 | 51,5423 |
| K10. VoxelCentric2NodeCentric [8] (0,76%) | 1,9715 | 0,0342 | 1,0000 | 57,5614 |
| Version 2.0 | 1,1764 | 0,0180 | 1,9000 | 65,3565 |
| Version 2.1 | 0,8141 | 0,1830 | 0,1869 | 44,4610 |
| Version 3 | 2,0374 | 0,0378 | 0,9048 | 53,8282 |
| Version 4 | 1,3829 | 0,0220 | 1,5545 | 62,8869 |

For a final characterization of every effort we have made for tuning the CUDA code of NiftyReg in terms of both performance and energy, we summarize in Table 6 the average acceleration and energy required by each version of NiftyReg, along with the subset of kernels contributing to that average. The penultimate column divides energy spent by acceleration attained, and therefore, the lower this number, the better behaviour concerning watts consumed for every speed-up factor attained. We can see how the best versions of the code are those

**Table 6.** Summary of NiftyReg optimizations deployed along our different versions implemented, with the set of kernels involved on each case. We include the average of acceleration factors and energy spent (Jules/second) by each version, and a ratio, energy/acceleration, to be minimized as overall goal. A final ranking is given for each NiftyReg version, proclaiming v1.3 as overall winner.

| CUDA code version | K1 | K2 | K3 | K4 | K5 | K6 | K7 | K8 | K9 | K10 | Acceleration versus v1.0 (on average) | Power/ time (on avg.) | Energy/ accel. ratio | Ran- king |
|---|---|---|---|---|---|---|---|---|---|---|---|---|---|---|
| 1.0 | ✓ | ✓ | ✓ | ✓ | ✓ | ✓ | ✓ | ✓ | ✓ | ✓ | 1.0000 | 39.9860 | 39.9860 | #7 |
| 1.1 | | | | | ✓ | ✓ | ✓ | | | | 1.0445 | 34.7627 | 33.2816 | #2 |
| 1.2 | | | | | ✓ | ✓ | ✓ | | | | 1.0339 | 34.8556 | 33.7127 | #3 |
| 1.3 | | | | | ✓ | ✓ | ✓ | | | | 1.0609 | 33.3325 | 31.4190 | #1 |
| 2.0 | ✓ | ✓ | ✓ | ✓ | ✓ | ✓ | ✓ | ✓ | ✓ | ✓ | 1.2435 | 45.9316 | 36.9373 | #4 |
| 2.1 | ✓ | ✓ | ✓ | ✓ | ✓ | ✓ | ✓ | ✓ | ✓ | ✓ | 1.1129 | 42.1782 | 37.8993 | #5 |
| 3 | ✓ | ✓ | ✓ | ✓ | ✓ | ✓ | ✓ | ✓ | ✓ | ✓ | 1.0310 | 41.0599 | 39.8253 | #6 |
| 4 | | ✓ | | ✓ | ✓ | ✓ | | | ✓ | ✓ | 0.7079 | 40.9554 | 57.8547 | #8 |

representing enhancements in parallelism, with version 1.3 as overall winner. On the other hand, version 4.0 corresponding to unified memory is the only effort in vain with respect to our departure point, but we are confident that upcoming GPU generations will soon lift this implementation to be highly competitive (and likely a predicted winner).

# 6  Conclusions

This work shows different ways to optimize a well-known biomedical library, NiftyReg, which we use to perform the non-rigid registration of brain images in clinical practice. If this work were solely focused on performance, we would have chosen a benchmark composed of hundreds of high-resolution images to lift speed-up factors. But this time energy is our foreground, and we have conducted our analysis to know more in deep how the energy is spent within the GPU.

Maxwell, our target GPU, has been optimized by Nvidia to minimize the energy consumed, and therefore, our departure point is already competitive versus previous executions on Kepler and Fermi GPUs. We have deployed a wide number of optimizations to accelerate the NiftyReg library, but as an instrument to see how power follows performance. And for our experimental cases, doubling performance increases the energy spent by a smaller factor, with a representative 2.87x peak performance which is translated into just 1.62x power increment. We save energy if we run faster, a remarkable finding.

As secondary achievements, we demonstrate that optimizations for enhancing arithmetic intensity (like using FMADDs) and exploiting more parallelism (via ILP or vector processing) are more energy efficient than those focused on the memory hierarchy, at least for the set of experiments that we have included here involving textures, constant, global and shared memory.

But in general, we expect the benefits of our memory optimizations to increase in future GPU generations, particularly with the imminent arrival of 3D memory. Textures, global and, overall, unified memory will benefit more than the first layers in memory hierarchy where CUDA programmer used to work extensively on early GPU generations.

**Acknowledgments.** This work was supported by the Ministry of Education of Spain under Project TIN2013-42253-P and by the Junta de Andalucia under Project of Excellence P12-TIC-1741. We thank Javier Cabero and Pablo Sánchez for their work on preliminary versions of these CUDA implementations. We also thank Marc Modat from University College London, for his support when using the NiftyReg library. We also thank Nvidia for hardware donations within GPU Education Center 2011–2016 and GPU Research Center 2012–2016 awards.

# References

1. Modat, M.: NIFTYREG - a library to perform rigid, affine and non-linear registration of NIfTI images. http://sourceforge.net/projects/niftyreg/
2. NIfTI: the NIfTI format home page. http://nifti.nimh.nih.gov
3. Clay, R.: Functional magnetic resonance imaging: a new research tool (2007). www.apa.org/research/tools/fmri-adult.pdf
4. DICOMNIFTI: a tool for converting DICOM files into the NIfTI data format. http://cbi.nyu.edu/software/dinifti.php. Accessed October 2013
5. NIFTILIB: input/output libraries for NIfTI-1 neuroimaging data format. http://niftilib.sourceforge.net
6. CMIC: the NifTK software platform. http://www.niftk.org
7. Modat, M., Ridgway, G.R., Taylor, Z.A., Lehmann, M., Barnes, J., Hawkes, D.J., Fox, N.C., Ourselin, S.: Fast free-form deformation using graphics processing units. Comput. Methods Program. Biomed. **98**(3), 278–284 (2010)
8. McNamee, J.: A comparison of methods for accurate summation. ACM SIGSAM Bull. **38**, 1–7 (2004)
9. Arduino: an open-source electronics platform based on easy-to-use hardware and software. https://www.arduino.cc/en/Main/ArduinoBoardMega2560
10. Adafruit: INA219 current sensor breakout. https://learn.adafruit.com/adafruit-ina219-current-sensor-breakout
11. Ziegler, S., Woodward, R., Iu, H., Borle, L.: Current sensing techniques: a review. IEEE Sens. J. **9**(4), 354–376 (2009)
12. Philips: I2C-bus specification and user manual. Philips Semiconductors (2014)
13. Igual, F., Jara, L., Gómez, J., Piñuel, L., Prieto, M.: A power measurement environment for PCIe accelerators. Comput. Sci. Res. Dev. **30**, 115–124 (2015)

# Unleashing the Graphic Processing Units-Based Version of NAMD

Yamandú González[1], Pablo Ezzatti[1(✉)], and Margot Paulino[2]

[1] Instituto de Computación,
Universidad de la República (UdelaR), Montevideo, Uruguay
{yamandug,pezzatti}@fing.edu.uy
[2] Faculty of Chemistry, Center of Structural Bioinformatic, DETEMA,
UdelaR, Montevideo, Uruguay
margot@fq.edu.uy

**Abstract.** NAMD is a parallel molecular dynamics software designed for high-performance simulations of large biomolecular systems. It scales from single computer up to hundreds of processors as high-end parallel platforms. Additionally, considering the evolution of Graphics Processing Units (GPUs) as a general purpose massively parallel co-processors, NAMD has included this kind of devices to leverage its computational power. In this work we analyze current NAMD GPU solution and develop an alternative based on Newton's third law. The results shows a significant reduction of the execution time of GPU computations, of up to 20 % when compared with a highly tuned version of the original GPU-enabled NAMD.

**Keywords:** NAMD · GPUs · Non-bonded forces · Newton's third law

## 1 Introduction

Molecular Dynamics (MD) has emerged as one of the most powerful chemistry computational tools, as it is capable of simulating a huge variety of systems [2]. Several software tools have been developed to perform molecular dynamics simulations. Suites as AMBER [4], GROMACS [1] and NAMD [12] are, among others, main references on the scientific community. Specifically, this work is focused on NAMD.

The increase on biosystem size and/or time period of the simulation has promoted the use of High Performance Computing (HPC) techniques, as new affordable alternatives have been developed on this field. In particular, the use of multi core platforms, in combination with GPUs to tackle this kind of problems, have become increasingly popular [14].

GPUs were originally designed to perform graphics processing in computers. However, in the last ten years GPUs have been used as a powerful parallel hardware architecture to achieve efficiency in the execution of different applications, specially, since 2007, when NVIDIA released CUDA, a framework for general purpose computing in GPUs [10].

© Springer International Publishing Switzerland 2016
F. Ortuño and I. Rojas (Eds.): IWBBIO 2016, LNBI 9656, pp. 639–650, 2016.
DOI: 10.1007/978-3-319-31744-1_56

Several software packages for MD have included GPU support for a subset or all the of required calculation steps on a conventional molecular dynamics simulation. To take advantage of this low-cost hardware is not easy, since CUDA architecture differs in many aspects from a traditional CPU, and an important re-engineering process is needed. This can be stated on Phillips et al. [13], that described the migration of the non-bonded interactions calculation in NAMD to GPU-based computations. Nowadays, NAMD includes the option of GPU-based computation to accelerate the MD simulations.

In this article we study and evaluate an alternative algorithm for the GPU-based method of NAMD (version 2.9), discussing a variant of the original proposal that is able to leverage the concurrency offered by GPUs in conjunction with Newton's third law. To evaluate this proposal the crystallographic structure of a ApoLipoprotein-A1 (APOA1) have been used. The obtained results demonstrate that this new proposal can achieve significant accelerations compared with a highly tuned version of the original NAMD.

The paper is structured as follows. In Sect. 2, we present an overview of the NAMD package. In Sect. 3 a brief explanation of basic concepts of CUDA, the use of GPU in NAMD, and a revision of the related work are given. The description of our proposal is described in Sect. 4, and in Sect. 5 the experimental results are presented. Concluding remarks and open questions are discussed in Sect. 6.

# 2   NAMD

NAMD is a parallel molecular dynamics software tool, designed for high performance simulation of large bio-molecular systems in realistic environments. One important feature of this tool is the ability to scale from individual desktop and laptop computers to hundreds of processors on high-end parallel hardware platforms [12]. Based on Charm++ and implemented in C++, NAMD works with AMBER and CHARMM potential functions, parameters, and file formats. It offers several features like Particle Mesh Ewald (PME) method, temperature and pressure controls, among others.

## 2.1   NAMD Architecture

Classical molecular dynamics requires computation of bonded and non-bonded forces. The later includes the calculation of electrostatic and van der Waal's forces. Which are frequently calculated, for performance reasons, using a *cut-off* radius $r_c$. With this strategy, forces between atoms within $r_c$ are calculated explicitly and those atoms beyond $r_c$ are not considered, reducing computation time to $O(N)$, where $N$ is the number of simulated atoms.

If a more accurate approximation is required, Particle-Mesh Ewald method for long range electrostatic forces can be applied [3]. This method translates the electric charge of each atom to electric potential on a grid and uses a Fast Fourier Transform (FFT) to calculate the influence of all atoms on each atom in $O(N \log(N))$ computation time.

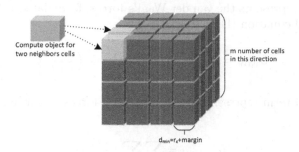

**Fig. 1.** NAMD spatial decomposition into cells and assignation to compute object.

NAMD uses a hybrid strategy performing both, spatial and force decompositions. The first is performed by assigning atoms to work units called "patches" or "cells" (see Fig. 1). These patches will be evenly distributed across all processors. The size of each cell should be $d_{min} \cong r_c + margin$ along every dimension, with $margin$ as a small constant. This ensures that atoms which are within a *cut-off* radius are placed in neighbouring boxes.

Then for every pair of interacting cells (or patches), a "compute object" is created, responsible for the computation of the pair-wise forces between the two cells, which is the force decomposition component. Each cell has 26 neighbouring cells, therefore $14 \times p$ compute objects $(26/2 + 1 = 14)$ are created, where $p$ is the number of patches. These compute objects are assigned to processors following a load balancing strategy (supported by Charm++ architecture that measures the work load of each processor).

## 2.2   NAMD Computations

Force in MD is defined as the negative gradient of the potential function:

$$\overrightarrow{F}(\overrightarrow{r}) = -\nabla U(\overrightarrow{r}) \tag{1}$$

where the potential function involves the following sum:

$$U(\overrightarrow{r}) = \sum U_{bonded}(\overrightarrow{r}) + \sum U_{non-bonded}(\overrightarrow{r}) \tag{2}$$

Equation (2) shows the criteria used in NAMD for the force decomposition. This effort is focused on non-bonded terms calculation, which are composed by the addition of two terms:

$$U_{non-bonded} = U_{vdW} + U_{Coulomb} \tag{3}$$

The first term represents the van der Waal's forces, formulated by the Lennard-Jones potential equation [11]:

$$U_{vdW} = \sum_i \sum_{j>i} 4\epsilon_{ij} \left[ \left( \frac{\sigma_{ij}}{r_{ij}} \right)^{12} - \left( \frac{\sigma_{ij}}{r_{ij}} \right)^6 \right] \tag{4}$$

and the second term represents the electrostatic forces, according to Coulombs law [5]:

$$U_{Coulomb} = \sum_i \sum_{j>i} \frac{q_i q_j}{4\pi\epsilon_0 r_{ij}} \tag{5}$$

Ideally, these two terms imply the calculation of Eqs. 4 and 5 for each pair of atoms in the simulation box, yielding an $O(N^2)$ computational cost. Considering the *cut-off* radius, the computational cost of simulations for both terms of Eq. 4 can be reduced to $O(N)$ and to $O(N\log(N))$ via Particle Mesh Ewald (PME) algorithms for Coulomb forces in Eq. 5, The selection of a *cut-off* value is related with the idea of discarding those charges whose contribution for potential are negligible, and usually a $8 - 10\mathring{A}$ *cut-off* radius should be enough.

As other molecular dynamics software, NAMD applies the pairlist concept on a specific simulation. During a simulation, the atoms change their positions and, in consequence, they get closer to some atoms and further from others. In the spatial decomposition applied by NAMD, this implies the migration from one cell to another. Computing a real-time mapping between atoms and cells is prohibitive. Consequently, a strategy commonly used to save time is to update the atom-cell mapping with a certain frecuency (a number of simulations steps), which must be carefully selected so the numerical errors introduced in the simulation are not significant.

## 3   GPGPU on MD

In this section we offer firstly an overview of CUDA infrastructure to use the GPUs for solving general purpose problems. After that, we study the GPU-based version of NAMD. Finally, we summarize the most important related works about the use of GPUs to accelerate MD.

### 3.1   Graphics Processing Unit Overview

Conventional CPUs are designed for sequential code execution. While running a program, it fetches instructions and associated data from the random access memory (RAM), decodes them, executes them, and then writes the result back to the RAM. This would be classified as single instruction – single data architecture (SISD). In contrast GPU can execute the same operation on different data at the same time, by means of a group (defined as block) of concurrent threads, which belongs to single instruction – multiple data category architecture (SIMD). Based on the facilities provided by CUDA [10] for GPU programming,

GPUs can be viewed as a set of shared memory multicore processors. GPUs follow the single-program multiple-thread (SPMT) parallel programming paradigm, in which many cores execute the same program on multiple parts of the data using different threads, but do not have to be executing the same instruction at the same time [6]. The number of threads that graphics cards can execute in parallel is in the order of hundreds and it is expected that this number will continue rapidly increasing, which makes these devices a powerful and low cost platform for implementing parallel algorithms.

The CUDA [10] consists of a stack of software layers including: a hardware driver, a C language application programming interface and the CUDA driver that is dedicated to transfer data between the GPU and CPU. It is available for NVIDIA's GeForce 8 and superior series of graphics cards.

CUDA architecture is built around a scalable multiprocessor array. Each multiprocessor on the GPU, consists of several (8, 32, 192, etc.) scalar processors as well as additional units like the multithreading instruction unit and the shared memory chip. When a part of an application can be run many times but independently on different data, it can be isolated in a function, called "kernel", to be executed on the device through many different threads. For this purpose, the kernel routine is compiled using the device instruction set and the resulting program is transferred to the device.

When a kernel is invoked, a large number of threads is generated on the GPU. All the threads generated by a kernel invocation form a "grid", which is partitioned in many blocks. These blocks comprise threads that are executed concurrently on a single multiprocessor. There is no fixed order of execution between blocks. If there are enough multiprocessors available on the card, they are executed in parallel. Otherwise, a time-sharing strategy is used.

Threads can access data across multiple memory spaces during their execution. Nowadays, GPUs have six different memory spaces: registers, local memory, shared block memory, global memory, constant memory and texture memory, that are briefly commented next.

The global memory is one of the slowest memories on the card and is not cached. On the other hand, constant memory is fast although it is located in the device memory. It can be seen as a read-only cache of the global memory. Finally, the texture memory has similar characteristics that constant memory.

Registers, that are located in the chip, are the fastest memory on the card and are only accessible by each thread. In addition to this, each thread has its own local memory, but is one of the slowest memories on the device because it is phisically located in the (off–chip) global memory. Both memory spaces are entirely managed by the compiler. Each block has a shared memory space that is almost as fast as registers and could be accessed by any thread of the block. The shared memory is located on the chip and its lifetime is equal to the lifetime of the block.

The data transferences can be performed both synchronously or asynchronously, making possible to overlap data transferences with computation on the device, for which is necessary the use of page-locked memory, since the GPU can fetch it without the help of the host ensuring faster memory transfer.

## 3.2    GPU Computations in NAMD

Original NAMD behaviour, when run on a distributed architecture (i.e. cluster), is to assign patches and compute objects over all available processors, leaving those compute objects near to its involved patches. An important aspect to take into consideration when including GPU computing in NAMD is that each interaction between patch pairs implies a compute object call. Thus, the direct use of this paradigm over GPU would imply one kernel call (compute object) per patch pair interaction [13]. For this reason, a new kind of compute objects was created, that instead of containing a single patch pair, contains all patch pairs scheduled for compute in GPU. This alternative reduces the kernel calls and the amount of memory transfers.

Force calculation on a kernel is computed by assigning one patch pair per thread block, considering that a patch pair can be composed by two copies of the same patch, as non-bonded forces inside a cell must be calculated too.

On NAMD when a patch pair is composed by two different patches, the reciprocal pair is scheduled for processing too. For example if we have patches A and B, non-bonded forces calculation over those patches imply the scheduling of pairs $A \rightarrow A$, $A \rightarrow B$, $B \rightarrow A$, $B \rightarrow B$. Forces calculation will be performed over the first element on the patch pair.

When a kernel is launched, each thread of a block copies an atom of the first patch into local-registers, while atoms from the second patch are loaded into shared memory. After that, each thread will iterate over atoms (on registers) accumulating force interactions between them (on shared memory). This task is performed for all at the same time. Finally, accumulated forces and energies are written into global memory, which is mapped to page-locked (pinned) memory.

However, the effort to take advantage of the GPU must not be focused only on how to use the GPU resources to get an acceptable CUDA routine. A considerable effort has to be targeted on how to use this routine from a host. A naive use of this kernel could serialize CPU evaluation and GPU invocation, which goes against both, CUDA suggested practices and NAMD original ideas, becoming a bottleneck on execution. Therefore, as NAMD is regarded as a parallel application, two strategies are proposed in [13]:

- The first is to overlap GPU and CPU execution by changing some priorities of several messages from the *Message Driven Architecture* policy of NAMD, such as PME related messages.
- The second is to overlap GPU execution and communication. A CUDA compute object may require coordinates from both, local processor and remote processor, and both calculated forces would be returned at the same time regardless if they were solved in different order inside kernel execution. Obviously this causes an unnecessary overhead as a consequence of the idle time of those processes waiting for force evaluation results. In order to optimize force calculation, each patch pair is classified as local or remote considering the location of its first element. Now, two kernel invocations will be triggered. The first one involves patch pairs whose first element does not reside on the local processor, and the second one involves patch pairs whose first element

resides on the local processor. this workaround allows overlapping the computation of remote patches and the return of these results with the computation of local patches.

GPU resources can be assigned in several forms when NAMD is launched. It is possible to share a GPU over threads running on different units of a multicore processor, as also a GPU can be assigned to a master thread that handles this resource and schedules the whole work. If a single node with multiple cores and multiple GPUs is used, different threads could share the GPUs with a previously configured assignation criteria, i.e. thread 0 and 1 uses GPU 0, thread 1 and 2 uses GPU 1, etc.

### 3.3 Related Work

NAMD is an important reference about the use of GPU to accelerate molecular dynamics simulations but is not the only one. Other initiatives on GPU integration for improvement of MD have been done. Next, we only comment a couple of works closely related to this one.

GROMACS is a software that includes GPUs support through a 3rd party library called OpenMM [7], providing an entire GPU implementation for molecular dynamics. On its last version, a native GPU acceleration is supported for non-bonded forces calculation. LAMMPS [15] is another software that proposes GPU use on heterogeneus clusters using MPI as communication framework. This work also proposes a strategy for taking advantage of Newton's third law, modifying the computations order. In particular it allows that some cells compute forces while others write the reciprocal forces in memory.

Götz et al. [8] proposed an implementation of AMBER that runs entirely on CUDA enabled NVIDIA GPUs. A strategy to reuse already calculated forces based on Newton's third law, is proposed in a similar way as in LAMMPS, controlling the order in which the atoms are read on each cell. This last idea will be explored on the next section. Other related works are focused on non-bonded interactions, for example, Imberón et al. [9].

## 4 Proposal

In this section we study and present a novel extension of the GPU-based NAMD able to leverage Newton's third law on CUDA kernels, based on the solution proposed by Götz et al. for AMBER [8].

Considering the distributed architecture of NAMD, any improvement in the implementation of the one node version will enhance the whole application. In consequence, this work is focused only in the use of a GPU in a single node.

Original NAMD GPU kernel is intended to exploit all the benefits offered by these architectures, taking advantage of the implicit high parallelism provided by CUDA. When a thread takes an atom $a$ from $A$, then the loop over each $b$ atom from $B$ will start always from the first atom ($b$) loaded on shared memory.

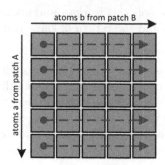

**Fig. 2.** NAMD current strategy.

This is the behaviour for all threads on one block. In other words, all threads of a block (where patch $A$ is processed) will take different atoms "a" from this patch and will read atoms "b" from $B$ in the same order (Fig. 2).

The previous strategy implies a simple idea for looping over elements of $A$, but causes that elements from $B$ can not accumulate reciprocal forces as race conditions could happen, i.e. several threads are accessing to the same memory address at certain moment in time.

Our proposal to address this issue is similar to GPU AMBER solution [8] for solving non-bonded forces calculation. The basic idea is to modify each thread-loop starting point over shared memory when $b$ contribution over $a$ is calculated. In other words, lets take a simple case, two patches $A$ and $B$, each composed by 5 atoms. On the original NAMD solution thread 0 loads atom $a_0$ from $A$, thread 1 loads atom $a_1$, and so on. Then each thread $i$ will loop over all atoms of patch $B$ starting on atom $b_0$, then $b_1$, until the last atom $b_5$. Our solution will use the same load policy for atoms "a" from patch $A$ but will change the loop over atoms "b" from $B$. Each thread will start looping over atoms from $B$ by starting on $b_i$ being $i$ the thread id. This causes that each thread will start on the next available atom, considering that previous atoms $b_i$ have been processed by thread id $i$.

Figure 3 shows how atoms $b$ from $B$ are visited. By this way each $b$ atom from $B$ will be processed by only one thread at each time. Hence race conditions are not possible, allowing reciprocal forces accumulation over each of these atoms.

Accordingly, from this point we are able to process both $A \rightarrow B$ and $B \rightarrow A$ at the same time, therefore instead of scheduling the same patch pair two times, one for each direction, now only one patch pair should be scheduled, $A \leftrightarrow B$.

*Changes on Communication-Execution Ovelapping:*
On each time step of a MD simulation a NAMD CUDA Kernel is invoked two times: the first time for solving remote patches, and the second time for solving local patches. Although this mechanism is conserved, the number of patch pairs scheduled on each stage is modified. We preserve the idea of dispatch remote (R) patches first and local (L) patches later. So, we should have four kinds of patch

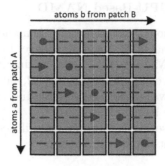

**Fig. 3.** AMBER strategy applied on NAMD.

pairs: $R \leftrightarrow R$, $R \leftrightarrow L$, $L \leftrightarrow R$ and $L \leftrightarrow L$. On the original NAMD kernel, each pair, except those that represent their own interaction, are scheduled two times. It should be noted that, depending on its location, the atom will be scheduled in the first or in the second kernel invocation. Conversely, when the proposed kernel is used, patch pairs $R \leftrightarrow R$, $R \leftrightarrow L$, $L \leftrightarrow R$ will be scheduled one time in the first invocation, and only $L \leftrightarrow L$ will be scheduled, also one time, but in second invocation. Local patch evaluations will use reciprocal values already left on global memory of the device on past kernel invocations (for remote patch values).

## 5   Experimental Evaluation

In this section, we evaluate the performance of the proposed method (Sect. 4). With this aim we choose as a test case Apolipoprotein A-I (APOA1) since it has become one of the main test subjects for molecular dynamics software evaluation. Simulation box of this biomolecule is composed by 92224 atoms.

Testing has been performed over a workstation equipped with a 3.10 GHz Intel Quad-Core i5-2400 CPU, with a GPU card GTX480 (CUDA 6.0), with a Linux Fedora 19 Operative System installed. NAMD Compilation has been performed with default compiler options for C code, but CUDA kernel has been compiled with two different maximum registry values (50 and 63) apart from the default value provided by NAMD (32) by changing *maxrregcount* flag of CUDA compiler NVCC. Other NVCC options remain unchanged, as optimization flag *O3*.

Before starting the evaluation of the proposal we try to highly tune up the original version of GPU-based NAMD (version 2.9 was used). All the runtimes reported in this section are the average of 10 executions. Additionally, the attained numerical results are in all cases very similar, i.e. the numerical differences are of a magnitude that can be easily explained by the unavoidable introduction of floating-point errors.

## 5.1   Tuning Original GPU-Based NAMD

To perform a fair comparison of our proposal it is mandatory to provide a highly tuned version for the original GPU NAMD package. Default compiler options on current distribution of NAMD are set to use a maximum of 32 register per thread on GPU. Three different compilations, using 32, 50 and 63 (max.) registers were evaluated. The test was performed considering both, remote patch kernel invocation and local patch kernel invocation.

In this line, Table 1 presents the runtimes (in milliseconds) for the GPU-based NAMD employing different number of registers.

**Table 1.** CUDA register use behaviour.

| Max. registers | GPU time (milliseconds) | Evaluated patches | Location |
|---|---|---|---|
| 32 | 46.506 | 1944 | Remote |
| 32 | 45.954 | 1944 | Local |
| 50 | **43.407** | **1944** | **Remote** |
| 50 | **42.945** | **1944** | **Local** |
| 63 | 47.161 | 1944 | Remote |
| 63 | 46.575 | 1944 | Local |

The results summarized on the previous table reveals that the best performance is achieved with 50 registers, gaining an improvement near to 7%.

## 5.2   Proposal Evaluation

After obtaining a tuned version of the original NAMD we advance on the evaluation of our proposal, whose runtimes are reported in Table 2.

**Table 2.** Proposed solution kernel performance.

| Kernel Used | GPU time (milliseconds) | Evaluated patches | Patch location |
|---|---|---|---|
| Tuned | 43.407 | 1944 | Remote |
| Proposed | 52.115 | 1512 | Remote |
| Tuned | 42.945 | 1944 | Local |
| Proposed | 19.574 | 504 | Local |

Two main facts can be observed on Table 2. While on original NAMD kernel both invocations (local and remote) take approximately the same time, employing our proposal these times become strongly different. The higher time on first call can be explained by the fact that (although the number of patch pairs is less than the original) this kernel performs more global memory accesses than the original one due to the saving of the reciprocal number. Also another factor is the additional synchronization instructions for memory writing and changes on how atoms are read, causing an additional overhead on memory reading. In the original kernel, the same atom is read by all threads at same time, while on the proposed version consecutive but different atoms are read by all threads.

In the other hand the lower time on the second kernel call is explained by a direct factor: the third part of the the patch pairs is processed (only those pairs with both elements locally hosted), i.e. second kernel calls are almost the third part of the first kernel calls. Additionally, because of how patch pairs are scheduled, on the first call any pair with a remote element is included, while on second call only those pairs with both elements locally hosted are scheduled, representing a quarter of all the patch pairs.

**Table 3.** Overall kernel performance.

| Kernel Used | Total GPU Time (milliseconds) | Total Evaluated Patches |
|---|---|---|
| Tuned | 86.352 | 3888 |
| **Proposed** | **71.689** | **2016** |

When total performance is evaluated (Table 3), the sum of both tuned NAMD kernel executions for APOA1 is around 86.352 milliseconds, while in our proposed version is 71.689 milliseconds, giving an improvement of approximately 20 %.

## 6    Concluding Remarks and Future Work

In this article, we have studied from an empirical perspective, the GPU-based version of NAMD tool and we have presented a novel variant. Our proposal is focused on leveraging Newton's third law and is based on other related works, such as the ideas used on AMBER package [8].

The experimental results obtained using APOA1 in a platform equipped with a GTX 480 GPU and 3.10 GHz Intel Quad-Core i5-2400 CPU, report improvements close to 20 % over a highly tuned version of the original GPU-based NAMD.

As part of future work, we plan to evaluate our proposal on several test cases as also on different hardware configurations. In particular, on distributed platforms and including nodes with newer GPUs (i.e. Kepler family).

# References

1. Berendsen, H.J.C., van der Spoel, D., van Drunen, R.: Gromacs: a message-passing parallel molecular dynamics implementation. Comput. Phys. Commun. **91**(1–3), 43–56 (1995)
2. Anderson, J.A., Lorenz, C.D., Travesset, A.: General purpose molecular dynamics simulations fully implemented on graphics processing units. J. Comp. Phys. **227**(10), 5342–5359 (2008)
3. Bhatele, A., Kumar, S., Mei, C., Phillips, J.C., Zheng, G., Kale, L.V.: NAMD: A portable and highly scalable program for biomolecular simulations. Technical report UIUCDCS-R-2009-3034, Department of Computer Science, University of Illinois, April 2009
4. Case, D.A., Cheatham, T.E., Darden, T., Gohlke, H., Luo, R., Merz, K.M., Onufriev, A., Simmerling, C., Wang, B., Woods, R.J.: The amber biomolecular simulation programs. J. Comput. Chem. **26**(16), 1668–1688 (2005)
5. Coulomb, C.A.: Premier-[troisième] mémoire sur l'electricité et le magnétisme. Académie Royale des sciences (1785)
6. Darema, F.: *The SPMD model*: past, present and future. In: Cotronis, Y., Dongarra, J. (eds.) PVM/MPI 2001. LNCS, vol. 2131, p. 1. Springer, Heidelberg (2001)
7. Friedrichs, M.S., Eastman, P., Vaidyanathan, V., Houston, M., Legrand, S., Beberg, A.L., Ensign, D.L., Bruns, C.M., Pande, V.S.: Accelerating molecular dynamic simulation on graphics processing units. J. Comp. Chem. **30**, 864–872 (2009)
8. Götz, A.W., Williamson, M.J., Xu, D., Poole, D., Grand, S.L., Walker, R.C.: Routine microsecond molecular dynamics simulations with amber on GPUs. 1. generalized born. J. Chem. Theor. Comput. **8**, 1542–1555 (2012)
9. Imbernón, B., Llanes, A., Peña-García, J., Abellán, J.L., Pérez-Sánchez, H., Cecilia, J.M.: Enhancing the parallelization of non-bonded interactions kernel for virtual screening on GPUs. In: Ortuño, F., Rojas, I. (eds.) IWBBIO 2015, Part II. LNCS, vol. 9044, pp. 620–626. Springer, Heidelberg (2015)
10. Kirk, D., Hwu, W.: Programming Massively Parallel Processors: a Hands-on Approach. Morgan Kaufmann, San Francisco (2012)
11. Lennard-Jones, J.E.: Cohesion. Proc. Phys. Soc. **43**, 461–482 (1931)
12. Phillips, J.C., Braun, R., Wang, W., Gumbart, J., Tajkhorshid, E., Villa, E., Chipot, C., Skeel, R.D., Kale, L., Schulten, K.: Scalable molecular dynamics with NAMD. J. Comp. Chem. **26**, 1781–1802 (2005)
13. Phillips, J.C., Stone, J.E., Schulten, K.: Adapting a message-driven parallel application to GPU-accelerated clusters. In: Proceedings of the 2008 ACM/IEEE Conference on Supercomputing, SC 2008, pp. 8:1–8:9. IEEE Press, Piscataway, NJ, USA (2008)
14. Stone, J.E., Phillips, J.C., Freddolino, P.L., Hardy, D.J., Trabuco, L.G., Schulten, K.: Accelerating molecular modeling applications with graphics processors. J. Comp. Chem. **28**, 2618–2640 (2007)
15. Trott, C.R., Winterfeld, L., Crozier, P.S., General-purpose molecular dynamics simulations on gpu-based clusters (2010). arXiv preprint arXiv:1009.4330

# Human Behavior Monitoring, Analysis and Understanding

# Consistency Verification of Marker-Less Gait Assessment System for Stair Walking

Ami Ogawa[1(✉)], Ayanori Yorozu[1], Akira Mita[1], Masaki Takahashi[1], Christos Georgoulas[2], and Thomas Bock[2]

[1] School of Science for Open and Environmental Systems, Keio University, Tokyo, Japan
ami_ogawa@keio.jp
[2] Chair of Building Realization and Robotics, Technical University of Munich, Munich, Germany

**Abstract.** The number of elderly people is drastically increasing. To support them, the gait information is under the spotlight since it has the relationship between the fall risk and dementia. Among other scenarios, a relatively higher level of ability is needed for the stair walking as it requires balancing and loading. Conventionally, 3D motion capture devices have been used to acquire the parameters of stair walking. However, it is difficult to acquire daily parameters as the equipment needs complicated preparation and body-worn markers. In this study, we propose a system which can acquire daily stair walking parameters using only depth data obtained by Kinect v2 without restraining by markers. We confirmed the accuracy of our proposed system compared with a 3D motion capture system.

**Keywords:** Stair walking · Marker-less gait measurement · Depth data · Kinect v2 · VICON

## 1    Introduction

Eight percent of the world's population was elderly people who were aged more than 65 in 2010, and this number is expected to rise up to sixteen percent by 2050. Moreover, the number of children is decreasing in worldwide. [1] This means the number of solitary aged person have been and will be increasing in the future. As they tend to spend their time at house, the prevention of accident in house is needed. And it is important to make them notice their own physical abilities. One of the most frequent activities is walking, and the correlation with fall risk [2–7] and dementia risk [8, 9] have been reported. In particular, stair walking is known to be a hard movement, we need larger contact force of hip [10] and knee [11] for stair walking than level walking. Thus, analysis of stair walking is more significant than level walking to recognize the change of resident's physical ability.

## 2    Related Work

The shape of stairs are related to several stair walking parameters [12, 13]. For example, the correlation between joint force, joint angle, joint moment and inclination of stairs is

© Springer International Publishing Switzerland 2016
F. Ortuño and I. Rojas (Eds.): IWBBIO 2016, LNBI 9656, pp. 653–663, 2016.
DOI: 10.1007/978-3-319-31744-1_57

reported [12], the longer run length provides high stability, and increase of run length and decrease of riser height makes the tilt of upper body small [13].

Also the differences between older and younger [13–18], and man and woman [17–20] were reported. The older group walked with larger lateral upper body movement than younger group [13], the ground reaction force was smaller, the electromyography (EMG) value and muscle co-activation was larger [14, 15], leg stiffness was larger at the impact phase [15], the support time was longer in performing push-off [15], and the elderly has less independent control than younger [17]. And there were significant differences in the magnitudes of the moment contributions, cadence, and stance time between the age groups [18]. The classification system of age group using knee joint force, moment, and angle applying PCA is suggested [16]. About the difference between genders, women performed greater change of angular momentum [19], larger movement of ankle angle and smaller movement of knee angle [20].

Moreover, the features of the knee osteoarthritis (OA) [21–23], and meniscal tear [24] patient's stair walking were reported. The EMG onset knee OA patients showed delayed onset of vastus lateralis activity during descending, and performed small knee flexion in the early stance phase [21]. Also the significant differences between knee OA patients and normal controls on knee, ankle and hip flexion angle are reported [22], and the knee OA patients perform less single support time, shorter step length, longer step width, shorter stride, and slower walking speed than controls [23]. The significant differences between meniscal tear patients and controls on maximum knee flexion angle, knee flexion and extension range of motion, and knee abduction and adduction are concluded [24].

These studies show that it is possible to evaluate the physical ability based on the stair walking performances. The 3D motion capture system using IR cameras and markers [13, 16–20, 22–24], force plate [11, 14–16, 18–20], EMG measurement device [14, 15], and electrogoniometer [15] were widely used in conventional studies.

In addition, the system which can detect the resident's abnormal event on stairs using depth data and skeleton data provided by Kinect by Microsoft, applying machine learning is suggested [25]. As the system is focused on the accident detection, any particular parameters were not measured. However, the suggestion indicated the possibility of acquisition of gait without putting any sensors and markers on the subject's body.

## 3  Problem

According to the related works, the 3D motion capture system, force plate, EMG measurement device, and electrogoniometer were widely used. However, the preparation of these devices is not simple, as we have to put them at several position of stairs and body parts for the measurement. Moreover, the 3D motion capture system, EMG measurement device, and electrogoniometer are always restraining the subject putting markers and electrodes on their body. Thus, these suggestions cannot be applied to daily parameter acquisition on stair walking. The Kinect which has depth sensor has the possibility of body position acquisition, however, EMG values cannot be acquired by body position, the others such as 3D motion capture, force plate, and electrogoniometer can be replaced with.

The purpose of this paper is to suggest the marker-less gait assessment system for stair walking and to validate the system with 3D motion capture which had been used in most of conventional researches. In this paper, we established the gait parameter acquisition system on stair walking using only depth data by Kinect v2, not to restrain the subject. In addition, we conducted validation experiment using 3D motion capture system.

# 4 Method

## 4.1 Parameter

According to the related works, we can conclude that the knee flexion angle is most important parameter, since it is used to evaluate the difference between elderly and younger [16], men and women [20], and knee OA patients and normal controls [21, 22, 24].

## 4.2 Depth Data of Kinect v2

Kinect v2 (Fig. 1), released in 2014 by Microsoft, has RGB camera, IR-depth camera, and microphone array, provides RGB, depth, and IR images as an animated film, also has a skeleton and face tracking functions (Fig. 2). As shown in Fig. 2b, we have depth data of 512 by 424 pixels at 30 fps.

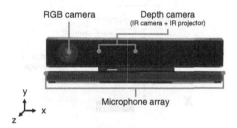

**Fig. 1.** Kinect v2 by microsoft

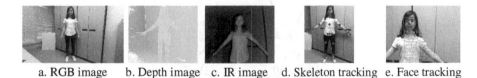

a. RGB image    b. Depth image    c. IR image    d. Skeleton tracking    e. Face tracking

**Fig. 2.** Main functions of Kinect v2

## 4.3 Preprocessing

512 by 424 depth matrix, $\mathbf{Z}$, is acquired by Kinect v2 in every frame, and it is save as CSV file. The program was written in C ++.

First of all, the calibration data without subject is subtracted from $\mathbf{Z}$.

According to the range of Kinect view, X and Y position matrix, $\mathbf{X}$ and $\mathbf{Y}$, can be calculated. The method for X position is shown in Fig. 3. For the calculation, the Eqs. (1) and (2) are used respectively, as the range of horizontal angle is 70 degree and vertical angle is 60 degree.

$$X_K = \frac{\left(\left(512 \div 2 - X_{pix}\right) \times Z_K \times \tan 35°\right)}{512 \div 2} \tag{1}$$

$$Y_K = \frac{\left(\left(424 \div 2 - Y_{pix}\right) \times Z_K \times \tan 30°\right)}{424 \div 2} \tag{2}$$

Now we have 512 by 424 **X**, **Y**, and **Z**, then the correction of angle for Y and Z position is needed. The method follows the Fig. 4, and the conversion Eqs. (3) and (4) are applied respectively.

$$Y_R = Z_K \times \sin \theta + Y_K \times \cos \theta \tag{3}$$

$$Z_R = Z_K \times \cos \theta - Y_K \times \sin \theta \tag{4}$$

After that, the body position is extracted in X-Z plane by using histogram. The data which is more than 1000 pixels away from body position is deleted.

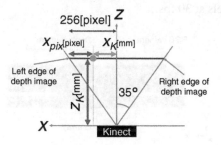

**Fig. 3.** X and Y position calculation method

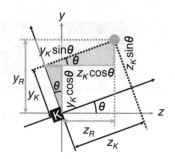

**Fig. 4.** Y and Z position correction method

### 4.4  Position Detection

Firstly, body is extracted and body height are defined. Fixing the column = j, consecutive row of **Z** which has non-zero value are referred as **Z** (non-zero start to non-zero end, j). Then the distance between **Y** (non-zero start, j) and **Y** (non-zero end, j) are calculated.

After that, if the distance is within 1300 mm to 2000 mm, the range of body height, the column is extracted. This process is repeated along all column, then whole body is extracted (Fig. 5). Finally, the body height is defined as the cell number between the average of each column's first and last non-zero row.

Then, the feet are founded. Firstly the 70 % height of body height row shown in green line in Fig. 5 is referred in the X-Z section (Fig. 6). There are two peaks which express both foot. Then the columns are fixed as the foot positions. Next, the 50 % to 80 % of body height row is defined as foot range, and both foot position columns are referred in Y-Z sections (Fig. 7). One peak is found respectively, and we get left and right knee positions.

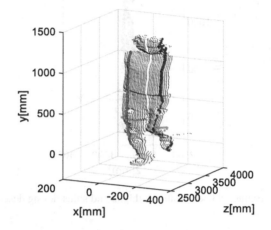

**Fig. 5.** Extracted body

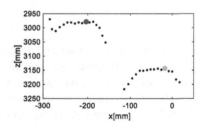

**Fig. 6.** X-Z section for feet detection

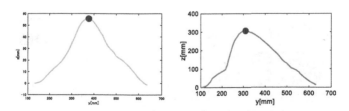

**Fig. 7.** Y-Z sections of left and right foot for knee position detection

### 4.5   Knee Flexion Angle Calculation

Extracted knee positions are defined as boundaries of thighs and shins. The range of foot is used again, from the start positions of foot range to the knee positions is defined as thighs, and from the knee position to the end position of foot range is defined as shins, respectively. Approximate lines of left and right thighs and shins are drawn, the left and right knee flexion angle are calculated (Fig. 8).

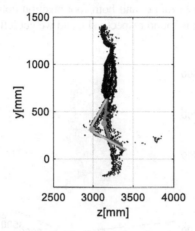

**Fig. 8.**  Extracted knee positions, thighs and shins during descending

## 5   Validation Experiment

### 5.1   Experiment Setup

Five subjects, named A to E, performed ascending and descending the six-step stairs (run length is 304 mm and riser height is 166 mm) for three times with usual speed. We prepared seven VICON cameras around stairs, and a Kinect v2 was set on the floor (Fig. 9). The Kinect was installed horizontally, so there was no need to correct the coordinates. The 16 markers of VICON were put on the subjects lower limbs (part of Plug-In-Gait marker set, LASI, LPSI, LTHI, LKNE, LTIB, LANK, LHEE, LTOE, RASI, RPSI, RTHI, RKNE, RTIB, RANK, RHEE, and RTOE).

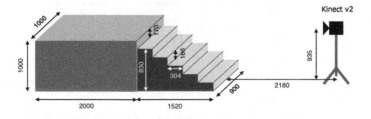

**Fig. 9.**  Experiment setup

## 5.2   Result

We got left and right knee flexion angle of each frame. The subject C's left knee flexion angle on first descent is shown in Fig. 10. The peak of knee flexion was extracted, and the example of results is shown in Table 1. For the comparison of Kinect and VICON data, the average and standard deviation of the difference of peak values were calculated and shown in Table 2.

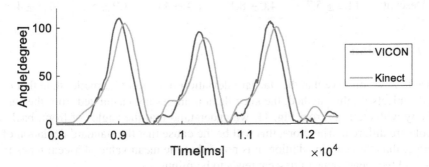

**Fig. 10.**   The left knee flexion angle of subject C's first descent

**Table 1.**   The peaks of left knee flexion angle of subject C [degree]

| Step | Ascent | | | Descent | | |
|---|---|---|---|---|---|---|
| | Kinect | VICON | Difference | Kinect | VICON | Difference |
| 1 | 111.2 | 102.3 | −8.9 | 111.5 | 109.7 | −1.8 |
| | 107.8 | 101.8 | −5.9 | 91.3 | 95.5 | 4.2 |
| | 112.3 | 100.9 | −11.4 | 111.7 | 106.7 | −5.0 |
| 2 | 111.2 | 109.8 | −1.5 | 99.2 | 104.0 | 4.8 |
| | 115.9 | 87.6 | −28.4 | 89.6 | 91.8 | 2.2 |
| | 113.5 | 106.9 | −6.6 | 89.9 | 94.2 | 4.3 |
| 3 | 111.0 | 107.2 | −3.8 | 97.4 | 105.2 | 7.8 |
| | 113.1 | 104.7 | −8.5 | 90.4 | 93.6 | 3.3 |
| | 100.5 | 106.0 | 5.5 | 89.2 | 98.2 | 8.9 |

**Table 2.** Mean and standard deviation of the differences between knee flexion angle acquired by VICON and Kinect (Mean ± SD) [degree]

| Subject | A | B | C | D | E |
|---|---|---|---|---|---|
| All | 8.7 ± 10.0 | −0.8 ± 9.9 | −2.1 ± 7.0 | −5.5 ± 6.7 | −1.3 ± 9.1 |
| Ascent | 6.3 ± 12.4 | −6.3 ± 8.8 | −5.6 ± 7.7 | −9.1 ± 5.5 | −8.6 ± 4.5 |
| Descent | 11.2 ± 5.7 | 4.0 ± 8.1 | 1.4 ± 4.0 | −1.9 ± 5.8 | 7.3 ± 4.5 |

## 6 Discussion

In Table 2, we observe that the standard deviation of descent is smaller than that of all in all subjects. In this method, the knee flexion angles are calculated from the surface of body with clothes. From Fig. 11, we understand that the angles of front, back and middle are different. Therefore, this will be the cause that the standard deviation of all is larger than descent. In addition, it is noted that the mean value of ascent tends to be plus, and the mean value of descent tends to be minus.

On the other hand, the standard deviation of ascent is large. In Fig. 12 with, we observe many dots associated with stairs. The number of these dots is much larger than those in Fig. 8. This fact tells us that the influence of stairs are larger in ascent than descent. In the method, we eliminated the background by simply subtracting a calibration data. To improve the accuracy, the calibration data should be obtained every subject.

Comparing with the conventional research, the difference of the mean peak angle during swing between knee OA patients and normal controls was 5.5 degree at ascent and 5.6 degree at descent [22]. From this result, the accuracy of proposed method is not enough to be used in the realistic environment. We will explore in the next stage to relate the acquired angle with the true angle. We will also seek for a method to reduce the noise.

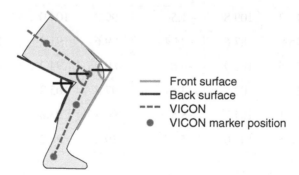

——— Front surface
——— Back surface
- - - VICON
● VICON marker position

**Fig. 11.** The angle obtained during ascent and descend is different

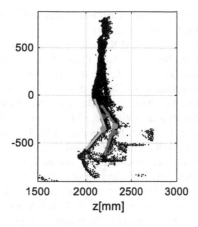

**Fig. 12.** Extracted knee positions, thighs and shins during ascending

## 7  Conclusion

In this paper, we proposed a marker-less gait acquisition system on stair walking using Kinect v2 by Microsoft, and conducted the experiment to confirm the accuracy of acquired knee flexion angle with the data acquired by 3D motion capture. We found that the proposed method could acquire the necessary information but with relatively large standard deviation. The sources of this large standard deviation will be studied in our future research.

**Acknowledgements.**  This work was supported by JSPS KAKENHI Grant-in-Aid for Scientific Research Grant Number 15H02890, and in part by MEXT Grant-in-Aid for the "Program for Leading Graduate Schools".

## References

1. World Health Organization: Global health and ageing (2011)
2. Murata, S., Tsuda, A., Tanaka, Y.: Physical and cognitive factors associated with falls among the elderly with disability at home. J. Jpn. Phys. Ther. Assoc. **32**(2), 88–95 (2005)
3. Suzuki, T., Sugiura, M., Furuna, T., Nishizawa, S., Yoshida, H., Ishizaki, T., Kim, H., Yukawa, H., Shibata, H.: Association of physical performance and falls among the community elderly in japan in a five year follow-up study. Jpn. J. Geriatr. **36**(7), 472–478 (1999)
4. Arai, T., Shiba, Y., Watanabe, S., Shibata, H.: The relationship between the stride time variability, motor ability and fall in community-dwelling elderly people. J. Jpn. Phys. Ther. Assoc. **38**(3), 165–172 (2011)
5. Kim, H., Suzuki, T., Yoshida, H., Shimada, H., Yamamoto, Y., Sudo, M., Niki, Y.: Are gait parameters related to knee pain, urinary incontinence and a history of falls in community-dwelling elderly women? Jpn. J. Geriatr. **50**(4), 528–535 (2013)

6. Tainaka, K., Aoki, J.: Fitness-related factors associated with falling in older women. Jpn. J. Phys. Fit. Sports Med. **56**(2), 279–285 (2007)

7. Hausdorff, J.M., Rios, D.A., Edelberg, H.K.: Gait variability and fall risk in community-living older adults: a 1-year prospective study. Arch. Phys. Med. Rehabil. **82**(8), 1050–1056 (2001)

8. Camicioli, R., Howieson, D., Oken, B., Sexton, G., Kaye, J.: Motor slowing precedes cognitive impairment in the oldest old. Neurol. **50**(5), 1496–1498 (1998)

9. Verghese, J., Annweiler, C., Ayers, E., Barzilai, N., Beauchet, O., Bennett, D.A., Bridenbaugh, S.A., Buchman, A.S., Callisaya, M.L., Camicioli, R., Capistrant, B., Chatterji, S., De Cock, A.M., Ferrucci, L., Giladi, N., Guralnik, J.M., Hausdorff, J.M., Holtzer, R., Kim, K.W., Kowal, P., Kressig, W.K., Lim, J.Y., Lord, S., Meguro, K., Montero-Odasso, M., Muir-Hunter, S.W., Noone, M.L., Rochester, L., Srikanth, V., Wang, C.: Motoric cognitive risk syndrome Multicountry prevalence and dementia risk. Neurol. **83**(8), 718–726 (2014)

10. Bergmann, G., Deuretzbacher, G., Heller, M., Graichen, F., Rohlamann, A.: Hip contact forces and gait patterns from routine activities. J. Biomech. **34**, 859–871 (2001)

11. Liikavainio, T., Isolehto, J., Helminen, H.J., Perttunen, J., Lepola, V., Kiviranta, I., Arokoski, J.P., Komi, P.V.: Loading and gait symmetry during level and stair walking in asymptomatic subjects with knee osteoarthritis: importance of quadriceps femoris in reducing impact force during heel strike? Knee **14**(3), 231–238 (2007)

12. Riener, R., Rabuffetti, M., Frigo, C.: Stair Ascent and Descent at Different Inclinations. Gait Posture **15**, 32–34 (2002)

13. Novak, A.C., Komisar, V., Maki, B.E., Fernie, G.R.: Age-related differences in dynamic balance control during stair descent and effect of varying step geometry. Appl. Ergonomics **52**, 275–284 (2016)

14. Larsen, A.H., Puggaard, L., Hämäläinen, U., Aagaard, P.: Comparison of ground reaction forces and antagonist muscle coactivation during stair walking with ageing. J. Electromyogr. Kinesiol. **18**, 568–580 (2008)

15. Hsu, M.J., Wei, S.H., Yu, Y.H., Chang, Y.J.: Leg stiffness and electromyography of knee extensors/flexors: comparison between older and younger adults during stair descent. J. Rehabil. Res. Dev. **44**(3), 429–436 (2007)

16. Reid, S.M., Graham, R.B., Costigan, P.A.: Differentiation of young and older adult stair climbing gait using principal component analysis. Gait Posture **31**, 197–203 (2010)

17. Chiu, S.L., Chang, C.C., Dennerlein, J.T., Xu, X.: Age-related differences in inter-joint coordination during stair walking transitions. Gait posture **42**, 152–157 (2015)

18. Novak, A.C., Brouwer, B.: Sagittal and frontal lower limb joint moments during stair ascent and descent in young and older adults. Gait Posture **33**, 54–60 (2011)

19. Singhal, K., Kim, J., Casebolt, J., Lee, S., Han, K.H., Kwon, Y.H.: Gender difference in older adult's utilization of gravitational and ground reaction force in regulation of angular momentum during stair descent. Hum. Mov. Sci. **41**, 230–239 (2015)

20. Hong, Y.N.G., Shin, C.S.: Gender differences of sagittal knee and ankle biomechanics during stair-to-ground descent transition. Clin. Biomech. **30**(10), 1210–1217 (2015)

21. Hinman, R.S., Bennell, K.L., Metcalf, B.R., Crossley, K.M.: Delayed onset of quadriceps activity and altered knee joint kinematics during stair stepping in individuals with knee osteoarthritis. Arch. Phys. Med. Rehabil. **83**(8), 1080–1086 (2002)

22. Hicks-Little, C.A., Peindl, R.D., Hubbard, T.J., Scannell, B.P., Springer, B.D., Odum, S.M., Fehring, T.K., Cordova, M.L.: Lower extremity joint kinecatics during stair climbing in knee osteoarthritis. Med. Sci. Sports Exerc. **43**(3), 516–524 (2011)

23. Hicks-Little, C.A., Peindl, R.D., Fehring, T.K., Odum, S.M., Hubbard, T.J., Cordova, M.L.: Temporal-spatial gait adaptations during stair ascent and descent in patients with knee osteoarthritis. J. Arthroplasty **27**(6), 1183–1189 (2012)

24. Roy, N., Boudreau, N., Vezina, F., Tousignant, M., Gaudreault, N.: Comparison of knee kinectics between meniscal tear and normal control during a step-down task. Clin. Biomech. **30**, 762–764 (2015)
25. Parra-Dominguez, G. S., Taati, B. and Mihailidis, A.: 3D human motion analysis to detect abnormal events on stairs. In: International Conference on 3D Imaging, Modeling, Processing, Visualization and Transmission, pp. 97–103 (2012)

# Full Body Gesture Recognition
# for Human-Machine Interaction
# in Intelligent Spaces

David Casillas-Perez[1,2]($\boxtimes$), Javier Macias-Guarasa[1,2],
Marta Marron-Romera[1,2], David Fuentes-Jimenez[1,2],
and Alvaro Fernandez-Rincon[1,2]

[1] Universidad de Alcalá, Madrid, Spain
{david.casillas,macias,marta,david.fuentes,
alvaro.fernandez}@depeca.uah.es
[2] Escuela Politécnica Superior - Campus externo, Universidad de Alcalá –
GEINTRA Research Group, Ctra. Madrid-Barcelona km. 33,600,
28805 Alcala de Henares, Spain
http://www.geintra-uah.org/en

**Abstract.** This paper describes a proposal for a full body gesture recognition system to be used in an intelligent space to allow users to control their environment. We describe a successful adaptation of the traditional strategy applied in the design of spoken language recognition systems, to the new domain of full body gesture recognition. The experimental evaluation has been done on a realistic task where different elements in the environment can be controlled by the users using gesture sequences. The evaluation results have been obtained applying a rigorous experimental procedure, evaluating different feature extraction strategies. The average recognition rates achieved are around 97 % for the gestural sentence level, and over 98 % at the gesture level, thus experimentally validating the proposal.

**Keywords:** Full body gesture recognition · Intelligent spaces · Human-machine interaction · Spoken language recognition strategies

## 1 Introduction

Current trends in human-machine interfaces depart from traditional interaction schemes that require a physical contact, and are based on the use of pointing devices (mouse, joystick) and keyboard. The new interface paradigms are oriented to more natural *human-like* strategies, including spoken dialog systems, which are reasonably accurate in controlled environments; touch interfaces, present in almost all current mobile devices; and human motion capture systems (that could further exploit the fact that most of our communication if nonverbal, through facial expressions and body gestures [1,2]), well represented in gaming scenarios by means of specific sensors such as the Wii mote [3] and kinect [4]

© Springer International Publishing Switzerland 2016
F. Ortuño and I. Rojas (Eds.): IWBBIO 2016, LNBI 9656, pp. 664–676, 2016.
DOI: 10.1007/978-3-319-31744-1_58

controllers, with more recent examples specifically oriented to detecting human (hand) gestures, such as the Google Glass system [5].

In everyday scenarios, a more natural interaction would allow a broader range of users to benefit from computing facilities or systems providing services in intelligent environments. In particular, we are mainly interested in users that are not able to provide spoken instructions, are not willing to use a standard mouse or keyboard access, and are not willing to use a touch interface, that would prevent them from moving freely around their normal environments without carrying a (probably cumbersome) mobile device or provide interaction in predefined locations.

In this context, gestural interfaces seem to be the way to go, as the users can use their full natural human motion capabilities to provide input to the computing systems to be further analyzed. Gestures include body movements and postures, and are able to communicate an intention or feeling, and can even be used to extend our communication capabilities by using a given code. Additionally, gestures are less prone to disfluency errors found in other communication modalities such as speech [6].

## 2    Related Work

In the literature there are a number of attempts describing solutions to gestural interfaces. [1] provide a review on selected proposals for face and body gesture recognition, oriented to be used in a multimodal analyzer for non-verbal behaviour analysis and recognition. They do not provide experimental results and their solution is based on visual processing. In [7] describes a full body tracking system using a two dimensional (2D) link-joint model, but with severe limitations on the movements allowed for the person (e.g. it was not allowed to move forward or backward), model limitations related to using only 2D information (tightly coupled to the available gesture vocabulary), and by the adoption of a rule based strategy for classification. More recent work, such as that described in [8] make use of a depth sensor, as the one used in this paper, but they focus on recognizing hand gestures to control a robot navigation, using Hidden Markov Models (HMMs [9]) for classification. The combination of a kinect depth sensor with the HMM paradigm is also used in [10], but their core features are just based on the 3D body joint positions, but no provision is given for exploiting redundancy in gestures with partially common sub-gestures or trajectories. Finally, [11] extends the HMM framework to combine gesture path and pose, allowing for the identification of different gesture phases, with application scenarios broader than the one presented in this work. Our main contribution in this work is adapting the traditional strategy in spoken language recognition systems (in what respect to the modeling paradigm: and the acoustic, lexical and grammatical levels) to the full body gesture recognition system.

The proposal we describe in this paper follows an original approach that mimics the traditional strategy used in spoken language recognition systems and adapts it to a full body gesture recognition system. The use of a depth

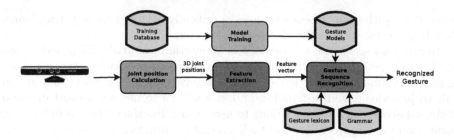

**Fig. 1.** Gesture recognition system architecture.

camera allows us to generate discriminative and relevant features based on measuring the human skeleton pose (evaluated from angle measurements) that are fed to a Hidden Markov Model to model the dynamic and static behaviour of the full body gestures. The use of specific gesture models composed of sub-gesture provide grammatical coherence to recognize full gesture sequences, and increases robustness when facing medium to large gestural vocabularies, although it also increases the alphabet size (that can be counterproductive for very small vocabularies and reduced training datasets).

The paper is structured as follows. In Sect. 1 we provide the introduction and some references on previous work. Section 3 describes the proposed system model. The experimental setup is detailed in Sect. 4 and the experimental results are discussed in Sect. 5. Finally, Sect. 6 summarizes the main conclusions and contributions of the paper and gives some ideas for future work.

## 3   System Description

### 3.1   General Strategy

The full body gesture recognition system proposed in this paper is adapted from the traditional processing paradigm used in spoken language recognition systems, and following the general architecture shown in Fig. 1, namely:

- Feature extraction: deriving relevant features for the task, based on angular measurements of the human body skeleton segments.
- Gesture modeling strategy: based on continuous densities Hidden Markov Models (CD HMMs [12]).
- Lexical and grammatical modeling: In our approach, any command or request by the users represent a *sentence* (as in speech recognition systems), which is composed by a sequence of gestures (that mimics the sequence of *words* concept), which are further composed by sequences of sub-gestures (mimicking the phone sequences within words). The grammatical structure within gesture sequences is modeled using a predefined grammar, and the lexical structure within gestures is modelled by a dictionary in which each gesture is associated to a sequence of sub-gestures (that can be shared among different gestures).
- Continuous gesture recognition: Using a standard Viterbi algorithm.

Our approach allows an easy expansion of the gesture *vocabulary* and *grammar*, as the training procedures are fully automatic and the CD HMM modeling is able to efficiently cope with both the static and dynamic behaviour expected in the human gesture production process.

### 3.2   Working Environment and Sensing Equipment

The working environment used in this work for our gesture recognizer will be located in an intelligent space [13], which is equipped with adequate sensors to provide relevant information to a computing infrastructure to extract significative information on the activities being carried out in the environment, to be able to provide intelligent responses to the user needs and requests.

In our case we want the users to generate full body gestures to be recognized by the system. With this requirement, we need to accurately estimate the three-dimensional coordinates of relevant body joints. In the literature there are approaches for 3D body joint estimations based on standard RGB cameras, but recent work using depth cameras [14] and it's implementation on low cost hardware [4] made us to select it.

We equipped our intelligent space with Kinect Sensors [4], which are capture devices created by Alex Kipman and developed by Microsoft for Xbox 360, featuring an RGB camera and a depth camera (both with a resolution of 640 × 480 pixels), and a microphone array composed of 4 microphones (used for beamforming and acoustic localization purposes). Only one Kinect Sensors is used to capture the user's movements and it's located in front of the scene.

The decision to use Kinect devices was also taken due to its extremely good quality to price ratio, and, more important, the fact that it has been widely used in scientific community, with lots of developing works that use or have used this specific sensor, thus generating a broad knowledge base to grow from, and sophisticated high level libraries freely available to handle the kinect information streams. Besides, Kinect Sensor allows to integrate voice and gesture information for a future voice-gesture recognizer.

Additionally, and in order to provide the system with elements to be controlled by the user, we created a virtual environment in which several appliances were deployed. That way, the system could allow for easy expansion and the tests could be done providing visual feedback to the users. More precisely, the following *fixed* (not moving) elements can be controlled in the environment: The room door (open/close), two windows (open/close) with the corresponding blinds (up/down), ceiling lamps (switch on/off), table lamps (switch on/off), a TV set (switch on/off), a radio set (switch on/off) and a telephone (to place phone calls). Additionally, we include a moving element, a robot, that could be used for assistance purposed. Figure 2 shows a front and top view of the virtual environment, displaying the main elements to be controlled, plus the skeleton of the user, furniture and building structure.

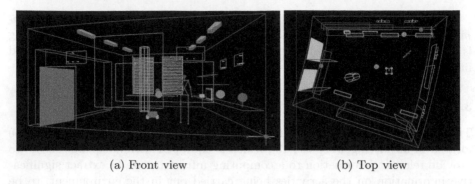

| (a) Front view | (b) Top view |

**Fig. 2.** Virtual intelligent space showing elements to be controlled: lamps (represented by spheres), ceiling lamps, door and two windows (in red), two window blinds (dashed light blue area), TV set (at the back of the right wall in the front view), radio set (next to the TV), and telephone (not seen) (Color figure online).

### 3.3 Grammar and Vocabulary

The specific vocabulary and grammar to be used, have to be specified depending on the application in which the recognizer is going to work. In our case, we had to define a gestural vocabulary to account for all the interactions expected in our virtual environment.

So, the gestures have been mapped either to words related to the element on which the user is willing to actuate (e.g. the gesture *stand still with your arms facing forward and forming a 90° angle* will mean *door*, as shown in Fig. 3), to actions (e.g. the gesture *place both arms straight facing forward and move them to a T position* will mean *open*, as shown in Fig. 4), or to full actions on elements (e.g. the gesture *touching your head with the right hand* will mean *I want to do a phone call*). Also, the gestures can be static (such as the one referring to the *door*, or dynamic (such as the one referring to the *open* action).

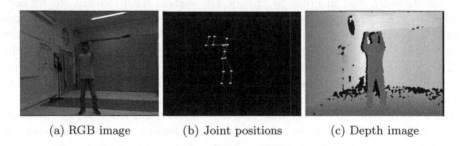

| (a) RGB image | (b) Joint positions | (c) Depth image |

**Fig. 3.** Gesture representing the *door*.

So, we defined a gestural vocabulary (equivalent to a word dictionary in a speech recognizer) composed of 22 gestures (11 corresponding to actions,

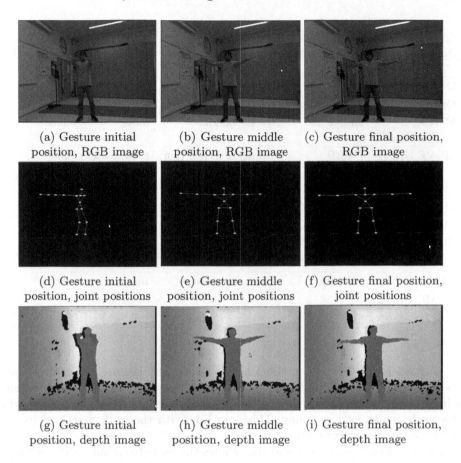

(a) Gesture initial position, RGB image

(b) Gesture middle position, RGB image

(c) Gesture final position, RGB image

(d) Gesture initial position, joint positions

(e) Gesture middle position, joint positions

(f) Gesture final position, joint positions

(g) Gesture initial position, depth image

(h) Gesture middle position, depth image

(i) Gesture final position, depth image

**Fig. 4.** Gesture representing the *open* action.

1 corresponding to a full action on an element, and 10 corresponding to environment elements), as shown in the first two columns of Table 1[1].

The dictionary also describes the sequence of sub-gestures (phones belonging to an alphabet in a speech recognizer) that form a gesture, so that we can exploit the redundancy of gestures sharing common positions or paths. In our case, we defined sub-gestures that are shared among several gestures (column Action sub-gestures (shared) in Table 1), and sub-gestures hat are assigned to individual gestures (column ... (not shared) in Table 1). To provide an example, gesture *OPEN* would be composed by subsets *arms front + arms straight + arms oblique + arms in T position*. Using sub-gestures will allow for a more robust training of the HMMs, as more training material will be available for each sub-gesture.

---

[1] Gestures and actions in the same table row are not necessarily related, the columns represent just a list of the different elements considered.

**Table 1.** List of gestures and sub-gestures considered

| Action gestures | Element gestures | Action sub-gestures (shared) | Action and element sub-gestures (not shared) | |
|---|---|---|---|---|
| OPEN | DOOR | arms up | | door |
| CLOSE | WINDOW | arms front | | window |
| MOVE FORWARD | BLINDS | arms oblique | move forward | blinds |
| MOVE BACK-WARD | LAMP | arms in T position | move back-ward | lamp |
| MOVE DOWN | RIGHT LAMP | right arm up | | right lamp |
| MOVE UP | LEFT LAMP | right arm right | | left lamp |
| SWITCH ON | TELEVISION | right arm front | | television |
| SWITCH OFF | RADIO | left arm up | | radio |
| TURN RIGHT | TELEPHONE | left arm right | turn right | telephone |
| TURN LEFT | ROBOT | left arm front | turn left | robot |
| COME HERE | | | come here | |
| CALL | | | call | |

Our approach relating an *alphabet* to given sub-gestures resembles that described in [7], but being more general, as there is no hard-coded relationship between gestures and their decomposition, and the HMM paradigm will effectively model such coupling. This is in line with the work described in [8], but for full body gesture recognition.

Finally, we defined a grammar that allows the users to combine different gestures to generate a request to the recognition system with complete meaning. In our case, we described the grammar in extended Backs-Naur Form (EBNF [15]) that is then converted to a finite state recognition network at run time. The grammar allows for producing pairs of actions and elements in any order.

## 3.4    Feature Extraction

The feature extraction module will be in charge of generating discriminative and significant features to be used by the modeling paradigm to classify different gestures. Choosing the adequate feature vector is a key step in the development of a robust recognizer, as the classification accuracy will heavily rely on the quality of the chosen features.

For our purposes, a gesture is defined as a movement of some joints or an specific static position, of both of them, with a predefined meaning.

We have made use of the Kinect Sensor and the OpenNI [16] library to extract the three dimensional body joint positions (related to the device coordinate system). In this work, we have evaluated four different feature extraction strategies. One of them uses boolean parameters (as described in [17], and was selected due to their good results in a related application) and the others are based on different strategies to provide the measurement of angles and distances (by changing the reference system used).

1. Boolean feature vectors [17]: Each feature vector is formed by an array of boolean. These parameters are calculated testing if a specific joint is at either side of a half space defined by a given plane. The planes are calculated considering the gestures the recognizer have to understand. For example, if the recognizer want to detect if a user has his arms straight up, it has to test if the user's hands are above a plane which contain the head position, being its normal vector defined by the position of the head and neck. Every plane has been built related to the user position so that the planes follow the user position and orientation. Boolean parameters allow for fast processing, but the information provided to the recognizer will be relatively poor (we will refer to this as BOOLEAN in the experimental discussion).

2. Distance and angles based feature vectors: This scheme of feature extracting keeps all the geometrical information that is relevant for identifying gestures: angles between skeleton segments and distances between skeleton joints. For example, if the recognizer wants to know if the user in scene has his right arm stretched or bent, it only needs to look at the angle formed by the right hand and the corresponding elbow and this elbow with the shoulder. The distance between a hand and the head can be a good parameter to measure if the user is self touching her/his head. In this work we tested three spherical coordinate reference systems, as angles provide more useful information for gesture identification:

    (a) Single reference system located in the Kinect sensor. This scheme produces variations in the features depending on the user position and orientation but is used for comparison purposes (we will refer to this as REFFIXED-KINECT in the experimental discussion).

    (b) Single reference system with the origin in the user's chest and the single basis set is built by vectors of the user's body. This system is invariant under rotation and translation changes of the user (we will refer to this as REFFIXED-BODY in the experimental discussion).

    (c) Multiple reference systems with multiple basis sets. The reference systems are located in the joint from which the angle is calculated, and the associated basis set is built containing one vector aligned with one of the skeleton segments used for the angle calculation (we will refer to this as REFVARIABLE in the experimental discussion).

## 3.5  Gesture Modeling

In our proposal, gestures (by means of their associated sub-gestures) are modelled using continuous densities HMMs which are able to efficiently model both the static and dynamic behaviour of the gestures.

In our HMM implementation (based on the HMM Toolkit HTK [18]) we again reproduce the successful approach typically used in speech recognition: we are using a three state left-to-right topology. A standard Baum Welch algorithm is applied to train *trigesture* models (following the *triphone* modeling strategy in speech recognition [19]), so that there is one distinct sub-gesture model for every different left and right sub-gesture contexts, to efficiently deal with context sensitivity. To alleviate the problem of lack of data for robust training, we apply parameter tying using a (*sub-gesture*) decision tree strategy (mimicking the *phonetic* decision trees adopted in HTK [20]).

### 3.6   Gesture Sequence Recognition

To finally generate a classification decision, we apply Viterbi decoding on a precompiled static recognition network that includes the lexical and grammatical information for the task. The Viterbi algorithm works with a trellis tree to obtain the sequence of states which maximize the observation probability $P(\mathbf{q}|\mathbf{O}, \lambda)$ giving the observation sequence of feature vectors $\mathbf{O}$ and the model $\lambda$, also deciding on the optimal state sequence $\mathbf{q}$ [9].

## 4   Experimental Setup

### 4.1   Database Description

In order to evaluate our proposal we recorded a multimodal and multiuser database, as there was no publicly available dataset meeting our requirements (comprising a gesture vocabulary for appliance control and including body joints positions). Two male speakers were recorded producing spoken and gesture sequences (in English and Spanish for the spoken part). A kinect sensor was used to record the audio, RGB video, depth and skeleton joint positions. Our database is being prepared to be made publicly available in a near future.

The database is composed by 750 sentences (375 per speaker): 250 English sentences, 250 Spanish sentences and 250 gesture sentences, and it has been manually labelled. Speech sentences have been recorded by a wireless close-talk using a sampling frequency of 48000 Hz in WAV format and by the Kinect Sensors using 16000 Hz in WAV format too. Gesture sentences have been recorded in RGB(24bits) video, depth(11bits) video and joints position from Kinect sensor. Speech sentences. Speech sentences have a mean duration of 3s and gesture sentences have a mean duration of 5 s. In this work, only the gestural part has been used.

### 4.2   Experimental Methodology

The experimental evaluation follows a rigorous approach in which the training and testing subsets are fully disjoint. In order to increase the statistical significance of the differences observed, we have applied a $k$-fold cross validation strategy, with $k = 10$ in our case.

The system was evaluated in user independent mode, that is, the system is trained with the material from one user, and evaluated on the testing material of the other user.

Finally, the evaluation will provide results for the four feature extraction strategies described in Sect. 3.4.

## 4.3   Evaluation Metrics

To provide the full picture of the system capabilities, we will provide metrics for full gestural sentence recognition performance (sentences recognized exactly as the reference ground truth, referred to as $CS$ in the tables below, being $NS$ the number of sentences in the evaluation database, so that the correct sentence rate will be $CSR(\%) = \frac{CS}{NS}$), and for individual gesture recognition performance.

Considering that $N$ = Total number of gestures in the evaluation database, $I$ = gesture insertions, $S$ = gesture substitutions and $D$ = gesture deletions, the number of correct gestures is $CG = N - S - D$, and the correct gestures rate is $CGR(\%) = \frac{N-S-D}{N} \times 100$.

Additionally, 95 % confidence intervals will also be provided (for $CSR$ and $CGR$, shown as $\pm \cdot \%$ accompanying the rates) to objectively assess the actual statistical relevance of the observed differences.

## 5   Results and Discussion

Table 2 shows the system performance at the gestural sentence level, and Table 3 at the gesture level. In both tables results are included for the four feature extraction strategies described in Sect. 3.4.

Table 2. Gestural sentence level recognition results.

| Strategy | User | $CSR(\%)$ | $CS$ | $NS - CS$ | $NS$ |
|---|---|---|---|---|---|
| BOOLEAN | USR1 | $2.80 \pm 2.05$ | 7 | 243 | 250 |
| | USR2 | $4.80 \pm 2.65$ | 12 | 238 | 250 |
| REFFIXED-KINECT | USR1 | $97.20 \pm 2.05$ | 243 | 7 | 250 |
| | USR2 | $91.60 \pm 3.44$ | 229 | 21 | 250 |
| REFFIXED-BODY | USR1 | $96.40 \pm 2.31$ | 241 | 9 | 250 |
| | USR2 | $93.60 \pm 3.03$ | 234 | 16 | 250 |
| REFVARIABLE | USR1 | $\mathbf{98.00 \pm 1.74}$ | 245 | 5 | 250 |
| | USR2 | $\mathbf{96.00 \pm 2.43}$ | 240 | 10 | 250 |

From both tables, it's clear that the boolean feature strategy (BOOLEAN) is unable to effectively cope with the task. Even though this strategy was successfully applied in [17], the HMM paradigm seem to be not well suited to handle boolean parameters (in [17] a SVM classifier is used instead).

**Table 3.** Gesture level recognition results

| Strategy | User | $CGR(\%)$ | $CG$ | $D$ | $S$ | $I$ | $N$ |
|---|---|---|---|---|---|---|---|
| BOOLEAN | USR1 | $13.20 \pm 2.97$ | 66 | 48 | 386 | 48 | 500 |
| | USR2 | $20.00 \pm 3.51$ | 100 | 44 | 356 | 44 | 500 |
| REFFIXED-KINECT | USR1 | $98.60 \pm 1.03$ | 493 | 5 | 2 | 5 | 500 |
| | USR2 | $94.20 \pm 2.05$ | 471 | 8 | 21 | 8 | 500 |
| REFFIXED-BODY | USR1 | $98.00 \pm 1.23$ | 490 | 6 | 4 | 6 | 500 |
| | USR2 | $96.20 \pm 1.68$ | 481 | 6 | 13 | 6 | 500 |
| REFVARIABLE | USR1 | $\mathbf{98.80 \pm 0.95}$ | 494 | 4 | 2 | 4 | 500 |
| | USR2 | $\mathbf{97.60 \pm 1.34}$ | 488 | 4 | 8 | 4 | 500 |

Regarding the different strategies to measure angles and distances, the one using a reference system tightly coupled to the corresponding skeleton segment exhibits the best performance. This is due to the fact that the angle and distance measurements are independent of the position and orientation of the user within the scene.

## 6    Conclusions and Future Work

In this document, we have describe a proposal for a full body gesture recognition system to be used in an intelligent space to allow users unable or unwilling to use standard communication channels, to control their environment.

Our main contribution is successfully adapting the traditional strategy applied in the design of spoken language recognition systems, to the new domain of full body gesture recognition. The adaptation has been done at every level, including the lexical and grammatical ones, with the HMM paradigm being the selected modeling (phones become sub-gestures, words become gestures, and finally gestures are combined in sentences, according to a properly defined grammar). The adoption of this strategy will ease future developments, both in the application of additional improvements, and the extension of the system within the same domain or the porting to new ones.

The evaluation has been done on a realistic task, defining a virtual environment where different elements can be controlled. In this scenario, gestures have been defined, including their segmentation in sub-gestures (building a dictionary) and a simplified grammar that is able to cope with the task and show the feasibility of the proposal.

The performance results have been obtained applying a rigorous experimental procedure, using cross-validation to increase the statistical significance of the results. Several feature extraction strategies have been evaluated, and the average recognition rates achieved are around 97 % for the gestural sentence level, and over 98 % at the gesture level, thus experimentally validating the proposal.

Additional work will be carried out to further validate the proposal with a bigger database, also increasing its variability so that more complex models can be used (e,g, increasing the number of mixtures), adaptation strategies can be applied and so on, all benefiting from the adoption of the speech recognition paradigm. Dealing with occlusions is also a pending issue and will be addressed in the future. Also, a gesture understanding and dialog management module is being developed, in order to allow the system to provide a more natural gestural interaction with the users.

**Acknowledgements.** This work has been supported by the Spanish Ministry of Economy and Competitiveness under project SPACES-UAH (TIN2013-47630-C2-1-R), and by the University of Alcalá under projects DETECTOR and ARMIS.

# References

1. Gunes, H., Piccardi, M., Jan, T.: Face and body gesture recognition for a vision-based multimodal analyzer. In: Piccardi, M., Hintz, T., He, S., Huang, M.L., Feng, D.D. (eds.) VIP, vol. 36 of CRPIT, pp. 19–28. Australian Computer Society (2003)
2. Mehrabian, A.: Communication without words. Psychol. Today **2**(9), 52–55 (1968)
3. Wikipedia: Wiimote sensor general information. https://en.wikipedia.org/wiki/Wii_Remote. Accessed January 2016
4. Wikipedia: Kinect sensor general information. http://en.wikipedia.org/wiki/Kinect . Accessed January 2016
5. Wikipedia: Google glass general information. https://en.wikipedia.org/wiki/Google_Glass. Accessed January 2016
6. Cassell, J.: A framework for gesture generation and interpretation. In: Computer Vision in Human-Machine Interaction, pp. 191–215. Cambridge University Press, Cambridge (2000)
7. Puranam, M.B.: Towards full-body gesture analysis and recognition. PhD thesis, University of Kentucky (2005)
8. Xu, D., Chen, Y., Lin, C., Kong, X., Wu, X.: Real-time dynamic gesture recognition system based on depth perception for robot navigation. In: 2012 IEEE International Conference on Robotics and Biomimetics, ROBIO 2012, Guangzhou, China, pp. 689–694, 11–14 December 2012
9. Rabiner, L.R.: A tutorial on hidden markov models and selected applications in speech recognition. Proc. IEEE **77**(2), 257–286 (1989)
10. Song, Y., Gu, Y., Wang, P., Liu, Y., Li, A.: A kinect based gesture recognition algorithm using GMM and HMM. In: 6th International Conference on Biomedical Engineering and Informatics, BMEI 2013, Hangzhou, China, pp. 750–754, 16–18 December 2013
11. Yin, Y.: Real-time continuous gesture recognition for natural multimodal interaction. PhD thesis, Massachusetts Institute of Technology, Cambridge, MA (2014)
12. Jurafsky, D., Martin, J.H.: Speech and Language Processing: An Introduction to Natural Language Processing, Computational Linguistics, and Speech Recognition, 1st edn. Prentice Hall PTR, Upper Saddle River (2000)
13. Lee, J.H., Hashimoto, H.: Intelligent space concept and contents. Adv. Rob. **16**(3), 265–280 (2002)

14. Shotton, J., Fitzgibbon, A.W., Cook, M., Sharp, T., Finocchio, M., Moore, R., Kipman, A., Blake, A.: Real-time human pose recognition in parts from single depth images. In: The 24th IEEE Conference on Computer Vision and Pattern Recognition, CVPR 2011, Colorado Springs, CO, USA, pp. 1297–1304, 20–25 June 2011
15. Pattis, R.E.: Ebnf: A notation to describe syntax. While developing a manuscript for a textbook on the Ada programming language in the late 1980s, I wrote a chapter on EBNF (1980)
16. OpenNI: https://github.com/OpenNI/OpenNI. Accessed January 2016
17. Müller, M., Röder, T., Clausen, M.: Efficient content-based retrieval of motion capture data. In: ACM Trans. Graph. (TOG), vol. 24, pp. 677–685. ACM (2005)
18. Young, S., Evermann, G., Kershaw, D., Moore, G., Odell, J., Ollason, D., Valtchev, V., Woodland, P.: The HTK Book, vol. 3. Cambridge University Engineering Department, Cambridge (2002)
19. Lee, K.F.: Context-dependent phonetic hidden markov models for speaker-independent continuous speech recognition. Acoust. Speech Signal Proc. IEEE Trans. **38**(4), 599–609 (1990)
20. Young, S.J., Odell, J.J., Woodland, P.C.: Tree-based state tying for high accuracy acoustic modelling. In: Proceedings of the Workshop on Human Language Technology, Association for Computational Linguistics, pp. 307–312 (1994)

# A Web System for Managing and Monitoring Smart Environments

Daniel Zafra[1], Javier Medina[1], Luis Martinez[1], Chris Nugent[2],
and Macarena Espinilla[1(✉)]

[1] Department of Computer Science, University of Jaén, Jaén, Spain
dzr00003@red.ujaen.es, {jmquero,martin,mestevez}@ujaen.es
[2] School of Computing and Mathematics, University of Ulster,
Jordanstown BT37 0QB, UK
cd.nugent@ulster.ac.uk

**Abstract.** Smart environments have the ability to record information about the behavior of the people by means of their interactions with the objects within an environment. This kind of environments are providing solutions to address some of the problems associated with the growing size and ageing of the population by means of the recognition of activities, monitoring activities of daily living and adapting the environment. In this contribution, a Web system for managing and monitoring smart environments is introduced as an useful tool to activity recognition. The Web system has the advantages to process the information, accessible services and analytic capabilities. Furthermore, a case study monitored by the proposed Web System is illustrated in order to show its performance, usefulness and effectiveness.

**Keywords:** Smart environments · Behavioral detection · Monitoring smart environments · Managing smart environments · Sensor-based activity recognition

## 1 Introduction

The number of elderly will reach 2 billion by the year 2050 and a key issue for this people is to stay as long as possible in their own homes in order to have a healthy ageing and wellbeing [1]. One of the most common diseases in this group is related to cognitive processes such as dementia. These illnesses are currently incurable, hence efforts are focused towards delaying their progression.

The dementia has a number of symptoms including memory loss, mood changes, communication problems and eventually results in problems with the completion of everyday tasks once the condition reaches the later stages, including activities of daily living (ADLs) and instrumental ADLs (IADLs) [2,3]. These alterations require long term monitoring and support in order to maximize quality of life and minimize progression.

© Springer International Publishing Switzerland 2016
F. Ortuño and I. Rojas (Eds.): IWBBIO 2016, LNBI 9656, pp. 677–688, 2016.
DOI: 10.1007/978-3-319-31744-1_59

In the early stages of dementia, it is useful to provide support in the form of prompting through the completion of ADLs and IADLs, offering a series of reminders for tasks such as medication management, eating or grooming [4–6].

Smart environments have the ability to record information about the behavior of the person by means of his/her interaction with the objects within an environment [7]. So, smart environments are residences with sensor technology in which sensors are connected to a range of objects or locations and networked in order to be used to identify people in the environment and their actions [8].

This kind of environments are providing solutions to address some of the problems associated with the growing size and ageing of the population by means of the recognition of activities, monitoring activities of daily living in order to adapt the environment.

Regarding a software tool for monitoring smart environments, we can find the tool presented in [9] in order to visualizate the data generated in real-time or summarized, using a density ring visualization format. However, this tool is a desktop application that must be installed on a personal computer to manage and monitor a smart environment. So, it does not allow the management of smart environments in a Web system.

In this contribution, we introduce a Web system for managing and monitoring smart environments in order to provide a useful tool to activity recognition. The Web system illustrates information about the behavior of the person by means of his/her interaction with the objects within a smart environment. The proposed Web system has been developed to be intuitive and flexible, providing the advantages to process the information with accessible services and analytic capabilities. To do so, the proposed Web system allows managing the components involved in each smart environment registered in the system: plane of the smart environment, objects, sensors, events, etc. The data generated within the environment can be visualized by the proposed Web system in real-time, off-line, and, finally, consulted using different filters.

In order to show the usefulness of the proposed tool, a case study has been designed for monitoring the smart lab of University of Jaen with six contact sensors (open/close) and three multisensors by using the introduced Web system.

The rest of the contribution is set out as follows: Sect. 2 introduces some preliminaries regarding sensor-based activity recognition. Section 3 presents our proposed Web system for managing and monitoring smart environments. Section 4 shows a case study monitored in the University of Jaen by the proposed Web System. Finally, in Sect. 5, conclusions are drawn.

## 2   Related Works

Sensor-based activity recognition is an important research topic that involves multiple fields of research including pervasive and mobile computing [10,11], context-aware computing [12–14] and ambient assisted living [15,16].

This area of research has witnessed a significant level interest mainly as a result of the rapid advances in the sensor technology development coupled with demands from an application perspective [8]. Advances in technology developments have mainly focused on providing a wide range of low cost sensors, with low-power requirements and decreased form factor.

These sensors can be connected to a range of objects or locations and networked and used to identify people in the environment, their actions, their emotions and to provide personalized assistance with their everyday tasks.

The process of activity recognition aims to recognise the actions and goals of inhabitants within the environment based on a series of observations of actions and environmental conditions. It can therefore be deemed as a complex process that involves the following steps: (i) to choose and deploy the appropriate sensors to objects within the environment in order to effectively monitor and capture a user's behavior along with the state change of the environment; (ii) to collect, store and process information and, finally, (iii) to infer/classify activities from sensor data through the use of computational activity models.

Traditionally, approaches used for sensor-based activity recognition have been divided into two main categories: Data-Driven Approaches (DDA) and Knowledge-Driven Approaches (KDA). The former, DDA, are based on machine learning techniques in which a preexistent dataset of user behaviors is required. A training process is carried out, usually, to build an activity model which is followed by a testing processes to evaluate the generalization of the model in classifying unseen activities [17–19]. With KDA, an activity model is built through the incorporation of rich prior knowledge gleaned from the application domain, using knowledge engineering and knowledge management techniques [20, 21].

It is necessary to train and test in depth activity models in order to check their adaptation with information about the behavior of the people by means of their interactions with the objects within an environment. In the following section, we introduce a Web system for managing and monitoring smart environments.

## 3    A Web System for Managing and Monitoring Smart Environments

In this section, we present a Web system for managing and monitoring Smart Environments as an useful tool to activity recognition. To do so, we pay attention to the architecture of the system, the database model and, finally, its functionality.

At the following URL: http://sinbad2.ujaen.es:8094/Smartlab, the proposed Web system is located.

### 3.1    Web System Architecture

Data generated by smart environments requires fusion information, accessibility services and analytic capabilities. The proposed Web system is focused on providing these requirements to friendly collect and handle the information offered by the set of sensors from a smart environment, suitable for using by non-technical users.

The proposed Web system processes the data generated from heterogeneous sensors to a homogeneous structure which is unique for any set of sensors within smart environments. In order to include a persistence of sensor data, we have defined an appropriate relational model, which has been focused on handling large volumes of data [22], whose persistence follows the next structure:

- *Objects*, which are related to physical sensor which are deployed on the environments.
- *Environments* to group the objects and sensors deployed in the environment. In this way, this approach enables the monitoring of multiple smart environments for scientific and analytic purposes.
- *Location.* Due to the fact that the object location usually changes continuously, we relate a dynamic set of locations to each object with its smart environment.
- *Property-Value.* We collect the sensor values adding a type which increases the interpretability of raw data.

In the context of smart environments, usually, an object is the abstraction of a sensor. So, when an object is created, a set of rules should be provided in order to define the interpretation of the values provided by its sensor. Therefore, the set of objects defined in each smart environment registered in our system should have three values: Object-property-value. Furthermore, in order to locate the object in the smart environment, its location must also be provided. We highlight we have defined an abstract schema where any mobile, wearable or ambient sensor can be described.

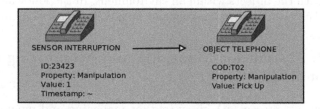

**Fig. 1.** Switch sensor associated to phone with the interpretation of values

The proposed Web system is transparent to all devices that interact with it. To create this interoperability between sensors, our system makes use of a central API provided by REST, where only the communication protocol, Hypertext Transfer Protocol (HTTP), is shared. With this solution, we can consume or produce content of our API on any platform, regardless of operating system or system architecture, thus having multiple heterogeneous clients. Using the REST services, any mobile or ambient computer is enable to send the data collected by their sensors.

The components of our architecture are described in Fig. 2:

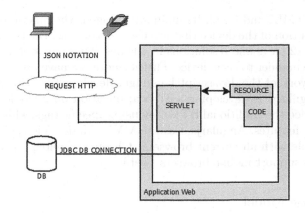

**Fig. 2.** System architecture based on REST service

- **Web Server**, being its aim to develop core services in order to provide a abstraction layer where different clients can communicate with the system. The remote services have been developed using Web Services, increasing the transparency and accessibility of mobile, web or desktop clients.

  In our Web system, we have implemented Web Service based on Representational State Transfer (REST) that has a client server architecture in which each service is identified by a URL. These services communicate with the server via servlet, providing a multi-platform architecture that uses the HTTP protocol. Each servlet is responsible for translating HTTP requests (GET, PUT, POST and DELETE), traveling accompanied by information in a language with a specific structure in JavaScript Object Notation (JSON) that is a simple and lightweight text-based format, which arises as XML alternative. So, the structure object-property-value is carried out using the notation JSON that are implemented by REST technology. Therefore, both the client and the server may make the exchange of information and both know the protocol and format of the information. Following, two main remote services are described:

**Environment Sensors:** Return sensors of an environment registered in the system
*/service/sensors/{environmentID}* "idSensor": "00144F0100006C62","idType": "Sunspot","active": true, "idSensor": "34037272","idType": "Tynetec","active": true

  **Current Status:** Return current status of environment and position map.
*/service/object/environment/{environmentID}/currenstatus* [JSON] "idObject": 21,"idSensor": "34037288","codObject": "M01", "description": "Puerta", "timestamp": "2015-11-27 11:32:43.0", "valueDescription": "Abierta", "posX": 641.0, "posY": 226.0,

- **Client**. A web client to consult the data from a Web browser. To provide a friendly user interface, we have integrated Bootstrap in the web view due to is one of the most important frameworks for the design of web applications

based on HTML5 and CSS3. Its main advantage is the functionality to adapt
to the resolution of the device that are used. To do this, Bootstrap divides the
screen into 12 parts and resolutions in 4 types. Thus, the web is adjusted in a
very simple manner to each device. Furthermore, we have used Angularjs that
is a framework of the Javascript language, developed by Google. The main
idea of Angularjs is developing a Web application on a single page. So, we
have extended the traditional HTML by using specific tags, which will provide
it with new features. Angularjs uses the MVC (Model, View, Controller) and
is compatible with all current browsers (Chrome, Firefox, Opera, Safari), in
addition to support mobile browsers (webkits).

## 3.2 Database Model

Modeling of Spatio-Temporal dimension is key in the analysis of data [23] but it is
necessary designing scalable data storages that provide efficient data mining [24].

To do so, we have defined a relational model based on an entity relationship
schema that are illustrated in Fig. 3. The general schema is divided into two
packages. These packages work together to provide meaning to our Web system.
Following, we describe the main components:

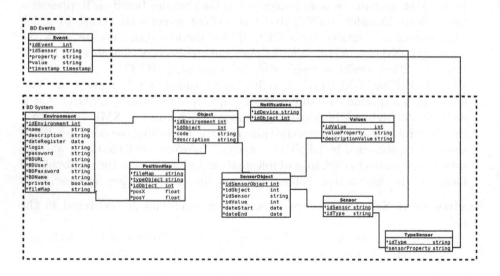

**Fig. 3.** Entity relationship schema

- *Events*: It is the most important component of package. This package contains one single table to store raw information of each sensor for each smart
  environment.
- *Web System*: This package translates the information from the sensor to a
  natural language. It contains the following components:

- *Environment*: This component describes different properties. One of them is the possibility to provide privacy to the environment. So, users could login to service for managing and monitoring the environment. Other property is the option of give a external database to the system to translate raw information of the sensors.
- *PositionMap*: This component collects the position of each sensor in the system dynamically.
- *Object*: It defines the object and its description, associating the object with its environment.
- *SensorObject*: This component can associate a sensor with an object, considering the time of association. So, this component has the following values: dateStart and dateEnd in order to indicate whether the sensor is currently located at object.
- *Value*: This component contains the value of a sensor with an interpretable description of the measure.

### 3.3   Web System Functionality

At this point, we present the functionality of the system that is shown from the point of view of the role that performs it.

There are 2 types of roles in the system: *administration* and *public*. The *administration* role is focused on manage smart environments and the *public* role is oriented to monitor smart environments. Following, we briefly summarize the actions that can be carried out for each role.

**Administration role:** The actions that can be performanced with this role are:
  - **Manage smart environments**. To add, edit and delete smart environments and its properties: description, plane, URL of database, etc.
  - **Manage sensors**. To add, edit and delete sensors in the Web system. Currently, our approach includes two types of sensors: contact sensors (open/close) and multisensors that can measure acceleration, light and temperature.
  - **Manage objects**. In order to add, edit and delete objects in a smart environment and associate with a sensor.
  - **Manage object location**. In order to locate objects in the plane of the smart environment.

**Public role:** The actions that can be performanced with this role are:
  - **Visualization of current status**. The plane of a smart environment is illustrated with the set of sensors and the last values provided for each property of each sensor.
  - **Visualization of past states (Historic)**. For a given time (day and hour), the plane of a smart environment is illustrated with the set of sensors and the set of values provided for each property in this moment for each sensor. Furthermore, the system can play the events that have occurred in one day through a timeline (off-line) in the plane of a smart environment.

- **Consultation past events (Historic)**. Reports about events within smart environment are provided. These reports can consider a date, a period and a number of recent changes. Furthermore, graphics of event frequency in each sensor per hour are illustrated.

## 4  Case Study

In order to highlight the effectiveness of the proposed Web system, a case study has been designed for monitoring the smart-lab of University of Jaen.

The smart lab is located in a room of 25 square meters in the CEATIC[1] (Center for Advanced Studies in Information Technology and Communication) of University of Jaen. In the smart lab are distributed the following four areas: a lobby, a living room, a kitchen and a bedroom with an integrated bathroom. Figure 4 illustrates the areas of the smart lab of the University of Jaen from different points of view.

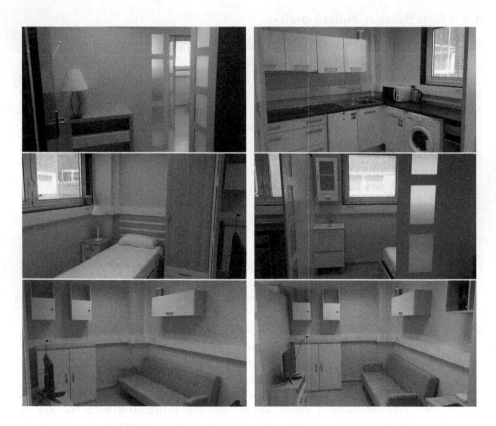

**Fig. 4.** Smart lab of the University of Jaen

---

[1] http://ceatic.ujaen.es/es/smart-lab-0.

Several sensors were located in different places of the smart lab, in order to gather the data from the interactions between the people within environment.

Regarding the sensors, two type of sensors were deployed: contact sensors (open/close)[2] and multisensors[3] that can measure acceleration, light and temperature that are illustrated in Fig. 5.

The sensors' network was composed by six contact sensors (open/close) and three multisensors. Contact sensors were located in the following objects: Front door, TV remote, bathroom faucet, refrigerator, microwave, bed room closet door. Multisensor was located in the kitchen, living room and in the workplace. The plane of the smart lab is illustrated in Fig. 6 in which the location of each sensor is indicated.

**Fig. 5.** Contact sensors and multisensors

---

[2] www.tynetec.co.uk.

[3] http://sunspotdev.org/.

686    D. Zafra et al.

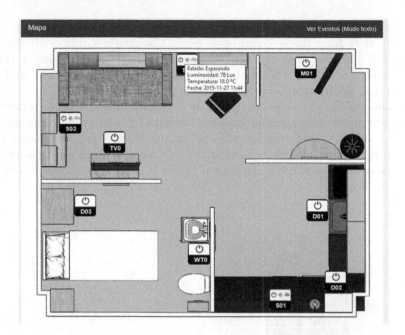

**Fig. 6.** Plane of the smart lab and deployed sensors

The proposed Web system was capable of visualizing the real-time state of the environment, showing events occurring in the environment. Figure 7 illustrates the last ten events generated in the smart lab.

| Objeto | Fecha Evento | Estado | Descripción |
|--------|--------------|--------|-------------|
| TV0 | 2015-10-15 17:40 | Encendido | Televisión |
| TV0 | 2015-10-15 20:49 | Encendido | Televisión |
| TV0 | 2015-10-16 09:00 | Encendido | Televisión |
| TV0 | 2015-10-16 09:00 | Encendido | Televisión |
| D01 | 2015-10-16 09:01 | Abierta | Frigorífico |
| WT0 | 2015-10-16 09:20 | Encendido | Grifo |
| D01 | 2015-10-16 09:54 | Abierta | Frigorífico |
| TV0 | 2015-10-19 09:40 | Encendido | Televisión |
| WT0 | 2015-10-19 09:40 | Encendido | Grifo |
| D01 | 2015-10-19 09:40 | Abierta | Frigorífico |
| WT0 | 2015-10-27 09:30 | Encendido | Grifo |

**Fig. 7.** Last ten events in the smart lab

## 5    Conclusions and Future Works

In this paper, a Web system for managing and monitoring smart environments has been introduced that aims to facilitate user interpretation of the data provided by the sensors generated by the interactions with the objects within an environment. The data generated within the environment can be visualized by the proposed Web system in real-time, off-line, and, finally, consulted using different filters. To do so, we have described the architecture of the system, the database model and, finally, its functionality. Furthermore, the introduced Web system has been tested for managing and monitoring a smart lab in the University of Jaen. However, the platform described in this paper has been developed for managing and monitoring of smart environments a scalable way. In future work, we focus on extending and deploying the Web System in the Smart Lab of University of Ulster.

One limitation of the introduced Web system is that includes only two types of sensors: contact sensors (open/close) and multisensors that can measure acceleration, light and temperature. Currently, we are working to incorporate other types of sensors such as motion or proximity.

On the data persistence, this first work includes a relational database. In future works, we will work on translating the relational schema to a BigTable Model, where tables are translated to family columns and attributes to single columns. It will provide redundancy and online analytical processing of huge amount of sensor records using several clusters.

On the data analysis, our future works are focused on including a sensor-based real-time activity recognition under a knowledge-driven approach based on fuzzy rules.

**Acknowledgements.** This contribution has been supported by research projects: UJA2014/06/14 and CEATIC-2013-001.

## References

1. Smith, G., Della Sala, S., Logie, R.H., Maylor, E.A.: Prospective and retrospective memory in normal aging and ementia: a questionnaire study. Memory **8**, 311–321 (2000)
2. Alzheimer's society. What is dementia? (2013). http://www.alzheimers.org.uk/site/scripts/documents_info.php?documentID=106
3. Von Strauss, E.: Aging and the occurrence of dementia: findings from a population-based cohort with a large sample of nonagenarians. Arch. Neurol. **56**(5), 587–592 (1999)
4. Holder, L.B., Cook, D.J.: Automated activity-aware prompting for activity initiation. Gerontechnology **11**(4), 534–544 (2013)
5. Feuz, K.D., Cook, D.J., Rosasco, C., Robertson, K., Schmitter-Edgecombe, M.: Automated detection of activity transitions for prompting. IEEE Trans. Hum. Mach. Syst. **45**(5), 575–585 (2014)

6.  Das, B., Cook, D.J., Schmitter-Edgecombe, M., Seelye, A.M.: Puck: an automated prompting system for smart environments: toward achieving automated prompting-challenges involved. Pers. Ubiquit. Comput. **16**(7), 859–873 (2012)
7.  Cook, D.J., Augusto, J.C., Jakkula, V.R.: Ambient intelligence: technologies, applications, and opportunities. Pervasive Mobile Comput. **5**(4), 277–298 (2009)
8.  Chen, L., Hoey, J., Nugent, C., Cook, D.J., Yu, Z.: Sensor-based activity recognition. IEEE Trans. Syst. Man Cybern. Part C Appl. Rev. **42**(6), 790–808 (2012)
9.  Synnott, J., Chen, L., Nugent, C.D., Moore, G.: Flexible and customizable visualization of data generated within intelligent environments. In: 2012 Annual International Conference of the IEEE Engineering in Medicine and Biology Society (EMBC), pp. 5819–5822, August 2012
10. Satyanarayanan, M.: Pervasive computing: vision and challenges. IEEE Pers. Commun. **8**(4), 10–17 (2001)
11. Varshney, U.: Pervasive healthcare and wireless health monitoring. Mobile Netw. Appl. **12**(2–3), 113–127 (2007)
12. Emmanouilidis, C., Koutsiamanis, R.-A., Tasidou, A.: Mobile guides: taxonomy of architectures, context awareness, technologies and applications. J. Netw. Comput. Appl. **36**(1), 103–125 (2013)
13. Makris, P., Skoutas, D.N., Skianis, C.: A survey on context-aware mobile and wireless networking: on networking and computing environments' integration. IEEE Commun. Surv. Tuts. **15**(1), 362–386 (2013)
14. Perera, C., Zaslavsky, A., Christen, P., Georgakopoulos, D.: Context aware computing for the internet of things: a survey. IEEE Commun. Surv. Tuts. **16**(1), 414–454 (2014)
15. Alam, M.M., Hamida, E.B.: Surveying wearable human assistive technology for life and safety critical applications: standards, challenges and opportunities. Sensors **14**(5), 9153–9209 (2014). (Switzerland)
16. Van Hoof, J., Wouters, E.J.M., Marston, H.R., Vanrumste, B., Overdiep, R.A.: Ambient assisted living and care in The Netherlands: the voice of the user. Int. J. Ambient Comput. Intell. **3**(4), 25–40 (2011)
17. Gu, T., Wang, L., Wu, Z., Tao, X., Lu, J.: A pattern mining approach to sensor-based human activity recognition. IEEE Trans. Knowl. Data Eng. **23**(9), 1359–1372 (2011)
18. Li, C., Lin, M., Yang, L.T., Ding, C.: Integrating the enriched feature with machine learning algorithms for human movement and fall detection. J. Supercomput. **67**(3), 854–865 (2014)
19. Martin, L.A., Pelaez, V.M., Gonzalez, R., Campos, A., Lobato, V.: Environmental user-preference learning for smart homes: an autonomous approach. J. Ambient. Intell. Smart. Environ. **2**(3), 327–342 (2010)
20. Chen, L., Nugent, C.: Ontology-based activity recognition in intelligent pervasive environments. Int. J. Web Inf. Syst. **5**(4), 410–430 (2009)
21. Chen, L., Nugent, C.D., Wang, H.: A knowledge-driven approach to activity recognition in smart homes. IEEE Trans. Knowl. Data Eng. **24**(6), 961–974 (2012)
22. Shah, M., Big data, the internet of things (2015). arXiv preprint arXiv:1503.07092
23. Maryvonne, M., Bédard, Y., Brisebois, A., Pouliot, J., Marchand, P., Brodeur, J.: Modeling multi-dimensional spatio-temporal data warehouses in a context of evolving specifications. Int. Arch. Photogrammetry Remote Sens. Spat. Inf. Sci. **34**(4), 142–147 (2002)
24. Zaslavsky, A.B., Perera, C., Georgakopoulos, D.: Sensing as a service and big data (2013). CoRR, abs/1301.0159

# A Study in Experimental Methods of Human-Computer Communication for Patients After Severe Brain Injuries

Andrzej Czyzewski[1(✉)] and Bozena Kostek[2]

[1] Faculty of Electronics, Telecommunication and Informatics,
Multimedia Systems Department, Gdansk University of Technology,
ul. Narutowicza 11/12, 80-233 Gdansk, Poland
ac@pg.gda.pl
[2] Faculty of Electronics, Telecommunication and Informatics,
Audio Acoustics Laboratory, Gdansk University of Technology,
ul. Narutowicza 11/12, 80-233 Gdansk, Poland

**Abstract.** Experimental research in the domain of multimedia technology applied to medical practice is discussed, employing a prototype of integrated multimodal system to assist diagnosis and polysensory stimulation of patients after severe brain injury. The system being developed includes among others: eye gaze tracker, and EEG monitoring of non-communicating patients after severe brain injuries. The proposed solutions are used for collecting and analyzing patients' responses and interactions induced by the multimodal stimulation, resulting in assessing the influence of stimuli on increase of patient's cognitive and communicative functions with the use of intelligent data analysis methods.

**Keywords:** CRS-R - coma recovery scale revised · DRS - disability rating scale · EEG– electroencephalography · EGT eye-gaze tracking · ERP - event related potentials EOG– electrooculography · VR - virtual reality · VS - vegetative state

## 1 Introduction

The increasing number of people who undergo brain damage is one of the most characteristic features of our contemporary society. Within the new research project we have been started in Poland in the year 2015, we understand brain damage in terms of severe disorders resulting from a traumatic injury, a cerebral stroke, or changes occurring in the brain that are consequent from cerebral hypoxia (e.g. due to sudden cardiac arrest (SCA) or suicidal strangulation). All of above mentioned causes of brain damage may lead to a coma [1]. A coma is a temporary state which may result in fast recovery or in a vegetative state (VS), a locked-in syndrome (LIS) or in brain death. Unfortunately, especially in case of patients in the VS, evaluation of their awareness is still problematic. It is also worth highlighting that there is a widespread erroneous belief that patients who recover from a coma without regaining consciousness, remain in a permanent vegetative state. In the vast majority of residential medical care facilities such patients are not diagnosed to a sufficiently detailed extent. Eastern European studies show that the ratio of incorrect diagnostic assessment of patients' consciousness

© Springer International Publishing Switzerland 2016
F. Ortuño and I. Rojas (Eds.): IWBBIO 2016, LNBI 9656, pp. 689–703, 2016.
DOI: 10.1007/978-3-319-31744-1_60

amounts to as much as 40 % [2, 38]. Therefore, within the project develop research experiments, employing a relatively inexpensive experimental setup of an integrated system that will become a diagnostic and polysensory stimulation tool and then to validate its usefulness in the management of patients belonging to the studied group.

It is planned within the project and it have been started already the validation of the following scientific hypothesis:

"Application of integrated technologies: eye-gaze tracking (EGT), Auditory Brainstem Response (ABR), electroencephalography (EEG), electrooculography (EOG), virtual reality and scent emitting for polysensory stimulation, along with results of subjective assessment in GCS (Glasgow Coma Scale), can support diagnosis and can positively influence cognitive and communicative functions of patients with brain injuries".

Above hypotheses is to be confirmed by:

1. Preparing experimental setups for multimodal polysensory stimulation and installing them in some selected patients' care centers (as is described in Sect. 3)
2. Utilizing technology of eye gaze tracking in an user interface to allow navigation by eyeball movements, enabling maximum possible patient's interaction with the multimodal virtual environment,
3. Applying EOG, EEG, ABR, for analysis of Event Related Potentials (ERP) for objectivation of patients' cognitive activity assessment,
4. Development of a combined metric of the patient state by an intelligent fusion of GCS (subjective Glasgow Coma Scale or its derivatives) with objective data acquired using ABR, EEG, EOG, ERP, and EGT. Employed data fusion to be based principally on inference methods originating from the rough set theory. It allows induction of rules describing patient consciousness state based on combined analysis of descriptive (subjective) data and objective results of measurements acquired with abovementioned techniques of multimodal data acquisition.

The motivation for conducting research in this area is based on three various scientific assumptions. First, as noted by Pąchalska [3], in most cases of patients with locomotor system damage it is possible to find a solution to the medical problems originating from the injury. However, it is much more difficult to prevent cognitive and emotional impairments. Therefore, we assume that the technological support of therapists working with such patients on an everyday basis is essential. We have acquired experience in designing and providing diagnostic and therapeutic tools which are now successfully being used in numerous facilities of medical care [4, 5]. The Gdansk University of Technology (Multimedia Systems Department) specializes in multimodal computer interfaces, which we have been developing within other type of projects (technology-oriented ones) involving also a cooperation with medical specialists and neurodegenerative patients care centers [6–8].

According to Edelman's theory of neural Darwinism, we assume that only those synaptic connections that are in use will be maintained, whereas those rarely used disintegrate themselves. Therefore, it is a definite necessity for a frequent stimulation and exercising cognitive function of patients with disorders of consciousness. Such an opportunity justifies our approach based on advanced technology [9]. Finally, as noted by Grüner et al. [10, 11], during the previous several years, polysensory stimulation has played a very significant role in the early rehabilitation stage of patients in a coma caused by brain damage. Thus, the effect of particular sense stimulation on the early rehabilitation of comatose patients is already known. Moreover, in our study we evaluate the effect of polysensory stimulation on the increase in the consciousness level of patients who have recovered from a coma, but who still remain in a vegetative or minimally conscious state, monitoring their performance by analysis of EEG/ERP and oculographic signals (eye gaze). This research brings a development of a new integrated metric for patient state involving intelligent fusion of GCS, ABR, EEG, ERP, and EGT results and employing some soft computing-based inference methods.

## 2 Study Rationale

The most commonly used scale for assessing the level of consciousness in patients with consciousness disturbances is the Glasgow Coma Scale (GCS) published in 1974 by Teasdale and Jennett [12]. The scale is composed of three tests: opening of the eyes (e.g. eyes open spontaneously, eyes open in response to painful stimuli), verbal response (e.g. converses logically or makes incomprehensible sounds) and motor functions (e.g. purposeful movements, reaction to pain). Each of these activities is scored according to some predefined values. The assessment of the level of a patient's consciousness with the GCS scale is relatively simple, thus it does not require much time. However, there are situations in which its implementation is difficult: for example, the scale does not allow for any differentiation between a patient in a persistent vegetative state and those remaining in minimal consciousness state. Since the Glasgow Coma Scale turned out to have limitations, in 1985 an improved version of it was proposed by Born et al.– GLS (Glasgow-Liège Scale) [13]. There have also been other tools designed for the same purposes, e.g.: the Disability Rating Scale (DRS) [14], the Sensory Modality Assessment and Rehabilitation Technique (SMART) [15], Wessex Head Injury Matrix (WHIM) [16], Coma Recovery Scale Revised (CRS-R) [17] or the Full Outline of Unresponsiveness (FOUR) [18]. In the last phase of our study, which includes tests conducted on patients with the use of our assessment system, we use three scales: GLS, DRS, CRS-R. The latter one is currently the most commonly used tool in the assessment of consciousness [19, 20]. The DRS scale supplements the evaluative abilities of CRS-R, and at the same time it verifies and confirms its results. As proven in the studies by Giacino et al. [21], such an approach to the assessment of consciousness level is fully justified. Apart from standard methods of neurological observation of the patient, there are also the techniques of dynamic neuroimaging, which can supplement the assessment of consciousness level. There have been numerous studies published, in which electroencephalography (EEG) [22, 23], positron emission tomography (PET) [24, 25] or functional magnetic resonance imaging (fMRI) [26, 27] were used. In addition to the above mentioned techniques, event-related potentials (ERP) [28] studies are becoming more and more common. ERP is associated with the electrical reactions of the brain to external stimulation of a selected sense. Nevertheless, PET or fMRI are in most cases used for scientific purposes only and they do not play any vital role in clinical practice. It is similar in the case of ERP, which has recently become extremely popular with the context of scientific studies. It is a popular belief [3] that it is sometimes difficult to analyze and to interpret the data acquired from modern brain imaging instruments. However, by focusing on EEG signals acquisition methods and processing algorithms employs an ERP in the diagnostic stage of our research for objectifying assessment of patient's consciousness level (mainly P300 signal), as well as in the evaluation stage in order to compare brain functioning before and after executing of the polysensory stimulation based on our approach.

The experimental setup of an integrated system designed within the project, is used in monitoring of patients' rehabilitation in 6 medical centers cooperating with Gdansk University of Technology. The method of treatment for patients with consciousness disturbances designed by us is based on polysensory stimulation of centers responsible for cognitive and linguistic processes. There have been studies of that direction described in literature. However, our approach is innovative, mainly because we integrate several different technologies into one advanced integrated tool. Moreover, our method allows for patients' interaction with the multimedia content shown on a computer screen and played back through loudspeakers. Such interaction is possible owing to the eye gaze tracking system (EGT), which analyses a patient's eyeballs movement and enables us to observe his/her visual activity and control the software and navigate the content accordingly. There has previously been no such use of eye gaze tracking systems, apart from our introductory studies described in the article: "Awareness evaluation of patients in vegetative state employing eye gaze tracking system" [4]. The article refers to the observation made by Giacino [29], according to which purposeful eye movement is usually connected with the patient's conscious action. Giacino also observed that minimally conscious state (MCS) patients are able to follow with their eyes a person moving in front of them, while vegetative state (VS) patients are not able to do that, either with moving people or any objects in their field of vision. Therefore, the main objective of Giacino's study was to establish a clear distinction between patients who show some signs of consciousness and those who remain in complete unconsciousness state. The eye gaze tracking system has played a crucial role in this study and provided stable and easily-controlled conditions during tests.

It should be highlighted that the eye gaze tracking system which was used in our pilot studies has been designed in the Multimedia Systems Department of Gdansk University of Technology and is recognized under the name of "CyberEye". A more detailed description of the interface designed by us can be found in our previous publications [4, 30].

Integrating the eye gaze tracking system with virtual reality of 3D visual and surround acoustic environment makes an important element of our system's design. Immersing patients in virtual reality is aimed at inducing polysensory stimulation on the one hand and, on the other, increasing their quality of life in a broad sense, by opening the whole world of multimedia content to patients interacting with it by eye gaze only. The approach based on intensive multimodal stimulation was in 2000 used by Grüner and Terhaag [10]. Our approach differs from the one proposed by Grüner and Terhaag among others by the fact that the two most crucial senses (vision and hearing) are in our study stimulated with the advanced multimodal system, enabling the patients to use their eyes to control the acoustic and visual content. The authors of the above mentioned article found significant differences in the values of heart and respiratory frequencies in patients caused mainly by tactile and auditory stimulation.

The most frequently observed result of tactile and visual stimulation was a change in the patient's facial expression and eye movement. In another article, Grüner et al. [11] observed that during the preceding several years, polysensory stimulation has played a very significant role in the early rehabilitation stage of patients in a coma caused by brain damage.

As mentioned above, the patient's sight and hearing are be mainly stimulated with the use of the virtual reality system. There are many examples of studies in the area of neurorehabilitation based on immersing patients in a virtual reality [31, 32]. However, most of these studies focus on motor rehabilitation. Systems of virtual reality enable cerebral stroke patients to operate their hands virtually (this is possible thanks to the so-called data gloves) and interact with the outside world by making various moves (the system is based on the analysis of the images provided by a camera tracking hand movement).

An interesting, but also controversial technique for visual (or visual-auditory) stimulation is the EEG biofeedback, also known as neurofeedback. The bibliography shows publications in which EEG biofeedback is thoroughly discussed, including its scientific aspects [33, 34]. Many of them also point to the positive effect of EEG biofeedback in the treatment of traumatic brain injury (TBI) patients. Even though Thornton [35] indicated that memory can be improved in patients with brain injury, Reddy et al. [36] observed a significant improvement in patients' verbal and visual memory, while Larsen et al. [37] investigated the influence of neurofeedback on the improvement of patient anxiety, mood disturbances, fatigue, pain, sleep problems, etc. Since the results are not fully convincing, thus we have decided not to include the EEG biofeedback into our study.

Nonetheless, the EEG is used in the project not in a neurofeedback approach, but for collecting evoked potentials– electrical potential recorded from the nervous system following the presentation of polysensory stimuli (event-related potentials– ERP). After presentation of a stimuli the P300 is expected to increase as it results from conscious decision making and categorization of stimulus. In EEG the P300 is acquired from parietal lobe responsible for processing sensory input. Reproducibility and ubiquity of this signal is widely acknowledged, what results in P300 being frequently used in clinical psychological tests [40]. Due to low signal-to-noise-ratio a time averaging and other dedicated processing of this signal is required, reducing the acquisition speed [41]. However, in the conducted research the speed is not an issue, as EEG is only used to confirm the long term cognitive activity of the patient. In the project a dedicated signal acquisition and preprocessing algorithms are being prepared to improve the quality of collected EEG measurements.

Moreover, the Electrooculography (EOG) technique allows measuring the corneo-retinal standing potential between the front and the back of the eye, associated with spontaneous or deliberate eye movements and contracting the muscles. EOG would be used in conjunction with optical EGT (eye gaze tracking), developed by the team of the Gdansk University of Technology. By precisely targeted measurements also other signals are collected, including: Auditory Brainstem Response (ABR) evoked by the presentation of a sound, recorded from the scalp but originating at brainstem level.

The project assumes development of a combined metric of patient state by intelligent fusion of GCS (subjective Glasgow Coma Scale or its derivatives) with objective data acquired using ABR, EEG, ERP, and EGT. Subjective values are not easily interpreted using typical arithmetic and Boolean logic, therefore an approach of soft computing is applied in this research, namely the rough set theory (RS) [42–45]. This property of RS makes this theory applicable for processing of medical data [46, 47, 48] and automatic inference of rules, and knowledge discovery from datasets.

As mentioned above, experiments assuming GCS + ABR + EEG + EOG + ERP + EGT fusion for inferring a combined metric of patient state are in progress. Each of these phases includes specific tasks determining the realization of subsequent tasks (Fig. 1).

The phase (1) includes following task: *Concept, development, and testing of the experimental setup and procedures for polysensory stimulation and monitoring of patient state*. It is dedicated to elaboration of a concept for an experimental test platform; design and studies of technological solutions integrating methods of gaze tracking and ABR, EEG, EOG, ERP, EGT measurements; development of the platform and data synchronization for presented stimuli and acquired signals; preparation of an experimental software and multimedia tests for implementing polysensory stimulations; elaboration of testing methods; reference tests of healthy persons; recruitment and training of staff of selected medical centers.

Phase (2) consists of following tasks: *Validation of the experimental setup of system in actual test research employing patients of selected medical centers*, involving sessions of patients interacting with and stimulated by the experimental setup, and synchronized collection of ABR, EEG, EOG, ERP, and EGT data, and subjective ratings in Glasgow Coma Scale (GCS), followed by an analysis of signals quality and optimization of elements of the experimental setup.

Phase (3): *Elaboration of objective metric of patient consciousness state by intelligent fusion of GCS, ABR, EEG, EOG, ERP, EGT results*. This fusion will be performed for inferring a combined metric of patient state. The input decision system for training of rough sets will be built from two collections: all values measured for healthy person interacting with the experimental setup, as a class of conscious and communicative persons, and all values for VS, UWS, LIS patients of selected medical centers; comprising the class of non-communicative patients. The boundaries of both sets are expected to be imprecise, therefore rough set processing is selected for the fusion and induction of rules describing patient's state. During the experiments of polysensory stimulation it is expected that registered parameters will deviate from initial values, positioning particular patient abilities closer to healthy persons.

Phase (4): *Assessment of the effectiveness of the developed methodology and influence of polysensory stimulation on cognitive functions and patient's communication ability*. During the numerous consultations and experiments in real conditions in the medical centers, therapists opinions will be collected as well and assessment of patient state will be correlated with the metrics elaborated within Task 3. The same metric will be collected for patients and for healthy persons using the same means of interfacing with the experimental setup. It will be verified further if results for those two groups reveal common trend & tendencies (e.g. if reaction on particular stimulus changes over time, is it closer to the reaction of healthy person, etc.).

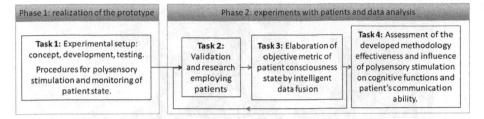

**Fig. 1.** Research and tasks dependencies

# 3   Results Obtained Hitherto in Studies

The first results of our pilot study has been published in several conferences and in the article "Awareness evaluation of patients in vegetative state employing eye gaze tracking system" [4] and they are documented in several reports obtained from 6 Polish medical care institution where pilot studies were organized. In the pilot experiments patients were presented among others with two photographs. One was associated with their past (presenting the patient himself or his relative ones), while the other one depicted a stranger. Patients showing some kinds of consciousness are able to cooperate with the eye gaze tracking system and to focus longer on an emotionally related photo (Fig. 2).

Moreover, results of "consciousness test" based on monitoring of patient's visual activity were also published [8]. Results expressing correctness of the task completion associated with a patient's visual activity can be presented as a box plot diagram (mean/standard error SE/standard deviation SD) (Fig. 3). The tasks included, for example, pointing to a particular picture or matching two pictures together. The tests were conducted over time span of 6 months. Best results were obtained by patients 01 and 02. The results reliability was confirmed by the fact that patient 02 finally recovered from a coma and preserved logical verbal contact, while patient 01 was considered to be a locked-in syndrome patient (therapists, working with her on everyday basis has confirmed this fact). The data obtained from patients are collected in Table 1.

For polysensory stimulations test materials with static content has been created including:

- 2D and 3D photographs
- 2D and 3D movies
- Spatial sound recordings with short sounds of nature, and speech, and long recordings of music
- Scents for aroma interface [6]

The comprehensive observation of patients should last for 52 weeks, with a 1 h long session once a week. During that time, changes in central nervous system function or the level of consciousness (by means of EEG/ERP) may be observed. It is expected that the developed combined metric taking into account subjective GCS scale and

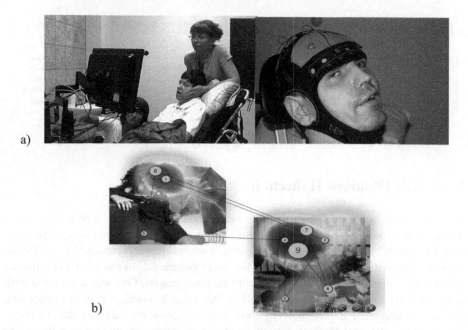

a)

b)

**Fig. 2.** Example pictures: (a) pilot study experimental stands and (b) testing pictures layout: the picture related emotionally to the patient is located in the right part of the screen (with gaze tracking results plotted)– large circles (e.g. no 9) reflect longer focus times

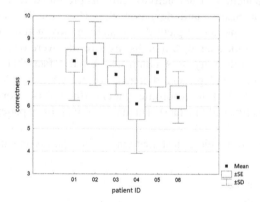

**Fig. 3.** The number of correctly performed tasks with regards to patient's visual activity (result of pilot studies [7])

objective measurements will reflect changes of patient state. We examine a variety of patients from 6 cooperating medical care centers, taking into account a certain margin of time for some unexpected obstacles which may disturb the schedule of the study (Fig. 4).

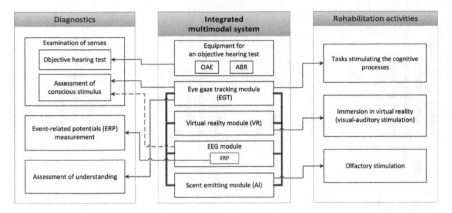

**Fig. 4.** The scheme of the integrated approach to patient diagnosis, stimulation and assessment

SCA– sudden cardiac arrest; CT– cerebellar tumor (after neurosurgical operation); SA– suicide attempt

WTB– TEOAE component present at the whole tested band

\* according to CRS-R criteria

NA– not available

The proposed diagnosis procedure is visualized on Fig. 5 and is described below.

The **rehabilitation phase** is a crucial part of the proposed methodology. As has already been stated, the main aim of the rehabilitation is to stimulate the centers responsible for cognitive processes by polysensory stimulation. Figure 6 presents the idea of integrated multimodal approach to neurorehabilitation of studied patients.

Apart from the rehabilitation function, the eye gaze tracking system also serves as a tool for monitoring the patient's condition throughout the time, as the results are stored in a database.

After consultations with therapists several scenarios of rehabilitation are being developed, exploiting the same experimental setup and polysensory stimulation. Among the ones that are considered are:

1. reading– matching a caption to the picture,
2. dialogue speech,
3. sense of time,
4. emotional states recognition,
5. memory,
6. classification,
7. arithmetic,
8. ability to synthesize,
9. logical thinking,
10. reading.

**Table 1.** Demographic and clinical data of studied patients

| Patient ID | Sex | Age (years) | Cause | Interval post-ictus (y-years, m-months) | Hearing test: OAE left ear/right ear | Hearing test: ABR left ear/right ear [dB nH] | Visual function before/after (interval of rating scale: 0-5)* | Communication before/after (interval of rating scale: 0-2)* | Arousal before/after (interval of rating scale: 0-3)* | Employing of: eye-gaze tracking system (Cyber-Eye) | EEG interface | Patient's state evaluation (diagnosis after the examination utilizing gaze tracking and EEG) |
|---|---|---|---|---|---|---|---|---|---|---|---|---|
| 01 | Female | 40 | trauma | 5 y | WTB/3 kHz, 4 kHz | 50/35 | 3/4 | 0/1 | 1/2 | + | + | LIS |
| 02 | Male | 21 | trauma | 8 m | 4 kHz/4 kHz | 20/30 | 3/5 | 0/2 | 1/3 | + | NA | awakened |
| 03 | Male | 26 | trauma | 5 y | WTB/1 kHz, 4 kHz | 30/20 | 3/4 | 0/0 | 1/2 | + | + | MCS |
| 04 | Female | 40 | SCA | 4 y | NA | NA | 1/2 | 0/0 | 1/1 | + | NA | MCS |
| 05 | Male | 32 | trauma | 5 y | NA | NA | 1/3 | 0/0 | 1/2 | + | NA | MCS |
| 06 | Female | 36 | SCA | 3 y | 1 kHz, 3 kHz/1 kHz, 4 kHz | NA/20 | 1/1 | 0/0 | 1/2 | NA | + | VS |
| 07 | Male | 56 | CT | 1 y | WTB/WTB | 20/30 | 1/3 | 1/2 | 1/3 | NA | + | MCS |
| 08 | Male | 22 | trauma | 8 m | 3 kHz, 4 kHz/4 kHz | 30/20 | 2/4 | 0/0 | 1/3 | + | + | awakened (without verbal logic contact) |
| 09 | Male | 22 | trauma | 3 m | WTB/3 kHz, 4 kHz | 25/45 | 1/3 | 0/0 | 1/1 | NA | + | VS |
| 10 | Male | 54 | CT | 5 m | WTB/WTB | 20/20 | 1/3 | 0/1 | 1/2 | NA | + | MCS |
| 11 | Male | 45 | SCA | 1 y | WTB/WTB | 30/40 | 1/1 | 1/0 | 1/1 | NA | + | VS |
| 12 | Female | 36 | SA | 4 m | 3 kHz, 4 kHz/3 kHz, 4 kHz | 30/50 | 0/1 | 0/0 | 1/1 | NA | + | VS |
| 13 | Female | 55 | trauma | 5 m | NA | NA | 1/5 | 0/2 | 1/3 | NA | + | awakened |
| 14 | Male | 53 | SCA | 2 y | WTB/WTB | NA/30 | 0/1 | 0/0 | 0/1 | NA | + | awakened (without verbal contact) |
| 15 | Male | 34 | trauma | 7 y | WTB/1 kHz, 3 kHz, 4 kHz | NA/30 | 3/3 | 0/0 | 0/1 | NA | + | MCS |

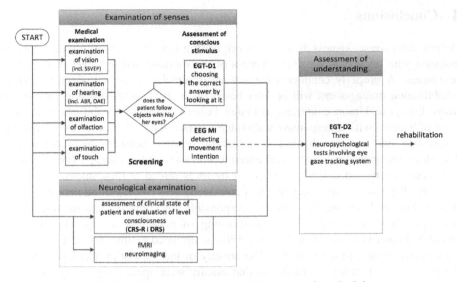

**Fig. 5.** The diagnostic procedure of the proposed methodology

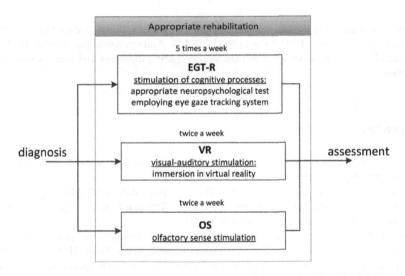

**Fig. 6.** Rehabilitation procedure being a part of the proposed methodology

For example "logical thinking" task requires finishing a story presented on the monitor in a logical way. The patients are encouraged to choose one the three pictures presented on the monitor screen which can make a logical ending to the story. The task may then become gradually more difficult.

Olfaction will be stimulated as well, in patients with confirmed olfactory consciousness test. A more detailed description of the interface designed by us can be found in our previous publications [6, 39].

# 4 Conclusions

Patients with consciousness disturbances originating from brain injury should be firstly assessed with regards to the extensiveness of the damage and the level of their consciousness. A properly conducted diagnostics is absolutely crucial, since the whole rehabilitation management will be later based on it. It is worth pointing out, that in many European Union countries, including Poland, standard rehabilitation management of patients with unresponsive wakefulness syndrome (UWS) patients is limited to a massage and simple physical exercises preventing bedsores and contractures. Therefore, the approach based on advanced technology for the management of patients with consciousness disturbances proposed by us is aimed at their intensive rehabilitation by diagnosing and stimulating particular structures of the central nervous system (CNS). This includes mainly structures responsible for cognitive and linguistic functions and those responsible for the functioning of particular senses. The presented research project is in progress, however hitherto obtained results in 6 care centers are fully encouraging and motivational. The novelty of the present project lays mainly in the integration of various technologies that usually were applied separately, as well as linking the following areas within a common setup, namely: diagnosis, stimulation and communication.

**Acknowledgments.** The project was funded partially by the National Science Centre of Poland on the basis of the decision DEC-2014/15/B/ST7/04724. Authors would like to thank Dr. Bartosz Kunka and Dr. Piotr Szczuko for their help with preparing drawings and some portions of the description.

# References

1. Laureys, S., Tononi, G.: The Neurology of Consciousness: Cognitive Neuroscience and Neuropathology. Academic Press, London (2009)
2. Monti, M.M., Laureys, S., Owen, A.M.: The vegetative state. BMJ **341**, 292–296 (2010)
3. Pąchalska, M.: The clinical neuropsychology. Traumatic brain injuries (in Polish). Warszawa: Wydawnictwo Naukowe PWN (2012)
4. Kunka, B., Czyzewski, A., Kwiatkowska, A.: Awareness evaluation of patients in vegetative state employing eye-gaze tracking system. Int. J. Artif. Intell. Tools **21**(02), 1–11 (2012)
5. Kupryjanow, A., Czyzewski, A.: Improved method for real-time speech stretching. Intell. Decis. Technol. **6**(2), 177–185 (2012)
6. Czyżewski, A., Odya, P., Smulko, J., Lentka, G., Kostek, B., Kotarski, M.: Scent emitting multimodal computer interface for learning enhancement. In: 21th International Workshops on Databse and Expert System Applications (DEXA), pp. 142–146 (2010)
7. Kunka, B., Sanner, T., Czyżewski, A., Kwiatkowska, A.: Consciousness study of subjects with unresponsive wakefulness syndrome employing multimodal interfaces. In: Ślęzak, D., Tan, A.-H., Peters, J.F., Schwabe, L. (eds.) BIH 2014. LNCS, vol. 8609, pp. 57–67. Springer, Heidelberg (2014)

8. Kwiatkowska, A., Izdebski, P., Kunka, B., Czyżewski, A.: Disordered reading and writing in post-comatose patients employing a video-based eye-gaze tracking system. In: Liberska, H., (ed.) Current psychosocial problems in traditional and novel approaches: the multiplicity of roles and difficulties experienced by human in life as determinants of well-being. Wydawnictwo Uniwersytetu Kazimierza Wielkiego, pp: 141–157, (2015). ISBN: 978-83-8018-010-9

9. Edelman, G.M.: Neural Darwinism– The Theory of Neuronal Group Selection. Basic Books, New York (1987)

10. Grüner, M.L., Terhaag, D.: Multimodal early onset stimulation (MEOS) in rehabilitation after brain injury. Brain Inj. 14(6), 585–594 (2000)

11. Grüner, M.L., Wedekind, C., Ernestus, R.I., Klug, N.: Early rehabilitative concepts in therapy of the comatose brain injured patients. Acta Neurochir. Suppl. 79, 21–23 (2002)

12. Teasdale, G., Jennett, B.: Assessment of coma and impaired consciousness. Lancet 304 (7872), 81–84 (1974)

13. Born, J.D., Albert, A., Hans, P., Bonnal, J.: Relative prognostic value of best motor response and brain stem reflexes in patients with severe head injury. Neurosurg. 16(5), 595–601 (1985)

14. Rappaport, M., Hall, K.M., Hopkins, H.K.: Disability rating scale for severe head trauma: coma to community. Arch. Phys. Med. Rehabil. 63(3), 118–123 (1982)

15. Gill-Thwaites, H.: The sensory modality assessment and rehabilitation technique— a tool for the assessment and treatment of patients with severe brain injury in a vegetative state. Brain Inj. 11(10), 723–724 (1997)

16. Shiel, A., Horn, S.A., Wilson, B.A., Watson, M.J., Campbell, M.J., McLellan, D.L.: The Wessex Head Injury Matrix (WHIM) main scale: a preliminary report on a scale to assess and monitor patient recovery after severe head injury. Clin. Rehabil. 14(4), 408–416 (2000)

17. Giacino, J.T., Kalmar, K., Whyte, J.: The JFK coma recovery scale - revised: measurement characteristics and diagnostic utility. Arch. Phys. Med. Rehabil. 85(12), 2020–2029 (2004)

18. Wijdicks, E., Bamlet, W., Maramattom, B., Manno, E., McClelland, R.: Validation of a new coma scale: The FOUR score. Ann. Neurol. 58(4), 585–593 (2005)

19. Bruno, M.A., Majerus, S., Boly, M., Vanhaudenhuyse, A., Schnakers, C., Gosseries, O., Boveroux, P., Kirsch, M., Demertzi, A., Bernard, C., Hustinx, R., Moonen, G., Laureys, S.: Functional neuroanatomy underlying the clinical subcategorization of minimally conscious state patients. J. Neurol. 259(6), 1087–1098 (2012)

20. Rosanova, M., Gosseries, O., Casarotto, S., Boly, M., Casali, A.G., Bruno, M.A., Mariotti, M., Boveroux, P., Tononi, G., Laureys, S., Massimini, M.: Recovery of cortical effective connectivity and recovery of consciousness in vegetative patients. Brain 135, 1308–1320 (2012)

21. Giacino, J.T., Whyte, J., Bagiella, E., Kalmar, K., Childs, N., Khademi, A., Eifert, B., Long, D., Katz, D.I., Cho, S., Yablon, S.A., Luther, M., Hammond, F.M., Nordenbo, A., Novak, P., Mercer, W., Maurer-Karattup, P., Sherer, M.: Placebo-controlled trial of amantadine for severe traumatic brain injury. N. Engl. J. Med. 366(9), 819–826 (2012)

22. Landsness, E., Bruno, M.A., Noirhomme, Q., Riedner, B., Gosseries, O., Schnakers, C., Massimini, M., Laureys, S., Tononi, G., Boly, M.: Electrophysiological correlates of behavioural changes in vigilance in vegetative state and minimally conscious state. Brain 134(Pt 8), 2222–2232 (2011)

23. Chennu, S., Finoia, P., Kamau, E., Monti, M.M., Allanson, J., Pickard, J.D., Owen, A.M., Bekinschtein, T.A.: Dissociable endogenous and exogenous attention in disorders of consciousness. NeuroImage. Clin. 3, 450–461 (2013)

24. Owen, A.M.: Detecting consciousness: a unique role for neuroimaging. Annu. Rev. Psychol. 64, 109–133 (2013)

25. Bruno, M.A., Fernández-Espejo, D., Lehembre, R., Tshibanda, L., Vanhaudenhuyse, A., Gosseries, O., Lommers, E., Napolitani, M., Noirhomme, Q., Boly, M., Papa, M., Owen, A., Maquet, P., Laureys, S., Soddu, A.: Multimodal neuroimaging in patients with disorders of consciousness showing 'functional hemispherectomy'. Prog. Brain Res. **193**, 323–333 (2011)
26. Tagliazucchi, E., Behrens, M., Laufs, H.: Sleep neuroimaging and models of consciousness. Front. Psychol. **4**, 256 (2013)
27. Di Perri, C., Stender, J., Laureys, S., Gosseries, O.: Functional neuroanatomy of disorders of consciousness. Epilepsy Behav. **30**, 28–32 (2014)
28. Risetti, M., Formisano, R., Toppi, J., Quitadamo, L.R., Bianchi, L., Astolfi, L., Cincotti, F., Mattia, D.: On ERPs detection in disorders of consciousness rehabilitation. Front. Hum. Neurosci. **7**(775), 1–10 (2013)
29. Giacino, J.T., Ashwal, S., Childs, N., Cranford, R., Jennett, B., Katz, D.I., Kelly, J.P., Rosenberg, J.H., Whyte, J., Zafonte, R.D., Zasler, N.D.: The minimally conscious state: Definition and diagnostic criteria. Neurol. **58**(3), 349–353 (2002)
30. Kunka, B., Kostek, B., Kulesza, M., Szczuko, P., Czyzewski, A.: Gaze-tracking-based audio-visual correlation analysis employing quality of experience methodology. Intell. Decis. Technol. **4**, 217–227 (2010)
31. De Mauro, A.: Virtual reality based rehabilitation and game technology. In: EICS4Med 2011, pp. 48–52 (2011)
32. Cameirão, M.S., Badia, S.B.I., Oller, E.D., Verschure, P.F.M.J.: Neurorehabilitation using the virtual reality based Rehabilitation Gaming System: methodology, design, psychometrics, usability and validation. J. Neuroeng. Rehabil. **7**(48), 1 (2010)
33. Evans, J.R., Abarbanel, A.: Introduction to Quantitative EEG and Neurofeedback. Academic Press, San Diego (1999)
34. Thompson, M., Thompson, L.: The Neurofeedback Book. W. W. Norton & Company, New York (2004)
35. Thornton, K.: Improvement/rehabilitation of memory functioning with neurotherapy/QEEG biofeedback. J. Head Trauma Rehabil. **15**(6), 1285–1296 (2000)
36. Reddy, R.P., Jamuna, N., Indira, D.B., Thennarasu, K.: Neurofeedback training to enhance learning and memory in patient with traumatic brain injury: A single case study. Indian J. Neurotrauma **6**(1), 87–90 (2009)
37. Larsen, S., Harrington, K., Hicks, S.: The LENS (low energy neurofeedback system): a clinical outcomes study on one hundred patients at stone mountain center, new york. J. Neurother. **10**(2–3), 69–78 (2006)
38. Tagliaferri, F., Compagnone, C., Korsic, M., Servadei, F., Kraus, J.: A systematic review of brain injury epidemiology in Europe. Acta Neurochirugica **148**(3), 255–268 (2006)
39. Kotarski, M., Smulko, J., Czyżewski, A., Melkonyan, S.: Fluctuation-enhanced scent sensing using a single gas sensor. Sens. Actuators B Chem. **157**(1), 85–91 (2011)
40. Comerchero, M.D., Polich, J.: P3a and P3b from typical auditory and visual stimuli. Clin. Neurophysiol. **110**(1), 24–30 (1999)
41. Donchin, E., Spencer, K.M., Wijesinghe, R.: The Mental Prosthesis: Assessing the Speed of a P300-based brain computer interface. IEEE Trans. Rehabil. Eng. **8**, 174–179 (2000)
42. Pawlak, Z.: Rough sets. Int. J. Parallel Prog. **11**(5), 341–356 (1982)
43. Kryszkiewicz, M.: Rules in incomplete systems. Inf. Sci. **113**, 271–292 (1999)
44. Stefanowski, J., Tsoukias, A.: Incomplete information tables and rough classification. Comput. Intell. **17**, 545–566 (2001)
45. Skowron, A.: Rough sets and vague concepts. Fundamenta informaticea **64**, 417–431 (1996)

46. Slowinski, K., Slowinski, R., Stefanowski, J.: Rough set approach to analysis of data from peritoneal lavage in acute pancreatitis. Med. Inform. **13**, 143–159 (1988)
47. Düntsch, I., Gediga, G.: Rough Set Dependency Analysis in Evaluation Studies– An Application in the Study of Repeated Heart Attacks. University of Ulster, Informatics Research Reports No. 10 (1995)
48. Pawlak, Z.: Rough set approach to knowledge-based decision support. Eur. J. Oper. Res. **99**, 48–57 (1997)

# Pattern Recognition and Machine Learning in the -omics Sciences

# Random Forests for Quality Control in G-Protein Coupled Receptor Databases

Aleksei Shkurin and Alfredo Vellido[✉]

Computer Science Department, Universitat Politècnica de Catalunya,
C. Jordi Girona, 1-3, Campus Nord, 08034 Barcelona, Spain
avellido@cs.upc.edu

**Abstract.** G protein-coupled receptors are a large and heterogeneous super-family of cell membrane proteins of interest to biology in general. One of its families, class C, is of particular interest to pharmacology and drug design. This family is quite heterogeneous on its own, and the discrimination of its several sub-families is a challenging problem. In the absence of known crystal structure, such discrimination must rely on their primary amino acid sequences. In this study, we are interested not as much in achieving maximum sub-family discrimination accuracy, but in exploring sequence misclassification behaviour. Specifically, we are interested in isolating those sequences showing consistent misclassification, that is, sequences that are very often misclassified and almost always to the same wrong sub-family. This analysis should assist database curators in receptor quality control tasks. Random Forests are used for this analysis due their ensemble nature, which makes them naturally suited to gauge the consistency of misclassification.

**Keywords:** G-protein coupled receptors · Random forest · Classification · Protein database curation

## 1 Introduction

The large super-family of G protein-coupled receptors (GPCRs) are eukaryotic cell membrane proteins of interest to biology in general. One of its families, class C, is particularly interesting for pharmacology and as therapeutic drug targets to treat specific neuro-degenerative diseases [1].

Very little is known of the full crystal 3-D structure of GPCRs. Only in recent years, some partial GPCR structures have been solved, mostly from class A [2]. For class C, instead, no full crystal structure has yet been solved; only two transmembrane domains and several extracellular domains have been described in [3,4]. This means that, in order to investigate the functionality of these receptors, we mostly have to rely on the analysis of their primary structure, expressed as an amino acid symbolic sequence.

The class C family is quite heterogeneous on its own and has a rich taxonomy of sub-families. Most of them are further sub-divided into types at different

© Springer International Publishing Switzerland 2016
F. Ortuño and I. Rojas (Eds.): IWBBIO 2016, LNBI 9656, pp. 707–718, 2016.
DOI: 10.1007/978-3-319-31744-1_61

levels. The automatic discrimination and classification of these sub-families thus becomes a challenging problem [5] for which machine learning techniques can provide well-founded solutions.

In this study, we are interested not as much in achieving maximum sub-family discrimination accuracy, but in exploring sequence misclassification behaviour. It has previously been reported that this discriminatory classification, from transformations of the primary sequences, has clear limits [6,7].

Specifically, we are interested in isolating those sequences showing consistent misclassification; that is, sequences that are very often misclassified and almost always to the same wrong sub-family. This is in contrast with sequences that might be misclassified for their partial similarity with different sub-families (borderline cases) which should show far less consistent behaviour. The results of this analysis should hopefully assist database curators in receptor quality control tasks by identifying and shortlisting cases whose original sub-family assignment is highly questionable from the data modelling results.

Random Forests (RF) are our machine learning tool of choice for the task of assessing the consistency of (mis)classification. This is because of their ensemble nature, in which the many base classifiers they consist of coalesce in the assignment of a class to each given sequence. This process makes them naturally suited to gauge the consistency of the classification decisions.

Given that our analyses are based on GPCR primary sequences, a first problem obviously arises, which is the choice of transformation of varying-length sequential symbolic data into formats that are suitable for multivariate data analysis. Such transformations might use the complete unaligned sequences, or methods of multiple sequence alignment. Here, we use a subset of unaligned sequence transformations.

The remainder of the paper is structured as follows. Section 2 describes the analysed class C GPCR data and the RF methods used to analyse them. Section 3 then reports and discusses the experimental results and some conclusions are gathered in Sect. 4.

## 2    Materials and Methods

### 2.1    Class C GPCR Data

The reported research is based on data extracted from GPCRDB, a publicly accessible molecular-class information repository for GPCRs [8]. In this repository, the GPCR super-family is divided into 6 major classes, namely: A (rhodopsin like), B (secretin like), C (metabotropic glutamate/pheromone), cAMP receptors, vomeronasal receptors (V1R and V2R) and Taste receptors T2R, based on the ligand types, functions and sequence similarities. As previously stated, this study focuses on class C GPCRs, a quite heterogeneous family that includes seven main sub-families: Metabotropic Glutamate (mG) receptors, Calcium sensing (CS), GABA$_B$ (GB), Vomeronasal (VN), Pheromone (Ph), Odorant (Od) and Taste (Ta).

mG receptors are activated by glutamate, a major excitatory neurotransmitter in the brain. These receptors are involved in neurological disorders including Alzheimers and Parkinsons diseases, Fragile X syndrome, depression, schizophrenia, anxiety, and pain. The CS receptor is activated by the calcium ion and it is known to play a key role in extra-cellular calcium homeostasis regulation. GB is a neurotransmitter that mediates most inhibitory actions in the central nervous system; it is involved in chronic pain, anxiety, depression and addiction pathologies.

A total of 1,510 class C GPCR sequences (from version 11.3.4, March 2011) belonging to these seven sub-families were analysed. Their distribution of cases by sub-family is summarized in Table 1.

**Table 1.** Number of available sequences in each of GPCR class C sub-families [9].

| Sub-family | Acronym | ♯ sequences |
|---|---|---|
| Metabotropic glutamate | mG | 351 |
| Calcium sensing | CS | 48 |
| GABA$_B$ | GB | 208 |
| Vomeronasal | Vn | 344 |
| Pheromone | Ph | 392 |
| Odorant | Od | 102 |
| Taste | Ta | 65 |

Given that the primary sequences cannot be analysed as symbolic arrays using standard statistical, pattern recognition and machine learning methods, they have to be transformed for subsequent investigation. Several transformations were considered in our experiments.

The first one uses directly the 20 amino acids (see Table 2) of which the sequence *alphabet* consists. An example of this type of transformations is amino acid composition (AAC) [10], which consists in calculating the frequencies of appearance of the amino acids. Despite its simplicity, its use has previously yielded surprisingly solid results [10,11].

Subsets of amino acids may share similar physico-chemical properties, which makes them equivalent at a functional level [12]. Amino acid grouping also helps computations by reducing the dimensionality of the analysed data set. For this study, two alternative groupings were used, in the form of sub-sequence frequencies (see Table 3): the Sezerman (SEZ) alphabet (11 groups) and the Davies Random (DAV) alphabet (9 groups).

Amino acids and their groupings were not just used as such in this study, but in the form of *n*-grams, which are subsequences of length *n*. The concept of *n*-grams is well-known in protein analysis [13,14]. Here, we used the relative frequencies of the *n*-grams. Therefore, the *n*-gram representation consists of the relative frequency of each *n*-gram in a sequence (note that for Sezerman and

**Table 2.** List of the 20 amino acids in the protein symbolic *alphabet*.

| AA name | Symbol | AA name | Symbol | AA name | Symbol |
|---------|--------|---------|--------|---------|--------|
| Alanine | A | Glycine | G | Proline | P |
| Arginine | R | Histidine | H | Serine | S |
| Asparagine | N | Isoleucine | I | Threonine | T |
| Aspartate | D | Leucine | L | Tryptophan | W |
| Cysteine | C | Lysine | K | Tyrosine | Y |
| Glutamate | E | Methionine | M | Valine | V |
| Glutamine | Q | Phenylalanine | F | | |

**Table 3.** Amino acid grouping schemes

| Grouping | 1 | 2 | 3 | 4 | 5 | 6 | 7 | 8 | 9 | 0 | X |
|----------|---|---|---|---|---|---|---|---|---|---|---|
| SEZ | IVLM | RKH | DE | QN | ST | A | GT | W | C | YF | P |
| DAV | SG | DVIA | RQN | KP | WHY | C | LE | MF | T | | |

Davies, the length of the $n$-gram is not taken in number of amino acids, but in number of groupings). Due to the exponential growth of the size of $n$-grams, experiments were limited to $n$-grams of size 1, 2 and 3.

## 2.2    Random Forests

Since their description on the first years of the century, RFs [15] have become a popular and widely-used machine learning tool for classification and regression tasks. This is particularly true in the areas of computational biology and bioinformatics [16]. In these fields, RFs have *de facto* become standard methods, especially adequate in settings with poor observations-to-variables ratios (of which the current study is a mild case). They also scale nicely to multi-class problems such as the one investigated here, avoiding more complex *one-vs-one* or *one-vs-all* classification schemes.

The general graphical scheme of the RF algorithm is sketched in Fig. 1. At each split of the observed sample data, a random subset of variables is selected and the process is repeated until the specified number of decision trees is generated. Each tree is built from a bootstrap sample drawn with replacement from the observed data, and the predictions of all trees are finally aggregated through majority voting. A feature of RFs is the definition of an out-of-bag (OOB) error, which is calculated from observations that were not used to build a particular tree; it can thus be considered as an internal cross-validation error measure [16,17]. This is an important feature for the type of experiments carried out in this study, because it simplifies the otherwise cumbersome cross-validation procedures that would be required if alternative classification methods such as, for instance, support vector machines or artificial neural networks were used [15].

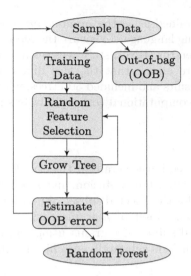

**Fig. 1.** General graphical scheme of the RF algorithm.

The fact that RFs are defined as ensemble-of-trees classifiers also means that these models are naturally suited to the task of analysing sequence misclassification behaviour; this is because we are interested in methods that naturally work according to a voting scheme in order to assign a sequence observation to a given class (sub-family). These votes from each individual decision tree allow us to more closely inspect the performance of each GPCR sequence; pooling the votes also allows us to gauge the consistency of the (mis)classification results for the sequences of any given sub-family.

## 3    Experiments

### 3.1    Settings

As mentioned in Sect. 2.1, the acquired GPCRDB sequences were used in three different versions: the 20 amino acid frequencies, Sezerman (11 groupings) and Davies (9 groupings). All possible $n$-grams of sizes 1 to 3 were built for each of them. In previous research [9], feature selection was performed on these transformed datasets using statistical $t$-test filtering to establish a ranking of the features. This led to the choice of subsets of features whose test was significant for different numbers of binary classifiers (note that, taking into account that we analyze 7 sub-families, we would have 21 different binary classifiers and, at most, we might get a subset of features that was significant for all 21).

The subsets of features that achieved the best classification performance in [9] become the starting point for our RF models. Their number of features is 585 for $n$-grams using all amino acids, 386 for $n$-grams using Sezerman groupings and 238 for $n$-grams using Davies groupings.

The RF model was trained using the *randomForest* and *matrixStatsR* packages of the R programming language. To ensure the reproducibility of the results, the random number generator (RNG) state was set to a value of 42. The model was stratified in order to ensure that the size difference of classes does not significantly affect the results and included 500 trees, which yields sufficient performance while keeping computational costs relatively small.

## 3.2    Results

Given that we are interested in the details of the class C GPCR misclassification behaviour, we first calculated the confusion matrices for our RFs. The results are presented in Table 4 for the selected subset of $n$-grams using all amino acids; in Table 5 for the selected subset of $n$-grams using Sezerman groupings; and in Table 6 for the selected subset of $n$-grams using Davies groupings. All these confusion matrices also include sub-family-specific classification errors.

**Table 4.** Confusion matrix corresponding to the RF model for the selected subset of $n$-grams using all amino acids. TC stands for *TrueClass* and PC stands for *PredictedClass*

|    |    | PC | | | | | | | |
|----|----|-----|----|-----|-----|-----|----|----|------------|
|    |    | mG  | CS | GB  | VN  | Ph  | Od | Ta | Class.error |
| TC | mG | 346 | 0  | 1   | 0   | 2   | 2  | 0  | 0.014      |
|    | CS | 2   | 45 | 1   | 0   | 0   | 0  | 0  | 0.062      |
|    | GB | 2   | 0  | 206 | 0   | 0   | 0  | 0  | 0.009      |
|    | VN | 3   | 1  | 0   | 314 | 22  | 4  | 0  | 0.087      |
|    | Ph | 12  | 0  | 1   | 19  | 358 | 2  | 0  | 0.087      |
|    | Od | 4   | 0  | 0   | 9   | 20  | 69 | 0  | 0.324      |
|    | Ta | 0   | 0  | 0   | 0   | 0   | 0  | 65 | 0.000      |

The overall accuracies of the models were as follows: 0.91 for the selected subset of $n$-grams using all amino acids; 0.90 for the selected subset of $n$-grams using Sezerman groupings; and 0.87 for the selected subset of $n$-grams using Davies groupings. Since the primary aim of this study is not to achieve the best possible accuracy, these results are presented only as a guarantee of the suitable performance of the models.

On our way to inspecting the details of sequences' misclassification behaviour, we now recall the fact that the sub-family assignments made by the RF model are the result of the partial voting of 500 individual base trees. On a first approximation, the votes of these trees for each sub-family were extracted and mean values of their ratios calculated as a measure of sub-family classification consistency. Hereafter, and to comply with paper extension limits, only the results for mG and Ph will be presented, as in Table 7. These two sub-families

**Table 5.** Confusion matrix corresponding to the RF model for the selected subset of $n$-grams using Sezerman groupings. TC stands for *TrueClass* and PC stands for *PredictedClass*.

| | | PC | | | | | | | |
|---|---|---|---|---|---|---|---|---|---|
| | | mG | CS | GB | VN | Ph | Od | Ta | Class.error |
| **TC** | mG | 345 | 0 | 2 | 0 | 2 | 2 | 0 | 0.017 |
| | CS | 1 | 44 | 1 | 0 | 2 | 0 | 0 | 0.083 |
| | GB | 3 | 0 | 205 | 0 | 0 | 0 | 0 | 0.014 |
| | Vn | 2 | 0 | 0 | 314 | 25 | 3 | 0 | 0.087 |
| | Ph | 11 | 0 | 1 | 17 | 362 | 1 | 0 | 0.076 |
| | Od | 2 | 0 | 0 | 7 | 25 | 68 | 0 | 0.333 |
| | Ta | 0 | 0 | 0 | 0 | 2 | 0 | 63 | 0.031 |

**Table 6.** Confusion matrix corresponding to the RF model for the selected subset of $n$-grams using Davies groupings. TC stands for *TrueClass* and PC stands for *PredictedClass*.

| | | PC | | | | | | | |
|---|---|---|---|---|---|---|---|---|---|
| | | mG | CS | GB | VN | Ph | Od | Ta | Class.error |
| **TC** | mG | 337 | 0 | 3 | 2 | 7 | 2 | 0 | 0.040 |
| | CS | 1 | 43 | 1 | 1 | 2 | 0 | 0 | 0.104 |
| | GB | 4 | 0 | 204 | 0 | 0 | 0 | 0 | 0.019 |
| | Vn | 1 | 1 | 0 | 309 | 29 | 4 | 0 | 0.102 |
| | Ph | 12 | 0 | 1 | 17 | 361 | 1 | 0 | 0.079 |
| | Od | 2 | 1 | 0 | 8 | 27 | 64 | 0 | 0.373 |
| | Ta | 3 | 0 | 0 | 1 | 6 | 0 | 55 | 0.154 |

were selected as opposite examples of a rather well discriminated sub-family (mG), *vs* a comparatively poorly discriminated one (Ph).

A consistency of 100 % for a given sequence is reached when all trees in the RF agree in their sub-family assignment. The maximum consistency achieved by an mG sequence was 100 %, whereas the maximum for Ph was 98 %. The detailed consistencies for all sequences in these two sub-families, for the selected subset of $n$-grams using all amino acids, are depicted on Figs. 2 and 3. In addition to that, histograms of these consistency values are also presented in Fig. 4. Similar depictions for the selected subset of $n$-grams using Sezerman groupings and for the selected subset of $n$-grams using Davies groupings were also produced but they are not shown here for the sake of brevity.

Some of the misclassified sequences may have a significantly low rate of RF tree votes to their *true* class, while, at the same time, a high rate of votes to some specific other sub-family. These cases are of special interest for database quality assessment due to the consistency of their misclassification. Table 8 lists

**Table 7.** Consistencies of the mG and Ph sub-families for the different data sets.

**Original**

|  | mG | CS | GB | VN | Ph | Od | Ta |
|---|---|---|---|---|---|---|---|
| mG | **0.832** | 0.010 | 0.054 | 0.025 | 0.048 | 0.016 | 0.016 |
| Ph | 0.044 | 0.010 | 0.018 | 0.163 | **0.697** | 0.053 | 0.014 |

**Sezerman**

|  | mG | CS | GB | VN | Ph | Od | Ta |
|---|---|---|---|---|---|---|---|
| mG | **0.811** | 0.012 | 0.052 | 0.033 | 0.055 | 0.021 | 0.017 |
| Ph | 0.049 | 0.013 | 0.019 | 0.180 | **0.660** | 0.058 | 0.020 |

**Davies**

|  | mG | CS | GB | VN | Ph | Od | Ta |
|---|---|---|---|---|---|---|---|
| mG | **0.796** | 0.010 | 0.046 | 0.044 | 0.067 | 0.019 | 0.019 |
| Ph | 0.059 | 0.015 | 0.025 | 0.182 | **0.631** | 0.062 | 0.025 |

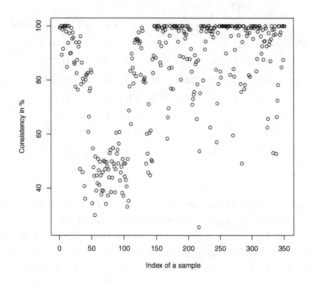

**Fig. 2.** mG consistency values for sequences described by the selected subset of $n$-grams using all amino acids.

those mG and Ph sequences that were misclassified (those whose rate of votes for their *true* sub-family was not the highest) and whose standard deviation of votes was higher than the average standard deviation of votes for sequences of their *true* sub-family.

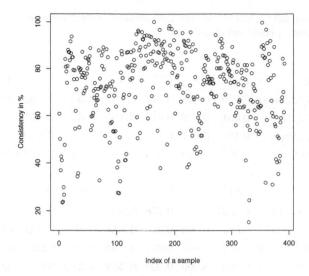

**Fig. 3.** Ph consistency values for sequences described by the selected subset of $n$-grams using all amino acids.

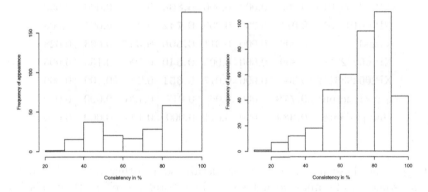

**Fig. 4.** Histograms of the mG (left) and Ph (right) consistency values depicted, in turn, in Figs. 2 and 3.

### 3.3   Discussion

The confusion matrices in Tables 4, 5 and 6 yield several clear messages. First, the RF model is shown to have adequate overall discrimination capabilities (as also reflected by the global accuracies reported in the previous section).

This classification, though, is not homogeneous across sub-families; mG and GB are very well discriminated in all data transformations, while Ta is well-discriminated in all data transformations but Davies'. On the opposite extreme, Od is poorly classified in all cases. This patterns are similar for all data transformations, but the selection of $n$-grams using all amino acids, perhaps unsurprisingly, shows some advantage over the more parsimonious transformations.

**Table 8.** mG (top list) and Ph (bottom list) sequences that were consistently misclassified to a different specific sub-family.

**mG**

| Name | mG | CS | GB | VN | Ph | Od | Ta |
|---|---|---|---|---|---|---|---|
| a8dz71_danre | 0.039 | 0.028 | 0.017 | 0.178 | **0.372** | **0.333** | 0.033 |
| a8dz72_danre | 0.091 | 0.048 | 0.012 | **0.188** | **0.370** | **0.267** | 0.024 |
| q5i5d4_9tele | 0.123 | 0.017 | 0.006 | 0.095 | 0.089 | **0.654** | 0.017 |
| q5i5c3_9tele | 0.067 | 0.000 | 0.010 | 0.062 | **0.144** | **0.701** | 0.015 |

**Ph**

| Name | mG | CS | GB | VN | **Ph** | Od | Ta |
|---|---|---|---|---|---|---|---|
| a7sdg9_nemve | **0.633** | 0.024 | 0.178 | 0.059 | 0.083 | 0.006 | 0.018 |
| XP_001368172 | 0.011 | 0.005 | 0.070 | **0.686** | 0.211 | 0.016 | 0.000 |
| NP_001096052 | 0.034 | 0.000 | 0.011 | **0.598** | 0.293 | 0.046 | 0.017 |
| XP_002723672 | 0.000 | 0.000 | 0.000 | **0.567** | **0.387** | 0.041 | 0.005 |
| NP_001093018 | 0.000 | 0.000 | 0.000 | **0.741** | 0.253 | 0.006 | 0.000 |
| NP_001093020 | 0.000 | 0.000 | 0.056 | **0.736** | 0.174 | 0.034 | 0.000 |
| NP_001093016 | 0.006 | 0.000 | 0.000 | **0.806** | 0.188 | 0.000 | 0.000 |
| NP_001093017 | 0.011 | 0.000 | 0.000 | **0.742** | 0.237 | 0.005 | 0.005 |
| d3zgs1_rat | 0.006 | 0.000 | 0.011 | **0.506** | **0.417** | 0.033 | 0.028 |
| XP_002723938 | 0.000 | 0.000 | 0.005 | **0.640** | 0.196 | 0.153 | 0.005 |
| XP_002936172 | 0.006 | 0.006 | 0.017 | **0.651** | 0.206 | 0.109 | 0.006 |
| q9pwe1_ictpu | **0.779** | 0.006 | 0.066 | 0.017 | 0.133 | 0.000 | 0.000 |
| b0uyj3_danre | **0.884** | 0.000 | 0.011 | 0.000 | 0.105 | 0.000 | 0.000 |

Inspecting these matrices in more detail, some other interesting patterns emerge: some sub-families do not show clear "preference" in their misclassification, namely mG, CS, GB and Ta; whereas VN, Ph and Od seem to mostly restrict misclassifications to happen between them. VN is mostly misclassified as Ph, while Ph is mostly misclassified as VN (and to a lesser extent to mG); in turn, Od is mostly misclassified as Ph and, to a lesser extent, as VN. Overall, the Ph sub-family seems to take a central role in this misclassification pattern, overlapping the other two sub-families.

The latter results are corroborated by the per-sub-family consistency means reported in Table 7 (and by the means corresponding to the rest of sub-families, not reported here), which show a sizeable overlapping between the Ph and VN sub-families.

Yet again, even if the overall results for each of the sub-families are relevant on their own right, the detailed consistency values per sequence for these sub-families are not less interesting. Corroborating the overall findings, a sizeable proportion of the consistencies of the individual sequences belonging to mG have either 100 % or near 100 % consistency values, as seen in Fig. 2. Nonetheless, the consistencies of many sequences with indices roughly between 50 and 120 fall sharply to values under 50 %, perhaps indicating the existence of a differentiated sub-group within this sub-family that would require specific inspection. This is also clearly reflected as a "lump" for consistencies between 40 and 50 % in the histogram distribution depicted in Fig. 4 (left); otherwise the histogram clearly reflects the high concentration of sequences with consistencies valued between 90 and 100 %. Also, corroborating overall findings, most of the individual consistencies of the sequences belonging to Ph have lower consistency values, as seen in Fig. 3, mostly with values between 60 and 90 %. In this case, sequences of very low consistencies seem to be quite evenly distributed. This is reflected by the negative-skewed histogram in Fig. 4 (right).

Finally, the detailed analysis of individual misclassification patterns reveals that some sequences of all sub-families but Ta were misclassified by the vast majority of RF trees to another particular class. Most of such cases, again, belong to the worst classified sub-families. The list of sequences for mG and Ph shown in Table 8 reveals some extreme cases. Note that the criterion used to create this shortlist is somehow subjective and that less or more restrictive criteria could have been defined. Ultimately, the strictness of this criterion should be set by the database curator. The characteristics of these individual sequences should be further investigated in the main international protein databases.

## 4   Conclusion

Class C of GPCRs is of great interest in pharmacology. This protein family has a heterogeneous sub-family structure, whose investigation has to be carried out from their primary structure. The automatic discrimination of these sub-families is challenging and it has been shown to have clear limits. This study has investigated in some detail the consistency of misclassification using RF techniques, whose ensemble nature is ideally suited to the analysis of such problem. Different sub-families have been shown to display very different discrimination consistency behaviours. Specific attention has been paid to the individual identification of sequences that are consistently assigned by the RF base classifiers to sub-families other than their *true* one. This type of analysis is meant to provide a quality control tool for GPCR database curators.

## References

1. Kniazeff, J., Prézeau, L., Rondard, P., Pin, J.P., Goudet, C.: Dimers and beyond: the functional puzzles of class C GPCRs. Pharmacol. Ther. **130**(1), 9–25 (2011)
2. Katritch, V., Cherezov, V., Stevens, R.C.: Structure-function of the G protein-coupled receptor superfamily. Ann. Rev. Pharmacol. Toxicol. **53**, 531–556 (2013)
3. Wu, H., Wang, C., Gregory, K.J., Han, G.W., Cho, K.P., Xia, Y., et al.: Structure of a class C GPCR metabotropic glutamate receptor 1 bound to an allosteric modulator. Science **344**(6179), 58–64 (2014)

4. Doré, A.S., Okrasa, K., Patel, J.C., Serrano-Vega, M., Bennett, K., Cooke, R.M., et al.: Structure of class C GPCR metabotropic glutamate receptor 5 transmembrane domain. Nature **551**, 557–562 (2014)
5. Gao, Q.B., Ye, X.F., He, J.: Classifying G-protein-coupled receptors to the finest subtype level. Biochem. Biophys. Res. Commun. **439**(2), 303–308 (2013)
6. König, K., Cárdenas, M., Giraldo, J., Alquézar, R., Vellido, A.: Label noise in subtype discrimination of class C G-protein coupled receptors: a systematic approach to the analysis of classification errors. BMC Bioinf. **16**(1), 314 (2015)
7. Cruz-Barbosa, R., Vellido, A., Giraldo, J.: The influence of alignment-free sequence representations on the semi-supervised classification of Class C G protein-coupled receptors. Med. Biol. Eng. Comput. **53**(2), 137–149 (2015)
8. Isberg, V., Vroling, B., van der Kant, R., Li, K., Vriend, G., Gloriam, D.: GPCRDB: an information system for G protein-coupled receptors. Nucleic Acids Res. **42**(Database issue), D4225 (2014)
9. König, K., Alquézar, R., Vellido, A., Giraldo, J.: Finding class C GPCR subtype-discriminating n-grams through feature selection. In: Proceedings of the 8th International Conference on Practical Applications of Computational Biology and Bioinformatics (PACBB 2014), pp. 89–96 (2014)
10. Sandberg, M., Eriksson, L., Jonsson, J., Sjöström, M., Wold, S.: New chemical descriptors relevant for the design of biologically active peptides. A multivariate characterization of 87 amino acids. J. Med. Chem. **41**, 2481–2491 (1998)
11. Cárdenas, M.I., Vellido, A., Giraldo, J.: Visual interpretation of class C GPCR subtype overlapping from the nonlinear mapping of transformed primary sequences. In: Proceedings of the International Conference on Biomedical and Health Informatics (IEEE BHI 2014), pp. 764–767 (2014)
12. Davies, M.N., Secker, A., Freitas, A., Clark, E., Timmis, J., Flower, D.R.: Optimizing amino acid groupings for GPCR classification. Bioinformatics **24**(18), 1980–1986 (2008)
13. Caragea, C., Silvescu, A., Mitra, P.: Protein sequence classification using feature hashing. In: Proceedings of the IEEE International Conference on Bioinformatics and Biomedicine (BIBM 2011), pp. 538–543 (2011)
14. Mhamdi, F., Elloumi, M., Rakotomalala, R.: Textmining, features selection and datamining for proteins classification. In: Proceedings of the IEEE International Conference on Information and Communication Technologies: From Theory to Applications, IEEE/ICTTA, pp. 457–458 (2004)
15. Breiman, L.: Random forests. Mach. Learn. **45**(1), 5–32 (2001)
16. Boulesteix, A.-L., Kruppa, J., Konig, I.: Overview of random forest methodology and practical guidance with emphasis on computational biology and bioinformatics. Wiley Interdiscip. Rev. Data Min. Knowl. Disc. **2**(6), 493–507 (2012)
17. Strobl, C., Boulesteix, A.-L., Kneib, T., Augustin, T., Zeileis, A.: Conditional variable importance for random forests. BMC Bioinf. **9**, 307 (2008)

# Automated Quality Control for Proton Magnetic Resonance Spectroscopy Data Using Convex Non-negative Matrix Factorization

Victor Mocioiu[1,4], Sreenath P. Kyathanahally[2], Carles Arús[1,4],
Alfredo Vellido[3,4], and Margarida Julià-Sapé[1,4(✉)]

[1] Departament de Bioquímica i Biologia Molecular,
Universitat Autònoma de Barcelona, Cerdanyola del Vallès, 08193 Barcelona, Spain
{victor.mocioiu,carles.arus,Margarita.Julia}@uab.cat
[2] Departments Radiology and Clinical Research, University of Bern, Bern, Switzerland
s.p.kyathanahally@insel.ch
[3] Departament de Ciències de la Computació, Universitat Politècnica de Catalunya,
BarcelonaTech, Campus Nord, 08034 Barcelona, Spain
avellido@cs.upc.edu
[4] Centro de Investigación Biomédica en Red en Bioingeniería,
Biomaterials y Nanomedicina CIBER-BBN, Cerdanyola del Vallès, Barcelona, Spain

**Abstract.** Proton Magnetic Resonance Spectroscopy ($^1$H MRS) has proven its diagnostic potential in a variety of conditions. However, MRS is not yet widely used in clinical routine because of the lack of experts on its diagnostic interpretation. Although data-based decision support systems exist to aid diagnosis, they often take for granted that the data is of good quality, which is not always the case in a real application context. Systems based on models built with bad quality data are likely to underperform in their decision support tasks. In this study, we propose a system to filter out such bad quality data. It is based on convex Non-Negative Matrix Factorization models, used as a dimensionality reduction procedure, and on the use of several classifiers to discriminate between good and bad quality data.

**Keywords:** Brain tumors · Magnetic resonance spectroscopy · Convex non-negative matrix factorization · Pattern recognition · Quality control · Machine learning

## 1    Introduction

Proton magnetic resonance spectroscopy ($^1$H MRS, henceforth only referred to as MRS) is a magnetic resonance modality that provides metabolic information about an investigated tissue volume, thus becoming a tool for metabolomics. MRS is inherently non-invasive and can be used either on its own, or in conjunction with other MR modalities [1] with the aim to improve diagnostic accuracy. Although MRS can be used to investigate a wide range of tissue types [2], it is mainly used for diseases of the central nervous system (CNS), and has proven a powerful tool in assessing a broad spectrum of diseases such as metabolic disorders, epilepsy, Alzheimer and Parkinson, amongst others. But, by far, its most common application is on brain tumors diagnostic assistance [2].

© Springer International Publishing Switzerland 2016
F. Ortuño and I. Rojas (Eds.): IWBBIO 2016, LNBI 9656, pp. 719–727, 2016.
DOI: 10.1007/978-3-319-31744-1_62

MRS can be single-voxel (SV), where the signal comes from a volume of interest, or multi-voxel, using a grid/matrix of many contiguous SVs (actually, SV-like spectral vectors). MRS has several parameters of importance that should be mentioned. First of all, echo time, which can be either short (STE, lower than 40 ms), or long (LTE, higher than 40 ms), controlling what metabolites can be better seen in the spectrum. STE only allows for positive peaks in the spectrum whereas LTE can also have negative peaks. Another important parameter is field strength (measured in Tesla, T) which, in laymen terms, determines how far apart the peaks of the metabolites are, and how many metabolites can be detected. A typical field strength used in clinical routine is 1.5 T. Figure 1 shows an example of an MR spectrum.

Although MRS is a promising technique, it is not yet widely implemented in clinical routine. It can be argued that the main reason for this is the explicit need for an expert (a radiologist) to interpret the spectrum and reach a diagnostic. Previous work has been carried out to compensate for the lack of experts in MRS interpretation by using (semi-)automated classifiers or decision support systems [3–5]. The main limitation of such systems is that they assume that the spectra are of consistently good quality. Unfortunately, the definition of 'good quality' has yet to be clearly established, and by this we mean that the gold standard is human-dependent and, although some guidelines have been proposed [3, 6], it may vary from expert to expert. Furthermore, multiple types of artifacts can contaminate the signal; an extensive gallery of such artifacts is presented in [7]. Figure 2 shows an example of a bad quality spectrum (compare to Fig. 1, which corresponds to a good quality spectrum).

In a previous work by van der Graaf et al. [8], a semi-automatic filtering procedure, based on the signal-to-noise ratio (SNR) of the spectrum and the water bandwidth (WBW - defined as the line width at half of the maximum intensity of the water peak), was proposed. However, this system relies on a board of experts for validating the final decision.

A fully automated system, trained on a subset of the eTUMOR [9, 10] and INTER-PRET [11, 12] SV spectra (144 SVs, 72 acceptable and 72 unacceptable), was proposed by Wright et al. [13]. The system consists of a least squares support vector machine [14] with a radial basis function kernel and fastICA [15], which was used for dimensionality reduction. The test set comprised 98 SVs (58 acceptable and 40 unacceptable) from the eTUMOR database. The results of this study were encouraging (an accuracy of 88 %), but we argue that this might be due to several reasons that might not hold up in a real clinical setting. First of all, because the number of cases analyzed was relatively low, something that can be seen as a detrimental for the generalization power of the system. Another more important aspect to consider is that the two classes are fairly balanced, for both training and testing – again affecting generalization and also performance metrics.

In this article, we propose a bad quality data filtering system based on convex non-negative matrix factorization (cNMF), used as a dimensionality reduction model as well as an artifact identification tool. The results from cNMF, specifically the mixing matrix (details in the Methods section), are then used as features for the following classifiers: Naïve Bayes (NB), Logistic Regression (LR), Linear Discriminant Analysis (LDA), AdaBoost, and Random Forests (RF). We evaluate these classifiers based on accuracy, area under the receiver operator characteristic curve (AUC), sensitivity, specificity, $F_1$ score, and balanced error rate. Performance metrics are averaged over a 10-fold stratified cross validation.

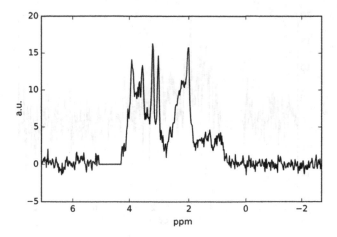

**Fig. 1.** Example of an MR spectrum from the eTumour database [10], considered of good quality. The acquisition parameters are described in the Materials section. Note that it has already undergone specific preprocessing procedures as described in the materials section. The x-axis of the spectrum is in parts per million (ppm) and the y-axis usually in arbitrary units. Note also that, traditionally in spectroscopy, the x-axis is reversed when compared to the Cartesian system [11].

## 2  Materials and Methods

### 2.1  Materials

The investigated dataset comes from the INTERPRET [11, 12] and eTUMOUR [10] multicentre databases and comprises of 1,196 SV STE spectra, out of which 982 were labeled as good quality and 198 as bad quality spectra, acquired at 1.5 T. The INTER-PRET data came from seven clinical centers, whereas the eTUMOUR data came from eleven different centers– in addition some of the centers were working with more than one MR scanner from different manufacturers. The spectrum-labeling panel consisted of three experts and the labeling procedure was as follows: if at least two experts accepted a spectrum, then it was labeled as 'good' and if at least two rejected the spectrum, then it was labeled as 'bad'. Since spectra of each multi-center project came from different clinical centers, and in order to avoid expert bias, the judging system was set so that experts from one center did not judge spectra from their own center. Therefore, different combinations of experts were set to judge spectra from each clinical center [10, 12].

Prior to building our system, the data underwent a well-defined pre-processing procedure [4]. Residual water filtering using the HLSVD algorithm with 10 Lorentzians was initially applied, followed by apodization using a 1 Hz bandwidth with a Lorentzian line-shape. Then the values in the [5.11, 4.31] interval were set to 0. Afterwards, baseline offset was corrected using two ranges– [−2 -1] ppm and [9 11] ppm. The spectrum was then normalized to unit length and multiplied by 100. Finally, the spectrum was aligned according to the algorithm presented in [4]. After preprocessing, each spectrum had 512 points in the 7.1 to −2.7 ppm range; this range was used for the remainder of our study. As a result, our analyzed data matrix was 1,196 × 512.

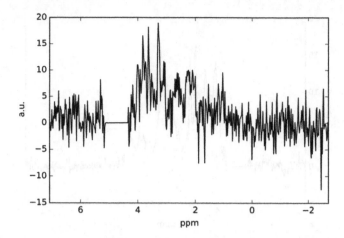

**Fig. 2.** An example of a bad quality spectrum due to low SNR from the eTumour database [10]. The acquisition parameters are described in the Materials section. Note that it has already undergone specific preprocessing procedures as described in the Materials section.

### 2.2  Methods

**Convex Non-negative Matrix Factorization.**  Convex non-negative matrix factorization (cNMF) [16] is a blind source separation method that, among other NMF algorithms [17], has been extensively used for MRS data analysis [17, 18]. NMF methods rely on factorizing an initial data matrix, $X$ ($m$ rows -data dimensionality, and $n$ columns - samples/spectra) into two matrices $F$ ($m$ rows and $k$ columns– sources/basis vectors) and $G$ ($k$ rows, $n$ columns). Furthermore, in cNMF, $F$ is constrained to lie in the column space of the input data $X$, so that the cNMF formula can be written as:

$$X \approx AXG \tag{1}$$

where $A$ fully determines $F$. $G$ is also called the mixing matrix, as it holds the coefficients to recompose a specific data sample. It is a well-known fact that the results of the cNMF algorithm are dependent on the initialization scheme; in our implementation, we use the k-means ++ algorithm [19]. It should be mentioned that there is no fully established method for choosing the optimal number of sources/basis vectors.

**Implementation Details.**  Some of the classifiers used in this study are not parameterless. LR was used in its basic variant, as well as with regularization, namely l2– LRCV. AdaBoost was built using fifty decision trees as estimators. In the case of RF, fifty estimators were used; bootstrap samples of the training set were used to build the trees and the maximum number of features considered when looking for the best split was the square root of the total number of features. All nodes were expanded until all leaves were rendered pure. LR, LRCV and RF versions that take into account class imbalance by assigning a proportional weight to the less represented class were also built– they will be in turn be named LRA, LRCVA and RFA.

Regarding the details of our proposed system, we extracted from three up to eleven sources. A maximum of eleven sources was extracted because, as reported in [6], one should account usually for up to 9 technical requirements to make the spectrum clinically interpretable. The 2 remaining sources should account for the variability that is present in the 'good' class. We then used a 10-fold stratified cross-validation loop to split the mixing matrix in training and test subsets; trained the classifiers; and computed the accuracy, sensitivity, specificity, F1 score, and balanced error rate. Final results are reported as averaged performance metrics.

## 3   Results

We start by presenting the eleven extracted sources in Fig. 3. Several aspects should be mentioned: first, that, even though the extracted sources are akin to the input space, they

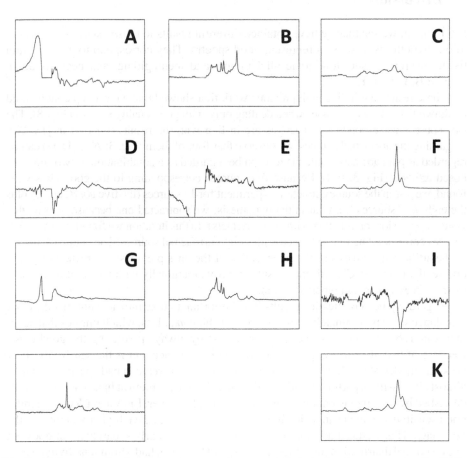

**Fig. 3.** Gallery of extracted sources: source A corresponds to bad water suppression (bad), B corresponds to normal brain tissue (good), C is bad homogeneity (bad), D is badly phased (bad), E is a combination of bad water suppression and low SNR (bad), F is necrotic tissue (good), G corresponds to bad water suppression and bad homogeneity(bad), H would correspond to the spectral pattern of a low grade glial tumour (good), I is bad phasing and other artifacts (bad), J represents the spectral pattern of meningioma (good), and K represents a source necrosis (good).

are not true spectra as such and, therefore, the $x$ and $y$ axes are left unlabeled (however the $x$ axis would correspond to the [7.1−2.7] ppm range and the $y$ axis to the [−20,40] a.u. range). Another thing to stress is that some of the negative peaks are not completely shown in this figure, for two reasons: first, because in STE, negative peaks are not possible and are directly regarded as artefacts; second, a more practical reason to not undermine the amplitude of the positive peaks. An important aspect to note is that we have found five sources pertaining to good spectra, and only six artifactual sources.

The classification results for the investigated classifiers are presented in Table 1. It is worth highlighting that RF shows the best performance in terms of accuracy, AUC, BER, and F1 score, whereas LR achieves the best specificity and LRA the best sensitivity.

# 4   Discussion

From Fig. 3, we see that we have obtained different results to those expected according to our hypothesis; 5 sources represent good spectra. They correspond to normal brain tissue, necrotic tissue, low grade glial tumor, and meningioma, and necrosis again (Fig. 3: B, F, H, J, K).

These results are in line with previous work that showed that a clear separation could be drawn between these classes when dealing only with good quality spectra [11, 18]. The two sources corresponding to necrosis are in line with the results reported in [18, 20]. Regarding the artifactual sources, we observe that four of them (Fig. 3: A, C, D, G) can be regarded as pure artifacts, while the rest can be regarded as a combination of two superimposed artifacts (Fig. 3: E, I). Because five sources corresponding to the good class were found, we re-run the source extraction experiment for 13 sources (the five *good* sources plus the eight *bad* sources from our initial hypothesis; we subtracted one because we saw that water suppression tends to mix with other artefacts). In this iteration we found the same five sources corresponding to good spectra, plus eight artifactual sources. The nature of the rest of the artifactual sources was the same as that of the ones previously extracted; they mix differently, however. Classification results improve marginally with the addition of the 2 extra sources and are not reported herein.

Regarding the classification results, it is important to take into account that the class imbalance was approximately 8 to 2 in favor of the *good* class, which implies that accuracies in Table 1 could have been achieved just by always predicting the good class. Because of this, the most important metric to take into account is the sensitivity (how good the model is at telling if a predicted bad spectrum is really a bad spectrum), where almost all classifiers perform poorly– indicating that the problem at hand cannot be well-described by our dimensionality reduction scheme. However, LRA and LRCVA exhibit good sensitivity, which can be justified by the fact that they give higher weights for the *bad* class. This would imply that our feature space is linearly separable, but does not obey a normal distribution (otherwise LDA would have also had a high sensitivity, which is not the case).

**Table 1.** Performance metrics for the investigated classifiers. Best performances are shown in bold.

|        | Accuracy | AUC  | Sensitvity | Specificity | BER  | F1 score |
|--------|----------|------|------------|-------------|------|----------|
| LR     | 0.85     | 0.84 | 0.25       | **0.97**    | 0.21 | 0.91     |
| LRA    | 0.77     | 0.84 | **0.72**   | 0.78        | 0.32 | 0.85     |
| LRCV   | 0.85     | 0.85 | 0.27       | 0.96        | 0.22 | 0.91     |
| LRCVA  | 0.78     | 0.85 | 0.68       | 0.80        | 0.30 | 0.85     |
| RF     | 0.85     | **0.86** | 0.37   | 0.95        | 0.21 | 0.91     |
| RFA    | **0.86** | **0.86** | 0.34   | 0.96        | **0.20** | **0.92** |
| LDA    | 0.85     | **0.86** | 0.35   | 0.95        | 0.21 | 0.91     |
| ADA    | 0.85     | 0.85 | 0.42       | 0.94        | 0.22 | 0.91     |
| NB     | 0.83     | 0.82 | 0.45       | 0.91        | 0.27 | 0.90     |

**Table 2.** Classification metrics after removing the samples that were labeled good by one expert spectrosopist and bad by the other two.

|        | Accuracy | AUC  | Sensitvity | Specificity | BER  | F1 score |
|--------|----------|------|------------|-------------|------|----------|
| LR     | 0.88     | 0.86 | 0.30       | **0.97**    | **0.20** | **0.93** |
| LRA    | 0.78     | 0.86 | **0.76**   | 0.78        | 0.33 | 0.86     |
| LRCV   | 0.88     | **0.88** | 0.32   | 0.96        | 0.21 | **0.93** |
| LRCVA  | 0.79     | 0.87 | 0.73       | 0.80        | 0.32 | 0.87     |
| RF     | 0.88     | **0.88** | 0.44   | 0.95        | 0.21 | **0.93** |
| RFA    | **0.89** | **0.88** | 0.44   | 0.92        | 0.21 | **0.93** |
| LDA    | 0.88     | **0.88** | 0.40   | 0.92        | 0.21 | **0.93** |
| ADA    | 0.88     | 0.86 | 0.46       | 0.94        | 0.22 | **0.93** |
| NB     | 0.86     | 0.84 | 0.50       | 0.91        | 0.27 | 0.91     |

As we previously mentioned, the original labeling of the analyzed data was performed by two or three expert spectroscopists; some cases were labeled as *bad*, but still one of the experts deemed it to be of acceptable quality. As such, we removed those 'borderline' spectra from the testing phase of our system, in order to see if the performance metrics would improve. The results, presented in Table 2, show that sensitivity improved for all classifiers, while other metrics did not change significantly.

One limitation of our system is that the cNMF optimization process is known to fall into local minima and, up to date, there is no way to assess whether the algorithm converged to the optimal minimum or not. In our case, this would translate into obtaining slightly different sources depending on the run of the algorithm (and thus on initialization). This issue is meant to be addressed in future work.

## 5    Conclusion

We have presented a system that tries to address an important issue in MRS for brain tumour analysis as a metabolomics problem: data quality control. Our system used cNMF as a dimensionality reduction and artifact-identification scheme and then investigated a range of classifiers in the task of discriminating between good and bad quality spectra.

By using LRA, a sensitivity of 0.72 and a specificity of 0.78 was achieved, and by taking out the 'borderline' cases, sensitivity was increased to a value of 0.76. Our results indicate that proper separation between the two classes can be achieved, but further investigation is needed.

**Acknowledgements.** This work was funded by the European Union's Seventh Framework Programme (FP7/2007-2013) under grant agreement n° ITN-GA-2012-316679 – TRANSACT. This work was also partially funded by CIBER-BBN, which is an initiative of the VI National R&D&i Plan 2008–2011, CIBER Actions and financed by the Instituto de Salud Carlos III with assistance from the European Regional Development Fund.

## References

1. Julia-Sape, M., Coronel, I., Majos, C., Candiota, A.P., Serrallonga, M., Cos, M., Aguilera, C., Acebes, J.J., Griffiths, J.R., Arus, C.: Prospective diagnostic performance evaluation of single-voxel 1H MRS for typing and grading of brain tumours. NMR Biomed. **25**, 661–673 (2012)
2. Stagg, C., Rothman, D.L.: Magnetic resonance spectroscopy: Tools for neuroscience research and emerging clinical applications. Academic Press, New York (2013)
3. Julia-Sape, M., Acosta, D., Mier, M., Arus, C., Watson, D., Consortium, I.: A multi-centre, web-accessible and quality control-checked database of in vivo MR spectra of brain tumour patients. Magma **19**, 192–233 (2006)
4. Pérez-Ruiz, A., Julià-Sapé, M., Mercadal, G., Olier, I., Majós, C., Arús, C.: The INTERPRET Decision-Support System version 3.0 for evaluation of Magnetic Resonance Spectroscopy data from human brain tumours and other abnormal brain masses. BMC Bioinform. **11**, 581 (2010)
5. Ortega-Martorell, S., Olier, I., Julia-Sape, M., Arus, C.: SpectraClassifier 1.0: A user friendly, automated MRS-based classifier-development system. BMC Bioinform. **11**, 106 (2010)
6. Oz, G., Alger, J.R., Barker, P.B., Bartha, R., Bizzi, A., Boesch, C., Bolan, P.J., Brindle, K.M., Cudalbu, C., Dincer, A., Dydak, U., Emir, U.E., Frahm, J., Gonzalez, R.G., Gruber, S., Gruetter, R., Gupta, R.K., Heerschap, A., Henning, A., Hetherington, H.P., Howe, F.A., Huppi, P.S., Hurd, R.E., Kantarci, K., Klomp, D.W., Kreis, R., Kruiskamp, M.J., Leach, M.O., Lin, A.P., Luijten, P.R., Marjanska, M., Maudsley, A.A., Meyerhoff, D.J., Mountford, C.E., Nelson, S.J., Pamir, M.N., Pan, J.W., Peet, A.C., Poptani, H., Posse, S., Pouwels, P.J., Ratai, E.M., Ross, B.D., Scheenen, T.W., Schuster, C., Smith, I.C., Soher, B.J., Tkac, I., Vigneron, D.B., Kauppinen, R.A.: Group, M.R.S.C.: Clinical proton MR spectroscopy in central nervous system disorders. Radiology **270**, 658–679 (2014)

7. Kreis, R.: Issues of spectral quality in clinical 1H-magnetic resonance spectroscopy and a gallery of artifacts. NMR Biomed. **17**, 361–381 (2004)
8. van der Graaf, M., Julia-Sape, M., Howe, F.A., Ziegler, A., Majos, C., Moreno-Torres, A., Rijpkema, M., Acosta, D., Opstad, K.S., van der Meulen, Y.M., Arus, C., Heerschap, A.: MRS quality assessment in a multicentre study on MRS-based classification of brain tumours. NMR Biomed. **21**, 148–158 (2008)
9. García-Gómez, J.M., Luts, J., Julià-Sapé, M., Krooshof, P., Tortajada, S., Robledo, J.V., Melssen, W., Fuster-García, E., Olier, I., Postma, G.: Multiproject–multicenter evaluation of automatic brain tumor classification by magnetic resonance spectroscopy. Magn. Reson. Mater. Phys., Biol. Med. **22**, 5–18 (2009)
10. Julia-Sape, M., Lurgi, M., Mier, M., Estanyol, F., Rafael, X., Candiota, A.P., Barcelo, A., Garcia, A., Martinez-Bisbal, M.C., Ferrer-Luna, R., Moreno-Torres, A., Celda, B., Arus, C.: Strategies for annotation and curation of translational databases: The eTUMOUR project. Database: The journal of biological databases and curation 2012, bas035 (2012)
11. Tate, A.R., Underwood, J., Acosta, D.M., Julia-Sape, M., Majos, C., Moreno-Torres, A., Howe, F.A., van der Graaf, M., Lefournier, V., Murphy, M.M., Loosemore, A., Ladroue, C., Wesseling, P., Bosson, J.L., Cabanas, M.E., Simonetti, A.W., Gajewicz, W., Calvar, J., Capdevila, A., Wilkins, P.R., Bell, B.A., Remy, C., Heerschap, A., Watson, D., Griffiths, J.R., Arus, C.: Development of a decision support system for diagnosis and grading of brain tumours using in vivo magnetic resonance single voxel spectra. NMR Biomed. **19**, 411–434 (2006)
12. Julià-Sapé, M., Griffiths, J.R., Tate, R.A., Howe, F.A., Acosta, D., Postma, G., Underwood, J., Majós, C., Arús, C.: Classification of brain tumours from MR spectra: The INTERPRET collaboration and its outcomes. NMR Biomed. **28**, 1772–1787 (2015)
13. Wright, A.J., Arus, C., Wijnen, J.P., Moreno-Torres, A., Griffiths, J.R., Celda, B., Howe, F.A.: Automated quality control protocol for MR spectra of brain tumors. Magn. Reson. Med.: Official J. Soc. Magn. Reson. Medi./Soc. Magn. Reson. Med. **59**, 1274–1281 (2008)
14. Van Gestel, T., Suykens, J.A., Lanckriet, G., Lambrechts, A., De Moor, B., Vandewalle, J.: Bayesian framework for least-squares support vector machine classifiers, gaussian processes, and kernel Fisher discriminant analysis. Neural Comput. **14**, 1115–1147 (2002)
15. Hyvarinen, A.: Fast and robust fixed-point algorithms for independent component analysis. IEEE Trans. Neural Networks/ a Publ. IEEE Neural Netw. Coun. **10**, 626–634 (1999)
16. Ding, C., Li, T., Jordan, M.I.: Convex and semi-nonnegative matrix factorizations. IEEE Trans. Pattern Anal. Mach. Intell. **32**, 45–55 (2010)
17. Sauwen, N., Sima, D.M., Van Cauter, S., Veraart, J., Leemans, A., Maes, F., Himmelreich, U., Van Huffel, S.: Hierarchical non-negative matrix factorization to characterize brain tumor heterogeneity using multi-parametric MRI. NMR Biomed. **28**, 1599–1624 (2015)
18. Ortega-Martorell, S., Lisboa, P.J., Vellido, A., Julia-Sape, M., Arus, C.: Non-negative matrix factorisation methods for the spectral decomposition of MRS data from human brain tumours. BMC Bioinform. **13**, 38 (2012)
19. Arthur, D., Vassilvitskii, S.: k-means ++: The advantages of careful seeding. In: Proceedings of the Eighteenth annual ACM-SIAM Symposium on Discrete Algorithms, pp. 1027–1035. Society for Industrial and Applied Mathematics (Year)
20. Tate, A.R., Griffiths, J.R., Martinez-Perez, I., Moreno, A., Barba, I., Cabanas, M.E., Watson, D., Alonso, J., Bartumeus, F., Isamat, F., Ferrer, I., Vila, F., Ferrer, E., Capdevila, A., Arus, C.: Towards a method for automated classification of 1H MRS spectra from brain tumours. NMR Biomed. **11**, 177–191 (1998)

# A Machine Learning Methodology for Enzyme Functional Classification Combining Structural and Protein Sequence Descriptors

Afshine Amidi[1], Shervine Amidi[1], Dimitrios Vlachakis[2], Nikos Paragios[1,3], and Evangelia I. Zacharaki[1,3(✉)]

[1] Center for Visual Computing, Department of Applied Mathematics, École Centrale de Paris, 92295 Châtenay-Malabry, France
evangelia.zacharaki@centralesupelec.fr
[2] Bioinformatics and Medical Informatics Laboratory, Biomedical Research Foundation of the Academy of Athens, Athens, Greece
[3] Equipe GALEN, INRIA Saclay, Île-de-France, Orsay, France

**Abstract.** The massive expansion of the worldwide Protein Data Bank (PDB) provides new opportunities for computational approaches which can learn from available data and extrapolate the knowledge into new coming instances. The aim of this work is to apply machine learning in order to train prediction models using data acquired by costly experimental procedures and perform enzyme functional classification. Enzymes constitute key pharmacological targets and the knowledge on the chemical reactions they catalyze is very important for the development of potent molecular agents that will either suppress or enhance the function of the given enzyme, thus modulating a pathogenicity, an illness or even the phenotype. Classification is performed on two levels: (i) using structural information into a Support Vector Machines (SVM) classifier and (ii) based on amino acid sequence alignment and Nearest Neighbor (NN) classification. The classification accuracy is increased by fusing the two classifiers and reaches 93.4 % on a large dataset of 39,251 proteins from the PDB database. The method is very competitive with respect to accuracy of classification into the 6 enzymatic classes, while at the same time its computational cost during prediction is very small.

**Keywords:** Enzyme classification · Protein structure · Amino acid sequence alignment · Multi-class SVM · PDB database

## 1 Introduction

Proteins are macromolecules which are made of amino acids. Although many distinct groups of proteins and protein families exist, enzymes constitute key pharmacological targets as their primary role is to catalyze chemical reactions. In contrast to most chemical catalysts which catalyze a wide range of reactions, enzymes are usually highly selective, catalyzing specific reactions only. The latter

© Springer International Publishing Switzerland 2016
F. Ortuño and I. Rojas (Eds.): IWBBIO 2016, LNBI 9656, pp. 728–738, 2016.
DOI: 10.1007/978-3-319-31744-1_63

are classified into 6 standard categories, Oxidoreductases, Transferases, Hydrolases, Lyases, Isomerases, and Ligases, which are identified by their Enzyme Classification (EC) number.

Knowing the EC number of a given enzyme is necessary for the development of potent molecular agents. Having a large dataset of uniquely annotated (by experimental procedures) enzymes at disposal, the goal of this work is to build classification models that are able to predict the EC number of new enzymes with high precision, repeatability and small computational time.

Previous work has been done on different datasets of enzymes. Dobson and Doig [1] used only structural information and achieved an accuracy of 35 % for top-ranked prediction using one-class versus one-class SVM on 498 enzymes from the PDB database. Others used only gene or amino acid sequences and achieved an accuracy stemming from 72.94 % on the PDB database using neural network [2] to 96 % using neural network on enzyme database [3], but also accuracies between 74 % [4] and 88.2 % using the Swiss-Prot database [5,6]. A systematic review on the various approaches used by different research groups, their utility and inference is presented in [7]. The methodologies have been classified according to the type of information used for descriptor generation into bioinformatics approaches and chemoinformatics approaches.

In this paper we present a bioinformatics approach that exploits both structural representation and protein sequence similarity in order to predict *in silico* the EC number of an enzyme using machine learning techniques. The structure is encoded by the torsion angles distribution, whereas the protein sequence is characterized by its alignment error to training sequences in which the class label (EC number) is known. Structural information has been previously used either as validation criteria for newly generated models [8] or during structure calculation to reproduce physically realistic conformational features [9]. In the following we present the method, the results achieved, some discussion and future work.

## 2    Materials and Methods

The outline of the method is shown in Fig. 1. Briefly, two supplementary descriptors are extracted from each protein model based on the structural information (SI) and the amino acid sequence (AA). Each descriptor is introduced into a classifier trained previously on annotated data and then the classifier outputs are fused into a single set of final class probabilities.

The method has been trained and tested on proteins from the PDB database. Enzymes that were found experimentally to catalyze more than one chemical reaction and were assigned multiple labels in the first level of the Enzyme Classification, were excluded from the analysis due to the uncertainty they introduce in both training and testing phase. Also PDB entries containing amino acids other than the 20 natural ones, were excluded from the AA analysis, as proper physicochemical parameterisation of Selenocysteine (U) and Pyrrolysine (O) was

# Enzymes

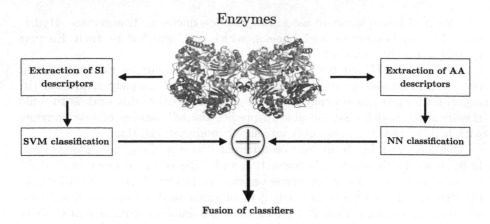

**Fig. 1.** Overview of the method

not part of this study. Same goes for ambiguous amino acids that are represented with the letters B, Z, J and X. We concluded that since the X-ray crystallography phases were incapable of giving a clear answer on which amino acid it is and provided that there were thousands of such cases in the full PDB dataset, we would introduce non-quantifiable "noise" to our dataset, which would inevitably sabotage the reliability of our findings.

The functional classes and number of enzymes obtained per class is shown in Table 1.

**Table 1.** Enzyme classification of the 39,251 enzymes

| ID | EC 1 | EC 2 | EC 3 | EC 4 | EC 5 | EC 6 |
|---|---|---|---|---|---|---|
| Name | Oxidoreductase | Transferase | Hydrolase | Lyase | Isomerase | Ligase |
| Number | 7,256 | 10,665 | 15,451 | 2,694 | 1,642 | 1,543 |

## 2.1  Feature Extraction

Proteins are chains of amino acids joined together by peptide bonds. The three-dimensional (3D) configuration of the amino acids chain is a very good predictor of protein function, thus there has been many efforts in extracting an appropriate representation of the 3D structure [10]. Since many conformations of this chain are possible due to the possible rotation of the chain about each carbon ($C_\alpha$) atom, the use of rotation invariant features is preferred over features based on cartesian coordinates of the atoms. In this study the two torsion angles of the polypeptide chain were used as structural features. The two torsion angles describe the rotation of the polypeptide backbone around the bonds between

N-$C_\alpha$ (angle $\phi$) and $C_\alpha$-C (angle $\psi$). The probability density of the torsion angles $\phi$ and $\psi \in [-180°, 180°]$ was estimated by calculating the 2D sample histogram of the angles of all residues in the protein using equal sized bins. When the protein consisted of more than one chain, the torsion angles of all chains were included into the feature vector. Smoothness in the density function was achieved by moving average filtering, i.e. by convoluting the 2D histogram with a uniform kernel. The range of angles was discretized using $19 \times 19$ bins centered at $0°$ (with bin size equal to $20°$ for all bins except the $1^{st}$ and last) and the obtained matrix of structural features was linearized to a 361-dimensional feature vector for each enzyme.

The structural description based on the probability density of the torsion angles does not provide any information about the spatial location of the amino acids in the chain, as well as their sequence. The connectivity patterns (protein sequence) reflects the intrinsic composition of the macromolecule and is an important descriptor of spatial composition. Many methods have been developed to quantify the similarity between two protein sequences which are either based on sequence alignment [11] or provide a similarity score without performing prior alignment [1]. A common algorithm that provides a similarity score to each pair of sequences is the Smith-Waterman algorithm [12]. The Smith-Waterman algorithm calculates the optimal local alignment of two sequences by computing a similarity matrix that takes into account matches, mismatches, substitutions, insertions and deletions between the two sequences [13]. Based on this algorithm, a score $\mathcal{S}(i, j)$ between each pair of sequences $i$ and $j$ was obtained and used to calculate the class probability of a sequence based on nearest neighbor classification rule.

## 2.2   Classification Using Structural and Amino Acid Sequence Information

For a given enzyme, the 361 obtained features representing the protein's structural conformation were introduced into a multi-class SVM to obtain 6 probabilistic output features. A one-versus-all classification scheme was applied in which 6 binary classifications are performed (class $i$ versus not class $i$) and combined by majority voting rule to decide which of the 6 classes is the most probable. Since the SVM's decision scores are not reflecting probabilities, an additional sigmoid function was fitted to the data in order to map the SVM outputs into pseudo-probabilities [14]. The latter are noted $(p_i^j)_{i \in [1,6]}$, with $p_i^j$ reflecting the probability for the enzyme $j$ to belong to the class EC $i$.

In order to allow the fusion of classifiers, the scores of amino acid sequence alignment (matrix $\mathcal{S}$) were also converted to pseudo-probabilistic output for the second classifier. Specifically, for a given enzyme $j$, the nearest neighbor of each class was found using as distance measure the alignment scores with the training samples. The alignment scores between the enzyme $j$ and the 6 neighbors were normalized by the sum of their scores:

$$\text{For each enzyme } j, \quad q_i^j = \frac{\displaystyle\max_{\substack{k \in \text{training} \cap \text{EC } i \\ k \neq j}} \mathcal{S}(k,j)}{\displaystyle\sum_{l=1}^{6} \max_{\substack{k \in \text{training} \cap \text{EC } l \\ k \neq j}} \mathcal{S}(k,j)}$$

Thus the classifier decision scores for both descriptors were appropriately scaled, such that $\sum_{i=1}^{6} p_i^j = \sum_{i=1}^{6} q_i^j = 1$, and allowed to be combined within a fusion scheme.

## 2.3 Fusion of Classifiers

It has been shown that fusion techniques that combine multiple machine learning methods achieve better predictive performance than any of the constituent methods [15]. In this work we combine the probabilistic output of each classifier, i.e. the SVM based on structural information and the nearest neighbor based on protein sequence alignment. The classification fusion is constructed by fusing the two probabilistic outputs $(p_i^j)$ and $(q_i^j)$ for each class $i$ through a linear combination into a final probability $(z_i^j)$:

$$\text{For each enzyme } j, \quad \forall i \in [\![1,6]\!], z_i^j = (1-\alpha)p_i^j + \alpha q_i^j$$

where $\alpha \in [0,1]$ is a weight that regulates the contribution of each classifier. Thus, the problem is to find the optimal parameter $\alpha$ that maximizes the overall classification accuracy.

## 2.4 Validation Scheme

The dataset was randomly split into 80 % for training and 20 % for testing. In order to determine the optimum $\alpha$, 20 % of the training set has been hold out and used as validation set, while the remaining part was used to train the classifier. Upon selection of $\alpha$ based on the validation set, the classifier was retrained using the whole training set, and the optimum $\alpha$ was used to assess accuracy of the fusion on the testing set. The assessment of the multi-class system was based on the confusion matrix. However, the performance of the system was also evaluated in respect to the cumulative accuracy after $i$ guesses, noted $\text{CA}_i (i \in [\![1,5]\!])$, which represents the classification accuracy after looking at the $i$ highest class probabilities.

# 3 Results

## 3.1 Structural Information and Amino Acid Sequences

Table 2 shows the cumulative accuracy (CA1 to CA5) for each enzymatic class and each descriptor separately. Classification with amino acid sequences yields from 12,2 % (EC 3) to 49,2 % (EC 6) better accuracy than classification with

**Table 2.** CA using structural information and amino-acid sequences separately

| Category | Structural information | | | | | Amino acid sequence | | | | |
|---|---|---|---|---|---|---|---|---|---|---|
| | CA1 | CA2 | CA3 | CA4 | CA5 | CA1 | CA2 | CA3 | CA4 | CA5 |
| EC 1 | 0.746 | 0.884 | 0.943 | 0.977 | 0.992 | 0.967 | 0.988 | 0.996 | 0.999 | 1.000 |
| EC 2 | 0.762 | 0.920 | 0.970 | 0.990 | 0.997 | 0.937 | 0.964 | 0.977 | 0.996 | 1.000 |
| EC 3 | 0.798 | 0.909 | 0.961 | 0.982 | 0.994 | 0.920 | 0.958 | 0.975 | 0.992 | 0.999 |
| EC 4 | 0.596 | 0.685 | 0.790 | 0.900 | 0.950 | 0.892 | 0.922 | 0.944 | 0.983 | 0.993 |
| EC 5 | 0.547 | 0.635 | 0.730 | 0.827 | 0.939 | 0.900 | 0.954 | 0.970 | 0.976 | 0.979 |
| EC 6 | 0.304 | 0.476 | 0.605 | 0.822 | 0.926 | 0.796 | 0.822 | 0.841 | 0.871 | 0.964 |
| Overall | 0.735 | 0.864 | 0.924 | 0.965 | 0.986 | 0.934 | 0.966 | 0.979 | 0.993 | 0.998 |

structural information. The accuracy difference between those two approaches might indicate that amino acid sequence content (with 93.4 % overall accuracy) is a better predictor of enzyme function than structural information (with 73.5 % overall accuracy).

### 3.2   Fusion of Both Information

Table 3 shows the most often predicted class by each method after each guess. It can be seen that the order of predicted classes varies when different features (SI and AA) are used for classification denoting that each descriptor captures different aspects of enzymatic function.

**Table 3.** Most representative guesses for each true class

| True class | Method | Guesses | | | | | |
|---|---|---|---|---|---|---|---|
| | | $1^{st}$ | $2^{nd}$ | $3^{rd}$ | $4^{th}$ | $5^{th}$ | $6^{th}$ |
| EC 1 | SI | EC 1 | EC 2 | EC 2 | EC 4 | EC 5 | EC 6 |
| | AA | EC 1 | EC 2 | EC 3 | EC 4 | EC 6 | EC 5 |
| EC 2 | SI | EC 2 | EC 3 | EC 1 | EC 6 | EC 5 | EC 5 |
| | AA | EC 2 | EC 3 | EC 1 | EC 1 | EC 5 | EC 5 |
| EC 3 | SI | EC 3 | EC 2 | EC 1 | EC 6 | EC 5 | EC 5 |
| | AA | EC 3 | EC 2 | EC 1 | EC 1 | EC 6 | EC 5 |
| EC 4 | SI | EC 4 | EC 2 | EC 1 | EC 1 | EC 5 | EC 5 |
| | AA | EC 4 | EC 3 | EC 2 | EC 2 | EC 1 | EC 5 |
| EC 5 | SI | EC 5 | EC 3 | EC 2 | EC 1 | EC 6 | EC 4 |
| | AA | EC 5 | EC 2 | EC 1 | EC 3 | EC 6 | EC 6 |
| EC 6 | SI | EC 2 | EC 2 | EC 1 | EC 1 | EC 5 | EC 5 |
| | AA | EC 6 | EC 1 | EC 2 | EC 1 | EC 4 | EC 5 |

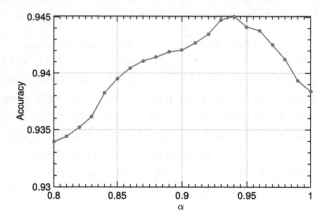

**Fig. 2.** Classification accuracy on the validation dataset as a function of the weight $\alpha$

The fusion of information relies on the study of the function $\alpha \mapsto \text{Acc}(\alpha)$, where Acc is the accuracy on the validation set obtained by weighting the probabilities obtained by the SI-based and AA-based classifiers. Fig. 2 illustrates the classification accuracy in the interval $\alpha \in [0.8, 1]$. The value that maximizes the accuracy is $\alpha = 0.94$ and was chosen for the assessment of the method on the testing dataset. As expected, the optimum $\alpha$ gives a more significant weight to amino acid sequences descriptors than to structural information descriptors.

The total cumulative accuracy (for multiple guesses) is shown in Fig. 3 for each of the three methods, while Fig. 4 illustrates the cumulative accuracy for each enzymatic class separately. For Transferases and Hydrolases the fusion of features induced an average increase of accuracy by 1 % on each guess compared to the accuracy obtained by amino acid sequences only. For the other classes, the accuracy of the fusion is comparable to the accuracy by AA indicating that structural information does not additionaly contribute to classification.

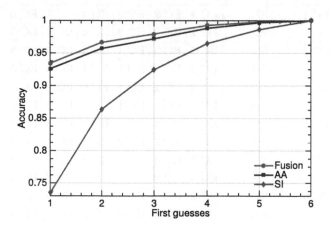

**Fig. 3.** Comparison of the overall CA between the 3 methods

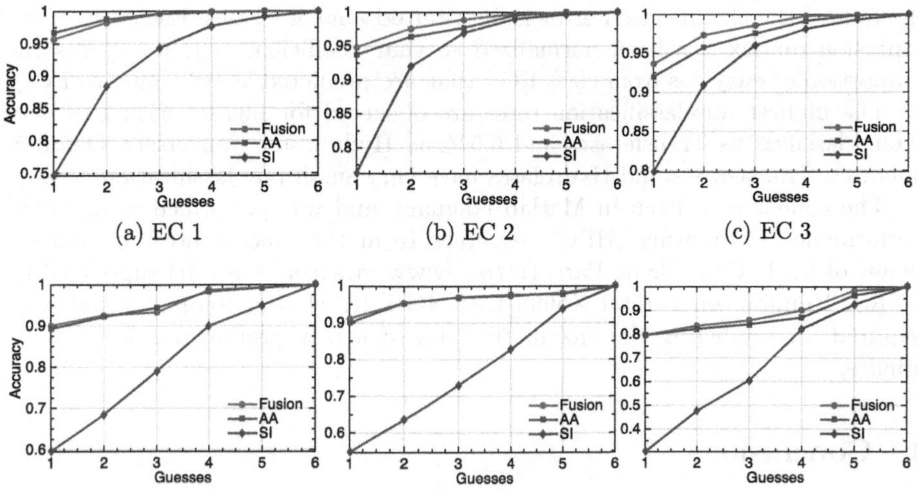


(a) EC 1 (b) EC 2 (c) EC 3

(d) EC 4 (e) EC 5 (f) EC 6

**Fig. 4.** Comparison of the CA for each true class between the 3 methods

**Predicted** (columns) / **Actual** (rows)

| Actual \ Predicted | EC 1 | EC 2 | EC 3 | EC 4 | EC 5 | EC 6 |
|---|---|---|---|---|---|---|
| EC 1 | **0.957** | 0.012 | 0.027 | 0.002 | 0.001 | 0.001 |
| EC 2 | 0.011 | **0.948** | 0.033 | 0.002 | 0.001 | 0.004 |
| EC 3 | 0.017 | 0.041 | **0.936** | 0.003 | 0 | 0.001 |
| EC 4 | 0 | 0.059 | 0.032 | **0.900** | 0.006 | 0.004 |
| EC 5 | 0.040 | 0.015 | 0.018 | 0.015 | **0.912** | 0 |
| EC 6 | 0.026 | 0.104 | 0.065 | 0.003 | 0.007 | **0.796** |

**Fig. 5.** Confusion matrix

Fig. 5 shows the confusion matrix of the classification fusion. The confusion matrix allows to see whether an enzyme class is more or less predictable and shows how the classification error is distributed among classes. Each row of the confusion matrix has been normalized so that coefficient $(i,j)$ represents the proportion of enzymes from class EC $i$ that are predicted as belonging to EC $j$.

The highest misclassification rates are observed for Ligases with 10.4 % of them classified as Transferases and 6.5 % as Hydrolases. Conversely, Oxidoreductases, Transferases and Hydrolases have very small misclassification rates.

The code was written in Matlab language and was performed using High-Performance Computing (HPC) resources from the "mesocentre" computing center of École Centrale de Paris (http://www.mesocentre.ecp.fr) supported by CNRS. Running on 2 Intel X5650 with 4GB RAM, the computational time required for predicting the enzymatic class of a new protein was less than a minute.

## 4    Conclusions

The bioinformatics bottleneck is a major problem in the fields of Next Generation Sequencing (NGS) and Drug Design (DD). Data accumulation is becoming a challenge on its own as the handling and analysis of huge datasets is not a trivial task. Therefore there is dear need in the field of bioinformatics to establish and optimize new methodologies that are faster and much more efficient for routine tasks in the aforementioned fields. A major drawback in protein classification is the fact that protein similarity is still judged upon primary sequence comparisons rather than combining structural information too. Using structural information was quite challenging a while ago as significant computing power was required. However, nowadays it is not too difficult to use optimized workstations and supercomputers to analyze all structural information available. Remarkably, the RCSB PDB database has now more than 114,000 entries. In 1990 there were only 507 structures available and in 2000 only 13,591 structures. Quick and reliable classification of NGS generated sequences is of paramount importance in the field of DD, as the problem of pharmacological target identification still holds. Classifying quickly and reliably, the several thousands of proteins that are being identified and sequenced on a daily basis would lead to an optimized pipeline that could yield results of great interest to protein science, DD and NGS related fields such as anticancer and antiviral therapy.

The proposed approach is fully automated, reproducible and computationally very efficient during prediction phase. Also the achieved accuracy (73.5 %) by the SVM classifier based on structural information is significantly higher than the reported accuracy (35 %) of the SVM classifier by Dobson and Doig [1] based also on protein structure. We used the torsion angles distribution to represent structure, whereas in [1] 55 structural attributes were originally extracted, including residue fractions, surface fractions, secondary structure content, cofactors, metals and general properties. Then attribute subset selection was applied based on backwards elimination.

Furthermore, the weighted combination of classifiers in the proposed scheme led to a significant increase of accuracy (27.1 % for the first guess) relatively to the classification performed using only structural information, while the accuracy increase relatively to the AA-based classification was only marginal (0.9 %). The trend is similar for the cumulative accuracy with multiple guesses. As can be seen from Fig. 3, the classification fusion performs better in respect to the cumulative accuracy with multiple guesses than the single classifiers.

The two classes that showed the best results after fusion (Transferases and Hydrolases) are the ones which have the largest number of samples in the dataset. This could be related to the selection of $\alpha$ through maximization of overall accuracy which is biased towards the majority classes. It is well known that most machine learning techniques rely on balanced datasets in which each class includes the same number of samples. In the future, we plan to investigate boot-strapping and data aggregation techniques [16] which provide efficient solutions to unbalanced datasets.

Moreover, in the current study, we investigated two common supervised classification techniques, the nearest neighbor and SVM. The nearest neighbor is a very simple classifier often used as baseline to assess the necessity of more advanced tools. SVM is a very popular algorithm due to its ability to capture complex relationships between the datapoints and its high predictive power. Also the number of required hyper-parameters to be optimized in SVM is small (usually 2, i.e. the parameters controlling the misclassification penalty and smoothness), and thus quite easy to tune even by manual search or grid-search. Ongoing work includes the investigation of different classifiers, in terms of accuracy and performance.

**Acknowledgment.** This research was partially supported by European Research Council Grant Diocles (ERC-STG-259112).

# References

1. Dobson, P.D., Doig, A.J.: Predicting enzyme class from protein structure without alignments. J. Mol. Biol. **345**(1), 187–199 (2005)
2. Osman, M.H., Choong-Yeun Liong, I.H.: Hybrid Learning algorithm in neural network system for enzyme classification. ICSRS **2** (2010). ISSN 2074–8523
3. Volpato, V., Adelfio, A., Pollastri, G.: Accurate prediction of protein enzymatic class by N-to-1 neural networks. BMC Bioinformatics **14**(Suppl 1), S11 (2013). doi:10.1186/1471-2105-14-S1-S11, licensee BioMed Central Ltd. 2013
4. des Jardins, M., Karp, P.D., Krummenacker, M., Lee, T.J., Ouzounis, C.A. : Prediction of enzyme classification from protein sequence without the use of sequence similarity. ISMB (1997)
5. Kumar, C., Choudhary, A.: A top-down approach to classify enzyme functional classes and sub-classes using random forest. EURASIP J Bioinform Syst Biol. (1), 1 (2012). doi:10.1186/1687-4153-2012-1

6. Lee, B.J., Lee, H.G., Lee, J.Y., Ryu, K.H.: Classification of enzyme function from protein sequence based of feature representation. In: Proceedings of the 7th IEEE International Conference on Bioinformatics and Bioengineering, 2007, BIBE 2007, pp. 741–747 (2007)
7. Sharma, M., Garg, P.: Computational approaches for enzyme functional class prediction: a review. Curr. Proteomics **11**(1), 17–22 (2014)
8. Read, R., Adams, P., Arendall III, W., Brunger, A., Emsley, P., Joosten, R., Keyweft, G., Krissinel, E., Lütteke, T., Otwinowski, Z., Perrakis, A., Richardson, J., Sheffler, W., Smith, J., Tickle, I., Vriend, G., Zwart, P.: A new generation of crystallographic validation tools for the protein data bank. PubMed (2011). doi:10.1016/j.str.2011.08.006
9. Bermejo, G., Clore, G., Schwieters, C.: Smooth statistical torsion angle potential derived from a large conformational database via adaptive kernel density estimation improves the quality of NMR protein structures. Proteine Sci. (2012). doi:10.1002/pro.2163
10. Lie, J., Koehl, P.: 3D representations of amino acids-applications to protein sequence comparison and classification. Comput. Struct. Biotechnol. J. **11**, 47–58 (2014). doi:10.1016/j.csbj.2014.09.001
11. Sharif, M.M., Thrwat, A., Amin, I.I., Ella, A., Hefeny, H.A.: Enzyme function classification based on sequence alignment. In: Mandal, J.K., Satapathy, S.C., Sanyal, M.K., Sarkar, P.P., Mukhopadhyay, A. (eds.) Advances in Intelligent Systems and Computing. Advances in Intelligent Systems and Computing, vol. 340, pp. 409–418. Springer, India (2015)
12. Jensen, L.J., Gupta, R., Blom, N., Devos, D., Tamames, J., Kesmir, C., Nielsen, H., Stærfeldt, H.H., Rapacki, K., Workman, C., Andersen, C.A., Knudsen, S., Krogh, A., Valencia, A., Brunak, S.: Prediction of human protein function from post-translational modifications and localization features. J. Mol. Biol. **319**(5), 1257–1265 (2002)
13. Smith, T., Waterman, M.: Identification of common molecular subsequences. J. Mol. Biol. **147**, 195–197 (1981). doi:10.1016/0022-2836(81)90087-5
14. Platt, J.C.: Probabilistic outputs for support vector machines and comparison to regularized likelihood methods. In: Advances in Large Margin Classifiers, pp. 61–74. MIT Press, Cambridge (1999)
15. Mohammed, A., Guda, C.: Application of a hierarchical enzyme classification method reveals the role of gut microbiome in human metabolism. BMC Genomics **16** (2015). doi:10.1186/1471-2164-16-S7-S16
16. Chawla, N.V.: Data Mining for Imbalanced Datasets: an overview (chap. 40). In: Maimon, O., Rokach, L. (eds.) Data Mining and Knwoledge Discovery Handbook, pp. 853–867. Springer, New York (2000). doi:10.1007/0-387-25465-X40

# Gene-Disease Prioritization Through Cost-Sensitive Graph-Based Methodologies

Marco Frasca$^{(\boxtimes)}$ and Simone Bassis

Department of Computer Science, University of Milano,
Via Comelico 39/41, 20135 Milano, Italy
{frasca,bassis}@di.unimi.it

**Abstract.** Finding genes associated with human genetic disorders is one of the most challenging problems in bio-medicine. In this context, to guide researchers in detecting the most reliable candidate causative-genes for the disease of interest, gene prioritization methods represent a necessary support to automatically rank genes according to their involvement in the disease under study. This problem is characterized by highly unbalanced classes (few causative and much more non-causative genes) and requires the adoption of cost-sensitive techniques to achieve reliable solutions. In this work we propose a network-based methodology for disease-gene prioritization designed to expressly cope with the data imbalance. Its validation over a benchmark composed of 708 selected medical subject headings (MeSH) diseases, shows that our approach is competitive with state-of-art methodologies, and its reduced time complexity makes its application feasible on large-size datasets.

**Keywords:** Gene-disease prioritization · Graph-based node ranking · Cost-sensitive learning

## 1 Introduction

Linkage studies for determining relevant genes for specific human diseases can point to a genomic region containing hundreds of genes, while the high-throughput sequencing approach will often identify a great number of non-synonymous genetic variants. Although the detection of potentially deleterious variants can be easily automated, this can often result in the identification of thousands candidate disease genes. Since the experimental verification of an individual gene can be both difficult and time consuming, an efficient way to reduce the validation cost is to narrow down the large list of candidate genes to a small and manageable set of promising genes; a process called *gene prioritization* (GP).

As manual examination of biological databases in order to select the most promising causative genes for the disease of interest has been only partially successful, since the selection is based solely on the subjective impressions of the researcher and genetic disorders often involve several primarily responsible genes, various computational GP methods have been proposed for this purpose.

© Springer International Publishing Switzerland 2016
F. Ortuño and I. Rojas (Eds.): IWBBIO 2016, LNBI 9656, pp. 739–751, 2016.
DOI: 10.1007/978-3-319-31744-1_64

Earlier works [1] investigated gene-diseases associations based on *gene expression profiles* or *genome wide association studies* (GWAS). Genome-wide association studies identify genes involved in human disease by searching the genome for small variations, called *single nucleotide polymorphisms* (SNPs), that occur more frequently in people with a particular disease than in healthy people. Each study can look at hundreds or thousands of SNPs at the same time. However, this approach tends to produce many false positive results, and the experimental validation of these candidate genes, for instance through resequencing, pathway or expression analysis, is still expensive and time consuming [2].

For these reasons other GP approaches have been investigated, such as *guilt-by-association* (GBA), in which candidate disease genes are ranked by exploiting the assumption that similar genes tend to share similar diseases [3]. The input of these methods is represented by gene networks, in which nodes represent genes and connections encode precomputed functional relationships among genes, such as common functional annotations (e.g. Gene Ontology annotations [4]), transcriptional co-expression regulation, direct molecular interactions [5]. In this context, many approaches have been adopted to compute the GP ranking, ranging from protein-protein interaction network analysis and semi-supervised graph partitioning [5], to flow propagation [6], and random walks [7].

To improve the accuracy of GP methods, recent studies have investigated the advantage of integrating multiple data sources, including expression profiles, SNP genotype data, expression quantitative trait loci, functional profiles, and network-based sources, such as gene-chemical networks, protein complexes and genetics/physical interactions [8]. A general approach in data source integration ranks each candidate gene according to each individual data source using various metrics, and then combine ranks from all data sources by using order statistics to obtain an overall rank [3]. For network-based integration approaches, a consensus network is constructed by combining the structure and the characteristics of each network, through different network integrating strategies [9]. The consensus network tends to provide better signal-to-noise ratio and complementary information about genes, thus leading to an improvement in prediction accuracy in most of cases [9,10].

Apart from the disadvantages and the benefits discussed above for each different approach, the main drawback shared by the above-mentioned GP methods is that they completely neglect the class imbalance problem characterizing GP: there are much fewer causative genes (the positive instances) than non-causative ones (the negative instances). For instance, around 40% (10/09/15 update) of known genetic diseases in the OMIM (Online Mendelian Inheritance in Man) database have still fewer or almost none established gene-disease associations [11]. Computational methodologies usually suffer from a drastic performance deterioration in case of imbalance classes, since algorithms tend more to focus on the classification of major class samples while ignoring or misclassifying minority class samples [12]. Unfortunately, in our context the minority class carries almost all the information we have about the disease under study, and this makes necessary the adoption of specifically designed imbalance-aware machine learning algorithms, often referred to as *cost-sensitive*. For instance, cost-sensitive techniques obtained successful result in similar contexts, e.g. in the protein function prediction [13,14].

Here we propose a novel network-based approach for detecting disease-gene association which aims at coping with the label imbalance by 'transforming' the input network so as to effectively represent the label imbalance, and by applying cost-sensitive methodologies on the obtained network representation. In particular, our procedure can be summarized as follows: (1) by following the approach proposed in [15], the input network is projected onto a bidimensional space, where each labeled input node corresponds to a labeled point whose coordinates depend on its positive and negative neighborhood in the input network, respectively; (2) the obtained couple of coordinates/features for each point are given in input to a cost-sensitive family of regressors to learn an cost-sensitive model to rank the unlabeled nodes. The node projection at Step 1 embeds the imbalance between positive and negative genes at each neighborhood in the corresponding point position. Moreover, working with just two features makes the Step 2 of our procedure very fast, thus allowing our method to efficiently handle large data sets. Finally, the method in general enough to include strategies for integrating heterogeneous network sources in a dedicated preprocessing step, so as to exploit the benefit of working with more reliable and informative networks. We experimentally validated our method on a public benchmark data set for GP, including almost nine thousands of human genes and around seven hundreds diseases collected from the Medical Subject Headings database[1].

The paper is organized as follows: in Sect. 2 we formalize the problem, while Sect. 3 is devoted to describe both the gene networks and the network integration techniques adopted in the benchmark experimental setting. In Sect. 4 we introduce our proposed two-step procedure; then in Sect. 5 we check its effectiveness by comparing its performance with state-of-the-art methodologies. Finally, Sect. 6 concludes the paper.

## 2    Problem Setting

The disease-gene prioritization problem can be seen as a semi-supervised bipartite ranking problem on undirected graphs [16]. Specifically, a gene network can be represented through an undirected weighted graph $G = (V, \boldsymbol{W})$, where $V = \{1, 2, \ldots, n\}$ is the set of vertices corresponding to genes, and $\boldsymbol{W}$ is the $n \times n$ weight matrix, where each element $W_{ij} \in [0, 1]$ represents some notion of functional similarity between vertices $i$ and $j$. Vertices in $V$ can be partitioned into two subsets: $S \subset V$ containing instances labeled according to a specific MeSH subject heading, and its complement $U = V \setminus S$, including unlabeled instances and therefore representing the object of our inference. As for the former, the set of positive/negative instances are denoted respectively with $S_+$ and $S_-$.

The task we are called to solve consists in learning a ranking function $\phi : U \rightarrow \mathbb{R}$ that assigns values to future positive instances higher than to negative ones,

---

[1] http://www.nlm.nih.gov/mesh.

ranking therefore the former higher than the latter. From this standpoint, GP is cast as a semi-supervised learning problem on graphs, since gene ranking can be inferred by exploiting both labeled and unlabeled nodes (genes) and the connections among them.

To make the problem even harder, the family of graphs under investigation is subjected to a strong imbalance between negative and positive instances, presenting a strong disproportion in favor of negative labeled nodes.

## 3   Materials

The input connection matrix $W$ represents a complex set of interactions or similarities between genes and/or their products (such as proteins), obtained as integration of several heterogeneous data sources. We adopt the benchmark experimental setting proposed in [9], which is composed of nine human gene networks covering 8449 genes, and describing functional interactions, transcriptional co-expression/regulation and localization, gene expression profiles,

Table 1. Gene networks used in experimental campaign.

| | |
|---|---|
| *finet* | *Functional interaction network* – A network covering 8441 selected proteins and containing protein-protein interactions inferred by a Naive Bayes classifier [18]. |
| *hnnet* | *Human net* – Functional gene network integrating 21 large-scale genomics and proteomics datasets from four species [19], spanning diverse distinct lines of evidence. |
| *cmnet* | *Cancer module network* – Gene-gene network composed of 8849 genes, where two genes are connected if they share at least one of the 263 biological and clinical conditions considered in [20], collecting expression profiles in different tumors. |
| *gcnet* | *Gene chemical network* – A network of 7649 genes constructed starting from the genes-chemicals interactions available at the CTD database. |
| *dbnet* | *BioGRID database network* – A protein-protein interaction network of 8449 proteins based upon direct physical and genetic interactions obtained from BioGRID [21] |
| *bgnet* | *BioGRID projected network* – Network obtained by: (i) constructing a bipartite graph exploiting interactions between genes available in BioGRID; and (ii) inserting an edge between two genes if they share at least one neighbor in the bipartite graph. |
| *bpnet, mfnet, ccnet* | *Semantic similarity-based networks* – Three networks obtained by considering the Gene Ontology terms [4] in the three branches (biological process, molecular function, and cellular component). Each connection weight is the maximum Rensik semantic similarity between all the terms for which the two genes are GO annotated. |

genes-chemicals relationships, protein-protein physical and genetic interactions, and GO semantic similarity (see Table 1). The database also provides the associations of such genes with 708 selected MeSH (Medical Subject Headings) diseases, downloaded from the CTD database [17]. The selected MeSH disease terms include between 5 and 200 causative genes.

## 3.1   Network Integration

Since the various networks have different number of genes, before their combination we extend them to the union of genes in the single networks, by filling each network with zeros in the corresponding missing rows/columns. As done in [9], in a pre-processing step we delete smaller edges so as to remove too small (and putative noisy) similarities, and ensure at least one neighbor for each node.

As integration scheme we adopt the *unweighted integration* of single networks, which performed better among the unweighted schemes proposed in [9]. It is the simple average of the $m$ available network adjacency matrices, i.e. $\boldsymbol{W}^* = \sum_{d=1}^m \boldsymbol{W}^{(d)}/m$. Finally, we apply to $\boldsymbol{W}^*$ the Laplacian normalizazion $\boldsymbol{D}^{-\frac{1}{2}} \boldsymbol{W}^* \boldsymbol{D}^{-\frac{1}{2}}$, where $\boldsymbol{D}$ is a diagonal matrix $D_{ii}|_{i=1}^n$, with $D_{ii} = \sum_j \boldsymbol{W}_{ij}^*$.

We performed our experimentations on two networks: the first, called Net6 hereinafter, was obtained by integrating the six gene networks with the exclusion of the semantic similarity-based ones; the latter (Net9) is defined as unweighted integration of all the nine single networks reported in Table 1.

## 4   Methods

We decided to solve the bipartite ranking problem introduced in Sect. 2 in terms of a generalized linear model (GLM) where the response variable, suitably thresholded through the sign function, decrees the membership to either the positive or negative class, while the predictors have been chosen so as to exploit the nodes similarity coded in the weight matrix $\boldsymbol{W}$. In order to keep the computational burden low and to exploit the network topology, we extract from the input network two features[2], as follows: each node $i \in S$ is associated with a point $\Delta_i = (\Delta_i^+, \Delta_i^-)$ in the plane, where

$$\Delta_i^+ = \sum_{j \in S_+} w_{ij}, \qquad \Delta_i^- = \sum_{j \in S_-} w_{ij} \qquad (1)$$

Intuitively, the more node $i$ is functionally similar to positive nodes and the higher will be the value of its $\Delta_i^+$ coordinate; analogously for the contribution given by negative nodes to the second coordinate. Remembering the one-to-one correspondence between genes and vertices, with this projection we hope to find a bipartition of nodes in $S$ which concentrates positive nodes mostly toward the rightmost lower region of the first quadrant, and negative ones in the remaining

---

[2] Actually the number of predictors, including the two-way interaction term (i.e. the product of the two features), is equal to 3.

portion of it. This network projection onto the plane, already adopted in [15], also allows to both avoid the *curse of dimensionality* problem, since the projected space has just two dimensions, and deal with the class imbalance problem, since the projected positive and negative two-dimensional points can be associated with different misclassification costs during the learning of the GLM.

**Table 2.** GLMs adopted in the experimental campaign.

| | |
|---|---|
| LR | *Linear regression model.* Usually disregarded in case of dichotomous categorical dependent variables, mainly to avoid the risk of "impossible predictions" (i.e. results outside of the unit interval), we include it in our analysis in view of both the straightforward interpretability of its coefficients, the not negligible speed-up factor observed when large datasets are given in input to the model, and the groundlessness of the aforementioned risks when interactions terms are included in the model [22]. |
| LogR | *Logit regression model.* Together with *probit* model, it is one of the widely used regression models for binary response variables. Despite the different assumptions the two models make about the error distribution, results tend to be so similar each other that preference for one over the other model tends to vary by discipline. We opted to work with logistic regression (whose link function reads as $g(x) = \log(x/(1-x))$) mainly for the straightforward interpretation of the estimated coefficients. |
| CLogLR | *Cloglog regression model.* While logit and probit are symmetric link functions, the choice of a skewed link function provides a better fit to unbalanced data [23]. Binomial regression model with complementary log-log link function (defined as $g(x) = \log(-\log(1-x))$) is frequently used when the probability of events' occurrence is very small. |
| PR | *Poisson regression model.* As an alternative to an asymmetric link function, the choice of a discrete and skewed distribution for the response variable is often suggested [24]. Poisson regression with the canonical log link function is widely used in case of binary outcome variables to cope with rare events. Indeed, imbalance classification problems represent a typical scenario to apply Poisson regression, since the main assumption which the model relies on that expected value and variance of the response variable coincide, is always (at least approximately) satisfied. |

We adopted the four GLM models, described in Table 2. Within the various cost-sensitive schemes proposed to allow regression models handling imbalance classes [25], one of the most effective is *maximum weighted likelihood estimation* [26], which consists in maximizing the sum of the log-density of each sample item, suitably weighted by a coefficient $\omega \in \mathbb{R}_0^+$: the higher the coefficient and more influential will be the corresponding sample point in the overall optimization. Here we propose two variants of the above vanilla regression models, by introducing two weighting schemes $\omega^a$ and $\omega^b$, as follows. Having denoted with

$n_+$ and $n_-$ respectively the number of positive and negative instances:

$$\omega_i^a = \begin{cases} 1/n_+ & \text{if gene } i \in S_+ \\ 1/n_- & \text{otherwise} \end{cases} \qquad \omega_i^b = \begin{cases} \Delta_i^+ / \sum_{j \in S_+} \Delta_j^+ & \text{if gene } i \in S_+ \\ 1/n_- & \text{otherwise} \end{cases} \quad (2)$$

Intuitively, both schemes try to compensate the class imbalance by giving higher weights to infrequent instances. Scheme 'b' breaks the flatness of positive weights by assigning higher influence to positive nodes when they are functionally more similar to nodes belonging to the same class. In other words, the higher is the positive neighborhood of a positive node and the higher will be its influence in the overall maximization process.

Summing up all possible combinations of GLMs and weight schemas, we obtained a total of 12 models, which we refer to with the schema "[W]GP-*mod* [*ws*]", where WGP stands for Weighted Gene Prioritization, *mod* is one of the four GLM acronyms used in Table 2, the weights schema $ws \in \{\text{'a','b'}\}$, and square brackets are used to denote optional arguments.

## 5   Results and Discussion

In order to have a fair comparison, the experimental validation of the proposed models follows the setting adopted in [9]. We compared our method with the state-of-the-art techniques briefly described in Table 3, and estimated the generalization performances by averaging the performances observed through the classical $k$-fold cross-validation (CV), with $k = 5$. Performances have been assessed using both the *Area Under the ROC Curve* (AUC) and the *Precision* at different *Recall* levels (PXR). Concerning the experimental campaign, as performed in [9], firstly we run our methods on the network Net6 (see Sect. 3.1). Table 4 shows the corresponding average AUC sorted in decreasing order. Apart from GP-PR and GP-CLogLR, all our methods outperform the top-performing benchmark algorithm ($S_{AV}$ $t = 5$). This witnesses the high informativeness of the two projected features defined in Eq. (1) and the effectiveness of the GLM to cope with the label imbalance at each node neighborhood. Moreover, to appreciate the benefit of the cost-sensitive models w.r.t. the corresponding vanilla versions, we performed the one-side Wilcoxon Signed Rank test between all couples of methods within the same family to assess whether their population mean ranks differ. As a results, we observed a meaningful increase in performance of the 'b' scheme over the 'a' variant – confirming our initial assumption that positives, carrying more information than negatives, should be taken into account when learning the predictive model – and singularly of both cost-sensitive models w.r.t. their vanilla version ($p$-value $< 0.001$). This regularity breaks down in both linear and Poisson regressions, where scheme 'a' outperforms variant 'b' ($p$-value $= 0.025$). We conjecture that such results are due to the convergence of GLM fitting procedures toward spurious optima in rare instances which, in turn, may be caused by the peaked landscape of weights distribution in 'b' scheme. Finally, due to the fast convergence of regression performed in the 2-dimensional projected space,

**Table 3.** Competitor benchmark methods.

| |
|---|
| *Kernelized score functions.* This kernel based ranking method adopts a suitable kernel matrix so as to extend the similarity between two nodes $i$ and $j$ also to non neighboring nodes [9]. The score of each gene $i$ for a given MeSH disease $M$ is defined according to a suitable metric $d(i, V_M)$, which is specified in terms of a distance $d_{\mathcal{H}}$ between the images in a suitably chosen Hilbert space $\mathcal{H}$ of $i$ and the subset of genes $V_M \subset V$ associated with $M$. By varying the definition of $d(i, V_M)$, authors obtained different scoring methods: <br> – $S_{NN}$, when $d(i, V_M)$ is the minimum distance (in $\mathcal{H}$) between $i$ and $V_M$; <br> – $S_{kNN}$, when $d(i, V_M)$ is defined by considering the closest $k$ neighbors in $V_M$; <br> – $S_{AV}$, when $d(i, V_M)$ is the average distance between $i$ and $V_M$. <br> As kernel matrix, the $t$-step ($t = 1, 2, \ldots$) random walk kernel $\boldsymbol{K}^t$ is adopted, where <br> $\boldsymbol{K} = \gamma \boldsymbol{I} + \boldsymbol{D}^{-\frac{1}{2}} \boldsymbol{W} \boldsymbol{D}^{-\frac{1}{2}}$, $\boldsymbol{I}$ is the $n \times n$ identity matrix, and $\gamma > 0$. |
| *Random walks.* The classical $t$-step random walk (RW) algorithm [27] assigns to a node $i \in V$ a score corresponding to the probability that a $t$-step random walk in $G$, starting from positive nodes ends at node $i$. The transition matrix $\boldsymbol{T}$ adopted by the random walker is obtained from $\boldsymbol{W}$ by row normalization, that is $\boldsymbol{T} = \boldsymbol{D}^{-1}\boldsymbol{W}$. |
| *Random walks with restart.* The rationale behind the random walk with restart (RWR) algorithm is that after many steps the walker may forget the prior information coded in the initial probability vector (0 for nodes in $V \setminus V_M$ and $1/|V_M|$ for nodes in $V_M$, for MeSH term $M$). Thus, the algorithm allows the walker to move another random walk step with probability $1 - \theta$, or to restart from its initial condition with probability $\theta$. |
| *Guilt-by-association methods.* Algorithms relying upon the GBA rule make predictions based on the interacting genes, which are assumed to share more likely similar functions [28]. Usually, the discriminant score for a gene $i$ w.r.t. a given MeSH disease $M$ is obtained as sum of the weights connecting $i$ to neighboring genes associated with the disease $M$, or as the maximum of these weights. The benchmark results adopt the latter version. |

our method is also scalable, taking around 5 seconds to perform the entire 5-fold CV procedure for a single MeSH disease on a Intel i7-860 CPU 2.80 GHz machine with 16 GB of RAM.

To better investigate the improvements achieved by cost-sensitive methods, we show in Fig. 1 the paired AUC obtained by vanilla regression and the corresponding cost-sensitive 'b' version for cloglog and logit link functions (similar trends were obtained for all other paired comparisons – results not shown). It is immediate to observe a large majority of bullets lying above the bisector, showing that cost-sensitive variants achieve higher AUC values for most of the considered MeSH diseases. Indeed, in the first two columns of Table 5 we report the proportion of MeSH terms where cost-sensitive methods outperform the corresponding vanilla ones. Such proportion ranges from 70.1 % to 85.9 %.

Similar AUC results are obtained when running the proposed methods over the network Net9, as reported in Table 6. Results obtained by GBA, RW and RWR methods are not reported in the referenced papers due to their low performances. All our methods (except for GP-CLogLR) perform better than the top-performing benchmark method ($S_{AV}$ $t = 5$). Note how the best method (GP-CLogLR b) makes more noticeable the gain due to the adopted cost-sensitive approach, ranking the correspondent vanilla version at thirteenth place. The Wilcoxon Signed Rank test confirms the results observed for the network Net6,

**Table 4.** Average AUC across MeSH terms on the network Net6.

| Method | AUC | Method | AUC | Method | AUC |
|---|---|---|---|---|---|
| WGP-CLogLR b | 0.8777 | RWR $\theta = 0.6$ | 0.8565 | $S_{kNN}$ $t = 1, k = 19$ | 0.8138 |
| WGP-LogR b | 0.8767 | GP-PR | 0.8563 | RW $t = 3$ | 0.7937 |
| WGP-CLogLR a | 0.8762 | $S_{AV}$ $t = 2$ | 0.8562 | RW $t = 5$ | 0.7773 |
| WGP-LR b | 0.8757 | $S_{AV}$ $t = 10$ | 0.8548 | RW $t = 10$ | 0.7720 |
| WGP-LR a | 0.8748 | $S_{AV}$ $t = 1$ | 0.8538 | $S_{kNN}$ $t = 10, k = 3$ | 0.7636 |
| WGP-LogR a | 0.8737 | RWR $\theta = 0.6$ | 0.8533 | $S_{kNN}$ $t = 5, k = 3$ | 0.7405 |
| GP-LR | 0.8705 | $S_{kNN}$ $t = 10, k = 19$ | 0.8374 | $S_{kNN}$ $t = 3, k = 3$ | 0.7332 |
| WGP-PR a | 0.8680 | GP-CLogLR | 0.8365 | $S_{kNN}$ $t = 2, k = 3$ | 0.7304 |
| WGP-PR b | 0.8665 | GBA | 0.8313 | $S_{kNN}$ $t = 1, k = 3$ | 0.7280 |
| GP-LogR | 0.8648 | $S_{kNN}$ $t = 5, k = 19$ | 0.8251 | $S_{NN}$ $t = 10$ | 0.7251 |
| $S_{AV}$ $t = 5$ | 0.8596 | $S_{kNN}$ $t = 3, k = 19$ | 0.8199 | $S_{NN}$ $t = 5$ | 0.7020 |
| $S_{AV}$ $t = 3$ | 0.8580 | RW $t = 2$ | 0.8186 | $S_{NN}$ $t = 3$ | 0.6968 |
| RW $t = 1$ | 0.8566 | $S_{kNN}$ $t = 2, k = 19$ | 0.8170 | $S_{NN}$ $t = 2$ | 0.6950 |
| | | | | $S_{NN}$ $t = 1$ | 0.6934 |

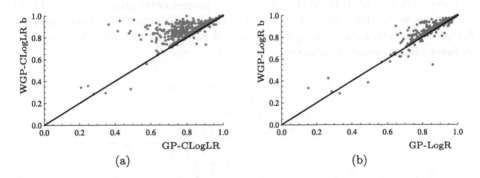

(a)          (b)

**Fig. 1.** Paired AUC comparison between cost-sensitive 'b' schema and the corresponding vanilla version for: (a) cloglog and (b) logit link functions.

**Table 5.** Proportion of wins (in terms of AUC) of cost-sensitive vs. cost-insensitive methods, observed over all the considered MeSH terms.

| | Network Net6 | | Network Net9 | |
|---|---|---|---|---|
| | WGP a | WGP b | WGP a | WGP b |
| GP-LR | 0.701 | 0.709 | 0.688 | 0.756 |
| GP-LogR | 0.743 | 0.804 | 0.441 | 0.579 |
| GP-CLogLR | 0.833 | 0.859 | 0.610 | 0.638 |
| GP-PR | 0.768 | 0.732 | 0.513 | 0.535 |

**Table 6.** Average AUC across MeSH terms on the network Net9.

| Method | AUC | Method | AUC | Method | AUC |
|---|---|---|---|---|---|
| WGP-CLogLR b | 0.8897 | GP-PR | 0.8877 | $S_{kNN}$ $t = 5, k = 19$ | 0.8500 |
| WGP-LR b | 0.8895 | GP-LR | 0.8860 | $S_{kNN}$ $t = 3, k = 19$ | 0.8413 |
| WGP-CLogLR a | 0.8894 | $S_{AV}$ $t = 5$ | 0.8831 | $S_{kNN}$ $t = 2, k = 19$ | 0.8368 |
| WGP-LogR b | 0.8890 | GP-CLogLR | 0.8827 | $S_{kNN}$ $t = 1, k = 19$ | 0.8322 |
| WGP-LR a | 0.8889 | $S_{AV}$ $t = 3$ | 0.8811 | $S_{NN}$ $t = 10$ | 0.7437 |
| GP-LogR | 0.8889 | $S_{AV}$ $t = 2$ | 0.8792 | $S_{NN}$ $t = 5$ | 0.7106 |
| WGP-PR b | 0.8884 | $S_{AV}$ $t = 1$ | 0.8765 | $S_{NN}$ $t = 3$ | 0.7014 |
| WGP-PR a | 0.8884 | $S_{AV}$ $t = 10$ | 0.8761 | $S_{NN}$ $t = 2$ | 0.698 |
| WGP-LogR a | 0.8881 | $S_{kNN}$ $t = 10, k = 19$ | 0.8665 | $S_{NN}$ $t = 1$ | 0.695 |

with some exceptions. Firstly, we observe no meaningful differences between both
the two cost-sensitive variants of the Poisson model, and 'a' schema with its naive
version ($p$-value $> 0.05$). Moreover, the only model privileging the vanilla variant w.r.t. its 'a' schema counterpart is the logistic one ($p$-value $< 0.001$). The
exceptional nature of such an event is confirmed by the entries reported in the
last two columns of Table 5: despite the less pronounced proportion of wins of
cost-sensitive methods over their cost-insensitive variants than those observed
for network Net6, six out of eight entries still shows a remarkable disproportion
in favor of cost-sensitive schemas.

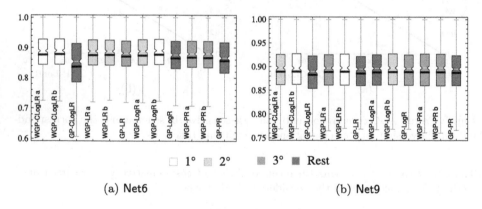

(a) Net6                    (b) Net9

**Fig. 2.** Performance distribution of the proposed methods across MeSH terms for:
(a) network Net6, and (b) network Net9. Box colors depict the performance ranking of
each method, as explained by the legend reported below the graphs. In such setting,
boxes sharing the same colors represent indistinguishable methods.

To better analyse the AUC distributions over MeSH diseases, in Fig. 2 we
report the box-and-whiskers plot of all proposed methods. Boxes are colored so

as to reflect the ranking of the methods, obtained by performing all pairwise comparisons under the one-side Wilcoxon Signed Rank test. Models sharing the same color represent maximal sets of indistinguishable methods under the above test with 0.05 significance level. The darker the color, the worst is the ranking, as shown in the legend under the picture. In particular, all methods ranking fourth downward are joined together in the same class, for the sake of visualization. Apart from the already discussed over-performance of cost-sensitive methods, we appreciate both a smaller variance and a reduced presence of outliers (not shown in the pictures). It is worth noting the marked skewness toward lower AUC values in all experiments, as confirmed by the fact that the means of AUC distributions (black markers in the pictures) are always lower than their medians (depicted with notches). Evaluating performances through means, as done in Tables 4 and 6, strongly penalizes all methods, being mean values strongly affected by the presence of outliers having low AUC values. To guarantee a fair comparison with benchmark results, we still make use of such estimator, noting that median values give a more informative and less biased view of the overall performances.

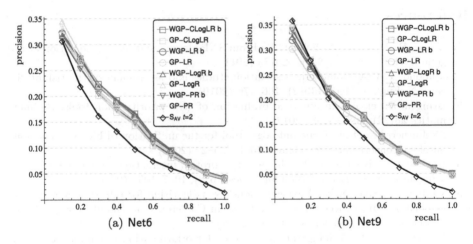

**Fig. 3.** PXR levels achieved on the network: (a) Net6, and (b) Net9.

We conclude this analysis by showing in Fig. 3 the PXR results for recall levels ranging from 0.1 to 1, with steps of 0.1. Undoubtedly, the performances of the proposed methods are very close each others; for this reason, for the sake of readability, we report just the results for vanilla and cost-sensitive 'b' scheme methods, since 'a' scheme achieves almost indistinguishable results. To better appreciate the advantage of working with cost-sensitive methods, we use a light gray level for all vanilla methods, and a dark gray one for their cost-sensitive variants. Apart from the slight but always remarkable behavior of the latter, we observe a noticeable gain w.r.t. $S_{AV}$ $t = 2$, the only method of which authors published the PXR performances, with the exception of precision value at a recall level of 0.1 in picture (b), where light and dark gray lines are almost overlapped, apart from GP-LR method which performs slightly worse. Note that in Fig. 3(a), vanilla methods tend to be more accurate for lower levels of recall. Nevertheless, for all the remaining recall values, in particular in the range [0.3, 1], cost-sensitive methods always outperform cost-insensitive ones.

# 6 Conclusions

In this work we propose a novel approach for gene-disease prioritization which is specifically designed to deal with imbalanced data, such as those character-izing databases of seed genes for known human diseases. We have shown that imbalance-aware methods can noticeably improve the performance in detecting gene-disease associations, evaluating the effectiveness of the proposed approach on a larged sized benchmark for gene prioritization problem. Future works will be devoted to exploit the hierarchical contribution coming from ontologically related gene classes.

# References

1. Lehne, B., Lewis, C.M., Schlitt, T.: From SNPs to genes: disease association at the gene level. PLoS ONE **6**(6), e20133 (2011)
2. Manolio, T.A.: Genomewide association studies and assessment of the risk of dis-ease. N. Engl. J. Med. **363**(2), 166–176 (2010)
3. Brnigen, D., et al.: An unbiased evaluation of gene prioritization tools. Bioinfor-matics **28**(23), 3081–3088 (2012)
4. Ashburner, M., et al.: Gene ontology: tool for the unification of biology. The gene ontology consortium. Nat. Genet. **25**(1), 25–29 (2000)
5. Navlakha, S., Kingsford, C.: The power of protein interaction networks for associ-ating genes with diseases. Bioinformatics **26**(8), 1057–1063 (2010)
6. Vanunu, O., Sharan, R.: A propagation-based algorithm for inferring gene-disease associations. In: Proceedings of the German Conference on Bioinformatics, GCB, September 9–12, Dresden, Germany (2008)
7. Kohler, S., et al.: Walking the interactome for prioritization of candidate disease genes. Am. J. Hum. Genet. **82**(4), 949–958 (2008)
8. Antanaviciute, A., et al.: Ova: integrating molecular and physical phenotype data from multiple biomedical domain ontologies with variant filtering for enhanced variant prioritization. Bioinformatics **31**(23), 3822–3829 (2015)
9. Valentini, G., et al.: An extensive analysis of disease-gene associations using net-work integration and fast kernel-based gene prioritization methods. Artif. Intell. Med. **61**(2), 63–78 (2014)
10. Frasca, M., et al.: UNIPred: unbalance-aware network integration and prediction of protein functions. J. Comput. Biol. **22**(12), 1057–1074 (2015)
11. Amberger, J., Bocchini, C., Hamosh, A.: A new face and new challenges for online mendelian inheritance in man (OMIM). Hum. Mutat. **32**(5), 564–567 (2011)
12. Elkan, C.: The foundations of cost-sensitive learning. In: Proceedings of the Seven-teenth International Joint Conference on Artificial Intelligence, pp. 973–978 (2001)
13. Frasca, M., et al.: A neural network algorithm for semi-supervised node label learn-ing from unbalanced data. Neural Netw. **43**, 84–98 (2013)

14. Frasca, M.: Automated gene function prediction through gene multifunctionality in biological networks. Neurocomputing **162**, 48–56 (2015)
15. Bertoni, A., Frasca, M., Valentini, G.: COSNet: a cost sensitive neural network for semi-supervised learning in graphs. In: Hofmann, T., Malerba, D., Vazirgiannis, M., Gunopulos, D. (eds.) ECML PKDD 2011, Part I. LNCS, vol. 6911, pp. 219–234. Springer, Heidelberg (2011)
16. Frasca, M., Pavesi, G.: A neural network based algorithm for gene expression prediction from chromatin structure. In: IEEE IJCNN, pp. 1–8 (2013). doi:10.1109/IJCNN.2013.6706954
17. Davis, A.P., et al.: Comparative toxicogenomics database: a knowledgebase and discovery tool for chemical-gene-disease networks. Nucleic Acids Res. **37**(Database issue), D786–D792 (2009)
18. Wu, G., Feng, X., Stein, L.: A human functional protein interaction network and its application to cancer data analysis. Genome Biol. **11**(5), R53+ (2010)
19. Lee, I., et al.: Prioritizing candidate disease genes by network-based boosting of genome-wide association data. Genome Res. **21**(7), 1109–1121 (2011)
20. Segal, E., et al.: A module map showing conditional activity of expression modules in cancer. Nat. Genet. **36**(3), 1090–1098 (2004)
21. Chatr-aryamontri, A., et al.: The biogrid interaction database: 2013 update. Nucleic Acids Res. **41**(Database-Issue), 816–823 (2013)
22. Hellevik, O.: Linear versus logistic regression when the dependent variable is a dichotomy. Qual. Quant. **43**(1), 59–74 (2009)
23. Van Del Paal, B.: A comparison of different methods for modelling rare events data. Master thesis in statistical data analysis, Department of Applied Mathematics, Computer Science and Statistics, Ghent University, Ghent, Belgium (2013–2014)
24. Derby, N.: An introduction to the analysis of rare events. In: SA16 Proceedings of the 2011 Midwest SAS Users Group Conference, Kansas City, KS (2011)
25. He, H., Garcia, E.A.: Learning from imbalanced data. IEEE Trans. Knowl. Data Eng. **21**(9), 1263–1284 (2009)
26. Dmochowski, J.P., Sajda, P., Parra, L.C.: Maximum likelihood in cost-sensitive learning: model specification, approximations, and upper bounds. J. Mach. Learn. Res. **11**, 3313–3332 (2010)
27. Lovász, L.: Random walks on graphs: a survey. In: Miklós, D., Sós, V.T., Szőnyi, T. (eds.) Combinatorics, Paul Erdős is Eighty, vol. 2, pp. 353–398. János Bolyai Mathematical Society, Budapest (1996)
28. Schwikowski, B., Uetz, P., Fields, S.: A network of protein-protein interactions in yeast. Nat. Biotechnol. **18**(12), 1257–1261 (2000)

# Network Ranking Assisted Semantic Data Mining

Jan Kralj[1,2(✉)], Anže Vavpetič[1,2], Michel Dumontier[4], and Nada Lavrač[1,2,3]

[1] Jožef Stefan Institute, Jamova 39, 1000 Ljubljana, Slovenia
{jan.kralj,anze.vavpetic,nada.lavrac}@ijs.si
[2] Jožef Stefan International Postgraduate School,
Jamova 39, 1000 Ljubljana, Slovenia
[3] University of Nova Gorica, Vipavska 13, 5000 Nova Gorica, Slovenia
[4] Stanford Center for Biomedical Informatics Research,
Stanford University, Stanford, USA
michel.dumontier@stanford.edu

**Abstract.** Semantic data mining (SDM) uses annotated data and inter-connected background knowledge to generate rules that are easily inter-preted by the end user. However, the complexity of SDM algorithms is high, resulting in long running times even when applied to relatively small data sets. On the other hand, network analysis algorithms are among the most scalable data mining algorithms. This paper proposes an effective SDM approach that combines semantic data mining and network analysis. The proposed approach uses network analysis to extract the most relevant part of the interconnected background knowledge, and then applies a semantic data mining algorithm on the pruned background knowledge. The application on acute lymphoblastic leukemia data set demonstrates that the approach is well motivated, is more efficient and results in rules that are comparable or better than the rules obtained by applying the incorporated SDM algorithm without network reduction in data preprocessing.

## 1 Introduction

Research into semantic data mining has so far focused on algorithms that pro-duce complex, high quality rules that describe the data they are applied to. The complexity of the outputs of SDM algorithms results in a severe performance bottleneck because the search space in which the algorithms look for rules is huge, and grows exponentially with the size of the background knowledge. On the other hand, network analysis is a research field with an abundance of research done to increase the performance and scalability of algorithms, resulting in algo-rithms that are capable of analyzing huge networks. The sizes of background knowledge data used in SDM approaches are usually several orders of magni-tude smaller than the problems typically encountered in network analysis. While, for example, our SDM algorithm Hedwig in [27] used a set of 337 examples and a background knowledge containing a total of 21,062 nodes, network analysis

© Springer International Publishing Switzerland 2016
F. Ortuño and I. Rojas (Eds.): IWBBIO 2016, LNBI 9656, pp. 752–764, 2016.
DOI: 10.1007/978-3-319-31744-1_65

algorithms are capable of handling much larger data sets, composed of hundreds of millions of nodes.

Despite the large difference in the sizes of data analyzed by network analysis compared to SDM, the two research fields are not fundamentally incompatible. In the most basic sense, both fields are interested in the question "Which part of the network structure is most important to my current interests?". This paper presents a method that is capable of utilizing aspects of network analysis, specifically the PageRank algorithm, along with the Hedwig semantic data mining algorithm, to produce high quality rules by only searching a fraction of the entire background knowledge space.

This paper is structured as follows. The related work is presented in Sect. 2. Section 3 presents Hedwig, the semantic data mining algorithm we used in the construction of our new algorithm. Section 4 presents how PageRank, a network ranking method, can be used to decrease the size of the background knowledge used by the Hedwig algorithm. Section 5 presents the setup and results of the experiments run with a method that merges both PageRank and Hedwig. Section 6 concludes the paper and describes further work that can be done to extend this research.

## 2   Related Work

The related work for this paper consists of research done in several different fields of research.

**Semantic pattern mining.** Rule learning, which was initially focused on building predictive models formed of sets of classification rules, has recently shifted its focus to descriptive pattern mining. Well-known pattern mining techniques in the literature are based on association rule learning [2,21]. While the initial studies in association rule mining have focused on finding interesting patterns from large data sets in an unsupervised setting, association rules have been used also in a supervised setting, to learn pattern descriptions from class-labeled data [17]. Building on top of the research in classification and association rule learning, subgroup discovery has emerged as a popular data mining methodology for finding patterns in class-labeled data, aiming to find interesting patterns as sets of individual rules that best describe the target variable [14,29].

Subgroup descriptions in the form of propositional rules are suitable descriptions of groups of instances. However, given the abundance of taxonomies and ontologies that are readily available, these can also be used to provide higher-level descriptors and explanations of discovered subgroups. Especially in the domain of systems biology the GO ontology [5], KEGG orthology [19] and Entrez gene–gene interaction data [18] are good examples of structured domain knowledge that can be used as additional higher-level descriptors in the induced rules.

The challenge of incorporating the domain ontologies in data mining was addressed in recent research on semantic data mining(SDM) [16,26,28].

In [28] an engineering ontology of Computer-Aided Design (CAD) elements and structures was used as background knowledge to extract frequent product design patterns in CAD repositories and discovering predictive rules from CAD data. Using ontologies, algorithm Fr–ONT for mining frequent concepts expressed in $\mathcal{EL}^{++}$ DL was introduced in [16]. In [26] we described and evaluated the SDM toolkit that includes two semantic data mining systems: SDM-SEGS and SDM-Aleph. SDM-SEGS is an extension of earlier domain-specific algorithm SEGS [24] which allows for semantic subgroup discovery in gene expression data. SEGS constructs gene sets as combinations of GO ontology [5] terms, KEGG orthology [19] terms, and terms describing gene–gene interactions obtained from the Entrez database [18]. SDM-SEGS extends and generalizes this approach by allowing the user to input any set of ontologies in the OWL ontology specification language and an empirical data collection which is annotated by domain ontology terms. SDM-SEGS employs ontologies to constrain and guide the top-down search of a hierarchically structured space of induced hypotheses. SDM-Aleph, which is built using the inductive logic programming system Aleph [23] does not have the limitations of SDM-SEGS, imposed by the domain-specific algorithm SEGS. Additionally, SDM-Aleph can accept any number of OWL ontologies as background knowledge which is then used in the learning process.

**Network node ranking.** The task of network node ranking in an information network provides means for assigning a *score* (or *rank*) to each node in the network, thus ranking the nodes from the highest to the lowest ranked node. The most famous ranking algorithm is the PageRank algorithm [20] used by the Google search engine, however several other network raking methods have been proposed such as a weighted version of the PageRank method called Weighted PageRank [30], as well as the related Hubs and Authorities method [13]. Another method to rank nodes in the network is to use centrality measures, for example using Freeman's network centrality [8], betweenness centrality [7], closeness centrality [4] and the Katz centrality measure [12].

**Previous work on acute lymphoblastic leukemia.** In the analysis we explored the acute lymphoblastic leukemia (ALL) data set used in a previous publication. We followed the steps used in [22] to obtain a set of $1,000$ enriched genes from a set of $10,000$ genes. The enriched genes were annotated by concepts from the Gene Ontology [3] which formed the background knowledge for our experiments. The original publication analyzing the ALL data set compared the performance of the DAVID [11] algorithm and the SegMine algorithm. In this work, however, we used the same data set to measure how we can improve the performance of the Hedwig algorithm, the algorithm which was already shown to perform well in a biological setting. The goal of this work is to examine whether network node ranking can decrease the runtime and improve the performance of the Hedwig algorithm.

# 3   Semantic Data Mining

This section describes the recently developed semantic subgroup discovery system Hedwig [27]. Compared to standard subgroup discovery algorithms, Hedwig uses domain ontologies to structure the search space and formulate generalized hypotheses [27]. Existing semantic subgroup discovery algorithms are either specialized for a specific domain [25] or adapted from systems that do not take into the account the hierarchical structure of background knowledge [26]. Hedwig overcomes these limitations as it is designed to be a general purpose semantic subgroup discovery system.

In addition to a financial use case [27], Hedwig was already shown to perform well in a biological setting, namely analyzing DNA aberration data for various cancer types [1], where it was part of a three-step methodology, together with mixture models and banded matrices. In the analysis, additional background knowledge was used in the form of several ontologies: hierarchical structure of multiresolution data, chromosomal location of fragile sites, virus integration sites, cancer genes, and amplification hotspots, obtained from various sources.

Semantic subgroup discovery, as addressed by the Hedwig system, results in relational descriptive rules. Hedwig uses ontologies as background knowledge and training examples in the form of Resource Description Framework (RDF) triples. The semantic data mining task addressed in this work takes as inputs the empirical data in the form of a set of training examples expressed as RDF triples, domain knowledge in the form of ontologies, and an object-to-ontology mapping which associates each object from the RDF triplets with appropriate ontological concepts, and finds a hypothesis (a predictive model or a set of descriptive patterns), expressed by domain ontology terms, explaining the given empirical data.

---

**Input**   : Input examples $E$, background knowledge $B$, target class value
              $c$, beam size $k$, $p$-value threshold $\alpha$
**Output**: Set of rules

1   $rules \leftarrow [\texttt{default\_rule}(E,\ c,\ B)]$

2   **while** $\texttt{improvement}(rules)$ **do**
3   $\quad$ // Add specializations of each rule to the beam
4   $\quad$ **for** $rule \in rules$ **do**
5   $\quad\quad$ | $\texttt{extend}(rules, \texttt{specialize}(rule,\ B))$
6   $\quad$ **end**
7   $\quad$ $rules \leftarrow \texttt{best}(rules,\ k)$ // Select the top $k$ rules
8   **end**
9   $rules \leftarrow \texttt{validate}(rules,\ \alpha)$ // Significance testing
10  **return** $rules$

---

**Algorithm 1:** Hedwig's $\texttt{induce}(E,\ B,\ c,\ k,\ \alpha)$ procedure.

---

**Input**    : Rule to specialize *rule*, background knowledge $B$
**Output**: Set of specializations of *rule*

1  *specializations* ← []
2  // Predicates that can be specialized
3  *eligible_preds* ← eligible(predicates(*rule*))

4  **for** *predicate* ∈ *eligible_preds* **do**
5  |    // Specialize by traversing the subClassOf hierarchy
6  |    **for** *subclass* ∈ subclasses(*predicate*, $B$) **do**
7  |    |    *new_rule* ← swap(*rule*, *predicate*, *subclass*)
8  |    |    **if** can_specialize(*new_rule*) **then**
9  |    |    |    append(*specializations*, *new_rule*)
10 |    |    **end**
11 |    **end**
12 |    // Specialize by negating
13 |    *new_rule* ← negate(*rule*, *predicate*)
14 |    **if** can_specialize(*new_rule*) **then**
15 |    |    append(*specializations*, *new_rule*)
16 |    **end**
17 **end**
18 **if** *rule* ≠ *default_rule* **then**
19 |    // Specialize by adding a new unary predicate
20 |    *new_predicate* ← next_non_ancestor(*eligible_preds*)
   |    *new_rule* ← append(*rule*, *new_predicate*)
21 |    **if** can_specialize(*new_rule*) *and* non_redundant(*new_rule*) **then**
22 |    |    append(*specializations*, *new_rule*)
23 |    **end**
24 **end**
25 **if** is_unary(last(predicates(*rule*))) **then**
26 |    // Specialize by adding new binary predicates
27 |    extend(*specializations*, specialize_binary(*new_rule*))
28 **end**
29 **return** *specializations*

**Algorithm 2:** Hedwig's specialize(*rule*, $B$) procedure.

Subgroup describing rules are first-order logical expressions. Consider the following rule, used to explain the format of induced subgroup describing rules, such as, for example: Class(X) ← C1(X), R(X,Y), C2(Y) with True Positives $(TP) = 80$ and False Positives $(FP) = 20$. Variables X, Y represent sets of input instances, $R$ is a binary relation between the examples and $C_1, C_2$ are ontological concepts. This rule is interpreted as follows. If an example $X$ is annotated with concept $C_1$, and is related with an example $Y$ via $R$, and $Y$ is annotated with concept $C_2$, then the conclusion $Class(X)$ holds. This rule condition is true for 100 input instances ($TP + FP$, also called *coverage*), 80 of which are of the target class (TP, also called *support*).

We implemented the Algorithms 1 and 2 to search for interesting subgroups. The Hedwig system, which implements this algorithm, supports ontologies and examples to be loaded as a collection of RDF triples (a graph). The system automatically parses the RDF graph for the subClassOf hierarchy, as well as any other user-defined binary relations. Hedwig also defines a namespace of classes and relations for specifying the training examples to which the input must adhere.

The algorithm uses beam search, where the beam contains the best N rules found so far. It starts with the default rule which covers all the input examples. In every iteration of the search, each rule from the beam is specialized via one of the four operations: (1) Replace predicate of a rule with a predicate that is a sub-class of the previous one, (2) negate predicate of a rule, (3) append a new unary predicate to the rule, or (4) append a new binary predicate, introducing a new existentially quantified variable.[1]

Rule induction via specializations is a well-established way of inducing rules, since every specialization either maintains or reduces the current number of covered examples. A rule will not be specialized once its coverage is zero or falls below some predetermined threshold. When adding a new conjunction, we check that if the extended rule does not improve the probability of the conclusion (we use the redundancy coefficient, as in [9]), then it is not added to the pool of specializations. After the specialization step is applied to each rule in the beam, we select new set of the best scoring N rules. If no improvement is made to the collection of rules, the search is stopped. In principle, our procedure supports any rule scoring function. Numerous rule scoring functions (for discrete targets) are available: $\chi^2$, precision, WRAcc [15], leverage and lift. The latter is the default choice and was also used in our experiments. After the induction phase, the significance of the findings is tested using the Fisher's exact test [6]. To cope with the multiple-hypothesis testing problem, we use Holm-Bonferroni [10] direct adjustment method with $\alpha = 0.05$.

## 4    Using Network Node Ranking to Decrease Background Knowledge Size

We used network ranking, in particular the personalized PageRank [20] algorithm, to asses the importance of each node in the background knowledge. The personalized PageRank of a set of nodes $S$ (P-PR$_S$) in a network is defined as the stationary distribution of the position of a random walker who starts the walk in a randomly chosen member of $S$ and then at each step either selects one of the outgoing connections or teleports back to a randomly selected member of $S$. The probability (denoted $p$) of continuing the walk is a parameter of the personalized PageRank algorithm and is usually set to 0.85.

---

[1] The new variable needs to be 'consumed' by a literal to be added as a conjunction to this clause in the next step of rule refinement.

The fundamental idea in our algorithm is that the PageRank method can be used to assess the relevance of a given background knowledge node for a particular experiment, and that Hedwig and other SDM algorithms are more likely to use highly relevant nodes when constructing rules. Therefore, if we allow the SDM algorithms to construct rules using only the most important nodes, the quality of the rules should increase. At the same time, because the background knowledge is decreased in size, the SDM algorithm we use to construct the rules will have to search through a significantly reduced space of possible rules and should therefore take much less time to conclude.

The algorithm (described in pseudo-code as Algorithm 3) consists of three steps. In the first step, we construct a network which we will use to assess the importance of background knowledge nodes. We begin with a background knowledge represented as a graph $G = (V, E)$, where $V$ is the set of nodes and $E$ a set of edges, and a data set $S$ we wish to analyze. The data set $S$ is split into a set of positive examples $S_+$ and a set of negative examples $S_-$, i.e. $S_+ \cup S_- = S, S_+ \cap S_- = \emptyset$. Each example $s \in S$ is *annotated* with some set of background knowledge nodes.

From $G$ and $S$, we construct a new network $G' = (V', E')$ by taking the original network $G$ and adding all positive examples to the set of background knowledge nodes (in other words, we set $V' = V \cup S_+$), connecting them to background knowledge nodes through the annotations ($E' = E \cup \{(e, a) \in S \times V |$annotation $a$ annotates example e$\}$).

---

**Data**: Background knowledge network $G$ and set of examples $S$
    annotated with nodes from $G$
**Result**: Rules describing the positive examples
1 **Parameters**: PageRank restart probability $p \in [0, 1]$, Cutoff percentage $c \in [0, 1]$ Set $G' = \{g \in G : \exists e \in S : e$ is annotated by $g\}$;
2 Calculate $r = PPR_G$;
3 **for** *node* $g \in G$ **do**
4    **if** $|\{g' \in G : r(g') > r(g)\}| > c \cdot |G|$ **then**
5      | remove node $g$ from $G$.
6    **end**
7 **end**
8 Run semantic data mining on $S$ using the pruned $G$ as background knowledge **return** *Rules, discovered by the SDM algorithm on the pruned background knowledge*

---

**Algorithm 3:** The proposed network ranking supported semantic data mining algorithm

In the second step of the algorithm, we decrease the size of the background knowledge network $G$ by removing less important nodes. We calculate the personalized PageRank values of the nodes in the expanded network $G'$, setting the starting nodes of for the iteration to all nodes in $S$. This allows the pagerank values to flow from the data set examples to the nodes that annotate them. The background knowledge network is then decreased by removing from it all but the top $t$ percent of nodes, where $t$ is the selected threshold and a parameter of our algorithm. We thus create a new background knowledge network $G_s$ whose nodes consist of a subset of $V$ and whose edges are induced by the edges in $E$.

In the final step, we use the Hedwig algorithm to construct rules, consisting of conjuncts of nodes in $G_s$, that best describe the set $S_+$.

## 5   Experiments

The experimental setup of our work consisted of two steps. In the first set of experiments, we ran the Hedwig algorithm on the data set to determine the baseline performance of the algorithm. We ran the algorithm with several settings of depth and beam width. The results of this round of experiments are shown in Table 1 and show that consistently, the gene ontology nodes that appear in the discovered rules have a PageRank value that is highly above normal.

The rules, discovered in this round of experiments, are also biologically significant. In all three settings when the search beam for the algorithm was set to 1, the only significant rule discovered was the gene ontology term GO:3674, a term denoting molecular function. This is a very broad term which offers little insight and shows that a larger search beam is necessary in order for Hedwig to make significant discoveries. The most interesting results are the results uncovered when the beam size is set to 10 and the support is set to 0.01. When the depth is set to 1, the most important term GO:50851 (antigen receptor-mediated signaling pathway) is interesting as it relates to the immune system related cell type. When searching with a depth of 10, we discovered a conjunct of four

**Table 1.** Best rules discovered by the Hedwig algorithm for the ALL data set. Each row presents the conjuncts (Gene Ontology terms) of the top ranking rule. The number in parentheses is the percentage of GO terms with a PageRank higher than the term in the rule. The numbers are remarkably low, showing that Hedwig consistently constructs rules with the top 1 % GO terms as ranked by the PageRank algorithm.

| Rule [ranking] | Beam | Depth | Support | Lift |
|---|---|---|---|---|
| GO:3674[0.0046] | 1 | 1 | 0.01 | 1 |
| GO:3674, [0.0046] | 1 | 10 | 0.01 | 1 |
| GO:50851, [0.7368] | 10 | 1 | 0.01 | 2.687 |
| GO:2376 [0.16047], GO:2429[0.6880], GO:5886 [0.0790], GO:5488[0.0070] | 10 | 10 | 0.01 | 3.42 |
| GO:3674, [0.0046] | 1 | 10 | 0.1 | 1 |
| GO:2376, [0.1604] | 10 | 1 | 0.1 | 1.292 |
| GO:2376 [0.1604], GO:5488 [0.0070] GO:48518[0.4277] | 10 | 10 | 0.1 | 1.414 |
| GO:2376 [0.1604], GO:5488 [0.0070] GO:48518 [0.4277] | 10 | 10 | 0.1 | 1.414 |

**Table 2.** The best rules discovered by the Hedwig using a truncated Gene Ontology as background knowledge. A cutoff value of 0.05 means that only ontology terms ranking in the top 5 % were used in rule construction. The rank was calculated by calculating the PageRank value and viewing relations as directed edges.

| Cutoff | Rules | Beam | Depth | Support | Lift |
|--------|-------|------|-------|---------|------|
| 0.05 | GO:50851 | 10 | 1 | 0.01 | 2.687 |
| 0.1 | GO:50851 | 10 | 1 | 0.01 | 2.687 |
| 0.2 | GO:50851 | 10 | 1 | 0.01 | 2.687 |
| 0.5 | GO:50851 | 10 | 1 | 0.01 | 2.687 |
| 1 | GO:50851 | 10 | 1 | 0.01 | 2.687 |
| 0.05 | GO:2376, GO:2694, GO:34110 | 10 | 10 | 0.01 | 3.235 |
| 0.1 | GO:2376, GO:2694, GO:44459 | 10 | 10 | 0.01 | 4.09 |
| 0.2 | GO:3824, GO:44283, GO:44444 | 10 | 10 | 0.01 | 4.257 |
| 0.5 | GO:2376, GO:2429 GO:5886 GO:5488, | 10 | 10 | 0.01 | 3.42 |
| 1 | GO:2376, GO:2429, GO:5886, GO:5488 | 10 | 10 | 0.01 | 3.42 |
| 0.05 | GO:43234 | 10 | 1 | 0.1 | 1.521 |
| 0.1 | GO:43234 | 10 | 1 | 0.1 | 1.543 |
| 0.2 | GO:43234 | 10 | 1 | 0.1 | 1.506 |
| 0.5 | GO:2376 | 10 | 1 | 0.1 | 1.292 |
| 1 | GO:2376 | 10 | 1 | 0.1 | 1.292 |
| 0.05 | GO:43234, GO:44464 | 10 | 10 | 0.1 | 1.537 |
| 0.1 | GO:65007, GO:43234, GO:44424 | 10 | 10 | 0.1 | 1.655 |
| 0.2 | GO:16020, GO:43234, GO:8150 | 10 | 10 | 0.1 | 1.709 |
| 0.5 | GO:2376, GO:5488, GO:48518 | 10 | 10 | 0.1 | 1.414 |
| 1 | GO:2376, GO:5488, GO:48518 | 10 | 10 | 0.1 | 1.414 |

terms: immune system process (GO:2376), immune response-activating cell surface receptor signaling pathway, (GO:2429), plasma membrane (GO:5886) and binding (GO:5488). This conjunct is interesting as it begins to provide some additional insight of the action (binding), effect (immune response signalling pathway), and location (plasma membrane).

In the second round of experiments, we decreased the size of the background data by removing low ranking nodes. We calculated the PageRank value of the GO nodes in two ways: in the first, we viewed is_a relations as directed edges pointing from the more specific GO term to the more general term. In the second, we viewed the relations as undirected edges. We ran the Hedwig algorithm on a gene ontology background data set containing only the 5 %, 10 %, 20 % and 50 % of nodes with the highest PageRank value. We also only focused on setting the size of the beam for the search to 10, as the results of the first round of experiments showed that the rules obtained by setting it to 1, are too vague to be of any biological interest.

**Table 3.** Table showing the second version of the experiment. The meaning of the rows is the same as in Table 2, but this time, the PageRank of terms was calculated by viewing relations as undirected edges.

| Cutoff | Rules | Beam | Depth | Support | Lift |
|--------|-------|------|-------|---------|------|
| 0.05 | GO:50851 | 10 | 1 | 0.01 | 2.687 |
| 0.1 | GO:50851 | 10 | 1 | 0.01 | 2.687 |
| 0.2 | GO:7584 | 10 | 1 | 0.01 | 1.811 |
| 0.5 | GO:50851 | 10 | 1 | 0.01 | 2.687 |
| 1 | GO:50851 | 10 | 1 | 0.01 | 2.687 |
| 0.05 | GO:3824, GO:44283, GO:44444, GO:44238 | 10 | 10 | 0.01 | 3.741 |
| 0.1 | GO:3824, GO:44283, GO:44444, GO:44238 | 10 | 10 | 0.01 | 3.769 |
| 0.2 | GO:45936, GO:3824, GO:9892 | 10 | 10 | 0.01 | 2.219 |
| 0.5 | GO:2376, GO:2429, GO:5886, GO:5488 | 10 | 10 | 0.01 | 3.42 |
| 1 | GO:2376, GO:2429, GO:5886, GO:5488 | 10 | 10 | 0.01 | 3.42 |
| 0.05 | GO:3824 | 10 | 1 | 0.1 | 1.322 |
| 0.1 | GO:43234 | 10 | 1 | 0.1 | 1.334 |
| 0.2 | GO:43234 | 10 | 1 | 0.1 | 1.524 |
| 0.5 | GO:2376 | 10 | 1 | 0.1 | 1.296 |
| 1 | GO:2376 | 10 | 1 | 0.1 | 1.292 |
| 0.05 | GO:3824, GO:44444, GO:44710, GO:44238 | 10 | 10 | 0.1 | 1.725 |
| 0.1 | GO:3824, GO:44444, GO:44710, GO:44238 | 10 | 10 | 0.1 | 1.744 |
| 0.2 | GO:48518, GO:43234, GO:5488 | 10 | 10 | 0.1 | 1.721 |
| 0.5 | GO:3824, GO:44444, GO:44710, GO:44238 | 10 | 10 | 0.1 | 1.639 |
| 1 | GO:2376, GO:5488, GO:48518 | 10 | 10 | 0.1 | 1.414 |

The results of the second round of experiments are shown in Tables 2 and 3. Both tables show a similar phenomenon: by decreasing the cutoff threshold the rules discovered by the Hedwig algorithm either stay the same or change to other rules with a higher lift value. For example, when searching for rules with depth set to 1, the same GO term, GO:50851, is discovered even when the size of the network is reduced to only 5 % of the original network. When searching for longer rules, decreasing the size of the network by 50 % still allows us to discover the same high quality conjunct of GO:2376, GO:2429, GO:5886 and GO:5488 as before, however decreasing the size further leaves a conjunct of GO:3824, GO:44283, GO:44444, GO:44238 which is more vague and less interesting.

While the increasing rule quality alone shows that using PageRank as a filter before applying the Hedwig algorithm can improve the performance of the algorithm, the results become even more promising if we also consider the fact that in the case when the cutoff threshold is low, the search space that Hedwig must analyze, and thus the computational complexity of the algorithm, is much smaller.

# 6  Conclusion and Further Work

The results show that network analysis method PageRank can be effectively used to reduce the size of the search space that needs to be examined by SDM methods without reducing their performance. Furthermore, the performance is in some cases even increased. This means that the proposed algorithm improvement approach shows great promise for future use of computationally expensive, but highly informative algorithms such as Hedwig, on data sets much larger than the ones used today.

In future work, we plan a more comprehensive examination of how the performance of Hedwig compares to existing enrichment methods like the SegMine method used in [22]. The comparison will be run on several biological data sets, including a data set of responses of rheumatoid arthritis patients to drug treatment.

Furthermore, we wish to perform further experiments with different methods of network reduction. For example, other network ranking methods or even other network analysis methods, such as community detection, can be used to identify the most relevant part of the background knowledge network. Also, network shrinking in our experiments was done in a basic way by simply removing all nodes whose PageRank value was too low and the edges that start or end in them. This method may cause some high ranking nodes to get "cut off" from the rest of the network, making them uninteresting for the Hedwig algorithm. In such a case, a better way may be to remove low ranking nodes, but keep the edges that start or end in them and simply extend these edges to the deleted node's neighbors. Furthermore, we will run other experiments testing the performance of our algorithm using different settings for the Hedwig algorithm.

# References

1. Adhikari, P.R., Vavpetič, A., Kralj, J., Lavrač, N., Hollmén, J.: Explaining mixture models through semantic pattern mining and banded matrix visualization. In: Džeroski, S., Panov, P., Kocev, D., Todorovski, L. (eds.) DS 2014. LNCS, vol. 8777, pp. 1–12. Springer, Heidelberg (2014)
2. Agrawal, R., Srikant, R.: Fast algorithms for mining association rules in large databases. In: Bocca, J.B., Jarke, M., Zaniolo, C. (eds.) Proceedings of the 20th International Conference on Very Large Data Bases, pp. 487–499. Morgan Kaufmann Publishers Inc., San Francisco (1994)
3. Ashburner, M., Ball, C.A., Blake, J.A., Botstein, D., Butler, H., Cherry, J.M., Davis, A.P., Dolinski, K., Dwight, S.S., Eppig, J.T., et al.: Gene ontology: tool for the unification of biology. Nat. Genet. 25(1), 25–29 (2000)
4. Bavelas, A.: Communication patterns in task-oriented groups. J. Acoust. Soc. Am. 22, 723–730 (1950)
5. Consortium, G.O.: The gene ontology project in 2008. Nucleic Acids Res. 36(Database–Issue), 440–444 (2008)
6. Fisher, R.A.: On the interpretation of $\chi^2$ from contingency tables, and the calculation of P. J. Roy. Stat. Soc. 85(1), 87–94 (1922)

7. Freeman, L.C.: A set of measures of centrality based on betweenness. Sociometry **40**(1), 35–41 (1977)
8. Freeman, L.C.: Centrality in social networks conceptual clarification. Soc. Netw. **1**(3), 215–239 (1979)
9. Hämäläinen, W.: Efficient search for statistically significant dependency rules in binary data. Ph.D. thesis, Department of Computer Science, University of Helsinki, Finland (2010)
10. Holm, S.: A simple sequentially rejective multiple test procedure. Scand. J. Stat. **6**(2), 65–70 (1979)
11. Huang, D.W., Sherman, B.T., Lempicki, R.A.: Systematic and integrative analysis of large gene lists using DAVID bioinformatics resources. Nat. Protoc. **4**(1), 44–57 (2008)
12. Katz, L.: A new status index derived from sociometric analysis. Psychometrika **18**(1), 39–43 (1953)
13. Kleinberg, J.M.: Authoritative sources in a hyperlinked environment. J. ACM **46**(5), 604–632 (1999)
14. Klösgen, W.: Explora: a multipattern and multistrategy discovery assistant. In: Fayyad, U.M., Piatetsky-Shapiro, G., Smyth, P., Uthurusamy, R. (eds.) Advances in Knowledge Discovery and Data Mining, pp. 249–271. American Association for Artificial Intelligence, Menlo Park (1996)
15. Lavrač, N., Kavšek, B., Flach, P.A., Todorovski, L.: Subgroup discovery with CN2-SD. J. Mach. Learn. Res. **5**, 153–188 (2004)
16. Ławrynowicz, A., Potoniec, J.: Fr-ONT: an algorithm for frequent concept mining with formal ontologies. In: Kryszkiewicz, M., Rybinski, H., Skowron, A., Raś, Z.W. (eds.) ISMIS 2011. LNCS, vol. 6804, pp. 428–437. Springer, Heidelberg (2011)
17. Liu, B., Hsu, W., Ma, Y.: Integrating classification and association rule mining. In: Proceedings of the 4th International Conference on Knowledge Discovery and Data mining (KDD 1998), pp. 80–86. AAAI Press (1998)
18. Maglott, D., Ostell, J., Pruitt, K.D., Tatusova, T.: Entrez gene: gene-centered information at NCBI. Nucleic Acids Res. **33**(Database issue), D54–D58 (2005)
19. Ogata, H., Goto, S., Sato, K., Fujibuchi, W., Bono, H., Kanehisa, M.: KEGG: kyoto encyclopedia of genes and genomes. Nucleic Acids Res. **27**(1), 29–34 (1999)
20. Page, L., Brin, S., Motwani, R., Winograd, T.: The PageRank citation ranking: bringing order to the web. Technical report, Stanford InfoLab (1999)
21. Piatetsky-Shapiro, G.: Discovery, analysis, and presentation of strong rules. In: Piatetsky-Shapiro, G., Frawley, W.J. (eds.) Knowledge Discovery in Databases, pp. 229–248. AAAI/MIT Press, Cambridge (1991)
22. Podpečan, V., Lavrač, N., Mozetič, I., Novak, P.K., Trajkovski, I., Langohr, L., Kulovesi, K., Toivonen, H., Petek, M., Motaln, H., et al.: SegMine workflows for semantic microarray data analysis in Orange4WS. BMC Bioinformatics **12**(1), 416 (2011)
23. Srinivasan, A.: Aleph Manual (2007)
24. Trajkovski, I., Lavrač, N., Tolar, J.: SEGS: search for enriched gene sets in microarray data. J. Biomed. Inform. **41**(4), 588–601 (2008a)
25. Trajkovski, I., Železný, F., Lavrač, N., Tolar, J.: Learning relational descriptions of differentially expressed gene groups. IEEE Trans. Syst. Man Cybern. Part C **38**(1), 16–25 (2008b)
26. Vavpetič, A., Lavrač, N.: Semantic subgroup discovery systems and workflows in the SDM-toolkit. Comput. J. **56**(3), 304–320 (2013)

27. Vavpetič, A., Novak, P.K., Grčar, M., Mozetič, I., Lavrač, N.: Semantic data mining of financial news articles. In: Fürnkranz, J., Hüllermeier, E., Higuchi, T. (eds.) DS 2013. LNCS, vol. 8140, pp. 294–307. Springer, Heidelberg (2013)

28. Žáková, M., Železný, F., Garcia-Sedano, J.A., Masia Tissot, C., Lavrač, N., Křemen, P., Molina, J.: Relational data mining applied to virtual engineering of product designs. In: Muggleton, S.H., Otero, R., Tamaddoni-Nezhad, A. (eds.) ILP 2006. LNCS (LNAI), vol. 4455, pp. 439–453. Springer, Heidelberg (2007)

29. Wrobel, S.: An algorithm for multi-relational discovery of subgroups. In: Komorowski, J., Żytkow, J.M. (eds.) PKDD 1997. LNCS, vol. 1263, pp. 78–87. Springer, Heidelberg (1997)

30. Xing, W., Ghorbani, A.: Weighted pagerank algorithm. In: 2nd Annual Conference on Communication Networks and Services Research, pp. 305–314. IEEE (2004)

# A Comprehensive Comparison of Two MEDLINE Annotators for Disease and Gene Linkage: Sometimes Less is More

Sarah ElShal[1,2(✉)], Jaak Simm[1,2], Adam Arany[1,2], Pooya Zakeri[1,2], Jesse Davis[3], and Yves Moreau[1,2]

[1] Department of Electrical Engineering (ESAT) STADIUS Center for Dynamical Systems, Signal Processing and Data Analytics Department, KU Leuven, 3001 Louvain, Belgium
sarah.elshal@esat.kuleuven.be
[2] iMinds Future Health Department, KU Leuven, 3001 Louvain, Belgium
[3] Department of Computer Science (DTAI), KU Leuven, 3001 Louvain, Belgium

**Abstract.** Text mining is popular in biomedical applications because it allows retrieving highly relevant information. Particularly for us, it is quite practical in linking diseases to the genes involved in them. However text mining involves multiple challenges, such as (1) recognizing named entities (*e.g.*, diseases and genes) inside the text, (2) constructing specific vocabularies that efficiently represent the available text, and (3) applying the correct statistical criteria to link biomedical entities with each other. We have previously developed Beegle, a tool that allows prioritizing genes for any search query of interest. The method starts with a search phase, where relevant genes are identified via the literature. Once known genes are identified, a second phase allows prioritizing novel candidate genes through a data fusion strategy. Many aspects of our method could be potentially improved. Here we evaluate two MEDLINE annotators that recognize biomedical entities inside a given abstract using different dictionaries and annotation strategies. We compare the contribution of each of the two annotators in associating genes with diseases under different vocabulary settings. Somewhat surprisingly, with fewer recognized entities and a more compact vocabulary, we obtain better associations between genes and diseases. We also propose a novel but simple association criterion to link genes with diseases, which relies on recognizing only gene entities inside the biomedical text. These refinements significantly improve the performance of our method.

## 1 Introduction

MEDLINE is a very large biomedical corpus containing over 25 million abstracts on life science and biomedical research [1]. This huge amount of text makes it challenging for genetic researchers to extract the desired information in a reasonable amount of time [2]. Hence, text mining has become a popular tool to help researchers extract relevant information more easily. One application of text mining is to identify links between biomedical entities of interest, such as genes and diseases. Multiple approaches have

© Springer International Publishing Switzerland 2016
F. Ortuño and I. Rojas (Eds.): IWBBIO 2016, LNBI 9656, pp. 765–778, 2016.
DOI: 10.1007/978-3-319-31744-1_66

been developed for this task, which rely on co-occurrence [3], concept profile similarity [4–6], or a combination of both [7]. These solutions address challenges including (1) recognizing the correct entity occurring in a given text, (2) selecting the correct set of concepts that defines a concept profile for a given entity, and (3) using the best criteria to link one entity with another. We introduce each challenge separately as follows.

Recognizing specific concepts (e.g., diseases and genes) within a given text is widely known as Named Entity Recognition (NER). NER is a basic step in text mining that involves (1) dividing the text into tokens that correspond to entities of interest, and (2) mapping the identified tokens to the correct entities [8, 9]. Different NER approaches exist to annotate a given text (e.g. MEDLINE abstracts) with biomedical entities [8–11]. Examples include MetaMap [8] and EXTRACT [9], which can be used to map MEDLINE abstracts to different sets of biomedical concepts. On the one hand, MetaMap maps the given text to the UMLS Metathesaurus [12]. On the other hand, EXTRACT maps the given text to a selection of biomedical ontologies (such as Gene Ontology [13] and Disease Ontology [14]). The resulting annotations can then be used to generate concept profiles for each MEDLINE abstract, and consequently concept profiles for any desired biomedical entity that is linked to MEDLINE abstracts.

A crucial aspect to building concept profiles is selecting the set of concepts, often called the vocabulary that describes a given profile. When the annotations for all MEDLINE abstracts are available, one can simply choose the vocabulary as the set of all unique concepts that are annotated. However, this is not always optimal computationally. For example, MetaMap extracts more than 500,000 unique concepts from all MEDLINE abstracts. Hence choosing this as the concept vocabulary for describing human genes requires using a structure (e.g., a matrix) whose dimensions is around $20,000 \times 500,000$. Loading such a data structure requires a lot of memory and doing any computation (e.g., matrix multiplication) on this data is expensive. A more practical choice would be to narrow down this vocabulary to a smaller one that covers the most important concepts for the task at hand.

Deciding on whether a gene is linked to a disease or not can be approached from many directions. For example, a gene that frequently occurs in the abstracts that are linked to a given disease has a high chance of getting annotated with that disease. This is related to co-occurrence. Also, if a gene is linked with a set of concepts that is similar to that of the disease, the chances are high that both the gene and the disease are linked together. This is related to concept profile similarity. Both directions require taking into account a background set of abstracts and concepts such that we only keep the links with a given gene that are specific to one disease and not to every other disease. For example, we do not want a gene that frequently occurs in all abstracts to get highly annotated with a given disease. Also, we do not want a concept that frequently occurs in all profiles to highly influence a disease or gene profile such that it erroneously suggests a strong link between both profiles. Hence, selecting a criterion or measure to link a gene with a given disease is challenging.

In our previous work in Beegle [7], we applied a combination of co-occurrence and concept profile similarity to associate genes with diseases, such that we selected the best rank that results from each approach separately as the final rank by Beegle for a gene given a certain disease. We used the Jaccard Similarity to measure co-occurrence, and the Cosine Similarity to measure the similarity between concept profiles. Also, we employed MetaMap to extract the biomedical concepts from the MEDLINE abstracts. For more details about Beegle, we refer the reader to our previous publication [7].

In this work, we compare the concept profiles generated by MetaMap to their counterparts generated by EXTRACT. We evaluate the influence of each concept profile setting in finding links between genes and diseases. We investigate different choices of vocabulary that we generated either manually or automatically. Our manual choices were related to choosing the starting set of unique concepts (e.g., the unique set that comes out from considering only gene-related abstracts) and the set of sources that each concept belongs to (e.g., MeSH or Ensembl). Our automatic approaches were related to combining similar concepts with each other as one united concept (e.g., via Latent Semantic Indexing (LSI)) and hence reduce our vocabulary set without losing much information. Finally, we propose an association criterion to associate genes with diseases that simplifies the concept profile similarity measure and improves its performance. We evaluate this criterion in comparison to co-occurrence and concept profile similarity as two reference criteria.

## 2   Materials and Methods

**Named Entity Recognition according to MetaMap and EXTRACT.** MetaMap is a tool that recognizes UMLS concepts inside a given text. It has been developed at the National Library of Medicine (NLM) to map biomedical text to the UMLS metathesaurus [8]. This corresponds to concepts recognized as MeSH terms, OMIM terms, Gene Ontology terms, SNOMED clinical terms, and many others. As of February 2014, MetaMap started to release its yearly-updated annotations for the MEDLINE baselines created November the year before. These baselines correspond to all the completed citations as of that date, which include the title and abstract texts for each included citation. MetaMap provides its annotations in the MetaMap Machine Output (MMO) format which is publicly available at their FTP website [15].

EXTRACT recognizes a collection of biomedical entities inside a given text, which corresponds to terms available in Gene Ontology (GO), Disease Ontology (DO), Ensembl, Brenda Tissue Ontology (BTO), NCBI Taxonomy, and others. It has been developed as a text mining pipeline at JensenLab [16] to serve many applications such as STRING [17]. It provides annotations for all MEDLINE titles and abstracts and it is updated every month. EXTRACT is available as a web service, and it can be downloaded as a tab separated file. The columns in this file correspond to information about the MEDLINE citation that is being annotated such as character positions and the annotated entities.

For more illustration, Table 1 provides a summary of the properties of each annotator. We also present the resulting annotations of MetaMap and EXTRACT given the same piece of text in Fig. 1. We observe that MetaMap provides more annotations given that it relies on UMLS, which includes a large number of sources for biomedical concepts. We also observe that EXTRACT provides the whole hierarchy of terms

**Table 1.** A summary of the MEDLINE annotators

|  | MetaMap | EXTRACT |
|---|---|---|
| Developed at | NLM | JensenLab |
| Annotations according to | GO, MeSH, OMIM, ... | GO, DO, BTO, Ensembl, ... |
| Format | MMO | TSV |
| Frequently updated | Yearly | monthly |

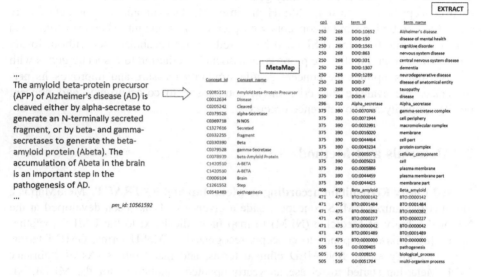

**Fig. 1.** The annotations of MetaMap vs. EXTRACT given PubMed record 10561592

(concepts) at a given character position, which is not the case for MetaMap that provides one concept at a given position. Note that we needed to parse the MMO of MetaMap to extract which concepts belong to which citation and construct the table as presented in Fig. 1, which was not the case for EXTRACT where we directly received the annotations in the presented format. However we needed to integrate data from GO and DO for example to find out which terms correspond to the given term ids.

**Imposing Manual and Automatic Vocabulary Settings.** Given that we could obtain the concept annotations for all MEDLINE citations (in terms of titles and abstracts) either through MetaMap or EXTRACT, the question then was how to make use of these annotations to build concept profiles for diseases and genes to find links between such entities. This translates to choosing the sets of vocabulary used to build the concept profiles. The simplest choice would be to choose the unique set of concepts extracted from all the MEDLINE annotations; however, this was not optimal as we briefly introduced given the size of such vocabulary set. Hence, we tried different choices as follows:

1. Choose the vocabulary to be the unique set of concepts that we could extract from all MEDLINE citations that are linked with genes according to PubMed. We call this subset the PubMed vocabulary.
2. Choose the vocabulary set to be the unique set of concepts that we could extract from all the MEDLINE citations that are linked with gene functions according to GeneRIF [18]. We call this subset the GeneRIF vocabulary.
3. Choose the vocabulary set to be the unique set of concepts that only belong to a selection of biomedical sources inside a subset setting (e.g. GO, DO, and Ensembl concepts inside the GeneRIF vocabulary).
4. Apply automatic techniques such as LSI to reduce one subset setting (e.g. GeneRIF vocabulary) into a more representative set with fewer concepts.

We think that narrowing down the vocabulary corpus into the set of abstracts that talk about genes is a reasonable choice, given that it results in profiles that are focused on concepts which are gene-related and hence perform better in our problem of associating genes with diseases. For the PubMed vocabulary we used PubMed to download the ids of all the MEDLINE citations that were found to be linked with all human genes. This corresponds to a unique set of over than 2 million citations and 283,507 concepts (according to EXTRACT). For the GeneRIF vocabulary, we downloaded the ids from GeneRIF which corresponds to a unique set of 349,274 citations and 73,027 concepts (again according to EXTRACT). We applied different selections of sources inside the GeneRIF vocabulary (according to each annotator). Given the MetaMap annotations, we chose the following sources: GO, MeSH, OMIM, HUGO, and the Disease Database. This resulted in 72,822 concepts. Given EXTRACT, we chose the following sources: GO, DO, and Ensembl. This resulted in 25,791 concepts. We selected these sources such that they are related to the two main entities in our text mining question (finding links between genes and diseases), and such that they are widely used within the annotation community [9, 19]. Finally we applied LSI via Singular Value Decomposition (SVD) to automatically reduce the GeneRIF vocabulary into a more representative subset where we could combine similar concepts together in one group. This group is called a dimension in an LSI context. We tried multiple dimension settings (starting from 2000 up to 10,000). We present a summary of the different vocabulary settings we just discussed in Table 2.

**Table 2.** A summary of the different vocabulary settings

|  | The PubMed vocabulary | The GeneRIF vocabulary | |
| --- | --- | --- | --- |
| # citations | 2,801,750 | 349,274 | |
| # concepts (complete set) | 283,507 (EXTRACT) | 73,027 (EXTRACT) | 119,336 (MetaMap) |
| # concepts (selecting sources) | n.a. | 25,791 (EXTRACT) | 72,822 (MetaMap) |
| # concepts (LSI) | n.a. | up to 10,000 | |

**Investigating Multiple Association Measures.** Different measures exist to associate genes with diseases inside text mining. Co-occurrence and concept profile similarity are two examples. In co-occurrence, we rely on the disease being linked with a set of MEDLINE citations that is similar to the set of the gene. Hence the disease and the gene frequently co-occur, which either can happen in the full citation level, the abstract level, or even the sentence level. In concept profile similarity, we rely on the fact that a disease is found to share a similar concept profile to that of the gene. Hence the disease and the gene are described by the same biomedical concepts, from which we could infer that there is a strong link between the disease and the gene. Here we used the Jaccard Similarity to measure co-occurrence, and we used the Cosine Similarity to measure concept profile similarity. For more information about each measure, we refer the reader to our previous work [7].

In this work we propose a novel measure to associate genes with diseases, which can be seen as a mix between co-occurrence and concept profile similarity. In concept profile similarity we represent each concept inside the profile by its TF-IDF (Term Frequency– Inverse Document Frequency) value. This representation gives higher weights to concepts that frequently occur with the entity they describe but don't frequently occur in general, and it gives lower weights to concepts that frequently occur in general even though they frequently occur inside a given profile. Hence we decided to use the TF-IDF values for "gene" concepts inside a disease profile to be used as the score (or measure) that ranks how well a gene is linked with a given disease. We call this measure the TF-IDF scores. We show an example of this measure in Fig. 2. On the left hand side we present the concept profile for Alzheimer's disease that is ranked by TF-IDF values in a descending order. We only show the top 13 concepts. We highlight the gene concepts in bold. On the right hand side we present the ranks of the genes against Alzheimer's disease according to their TF-IDF scores.

**The Datasets.** In our experiments, we used the 2014 release of MetaMap for the MEDLINE annotations. This corresponds to annotations for 22,076,054 MEDLINE citations. We used a version of EXTRACT that we downloaded in December 2014.

**Alzheimer's Disease concept profile (ranked by TF-IDF values)**

| concept_id | concept_count | Tf value | Tf_idf value | concept_name |
|---|---|---|---|---|
| DOID:10652 | 4570 | 0.152201425 | 0.806046297668408 | Alzheimer's disease |
| GO:0007613 | 1232 | 0.041031106 | 0.187570290105574 | memory |
| DOID:4 | 2388 | 0.079531073 | 0.17949582720061 | disease |
| **ENSP00000252486** | 658 | 0.02191434 | **0.144862663967702** | **APOE** |
| **ENSP00000284981** | 409 | 0.013621528 | **0.107623839053257** | **APP** |
| GO:0007568 | 650 | 0.021647905 | 0.100220470098624 | aging |
| **ENSP00000318585** | 306 | 0.010191167 | **0.0896536948264687** | **BACE1** |
| **ENSP00000326366** | 288 | 0.009591687 | **0.0833396769855539** | **PSEN1** |
| GO:0008219 | 380 | 0.012655698 | 0.0792002750792736 | cell death |
| GO:0050890 | 377 | 0.012555784 | 0.0745013316675462 | cognition |
| GO:0009405 | 500 | 0.016652234 | 0.0643358692114075 | pathogenesis |
| **ENSP00000340820** | 236 | 0.007859854 | **0.0629324239602534** | **MAPT** |
| GO:0007612 | 364 | 0.012122826 | 0.0549753017951828 | learning |
| ... | ... | ... | ... | ... |

⟹

**Alzheimer's Disease ranked genes**

| TF-IDF scores |
|---|
| APOE |
| APP |
| BACE1 |
| PSEN1 |
| MAPT |
| ... |

**Fig. 2.** An example for the TF-IDF scores

This corresponds to annotations for 20,686,757 MEDLINE citations. As for the validation set, we used a benchmark of experimentally validated disease–gene annotations that we extracted from the OMIM morbidmap (downloaded in May 2015). This corresponds to 330 diseases, 2214 genes, and 2789 disease–gene pairs. We downloaded our gene data (ids and symbols) from the Ensembl database (in March 2013). This corresponds to 17,116 gene records. We only consider human genes in our experiments. In order to generate the gene concept profiles, we used GeneRIF to download the ids of the MEDLINE citations that are functionally linked with our Ensembl genes (downloaded in March 2015). This corresponds to a unique set of 349,274 citations, which we used to generate the GeneRIF vocabulary. Additionally we used PubMed to download the ids of the more general list of MEDLINE citations that were found to be linked with our Ensembl genes, which we used to generate the PubMed vocabulary. As for the disease concept profiles, we similarly used PubMed to download the corresponding list of MEDLINE ids. This corresponds to a set of 936,668 unique citations. Note that on PubMed, we restrict the maximum number of ids retrieved per entity to 6500. This is the maximum number of ids that we found linked to a gene in GeneRIF.

**Boltzmann-Enhanced Discrimination (BEDROC) Evaluation.** The Area Under the Receiver Operating Characteristic (ROC) curve (AUC) has been widely used to evaluate and compare prioritization tools. It can be interpreted as the probability of a disease-associated gene being ranked earlier than a gene selected at random by a uniform distribution. To estimate the AUC value of a prioritization model, we can simply take the average of the ranks of disease-associated genes considered as the test set. However, the AUC score often leads to a misinterpretation of the model's performance in early discovery of disease-associated genes [20, 21], especially from a researcher's perspective who is normally interested in the top results for a given disease. As a result, Boltzmann-Enhanced Discrimination of ROC (BEDROC) has been proposed [20] as a proper and robust evaluation measurement for the early discovery.

For n disease-associated genes ranked $<r_i>_{i=1}^{n}$ among N genes, the BEDROC score is calculated as follows:

$$\text{BEDROC} = \frac{\sum_{i=1}^{n} \exp(-\alpha p_i)}{n \frac{1-\exp(-\alpha)}{N \exp(\frac{\alpha}{N}-1)}} + \frac{R_a \sinh(\frac{\alpha}{2})}{\cosh(\frac{\alpha}{2}) - \cosh(\frac{\alpha}{2} - \alpha R_a)} + \frac{1}{1 - (\exp(\alpha(1 - R_a)))} \quad (1)$$

where $p_i(\frac{r_i}{N})$ is the normalized rank of the $i^{\text{th}}$ disease-associated gene, $Ra = \frac{n}{N}$ is the ratio of the number of disease-associated genes to the total number of genes, and the parameter $\alpha$ tunes the importance given to early recognition. For example, when alpha equals to 275.5, 80 % of BEDROC score is assigned to the top 100 ranked genes. The BEDROC value can be interpreted as the probability that a disease-associated gene being ranked better than a gene selected at random from an exponential probability distribution function of parameter $\alpha$. In this study, we consider values of $\alpha$ equal to $\alpha = 160.9$, $\alpha = 275.5$ and $\alpha = 550.9$, which correspond to 80 % of the BEDROC being assigned to the top 1 %, top 100 and top 50 ranked genes, respectively.

**Setting up the Experiments.** In this work we had three objectives. First was to compare the contribution of MetaMap and EXTRACT as two MEDLINE annotators in generating concept profiles for diseases and genes; mainly in terms of how well each concept profile setting links the correct genes with their corresponding disease in our OMIM test set. Second was to check the impact of choosing the vocabulary on shaping the concept profiles and how that influences the disease–gene annotation process. Third was to compare the TF-IDF scores to concept profile similarity and co-occurrence as two traditional approaches. So we proceeded as follows:

1. Starting from the GeneRIF vocabulary, we used the complete annotations of MetaMap and EXTRACT to generate the concept profiles for our genes and diseases. Then we applied concept profile similarity and measured the cosine similarity on the TF-IDF representations of the profiles to score genes against diseases. We used these scores to rank the genes and calculate the BEDROC scores at the different $\alpha$ values. Note that for EXTRACT, we tried the complete annotations once using the whole hierarchy (including parent terms) and once using only the leaf terms at a given character position. We call this experiment the complete setup.
2. We used our manual selection of sources inside MetaMap and EXTRACT to generate a reduced version of the concept profiles we constructed in the first experiment. In parallel, we applied LSI. Again we measured the cosine similarity, computed the gene scores, and calculated the BEDROC score. Note that here for EXTRACT, we included only the leaf terms at a given position. We call this experiment the reduced setup.
3. We applied the TF-IDF scores measure on the (manually) reduced disease concept profiles of MetaMap and EXTRACT. We also tried a combination of TF-IDF scores and concept profile similarity by assigning the best rank that results from each approach as the gene's new score. Furthermore, we compared that to co-occurrence in which we applied the Jaccard-similarity to score a gene against a given disease.
4. We additionally applied TF-IDF scores on the disease concept profiles resulting from the complete annotations of EXTRACT according to the PubMed vocabulary. Given the size of this vocabulary the TF-IDF scores measure was the most convenient computationally.

## 3    Results

**MetaMap vs. EXTRACT (The Complete Setup).** We present the average BEDROC from the complete setup experiment in Fig. 3. We observe that by applying concept profile similarity and including only the leaf terms of EXTRACT, we achieve the best average score of 62 %, 57 %, and 51 % at $\alpha = 160.9$, $\alpha = 275.5$, and $\alpha = 550.9$ respectively. This compares to 54 %, 51 %, and 47 % when employing MetaMap, and 44 %, 40 %, and 36 % when considering the whole hierarchy of EXTRACT. Note that we highlight the black solid lines in the box plots correspond to the median value and not the average. This remark applies to the following box plots as well.

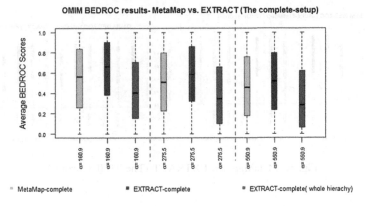

**Fig. 3.** MetaMap vs. EXTRACT (The complete setup)

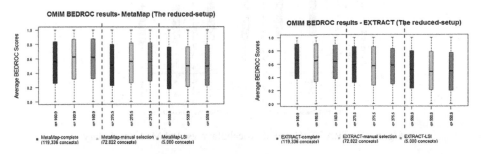

**Fig. 4.** MetaMap and EXTRACT (The reduced setup)

**MetaMap vs. EXTRACT (The Reduced Setup).** We present the average BEDROC from the reduced setup when applying concept profile similarity and employing MetaMap on the left hand side of Fig. 4. We observe that both the manual and the automatic reductions of the concept profiles result in an average score of 57 %, 53 %, and 49 % at $\alpha = 160.9$, $\alpha = 275.5$, and $\alpha = 550.9$, which slightly improves the complete setup when employing MetaMap (especially at $\alpha = 160.9$). We also present the results from the reduced setup when employing EXTRACT on the right hand side of Fig. 4. We observe that the reduced setup results in a comparable performance to the complete setup.

**TF-IDF Scores.** We present the average BEDROC when applying TF-IDF scores and employing MetaMap on the left hand side of Fig. 5. We observe that the TF-IDF scores measure improves the BEDROC results of concept profile similarity such that it reaches an average of 59 %, 56 %, and 52 % at $\alpha = 160.9$, $\alpha = 275.5$, and $\alpha = 550.9$. We also observe that when combining both TF-IDF scores and concept profile similarity, we achieve the best BEDROC results in this setting, which correspond to an average of 63 %, 59 %, and 55 % at $\alpha = 160.9$, $\alpha = 275.5$, and $\alpha = 550.9$. We also present the performance of TF-IDF scores when employing EXTRACT on the right hand side of Fig. 5. Again we observe that TF-IDF scores improve the results and when combined

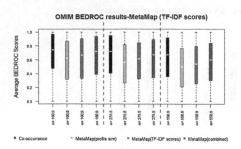
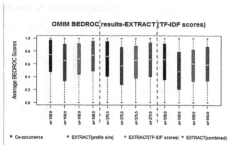

**Fig. 5.** MetaMap and EXTRACT (The TF-IDF scores)

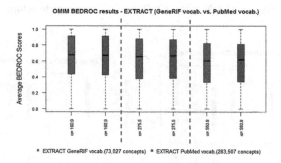

**Fig. 6.** EXTRACT (GeneRIF vocabulary vs. PubMed vocabulary)

with concept profile similarity, we achieve the best results of 68 %, 63 %, and 58 % at $\alpha = 160.9$, $\alpha = 275.5$, and $\alpha = 550.9$. We also observe that the improvement is more significant at the earlier discovery ($\alpha = 550.9$) in both models.

We additionally present the results when applying TF-IDF scores and employing EXTRACT while including the PubMed vocabulary in Fig. 6. We observe a comparable performance to the setting where we included GeneRIF as our vocabulary.

## 4  Discussion

In this work, we studied the contribution of MetaMap and EXTRACT as two different MEDLINE annotators in generating concept profiles for diseases and genes so that we could associate these entities with each other. We tried different vocabulary settings and compared different versions of the concept profiles generated by each annotator. We imposed these settings in manual and automatic fashions either by selecting the source vocabularies that generate the mapped concepts inside a given annotator or by applying LSI techniques. We also discussed TF-IDF scores as a criterion that we propose to associate genes with diseases.

We present a detailed summary of our results in Table 3. Our results show that EXTRACT outperforms MetaMap for disease-gene association in the complete setup experiment. This is achieved with more compact concept profiles and fewer concepts. We also show that when we further reduced the concept profiles generated from both annotators, either manually or automatically, we achieved at least as good performance with even fewer concepts. Furthermore, we showed that applying TF-IDF scores significantly improve the disease-gene associations especially when being combined with concept profile similarity. This combination approximates the performance of co-occurrence and it even improves it at the top 1 % threshold. We additionally applied the t-test to assess the significance between our results (*e.g.* comparing EXTRACT_-combined and MetaMap_combined at $\alpha$ = 160.9, we achieved t = 6.0629 and p-value = 3.665e-09). The application of TF-IDF scores as an association criterion is interesting because it is simpler than concept profile similarity. In TF-IDF scores, we only need concept profiles for diseases, unlike concept profile similarity where we need concept profiles for disease and gene entities. Also in TF-IDF scores, we directly use the scores as the TF-IDF values of the gene concepts inside a disease profile, while in concept profile similarity we need to calculate the scores according to some similarity

**Table 3.** A summary of the results

| Methods | Average BEDROC score | | |
|---|---|---|---|
| | $\alpha$ = 160.9 | $\alpha$ = 275.5 | $\alpha$ = 550.9 |
| 80 % of score given to the | **Top 1 %** | **Top 100** | **Top 50** |
| TF-IDF scores + concept profile similarity EXTRACT (manual selection) | **0.6800** | **0.6343** | **0.5756** |
| TF-IDF scores EXTRACT | 0.6453 | 0.6110 | 0.5649 |
| Concept profile similarity EXTRACT (LSI: 5000 dimensions) | 0.6037 | 0.5527 | 0.4891 |
| Concept profile similarity EXTRACT (manual selection: 25,791 concepts) | 0.6048 | 0.5572 | 0.4980 |
| Concept profile similarity EXTRACT (leaf terms: 73,027 concepts) | 0.6162 | 0.5704 | 0.5123 |
| Concept profile similarity EXTRACT (whole hierarchy: 108,392 concepts) | 0.4408 | 0.4006 | 0.3575 |
| Concept profile similarity MetaMap (119,336 concepts) | 0.5369 | 0.5069 | 0.4724 |
| Concept profile similarity MetaMap (manual selection: 72,822 concepts) | 0.5661 | 0.5329 | 0.4920 |
| Concept profile similarity MetaMap (LSI: 5000 dimensions) | 0.5752 | 0.5380 | 0.4907 |
| TF-IDF scores MetaMap | 0.5906 | 0.5570 | 0.5187 |
| TF-IDF scores + concept profile similarity MetaMap (manual selection) | 0.6313 | 0.5946 | 0.5464 |
| *Co-occurrence* | *0.6751* | *0.6504* | *0.6154* |

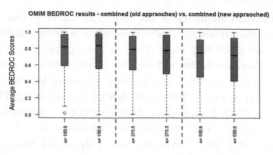

**Fig. 7.** In comparison to old Beegle

statistic (e.g., cosine similarity). Hence with TF-IDF scores we consume less space, do fewer computations, and arrive to better disease-gene associations.

In comparison to our previous approaches in Beegle, we combined co-occurrence with TF-IDF scores on the disease profiles according to EXTRACT using best rank, and then computed the BEDROC scores against our previous OMIM benchmark. We compared this to our previous best approach where we combined co-occurrence with concept profile similarity according to MetaMap. We present the results in Fig. 7. We observe comparable BEDROC results.

We wanted to get an additional insight on the performance of each annotator and whether one works better on some disease queries that are different from the other or not. Hence we checked the diseases that achieved minimum recall (recall = 0) in the top 100 ranked genes when applying TF-IDF scores given each annotator. We found out that the zero-recall set resulting from applying TF-IDF scores on the disease profiles according to EXTRACT is simply a subset of its counterpart according to MetaMap. It is also a subset of the zero-recall set when applying co-occurrence. We further investigated these disease queries and checked why they consistently lead to very poor recall. We present them in Table 4. We observe two things. First, most of the diseases are linked to very few citations, hence text mining cannot do much here and annotation with the correct genes fails. This is further confirmed when we checked the average number of citations for the one-recall set in the top 10 ranked genes, which is 2208.4 citations. Second, when enough text is available for the disease query, the corresponding top ranking genes are not random, however they share a fair number of citations with their corresponding disease query but they are not annotated in OMIM. Hence, text mining still returns some true biology here however the benchmark is probably not complete.

As for future work, we are currently integrating the annotations from EXTRACT to generate our concept profiles for genes and queries inside Beegle. We also plan to apply the TF-IDF scores measure in combination with the current approaches we apply there. Finally, we plan to study more automatic techniques (e.g., Latent Dirichlet Allocation and Logistic Regression) to sort out the most important concepts inside a concept profile and construct more relevant vocabularies.

**Table 4.** Zero-recall diseases

| Disease name | #citations | Remarks |
|---|---|---|
| Barrett esophageal adenocarcinom | 1 | Very few text available |
| Cerebrooculofacioskeletal syndrome | 4 | Very few text available |
| Cirrhosis | 6500 | Enough text available however, top 3 genes are: CFTR: 7844 common citations CTGF: 1229 common citations SMAD2: 861 common citations |
| Heinz body anemias | 5 | Very few text available |
| Microcephaly and chorioretinopathy | 27 | Very few text available |
| Coronary artery disease | 6500 | Enough text available however, top 3 genes are: CRP: 1810 common citations IL6: 138 common citations MPO: 143 common citations |
| Lymphoma | 6500 | Enough text available however, top 3 genes are: ALK: 2108 common citations BCL6: 890 common citations CD4: 5015 common citations |
| Major depressive disorder and accelerated response to antidepressant drug treatment | 21 | Very few text available |
| Renal plasia | 2 | Very few text available |

**Acknowledgements.** This work was supported by the Research Council KU Leuven [CoE PFV/10/016 SymBioSys, OT/11/051] to Y.M. and J.D.; the government agency for Innovation by Science and Technology to Y.M.; Industrial Research fund to Y.M.; Hercules Stichting to Y.M.; iMinds Medical Information Technologies [SBO 2015] to Y.M.; EU FP7 Marie Curie Career Integration Grant [\#294068] to J.D.; FWO-Vlaanderen [G.0356.12] to J.D.; and IMEC mandaat - Ph.D mandaat to A.A.. Funding for open access charge: Research Council KU Leuven.

# References

1. United States National Library of Medicine (2002) PubMed: MEDLINE Retrieval on the World Wide Web. Fact Sheet
2. Jensen, L.J., Saric, J., Bork, P.: Literature mining for the biologist: from information retrieval to biological discovery. Nat. Rev. Genet. **7**(2), 119–129 (2006)
3. Fleuren, W.W., Verhoeven, S., Frijters, R., Heupers, B., Polman, J., van Schaik, R., de Vlieg, J., Alkema, W.: CoPub update: CoPub 5.0 a text mining system to answer biological questions. Nucleic Acids Res. **39**, 450–454 (2011)
4. Jelier, R., et al.: Text-derived concept profiles support assessment of DNA microarray data for acute myeloid leukemia and for androgen receptor stimulation. BMC Bioinform. **18**, 8–14 (2007)
5. Jelier, R., Schuemie, M.J., Roes, P.J., van Mulligen, E.M., Kors, J.A.: Literature-based concept profiles for gene annotation: The issue of weighting. Int. J. Med. Inform. **77**, 354–362 (2008)
6. Jelier, R., Schuemie, M.J., Veldhoven, A., Dorssers, L.C., Jenster, G., Kors, J.A.: Anni 2.0: A multipurpose text-mining tool for the life sciences. Genome Biol. **9**(6), R96 (2008)
7. ElShal, S., Tranchevent, L.-C., Sifrim, A., Ardeshirdavani, A., Davis, J., Moreau, Y.: Beegle: from literature mining to disease-gene discovery. Nucleic Acids Res. **44**(2), e18 (2015)
8. Aronson, A.R., Lang, F.-M.: An overview of MetaMap: historical perspective and recent advances. J. Am. Med. Inform. Assoc. **17**(3), 229–236 (2010)
9. Pafilis, E., et al.: EXTRACT: Interactive extraction of environment metadata and term suggestion for metagenomics sample annotation. To appear in Database (2015)
10. Netherlands Bioinformatics Centre. Peregrine literature indexing service
11. United States National Library of Medicine. PubMed MeSH indexing
12. Bodenreider, O.: The Unified Medical Language System (UMLS): integrating biomedical terminology. Nucleic Acids Res. **32**, D267–D270 (2004)
13. Ashburner, M., Ball, C.A., Blake, J.A., et al.: Gene ontology: Tool for the unification of biology. Gene Ontology Consortium. Nat. Genet. **25**, 25–29 (2000)
14. Kibbe, W.A., Arze, C., Felix, V., et al.: Disease Ontology 2015 update: an expanded and updated database of human diseases for linking biomedical knowledge through disease data. Nucleic Acids Res. **43**, D1071–D1078 (2015)
15. United States National Library of Medicine. MetaMapped MEDLINE Baseline Results. http://ii.nlm.nih.gov/MMBaseline/index.shtml
16. Lars Juhl Jensen from the Novo Nordisk Foundation Center for Protein Research. JensenLab: Cellular Network Biology. http://jensenlab.org/
17. Szklarczyk, D., et al.: STRING v10: protein-protein interaction networks, integrated over the tree of life. Nucleic Acids Research, 43(Database issue), D447–452 (2015)
18. Mitchell, J.A., Aronson, A.R., Mork, J.G., Folk, L.C., Humphrey, S.M., Ward, J.M.: Gene indexing: Characterization and analysis of NLM's GeneRIFs. In: AMIA Annual Symposium Proceedings, pp. 460–464 (2003)
19. Cheung, W.A., Ouellette, B.F., Wasserman, W.W.: Inferring novel gene-disease associations using medical subject heading over-representation profiles. Genome Med. **4**(9), 75 (2012)
20. Truchon, J.F., Bayly, C.I.: Evaluating virtual screening methods: Good and bad metrics for the "early recognition" problem. J. Chem. Inf. Model. **47**, 488–508 (2007)
21. Zhao, W., Hevener, K., White, S., Lee, R., Boyett, J.: A statistical framework to evaluate virtual screening. BMC Bioinformatics **10**, 225 (2009)

# Resources for Bioinformatics

# A Mechanistic Study of lncRNA Fendrr Regulation of FoxF1 Lung Cancer Tumor Supressor

Carmen Navarro[1](✉), Carlos Cano[1], Marta Cuadros[2],
Antonio Herrera-Merchan[3], Miguel Molina[4], and Armando Blanco[1]

[1] Department of Computer Science and AI, University of Granada, Granada, Spain
{cnluzon,ccano,armando}@decsai.ugr.es
[2] Department of Biochemistry and Molecular Biology,
University of Granada, Granada, Spain
mcuadros@ugr.es
[3] GENyO, Centre for Genomics and Oncological Research: Pfizer,
University of Granada, Andalusian Regional Government, Granada, Spain
antonio.herrera@genyo.es
[4] Data Science Institute, Imperial College London, London, UK
m.molina-solana@imperial.ac.uk

**Abstract.** Long non-coding RNAs are known to play multiple roles in the complex machinery of the cell. However, their recent addition to genomic research has increased the complexity of gene expression analyses. In this work, we perform a computational study that aims to contribute to the current understanding of the mechanisms that underlie the experimentally suggested interaction between the lncRNA Fendrr and FoxF1 lung cancer tumor suppressor in carcinogenesis. Results suggest that there exists indeed a multi-level interaction between Fendrr and FoxF1 promoter region, both direct via RNA-DNA:DNA triplex domain formation or mediated by proteins that interact simultaneously with the promoter region of FoxF1 and Fendrr transcripts. Moreover, the applied computational methodology can serve as a pipeline to process any candidate lncRNA-gene pair of interest and obtain putative sources of lncRNA-gene interaction.

**Keywords:** lncRNAs · Gene regulation · Computational motif finding · Data integration · DNA-lncRNA interaction

## 1 Background

Carcinogenesis has been extensively studied at a molecular point of view, and has recently entered the era of long non-coding RNAs (lncRNAs), a class of noncoding RNAs that regulate gene expression and control various cellular mechanisms [1,2]. The lncRNA gene Fendrr is located on chromosome 16 (16q24.1). This gene produces a long non-coding RNA transcribed bidirectionally with the forkhead box-F1 (FoxF1) transcription factor on the opposite strand.

© Springer International Publishing Switzerland 2016
F. Ortuño and I. Rojas (Eds.): IWBBIO 2016, LNBI 9656, pp. 781–789, 2016.
DOI: 10.1007/978-3-319-31744-1_67

There is contradictory evidence to support a correlation between Fendrr and FoxF1. Szafranski et al. identified that Fendrr regulates negatively FoxF1 [3], Cabili et al. showed that expression and subcellular localization of Fendrr and FoxF1 mRNA are in the same proportion and compartmented (nucleus and cytoplasm) [4]; Grote and Herrmann indicated that Fendrr interacts with PRC2 complex that inhibit expression of FoxF1 [5].

In the last years, several experimental approaches have emerged that are suitable and necessary for the confirmation or rebuttal of this kind of hypothesis. These include approaches like iClip [6], CLASH [7], CLIP [8], the very recent PAR-CLIP [9] and even the lncRNA-specific CHART [10], which allows to obtain information on lncRNA-protein and DNA interaction. Nonetheless, such approaches involve time and economical costs that could be reduced if a computational prediction that could narrow down the amount of hypotheses to test or discard the unlikely ones was performed beforehand.

To the extent of our knowledge, there are no integrative computational approaches oriented to study this specific case where a lncRNA interacts with a promoter region of a gene. To this end, we applied a two-level based pipeline to assess interaction between a DNA promoter region of interest and a customized set of lncRNA transcripts: both direct and indirect, mediated by transcription factors (TFs) or RNA-binding proteins (RBPs).

## 2    Methods and Data

In order to better understand the molecular mechanisms by which Fendrr may regulate FoxF1 expression in lung cancer, we have performed a sequence-based *in-silico* analysis to assess possible direct and indirect interactions between Fendrr transcripts and the promoter region of FoxF1.

LncRNAs can interact with DNA forming RNA-DNA:DNA triplexes [11], but they can also bind to proteins such as RNA-binding proteins (RBPs) [2] or Transcription Factors (TFs) [12] to perform guidance, scaffolding to higher level structures or acting as a decoy, among others [13]. Furthermore, TFs are known for their role on transcription regulation [14]. Based on this knowledge about how proteins, RNA and DNA may cooperate, we tested computationally two levels of interaction between Fendrr transcripts and the promoter region of FoxF1 to search for plausible hypotheses that can explain the the mechanisms underlying Fendrr putative regulation of FoxF1 gene.

1. **Direct Interaction.** Fendrr transcripts may interact with the promoter region of FoxF1 directly, forming a RNA-DNA:DNA triplex.
2. **Indirect Interaction.** Fendrr transcripts interact with the promoter region of FoxF1 via protein mediators such as RBPs or TFs.

Note that these two levels of interaction are also not mutually exclusive, being possible that triplex-forming transcripts are also interacting with TFs or RBPs working as decoy, sacaffolding or guidance.

## 2.1  Direct Interaction Between Fendrr Transcripts and FoxF1 Promoter

RNA transcripts can interact with DNA forming RNA-DNA:DNA triplexes [11]. These triplexes are bound according to triplex formation rules based on nucleic acid chemical properties [15]. In the past years, a few *in-silico* tools that predict such interactions have been made available, such as Triplexator [16], a method that computationally predicts sites of a given length where a RNA transcript might bind, and Triplex Domain Finder (TDF) [11], a tool that takes as an input a set of DNA promoters candidate to be regulated by a certain RNA transcript and searches for DNA binding domains.

In order to provide further information about whether Fendrr lncRNA transcripts bind directly to the promoter region of FoxF1 and in which manner, we applied Triplexator [16], since it focuses on single sequence analysis. TDF uses it internally among other tools in order to find triplexes in a more genome-wide fashion and uses Fendrr transcripts as input for a positive control, since Fendrr is known to form RNA-DNA:DNA triplexes on several genome locations [5]. In their TDF analysis, the authors claim that FoxF1 shows up as a candidate promoter region to form a triplex with a Fendrr transcript [11], although no explicit information about where or which sequences did match was provided. To this end, we applied Triplexator. Parameters used to this end are very similar to the parameters found in Triplexator's user guide: minimum length of 15 nucleotides, having at most 20 % errors where no consecutive mismatches are allowed, disabling low-complexity filtering (since Fendrr has been suggested to bind to low-complexity repeats [17]) and requiring a guanine percentage of at least 20 %. Table 1 shows matches found for the promoter region of FoxF1. In addition, the same test was performed with the full length of the FoxF1 gene including promoter region (5365 bp sequence) and no further *spurious* results were obtained. Furthermore, these matches overlap partially or totally with conserved regions, suggesting there might indeed exist functionality for this triplex binding.

**Table 1. Triplexator matches for promoter sequence of FoxF1.** Two transcripts of Fendrr show affinity to form a triplex at the promoter region of FoxF1 according to Triplexator. Positions are relative to transcript and promoter start. All four matches form two regions that overlap with conserved regions.

| Transc. ID | Transc. position | DNA position | DNA strand |
|---|---|---|---|
| ENST00000598996.2 | 1928–1943 | 1160–1175 | + |
| ENST00000599749.5 | 146–161 | 1170–1185 | + |
| ENST00000599749.5 | 144–159 | 1253–1268 | + |
| ENST00000599749.5 | 1967–1982 | 1160–1175 | + |

## 2.2    Protein-Mediated Interaction Between Fendrr Transcripts and FoxF1 Promoter

On a slightly more indirect level of interaction, RNA transcripts may interplay with DNA through proteins that act as mediators [1]. There are many ways in which lncRNAs may cooperate with proteins in order to achieve multiple functions. lncRNAs can bind to proteins meant to be bound to a DNA sequence and take them away, acting as a decoy, or they can function as a guidance to such proteins [13]. Moreover, since there is a suggested interaction of lncRNA and the DNA sequence of FoxF1 promoter, it is also of interest to see whether there can be interaction between TFs that might bind to such promoter regions and lncRNAs.

**Transcription Factors as Mediators.** In order to predict likely Transcription Factor candidates to act as mediator between Fendrr and FoxF1, scans with FIMO, a tool for DNA-Motif sequence scoring [18], were performed. FIMO is a statistically robust tool that uses a log-likelihood ratio score, which is afterwards converted to a p-value assuming a zero-order null model. All matrices in JASPAR Core 2016 Vertebrata [19] and HOCOMOCOv9 [20] have been used for this purpose. The parameters used for FIMO scan are default, but kept single stranded for the lncRNA transcripts. Fendrr transcripts have also been scanned for motifs in the hypothesis that a TF might act as a mediator to the interaction between FoxF1 promoter and a Fendrr transcript. In this case, RNA secondary structure has been integrated with this motif scan and Fendrr transcripts secondary structure has been predicted using RNAfold from Vienna Package 2.0 [21] to compare the results for folded and unfolded transcripts.

Once folded, only parts of the transcripts left single-stranded are used for prediction, since there is experimental evidence that proteins can show binding preference to single-stranded regions of folded RNA [22].

Results in the folded transcripts tend to be scarce, whereas result sets are very large when RNA secondary structure is not considered. In order to reduce false positives, we sought a more accurate representation on how proteins may bind to folded RNA, and implemented an intermediate approach, where the preference for single-stranded folded RNA is met but not restricted to only those sequences, has been implemented. We increased the single-stranded RNA sequences provided by RNAfold in a small number of nucleotides around the single stranded regions, considering that a TF might bind partially to such single-stranded RNA sequences [12]. Since the average motif length of JASPAR Core 2016 Transcription Factor database is 10.64, and the average length of single-stranded RNA sequences left by RNAfold in our dataset was of 3.19, we have found that an extension of 1–2 nucleotides around each free single-stranded RNA sequence raises the average to 6.19 and 13, respectively, while still reducing the total length of sequence processed a 40.77 % and 23.51 %, respectively, narrowing down the search space. This helps finding TFBSs that would otherwise be overlooked while still putting the stress on the interest area around free single-stranded RNAs.

Table 2 shows the results in FoxF1 promoter region that are also significantly found in folded Fendrr transcripts (i.e., TFs that show high binding affinity to to both the RNA transcript and the DNA sequence, therefore candidates to lncRNA-promoter mediation). Single-stranded RNA sequences were increased in 2 nucleotides according to the average length of JASPAR motifs. Manual research in the literature has shown that the highlighted TFs, REST and E2F6, are thought to be related to gene repression by Polycomb-group proteins PRC2 and PRC1 recruiting [23]. This, in addition to the close proximity of their putative binding sites (as shown in Fig. 1) in the DNA sequence and the direct RNA-DNA interaction predicted by Triplexator might point to them as main actors in the Fendrr-FoxF1 regulation machinery.

**RNA-Binding Proteins as Mediators.** RNA transcripts often interact with RNA-binding proteins (RBPs) [26], proteins that can take part on many key cellular functions such as RNA transport and cellular localization or biogenesis [27]. The same approach performed to find TFs that might bind to both Fendrr transcript and the FoxF1 promoter region has been applied to RBP motifs present in the CISBP-RNA [14] database, the results were compared to RNA-binding tables. As shown in Table 3, manual search of these results through the literature showed that most of the RBP hits are directly related to RNA-specific function.

**Table 2. Putative mediator TFs between Fendrr transcripts and FoxF1 promoter region.** Top 15 FIMO significant (p-value under $10^{-4}$) results (of 26 total results) that match in both Fendrr transcripts and FoxF1 promoter, for 519 JASPAR Core 2016 Vertebrata matrices [19]. Highlighted results have been found in the literature to be related to Polycomb-group proteins PRC2 and PRC1 recruiting [23].

| Matrix ID | TF Name | RNA pos | DNA pos | RNA p-val | DNA p-val |
|---|---|---|---|---|---|
| MA0146.2 | Zfx | 178–191 + | 1119–1132 + | $1.34 \times 10^{-6}$ | $1.10 \times 10^{-5}$ |
| MA0073.1 | RREB1 | 474–493 + | 1066–1085 + | $2.77 \times 10^{-6}$ | $3.54 \times 10^{-5}$ |
| MA0119.1 | NFIC::TLX1 | 261–274 − | 1213–1226 + | $3.13 \times 10^{-6}$ | $3.81 \times 10^{-5}$ |
| MA0057.1 | MZF1(var.2) | 151–160 + | 883–892 + | $3.59 \times 10^{-6}$ | $9.27 \times 10^{-5}$ |
| **MA0138.2** | **REST** | **2177–2197 −** | **1241–1261 +** | $1.33 \times 10^{-5}$ | $5.70 \times 10^{-5}$ |
| MA0506.1 | NRF1 | 175–185 + | 471–481 + | $2.44 \times 10^{-5}$ | $2.35 \times 10^{-5}$ |
| MA0516.1 | SP2 | 53–67 + | 1352–1366 + | $2.47 \times 10^{-5}$ | $3.02 \times 10^{-6}$ |
| MA0597.1 | THAP1 | 52–60 + | 475–483 + | $2.83 \times 10^{-5}$ | $2.83 \times 10^{-5}$ |
| MA0149.1 | EWSR1-FLI1 | 424–441 + | 895–912 + | $3.13 \times 10^{-5}$ | $3.46 \times 10^{-5}$ |
| MA0162.2 | EGR1 | 84–97 − | 1391–1404 + | $3.18 \times 10^{-5}$ | $3.92 \times 10^{-6}$ |
| MA0747.1 | SP8 | 154–165 − | 1352–1363 + | $3.92 \times 10^{-5}$ | $6.02 \times 10^{-5}$ |
| MA0517.1 | STAT1::STAT2 | 227–241 − | 439–453 + | $4.18 \times 10^{-5}$ | $1.50 \times 10^{-6}$ |
| MA0872.1 | TFAP2A(var.3) | 172–184 + | 464–476 + | $4.47 \times 10^{-5}$ | $8.57 \times 10^{-6}$ |
| MA0039.2 | Klf4 | 706–715 − | 25–34 + | $4.49 \times 10^{-5}$ | $2.12 \times 10^{-5}$ |
| **MA0471.1** | **E2F6** | **2268–2278 +** | **1261–1271 +** | $4.53 \times 10^{-5}$ | $2.66 \times 10^{-6}$ |

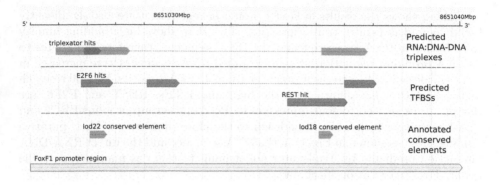

**Fig. 1.** Relative positions between putative regulatory elements involved in Fendrr-FoxF1 promoter interaction. The region depicted corresponds to a 150 bp long section of the FoxF1 promoter located at coordinates chr16:86510250-86510400 of the genome. First track includes triplexator putative RNA:DNA-DNA triplex matches between Fendrr transcripts and FoxF1 promoter region. Second track shows TFBSs E2F6 and REST predicted by FIMO in the mentioned region. These two proximal regions of the genome overlap with two conserved elements on 100 vertebrates conserved track at Genome Browser [24]. The annotations were drawn using AnnotationSketch [25] and manually vectorized.

**Table 3. Putative mediator RBPs between Fendrr transcripts and FoxF1 promoter region.** Significant RNA-binding protein motifs found by FIMO on both the promoter region of FoxF1 and Fendrr transcripts (folded by RNAfold and extended by 2 nt). Repeated matches were filtered. Results highlighted have not been found to be directly related to RNA-specific function.

| ID | RBP name | DNA position | RNA position | DNA p-val | RNA p-val |
|---|---|---|---|---|---|
| M050_0.6 | RBM4 | 1269–1275 + | 9–15 + | $2.96 \times 10^{-5}$ | $2.96 \times 10^{-5}$ |
| M109_0.6 | RBM4B | 1269–1275 + | 9–15 + | $2.96 \times 10^{-5}$ | $2.96 \times 10^{-5}$ |
| M044_0.6 | **PPRC1** | 1342–1348 + | 217–223 + | $2.96 \times 10^{-5}$ | $8.88 \times 10^{-5}$ |
| M151_0.6 | HNRNPH2 | 32–38 + | 89–95 + | $3.60 \times 10^{-5}$ | $7.21 \times 10^{-5}$ |
| M126_0.6 | SRSF4 | 1353–1359 − | 68–74 + | $3.60 \times 10^{-5}$ | $3.60 \times 10^{-5}$ |
| M043_0.6 | **PCBP2** | 682–688 − | 650–656 + | $4.39 \times 10^{-5}$ | $4.39 \times 10^{-5}$ |
| M086_0.6 | SRSF12 | 1179–1185 + | 44–50 + | $5.35 \times 10^{-5}$ | $5.35 \times 10^{-5}$ |
| M177_0.6 | **PCBP1** | 1142–1148 + | 2850–2856 + | $5.35 \times 10^{-5}$ | $5.35 \times 10^{-5}$ |
| M027_0.6 | HNRNPL | 16–22 − | 499–505 + | $6.51 \times 10^{-5}$ | $6.51 \times 10^{-5}$ |
| M169_0.6 | hnRNPLL | 16–22 − | 499–505 + | $6.51 \times 10^{-5}$ | $6.51 \times 10^{-5}$ |

## 2.3  Data

FENDRR annotation and its 7 corresponding transcripts were taken from the last version to date of GENCODE (v23) [28]. The full RNA sequence of these 7 transcripts was used. GENCODE v23 maps to genome version GRCh38, reference from which the 1428 bp long promoter region sequence of FoxF1 at positions chr16:86509099-86510527 was extracted. 1082 TFBS motifs were extracted from JASPAR Core 2016 [19], from which 519 vertebrate motifs are used. 98 Human RBP motifs from CISBP-RNA database [29], and 426 motifs from HOCOMOCOv9, downloaded from their MEME version at MEME-suite website http://meme-suite.org (needed to run with FIMO tool).

# 3  Conclusions and Further Work

In this work, we have applied a computational methodology to explore two levels of possible molecular interaction between a pair of biological entities of interest, transcripts of the lncRNA Fendrr and FoxF1, a near-antisense gene for which there is controversial evidence of interaction with Fendrr, reaching the conclusion that such interaction is likely to occur in two levels, both direct and indirect. Knowing the molecular mechanisms in which lncRNA transcripts can regulate genes is key to understand gene regulation. The abundance of biological entities that usually interact in the cell machinery poses a challenge when detailed mechanistical studies are the goal. In this sense, computational approaches can be of help, reducing the search space and leaving the researchers the development of hypotheses based on the provided multi-level information. Interaction between Fendrr and FoxF1 has been suggested by our proposed pipeline in both direct and indirect levels and our methodology is ready to be applied to any lncRNA-promoter pairs of interest. However, further filtering for the results in the protein-binding approach would be of interest, since many mediator candidates are likely to bind to both DNA sequence and lncRNA transcript. Additional layers of information, such as across-species conservation, histone marks and other epigenetic features can be added to screen for the most interesting putative protein mediators. In addition, it is also known that RNA can act as a scaffold to more complex protein structures [1,30]. The single-protein approach suggested in this work may be extended with protein-protein interaction networks in order to include such higher order mechanisms. Further experimental research is needed to corroborate these results. Nonetheless, they are promising as they are aligned with current general knowledge about RNA function and specific information about Fendrr behaviour. Moreover, the proposed computational approach can be of use for research on any lncRNA-gene pair of interest. Although our approach should work best when there is some external evidence that links a lncRNA to a gene, results can be interesting for any lncRNA proximal to a promoter region of a gene.

**Acknowledgements.** This work has been funded as part of projects PI-0710-2013 of J. A., Sevilla and TIN2013-41990-R of DGICT, Madrid and from FEDER. C. Navarro's work is funded as part of a FPU grant by the Spanish Ministry of Education, Culture and Sports.

# References

1. Geisler, S., Coller, J.: RNA in unexpected places: long non-coding RNA functions in diverse cellular contexts. Nat. Rev. Mol. Cell Biol. **14**(11), 699–712 (2013)
2. Wapinski, O., Chang, H.Y.: Long noncoding RNAs and human disease. Trends Cell Biol. **21**(6), 354–361 (2011)
3. Szafranski, P., Dharmadhikari, A.V., Brosens, E., Gurha, P., Kołodziejska, K.E., Zhishuo, O., Dittwald, P., Majewski, T., Mohan, K.N., Chen, B., et al.: Small noncoding differentially methylated copy-number variants, including lncRNA genes, cause a lethal lung developmental disorder. Genome Res. **23**(1), 23–33 (2013)
4. Cabili, M.N., Dunagin, M.C., McClanahan, P.D., Biaesch, A., Padovan-Merhar, O., Regev, A., Rinn, J.L., Raj, A.: Localization and abundance analysis of human lncR-NAs at single-cell and single-molecule resolution. Genome Biol. **16**(1), 20 (2015)
5. Grote, P., Herrmann, B.G.: The long non-coding RNA Fendrr links epigenetic control mechanisms to gene regulatory networks in mammalian embryogenesis. RNA Biol. **10**(10), 1579–1585 (2013)
6. Konig, J., Zarnack, K., Rot, G., Curk, T., Kayikci, M., Zupan, B., Turner, D.J., Luscombe, N.M., Ule, J.: iCLIP-transcriptome-wide mapping of protein-RNA interactions with individual nucleotide resolution. J. Vis. Exp.: JoVE **50**, e2638 (2011)
7. Kudla, G., Granneman, S., Hahn, D., Beggs, J.D., Tollervey, D.: Cross-linking, ligation, and sequencing of hybrids reveals RNA-RNA interactions in yeast. Proc. Nat. Acad. Sci. **108**(24), 10010–10015 (2011)
8. Darnell, R.: CLIP (Cross-Linking and Immunoprecipitation) identification of RNAs bound by a specific protein. Cold Spring Harb. Protoc. **2012**(11), pdb-prot072132 (2012)
9. Danan, C., Manickavel, S., Hafner, M.: A method for Transcriptome-wide identification of RNA binding protein interaction sites. In: Dassi, E. (ed.) Post-Transcriptional Gene Regulation. MMB, vol. 1358, pp. 153–173. Springer, New York (2016)
10. Simon, M.D., Wang, C.I., Kharchenko, P.V., West, J.A., Chapman, B.A., Alekseyenko, A.A., Borowsky, M.L., Kuroda, M.I., Kingston, R.E.: The genomic binding sites of a noncoding RNA. Proc. Nat. Acad. Sci. **108**(51), 20497–20502 (2011)
11. Hanzelmann, S., Kuo, C.C., Kalwa, M., Wagner, W., Costa, I.G.: Triplex domain finder: detection of triple helix binding domains in long non-coding RNAs (2015). bioRxiv 020297
12. Lee, N., Steitz, J.A.: Noncoding RNA-guided recruitment of transcription factors: a prevalent but undocumented mechanism? BioEssays **37**(9), 936–941 (2015)
13. Wang, K.C., Chang, H.Y.: Molecular mechanisms of long noncoding RNAs. Mol. Cell **43**(6), 904–914 (2011)
14. Vaquerizas, J.M., Kummerfeld, S.K., Teichmann, S.A., Luscombe, N.M.: A census of human transcription factors: function, expression and evolution. Nat. Rev. Genet. **10**(4), 252–263 (2009)
15. Buske, F.A., Mattick, J.S., Bailey, T.L.: Potential in vivo roles of nucleic acid triple-helices. RNA Biol. **8**(3), 427–439 (2011)

16. Buske, F.A., Bauer, D.C., Mattick, J.S., Bailey, T.L.: Triplexator: detecting nucleic acid triple helices in genomic and transcriptomic data. Genome Res. **22**(7), 1372–1381 (2012)
17. Johnson, R., Guigó, R.: The RIDL hypothesis: transposable elements as functional domains of long noncoding RNAs. RNA **20**(7), 959–976 (2014)
18. Grant, C.E., Bailey, T.L., Noble, W.S.: FIMO: scanning for occurrences of a given motif. Bioinformatics **27**(7), 1017–1018 (2011)
19. Mathelier, A., Fornes, O., Arenillas, D.J., Chen, C.Y., Denay, G., Lee, J., Shi, W., Shyr, C., Tan, G., Worsley-Hunt, R., Zhang, A.W., Parcy, F., Lenhard, B., Sandelin, A., Wasserman, W.W.: JASPAR 2016: a major expansion and update of the open-access database of transcription factor binding profiles. Nucleic Acids Res. **44**(Database issue), D110 (2015)
20. Kulakovskiy, I.V., Medvedeva, Y.A., Schaefer, U., Kasianov, A.S., Vorontsov, I.E., Bajic, V.B., Makeev, V.J.: HOCOMOCO: a comprehensive collection of human transcription factor binding sites models. Nucleic Acids Res. **41**(D1), D195–D202 (2013)
21. Lorenz, R., Bernhart, S.H., Zu Siederdissen, C.H., Tafer, H., Flamm, C., Stadler, P.F., Hofacker, I.L., et al.: Viennarna package 2.0. Algorithms Mol. Biol. **6**(1), 26 (2011)
22. Cook, K.B., Kazan, H., Zuberi, K., Morris, Q., Hughes, T.R.: RBPDB: a database of RNA-binding specificities. Nucleic Acids Res. **39**(suppl 1), D301–D308 (2011)
23. Bauer, M., Trupke, J., Ringrose, L.: The quest for mammalian Polycomb response elements: are we there yet? Chromosoma, 1–26 (2015). doi:10.1007/s00412-015-0539-4
24. Rosenbloom, K.R., Armstrong, J., Barber, G.P., Casper, J., Clawson, H., Diekhans, M., Dreszer, T.R., Fujita, P.A., Guruvadoo, L., Haeussler, M., et al.: The UCSC genome browser database: 2015 update. Nucleic Acids Res. **43**(D1), D670–D681 (2015)
25. Steinbiss, S., Gremme, G., Schärfer, C., Mader, M., Kurtz, S.: Annotationsketch: a genome annotation drawing library. Bioinformatics **25**(4), 533–534 (2009)
26. Lunde, B.M., Moore, C., Varani, G.: RNA-binding proteins: modular design for efficient function. Nat. Rev. Mol. Cell Biol. **8**(6), 479–490 (2007)
27. Glisovic, T., Bachorik, J.L., Yong, J., Dreyfuss, G.: RNA-binding proteins and post-transcriptional gene regulation. FEBS Lett. **582**(14), 1977–1986 (2008)
28. Harrow, J., Frankish, A., Gonzalez, J.M., Tapanari, E., Diekhans, M., Kokocinski, F., Aken, B.L., Barrell, D., Zadissa, A., Searle, S., et al.: Gencode: the reference human genome annotation for the encode project. Genome Res. **22**(9), 1760–1774 (2012)
29. Ray, D., Kazan, H., Cook, K.B., Weirauch, M.T., Najafabadi, H.S., Li, X., Guer-oussov, S., Albu, M., Zheng, H., Yang, A., et al.: A compendium of RNA-binding motifs for decoding gene regulation. Nature **499**(7457), 172–177 (2013)
30. Tsai, M.C., Manor, O., Wan, Y., Mosammaparast, N., Wang, J.K., Lan, F., Shi, Y., Segal, E., Chang, H.Y.: Long noncoding RNA as modular scaffold of histone modification complexes. Science **329**(5992), 689–693 (2010)

# Evaluation of Disambiguation Strategies on Biomedical Text Categorization

Mohammed Rais[(✉)] and Abdelmonaime Lachkar

L.I.S.A, Department of Electrical and Computer Engineering,
ENSA, USMBA, Fez, Morocco
{mohammed.rais,abdelmonaime.lachkar}@usmba.ac.ma

**Abstract.** A common and ordinary way of representing a text is as a Bag of its component Words BoW. This Representation suffers from the lack of sense in resulting representations ignoring all semantics that reside in the original text, instead of, the Conceptualization using background knowledge enriches document representation models. While searching polysemic term corresponding senses in semantic resources, multiple matches are detected then introduce some ambiguities in the final document representation, three strategies for Disambiguation can be used: First Concept, All Concepts and Context-Based. SenseRelate is a well-known Context-Based algorithm, which use a fixed window size and taking into consideration the distance weight on how far the terms in the context are from the target word. This may impact negatively on the yielded concepts or senses.

To overcome this problem, and therefore to enhance the process of Biomedical WSD, in this paper we propose a simple modified versions of SenseRelate algorithm named NoDistanceSenseRelate which simply ignore the distance, that is the terms in the context will have the same distance weight.

To illustrate the efficiency of both SenseRelate algorithm and NoDistanceSenseRelate one over the others methods, in this study, several experiments have been conducted using OHSUMED corpus. The obtained results using Biomedical Text Categorization system based on three machine learning models: Support Vector Machine (SVM), Naïve Bayes (NB) and Maximum Entropy (ME) show that the Context-Based methods (SenseRelate and NoDistanceSenseRelate) outperform the others ones.

**Keywords:** Biomedical text-processing · Word sense disambiguation · WSD · Biomedical text categorization · SVM · NB · ME · Bag of words · Bag of concepts · Bow · Boc

## 1 Introduction

A word is ambiguous when it has more than one sense, for example, the word "cold" may refer both to a respiratory disorder and to the absence of heat. Humans can disambiguate the proper sense of a term from its context. Word Sense Disambiguation (WSD) systems use the context information in order to assign a unique sense or concept to an ambiguous term. WSD play a vital role in many biomedical text mining applications such as categorization.

© Springer International Publishing Switzerland 2016
F. Ortuño and I. Rojas (Eds.): IWBBIO 2016, LNBI 9656, pp. 790–801, 2016.
DOI: 10.1007/978-3-319-31744-1_68

Text Categorization is a challenging research area where documents are categorized based on a predefined categories using the content of documents.

Managing the growing number of articles in biomedical databases such as MedLine biomedical digital library has become critical. Applying Text Categorization techniques is essential to organize, facilitate indexing, searching and filtering articles in those databases.

Generally, Text Categorization methods use syntactical and statistical models to represent a text document, and this mainly applies to text classification methods such as: Naïve Bayes (NB), Support Vector Machines (SVMs) and Maximum Entropy (ME).

These syntactical and statistical models suffer from the lack of sense in final document representation due to the absence of all semantics that reside in the original text, instead of, the Conceptualization using background knowledge enriches document representation models.

Nevertheless, in order to conceptualize biomedical documents, assigning a term to its concepts in an ontology can imply in certain cases the risk of the loss of information, because of the natural ambiguity of the Unified Medical Language System (UMLS) thesaurus, then introduce some ambiguities in the final document representation. We distinguish three strategies for Disambiguation: *All Concepts, First Concept and Based-Context*.

In this work, we try to estimate the impact of the three Disambiguation strategies to Text Categorization, and evaluated the contribution of Word Sense Disambiguation (WSD) algorithms to Biomedical Text Categorization using three machine learning models: Support Vector Machine (SVM), Naïve Bayes (NB) and Maximum Entropy (ME). These experiments are realized particularly in the biomedical domain on the OHSUMED corpus, using domain specific knowledge base UMLS.

The remainder of this document is structured as follows: some related works are presented in Sect. 2. The task of Text Conceptualization are presented in Sect. 3. In Sect. 4 we give the architecture used for our evaluation system. The results and discussion for our evaluation system are in Sect. 4. At the end, a conclusion and future works are in Sect. 5.

## 2    Related Work

Document representation is typically based on the traditional approach 'Bag-of-words' (BoW) [1]. Nevertheless the Bag-of-Words or simply Term-Based representation suffers from the lack of sense in final document representation due to the absence of all semantics that reside in the original text. However, Concept-Based (Bag of Concepts) representation permit semantic integration namely Conceptualization that enriches document representation using background knowledge resources.

Amine et al. [2] integrate an ontology on the process of document clustering and conclude that representation based on concepts, provides the best results.

Sanchez et al. [5] present a methodology to build automatically an ontology, extracting information from the World Wide Web from an initial keyword, based on the hypothesis that building a robust ontology design requires the integration of semantics.

Litvak et al. [4] introduce the ontology-based web content mining application for analyzing and classifying web documents in a given domain. The main contribution of this work is using domain-based Multi-Lingual Ontology in the conceptual representation of documents.

Guyot et al. [3] evaluated a multilingual ontology-based approach for multilingual information retrieval.

Mu-Hee et al. [6] suggests an automated method for document classification using an ontology, which expresses terminology information and vocabulary contained in Web documents by way of a hierarchical structure.

In this paper we use UMLS thesaurus as domain specific resources in order to realize the Conceptualization process.

# 3   Text Conceptualization Task

Knowledge resources such as thesaurus or ontologies can be used to resolve the issue of term based representation by replacing with concept-based one. This section presents text Conceptualization task, introducing text preprocessing step, then different possible Conceptualization and Disambiguation strategies.

## 3.1   Text Preprocessing

The first step in text representation is to convert the documents, which its data is strings of characters and words, into a format suitable for the Conceptualization process. Then, extract all the words from the documents and using preprocessing treatments and content extraction to help prepare these words for the Conceptualization process.

## 3.2   Conceptualization

Conceptualization is the process of mapping literally occurring terms detected in text to semantically related concepts, and then the integration of these concepts in text producing the final conceptualized text. In this paper we use UMLS thesaurus as domain specific resources in order to realize this Conceptualization.

Three different strategies can be used for text Conceptualization [1, 7]:

- *Adding Concepts*: this strategy consists in adding the corresponding concepts on each vector of terms.
- *Partial Conceptualization*: this strategy consist of replacing terms by corresponding concepts, indeed, the terms having related concepts in the ontology will be substituted by those concepts, and terms with no concepts are reserved in the document representation.
- *Complete Conceptualization (Concept only)*: similar to the second strategy, but it excludes all terms having no corresponding concepts from the final representation.

In our study, we chose to experiment with the second mapping strategy partial Conceptualization.

## 3.3    Disambiguation Strategies

To add or replace terms by concepts in an ontology can imply in certain cases the risk of the loss of information, because of the natural ambiguity of the Unified Medical Language System (UMLS) thesaurus, then introduce some ambiguities in the final document representation. Albitar et al. [7] distinguishes three Disambiguation strategies:

- *All concepts*: Consists of keeping all the concepts returned by the ontology, which enables an over generation of concepts.
- *First*: This Disambiguation strategy consists in taking only the *First Concept* of the list considered as the best sense.
- *Context*: This strategy assign to the term the preferred sense or concept based on the context in which the ambiguous word is used in the document.

Despite its complexity, the context based strategy is the most recommended one, this strategy has been introduced in the Word Sense Disambiguation task. Word Sense Disambiguation (WSD) methods automatically assign the proper concept to an ambiguous term based on context, next section, we are introducing the Word Sense Disambiguation task and describe the method used in this study.

## 3.4    Word Sense Disambiguation

A word is ambiguous when it has more than one sense, for example, the word "cold" may refer both to a respiratory disorder and to the absence of heat. It is the context in which the word is used that determines its correct meaning.

For Unsupervised Knowledge-Based Word Sense Disambiguation, to determinate the correct meaning (correct concept), generally we use semantic similarity or relatedness measures which attempt to quantify the semantic proximity between two concepts. In our study we use the *SenseRelate* algorithm, this later is an implementation of Patwardhan et al. method [8].

**SenseRelate Algorithm.**    This algorithm is mainly based on the Lesk one to measure gloss overlap [9]. The algorithm attempts to identify the most likely meaning for a word in a given context based on the semantic relations between the target word and its neighboring words in a sentence.

In order to implement *SenseRelate* algorithm, we process as follows: we start by identifying the terms surrounding the ambiguous word using the SPECIALIST Lexicon. Then, we calculate the relatedness between the possible concept of the ambiguous word and each of the surrounding terms using Perl package UMLS: Similarity [10].

The algorithm take into consideration the distance weight on how far the terms in the context are from the target word, by multiplying the reciprocal of this distance to the returned similarity score. This may impact negatively on the yielded concepts or senses.

To overcome this problem, and therefore to enhance the process of Biomedical WSD, in this paper we propose a simple modified versions of *SenseRelate* algorithm named *NoDistanceSenseRelate* which simply ignore the distance, that is the terms in the context will have the same distance weight.

**NoDistanceSenseRelate Algorithm.** This algorithm is a modified versions of *SenseRelate* one, on this algorithm we don't multiply the similarity score by the distance between the term surrounding the ambiguous word and the target word, in so doing the score is then not based on how far it is from the target word. In the next section we are providing an example by way of illustration of *SenseRelate* and *NoDistanceSenseRelate* methods.

**Illustration.** Still as an example, from the NLM's MSH-WSD dataset developed by Jimeno-Yepes et al. [11], consider the following sentence *"We measured the environmental radioactivity and isotope ratio, (87)Sr/(86)Sr ratio, (234)U/(238)U, delta (13)C, and (228) < e>Ra </e >/(226)Ra activity ratio, of adhesion that adhered to a wooden tubing vessel"* containing the Target Word "RA" which has the possible concepts: Refractory anemias [C0002893], Radium [C0034625] and Rheumatoid Arthritis [C0003873].

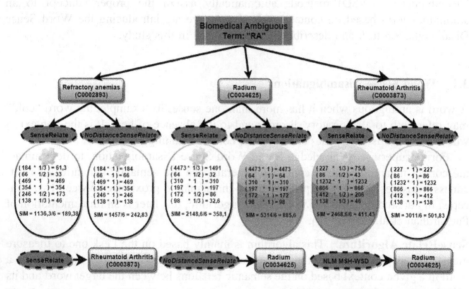

**Fig. 1.** Illustration of WSD Context-based methods: *SenseRelate* and *NoDistanceSenseRelate*

We use a window size of three which refers to three content terms to the right and to the left of the target word and attempt to map them to Concept Unique Identifier (CUI). In this example, the content words are: *"U", "Delta", "C", "right atrium", "Activity"* and *"Ratio"*.

The WSD algorithms (*SenseRelate* and *NoDistanceSenseRelate*) use similarity scores obtained between each of the possible concepts and the concepts of the content words in the window of context. For *SenseRelate* algorithm, each score is then multiplied by the reciprocal of its distance from the target words, for example, the term *"Ratio"* is three content words away from Malaria and therefore multiplied by 1/3.

**Table 1.** Example of adapted lesk relatedness measure

| a-lesk (measure) | Distance | Possible Concepts | | "RA" Possible Concepts | | |
|---|---|---|---|---|---|---|
| | | Label | CUI | C0002893 | C0034625 | C0003873 |
| U | -3 (Before) | Uranium | C0041928 | 94 | 4473 | 107 |
| | | Unit | C0439148 | 184 | 133 | 227 |
| Delta | -2 (Before) | YY1 gene | C1421565 | 66 | 64 | 86 |
| C | -1 (Before) | Catechin | C0007404 | 251 | 192 | 333 |
| | | Cocaine | C0009170 | 469 | 310 | 1232 |
| | | Maxillary right primary canine | C0227087 | 13 | 13 | 19 |
| RA | 0 | Refractory anemias | C0002893 | - | - | - |
| | | Radium | C0034625 | - | - | - |
| | | Rheumatoid Arthritis | C0003873 | - | - | - |
| right atrium | 1 (After) | Right atrial structure | C0225844 | 354 | 197 | 866 |
| | | Entire right atrium | C1269890 | 53 | 47 | 65 |
| Activity | 2 (After) | Activities | C0441655 | 246 | 172 | 412 |
| | | Endurance of Activity | C0600075 | 6 | 5 | 7 |
| | | Percent Activity | C0439167 | 26 | 22 | 27 |
| Ratio | 3 (After) | Ratio | C0456603 | 138 | 98 | 138 |
| | | data type - ratio | C1547037 | 40 | 36 | 45 |

But for *NoDistanceSenseRelate* we don't multiply by the reciprocal of distance. The scores are then summed to obtain a total score for each possible concept as shown in Fig. 1 and Table 1.

After calculating score for each methods we deduct the assigned concept for *SenseRelate* and *NoDistanceSenseRelate*, for this example, experts on MSH-WSD, have chosen the concept "C0034625", which are the concept selected by *NoDistanceSenseRelate*.

## 4    Biomedical Text Categorization System

In order to compare the Disambiguation strategies described above, we developed a Biomedical Text Categorization system to evaluate the contribution of Word Sense Disambiguation methods using three machine learning algorithms: Support Vector Machine (SVM), Naïve Bayes (NB) and Maximum Entropy (ME).

In this Section, we introduce the machine learning algorithms, then we provide dataset and flowchart of our evaluation system.

## 4.1 Machine Learning Algorithms

Once, the Conceptualization process is done and the disambiguation is performed, the final step of Text Categorization consists of document classification. As it is always done in machine learning, to classify a document, we build machine learning models and we expose the document descriptor vectors to be classified on the model created in the learning stage, in order to determine the adequate class or category of the new document. Of course, the document to classify will be processed in the same way as the other documents of the learning phase. In this paper, we propose to use Support Vector Machine (SVM), Naïve Bayes (NB) and Maximum Entropy (ME) machine learning models on our evaluation system. Next section presents the dataset used in our evaluation system.

## 4.2 Data

Ohsumed corpus or test collection [12], is a subset of the MEDLINE database, which is a bibliographic database of important, peer-reviewed medical literature maintained by the National Library of Medicine. We consider 20,000 documents from the 50,216 medical abstracts of the year 1991, devised into 23 Medical Subject Headings (MeSH) categories of cardiovascular diseases group. Of the 23 categories of the cardiovascular diseases group, we consider 10 categories for our evaluation system: Bacterial Infections and Mycoses, Virus Diseases, Parasitic Diseases, Neoplasms, Musculoskeletal Diseases, Digestive System Diseases, Stomatognathic Diseases, Respiratory Tract Diseases, Otorhinolaryngologic Diseases, Nervous System Diseases.

## 4.3 Flowchart System

As mentioned above, we use Ohsumed [12] as dataset to evaluate the contribution of Disambiguation strategies to Biomedical Text Categorization. Figure 2, presents the flowchart of the evaluation system.

In this system, we proceed as follows:

(a) Executing Term Extractor module for biomedical domain which attempt to extract biomedical terms from the corpus using UMLS as dictionary.

(a) Mapping all terms to their corresponding concepts using The UMLS Terminology Services (UTS) provide Web Services to search and retrieve UMLS data.
(b) The Conceptualization process is carried out, and for polysemic terms, we use the strategies for disambiguation: First Concept, All Concepts, Context-Based (*SenseRelate*, *NoDistanceSenseRelate*). So for each strategies, the document will be represented by a Bag of Concepts.
(c) Training vectors of document using machine learning models.
(d) Using these classifiers to categorize testing documents vectors.

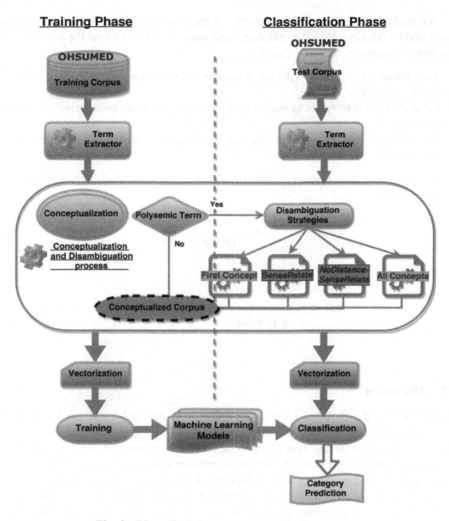

**Fig. 2.** Biomedical documents categorization system

# 5 Results and Discussion

Given the previously presented system, we have performed experiments in order to evaluate the effect of Disambiguation strategies on Biomedical Text Categorization using three machine learning algorithms: NB, SVMs and ME.

In this section, we introduce the performance metrics, classification results obtained and finally we analyze as well as discuss these results.

## 5.1 Results

In our experiments, we have tested all methods on the OHSUMED corpus after Conceptualization that is realized according to four different disambiguation strategies.

Three machine learning algorithms we used in experiments: SVMs, NB and ME.

We evaluated the category assignments of the classifier system by using precision, recall and F1-Measure [13]. To calculate these metrics, we first found the values of the following parameters for each classifier:

- True Positive (TP) refers to the set of documents which are correctly assigned to the given category.
- False Positive (FP) refers to the set of documents which are incorrectly assigned to the category.
- False Negative (FN) refers to the set of documents which are incorrectly not assigned to the category.

Then, we calculated precision and recall for each of the classifiers with the following formulas:

$$P = \frac{TP}{TP + FP} \tag{1}$$

$$R = \frac{TP}{TP + FN} \tag{2}$$

$$F1measure = 2\frac{P * R}{P + R} \tag{3}$$

## 5.2   Discussion

Table 2 shows the obtained F1-measure using ME, NB and SVM for OHSUMED Documents Categorization. As mentioned above we use ten categories from OHSUMED corpus for the Text Categorization System using first the three classification methods on original OHSUMED corpus without Conceptualization (Term Based), Then methods are tested using each of the three different Disambiguation strategies: All Concepts, First Concept, Context-Based (SenseRelate and NoDistanceSenseRelate).

As illustrated in the table, in all cases, we can observe that disambiguation using **SenseRelate** and **NoDistanceSenseRelate** improves the outcome.

Considering each category independently in the results, we can observe that **NoDistanceSenseRelate** outperforms other strategies in about 94 % of cases for the categories *"Bacterial Infections and Mycoses", "Virus Diseases", "Digestive System Diseases"* and *"Respiratory Tract Diseases"* using NB, ME and SVM models. Meanwhile **SenseRelate** outperforms other strategies in about 93 % for the categories *"Parasitic Diseases", "Neoplasms", "Musculoskeletal Diseases", "Stomatognathic Diseases"* and *"Otorhinolaryngologic Diseases Nervous"* using NB, ME and SVM models.

**NoDistanceSenseRelate** strategy outperforms traditional representation (Term Based) and also **First Concept** and **All Concepts**, achieving an F1-Measure of 81.40 % on *"Virus Diseases"* category, it presents a significant improvement of 3,80 % over **Term Based** representation, 4,47 % over the **First Concept** strategy and 5,36 % over **All Concepts** strategy when using **ME**.

**Table 2.** Ohsumed disambiguation strategies using machine learning models

| | | Bacterial Infections and Mycoses | | Virus Diseases | | Parasitic Diseases | | Neoplasms | | Musculoskeletal Diseases | |
|---|---|---|---|---|---|---|---|---|---|---|---|
| | | F1-Measure | Baseline Gap | F1-Measure | Baseline Gap | F1-Measure | Baseline Gap | F1-Measure | Baseline Gap | F1-Measure | Baseline Gap |
| ME Classifier | Term-Based (Baseline) | 72,90% | | 77,60% | | 87,71% | | 70,94% | | 74,35% | |
| | First Concept | 72.08% | -0,83% | 76.85% | -0,75% | 70.27% | +0,58% | 70,27% | -0,67% | 70,27% | -0,35% |
| | All Concepts | 71.18% | -1,72% | 75.31% | -2,28% | 70.77% | -1,08% | 70,77% | -0,16% | 70,77% | -1,70% |
| | SenseRelate | 76.53% | +3,63% | 81.26% | +3,66% | 74.85% | +3,24% | 74,85% | +3,91% | 74,85% | +4,08% |
| | NoDistanceSenseRelate | 76.55% | +3,64% | 81.40% | +3,80% | 74.65% | +2,97% | 74,65% | +3,71% | 74,65% | +4,01% |
| NB Classifier | Term-Based (Baseline) | 72,14% | | 76,91% | | 87,76% | | 69,66% | | 76,07% | |
| | First Concept | 71.35% | -0,78% | 77.14% | +0,23% | 88.08% | +0,32% | 69,45% | -0,21% | 76,24% | +0,16% |
| | All Concepts | 69.75% | -2,38% | 76.32% | -0,58% | 86.69% | -1,07% | 69,63% | -0,03% | 75,59% | -0,48% |
| | SenseRelate | 75.74% | +3,60% | 81.72% | +4,81% | 91.03% | +3,27% | 74,44% | +4,77% | 80,84% | +4,76% |
| | NoDistanceSenseRelate | 75.78% | +3,65% | 81.74% | +4,83% | 90.82% | +3,06% | 74,17% | +4,51% | 80,39% | +4,32% |
| SVM Classifier | Term-Based (Baseline) | 74,17% | | 78,43% | | 88,98% | | 71,13% | | 78,24% | |
| | First Concept | 73.68% | -0,49% | 78.36% | -0,07% | 89.09% | +0,11% | 71,79% | +0,66% | 76,78% | -1,46% |
| | All Concepts | 72.42% | -1,75% | 77.83% | -0,60% | 87.46% | -1,52% | 72,53% | +1,40% | 76,14% | -2,10% |
| | SenseRelate | 78.31% | +4,14% | 80.88% | +2,45% | 92.41% | +3,43% | 76,44% | +5,31% | 81,35% | +3,11% |
| | NoDistanceSenseRelate | 78.36% | +4,19% | 80.94% | +2,51% | 92.22% | +3,24% | 74,49% | +3,36% | 81,27% | +3,03% |

| | | Digestive System Diseases | | Stomatognathic Diseases | | Respiratory Tract Diseases | | Otorhinolaryngologic Diseases | | Nervous System Diseases | |
|---|---|---|---|---|---|---|---|---|---|---|---|
| ME Classifier | Term-Based (Baseline) | 74,63% | | 74,66% | | 68,18% | | 77,96% | | 69,44% | |
| | First Concept | 73,55% | -1,08% | 76,24% | +1,58% | 69,82% | +1,64% | 78,17% | +0,21% | 69,27% | -0,16% |
| | All Concepts | 72,74% | -1,89% | 75,81% | +1,15% | 69,03% | +0,85% | 77,09% | -0,88% | 69,76% | +0,32% |
| | SenseRelate | 77,88% | +3,25% | 80,82% | +6,16% | 74,24% | +6,07% | 82,47% | +4,51% | 73,92% | +4,48% |
| | NoDistanceSenseRelate | 78,18% | +3,54% | 80,75% | +6,09% | 74,42% | +6,25% | 82,43% | +4,47% | 73,83% | +4,39% |
| NB Classifier | Term-Based (Baseline) | 76,16% | | 80,12% | | 71,28% | | 78,21% | | 71,99% | |
| | First Concept | 76,14% | -0,02% | 80,00% | -0,13% | 71,18% | -0,09% | 77,48% | -0,73% | 72,04% | +0,05% |
| | All Concepts | 75,45% | -0,71% | 79,60% | -0,52% | 70,47% | -0,81% | 77,52% | -0,69% | 71,31% | -0,68% |
| | SenseRelate | 80,47% | +4,32% | 84,40% | +4,28% | 75,54% | +4,26% | 81,83% | +3,62% | 76,70% | +4,71% |
| | NoDistanceSenseRelate | 80,69% | +4,53% | 84,24% | +4,12% | 75,59% | +4,32% | 81,91% | +3,69% | 76,32% | +4,33% |
| SVM Classifier | Term-Based (Baseline) | 78,76% | | 80,93% | | 70,50% | | 81,59% | | 71,43% | |
| | First Concept | 77,26% | -1,50% | 81,91% | +0,98% | 71,55% | +1,06% | 80,65% | -0,95% | 71,26% | -0,17% |
| | All Concepts | 77,21% | -1,55% | 80,99% | +0,06% | 70,79% | +0,29% | 79,39% | -2,21% | 71,41% | -0,01% |
| | SenseRelate | 81,96% | +3,20% | 83,60% | +2,67% | 76,63% | +6,13% | 84,98% | +3,39% | 76,06% | +4,63% |
| | NoDistanceSenseRelate | 82,14% | +3,38% | 83,29% | +2,36% | 76,16% | +5,66% | 85,16% | +3,56% | 75,90% | +4,47% |

*SenseRelate* strategy outperforms **Term Based**, **First Concept** and **All Concepts**, achieving an F1-Measure of 76,70 % on *"Nervous System Diseases"* category, it presents a significant improvement of 4,66 % over the **First Concept** Strategy, 4,71 % over **Term Based** representation, and 5,39 % over **All Concepts** strategy when using **NB**.

The outcomes of **Term Based** representation in general is very close to the **First Concept** performance, while the two strategies outperforms **All concept** strategy in about 70 % of cases for all categories due to the increase of the representation space.

The overall results for all Disambiguation strategies show that classification using **SVM** model obtain higher score, achieving an F1-measure of 81,96 % on *"Digestive System Diseases"* using **SenseRelate**, it presents an improvement of 1,49 % over the second best model **NB**, compared to **ME**, **NB** presents an improvement of 2,59 %.

# 6 Conclusion and Perspectives

Word Sense Disambiguation (WSD) play a vital role in many biomedical text mining applications such as categorization. *SenseRelate* is a well-known WSD Context-Based algorithm, which use a fixed window size and taking into consideration the distance weight on how far the terms in the context are from the target word. This may impact negatively on the yielded concepts or senses. To overcome this problem, and therefore to enhance the process of Biomedical WSD, in this paper we have proposed a simple modified versions of *SenseRelate* algorithm named *NoDistanceSenseRelate* which simply ignore the distance, that is the terms in the context will have the same distance weight.

To illustrate the efficiency of our proposition, both *SenseRelate* and *NoDistanceSenseRelate* algorithms have been integrated in our Biomedical Text Categorization system. Several experiments have been conducted using three machine learning models: Support Vector Machine (SVM), Naïve Bayes (NB) and Maximum Entropy (ME).

The obtained results using OHSUMED corpus show that the Context-Based methods (*SenseRelate* and *NoDistanceSenseRelate*) outperform the others ones.

In future work, we propose to integrate the Context-Based strategy for Biomedical WSD in a new Biomedical Information Retrieval System (IRS).

# References

1. Elberrichi, Z., Taibi, M., Belaggoun, A.: Multilingual Medical Documents Classification Based on MesH Domain Ontology. CoRR abs/1206.4883 (2012)
2. Amine, A., Elberrichi, Z., Simonet, M.: Evaluation of text clustering methods using WordNet. Int. Arab J. Inf. Technol. **7**, 351 (2010)
3. Guyot, J., Radhoum, S., Falquet, G.: Ontology-based multilingual information retrieval. In: CLEF (2005)

4. Litvak, M., Last, M., Kisilevich, S.: Improving classification of multi-lingual web documents using domain ontologies. In: The Second International Workshop on Knowledge Discovery and Ontologies, KDO05, Porto, Portugal, October 7th 2006
5. Sanchez, D., Moreno, A.: Creating ontologies from Web documents. In: Recent Advances in Artificial Intelligence Research and Development, vol. 113, pp. 11–18. IOS Press (2004)
6. Song, M.-H., Lim, S.-Y., Park, S.-B., Kang, D.-J., Lee, S.-J.: An automatic approach to classify web documents using a domain ontology. In: Pal, S.K., Bandyopadhyay, S., Biswas, S. (eds.) PReMI 2005. LNCS, vol. 3776, pp. 666–671. Springer, Heidelberg (2005)
7. Albitar, S., Fournier, S., Espinasse, B.: The Impact of Conceptualization on Text Classification
8. Patwardhan, S., Banerjee, S., Pedersen, T.: Using measures of semantic relatedness for word sense disambiguation. In: Proceedings of the Fourth International Conference on Intelligent Text Processing and Computational Linguistics, pp. 241–57 (2003)
9. Lesk, M.: Automatic sense disambiguation using machine readable dictionaries: How to tell a pine cone from an ice cream cone. In: Proceedings of the 5th Annual International Conference on Systems Documentation, pp. 24–6 (1986)
10. McInnes, B.T., Pedersen, T., Pakhomov, S.V.S., Liu, Y., Melton-Meaux, G.: UMLS: Similarity: Measuring the relatedness and similarity of biomedical concepts. In: HLT-NAACL, pp. 28–31 (2013)
11. Jimeno-Yepes, A., McInnes, B., Aronson, A.: An unsupervised vector approach to biomedical term disambiguation: Integrating umls and medline. BMC Bioinform. 12(1), 223 (2011)
12. Hersh, W., et al.: OHSUMED: An interactive retrieval evaluation and new large test collection for research. In: 17th Annual International ACM SIGIR Conference on Research and Development in Information Retrieval, pp. 192–201. New York, Inc., Dublin (1994)
13. Sokolova, M., Lapalme, G.: A systematic analysis of performance measures for classification tasks. Inf. Process. Manag. 45(4), 427–437 (2009)

# Biomolecular Annotation Integration and Querying to Help Unveiling New Biomedical Knowledge

Arif Canakoglu, Stefano Ceri, and Marco Masseroli[✉]

Dipartimento di Elettronica, Informazione e Bioingegneria,
Politecnico di Milano, 20133 Milan, Italy
{arif.canakoglu,stefano.ceri,marco.masseroli}@polimi.it

**Abstract.** Targeting biological questions requires comprehensive evaluation of multiple types of annotations describing current biological knowledge; they are increasingly available, but their fast evolution, heterogeneity and dispersion in many different sources hamper their effective use. Leveraging on innovative flexible data schema and automatic software procedures that support the integration of data sources evolving in number, data content and structure, while assuring quality and provenance tracking of the integrated data, we created a multi-organism Genomic and Proteomic Knowledge Base (GPKB) and easily maintained it updated. From several well-known databases it imports and integrates very numerous gene and protein data, external references and annotations, expressed through multiple biomedical terminologies. To easily query such integrated data, we developed intuitive web interfaces and services for programmatic access to the GPKB; they are publicly available respectively at http://www.bioinformatics.deib.polimi.it/GPKB/ and http://www.bioinformatics.deib.polimi.it/GPKB-REST/. The created GPKB is a very valuable resource used in several projects by many users; the developed interfaces enhance its relevance to the community by allowing the seamlessly composition of queries, although complex, on all data integrated in the GPKB, which can help unveiling new biomedical knowledge.

**Keywords:** Heterogeneous and distributed biological data management and integration · Biomedical ontologies · Biomolecular annotations querying

## 1   Background

Several biological questions can be addressed only by comprehensively analyzing different types of data and available knowledge. For example, the identification of biomolecular phenomena involved in a specific biological condition requires the evaluation of several different structural, functional and phenotypic characteristics (i.e. annotations) of numerous genes, e.g. differentially expressed in that biological condition. Although spread into many data sources, such biomolecular data and knowledge are increasingly available in structured or semi-structured form [1]. Among them, controlled biomolecular annotations are of great value to describe structural, functional and phenotypic features of genes and proteins; they effectively help the biomedical

© Springer International Publishing Switzerland 2016
F. Ortuño and I. Rojas (Eds.): IWBBIO 2016, LNBI 9656, pp. 802–813, 2016.
DOI: 10.1007/978-3-319-31744-1_69

interpretation of biomolecular test results and the extraction of new biomedical knowledge. Combining information from multiple diverse sources and executing complex source-independent queries on them to derive knowledge are hence paramount tasks. Yet, the dispersion of biomolecular annotations in many complementary but partially overlapping resources, the data source evolution in number, content and structure, the great variety of data types, the high amount of data and the requirements for their efficient and effective use represent very important research challenges to face for the integration of these data [2, 3].

Several approaches and implementations have been proposed to integrate distributed heterogeneous biological data from multiple sources; they include BioMart [4], TAMBIS [5], BioDataServer [6], Biomediator [7], BioWarehouse [8], BIOZON [9] and Bio2RDF [10]. Among them, just as an example, BioMart adopts a federated database system approach, and provides a unified access to disparate, geographically distributed data sources by organizing data from individual sources into one query-optimized system. It has the positive feature to be data agnostic and platform independent, which makes easier to federate additional data sources; yet, its federated approach, although useful to query and retrieve distributed data, does not support full data integration, comprehensive processing, and quality checking. Similar comment applies to TAMBIS, which also adopts a federated approach, as well as to BioDataServer and Biomediator which instead use a mediator-based approach. Furthermore, both these approaches are not suitable when off-line processing is required to efficiently and comprehensively mine the integrated data. Conversely, the data warehousing architecture, which BioWarehouse and BIOZON implement, well supports it through an integrative data management approach consisting of an *a priory* integration and reconciliation of data extracted from multiple sources.

Despite many efforts, creating a public, high-quality consistent integration of reconciled distributed annotations and maintaining it updated remain a challenge not completely solved [11, 12]. When the data to be integrated are very numerous and off-line processing is required to efficiently and comprehensively mine the integrated data, data warehousing is the advisable approach. In a data warehouse, integration is performed off-line, according to a predefined data model that gives a unified global view of the data. Yet, when data in the original sources vary frequently, as often occurs in biomolecular databases, data warehousing requires a committed maintenance effort and automatic procedures to retrieve the new data from the integrated distributed sources and keep the data warehouse updated [13]. Some global schemas for biological data have been proposed [14, 15]; generally, they are very expressive, but quite complex, mainly suitable for organism or tissue specific research projects. They make difficult to face the above mentioned data integration challenges, both in maintaining the data warehouse that adopts one of such data schemas updated and in expanding it with additional data and data types from new sources. Furthermore, they usually do not provide good support for provenance and versioning [16].

Targeting the above issues and challenges by focusing on integration of genomic and proteomic controlled annotations of different species from multiple sources, we previously developed a software framework to create and easily maintain updated a Genomic and Proteomic Data Warehouse (GPDW) [17], i.e. a high-quality and consistent integration of reconciled heterogeneous and distributed annotations. Here, we present and discuss our work aimed at (1) leveraging the GPDW framework to create, keep updated and progressively extend a knowledge base of genomic and proteomic annotations from multiple sources, and (2) developing easy-to-use web interfaces and services to give easy access to such knowledge base and enable users to comprehensively query the integrated data for biomedical knowledge extraction and discovery. The results of these aims provide to the community easy access to a very valuable unique resource which proved very useful in multiple applications and projects.

## 2   Integrative Genomic and Proteomic Knowledge Base

By taking advantage of the recently developed GPDW software framework [17], we created an integrative Genomic and Proteomic Knowledge Base (GPKB) in a PostgreSQL relational data base management system. The used framework leverages the warehousing approach to thoroughly check quality and consistency of the data to be integrated; these are crucial aspects to be then able to reliably use the integrated data for relevant information extraction and knowledge discovery. Furthermore, the framework greatly limits warehousing difficulties in keeping the data warehouse up to date and expanding it with additional data and data types from new sources. Towards this goal, in GPKB we adopted a global integrative data schema composed of multiple interconnected modules [17], which eases maintenance of the integrated data and assures their provenance tracking. Briefly, each module represents a single feature, whose data are provided by one or more of the integrated data sources (Fig. 1). A feature can be a biomolecular entity (i.e. DNA sequence, gene, transcript, or protein), or a biomedical characteristic (e.g. biological function, pathway, genetic disorder, etc.). Each feature instance (e.g. a specific gene or biological function) is identified by the value of its Source ID and Source name attributes and contains the Reference attribute, representing the source that provided the data (i.e. their provenance), which enable users to assess their confidence in the data. Each feature module can also include history and/or similarity data (not shown in Fig. 1).

To ease maintenance and extension of the global data schema, each feature module is internally organized in two levels: an *import level* and an *aggregation level*. The former is composed of separated sub-schemas, each one for every single data source considered which provides data for that feature. These sub-schemas are individually structured as in the original data source, i.e. in a global-as-view (GAV) data integration manner. The import level allows positioning together originally dispersed data, while their consistency and quality is thoroughly checked [18], and identifying the feature they refer to and their main attributes. This lets automatic aggregation of the main attribute data of each feature source in the aggregation level, where they are associated with a unique OID, while duplicate instances are identified. Furthermore, it eases maintenance and expansion of the global data schema: if data schema variations occur in the original data sources, or as consequence of the integration of an additional data source, they can be easily managed, since they affect only the source specific part of the data global schema. Then, for all feature modules the aggregation level of the global schema is automatically derived from the modified import level.

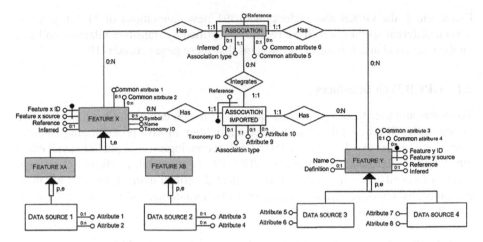

**Fig. 1. General Entity-Relationship diagram of two associated multi-level feature modules of the integrated global data schema.** White shapes represent import level data, whereas dark shapes represent aggregation level data. As usual, boxes represent entities, rhombuses represent relationships, empty circles represent attributes and full circles represent unique identifiers of entity instances; values on lines connecting boxes and rhombuses represent relationship cardinalities; values on lines connecting circles to boxes represent attribute cardinalities.

Feature modules are pair wise associated through association/annotation data, which are as well structured in an import and an aggregation level (Fig. 1). In the latter one, the association data, which are stored in the import level as feature instance ID pairs, are automatically translated into pairs of unique OIDs and matched to the feature instance OIDs of the two associated features. Leveraging the adopted data schema and the included history and similarity data, the used framework also supports reconciliation of unsynchronized data from different sources and identification of multiple data instances as representing the same data, providing remarkable assets to our GPKB that are rarely present in other integrative systems. They are paramount to ensure high quality and consistency of the data integrated in the GPKB and to be able to subsequently use them for reliable analysis. Additionally, the framework allows defining metadata that describe all data and their sources and relationships integrated in the GPKB; such metadata are stored in a metadata portion of the data warehouse data schema. Thanks to the modular nature of the used global data schema and the defined procedures for its creation, the data warehouse structure can be automatically derived easily from the defined metadata.

The GPKB ambitious goal is to provide an updated consistent integration of numerous heterogeneous annotations from multiple data sources. At the time of writing, the GPKB integrated very many, high-quality reconciled data (Table 1): they regard multiple features, including biomolecular entities and their associations and annotations with many different biomedical-molecular characteristics. All these data are down-loaded from several well-known databases, carefully selected according to their renowned relevance and reliability of the provide data, which include Entrez Gene, UniProt, IntAct, Expasy Enzyme, GO, GOA, BioCyc, KEGG, Reactome and OMIM.

Furthermore, the GPKB also includes 236,391 new annotations of 51,286 genes of several different species, which we recently identified by transitive relationship based on the integrated annotations of the proteins that these genes encode [19].

## 2.1 GPKB Web Interfaces

To enable any user to easily compose queries, although complex, on all the data integrated in the GPKB, we developed a web application in Java programming language using Java servlet and Java Server Page (JSP) technologies; it is publicly available at http://www.bioinformatics.deib.polimi.it/GPKB/. Its graphical interfaces are automatically composed according to the GPKB content described in the defined GPKB metadata; thus, the visualized features and their attributes automatically adapt to the specific GPKB instance, transparently to the user.

**Table 1.** Biomolecular entities and their associations and annotations with biomedical-molecular characteristics integrated in the Genomic and Proteomic Knowledge Base

| | Number of items (*Homo sapiens*) | Number of organisms | Total annotations (*Homo sapiens*) | Gene annotations (*Homo sapiens*) | Protein annotations (*Homo sapiens*) |
|---|---|---|---|---|---|
| DNA Sequences | 1,841,019 (37,311) | 13,724 | 44,039,462 (787,489) | 19,533,124 (270,083) | 12,294,437 (126,574) |
| Genes | 17,483,280 (47,760) | 14,992 | 61,909,745 (825,336) | 176,315 (623) | 12,057,465 (21,524) |
| Transcripts | 9,686,716 (106,957) | 490 | 25,218,329 (546,440) | 9,754,807 (107,697) | 3,251,621 (47,911) |
| Proteins | 13,400,602 (21,134) | 13,374 | 58,306,206 (734,317) | 12,057,465 (21,524) | 302,131 (66,584) |
| Enzymes | 5,862 | 7,904 | 511,094 (6,597) | 231,082 (629) | 273,130 (4,627) |
| Biological Functions (Gene Ontology Terms) | 41,829 | 1,454 | 42,164,940 (546,389) | 16,621,192 (289,260) | 25,469,166 (234,629) |
| Biochemical pathways | 359,500 | 1,892 | 2,578,089 (207,348) | 986,510 (99,993) | 1,562,746 (106,859) |
| Genetic disorders | 7,936 | 1 | 42,351 (42,351) | 12,326 (12,326) | 4,729 (4,729) |
| Clinical synopses | 99 | 1 | 26,338 (26,338) | 471 (471) | 571 (571) |

Through the developed *Easy search* interface (Fig. 2), to compose a query the user is only required to select, out of the list of the features integrated in the GPKB, the ones to be included in the query and, if necessary for the specific query, to specify the filtering conditions on the data values to be retrieved. Users can graphically compose queries involving any number and combination of features in the GPKB, according to the feature relationships described in the GPKB metadata and defined by the annotation data integrated in the GPKB; for any selected feature, the list of related features that can be further selected is shown in the interface.

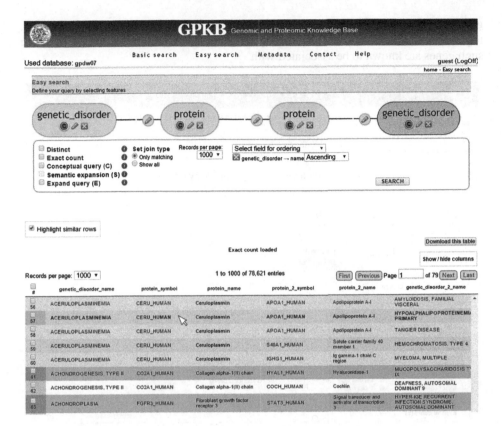

**Fig. 2. GPKB** *Easy search* **interface showing a graphically composed query and its obtained results.** The query shown extracts all the genetic disorders associated with interacting proteins and shows their names and the associated interacting proteins. Selecting the "*Highlight similar rows*" option and positioning the mouse pointer on a protein name highlights all the retrieved results which regard that protein.

Query results are displayed in tabular format in the web interface, where the user can interactively modify their ordering and the one of their attributes/columns in the result table for best viewing. Furthermore, by selecting the "*Highlight similar rows*" option and positioning the mouse pointer on the value of a result column, the user can highlight all the retrieved results sharing that same value of that attribute. Using the "*Download this table*" button, obtained results can be downloaded in tabular-separated file format for their subsequent easy use. Figure 2 shows the results (figure bottom part) of a graphically composed query (figure upper part) to extract from the GPKB the genetic disorders associated with interacting proteins. The default retrieved attributes of each feature/association can be modified by clicking on the "*pencil*" icon on the feature/association. This makes appearing a popup window where it is possible to select the attributes to be shown in the query result and to define their filtering values according to their values integrated in the GPKB. Multiple annotation extraction is also possible from a single query. As an example, in a single call, the query in Fig. 3 extracts all the genes and their known associated genetic disorders, with the related clinical synopses, together with all the enzymes encoded by the same genes and the pathways in which such genes are known to be associated with.

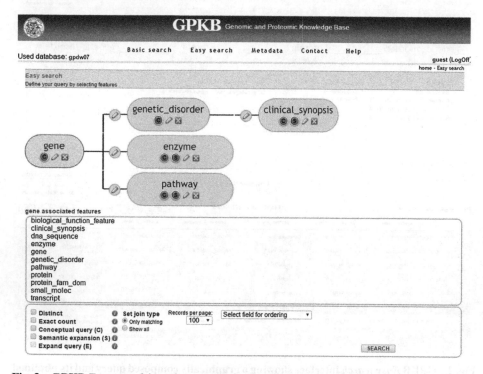

**Fig. 3.  GPKB** *Easy search* **interface: multiple annotation extraction.** It is shown a graphically composed query to extract from the GPKB all the integrated genes and their known associated genetic disorders, with the related clinical synopses, and also all the enzymes that the same genes encode and the pathways in which such genes are known to be associated with.

We also developed another web interface, named *Basic search*, which can be used to straightforwardly extract all the associations between a single specific feature instance and all the other feature instances integrated in the GPKB; for example, all the associations of a single gene, e.g. the *D-dopachrome tautomerase-like* (DDLT) human gene (with Entrez Gene ID 100037417), to all other features integrated in the GPKB, i.e. DNA sequences, transcripts, proteins, enzymes, biological function features, pathways, genetic disorders and clinical synopses. In particular, with this interface it is easy to visualize the new gene annotations that we identified by using a transitive relationship based algorithm [19] and stored in the data warehouse (Fig. 4).

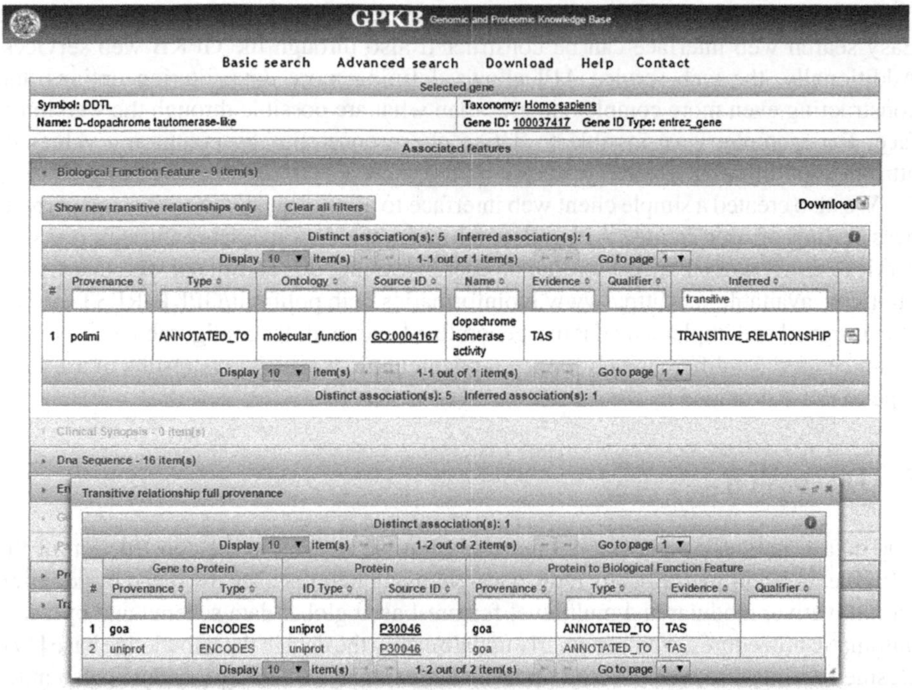

**Fig. 4. GPKB *Basic search* interface result page.** Results of the query for the *D-dopachrome tautomerase-like* (DDTL) human gene are shown; by clicking the "*Show new transitive relationships only*" button, out of the 5 retrieved biological function feature annotations of the gene, only the new one identified by transitive relationship, *dopachrome isomerase activity (GO: 0004167)*, is shown. In the dialog window with title "Transitive relationship full provenance", evidence of this inference is shown; DDTL human gene encodes the protein with UniProt code P30046 and the protein is annotated to the Gene Ontology (GO:0004167).

## 2.2 GPKB Web Services

In order to make possible also the programmatic access to the GPKB, we developed Representational State Transfer (REST) compliant web services. They can be used to perform queries over the data integrated in the GPKB, and extract them within well-defined eXtensible Markup Language (XML) files. This makes the GPKB interoperable with other computational systems and usable also within workflow management systems, such as Galaxy [20] or Taverna [21], through their simple XML parsing tools.

The developed web services, and a help page to use them, are publicly available at http://www.bioinformatics.deib.polimi.it/GPKB-REST/. They can be used first to query the GPKB metadata for the features available in the data warehouse and for the features associated with each of them. Then, through an Application Programming Interface (API) call, such extracted information can be used to query the data integrated in the GPKB, as in the developed GPKB web interfaces, and retrieve the query results in XML format. All queries, although complex, that can be performed through the developed Easy search web interface can be constructed also through the GPKB web services. Additionally, the web service API allows defining more data filtering options and constructing even more complex queries than what are possible through the web interface; for example, it is possible to define query constrains by comparing values of attributes of different selected features.

We also created a simple client web interface to allow easy composition and correctness testing of user-created API calls, and make the data extracted through our web services easily available also for human user browsing; this additional web interface is publicly available at http://www.bioinformatics.deib.polimi.it/GPKB-REST-client/. We designed and implemented it in a generalized and modular configurable way, so that it can be used for different web service implementations which give results of a similar type to those provided by the GPKB.

## 3   Discussion

The data warehousing integration approach that we applied in the construction of the GPKB enabled us to drastically reduce warehousing maintenance overhead by adopting an innovative, modular and multilevel feature-based global data schema and by using automatic procedures, which regularly update easily the data in the knowledge base [17]. Besides easing data warehousing updates and extensions, these procedures also automatically perform thorough data quality and consistency checking [18], as well as reconciliation of unsynchronized data, in order to integrate in the GPKB only high quality consistent data [17]; both these aspects are paramount to subsequently use the integrated data for reliable comprehensive evaluations. Furthermore, they also ensure provenance tracking of all the data integrated in the GPKB, which is fundamental for their proper subsequent processing and for assessing the user confidence in the interpretation of processing results.

Despite the huge and increasing number of data integrated in the GPKB, its design and the automatic procedures used for its maintenance ensure smooth performing of updates and data synchronization among the integrated sources. The great amount of high-quality heterogeneous genomic and proteomic annotation data integrated and maintained updated in the GPKB makes it a unique valuable resource to perform comprehensive queries and evaluations on all the integrated data that it contains. The developed intuitive web interfaces for easy human interaction and the services for seamless programmatic access to the GPKB offer to the community full access to this outstanding knowledge base; they empower users to comprehensively query the important integrated data and take freely advantage of them for biomedical knowledge extraction and discovery. We took advantage of it in several applications; among them, recently we used the GPKB as database where to apply and test our implemented transitive relationship based method to detect missing new biomedical-molecular annotations based on integrated available annotations [19]. This method is close to the Linked Open Data approach of the Semantic Web [22], which has been recently used to link various sources of drug data to answer interesting scientific and business questions [23].

Its application to the high-quality, consistent and reconciled data integrated in the GPKB allowed reliably detecting more than 236,000 new annotations of nearly 51,800 genes; we integrated them in the GPKB, where they are easily identifiable as such. The developed Basic search web interface allows easily extracting them, alone or together with the other annotations integrated in the GPKB from several well-known databases.

We previously used the GPKB as one of the valuable data sources and bioinformatics search services modeled, registered and integrated in the Bioinformatics Search Computing (Bio-SeCo) system [24], publicly available at http://www.bioinformatics.deib.polimi.it/bio-seco/seco/. Bio-SeCo provides an integrated environment that eases search, exploration and ranking-aware combination of heterogeneous data provided by the available registered services, and supplies global results that can support answering complex multi-topic biomedical questions. Thus, it offers an extremely useful automated support for exploratory bio-search, which is fundamental for Life Science data driven knowledge discovery.

Lately, we are using the developed web services for the programmatic access to the GPKB, and their client web interface, in a drug repurposing [25] project that we are carrying on in collaboration with the US National Library of Medicine (NLM). It involves the integration of the GPKB with other data sources and web services available from the NLM, including the SemMedDB [26]; we plan to open to the public a new system for drug repurposing searches leveraging such integrated sources.

All these applications and their provided results prove the relevance of the GPKB and the high utility of its easy access offered to the community through the newly developed web interfaces and services.

## 4 Conclusions

Thanks to the innovative global data schema and the software framework used for its creation and maintenance, and to ensure high quality of the integrated data, the created GPKB represents a very valuable resource; the many annotation data that it integrates from several well-known databases are highly reliable, which makes them suitable to be used for data-driven biomedical knowledge discoveries.

Besides its use within multiple research projects, the relevance of the GPKB is also proved by the more than 52,000 accesses received by nearly 1,600 visitors since when it opened on the web. The multiple web and service interfaces developed enhance the relevance of the GPKB by providing easy human user and programmatic access to the numerous integrated data; they enable the community to freely and seamlessly perform queries, although complex, on all the data integrated in the GPKB and can thus be profitably leveraged to help unveiling new biomedical knowledge.

**Acknowledgements.** The authors would like to thank the several students who co-worked on developing and making publicly available the GPKB and its web and service interfaces, particularly Maria Carucci, Vincenzo Di Girolamo, Stefano Gennaro, and Marta Morfina.

# References

1. Galperin, M.Y., Rigden, D.J., Fernández-Suárez, X.M.: The 2015 nucleic acids research database issue and molecular biology database collection. Nucleic Acids Res. **43**(Database issue), D1–D5 (2015)
2. Sujansky, W.: Heterogeneous database integration in biomedicine. J. Biomed. Inform. **34**(4), 285–298 (2001)
3. Hernandez, T., Kambhampati, S.: Integration of biological sources: current systems and challenges ahead. ACM Sigmod Rec. **33**(3), 51–60 (2004)
4. Smedley, D., Haider, S., Ballester, B., Holland, R., London, D., Thorisson, G., Kasprzyk, A.: BioMart - Biological queries made easy. BMC Genom. **10**(1), 22 (2009)
5. Stevens, R., Baker, P., Bechhofer, S., Ng, G., Jacoby, A., Paton, N.W., et al.: TAMBIS: Transparent Access to Multiple Bioinformatics Information Sources. Bioinform. **16**(2), 184–185 (2000)
6. Freier, A., Hofestädt, R., Lange, M., Scholz, U., Stephanik, A.: BioDataServer: a SQL-based service for the online integration of life science data. Silico Biol. **2**(2), 37–57 (2002)
7. Cadag, E., Louie, B., Myler, P.J., Tarczy-Hornoch, P.: Biomediator data integration and inference for functional annotation of anonymous sequences. In: Pacific Symposium on Biocomputing, pp. 343–354 (2007)
8. Lee, T.J., Pouliot, Y., Wagner, V., Gupta, P., Stringer-Calvert, D.W., Tenenbaum, J.D., Karp, P.D.: BioWarehouse: a bioinformatics database warehouse toolkit. BMC Bioinform. **7**, 170 (2006)
9. Birkland, A., Yona, G.: BIOZON: a system for unification, management and analysis of heterogeneous biological data. BMC Bioinform. **7**, 70 (2006)
10. Belleau, F., Nolin, M.A., Tourigny, N., Rigault, P., Morissette, J.: Bio2RDF: towards a mashup to build bioinformatics knowledge systems. J. Biomed. Inform. **41**(5), 706–716 (2008)
11. Louie, B., Mork, P., Martin-Sanchez, F., Halevy, A., Tarczy-Hornoch, P.: Data integration and genomic medicine. J. Biomed. Inform. **40**(1), 5–16 (2007)
12. Goble, C., Stevens, R.: State of the nation in data integration for bioinformatics. J. Biomed. Inform. **41**(5), 687–693 (2008)
13. Lapatas, V., Stefanidakis, M., Jimenez, R.C., Via, A., Schneider, M.V.: Data integration in biological research: an overview. J. Biol. Res. (Thessalon) **22**(1), 9 (2015)
14. Davidson, S.B., Crabtree, J., Brunk, B.P., Schug, J., Tannen, V., Overton, G.C., et al.: K2/Kleisli and GUS: Experiments in integrated access to genomic data sources. IBM Syst. J. **40**(2), 512–531 (2001)
15. Bornberg-Bauer, E., Paton, N.W.: Conceptual data modelling for bioinformatics. Brief. Bioinform. **3**(2), 166–180 (2002)
16. Masseroli, M., Ceri, S., Campi, A.: Integration and mining of genomic annotations: experiences and perspectives in GFINDer data warehousing. In: Paton, N.W., Missier, P., Hedeler, C. (eds.) DILS 2009. LNCS, vol. 5647, pp. 88–95. Springer, Heidelberg (2009)
17. Canakoglu, A., Masseroli, M., Ceri, S., Tettamanti, L., Ghisalberti, G., Campi, A.: Integrative warehousing of biomolecular information to support complex multi-topic queries for biomedical knowledge discovery. In: Nikita, S.K., Fotiadis, D.I., (eds.) Proceedings of Thirteenth IEEE International Conference Bioinformatics and Bioengineering, (BIBE 2013), vol. 159, pp.1–4. IEEE Computer Society, Los Alamitos, CA (2013)
18. Ghisalberti, G., Masseroli, M., Tettamanti, L.: Quality controls in integrative approaches to detect errors and inconsistencies in biological databases. J. Integr. Bioinform. **7**(3), 119, 1–13 (2010)

19. Masseroli, M., Canakoglu, A., Quigliatti, M.: Detection of gene annotations and protein-protein interaction associated disorders through transitive relationships between integrated annotations. BMC Genom. **16**(Suppl 6), S5 (2015)
20. Giardine, B., Riemer, C., Hardison, R.C., Burhans, R., Elnitski, L., Shah, P., et al.: Galaxy: a platform for interactive large-scale genome analysis. Genome Res. **15**(10), 1451–1455 (2005)
21. Wolstencroft, K., Haines, R., Fellows, D., Williams, A., Withers, D., Owen, S., et al.: The Taverna workflow suite: designing and executing workflows of Web Services on the desktop, web or in the cloud. Nucleic Acids Res. **41**(Web Server issue), W557–W561 (2013)
22. LinkingOpenData W3C SWEO community project. http://www.w3.org/wiki/SweoIG/TaskForces/CommunityProjects/LinkingOpenData. Accessed 01 December 2015
23. Samwald, M., Jentzsch, A., Bouton, C., Kallesøe, C.S., Willighagen, E., Hajagos, J., et al.: Linked open drug data for pharmaceutical research and development. J. Cheminform. **3**(1), 19 (2011)
24. Masseroli, M., Picozzi, M., Ghisalberti, G., Ceri, S.: Explorative search of distributed bio-data to answer complex biomedical questions. BMC Bioinform. **15**(Suppl 1): S3, 1–14 (2014)
25. Cohen, T., Widdows, D., Schvaneveldt, R.W., Davies, P., Rindflesch, T.C.: Discovering discovery patterns with predication-based semantic indexing. J. Biomed. Inform. **45**(6), 1049–1065 (2012)
26. Kilicoglu, H., Shin, D., Fiszman, M., Rosemblat, G., Rindflesch, T.C.: SemMedDB: a PubMed-scale repository of biomedical semantic predications. Bioinform. **28**(23), 3158–3160 (2012)

# Author Index

Printed in the United States
By Bookmasters